PRINCIPLES OF SEMICONDUCTOR DEVICES

The Oxford Series in Electrical and Computer Engineering

Adel S. Sedra, Series Editor

Allen and Holberg, *CMOS Analog Circuit Design, 2nd edition*
Bobrow, *Elementary Linear Circuit Analysis, 2nd edition*
Bobrow, *Fundamentals of Electrical Engineering, 2nd edition*
Burns and Roberts, *An Introduction to Mixed-Signal IC Test and Measurement*
Campbell, *Fabrication Engineering at the Micro- and Nanoscale, 3rd edition*
Chen, *Digital Signal Processing*
Chen, *Linear System Theory and Design, 3rd edition*
Chen, *Signals and Systems, 3rd edition*
Comer, *Digital Logic and State Machine Design, 3rd edition*
Comer, *Microprocessor-Based System Design*
Cooper and McGillem, *Probabilistic Methods of Signal and System Analysis, 3rd edition*
DeCarlo and Lin, *Linear Circuit Analysis, 2nd edition*
Dimitrijev, *Principles of Semiconductor Devices, 2nd edition*
Dimitrijev, *Understanding Semiconductor Devices*
Fortney, *Principles of Electronics: Analog & Digital*
Franco, *Electric Circuits Fundamentals*
Ghausi, *Electronic Devices and Circuits: Discrete and Integrated*
Guru and Hiziroğlu, *Electric Machinery and Transformers, 3rd edition*
Houts, *Signal Analysis in Linear Systems*
Jones, *Introduction to Optical Fiber Communication Systems*
Krein, *Elements of Power Electronics*
Kuo, *Digital Control Systems, 2nd edition*
Lathi, *Linear Systems and Signals, 2nd edition*
Lathi and Ding, *Modern Digital and Analog Communication Systems, 4th edition*
Lathi, *Signal Processing and Linear Systems*
Martin, *Digital Integrated Circuit Design*
Miner, *Lines and Electromagnetic Fields for Engineers*
Parhami, *Computer Architecture*
Parhami, *Computer Arithmetic, 2nd edition*
Roberts and Sedra, *SPICE, 2nd edition*
Roulston, *An Introduction to the Physics of Semiconductor Devices*
Sadiku, *Elements of Electromagnetics, 5th edition*
Santina, Stubberud, and Hostetter, *Digital Control System Design, 2nd edition*
Sarma, *Introduction to Electrical Engineering*
Schaumann, Xiao, and Van Valkenburg, *Design of Analog Filters, 2nd edition*
Schwarz and Oldham, *Electrical Engineering: An Introduction, 2nd edition*
Sedra and Smith, *Microelectronic Circuits, 6th edition*
Stefani, Shahian, Savant, and Hostetter, *Design of Feedback Control Systems, 4th edition*
Tsividis, *Operation and Modeling of the MOS Transistor, 3rd edition*
Van Valkenburg, *Analog Filter Design*
Warner and Grung, *Semiconductor Device Electronics*
Wolovich, *Automatic Control Systems*
Yariv and Yeh, *Photonics: Optical Electronics in Modern Communications, 6th edition*
Żak, *Systems and Control*

PRINCIPLES OF SEMICONDUCTOR DEVICES

SECOND EDITION

SIMA DIMITRIJEV
Griffith University

New York Oxford
OXFORD UNIVERSITY PRESS

Oxford University Press, Inc., publishes works that further Oxford University's objective of excellence in research, scholarship, and education.

Oxford New York
Auckland Cape Town Dar es Salaam Hong Kong Karachi
Kuala Lumpur Madrid Melbourne Mexico City Nairobi
New Delhi Shanghai Taipei Toronto

With offices in
Argentina Austria Brazil Chile Czech Republic France Greece
Guatemala Hungary Italy Japan Poland Portugal Singapore
South Korea Switzerland Thailand Turkey Ukraine Vietnam

For titles covered by Section 112 of the U.S. Higher Education Opportunity Act, please visit www.oup.com/us/he for the latest information about pricing and alternate formats.

Published by Oxford University Press, Inc.
198 Madison Avenue, New York, New York 10016
http://www.oup.com

Library of Congress Cataloging-in-Publication Data

Dimitrijev, Sima, 1958–
Principles of semiconductor devices/Sima Dimitrijev.—2nd ed.
p. cm.—(The Oxford series in electrical and computer engineering)
ISBN 978-0-19-538803-9 (hardback)
1. Semiconductors. I. Title.
TK7871.85.D54697 2011
621.3815′2—dc22 2010048450

Printing number: 9 8 7 6 5 4 3 2 1

Printed in the United States of America
on acid-free paper

BRIEF CONTENTS

CONTENTS

Note: Dagger (†) indicates a section that can be treated as a read-only section; asterisk (*) indicates a section that can be omitted without loss of continuity.

PREFACE

This edition of *Principles of Semiconductor Devices* maintains the main aims of the previous edition—to offer a student-friendly text for senior undergraduate and graduate students of electrical and computer engineering that provides a comprehensive introduction to semiconductor devices. Related to the **student-friendly** aspect, the aim is to provide the best explanations of the underlying physics, device operation principles, and mathematical models used for device and circuit design and to support these explanations by intuitive figures. The **comprehensive** character of the text emerges from the links that it establishes between the underlying principles and modern practical applications, including the link to the SPICE models and parameters that are commonly used during circuit design.

New to This Edition

The aim of linking device physics to **modern** applications sets the need for changes that are made in this edition. The dimensions of modern semiconductor devices are reduced to the point where electronic engineers have to question applicability of the basic concepts and models presented in semiconductor textbooks. For example, the *average number* of minority-current carriers is smaller than one carrier ($\overline{N} < 1$) in almost all modern semiconductor devices, which calls into question the concepts of continuous particle concentration and continuous current as fundamental elements of standard semiconductor theory. Further questions are due to increasing practical manifestations of quantum-mechanical effects in nanoscale devices and potential applications of nanowires and carbon nanotubes that exhibit one-dimensional transport. The answer to these questions should not be to simply disregard well-established standard semiconductor theory and as a consequence to disregard all the design tools and practices based on this theory that have been developed over several decades. This edition of *Principles of Semiconductor Devices* is the first textbook to address these questions by specifying the fundamental principles and by logical application of these principles to upgrade the standard theory for proper interpretation and modeling of the effects in modern devices.

Following is a summary of the new elements and main changes in this edition:

- A new chapter—the first in semiconductor textbooks—on the physics of nanoscale devices, including the physics of single-carrier events, two-dimensional transport in MOSFETs and HEMTs, and one-dimensional ohmic and ballistic transport in nanowires and carbon nanotubes.
- Fully revised and upgraded material on crystals to introduce graphene and carbon nanotubes as two-dimensional crystals and to link them to the standard three-dimensional crystals through the underlying atomic-bond concepts.
- Revised P–N junction chapter to emphasize the current mechanisms that are relevant in modern devices.

- JFETs and MESFETs presented in a separate chapter.
- Revised chapter on the energy-band model.
- 57 new problems and 11 new examples.

Course Organization

The organization of the text is such that the core material is presented in Part I (semiconductor physics) and Part II (fundamental device structures). This will be quite sufficient for introductory undergraduate courses. Selected sections from Part III can be used in courses that are focused on the core material in different ways: (1) as read-only material, (2) as material for assignments, (3) as reference material, and (4) as supporting material for computer and/or laboratory exercises. Many courses will require full integration of selected sections from Part III. The order of the sections in the book does not imply the sequence to be followed in these courses. Selected specific/advanced sections can be read immediately after the relevant fundamental material. A typical example is the electronics-oriented material, such as the equivalent circuits and SPICE parameter measurements, presented in Chapter 11. To integrate the equivalent circuits into a course, the sections on equivalent circuits of diodes can be used as extensions of the diode chapter, the sections on equivalent circuits of MOSFETs can be used as extensions of the MOSFET chapter, and the sections on equivalent circuits of BJTs can be used as extensions of the BJT chapter. Similarly, if the IC technology sections from Chapter 16 are to be integrated, they can be used as extensions of the diode, MOSFET, and BJT chapters. Another example relates to the JFET and MESFET sections in Chapter 13. If these devices are an integral part of a course, then these sections can be included after the MOSFET chapter to create integrated coverage of FET devices. For courses that integrate photonic devices, the specific material from Chapter 12 can be included after the diode chapter. Analogously, for courses that need to address issues that are specific for power electronics, the sections on power diodes and power MOSFETs from Chapter 14 can be used as extensions of the diode and MOSFET chapters, respectively.

Acknowledgments

I value very much the feedback I received about the first edition of the book, and I wish to thank everybody who sent me comments, suggestions, and errata. I am indebted to all anonymous reviewers from different stages of the development of this book, because they provided absolutely essential feedback and extremely valuable comments and suggestions. I am especially grateful to the reviewers, whose opinions and comments directly influenced the development of this edition:

Petru Andrei, Florida State University
Terence Brown, Michigan State University
Godi Fischer, Rhode Island University
Siddhartha Ghosh, University of Illinois at Chicago
Matthew Grayson, Northeastern University
Ying-Cheng Lai, Arizona State University

It is my pleasure to acknowledge that a large team from Oxford University Press is behind this project and I am most thankful for their continuing support. The project started with important decisions and involvement by John Challice, publisher, Danielle Christensen, development editor, and Rachael Zimmermann, associate editor, and crucially relied on the work and expertise of Claire Sullivan, editorial assistant, Barbara Mathieu, senior production editor, Claire Sullivan, editorial assistant, Brenda Griffing, copyeditor, and Jill Crosson, copywriter.

Finally, I would like to acknowledge the support of my family, in particular my wife, Vesna, because it provided me with a lot of necessary time and inspiration.

Sima Dimitrijev

1 Introduction to Crystals and Current Carriers in Semiconductors: The Atomic-Bond Model

Electric current in both metals and semiconductors is due to the flow of electrons, although many electrons are tied to the parent atoms and are unable to contribute to the electric current. The regular placement of atoms in metal and semiconductor crystals, shown in Fig. 1.1 for the case of silicon crystal, provides the conditions for some electrons to be shared by all the atoms in the crystal. It is these electrons that can make electric current and are referred to as current carriers. The effects of regular atom placement on the essential properties of the current carriers in semiconductors are progressively introduced in two steps: (1) at the level of the atomic-bond model in this chapter and (2) at the level of the energy-band model in the next chapter.

This chapter begins with a description of atomic bonds and then proceeds to the important concepts related to spatial placement of atoms in both three-dimensional and two-dimensional crystals (including graphene and carbon nanotubes). This chapter also

Figure 1.1 Image of silicon crystal obtained by transmission-electron microscopy.

introduces current carriers in semiconductors to the level that is possible with the atomic-bond model. Although lacking certain important details, this level is a very important initial step. The usual model of current conduction in metals has to be gradually upgraded to introduce the model of conduction by carriers of two types: free electrons and holes among bound electrons. It is the existence of two types of carrier that distinguishes semiconductors from metals. Given that semiconductor devices utilize combinations of layers with predominantly electron-based conduction (N-type layers) and layers with predominantly hole-based conduction (P-type layers), the concepts and effects of N-type and P-type doping are also introduced in this chapter. The effects of doping are essential because semiconductors are distinguished from insulators by the ability to achieve N-type and P-type layers. Finally, to round up the introduction to semiconductors at the atomic-bond level, the last section briefly presents the basic techniques of crystal growth and doping.

†1.1 INTRODUCTION TO CRYSTALS

1.1.1 Atomic Bonds

Certain atoms can pack spontaneously into an orderly pattern called a *crystal lattice*. If this were not so, it would be practically impossible to create even one cubic millimeter of a crystalline material, as this would require the placement of more than 10^{19} atoms in almost perfect order. Clearly, there are natural forces that hold the atoms of a crystalline material together. These forces are related to the stability of the electronic configuration of individual atoms. For example, there are atoms with quite stable electronic configurations; they are referred to as the noble gases, and they are chemically inert. Helium, the noble gas element with the smallest number of electrons, has two electrons with spherical symmetry and opposite spins. The next noble gas element, neon, has 10 electrons. In neon, two electrons are in the first *shell* (as in the case of helium), which is usually denoted by $1s^2$ (1 indicates the first shell and s^2 indicates the two electrons in the spherical *orbital* labeled by s). The remaining eight electrons fill the second shell—one pair with spherical symmetry (s^2) and three pairs at p orbitals with x-, y-, and z-symmetries ($p_x^2 p_y^2 p_z^2 = p^6$). The shapes and symmetries of $2s$ and $2p$ electron orbitals are illustrated in Fig. 1.2. Accordingly, the complete electronic configuration of neon is expressed as $1s^2 2s^2 2p^6$. Sodium is the eleventh element, with the eleventh electron placed in the s orbital of the third shell: $1s^2 2s^2 2p^6 3s^1$. To reach the stability of the electronic configuration found in neon, sodium tends to give the eleventh electron away. On the other hand, chlorine, the seventeenth, element, can reach the stability of argon (the next noble gas element) by accepting an extra electron. Therefore, sodium and chlorine atoms relatively easily exchange electrons, creating positive sodium ions and negative chlorine ions. The attractive forces between the positive and negative ions (*ionic bonds*) hold the atoms of NaCl crystals together.

---

†Sections marked by a dagger can be used as read-only sections.

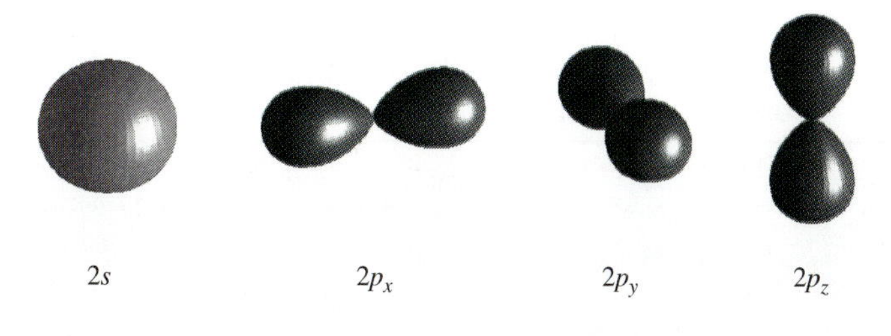

Figure 1.2 The shapes and symmetries of $2s$ and $2p$ electron orbitals.

The atoms in metal crystals are held together by another type of bond, the *metallic bond*. In this case, the atoms simply give the extra electrons away to reach stable electronic configurations. The extra electrons are shared by all the atoms (positive ions) in the crystal, so that we can think of ions submerged in a sea of electrons. The sea of electrons holds the crystal together; but because these electrons are shared by all the atoms, they move through the crystal when an electric field is applied. Consequently, metals are excellent conductors of electric current.

The atoms with half-filled shells can reach stable electronic configurations in two symmetric ways: (1) by giving the electrons from the half-filled shell away or (2) by accepting electrons from neighboring atoms to fill the half-empty shell. The first element that exhibits a half-filled shell is hydrogen: it has one electron in the first shell that can accommodate two electrons. Two hydrogen atoms form a hydrogen molecule, where one of the hydrogen atoms gives its electron away (to eliminate the unstable electronic configuration associated with a single electron) and the other hydrogen atom accepts the electron (to reach the stable electronic configuration of helium). Because of the symmetry of this situation, the hydrogen atom that gives and the hydrogen atom that accepts an electron are indistinguishable. This type of bond, which is clearly different from both the ionic and metallic bonds, is the *covalent bond*.

The next atom that has a half-filled shell is carbon. Carbon has six electrons: two electrons completely fill the first shell, with the remaining four electrons appearing in the second shell, which can accommodate eight electrons: $1s^2 2s^2 2p^2$. The four *valence* electrons in the second shell, which are the unstable or active electrons, make carbon the first element in the fourth group of the periodic table of elements (as shown in Table 1.1). A carbon atom can form four covalent bonds with four hydrogen atoms, which creates a stable methane molecule (CH_4). In a methane molecule, the carbon atom either gives away the four valence electrons (to reach the stable electronic configuration of helium) or accepts the four electrons from the four hydrogen atoms (to reach the stable electronic configuration of neon). The four covalent bonds in a methane molecule are indistinguishable because of the symmetry of this molecule, which is not consistent with the difference between the two s electrons and the two p electrons in the $2s^2 2p^2$ configuration of the second shell of a carbon atom. A carbon atom can form four symmetrical covalent bonds because the $2s$, $2p_x$, $2p_y$, and $2p_z$ orbitals (Fig. 1.2) can be transformed into the four symmetrical hybrid orbitals illustrated in Fig. 1.3a. This transformation is called *hybridization*, and the four symmetric hybrid orbitals are called sp^3 hybrid orbitals; the label sp^3 indicates that these orbitals are the result of hybridization of one s ($2s$) and three p ($2p_x$, $2p_y$, and $2p_z$) orbitals. Figure 1.3b illustrates the interaction between the four sp^3 orbitals of the carbon atom and the $1s$ orbitals of the four hydrogen atoms in a methane molecule,

TABLE 1.1 Semiconductor Related Elements in the Periodic Table (with Atomic Number and Atomic Weight)

III	IV	V
+3	+4	+5
5 B Boron 10.82	6 C Carbon 12.01	7 N Nitrogen 14.008
13 Al Aluminum 26.97	14 Si Silicon 28.09	15 P Phosphorus 31.02
31 Ga Gallium 69.72	32 Ge Germanium 72.60	33 As Arsenic 74.91
49 In Indium 114.8	50 Sn Tin 118.7	51 Sb Antimony 121.8

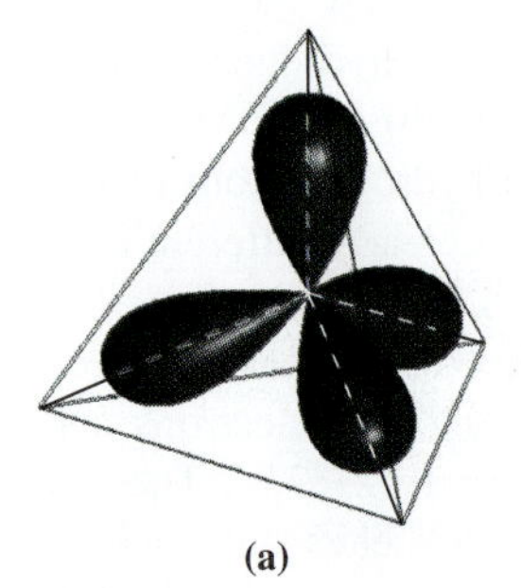

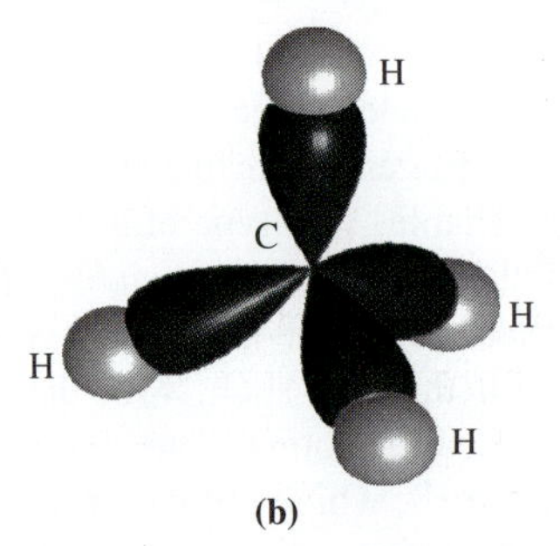

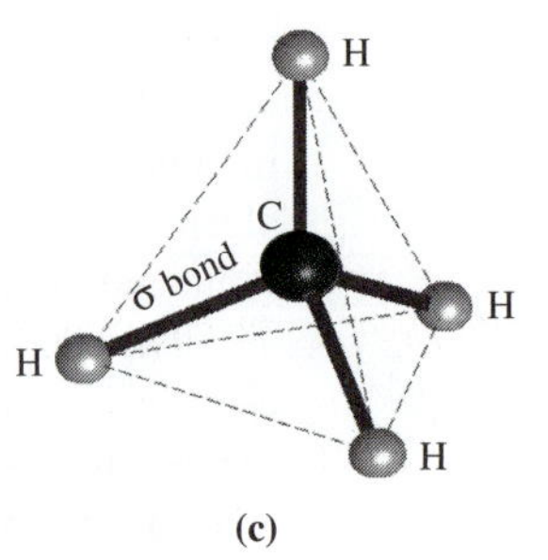

Figure 1.3 Hybrid electronic configuration that enables four symmetrical covalent bonds of a carbon atom: (a) the shape and tetrahedral symmetry of sp^3 hybrid orbitals, (b) the interactions of the four sp^3 orbitals with 1s orbitals of four hydrogen atoms that form the four covalent bonds of a CH_4 molecule, and (c) the three-dimensional atomic-bond model of the CH_4 molecule.

whereas Fig. 1.3c shows the atomic-bond model of this molecule in three dimensions. As indicated in Fig.1.3c, covalent atomic bonds of this type are called σ bonds.

Another important hybridization of the $2s$, $2p_x$, $2p_y$, and $2p_z$ orbitals in carbon is called sp^2 hybridization. In this case, the s orbital and two p orbitals, say p_x and p_y, are transformed into three sp^2 orbitals with triangular planar symmetry (Fig. 1.4a); the p_z orbital remains unhybridized. Carbon atoms with sp^2 hybrid orbitals form σ covalent bonds with planar (two-dimensional) structures. The simplest structure from this class is the ethylene molecule, C_2H_4. Figure 1.4b illustrates that the sp^2 orbitals form σ bonds (four bonds with the s orbitals of four hydrogen atoms and one bond between two sp^2

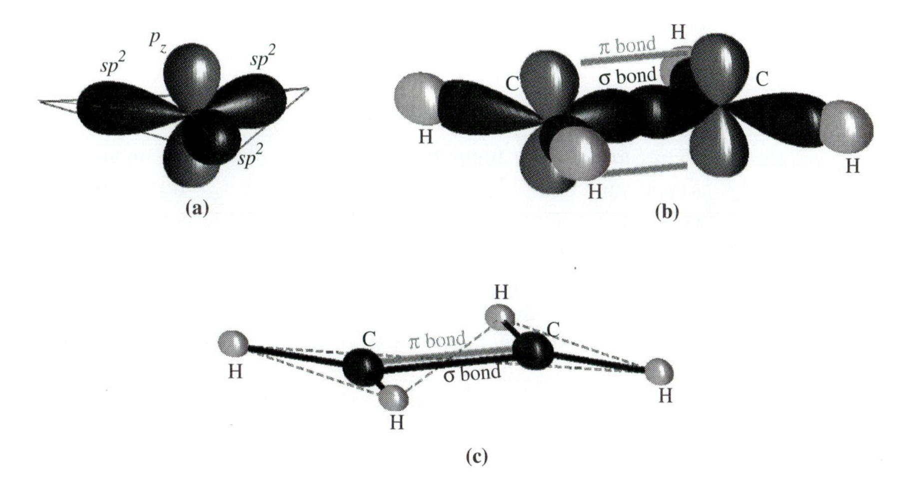

Figure 1.4 Hybrid electronic configuration of a carbon atom that results in triangular planar bonds: (a) the shape and triangular symmetry of sp^2 hybrid orbitals, (b) σ bond formed by two sp^2 orbitals and π bond formed by the unhybridized p orbitals of two carbon atoms in a C_2H_4 molecule, and (c) the atomic-bond model of the C_2H_4 molecule.

orbitals of two carbon atoms), whereas the unhybridized p orbitals of the two carbon atoms form a much weaker π bond.

Covalent bonds between carbon and hydrogen atoms result in molecules that form hydrocarbon gases of different types. However, the situation is very different when covalent bonding is limited to carbon atoms themselves. Consider sp^3-hybridized carbon atoms (as in Fig. 1.3a) and replace the four hydrogen atoms with identical, sp^3-hybridized carbon atoms. The carbon atom in the center of the tetrahedron will be stable owing to the four σ covalent bonds with the four neighboring carbon atoms. As distinct from hydrogen atoms in the corners of the tetrahedron, carbon atoms in the corners need to form three extra covalent bonds each to be stable. This means that each of the corner carbon atoms needs four carbon neighbors in analogous tetrahedral structures, so that each of these atoms appears in the center of its own tetrahedron. This arrangement requires continuous replication of the tetrahedral pattern, which is the basic or *primitive* cell, in all three dimensions in space. Accordingly, sp^3-hybridized carbon atoms form three-dimensional crystals with the tetrahedral primitive cell, which is the diamond version (*polytype*) of solid carbon. Silicon and germanium, as the second and third elements in the fourth column of the periodic table, respectively, also form three-dimensional crystals with the same diamond-type lattice. Section 1.1.2 considers the most important three-dimensional crystals for device applications in more detail.

Consider now sp^2-hybridized carbon atoms (as in Fig. 1.4a) and replace the four hydrogen atoms with identical sp^2-hybridized carbon atoms. The replacement of a hydrogen atom by an sp^2 carbon atom means that this atom will have to connect to two additional sp^2 carbon atoms to be stable, symmetrically to the carbon atoms in the ethylene molecule. In this case, the triangular planar structure is replicated to form a two-dimensional crystal known as *graphene*. Section 1.1.2 describes the two-dimensional structure of graphene and the related carbon nanotubes.

1.1.2 Three-Dimensional Crystals

Crystal Lattices

The atoms of a crystalline material are regularly placed in points that define a particular *crystal lattice*. The regularity of the atom placement means that the pattern of a *unit cell* is replicated to build the entire crystal. Figure 1.5 illustrates the unit cell of a *cubic lattice,* which is the simplest three-dimensional crystal lattice. Important parameters of any crystal lattice are the lengths of the unit cell edges. Cubic crystals have unit-cell edges of the same length (a), which is called the *crystal-lattice constant*. In general, unit-cell edges can be different ($a \neq b \neq c$). The unit cell fully defines the crystal lattice: a whole crystal can be created by shifting the unit cell along the cell edges in steps that are equal to the cell edges.

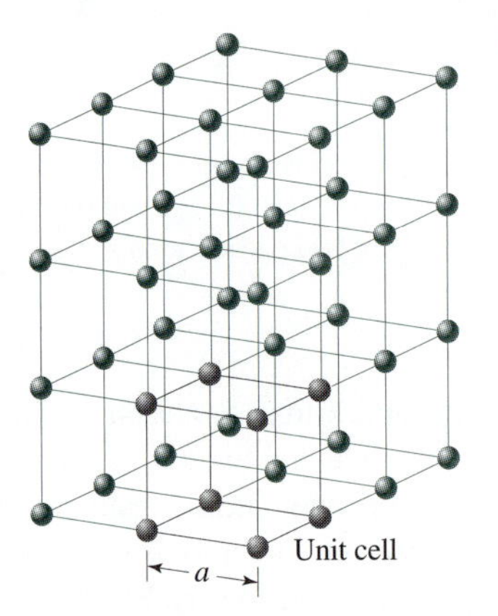

Figure 1.5 Simple cubic lattice and its unit cell (a is the lattice constant).

In addition to the simple cubic lattice, a number of more complex lattice structures have cubic unit cells. Figure 1.6 illustrates two additional cases: body-centered cubic (Fig. 1.6b) and face-centered cubic (Fig. 1.6c) unit cells. A body-centered cubic cell has

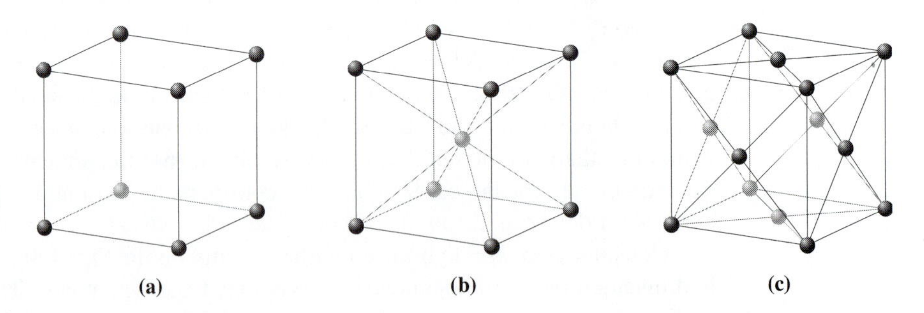

Figure 1.6 Three different types of cubic unit cell: (a) simple cubic (sc), (b) body-centered cubic (bcc), and (c) face-centered cubic (fcc).

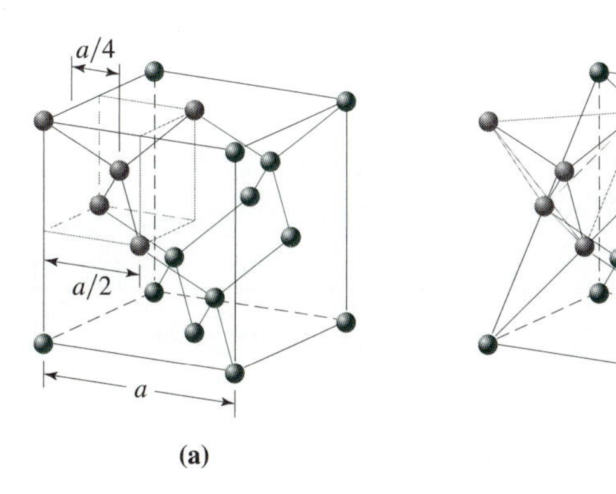

Figure 1.7 Diamond unit cell, illustrated in two ways to show (a) the cubic unit cell and (b) the inherent tetrahedral structure.

an additional atom in the center of the cube, whereas in a face-centered cubic cell, six additional atoms are centered on the six faces.

As described in the preceding section, carbon, silicon, and germanium (elements of the fourth column of the periodic table) form a tetrahedral primitive cell. The term *primitive* is used rather *unit* because a primitive cell requires rotations, in addition to shifts along the cell edges, to create the complete crystal lattice. Figure 1.7 shows the unit cell of diamond, silicon, and germanium crystals (it is commonly referred to as the *diamond unit cell*). It can be seen that each atom appears in the center of a tetrahedral primitive cell (analogous to the tetrahedral structure of the methane molecule, illustrated in Fig. 1.3).

In addition to semiconductors with single elements from the fourth column of the periodic table (diamond, silicon, and germanium), semiconductor crystals can be created from two different fourth-column elements. These materials, such as SiC and SiGe, are known as *compound semiconductors*. The diamond crystal lattice is the simplest crystal lattice for compound semiconductors. In the case of cubic SiC (also labeled as 3C-SiC and β-SiC), each silicon atom is connected to four carbon atoms in the corners of its tetrahedral primitive cell, and each carbon atom is connected to four silicon atoms. SiC appears in many other crystalline *polytypes* that are based on different hexagonal and rhomboidal unit cells.

Compound semiconductors can also be formed by elements from the third and the fifth columns of the periodic table (called III–V compound materials) and between the second and sixth columns (called II–VI compound semiconductors). Although the two different elements in compound materials contribute different number of electrons, a pair of these elements has eight valence electrons. This means that covalent bonds stabilize the electronic configurations of the atoms in these crystals. Examples of III–V semiconductors are GaAs, GaP, GaN, AlP, AlAs, InP, and InAs, whereas examples of II–VI semiconductors are ZnS, ZnSe, CdS, and CdTe. The unit cell of GaAs (and many other compound semiconductors) is also cubic with the tetrahedral primitive cell (Fig. 1.7), where gallium and arsenic atoms occupy alternating sites, so that each gallium atom is linked to four arsenic atoms and vice versa. This type of lattice is also referred to as a *zincblade lattice structure*.

EXAMPLE 1.1 Atom-Packing Fraction

Assuming that the atoms are hard spheres that just touch the nearest neighbors in the crystal, find the packing fraction (the fraction of unit-cell volume filled with atoms) of a simple cubic cell.

SOLUTION

The packing fraction can be obtained by dividing the volume of the atoms (spheres) that belong to a unit cell by the volume of the unit cell:

$$PF = \frac{N_{atoms} V_{atom}}{V_{cell}}$$

where V_{atom} is the volume of each atom, and N_{atoms} is the equivalent number of atoms inside each cell. The volume of the unit cell is $V_{cell} = a^3$, where a is the crystal-lattice constant (the length of the cube side). Two nearest atoms appear in the cube corners, and if they are just touching each other, the radius of each atom is $r = a/2$. This means the volume of each atom is $V_{atom} = \frac{4}{3}\pi r^3 = \pi a^3/6$. There are eight atoms (in each of the corners), but each of them is shared between 8 unit cells. Because 1/8 of each of the eight spheres belongs to the considered unit cell, $N_{atoms} = 8 \times \frac{1}{8} = 1$. Therefore,

$$PF = \frac{\pi a^3/6}{a^3} = \pi/6$$

which means that the packing fraction is $PF = 52.4\%$.

EXAMPLE 1.2 Crystal-Lattice Constant

The diameter of a silicon atom is $d = 0.235$ nm, under the assumption that silicon atoms are hard spheres that just touch each other in the silicon crystal. Determine the crystal-lattice constant (width of the unit cell, a).

SOLUTION

We saw in Fig. 1.7 that every silicon atom can be considered to be in the center of a tetrahedron, its closest neighbors being the atoms in the tetrahedron corners. Figure 1.7 also shows that each tetrahedron fits in a small cube with side of $a/2$, where the atom that is inside the tetrahedron appears in the center of the small cube, whereas the corner atoms are in four corners of the small cube. With this observation, we find that the distance between the centers of the atom in the cube center and any of the corner atoms is equal to the half of the small-cube diagonal. If the nearest atoms are just touching each other, their radii must be equal to a quarter of the small-cube diagonal. Because the small-cube diagonal is $D = \sqrt{3}a/2$, the atomic radius is $r = \sqrt{3}a/8$. From here, $a = 8r/\sqrt{3} = 0.543$ nm.

EXAMPLE 1.3 Volume, Area, and Mass Densities of Atoms

The crystal-lattice constant and atomic mass of silicon are 0.543 nm and 28.09 g/mol, respectively. Avogadro's number is 6.02×10^{23} atoms/mol. Determine:

(a) the atom concentration (volume density of atoms)
(b) the mass density
(c) the surface atom density if the crystal is terminated at the faces of unit cells defining a single plane.

SOLUTION

(a) Referring to Fig. 1.7, we see that there are eight equivalent atoms in each unit cell:

$$N_{atoms} = \underbrace{8 \times \frac{1}{8}}_{\text{corners}} + \underbrace{6 \times \frac{1}{2}}_{\text{faces}} + \underbrace{4}_{\text{inside}} = 8$$

The volume of each cell is $V_{cell} = a^3$. Therefore, the atom concentration is

$$N_{Si} = 8/a^3 = 49.97 \text{ nm}^{-3} = 4.991 \times 10^{22} \text{ cm}^{-3}$$

(b) Avogadro's number expresses the number of particles (atoms) in a mole, and the atomic mass is the total mass of all these particles (atoms). Therefore, the mass of each silicon atom is $m_{Si} = 28.09/6.02 \times 10^{23} = 4.67 \times 10^{-23}$ g. If there are N_{Si} atoms per unit volume, then the mass density is $\rho_{Si} = N_{Si} m_{Si} = 2.33$ g/cm^3.

(c) Again, referring to Fig. 1.7, we find that there are two equivalent silicon atoms in each face of the unit-cell cube:

$$N_{atoms} = \underbrace{4 \times \frac{1}{4}}_{\text{corners}} + \underbrace{1}_{\text{center}} = 2$$

The area of each of these square faces is a^2. Therefore, the number of atoms per unit area is

$$N_{\{100\}} = 2/a^2 = 6.78 \text{ nm}^{-2} = 6.78 \times 10^{14} \text{ cm}^{-2}$$

Planes and Directions

It is convenient to express the positions of atoms and different crystallographic planes in terms of the lengths of the unit-cell edges, a, b, and c. Any plane in space may be described by the equation

$$h\frac{x}{a} + k\frac{y}{b} + l\frac{z}{c} = 1 \tag{1.1}$$

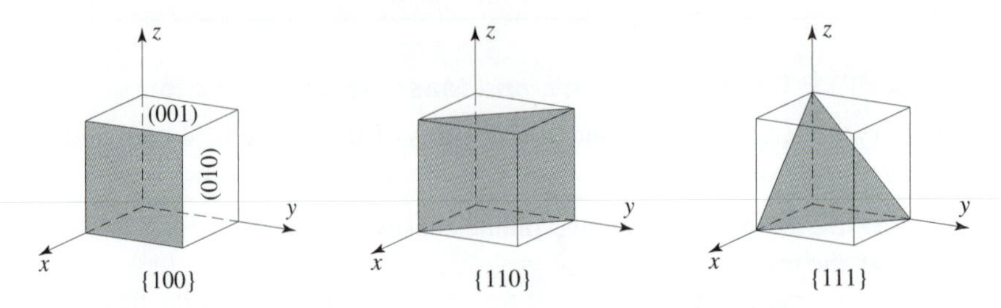

Figure 1.8 Miller indices for the three most important planes in cubic crystals.

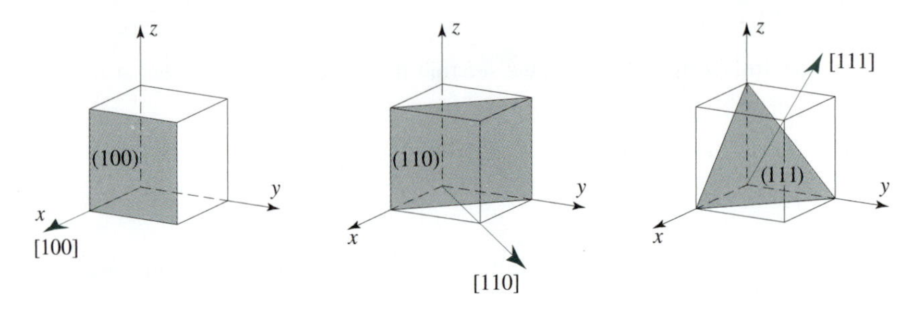

Figure 1.9 Important directions in cubic crystals.

where a/h, b/k, and c/l are intercepts of the x-, y-, and z-axes, respectively. In general, h, k, and l can take any value. For example, $h = k = l = 1/2$ defines the plane that intersects the x-, y-, and z-axes at $2a$, $2b$, and $2c$, respectively. However, this plane is crystallographically identical to the plane that intersects the axes at a, b, and c, in which case $h = k = l = 1$. For that reason, h, k, and l are defined as integers in the case of crystals. *Characteristic crystallographic planes are defined by a set of integers,* (hkl), *known as Miller indices.* In the case of negative intercepts, the minus sign is placed above the corresponding Miller index—for example, $(\bar{h}kl)$.

Figure 1.8 illustrates three important planes in cubic crystals. The shaded plane labeled as {100} intersects the x-axis at a, and it is parallel to the y- and z-axes (mathematically, it intersects the y- and z-axes at infinity, as $a/0 = \infty$). Analogously, the plane (010) intersects the y-axis at a and is parallel to the x- and z-axes. Because there is no crystallographic difference between the (100), (010), and (001) planes, they are uniquely labeled as {100}, where the braces {} indicate that the notation is for all the planes of equivalent symmetry.

The second plane shown in Fig. 1.8, labeled as {110}, intersects two axes at a and is parallel to the third axis. Finally, {111} plane intersects all the axes at a.

In addition to planes, it is necessary to describe directions in crystals. By convention, the direction perpendicular to (hkl) plane is labeled as $[hkl]$. A set of equivalent directions is labeled as $\langle hkl \rangle$; for example, $\langle 100 \rangle$ represents [100], [010], [001], $[\bar{1}00]$, and so on. Figure 1.9 illustrates the most important directions in cubic crystals.

EXAMPLE 1.4 Miller Indices

(a) Determine the Miller indices for the plane illustrated in Fig. 1.10.

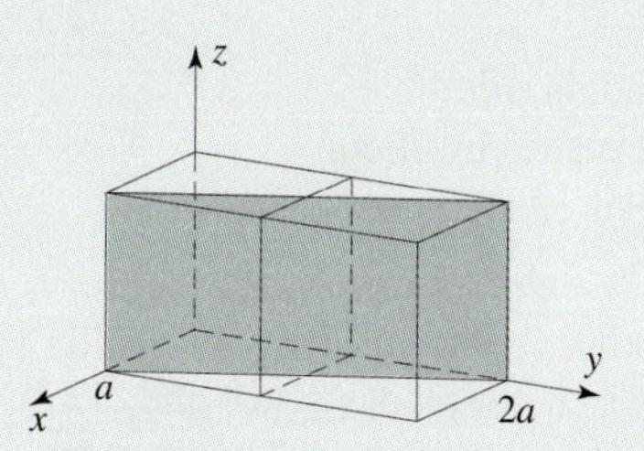

Figure 1.10 A plane in a cubic crystal.

(b) Looking along z-axis, which is the [001] direction, draw the intersection between the (001) plane and the plane indicated in Fig. 1.10. Draw the direction perpendicular to the indicated plane, and label it with the corresponding Miller indices.

SOLUTION

(a) The Miller indices can be determined by the following procedure:

1. *Note where the plane intercepts the axes:* a, $2a$, and ∞ in this case.
2. *Divide the intercepts by the unit-cell length(s) to normalize them, and invert the normalized values:* 1, $\frac{1}{2}$, and $\frac{1}{\infty} = 0$.
3. *Multiply the set of obtained numbers by an appropriate number to obtain the corresponding set of the smallest possible whole numbers:* multiplying by 2 in this case, we obtain 2, 1, and 0.

The numbers obtained after step 3 are the Miller indices, so we enclose the set of these numbers in brackets: (210).

(b) The drawing and the labels are shown in Fig. 1.11.

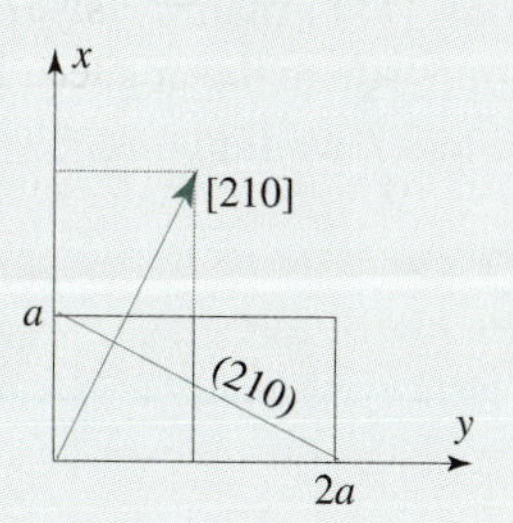

Figure 1.11 Solution of Example 1.4b.

EXAMPLE 1.5 Angle and Distance Between Crystal Planes

KOH-based etching of silicon, through an appropriately oriented square window, produces an inverted pyramid where the pyramid base is the {100} silicon surface and the four sides are {111} planes.

(a) What is the angle between the pyramid base and its side?
(b) What is the distance between the closest {111} planes in silicon?

SOLUTION

(a) Figure 1.12 illustrates this angle. The intersection between {100} and {111} planes is along the line labeled as $\overline{AB}$. Lines $\overline{DC}$ and $\overline{EC}$ lie in planes {100} and {111}, respectively, hitting the intersecting line $\overline{AB}$ at point C at right angles. The lines $\overline{DC}$ and $\overline{EC}$ define the requested angle: $\angle DCE = \alpha$. The points D, C, and E create a right-angled triangle with sides $\overline{DE} = a$ and $\overline{DC} = \sqrt{2}a/2$. Because the length of the hypotenuse is

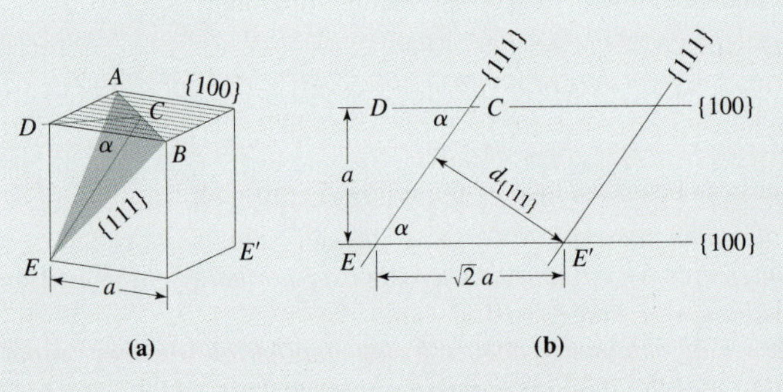

Figure 1.12 Illustrations of the angle between {100} and {111} planes (α) and the distance between neighboring {111} planes ($d_{\{111\}}$).

$\overline{EC} = \sqrt{a^2 + a^2/2} = \sqrt{3/2}a$, the angle α can be found as

$$\alpha = \arccos \frac{\overline{DC}}{\overline{CE}} = \arccos \frac{a\sqrt{2}/2}{a\sqrt{3/2}} = \arccos\left(1/\sqrt{3}\right) = 54.74^\circ$$

(b) Figure 1.12b illustrates the distance between neighboring {111} planes ($d_{\{111\}}$). It can be seen that $d_{\{111\}}$ is the side of a right-angled triangle, opposite to the angle α. Because the length of the hypotenuse is $\sqrt{2}a$ (diagonal of a cube face), we find that

$$d_{\{111\}} = \sqrt{2}a \sin(\alpha) = 0.627 \text{ nm}$$

Crystal Defects

It is possible to grow regions of defect-free crystals much larger than that shown in Fig. 1.1. Inevitably, however, the crystal region must be terminated and the atomic bonds disrupted at the crystal surface; this creates crystal defects that can have a significant impact on device operation. Frequently used materials in practice consist of multiple crystalline grains, as

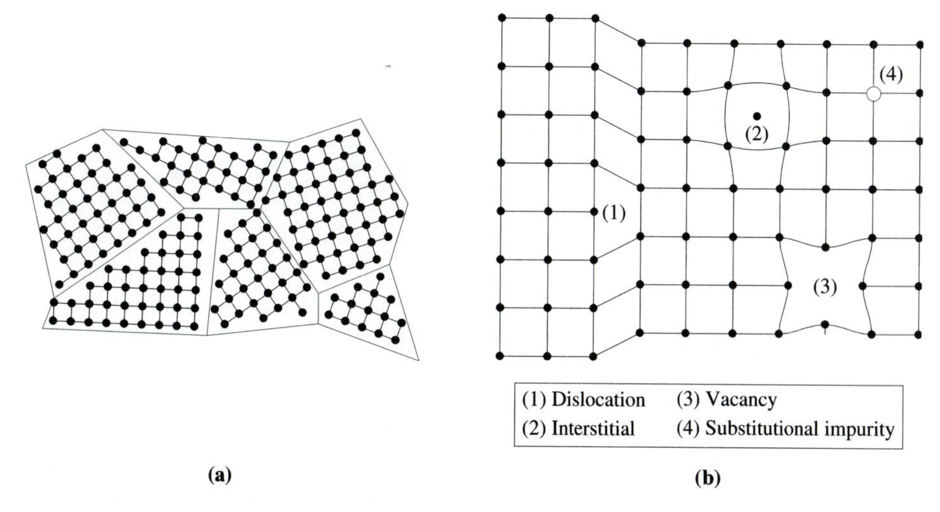

Figure 1.13 Schematic illustration of (a) polycrystalline material and (b) various crystal defects.

shown in Fig. 1.13a. This type of material is referred to as polycrystalline to distinguish it from the monocrystalline structure of the individual grain. The electrical properties of polycrystalline materials are dominated by the defects at the grain boundaries.

In addition to the inevitable surface termination of a three-dimensional crystal, a number of other crystal defects may appear inside a monocrystalline region. As illustrated in Fig. 1.13b, these defects can be *dislocations* of crystallographic planes, *vacancies* (missing atoms), *interstitial* atoms, and interstitial and substitutional *impurity* atoms. Although substitutional impurity atoms are crystal defects in principle, some are deliberately introduced as a way of controlling the fundamental electrical properties of semiconductor materials. This important concept, called semiconductor doping, is described in Section 1.2.2.

1.1.3 Two-Dimensional Crystals: Graphene and Carbon Nanotubes

As distinct from sp^3-hybridized carbon atoms that form three-dimensional diamond crystals, sp^2-hybridized carbon atoms form two-dimensional structures. As described in Section 1.1.1, sp^2 carbon atoms form the two-dimensional C_2H_4 molecule when combined with hydrogen. Consider what happens when the four hydrogen atoms in a C_2H_4 molecule (Fig. 1.4c) are replaced with identical sp^2 carbon atoms. The replacement of a hydrogen atom by an sp^2 carbon atom means that this atom will have to connect to two additional sp^2 carbon atoms to be stable, symmetrically to the carbon atoms in the C_2H_4 molecule. As shown in Fig. 1.14, this requirement leads to replication of the planar bond structure of the carbon atoms, forming a two-dimensional crystal with a honeycomb structure. This two-dimensional crystal is known as *graphene*.

The concept of graphene has been used for a long time, in particular to explain the properties of graphite, which consists of loosely connected parallel sheets of graphene. It is established that (1) the distance between the neighboring carbon atoms in graphene is

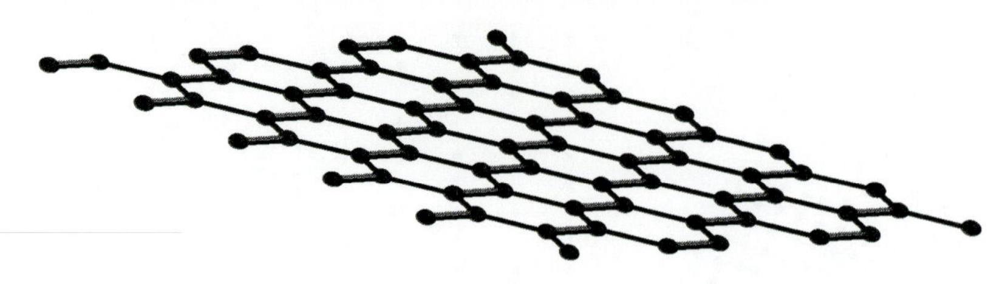

Figure 1.14 The crystal structure of graphene: a two-dimensional crystal formed by sp^2-hybridized carbon atoms.

0.1421 nm; (b) the σ bonds between the carbon atoms within each layer are stronger than those in diamond; and (3) the π bonds between the carbon atoms are distributed (similar to metallic bonding), and this property is responsible for the good electrical conductivity of graphite. Although it was believed that single layers of graphene could not be stable, this was recently proved wrong by a successful exfoliation of films consisting of single and few layers of graphene.[1] Study of these graphene films—the thinnest material that can be made out of atoms—revealed very high electrical conductivity and mechanical strength. In addition, it was shown that the films' electrical conductivity can be modulated by applying an electric field. This field effect, which underpins the operation of the most important semiconductor devices, is not possible in three-dimensional metals because the abundance of free electrons in metals screens the electric field to atomic distances. Although commercial graphene-based devices do not yet exist, the extraordinary electrical and mechanical properties of the material have unleashed extensive research activities with respect to many possible applications, including transistors, because of the very high electrical conductivity and the ability to modulate it by electric field; gas sensors, because electrical conductivity is affected by the smallest number of absorbed molecules; inert coatings, because of graphene's chemical resistance; and support membranes for microscopy, because of the material's mechanical strength.

Graphene sheets can spontaneously roll up to form microscopic tubes known as carbon nanotubes. Carbon nanotubes have been under intensive research since the 1991 paper by Iijima,[2] well before the demonstration of graphene sheets themselves. Figure 1.15 illustrates a graphene sheet that would be obtained if a carbon nanotube were cut and unfolded. The vectors $\boldsymbol{a}_1$ and $\boldsymbol{a}_2$ are the graphene lattice vectors: each node of the lattice can be presented by a vector $n\boldsymbol{a}_1+m\boldsymbol{a}_2 \equiv (n, m)$, where n and m are integers. The vector $\boldsymbol{c}_h$ shown in Fig. 1.15 uniquely determines the diameter and the helicity of a carbon nanotube; this vector connects crystallographically identical nodes in the carbon nanotube, meaning that the nodes (0,0) and (5,2) in this example are identical when the tube is rolled up.

The helicity of a carbon nanotube depends on the angle between the lattice vector $\boldsymbol{a}_1$ and the roll-up vector $\boldsymbol{c}_h$. The dashed lines show the two extreme values for this angle: 0° for $\boldsymbol{c}_h = (n, 0)$ and 30° for $\boldsymbol{c}_h = (n, n)$. Carbon nanotubes are called *zigzag* when $\boldsymbol{c}_h = (n, 0)$, *armchair* when $\boldsymbol{c}_h = (n, n)$, and *chiral* when the angle between $\boldsymbol{c}_h$ and $\boldsymbol{a}_1$ is between 0° and 30°, as illustrated in Fig. 1.16.

---

[1] K. S. Novoselov *et al.*, Electric field effect in atomically thin carbon films, *Science*, vol. 306, pp. 666–669 (2004).

[2] S. Iijima, Helical microtubules of graphitic carbon, *Nature*, vol. 354, pp. 56–58 (1991).

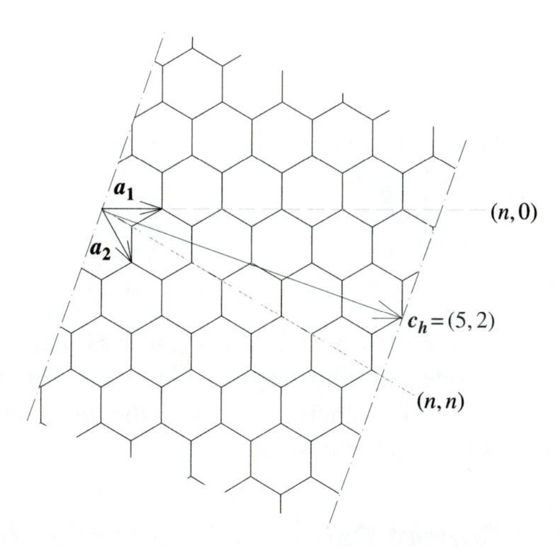

Figure 1.15 Schematic presentation of a two-dimensional graphene sheet obtained by cutting a chiral carbon nanotube along the dashed-dotted line.

Figure 1.16 Side and top views of (a) zig-zag, (b) armchair, and (c) chiral carbon nanotubes.

Carbon nanotubes can also be concentrically inserted into one another, like the rings of a tree trunk, in which case they are called multiwall nanotubes to distinguish them from the single-walled variety. The diameter of carbon nanotubes is usually in the order of nanometers, whereas the length can be in the order of centimeters.

An inherent and very interesting property of carbon nanotubes is that they can provide perfect crystalline structure between two electrical contacts at the tube ends without any defects due to surface termination of the atomic bonds. Similar to graphene, carbon nanotubes also exhibit very high electrical conductivity and mechanical strength, properties that make them attractive for a number of different applications. In addition, carbon nanotubes can appear as either metallic or semiconductive, depending on the helicity. Although this is a positive aspect in terms of possible applications, it has also been the biggest issue in terms of practical use of carbon nanotubes because there is no method as yet for controlling the helicity of nanotubes during their growth.

1.2 CURRENT CARRIERS

The general electronic properties of semiconductors can be explained by using a simplified two-dimensional crystal model, as illustrated in Fig. 1.17. This representation shows positively charged silicon cores (four charge units) with the four valence electrons forming the covalent bonds in a simple quadratic structure.

1.2.1 Two Types of Current Carrier in Semiconductors

An important thing to note from Fig. 1.17 is that all the electrons are bound through covalent bonds only at the temperature of 0 K. There can be a certain number of broken covalent bonds at temperatures higher than 0 K. This happens when the heat energy absorbed by a silicon atom is released through breakage of a covalent bond and release of a free electron that carries the energy away. The energy needed to break a covalent bond in silicon crystal is about 1.1 eV at room temperature, and it is slightly different at different temperatures. Obviously, there will be more broken covalent bonds at higher temperatures because the silicon atoms possess more thermal energy, which eventually destroys covalent

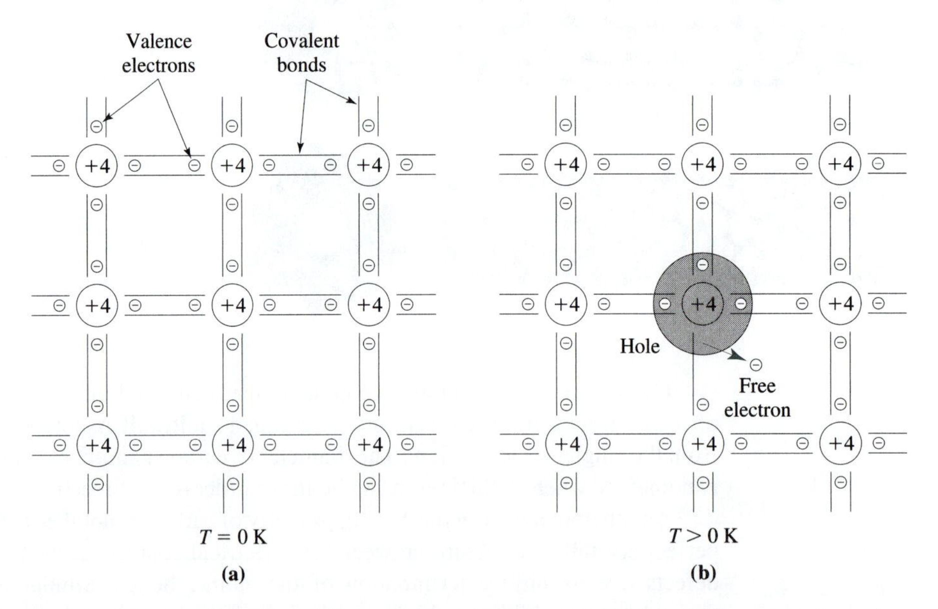

Figure 1.17 Two-dimensional representation of silicon crystal: (a) all the electrons are bound at 0 K, whereas at temperatures >0 K there are broken bonds, creating free electron–hole pairs (b).

bonds. When a silicon atom releases an electron, it becomes positively charged with a hole in its bond structure, as illustrated in Fig. 1.17b.

The electrons released from covalent bonds broken by thermal energy (Fig.1.17) are mobile charged particles, so they can flow through the crystal to create electric current. Accordingly, we define the *free electrons* as current carriers.

In addition to the free electrons, it is necessary to consider what happens with the holes in the bond structures of the silicon atoms that released the free electrons (Fig. 1.17). A silicon atom with a hole in its bond structure is unstable, and it will "use" any opportunity to "steal" an electron from a neighboring atom to rebuild its four covalent bonds. If there is an electric field applied in the crystal, as shown in Fig. 1.18, the field would help this atom to take an electron over from a neighboring atom, leaving the neighboring atom with a hole in its bond structure. We can say that the hole has moved to the neighboring atom, as illustrated in Fig. 1.18. Obviously, there is no reason why this new silicon atom should stay with the hole for much longer than the first atom; it can equally well use the field's help to take an electron over from its neighbor. As a consequence, the hole has moved one step further in the direction of the field direction, a process illustrated again in Fig. 1.18. As this process continues, we get an impression of the hole as a positive charge moving in the crystal in the direction of the electric field lines.

To simplify our presentation of current mechanisms in semiconductors, it is not necessary to constantly keep in mind the details of the process of hole motion as just described; we can simply treat the hole as a positively charged current carrier. Therefore, we introduce the model of two types of current carriers in semiconductors: (1) negatively charged free electrons and (2) positively charge holes, which represent the motion of the bound electrons.

Quite obviously, the electrical properties of a semiconductor material directly depend on the number of current carriers per unit volume, which is the *concentration* of carriers. It may seem that the concentration of free electrons (n_0) has to be always equal to the concentration of holes (p_0), given that the process of covalent-bond breakage creates the free electrons and holes in pairs (Fig. 1.17). This is certainly true for the case of an *intrinsic semiconductor,* which is a semiconductor crystal containing only the native atoms. Even though it is not possible in reality to obtain an ideally pure semiconductor, the intrinsic-semiconductor model is very useful for explaining semiconductor properties. The concentration of free electrons and holes in an intrinsic semiconductor is denoted by n_i and is called *intrinsic carrier concentration*. Therefore, in an intrinsic semiconductor we have

$$n_0 = p_0 = n_i \tag{1.2}$$

The intrinsic carrier concentration n_i is constant for a given semiconductor at a given temperature. The room-temperature values for Si, GaAs, and Ge are given in Table 1.2.[3] If the temperature is increased, the atoms possess more thermal energy, because of which more covalent bonds are destroyed; therefore, the concentration of free electrons and

---

[3]M. A. Green, Intrinsic concentration, effective densities of states, and effective mass in silicon, *J. Appl. Phys.*, vol. 67, pp. 2944–2954 (1990).

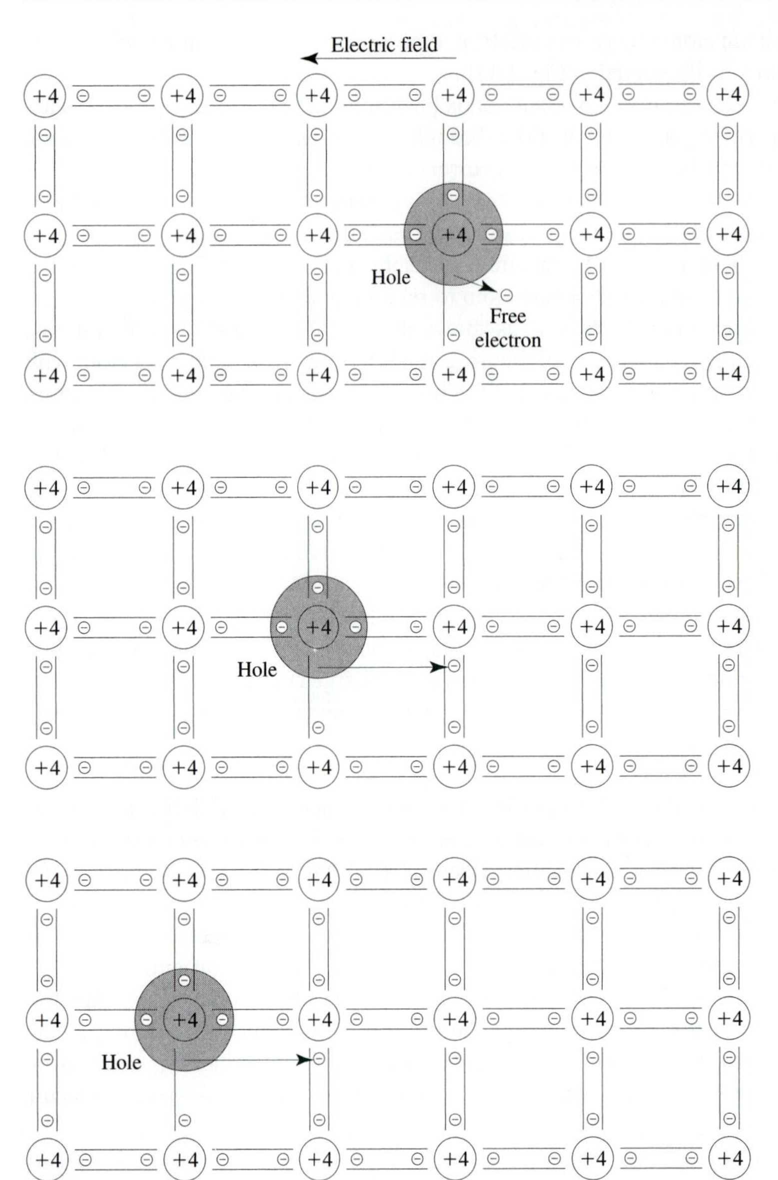

Figure 1.18 Model of hole as a mobile carrier of positive charge.

holes (n_i) is higher. More detailed consideration of the dependence of the intrinsic carrier concentration on temperature is a subject of Section 2.4. At this stage, it should be pointed out that not only the temperature can destroy the covalent bonds (thereby creating free electrons and holes), but other types of energy can do the job as well. An important example is *light*. If the surface of a semiconductor is illuminated, the absorbed photons of the light can transfer their energy to the electrons in the crystal, a process in which the covalent bonds can be destroyed and pairs of free electrons and holes can be generated. This is the

TABLE 1.2 Intrinsic Carrier Concentrations @ 300 K

Si	1.02×10^{10} cm^{-3}
GaAs	2.1×10^{6} cm^{-3}
Ge	2.4×10^{13} cm^{-3}

effect that makes the use of semiconductor-based devices as light detectors and solar cells possible.

1.2.2 N-Type and P-Type Doping

To create useful microelectronic devices, the carrier concentration in semiconductors is changed by technological means. Semiconductors with technologically changed concentrations of free electrons and/or holes are called *doped semiconductors*. In the process of semiconductor doping, some silicon atoms are replaced by different types of atoms, which are called *doping atoms* or *impurity atoms*. In a real silicon crystal there are many silicon atom places that are occupied by impurity atoms. Useful properties are obtained when some silicon atom places are taken by atoms from the third or the fifth column of the periodic table (Table 1.1).

Consider first the case of a silicon atom replaced by an atom from the fifth column of the periodic table, say phosphorus, as shown in Fig. 1.19a. The phosphorus atom has five electrons in the outer orbit. To replace a silicon atom in the lattice, it will use four electrons to form the four covalent bonds with the four neighboring silicon atoms, as

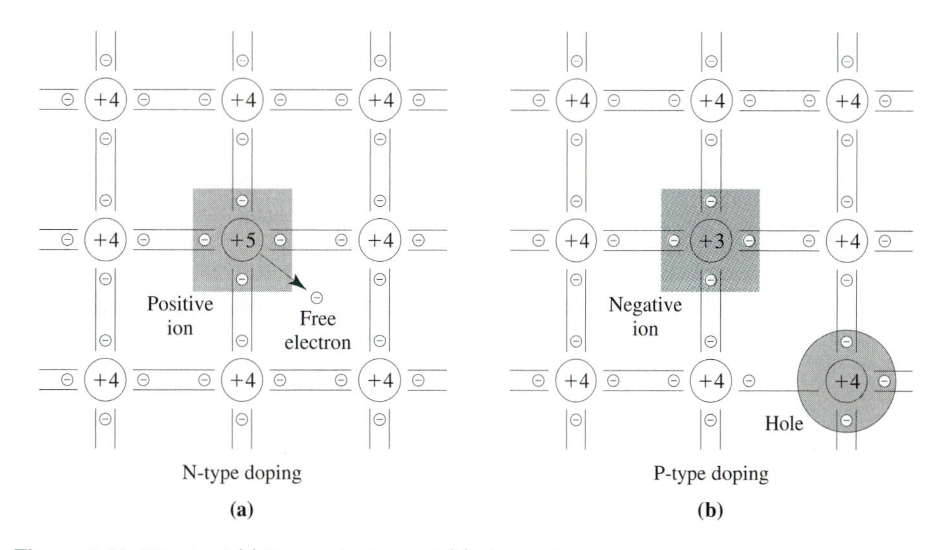

Figure 1.19 Effects of (a) N-type doping and (b) P-type doping.

shown in Fig. 1.19a. The fifth electron does not fit into this structure; it would not be able to find a comfortable place around the parent atom, and with a little help from thermal energy it would leave the phosphorus atom "looking for a better place." In other words, the fifth phosphorus electron is easily liberated by the thermal energy, by which it becomes a free electron. This electron cannot be distinguished from the free electrons produced by covalent-bond breakage. Therefore, a replacement of a silicon atom by a phosphorus atom produces a free electron, increasing the concentration of free electrons n_0. This process is called *N-type doping*. The elements from the fifth group of the periodic table that can produce free electrons when inserted into the silicon crystal lattice are phosphorus, arsenic, and antimony. These doping elements are called *donors* (they donate electrons). In silicon, almost every donor atom produces a free electron at room temperature, but at low temperatures and in the case of some semiconductors at room temperature, a significant fraction of the doping atoms may remain inactive.[4] The concentration of the donor atoms that do produce free electrons will be denoted by N_D.

Although the free electrons in N-type semiconductors are indistinguishable from each other, we know that they can come from two independent sources: thermal generation and doping atoms. As shown in Table 1.2, thermal generation produces about 10^{10} electrons/cm^3 in the intrinsic silicon. Doping can, however, introduce as many as 10^{21} electrons/cm^3. In fact, the lowest concentration of electrons that can be introduced by doping in a controllable way is not much smaller than 10^{14} electrons/cm^3. These numbers show that the concentration of thermally generated electrons is negligible when compared to the concentration of doping induced electrons in an N-type semiconductor.

N-type doping does not produce free electrons only—generation of every free electron in turn creates a positive ion. There is usually the same number of positive and negative charges in a crystal, to preserve its overall *electroneutrality*. When a phosphorus atom liberates its fifth electron, it remains positively charged, as illustrated in Fig.1.19a. There is an important difference between this positive charge and the positive charge created as a broken covalent bond (Fig. 1.17b). The positive charge of Fig. 1.17b is due to a missing electron, or a hole in the bond structure of a silicon atom. There is no hole in the bond structure that phosphorus (or another fifth-group element) creates with the neighboring silicon atoms; the structure is stable as all four bonds are satisfied. The positive charge created due to the absence of the fifth (extra) phosphorus electron will be neither neutralized nor moved to a neighboring atom. This charge appears as a fixed positive ion. The immobile positive ions are not current carriers because they do not contribute to a current flow. They cannot be forgotten, however, because they create important effects in semiconductor devices by their charge and the associated electric field. To distinguish from holes and electrons, squares are used as symbols for the immobile ion charge.

A summary of the types of charges appearing in a semiconductor is given in Table 1.3.

P-type doping is obtained when a number of silicon atoms are replaced by atoms from the third column of the periodic table, basically boron atoms. Boron atoms have three electrons in the outer shell, which are taken to create three covalent bonds with the

[4]This effect will be considered in more detail in Chapter 2 (Examples 2.15 and 2.16).

TABLE 1.3 Types of Charge in Doped Semiconductors

Current Carriers (Mobile Charge)	
Free electrons (n_0)	
Holes (p_0)	
Fixed Charge	
Positive donor ions (N_D)	
Negative acceptor ions (N_A)	
N Type	**P Type**
Majority carriers: electrons	Majority carriers: holes
Minority carriers: holes	Minority carriers: electrons
($N_D > N_A$)	($N_A > N_D$)

neighboring silicon atoms. Because an additional electron per boron atom is needed to satisfy the fourth covalent bond, the boron atoms will capture thermally generated free electrons (generated by breakage of covalent bonds of silicon atoms), creating in effect excess holes as positive mobile charge (Fig. 1.19b).

The boron atoms that accept electrons to complete their covalent-bond structure with the neighboring silicon atoms become negatively charged ions—that is, immobile charges analogous to the positively charged donor atoms in the case of N-type doping. Boron atoms, or, in general, P-type doping atoms, are called *acceptors*. Similarly to the case of N-type doping, almost all boron (acceptor) atoms are ionized at room temperature, which means they have taken electrons creating mobile holes. The concentration of acceptor ions (acceptor atoms that create holes) will be denoted by N_A.

1.2.3 Electroneutrality Equation

As Table 1.3 illustrates, the concentrations of the four types of charges in a semiconductor—electrons, holes, positive donor ions, and negative acceptor ions—are labeled by n_0, p_0, N_D, and N_A. In normal conditions, a semiconductor material is electroneutral, meaning that the concentrations of the negative and the positive charges are fully balanced. This is expressed by the electroneutrality equation:

$$p_0 - n_0 + N_D - N_A = 0 \tag{1.3}$$

EXAMPLE 1.6 Simplified Electroneutrality Equation

(a) Identifying the terms in the electroneutrality equation that can be neglected, determine the concentration of electrons in silicon doped with 10^{15} cm^{-3} donor atoms.

(b) Determine the concentration of holes if the silicon crystal considered in part (a) is additionally doped with 10^{17} cm^{-3} acceptor atoms.

SOLUTION

(a) As almost every donor atom is ionized (by giving an electron away), the concentration of the donor ions is $N_D \approx 10^{15}$ cm^{-3}. There are no acceptor ions, so $N_A = 0$. The concentration of thermally induced holes in the intrinsic silicon is about 10^{10} cm^{-3}. This is already negligible in comparison to the positive donor ions, and in N-type silicon the concentration of holes is below the intrinsic level (this effect will be explained in the next section). Therefore,

$$\underbrace{p_0}_{\ll N_D} - n_0 + N_D - \underbrace{N_A}_{=0} = 0$$

$$n \approx N_D = 10^{15} \text{cm}^{-3}$$

(b) In this case, $N_A \gg N_D$, N_D can be neglected. The concentration of electrons is at about the same level as N_D, so n_0 can be neglected as well:

$$p \approx N_A = 10^{17} \text{ cm}^{-3}$$

1.2.4 Electron and Hole Generation and Recombination in Thermal Equilibrium[5]

The concentration of minority carriers is much smaller than to the concentration of the majority carriers, and sometimes we can simply neglect their existence. However, the minority carriers have opposite polarity from the majority carriers, so they may produce effects that are different from the effects of the majority carriers. Because the effects of the minority and the majority carriers are not always merged, it is sometimes important to determine the concentrations of both types of carriers. We already know that the concentration of majority carriers is basically set by their generation from the doping atoms. A deeper consideration of the generation and the associated recombination processes is needed to be able to determine the concentration of minority carriers.

Generation of free electrons and holes is a process in which bound electrons are given enough energy to (1) liberate themselves from silicon atoms, creating electron–hole pairs (Fig. 1.17), (2) liberate themselves from donor-type atoms, creating free electrons and fixed positive charge (Fig. 1.19a), or (3) liberate themselves from silicon atoms to provide only the fourth bond of acceptor-type atoms, creating mobile holes and negative fixed charge (Fig. 1.19b). The doping-induced generation of electrons/holes is obviously limited to the level of doping—no more electrons/holes can be generated once all the doping atoms have been ionized. What does, however, limit the process of thermal generation

[5]For detailed description and modeling of recombination–generation processes, see Chapter 5.

of electron–hole pairs due to breakage of covalent bonds? Would such created electrons and holes accumulate in time to very high concentrations?

A free electron that carries energy taken from the crystal lattice can easily return this energy to the lattice and bond itself again when it finds a silicon atom with a hole in its bond structure. This process, called *recombination,* results in annihilation of free electron–hole pairs. Obviously, if the concentration of free electrons is higher, it is more likely that a hole will be "met" by an electron and recombined. Similarly, an increase in the hole concentration also increases the recombination rate—that is, the concentration of electron–hole pairs recombined per unit time. It is the recombination rate that balances the generation rate, limiting the concentration of free electrons and holes to a certain level. If the generation rate is increased—for example, by an increase in the temperature—the resulting increase in concentration of free electrons and holes will automatically make the recombination events more probable; therefore the recombination rate is automatically increased to the level of the generation rate. *In thermal equilibrium, the generation rate is equal to the recombination rate.* These rates are, however, different at different temperatures. At 0 K they are both equal to 0, as is the concentration of free electrons and holes; with increase in the temperature, the recombination and generation rates are increased, which results in increased concentrations of free electrons and holes.

Doping influences the recombination and generation rates as well. Take as an example an N-type semiconductor where the concentration of free electrons is increased by doping-induced electrons. We have already concluded that an increased concentration means it is more likely that a hole will be "met" by an electron and recombined. Consequently, the concentration of holes in an N-type semiconductor is smaller than in an intrinsic semiconductor, because the holes are recombined not only by the thermally generated electrons but also, and much more, by the doping-induced electrons. Therefore, with an increase in the concentration of electrons, the concentration of holes is reduced. This dependence can be expressed as

$$p_0 = C\frac{1}{n_0} \tag{1.4}$$

where the proportionality coefficient C is a temperature-dependent constant. This equation is written for N-type semiconductors, but it can also be applied to P-type semiconductors, as well as intrinsic semiconductors. It can be rewritten as $n_0 = C/p_0$ for P-type semiconductors, but it is essentially the same equation. The constant C can be determined if Eq. (1.4) is applied to the case of intrinsic semiconductor. Because $n_0 = p_0 = n_i$ in that case, we see that the constant C is given as

$$C = n_i^2 \tag{1.5}$$

Using the preceding value for the constant C, Eq. (1.4) becomes

$$n_0 p_0 = n_i^2 \tag{1.6}$$

Equation (1.6), when combined with the electroneutrality equation, enables calculation of the concentration of minority carriers in doped semiconductors.

EXAMPLE 1.7 Calculating the Concentration of Minority Carriers

The doping level of P-type silicon is $N_A = 5 \times 10^{14}$ cm^{-3}. Determine the concentrations of holes and electrons at

(a) room temperature ($n_i = 1.02 \times 10^{10}$ cm^{-3})
(b) $T = 273°$C ($n_i = 5 \times 10^{14}$ cm^{-3}, the same as the doping level)

SOLUTION

(a) As $n_0 \ll N_A$ at room temperature, the electroneutrality equation can be simplified to (refer to Example 1.6)

$$p_0 \approx N_A = 5 \times 10^{14} \text{ cm}^{-3}$$

Once the concentration of the majority carriers is known, the concentration of the minority carriers can be determined from the $n_0 p_0 = n_i^2$ relationship:

$$n_0 = \frac{n_i^2}{p_0} = 2.1 \times 10^5 \text{ cm}^{-3}$$

(b) In this case we cannot assume that $p_0 \approx N_A$, because the concentrations of the thermally generated electrons and holes cannot be neglected. Both n_0 and p_0 have to appear in the electroneutrality equation:

$$p_0 = N_A + n_0$$

We have two unknown quantities (p_0 and n_0), but we have an additional equation that relates them:

$$n_0 p_0 = n_i^2$$

Therefore,

$$p_0 = N_A + n_i^2 / p_0$$
$$p_0^2 - N_A p_0 - n_i^2 = 0$$
$$p_{0-1,2} = \frac{N_A \pm \sqrt{N_A^2 + 4n_i^2}}{2}$$

We will select the plus sign in the above equation, because $\sqrt{N_A^2 + 4n_i} > N_A$ and the minus sign would give physically meaningless $p_0 < 0$. Once the numerical value for p_0

has been determined, n_0 is calculated as $n_0 = n_i^2/p_0$. The following script can be used to perform the calculations in MATLAB®:

```
>>ni=5e14;
>>Na=5e14;
>>p=(Na+sqrt(Na^2+4*ni^2))/2
p =
    8.0902e+014
>>n=ni^2/p
n =
    3.0902e+014
```

Therefore, $p_0 = 8.09 \times 10^{14}$ cm^{-3}, $n_0 = 3.09 \times 10^{14}$ cm^{-3}.

EXAMPLE 1.8 A Question Related to the Number of Minority Carriers

(a) P-type silicon can be doped in the range from 5×10^{14} cm^{-3} to 10^{20} cm^{-3}. Determine the maximum possible number of minority electrons in a neutral P-type region if the device area is limited to $A_D = 1$ cm $\times$ 1 cm and the thickness of the P-type region is limited to $t_P = 100\ \mu$m. Assume room temperature and full acceptor ionization.

(b) Determine the number of minority electrons in the lowest- and highest-doped P-type region ($N_A = 5 \times 10^{14}$ cm^{-3} and $N_A = 10^{20}$ cm^{-3}, respectively) if the area of the P-type region is reduced to $A_D = 100\ \mu$m $\times$ 100 μm and the thickness is reduced to $t_P = 10\ \mu$m.

SOLUTION

(a) The maximum concentration of minority carriers is obtained for the minimum doping level:

$$n = \frac{n_i^2}{N_A} = \frac{(1.02 \times 10^{10})^2}{5 \times 10^{14}} = 2.1 \times 10^5 \text{ cm}^{-3} = 2.1 \times 10^{11} \text{ m}^{-3}$$

The maximum volume is

$$V = A_D t_P = (0.01)^2 \times 100 \times 10^{-6} = 10^{-8} \text{ m}^3$$

The maximum number of minority electrons in this volume is

$$N = nV = 2.1 \times 10^3$$

(b) The concentration of the minority electrons in the lowest-doped P-type region is the same: $n = 2.1 \times 10^5\ \mathrm{cm}^{-3} = 2.1 \times 10^{11}\ \mathrm{m}^{-3}$. The volume of the P-type region is

$$V = A_D t_P = (100 \times 10^{-6})^2 \times 10 \times 10^{-6} = 10^{-13}\ \mathrm{m}^3$$

Therefore, the number of minority electrons in the lowest-doped P-type region is

$$N = nV = 2.1 \times 10^{11} \times 10^{-13} = 0.021\ !?$$

The concentration of the minority electrons in the highest-doped P-type region is

$$n = \frac{n_i^2}{N_A} = \frac{(1.02 \times 10^{10})^2}{10^{20}} = 1.04\ \mathrm{cm}^{-3} = 1.04 \times 10^6\ \mathrm{m}^{-3}$$

Therefore, the number of minority electrons in the highest-doped P-type region is

$$N = nV = 1.04 \times 10^6 \times 10^{-13} = 1.04 \times 10^{-7}\ !?$$

What is the meaning of 0.021 **and** 10^{-7} **electron?** This question will be considered in Section 10.1.

*1.3 BASICS OF CRYSTAL GROWTH AND DOPING TECHNIQUES

This is a *read-only* section that completes this chapter by providing brief descriptions of the techniques for crystal growth and semiconductor doping.

1.3.1 Crystal-Growth Techniques

It was mentioned that certain atoms pack spontaneously into a regular pattern when proper conditions are met. The reason for this was described as a tendency toward stability. The same phenomenon is frequently explained in terms of energy, where the reason for creating regular patterns is described as a tendency toward the minimum energy of the system. As far as the necessary conditions are concerned, it is essential that many atoms of the desired element be brought together; ideally, there should be no alien (impurity) atoms. Further conditions relate to favorable ambient parameters, such as temperature. The necessary conditions for growth of the common semiconductors do not appear naturally. Therefore, crystal-growth techniques are needed to provide the conditions for the growth of semiconductor crystals for commercial use.

---

*Sections marked with an asterisk can be omitted without loss of continuity.

Bulk Crystals and Wafers

Silicon, the most used semiconductor material, is the second most abundant element in the earth's crust. However, to create silicon crystals, silicon has to be separated from its compounds, the most frequent of which is sand (impure SiO_2, called silica). The separation from oxygen (reduction) is achieved by heating the sand with carbon in an electric furnace. At very high temperatures ($\approx 1800°C$), SiO_2 reacts with carbon to create CO, leaving behind Si. This silicon is not pure enough for electronic applications, so it is reacted with HCl to convert it into $SiCl_4$ or $SiHCl_3$. Both these compounds are liquids, enabling distillation and other liquid purification procedures to be applied, to achieve ultrapure $SiCl_4$ or $SiHCl_3$. These compounds are then converted to high-purity silicon by reacting them with H_2:

$$2SiHCl_3 + 2H_2 \rightarrow 2Si + 6HCl$$

$$SiCl_4 + 2H_2 \rightarrow Si + 4HCl$$

Bulk silicon obtained in this way is in the polycrystalline form. To convert polycrystalline bulk silicon into single-crystal *ingots* or *boules,* the material is heated in an inert atmosphere in a graphite crucible to create a silicon melt. A seed crystal (a small single crystal) is carefully aligned along a desired direction (typically $\langle 100 \rangle$) and brought into contact with the melt. As the temperature of the melt that is in contact with the seed is reduced, the melt crystallizes following the pattern of the seed. The seed together with the formed crystal is then slowly rotated and pulled out, creating the conditions for continuing crystallization of the molten material. This technique for growing single-crystal materials is called the Czochralski method. It results in cylindrical ingots, which in the case of silicon can be 300 mm in diameter and over 1 m long.

If the concentration of the impurities in the ingot is too high, it can be further purified by *zone refining*. The technique utilizes an effect called *segregation*. The effect is that there is usually a difference between the concentrations of impurities in solid (C_S) and liquid (C_L) phases of certain material. If the ratio of these concentrations (called the *segregation coefficient*) is <1, the impurities will tend to accumulate in the liquid layer, leaving behind purified crystal.

To prepare a monocrystalline semiconductor for device and integrated-circuit fabrication, the ingot is mechanically shaped into a perfect cylinder and then sliced by a diamond saw into *wafers*. The thickness of the wafers has to be sufficient to allow wafer handling. The wafers are lapped and ground on both sides, and finally one side is mechanically and chemically polished to prepare it for the fabrication process.

Monocrystalline and Polycrystalline Layers

Thin crystal layers can be grown on a wafer of the same material or different material with a compatible crystal lattice. These layers are called *epitaxial layers*. There are different reasons for growing epitaxial layers. Perhaps the most obvious is the creation of structures consisting of layers of different semiconductor materials. Layers of different semiconductors can be grown on one another if the crystal-lattice structure is compatible and there is no large mismatch between the crystal-lattice constants. A typical example is growing GaAs and AlGaAs films on each other to achieve useful effects by combining the properties of GaAs and AlGaAs. Another reason for growing epitaxial layers is to create active layers with reduced concentration of defects or doping level. For many devices, it is desirable to have highly doped substrate material with low-doped layers on the surface. This is achieved by epitaxial growth.

The most frequent epitaxial technique is *chemical-vapor deposition* (CVD). In this technique, the atoms or molecules of the desired material are chemically created inside a processing chamber and then they fall down to coat the wafer(s). Obviously, this process relies on an appropriate chemical reaction that is initiated by providing the reacting gases and energy. In standard CVD, the energy is supplied in the form of heat, although the reaction energy may also be supplied by plasma or optical excitation, in which case the processes are referred to as *plasma* and *rapid-thermal processing,* respectively. Examples of chemical reactions used to create silicon are

$$SiH_4 \rightarrow Si + 2H_2 \tag{1.7}$$

$$SiCl_4 + 2H_2 \rightarrow Si + 4HCl \tag{1.8}$$

Another technique for growing epitaxial layers is called *molecular-beam epitaxy.* Molecular or atomic beams of different elements are created by evaporation from separate cells. The beams are then directed to the surface of the wafer, which is held in very high vacuum and at elevated temperature (400–800°C). By selecting desired beams (using shutters in front of the cells) and by controlling their intensity, abrupt changes in doping levels or sharp transitions from one material to another can be achieved. For example, a monolayer transition from GaAs to AlGaAs and vice versa can be achieved by molecular-beam epitaxy.

The layer-deposition techniques—in particular, CVD—are also used to deposit noncrystalline but still very useful layers. A frequent example is the deposition of insulating layers, such as SiO_2 and Si_3N_4. The following chemical reactions can be used to create SiO_2 and Si_3N_4 in the CVD chamber:

$$SiH_4 + O_2 \rightarrow SiO_2 + 2H_2 \tag{1.9}$$

$$3SiH_4 + 4NH_3 \rightarrow Si_3N_4 + 12H_2 \tag{1.10}$$

If the CVD process is used to deposit silicon on these noncrystalline films, the silicon layer takes polycrystalline form. Although this silicon film on an insulator is not monocrystalline, it still has very useful properties. It is a commonly used film, and it is usually called *polysilicon.*

1.3.2 Doping Techniques

Section 1.2.2 explained that doping is achieved by replacing some atoms of the native crystal by *impurity* (*doping*) atoms. This can be achieved during crystal growth by simply providing the impurity atoms to the melt or the gas that is used for the crystal growth. The following text briefly describes the techniques that can be used to dope a semiconductor crystal once it has been grown.

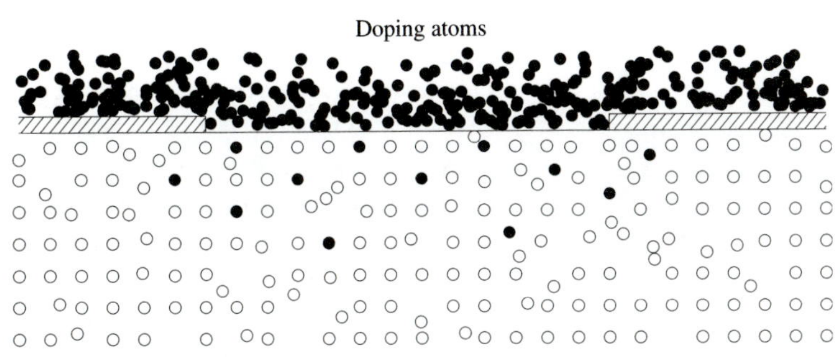

Figure 1.20 Doping of a semiconductor by diffusion.

Diffusion

As will be described in Chapter 4, the thermal random motion of particles causes them to effectively flow from the points of higher toward the points of lower particle concentrations. The tendency of diffusion is to reach uniform concentration of the particles. Diffusion is not limited to gases; it happens in liquids and solids as well, although a very high temperature is typically needed for diffusion of atoms to be clearly observed in solids. The semiconductor crystal has to be heated to about 1000°C, so that a sufficient number of semiconductor atoms are released from their crystal-lattice positions, as illustrated in Fig. 1.20. The semiconductor atoms that leave their crystal-lattice positions are called *interstitials,* whereas the empty positions left behind are called *vacancies.* The doping atoms that are provided at the surface of the semiconductor can diffuse into the semiconductor at this high temperature and place themselves into the created vacancies, thereby taking crystal-lattice positions. When the semiconductor is cooled down to room temperature, the thermal motion of semiconductor and doping atoms becomes insignificant; therefore, the doping atoms will stay "frozen" in their positions. The doped semiconductor layer created in this way expands only to a certain depth and appears roughly under the provided window in the diffusion-protective "wall." By using diffusion at a high temperature, a doped semiconductor layer having a desired depth, length, and width can be created.

Diffusion of acceptors into N-type substrate, or donors into P-type substrate, will create P–N junctions that surround the doped regions created by the diffusion. Assume that the concentration of donor atoms in an N-type substrate is $N_D = 10^{16}$ cm^{-3}, as illustrated in Fig. 1.21 by the solid line. The concentration of the donors is uniform throughout the substrate because this doping is performed while the crystal substrate is being grown. To create a P-type region, boron diffusion is performed. The boron diffusion will create a nonuniform doping profile, as most of the boron atoms incorporated by the diffusion in the silicon will remain at the surface; going deeper into the substrate, a lower concentration of boron atoms will be found (dashed line in Fig. 1.21). If a higher concentration of boron atoms is achieved at the surface (say $N_A = 10^{18}$ cm^{-3}), the acceptor-type doping atoms will prevail and the surface of the silicon will appear as P type. Figure 1.21 shows that there is a point where $N_A = N_D$. This point is called the *P–N junction*—on one side of the junction the semiconductor is a P type, whereas on the other side of the junction it is an N type.

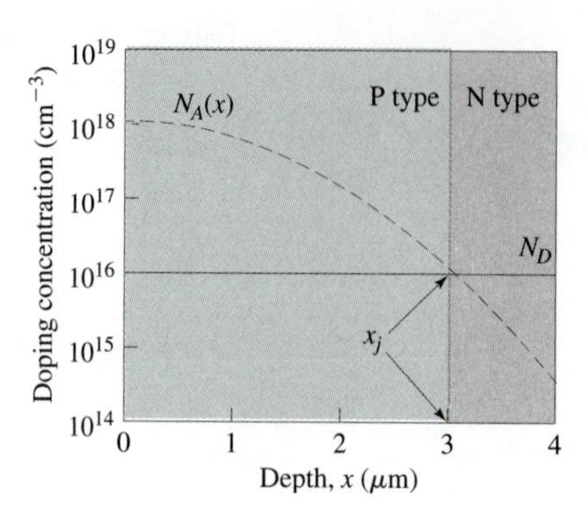

Figure 1.21 Diffusion of acceptors into N-type substrate creates a P–N junction: P-type semiconductor between the surface and x_j, and N-type semiconductor from x_j into the bulk of the substrate.

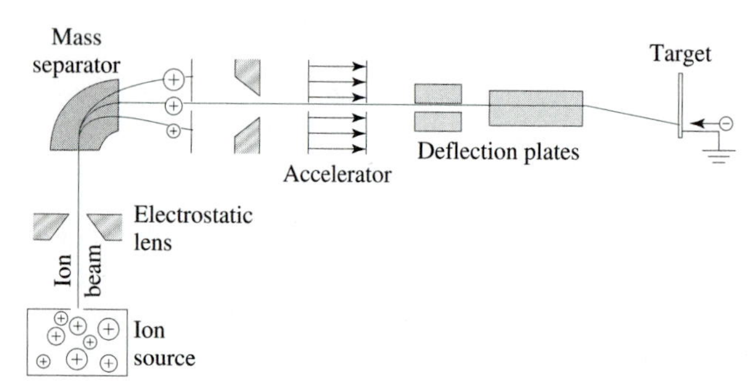

Figure 1.22 Ion implanter diagram.

Ion Implantation

The ion implanter diagram of Fig. 1.22 illustrates the ion-implantation process. The process begins with gas ionization, which creates an ion mixture containing the ions of the doping element. The desired ions are separated according to their atomic mass in a mass separator, the ion beam is then focused, and the ions are accelerated to the desired energy (typically between 10 keV and 200 keV). The ion beam is scanned across the wafer to achieve uniform doping. When the ions of the doping element hit the wafer, they suffer many collisions with the semiconductor atoms before eventually stopping at some depth beneath the surface. As the target is grounded, to complete the electric circuit, the implanted ions are neutralized by electrons flowing into the substrate.

Although all the beam ions have the same energy, they do not stop at the same distance, since the stopping process involves a series of random events. This is illustrated in Fig. 1.23a. Therefore, the ion-implantation process leads to a bell-shaped profile of the doping atoms, as shown in Fig. 1.23b.

As already explained, the doping atoms are electrically active only when they replace semiconductor atoms from their positions in the crystal lattice. The implanted atoms generally terminate in interstitial positions. Also, a number of semiconductor atoms will be displaced from their positions, which results in damage to the semiconductor crystal. Because of that, a post-implant anneal must be performed. This anneal should provide sufficient energy to enable the silicon and doping atoms to rearrange themselves back into the crystal structure. The change of the ion-implant profile during this annealing is illustrated in Fig. 1.23c.

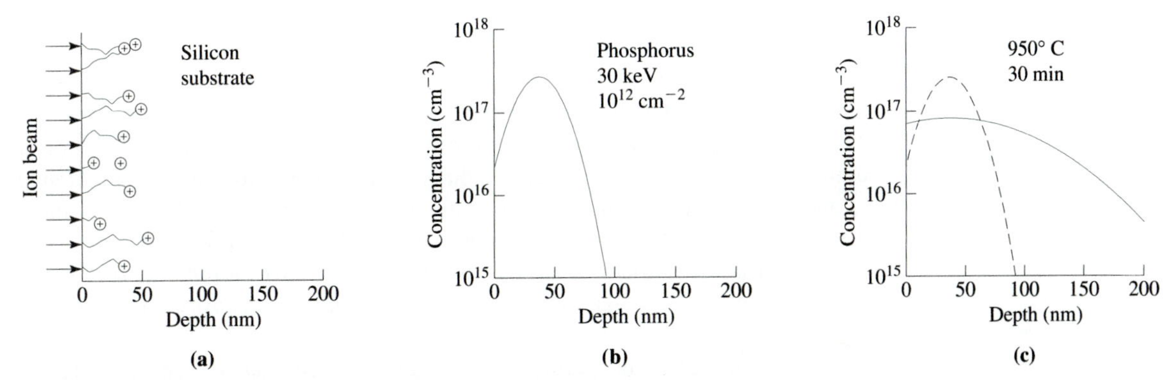

Figure 1.23 (a) Accelerated ions of the doping element collide inside the semiconductor, stopping at scattered depths. The doping profile immediately after ion implantation (b) and after annealing (c).

Two important parameters of the ion-implantation process are the implant energy and the density of implanted ions (dose). The implant energy determines the depth of the implanted ions. The dose (number of implanted ions per unit area) is determined by the implant time and the current of the ion beam. The current of the ion beam can be found by measuring the current of electrons flowing from the ground into the substrate to neutralize the implanted ions. Integrating the measured current in time gives the charge implanted into the substrate, which is divided by the unit charge q and the substrate area to obtain the dose. In this way, the doping level achieved by the ion implantation can very precisely be controlled. This has proved to be one of the most important advantages of ion implantation compared to the diffusion technique. Added flexibility for doping-profile engineering appears as an additional advantage of ion implantation. A disadvantage of ion implantation is the complex equipment, which is reflected in the cost of the doping process.

SUMMARY

1. *Crystals* are solid materials with atoms appearing in a regular pattern. (It is the regular placement of the atoms that enables electrons to flow in solid crystals.)
2. The *bonds* between atoms in a crystal enable increased stability of the electronic structure of individual atoms; that is, the bonds reduce the energy of the system of atoms. The most frequently encountered types of atomic bonds are *ionic*, *metallic*, and *covalent*. Covalent bonds appear in molecules (H_2, CH_4, C_2H_2, ...) in which atom stability can be reached by a small number of atoms, and in insulator/semiconductor crystals, where atom stability requires a periodically repeated atom or a group of atoms.
3. *Crystal lattice* defines the positions of the atoms; it can be obtained by replicating a *unit cell*. Many three-dimensional crystals have *cubic unit cells,* with the side of the cube called the *crystal-lattice constant*. The most common cubic cells are the simple, body-centered, face-centered, and diamond/zincblade unit cells. Each atom

in a diamond/zincblade crystal structure appears in the center of a tetrahedron and creates four identical bonds with the neighbors in the tetrahedron corners; there are four tetrahedra in each unit cell. This crystal structure is due to sp^3 hybridization, which is also observed in the CH_4 molecule.

4. *The atom-packing fraction* is the fraction of a unit-cell volume filled with atoms, assuming that the atoms are hard spheres that just touch the nearest neighbors. *Atomic mass* and *volume* are the mass and the volume of 1 mole of material, where 1 mole of material consists of 6.02×10^{23} atoms or molecules (Avogadro's number). *Atom concentration* is the number of atoms per unit volume; *surface atom density* is the number of atoms per unit area (of a specified plane); and *mass density* is the mass per unit volume.
5. The equation for a plane in a cubic crystal, with lattice constant a, can be written as $h(x/a) + k(y/a) + l(z/a) = 1$. The defining coefficients h, k, and l are called Miller indices: (hkl) denotes a specific plane, whereas $\{hkl\}$ denotes a set of equivalent planes. The direction perpendicular to a given plane is also labeled by the same Miller indices, where $[hkl]$ is used for a specific direction, perpendicular to (hkl), whereas $\langle hkl \rangle$ is used for a set of equivalent directions.
6. As distinct from the sp^3 hybridization, sp^2 hybridization (observed in the C_2H_2 molecule) leads to two-dimensional graphene crystal and carbon nanotubes. Carbon nanotubes can be thought of as cylinders rolled from two-dimensional graphene sheets. Both graphene and carbon nanotubes exhibit excellent mechanical strength owing to the strong atomic bonds and some unique electrical properties.
7. There are two types of current carriers in semiconductors: negatively charged *free electrons* and positively charged *holes* (missing electrons in the bonding structure). In an *intrinsic semiconductor,* the free electrons and holes are created in pairs, due to breakage of covalent bonds; the concentrations of free electrons and holes are equal, $n_0 = p_0 = n_i$, where n_i is the intrinsic-carrier concentration. The intrinsic-carrier concentration is temperature-dependent constant for a given semiconductor (n_i increases with temperature, because more covalent bonds are broken at a higher temperature).
8. *Donor atoms* have more valence electrons than are necessary to form the covalent bonds when these atoms replace native atoms in the crystal structure. Given that the extra electrons are easily released as free electrons, *doping* a semiconductor by donor atoms (N-type doping) increases the concentration of free electrons. By N-type doping, the concentration of electrons can technologically be set at $n_0 \approx N_D$, where N_D is the concentration of *ionized donors.*
9. *Acceptor atoms* have fewer electrons than are necessary to form the covalent bonds, so they create holes when they replace native atoms in a semiconductor crystal. This is called P-type doping, and it can be used to set the concentration of holes at $p_0 \approx N_A$, where N_A is the concentration of *ionized acceptor* atoms.
10. In an N-type semiconductor ($N_D > N_A$), the electrons are *majority* carriers, whereas the holes are *minority* carriers; usually, $n_0 \gg p_0$. In a P-type semiconductor ($N_A > N_D$), the holes are majority and the free electrons are the minority carriers.
11. In an electroneutral semiconductor material, the charge of the negative free electrons and positive holes is balanced by the charge of the fixed ions: positive donor ions and

negative acceptor ions. This is expressed by the *electroneutrality equation:*

$$p_0 - n_0 + N_D - N_A = 0$$

12. The process of *thermal generation* of free electrons and holes (thermally induced breakage of covalent bonds) is balanced by the opposite process of electron–hole *recombination*. The recombination process leads to reduction of hole concentration when the concentration of free electrons is increased by doping, and it leads to reduction of electron concentration when the concentration of holes is increased. In thermal equilibrium, the product of electron and hole concentrations is constant:

$$n_0 p_0 = n_i^2$$

13. The Czochralski method (growth from a melt) is a common technique for growing *ingots* (or *boules*) of monocrystalline silicon. The ingot is sliced into wafers that will be used for the fabrication of devices and integrated circuits. Chemical-vapor deposition (CVD) or molecular-beam epitaxy (MBE) can be used to deposit monocrystalline layers on semiconductor wafers (so-called *epitaxial* layers). When CVD is used to deposit silicon on a noncrystalline substrate (such as a film of SiO_2 or Si_3N_4), the deposited film takes polycrystalline form and is called *polysilicon*.

14. At high temperature, *diffusion* of donor and acceptor atoms into a semiconductor is significant and is used as a doping technique. Another doping technique is *ion implantation*.

PROBLEMS

1.1 Find the packing fractions for

(a) body-centered

(b) face-centered **A***

(c) diamond cubic cells

1.2 The crystal-lattice constant of GaAs is $a = 0.565$ nm. Calculate

(a) the distance between the centers of the nearest Ga and As atoms **A**

(b) the distance between the centers of the nearest Ga neighbors

1.3 The atom radius of copper is $r = 0.1278$ nm. Its crystal structure has face-centered cubic cells.

(a) What is the crystal-lattice constant?

(b) What is the concentration of copper atoms?

(c) What is the atomic volume (the volume of 1 mol of copper)? **A**

1.4 The crystal-lattice constant of GaAs is $a = 0.565$ nm, whereas the atomic masses of Ga and As are 69.72 g/mol and 74.91 g/mol, respectively. Determine the mass density of GaAs.

1.5 The atomic mass and the atomic volume of silver are 107.87 g/mol and 10.3 cm^3/mol, respectively. Its crystal structure has face-centered cubic cells.

(a) What is the crystal-lattice constant?

(b) What is the mass density?

1.6 Assuming that the radii of silicon and carbon atoms are $r_{Si} = 0.1175$ nm and $r_C = 0.0712$ nm, respectively, determine the concentration of silicon atoms in the zincblade SiC crystal (3C SiC). **A**

1.7 A 5-kg silicon ingot is doped with 10^{16} cm^{-3} phosphorus atoms. What is the total mass of phosphorus

*Answers to selected problems are provided beginning on page 610. The problems are marked with **A**.

in this ingot? The atomic masses of silicon and phosphorus are 28.09 g/mol and 31.02 g/mol, respectively.

1.8 Identify the correct statement in the following list:

(1) (100): [100], $[\bar{1}00]$, [010], $[0\bar{1}0]$, [001], $[00\bar{1}]$
(2) (100): {100}, {010}, {001}
(3) $\langle 100\rangle$: [100], $[\bar{1}00]$, [010], $[0\bar{1}0]$, [001], $[00\bar{1}]$
(4) $\langle 111\rangle$ is perpendicular to {111}
(5) (110) and (101) are parallel
(6) [100] and [011] are opposite directions

1.9 What is the density of atoms at the surface of a simple cubic crystal, if the crystal is terminated at

(a) {100} plane
(b) {110} plane
(c) {111} plane **A**

The crystal-lattice constant is $a = 0.5$ nm.

1.10 How many silicon atoms per unit area are found in

(a) {110} plane **A**
(b) {111} plane

1.11 Determine the Miller indices for a plane that intersects the x-, y-, and z-axes at $-a$, $2a$, and $-3a$, respectively (a is the crystal-lattice constant).

1.12 At what distances from the origin does the plane (012) intersect the x-, y-, and z-axes if the crystal-lattice constant is 0.5 nm?

1.13 An atom in a simple cubic crystal has six neighbors at distance a. Label the directions toward all the six neighbors by Miller indices. **A**

1.14 There are four equivalent $\langle 110\rangle$ directions that are perpendicular to the z-axis: [110], $[\bar{1}10]$, $[1\bar{1}0]$, and $[\bar{1}\bar{1}0]$. Looking along the z-axis, draw a simple cubic crystal and indicate the four directions that are perpendicular to the z-axis. What is the total number of equivalent $\langle 110\rangle$ directions?

1.15 How many equivalent {111} planes and how many equivalent $\langle 111\rangle$ directions exist in a cubic lattice?

1.16 Looking down the $[\bar{1}00]$ direction of a diamond crystal lattice, draw the two-dimensional position of the atoms. Indicate the crystal-lattice constant and the bonds.

1.17 What is the angle between {100} and {110} planes in silicon crystal? What is the distance between neighboring {110} planes? The crystal-lattice constant is $a = 0.543$ nm.

1.18 Determine the concentration of valence electrons in

(a) Si ($a = 0.543$ nm) **A**
(b) GaAs ($a = 0.565$ nm)

1.19 One electron per crystalline atom in silver is free to conduct electricity (it is shared by all the atoms in the crystal). Determine the concentration of free electrons, knowing that the atomic volume (volume of 1 mol) of silver is 10.3 cm^3/mol.

1.20 The doping level of a sample of N-type GaAs is $N_D = 10^{16}$ cm^{-3}.

(a) What is the concentration of electrons at 0 K?
(b) What is the concentration of holes at 0 K? **A**
(c) What is the net-charge concentration at 300 K?

The possible answers are

(1) 0
(2) 4.41×10^{-6} cm^{-3}
(3) 10, 404 cm^{-3}
(4) 2.1×10^{6} cm^{-3}
(5) 1.02×10^{10} cm^{-3}
(6) 10^{16} cm^{-3}

1.21 The concentration of donor atoms in an N-type semiconductor is $N_D = 10^{16}$ cm^{-3}. Calculate the concentration of minority carriers at room temperature if the semiconductor is

(a) Si
(b) GaAs **A**
(c) Ge

1.22 The doping level of a sample of N-type silicon is $N_D = 10^{16}$ cm^{-3}.

(a) List the types of charge that exist in this sample and determine their concentrations.
(b) What is the net charge concentration?

1.23 Calculate the concentration of holes in a heavily doped silicon having donor concentration $N_D = 10^{20}$ cm^{-3}.

1.24 In a silicon crystal, $N_D = 10^{17}$ cm^{-3} and $N_A = 10^{16}$ cm^{-3}. Find the concentrations of the minority and the majority carriers.

1.25 For N-type semiconductor, doped at the level of $N_D = 10^{16}$ cm^{-3}, determine the average distance between two holes if the semiconductor is

(a) Si
(b) 3C SiC ($n_i \approx 10^{-1}$ cm^{-3})

Assume that the holes are uniformly distributed in space.

1.26 N-type Si, GaAs, and 4H SiC are all doped to the level of $N_D = 10^{17}$ cm^{-3}. Determine the volumes of each of these semiconductors that will on average contain 1 hole. The intrinsic carrier concentrations are $n_i = 1.02 \times 10^{10}$ cm^{-3}, $n_i = 2.1 \times 10^6$ cm^{-3}, and $n_i \approx 10^{-7}$ cm^{-3}, respectively. (A for GaAs)

1.27 The substrate concentration of an N-type semiconductor is $N_D = 10^{15}$ cm^{-3}. The wafer is doped with $N_A = 1.1 \times 10^{15}$ cm^{-3}, so that a very lightly doped P-type region is created at the surface. What is the concentration of electrons in the P-type region if the semiconductor is

(a) Si

(b) Ge A

1.28 The concentration of thermally generated electrons in N-type silicon increases with the temperature. At certain temperature, it is equal to the concentration of the doping-induced electrons (the concentration of the thermally induced electrons is no longer negligible).

(a) If the doping level is $N_D = 10^{15}$ cm^{-3}, what is the concentration of holes? A

(b) What is the intrinsic concentration (n_i) at this temperature?

1.29 P-type silicon substrate, with $N_A = 5 \times 10^{14}$ cm^{-3}, is doped by phosphorus. Determine the concentration of electrons and holes at room temperature if the doping level is $N_D = 10^{15}$ cm^{-3}. What is the concentration of electrons and holes at $T = 273$°C, where the intrinsic concentration is $n_i = 5 \times 10^{14}$ cm^{-3}?

1.30 What concentration of acceptors should be added to an N-type silicon crystal to reduce the effective doping concentration from 10^{18} cm^{-3} to 10^{15} cm^{-3}? How much should the concentration of acceptor atoms change in order for the electron concentration to change by $|dn_0/n_0| = 0.01$? Based on this result, is it practical to reduce the doping concentration by adding doping atoms of the opposite type?

1.31 It is found that $n_0 \approx p_0 \approx n_i$ at very high temperatures and that $n_0 = 10^{16}$ cm^{-3}, $p_0 \ll n_0$, and $n_i \ll n_0$ at much lower temperatures. What are n_0 and p_0 at the medium temperature where $n_i = 10^{16}$ cm^{-3}?

1.32 Consider a silicon wafer doped with $N_D = 10^{17}$ cm^{-3} and thickness $t_W = 0.5$ mm. Calculate the side of a square that is needed to enclose 1000 holes at room temperature.

1.33 Exposure of a silicon sample to light increases the concentration of holes from 10^4 cm^{-3} to 10^{12} cm^{-3}. What are the concentrations of electrons before and after the exposure to light? Assume room temperature.

1.34 Silicon is doped by boron atoms to the maximum level of 4×10^{20} cm^{-3}. Determine the percentage of boron atoms in the silicon crystal.

1.35 Silicon is doped so that 1 in 10,000 Si atoms is replaced by a phosphorus atom.

(a) Knowing that only 30% of these phosporus atoms donate electrons at room temperature (30% ionization), determine the effective doping level and the electron concentration.

(b) Determine the average distances between the phosphorus atoms and between the mobile electrons.

REVIEW QUESTIONS

R-1.1 A simple cubic crystal lattice can be constructed by replicating two cubes with two sides joined. Do the two cubes satisfy the definition of a unit cell?

R-1.2 The smallest unit cell is called the *primitive cell*. Identify the primitive cell in the diamond lattice.

R-1.3 The primitive cell of the diamond lattice has to be shifted along the x-, y-, and z-axes and rotated to construct the conventional unit cell (the cubic cell whose side is defined as the crystal-lattice constant). Is there a need to rotate the conventional unit cell in order to construct the diamond lattice?

R-1.4 Is diamond (carbon in crystalline form) a semiconductor?

R-1.5 Both hydrogen and silicon atoms form covalent bonds to create hydrogen molecules and silicon crystal, respectively. Why do hydrogen atoms create just molecules and silicon atoms an entire crystal?

R-1.6 The valence electrons in a metal are shared by all atoms, whereas the valence electrons in a semiconductor are shared by neighboring atoms. Is this related to the fact that metals conduct electric current at 0 K and semiconductors do not?

R-1.7 Is the total concentration of free electrons (current carriers) in metals equal to the concentration of valence electrons?

R-1.8 Is the total concentration of current carriers (free electrons and holes) in a semiconductor equal to the concentration of valence electrons?

R-1.9 There are two types of current carrier in semiconductors (free electrons and holes) and only one type (free electrons) in metals, yet metals are much better conductors of electric current. Is this because the currents of electrons and holes oppose each other? If not, what is the reason?

R-1.10 The intrinsic carrier concentration is much higher in silicon than in GaAs. Does this mean that more energy is needed to break a covalent bond in GaAs?

R-1.11 Can doping-induced and thermally generated holes be distinguished from each other?

R-1.12 Is there any positive charge in N-type semiconductors? If so, how many types of positive charge are there?

R-1.13 Which type of positive charge dominates in terms of the concentration in N-type semiconductors?

R-1.14 Does the dominant positive charge in N-type semiconductors contribute to the current flow?

R-1.15 A semiconductor has equal concentrations of donor and acceptor ions. Are the concentrations of the current carriers equal? If so, are they equal to the intrinsic-carrier concentration?

R-1.16 Does an increase in the doping level influence the concentration of broken covalent bonds? If so, is the concentration of broken covalent bonds increased or decreased?

R-1.17 An increase in temperature increases the carrier-generation rate (the concentration of broken covalent bonds per unit time). Does it influence the recombination rate? If so, why?

R-1.18 Light can also break covalent bonds to generate current carriers (electron–hole pairs). Is the recombination rate in an illuminated sample higher than the recombination rate in thermal equilibrium?

R-1.19 Can the equation $n_0 p_0 = n_i^2$ be applied to the case of an illuminated sample?

R-1.20 After a period of illumination, the light is switched off and the generation rate drops immediately to the equilibrium level. Does the recombination rate drop *immediately* to the equilibrium level? If so, what reduces the carrier concentrations to their equilibrium levels?

R-1.21 Is a semiconductor crystal neutral after ion implantation?

2 The Energy-Band Model

It has been possible to use the atomic-bond model alone to define current carriers and to introduce some of their fundamental properties. However, the operation of an electronic device does not relate to electric *current* alone; the inseparable concept is *voltage,* that is, electric-potential difference. The electric potential (φ) is directly related to potential energy (E_{pot}) through the electron charge ($-q$), which is a constant: $E_{pot} = -q\varphi$. Since the simple atomic-bond model does not express the energy state of the current carriers, it is not sufficient for a proper understanding of semiconductor device operation. A commonly used tool for descriptions of the phenomena observed in semiconductor devices is referred to as the *energy-band model.* This powerful tool can provide intuition about and visualization of abstract and complex phenomena, but only when clearly understood. The aim of this chapter is to introduce the energy-band model, finalizing at the same time the introduction to the fundamental properties of current carriers.

Electrons exhibit wave properties that combine with the regular placement of atoms in crystals, which in turn leads to important effects in terms of the potential energy of current carriers. The theory that accounts for the wave properties of small particles is called *quantum mechanics.* The first two sections of this chapter introduce the quantum-mechanical effects necessary to link the wave properties of electrons to the energy bands in semiconductors. In the third section, the concepts of effective mass and density of states are introduced to complete the incorporation of quantum-mechanical effects into the particle model of electrons and holes as the current carriers in semiconductors. These three sections can be considered as read-only sections that provide an insight into the fundamental elements of the energy-band model: energy bands, energy gaps, effective mass, and density of states. The solved examples in these sections provide a tool for deeper study of selected effects. The last section introduces the essential link between the energy bands and the concentration of electrons and holes, through the Fermi–Dirac distribution and the Fermi level. The presentation of the energy-band model in this section is necessary for proper understanding of semiconductor device operation.

†2.1 ELECTRONS AS WAVES

2.1.1 De Broglie Relationship Between Particle and Wave Properties

We as humans have developed considerable knowledge that distinguishes particles from waves. This knowledge is important for many functions in human society, and it is taught as what is now described as *classical mechanics*. Building on this knowledge, the model of electrons as negatively charged particles was used in Chapter 1 to introduce the fundamental properties of current carriers. This is justified, as the electrons possess some undeniable particle properties. It is proved that an electron carries a unit of charge ($-q = -1.6 \times 10^{-19}$ C) and that it can be released from an atom or captured by an atom only as a whole—there is no such a thing as a half or a quarter of an electron. Also, each electron has a specific *mass,* which is another important property that we associate with particles. The mass of an electron at rest is $m_0 = 9.1 \times 10^{-31}$ kg.

Notwithstanding the above-mentioned particle properties, electrons also exhibit wave properties, such as diffraction and interference. The electron microscope quite successfully exploits these wave properties of electrons. This fact does not mean that the simple particle model used in the previous chapter was wrong. Also, it does not mean that the classical concept and description of waves could not be used to describe and model the operation of an electron microscope. However, it does mean that the concepts of particles and waves from classical mechanics are not general. In the framework of classical mechanics, we associate the wavelength (λ) with waves, whereas we associate the momentum (the product of mass and velocity, $p = mv$) with particles. In classical mechanics these quantities are unrelated to each other. However, there is a fundamental relationship between them:

$$\lambda p = h \tag{2.1}$$

where h is the Planck constant ($h = 6.626 \times 10^{-34}$ J · s). It was suggested by de Broglie that Eq. (2.1) could be applied to both waves and particles. This indicates that there are no fundamental differences between particles and waves, a concept referred to as *particle–wave duality*.

The existence of particle–wave duality does not mean something is wrong if we do not observe wave properties in the case of *large* particles. The wavelength of a 1-kg iron ball, moving with velocity $v = 1$ m/s, is $\lambda = 6.626 \times 10^{-34}$ m! This wavelength is so much smaller than the iron atom's that it loses any practical meaning. Likewise, waves with *large* wavelengths have meaningless momenta. In these cases, the classical particle and wave concepts work quite well. The particle–wave duality becomes important for *small* objects. As mentioned earlier, the theory that describes the properties of small objects is called *quantum mechanics,* the wave–particle duality being its central concept. Quantum mechanics successfully describes many effects associated with electrons and light (photons) that cannot be described by classical mechanics.

Our knowledge of many important principles is in terms of the classical concepts of particles and waves. The most efficient way of incorporating the quantum-mechanical

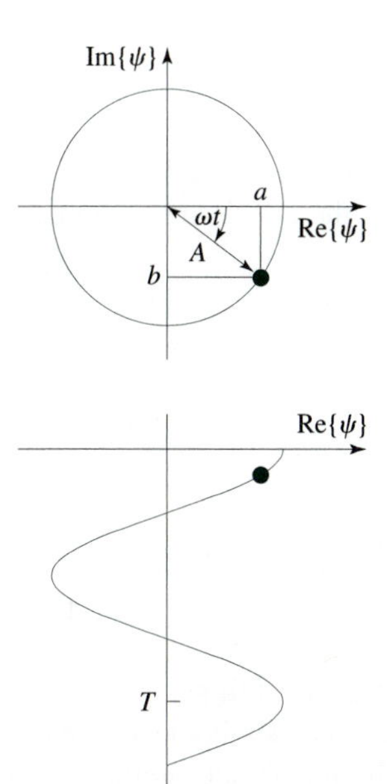

Figure 2.1 Illustration of the $\psi(t) = A\exp(-j\omega t)$ function, along with its relationship to waves.

effects is to upgrade our classical picture. The upgrade will set the electrons perceived as particles in a specific environment—energy bands.[1]

Consequently, we begin with a presentation of electrons as waves. This will lead to definitions of energy levels and energy bands in Section 2.2. A continuous transition from the wave presentation to the particle presentation will be made in Section 2.3 to define the density of electron states. The introduction of the energy-band model will be completed in Section 2.4 by the model for occupancy of the electron states, to link the energy bands to the already familiar concentrations of current carriers.

2.1.2 Wave Function and Wave Packet

The wave function, labeled by ψ, is the usual mathematical description of the wave properties of electrons. Figure 2.1 illustrates the simplest form of the wave function. Imagine a point circulating in the complex plane. The *rate of change of phase with time* (radians per second) is called *angular frequency* ω. Because the whole circle has 2π radians, the angular frequency is

$$\omega = \frac{2\pi}{T} \tag{2.2}$$

where the *period* T is the time that it takes to complete a circle.

[1]The energy bands will involve the wave properties of electrons, but they cannot separate the wave properties from the electrons, so we should never ignore them as we use the particle presentation.

The position of any point in the complex plane can be expressed by a complex number $a + jb$, where a and b are the real and imaginary parts, respectively, whereas $j = \sqrt{-1}$. Alternatively, it can be expressed as $A\exp(j\varphi)$, where the distance from the origin (A) is called *amplitude,* and the angle with respect to the real axis (φ) is called *phase*. In our case, $\varphi = -\omega t$. Therefore, the position of the circulating point can be expressed by the following complex function:

$$\psi(t) = Ae^{-j\omega t} \tag{2.3}$$

or, alternatively:

$$\psi(t) = \underbrace{A\cos(\omega t)}_{\mathrm{Re}\{\psi\}} - j\underbrace{A\sin(\omega t)}_{\mathrm{Im}\{\psi\}} \tag{2.4}$$

Obviously, the real part of ψ is a cosine function of time.

Now, imagine that the point is attached to a membrane, causing it to follow the cosine $\mathrm{Re}\{\psi\}$ oscillation. Imagine further that the membrane oscillations are transferred to air particles, or any other set of particles. The push and pull of the membrane causes peaks and valleys of particle concentration, which travel in the direction perpendicular to the membrane, say the x-direction. If λ is the distance that a peak (or a valley) travels as the oscillating point completes a whole circle (2π radians), then the *rate of change of phase with distance* (radians per meter) is

$$k = \frac{2\pi}{\lambda} \tag{2.5}$$

More frequently, k is referred to as the *wave number* or the *wave vector* in the case of three-dimensional presentation ($\boldsymbol{k}$).

In general, a traveled distance x is related to the change of phase as $\varphi = kx$ (or $\varphi = \boldsymbol{kr}$ in the three-dimensional case). Therefore, we obtain the following wave function of time t and distance x:

$$\psi(x, t) = Ae^{-j(\omega t - kx)} \tag{2.6}$$

The real part of this wave function,

$$\mathrm{Re}\{\psi(x, t)\} = A\cos(\omega t - kx) \tag{2.7}$$

is plotted in Fig. 2.2. Obviously, λ is the wavelength, and is related to the period T through the wave velocity v:

$$\lambda = vT \tag{2.8}$$

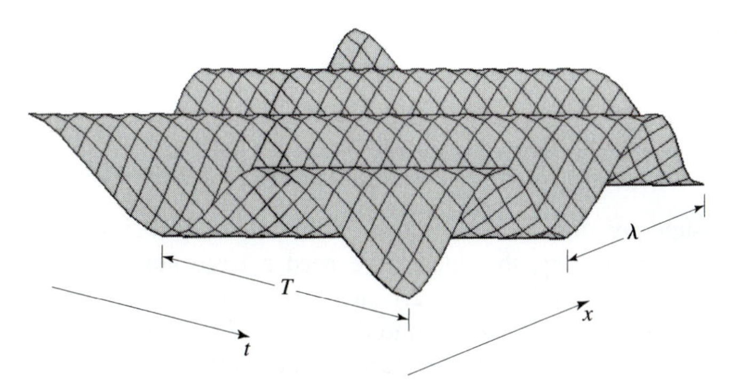

Figure 2.2 Plot of $\mathrm{Re}\{\psi(x,t)\} = A\cos(\omega t - kx)$. Illustration of a plane wave.

Equations (2.2), (2.5), and (2.8) show that the wave number and the angular frequency are also related to each other through the wave velocity:

$$\omega = vk \tag{2.9}$$

Equation (2.5) is analogous to Eq. (2.2). It shows that the wave number is related to inverse space analogously to the relationship between frequency and inverse time. Eliminating the wavelength from Eqs. (2.1) and (2.5), the fundamental particle–wave link can be expressed as a relationship between the momentum p and the wave number k:

$$p = \underbrace{\frac{h}{2\pi}}_{\hbar} k = \hbar k \tag{2.10}$$

Equation (2.6) describes a wave that is not localized in space (x-direction). To illustrate this point, let us say that the considered wave is *sound,* produced by the oscillating membrane. Because the wave is not decaying in the x-direction, we will hear a constant sound level at any point x. An analogous example can be made with light, which will appear with constant intensity at any point x. Mathematically, the wave intensity is expressed as $\psi(x,t)\psi^*(x,t)$, where ψ^* is the complex conjugate (the signs of j are reversed). It is easy to show that the intensity of the wave given by Eq. (2.6) is indeed time- and space-independent: $\psi(x,t)\psi^*(x,t) = A^2$.

There is no apparent problem in applying the preceding result to sound and light. However, how about electrons? It is correct that electron diffraction and interference demonstrate beyond any doubt that electrons can behave as waves as much as the light does. However, Eq. (2.6) does not properly describe electrons that are localized in space—for example, electrons in semiconductor devices. Equation (2.6) appears too simple to account for this case.

There is a way of making Eq. (2.6) complicated enough to obtain a wave function that could describe a localized electron. The wave function given by Eq. (2.6) can be considered as a single term (harmonic) of the Fourier series. With properly determined coefficients, the Fourier series can closely match any shape that the wave function may take. Therefore, the needed function $\psi(x,t)$ can be approximated by

$$\psi(x,t) \approx \sum_{n=-N}^{N} \underbrace{A\Delta k f(k_n)}_{A_n} e^{-j(\omega_n t - k_n x)} \tag{2.11}$$

What is happening here is a superposition of cosine waves of different frequencies (wave numbers) and amplitudes. In our analogy, this means we need a large number ($2N$) of "cosine wave machines" described earlier (Figs. 2.1 and 2.2), operating at different angular frequencies ω_n and amplitudes A_n. In addition to that, we need to place all these "machines" close to one another so that they have a perfectly blended effect on the air, or any other set of particles.

A problem with the wave function $\psi(x,t)$ obtained by Eq. (2.11) is that it remains periodic. To remove this problem, we have to push N to ∞ and Δk to dk.[2] In this case, the sum in Eq. (2.11) becomes an integral:

$$\psi(x,t) = A\int_{-\infty}^{\infty} f(k)e^{-j(\omega t - kx)}\, dk \tag{2.12}$$

The wave function $\psi(x,t)$ expressed in this way is called a *wave packet,* where $f(k)$ is a *spectral function* specifying the amplitude of the harmonic wave with wave number k and angular frequency $\omega = vk$.

There are so many things naturally distributed according to the normal, or Gaussian, distribution that it seems quite appropriate to take the normal distribution as an example of the spectral function $f(k)$:

$$f(k) = \frac{1}{\sqrt{2\pi}\sigma_k} \exp\left[-\frac{(k-k_0)^2}{2\sigma_k^2}\right] \tag{2.13}$$

This is a bell-shaped function centered at $k = k_0$ with width σ_k. In terms of our "cosine wave machines," this means that most of the "machines" would be producing waves with frequencies close to $\omega_0 = vk_0$, with the number of "machines" decaying as the frequency difference from ω_0 increases.

The integral in Eq. (2.12), with the normal distribution as the spectral function $f(k)$, can be explicitly performed, leading to the following form of the wave function:

$$\psi(x,t) = Ae^{-(\sigma_k^2/2)(vt-x)^2} e^{-j(\omega_0 t - k_0 x)} \tag{2.14}$$

The real part of this function,

$$\text{Re}\{\psi(x,t)\} = Ae^{-(\sigma_k^2/2)(vt-x)^2} \cos(\omega_0 t - k_0 x) \tag{2.15}$$

is plotted in Fig. 2.3. We can see that our normally distributed "cosine wave machines" produced a kind of localized wave packet that travels in the x-direction. If our membranes were generating sound, this time we would not hear a constant sound level at any x point.

[2] Both $N \to \infty$ and dk are mathematical abstractions with consequences discussed in Section 2.3.

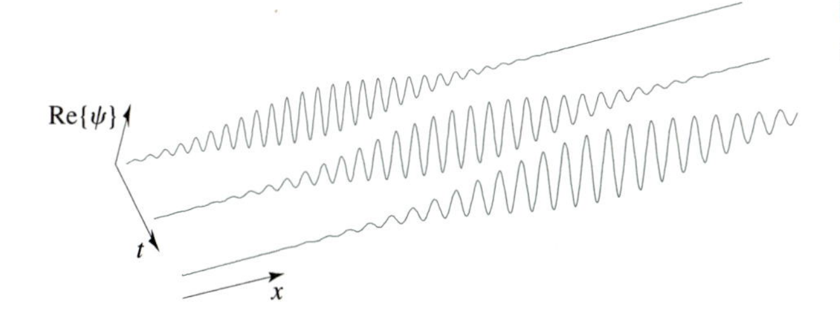

Figure 2.3 Illustration of a wave packet, traveling in the x-direction.

If the wave packet was light, we would not see the same light intensity at any x point. Standing at a single x point, we would hear or see the wave packet passing by us as a lump of sound or light, traveling with velocity v.

Obviously, the concept of wave packet is much more general than the single harmonic wave, and it appears as a tool that can model the wave–particle duality. We have used oscillating membranes ("wave packet machines") and sound to illustrate the wave packet. If applied to sound, everything looks clear: (1) Eq. (2.14) is the complex wave function of the sound packet, (2) Eq. (2.15) is the real instantaneous amplitude of the sound (plotted in Fig. 2.3), and (3) the intensity of the sound is

$$|\psi(x,t)|^2 = \psi(x,t)\psi^*(x,t) = A^2 \exp\left[-\sigma_k^2(vt-x)^2\right] = A^2 \exp\left[-\frac{(vt-x)^2}{2\sigma_x^2}\right] \tag{2.16}$$

where $\sigma_x^2 = 1/(2\sigma_k^2)$. It should be stressed that the complex wave function [Eq. (2.14)] is a convenient mathematical abstraction. The real instantaneous amplitude [Eq. (2.15)] can be related to physically real oscillations of air particles, but even this is an abstraction as far as our sense of hearing is concerned. What is real for our sense of hearing is the sound intensity [Eq. (2.16)]. Although the same mathematical apparatus can be applied to different types of waves, the links to reality may be different.

In the case of electrons, it is the wave intensity, $|\psi(x,t)|^2 = \psi(x,t)\psi^*(x,t)$, that is real. The question is what the intensity of electron waves means. In the analogy of a sound packet, the wave intensity is not too hard to understand: there are air particles that vibrate within a localized domain, producing a sound packet with sound intensity as in Fig. 2.4. In this example, $|\psi(x,t)|^2$ is applied to an ensemble of particles. However, what is the meaning of the intensity $|\psi(x,t)|^2$ when applied to a single particle, such as a single electron?

The answer to this question is in the relationship between the concepts of probability and statistics. It appears perfectly meaningful to say there are 2×10^{19} cm^{-3} air particles. Let us assume that an extremely good vacuum can be achieved, so good that there is only one particle left in a room of 200 cm × 300 cm × 167 cm $= 10^7$ cm^3. In this case, we find that there is $1/10^7 = 0.0000001$ particle/cm^3. What does this mean? The answer is: The *probability* of finding this particle in a specified volume of 1 cm^3. Let us assume that our membranes ("wave packet machines") are installed in this room with the single particle inside. This will affect the probability of finding the particle in our specified 1 cm^3, according to the specific wave function intensity $|\psi(x,t)|^2$.

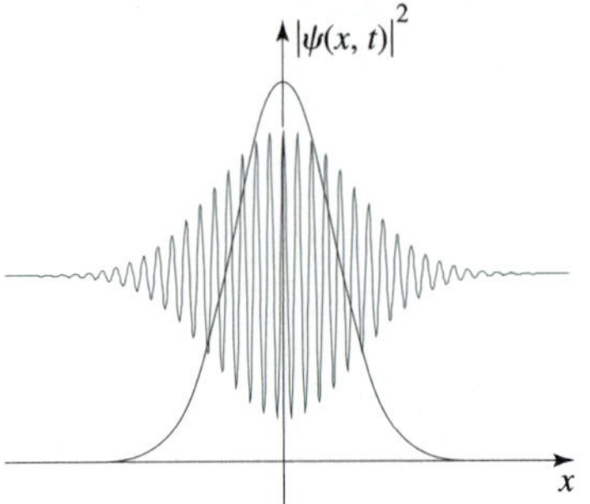

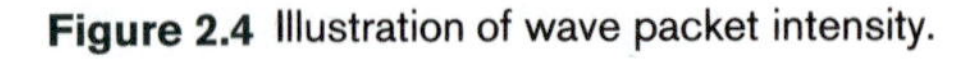
Figure 2.4 Illustration of wave packet intensity.

If the wave function $\psi(x, t)$ is to be a representation of a single particle (for example, an electron perceived as a particle), then the intensity $\psi(x, t)\psi^*(x, t)$ relates to the probability of finding this particle (electron) at point x and time t.

2.1.3 Schrödinger Equation

We have used a specific wave function $\psi(x, t)$ to introduce the concept of a wave packet. Clearly, this specifically selected wave function [Eq. (2.14)] cannot be used for any object in any situation. Obviously, we need to be able to find somehow the wave function that would specifically model a particular object (say an electron) in specific conditions.

In 1926, Schrödinger postulated a differential equation that results in the needed wave function when solved with appropriate boundary and/or initial conditions. The one-dimensional form of the time-dependent Schrödinger equation is given as follows:

$$j\hbar\frac{\partial\psi(x, t)}{\partial t} = -\frac{\hbar^2}{2m}\frac{\partial^2\psi(x, t)}{\partial x^2} + E_{pot}(x)\psi(x, t) \tag{2.17}$$

where $E_{pot}(x)$ is the potential-energy function incorporating any influence of the environment on the considered electron and $\hbar = h/2\pi$ is the reduced Planck constant. The wave function $\psi(x, t)$ and its first derivative are finite, continuous, and single-valued. The term $\hbar^2/2m$ involves the mass of electrons, m, so it relates the wave function to particle properties of electrons.

The three-dimensional form of the time-dependent Schrödinger equation can be written as

$$j\hbar\frac{\partial\psi}{\partial t} = -\frac{\hbar^2}{2m}\nabla^2\psi + E_{pot}\psi \tag{2.18}$$

where $\nabla^2 = \frac{\partial^2}{\partial x^2} + \frac{\partial^2}{\partial y^2} + \frac{\partial^2}{\partial z^2}$, $\psi = \psi(x, y, z, t)$, and $E_{pot} = E_{pot}(x, y, z)$. In the following text we will limit ourselves to the one-dimensional case.

The variables of the time-dependent Schrödinger equation [x and t in Eq. (2.17)] can be separated if the following form of the wave function is used: $\psi(x, t) = \psi(x)\chi(t)$. In this case, Eq. (2.17) can be transformed into

$$j\frac{\hbar}{\chi(t)}\frac{\partial\chi(t)}{\partial t} = -\frac{\hbar^2}{2m\psi(x)}\frac{\partial^2\psi(x)}{\partial x^2} + E_{pot}(x) \tag{2.19}$$

The left-hand side of this equation is a function of time alone, whereas the right-hand side is a function of position alone. This is possible only when the two sides are equal to a constant. The constant is in the units of energy, and it actually represents the total energy E:

$$j\frac{\hbar}{\chi(t)}\frac{\partial \chi(t)}{\partial t} = E$$

$$-\frac{\hbar^2}{2m\psi(x)}\frac{\partial^2 \psi(x)}{\partial x^2} + E_{pot}(x) = E \tag{2.20}$$

Obviously, the time-independent wave function $\psi(x)$ has to satisfy the following *time-independent Schrödinger equation:*

$$-\frac{\hbar^2}{2m}\frac{d^2\psi(x)}{dx^2} + E_{pot}(x)\psi(x) = E\psi(x) \tag{2.21}$$

EXAMPLE 2.1 *E–k* Diagram for a Free Electron

Solve the Schrödinger equation for a free electron to

(a) determine the general wave function for a free electron and
(b) determine and plot the E–k dependence, and relate this dependence to the classical equation for kinetic energy of a particle ($E_{kin} = mv^2/2$).

SOLUTION

(a) For the case of a free electron, $E_{pot} = 0$, so the Schrödinger equation is

$$\frac{d^2\psi(x)}{dx^2} + \frac{2m}{\hbar^2}E\psi(x) = 0$$

The solution of this type of differential equation can be expressed as

$$\psi(x) = A_+e^{s_1x} + A_-e^{s_2x}$$

where $s_{1,2}$ are the roots of its characteristic equation:[3]

$$s^2 + \underbrace{\frac{2m}{\hbar^2}E}_{k^2} = 0 \tag{2.22}$$

Given that the solutions of the characteristic equation $s^2 = -k^2$ are $s_{1,2} = \pm jk$, the general solution is expressed as

$$\psi(x) = A_+e^{jkx} + A_-e^{-jkx} \tag{2.23}$$

[3]Here s^2 represents the second derivative, $\frac{d^2\psi(x)}{dx^2}$, whereas $\psi(x)$ itself is represented by $s^0 = 1$.

Comparing this wave function to the time-independent part of Eq. (2.6), we can see that it consists of two plane waves traveling in the opposite directions: (1) $A_+ \exp(jkx)$, traveling in the positive x-direction, and (2) $A_- \exp(-jkx)$, traveling in the negative x-direction. Equation (2.23) represents the general solution for the electron wave function in free space ($E_{pot} = 0$), as the superposition of the two plane waves can account for any possible situation in terms of boundary conditions.

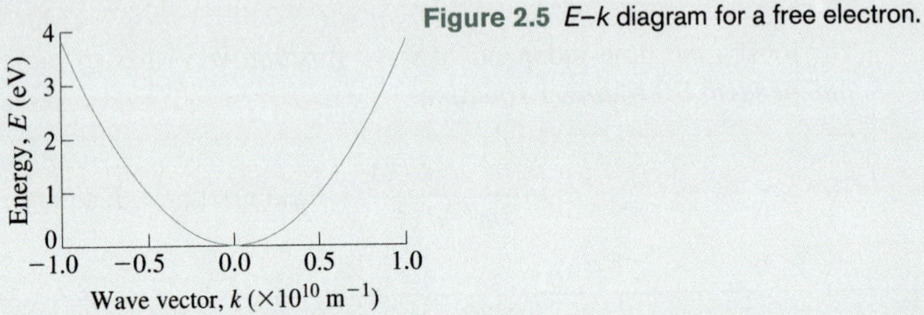

Figure 2.5 E–k diagram for a free electron.

(b) As Eq. (2.22) shows, the wave function given by Eq. (2.23) satisfies the Schrödinger equation when the dependence of the kinetic (and thus total) energy of a free electron is given by

$$E = \frac{\hbar^2}{2m} k^2 \tag{2.24}$$

This E–k dependence is plotted in Fig. 2.5.

To relate the obtained E–k dependence to the classical equation for E_{kin}, we notice first that $E = E_{kin}$ (given that a free electron has no potential energy). Then, we apply the fundamental link between the wave number and momentum ($p = \hbar k$) and convert the momentum into velocity ($p = mv$):

$$E_{kin} = \frac{\hbar^2}{2m} k^2 = \frac{p^2}{2m} = \frac{mv^2}{2}$$

This equation shows that the E–k dependence of a free electron is identical to the classical dependence of kinetic energy on velocity.

EXAMPLE 2.2 Electron in a Potential Well

The infinite potential well, illustrated in Fig. 2.6a, is mathematically defined as

$$E_{pot}(x) = \begin{cases} 0 & \text{for } 0 < x < W \\ \infty & \text{for } x \le 0 \text{ and } x \ge W \end{cases}$$

For an electron inside the potential well, determine

(a) the possible energy values and
(b) the wave functions corresponding to the possible energy levels.

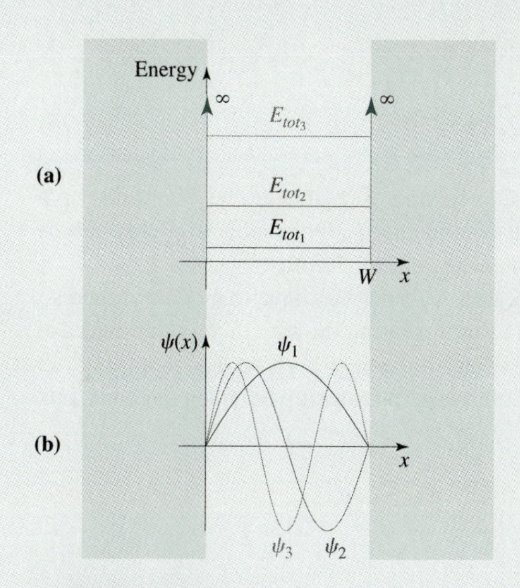

Figure 2.6 A particle in an infinite potential well. The possible energy levels (a) correspond to standing waves with integer half-wavelength multiples (b).

SOLUTION

The Schrödinger equation for this case is

$$\begin{aligned} \frac{d^2\psi(x)}{dx^2} + \frac{2m}{\hbar^2}E\psi(x) &= 0 \qquad \text{for } 0 < x < W \\ \psi(x) &= 0 \qquad \text{for } x \leq 0 \text{ and } x \geq W \end{aligned} \tag{2.25}$$

Mathematically, this problem is reduced to solving the Schrödinger equation for a free particle ($E_{pot} = 0$) with the following boundary conditions: $\psi(0) = 0$ and $\psi(W) = 0$. Therefore, the general solution is given by Eq. (2.23).

(a) In the case of an electron trapped between two walls, the plane waves moving in either direction are relevant as the electron is reflected backward and forward by the walls. Time-independent solutions (steady states) are still possible, but only in the form of standing waves, as illustrated in Fig. 2.6b. Obviously, the standing waves are formed only for specific values of $k = 2\pi/\lambda$ that correspond to integer multiples of half-wavelengths $\lambda/2$. Therefore,

$$n\frac{\lambda_n}{2} = W \qquad (n = 1, 2, 3, \ldots)$$

$$k_n = \frac{2\pi}{\lambda_n} = \frac{\pi}{W}n$$

Using the relationship between the total energy and the wave number from Eq. (2.22),

$$E_n = \frac{\hbar^2}{2m} k_n^2$$

we obtain

$$E_n = n^2 \frac{\pi^2 \hbar^2}{2mW^2} \qquad (n = 1, 2, 3, \ldots) \tag{2.26}$$

With this result the following important conclusion is reached: the electrons in a potential well cannot have an arbitrary value of total energy. Only specific energy values are possible (Fig. 2.6a). This effect is called *energy quantization*.

(b) The two constants in the general solution (2.23), A_+ and A_-, have to be determined so as to obtain the particular wave function $\psi(x)$ representing the specifically defined case (electrons inside a potential well of width W and infinitely high walls). Applying the boundary condition $\psi(0) = 0$, we find the following relationship between the constants A_+ and A_-:

$$A_+ e^0 + A_- e^0 = 0 \Rightarrow A_+ = -A_-$$

After this, Eq. (2.23) can be transformed as

$$\psi(x) = A_+ \cos(kx) + jA_+ \sin(kx) - A_+ \underbrace{\cos(-kx)}_{=\cos(kx)} - jA_+ \underbrace{\sin(-kx)}_{=-\sin(kx)} = \underbrace{2jA_+}_{A} \sin(kx) \tag{2.27}$$

where A is a new constant, involving the constant A_+.

The second boundary condition is $\psi(W) = 0$. With exception of the trivial case $A = 0$, no other value for A in Eq. (2.27) can satisfy this boundary condition. However, $\sin(kW)$ can be zero for a number of kW values, which means that the possible solutions are determined by specific (and discrete) k values. Because k is related to the total energy [Eq. (2.22)], this means that the possible solutions are determined by specific (and discrete) energy values. This is the same conclusion related to the *energy quantization* effect, only it is reached in a different way. Given that $\sin(kW) = 0$ for $kW = n\pi$, where $n = 1, 2, 3, \ldots$, the possible energy levels are according to Eq. (2.26).

There is an additional condition that has to be satisfied by the wave function that is to represent electrons in the potential well. It relates to the probability of finding an electron at x: $|\psi(x)|^2 = \psi(x)\psi^*(x)$. If an electron is trapped inside the potential well, then it has to be somewhere between 0 and W, which means

$$\int_0^W \psi(x)\psi^*(x)dx = 1$$

This is called the normalization condition. The normalization condition determines a specific value of the constant A, as

$$\int_0^W A^2\left(\sin\frac{n\pi}{W}x\right)^2 dx = 1 \Rightarrow A = \sqrt{\frac{2}{W}}$$

Therefore, the final solution is

$$\psi_n(x) = \sqrt{\frac{2}{W}}\sin\frac{n\pi}{W}x \qquad (n = 1, 2, 3, \ldots)$$

where $\psi_1(x)$, $\psi_2(x)$, $\psi_3(x), \ldots$ $(n = 1, 2, 3, \ldots)$ represent different electron states (electrons at different energy levels E_n). The wave functions $\psi_1(x)$, $\psi_2(x)$, and $\psi_3(x)$ are illustrated in Fig. 2.6b.

EXAMPLE 2.3 Tunneling

Another important quantum-mechanical effect, observed in semiconductor devices, is tunneling. To illustrate the effect of tunneling, determine the wave function $\psi(x)$ for the case when electrons are approaching a potential-energy barrier, as in Fig. 2.7. This potential barrier is mathematically defined as

$$E_{pot}(x) = \begin{cases} 0 & \text{for } x < 0 \\ E_{pot} & \text{for } 0 \leq x \leq W \\ 0 & \text{for } x > W \end{cases}$$

Consider both cases: (a) the electron energy is higher than the barrier height, $E > E_{pot}$, and (b) the electron energy is smaller than the barrier height, $E < E_{pot}$.

SOLUTION

Writing the Schrödinger equation in the following forms:

$$\frac{d^2\psi(x)}{dx^2} + \underbrace{\frac{2m}{\hbar^2}E}_{k^2}\,\psi(x) = 0 \qquad \text{for } x < 0 \text{ and } x > W$$

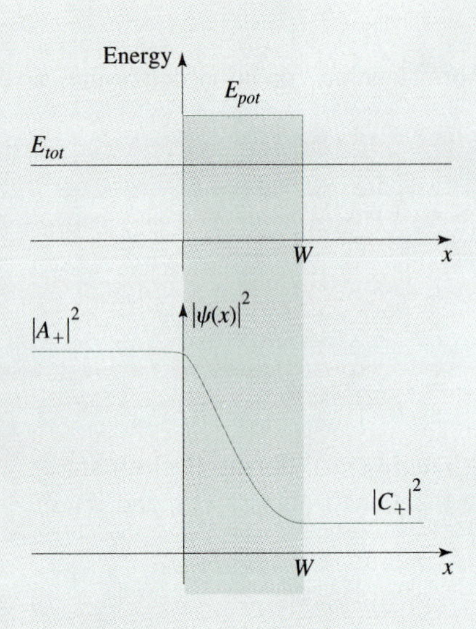

Figure 2.7 Illustration of tunneling.

and

$$\frac{d^2\psi(x)}{dx^2} - \underbrace{\frac{2m}{\hbar^2}(E_{pot} - E)}_{\kappa^2}\psi(x) = 0 \qquad \text{for } 0 \le x \le W$$

the general solution can be expressed as

$$\psi(x) = \begin{cases} A_+e^{jkx} + A_-e^{-jkx} & \text{for } x < 0 \\ B_+e^{\kappa x} + B_-e^{-\kappa x} & \text{for } 0 \le x \le W \\ C_+e^{jkx} + C_-e^{-jkx} & \text{for } x > W \end{cases} \tag{2.28}$$

Again, the electron wave function appears as a superposition of two plane waves traveling in opposite directions when $E_{pot} = 0$ [the first and third rows in Eq. (2.28)]. In the region where $E_{pot} \neq 0$, two different cases have to be considered: $E > E_{pot}$ and $E < E_{pot}$.

(a) $E > E_{pot}$: In this case, classical mechanics predicts that a particle with energy E should go over the lower barrier, E_{pot}, without any interference. However, the wave function inside the barrier region ($B_+e^{\kappa x}$ if we limit ourselves to the wave traveling in the positive x direction) is different from the incident wave (A_+e^{jkx}). Because $\kappa^2 = \frac{2m}{\hbar^2}(E_{pot} - E)$ is negative, κ can be expressed as $\kappa = jk_E$, where $k_E = \sqrt{\frac{2m}{\hbar^2}(E - E_{pot})}$ is the wave number inside the barrier region. This means that the incident plane wave continues to travel as a plane wave through the barrier region ($B_+e^{jk_Ex}$), but with an increased wavelength $\lambda_E = 2\pi/k_E$ ($k_E < k$). In addition to that, the intensity of the wave ($|B_+|^2$) is reduced because there is a finite probability ($|A_-|^2$) that the particle is reflected by the barrier.

(b) $E < E_{pot}$ (*Tunneling*): Classical mechanics predicts that a particle cannot go over a potential barrier that is higher than the total energy of the particle. However, the wave function of the particle does not suddenly vanish when it hits the barrier as $B_+e^{\kappa x} \neq 0$.[4] This time, $\kappa = -\sqrt{\frac{2m}{\hbar^2}(E_{pot} - E)}$ is a real number, and $\exp(\kappa x)$ causes an exponential decay of the wave-function intensity in the barrier region. This is illustrated in Fig. 2.7. If the barrier is wide, so that $W \gg 1/|\kappa|$, the wave function would practically drop to zero inside the potential barrier. This means that $C_+ \approx C_- \approx 0$. However, if the potential barrier is not as wide, the wave function will not drop to zero at the end of the potential barrier ($x = W$), which means that $C_+ \neq 0$. There is a finite probability ($|C_+|^2$) that the particle will be found beyond the barrier ($x > W$), which is the tunneling effect. Probability $|C_+|^2$, normalized by the probability of a particle hitting the barrier ($|A_+|^2$), defines the *tunneling probability*, also called the *tunneling coefficient*, which can be approximated as

$$T = \frac{|C_+|^2}{|A_+|^2} \approx \exp(2\kappa W) = \exp\left[-2W\sqrt{\frac{2m}{\hbar^2}(E_{pot} - E)}\right]$$

†2.2 ENERGY LEVELS IN ATOMS AND ENERGY BANDS IN CRYSTALS

2.2.1 Atomic Structure

The potential-well problem from the previous section introduced the *quantization concept*—the fact that electrons can have discrete values of energy only when appearing inside a potential well. This concept will be applied to the case of electrons in an atom and in a crystal in the following text. To be able to describe how electrons populate these levels, we need to know the fundamental properties of electrons. In addition to being negatively charged, electrons possess intrinsic angular momentum called *spin*. The spin can take two values, $s = \pm 1/2$ in the units of $\hbar$. Another fundamental property of electrons is that only two electrons, with different spins, can occupy the same energy level if their wave functions overlap.[5] This is known as the *Pauli exclusion principle*.

The electrons of an atom appear in a potential well created by the electric field of the positive core. The potential energy has spherical symmetry:

$$E_{pot}(r) = -\frac{Zq^2}{4\pi\varepsilon_0 r} \tag{2.29}$$

[4] In the specific case of $E_{pot} = \infty$ (infinite barrier height), the wave function is equal to zero inside the barrier region as $\kappa = -\sqrt{\frac{2m}{\hbar^2}(E_{pot} - E)} = -\infty$ and $\exp(-\infty x) = 0$.

[5] When there is an overlap in the wave functions, we say that the electrons belong to the same system (an atom, molecule, or crystal).

where Z is the number of the positively charged protons in the core, ε_0 is the permittivity of vacuum, and r is the distance from the center of the atom. If plotted along a line, the potential energy has a funnel shape. Although the shape of $E_{pot}(r)$ is different from the infinite potential well of Fig. 2.6, the effect of the energy-level quantization can be visualized in an analogous manner. An important difference is that we are now dealing with a three-dimensional case. In this case, electrons create standing waves when the orbital circumference is equal to an integral number n of the wavelength λ: $2\pi r = n\lambda$. The number n, expressing this quantization effect, is called the *principal quantum number*. The orbits that these numbers relate to are also referred to as *electron shells*. The wave functions corresponding to different values of n can be obtained by solving the Schrödinger equation with $E_{pot}(r)$ as given by Eq. (2.29). Although it has the spherical symmetry, this is a three-dimensional problem and the actual solving of the Schrödinger equation is beyond the scope of this book. It is quite sufficient to discuss the number of solutions that exist for $n = 1, 2, 3, \ldots$ so that we can relate the introduced quantum effects to the electron structure in atoms.

For the case of $n = 1$, the wave function is spherically symmetrical. Only two electrons in an atom, with different spins, can have this spherically symmetrical wave function. This means the first shell, labeled $1s$, is completely filled with two electrons. The element with two electrons is helium, and its electronic structure is represented as $1s^2$. For $n = 2$, one solution is also spherically symmetrical and is labeled as $2s$. However, there is an additional solution for $n = 2$ with x-, y-, or z-directional symmetry. This type of wave function is labeled by p and it can accommodate six electrons: three space times two spin directions. Therefore, the electronic structure of neon, $1s^2 2s^2 2p^6$, shows that the second shell is filled with eight electrons.

The value of the principal quantum number n shows how many different wave functions exist for that shell. The existence of these different wave functions causes splitting of the second shell into two subshells (s and p), splitting of the third shell into three subshells (s, p, and d), and so on. The subshells are represented by the second quantum number (also called *angular quantum number*). It is labeled by l, so we can write $l = s, p, d, f, g, \ldots$ to express the values (types of wave function) that this quantum number can take. The third quantum number is labeled by m and is called the *magnetic quantum number*. It shows the number of space directions associated with each type of wave function: one for s, three for p, five for d, seven for f, and so on. Finally, the *spin* is the fourth quantum number with two possible values, $\pm 1/2$. Because two electrons with different spins can have identical wave functions, each $s, p, d, f, \ldots$ subshell can accommodate 2, 6, 10, 14, ... electrons, respectively.

Silicon is the fourteenth element in the periodic table, with the following electronic configuration: $1s^2 2s^2 2p^6 3s^2 3p^2$. It has four electrons in the s and p subshells of the third shell. These are the valence electrons that combine through covalent bonding. As described in Section 1.1.1, covalent bonding enables individual atoms in a silicon crystal to share the four valence electrons with four neighbors so that they reach the stable electronic configuration of either neon (the four valence electrons given to the neighbors) or argon (four electrons taken from the neighbors). All other semiconductor elements also form analogous covalent bonding. The atoms in a metal crystal, however, are bound by a metallic bond. For example, copper is the twenty-ninth element in the periodic table, with the following electronic configuration: $1s^2 2s^2 2p^6 3s^2 3p^6 3d^{10} 4s^1$. The individual atoms in a copper crystal give away their valence ($4s^1$) electrons. These electrons form an "electronic cloud" shared by all the atoms in the crystal.

2.2.2 Energy Bands in Metals

The funnel-like shape of the potential energy in an isolated atom is illustrated in Fig. 2.8a, together with the energy level of a single valence electron. Although the potential well is different from the one-dimensional, infinite, and rectangular potential well considered in Example 2.2, the appearance of the discrete energy level for the valence electron is analogous to the standing waves in Fig. 2.6b. The situation is significantly different when there are atoms so close together that there is an overlap between the wave functions of electrons from different atoms. Figure 2.8b illustrates how this would occur in a two-atom molecule. We see that the shape of the potential-energy well is changed so that the standing waves of the valence electrons are no longer confined within the potential wells of the individual atoms. In this example, the two valence electrons are confined within the two-atom system and are shared by both atoms. According to the Pauli exclusion principle, both valence electrons can have the wave function corresponding to the lowest energy level, provided they have different spins. Approximating the funnel-like potential wells by square wells with infinite walls (Example 2.2 and Fig. 2.6), we can use Eq. (2.26) to *estimate* the lowest energy levels for the electrons in a separated atom (Fig. 2.8a) and in a molecule (Fig. 2.8b).[6] To do so, we express the width of the potential well (W) as $W = Na$, where a is the distance between the atom's centers (the potential-well width corresponding to one atom), $N = 1$ represents a single atom, and $N = 2$ represents a molecule. With this,

$$E_1 = \frac{\pi^2 \hbar^2}{2m(Na)^2} \tag{2.30}$$

Assuming $a = 1$ nm, the energy levels are $E_1(N = 1) = 0.377$ eV and $E_1(N = 2) = 0.094$ eV. We can see that there is a significant energy saving when two atoms share the valence electrons. The drop of the allowed energy for these electrons is due to the doubling of the potential-well width, which doubles the half-wavelength (λ_1), halves the wave number value ($k_1 = 2\pi/\lambda_1$), and reduces by factor of 4 the lowest energy level ($E_1 = \hbar^2 k_1^2/2m$).

If this consideration is extended to the case of a one-dimensional crystal with N atoms in the chain (Fig. 2.8c), Eq. (2.30) predicts that $E_1 \to 0$ as $N \to \infty$. It is also important to estimate the highest energy level [$n = N$ in Eq. (2.26)]:

$$E_N = N^2 \frac{\pi^2 \hbar^2}{2m(Na)^2} = \underbrace{\frac{\pi^2 \hbar^2}{2ma^2}}_{E_1(N=1)} \tag{2.31}$$

We can see that E_N does not depend on the number of atoms that are put together. For the example of $a = 1$ nm, E_N remains equal to 0.377 eV. Given that the number of energy levels between $E_1 \approx 0$ and E_N is equal to N, we can conclude that the differences between two subsequent energy levels are negligibly small in a large crystal. To support this conclusion, we note that the average difference between two subsequent energy levels in a

[6]The use of Eq. (2.26) also involves a shift from three-dimensional to one-dimensional potential well—these are only *estimates*, which can be used to conveniently illustrate important effects.

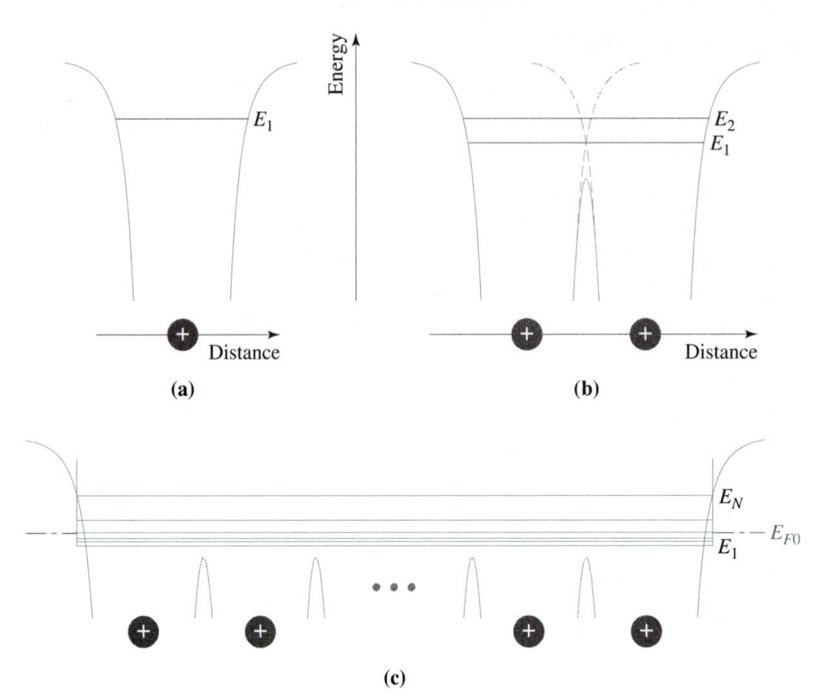

Figure 2.8 Splitting of an energy level, (a) corresponding to a single atom, into (b) two energy levels in a two-atom molecule and (c) N energy levels in a one-dimensional crystal with N atoms.

150-atom crystal (2.5 meV) is already much smaller than the thermal energy (26 meV at 300 K). Because these differences are so small, the energy quantization is practically lost. Accordingly, we can consider the set of energy levels between E_1 and E_N as an *energy band* of width $E_N - E_1$. Referring again to Fig. 2.8, we can say that the single energy level of the valence electron in an isolated atom corresponds to an energy band in a crystal.

The introduction of the concept of *energy band* does not remove the Pauli exclusion principle. There can still be only two electrons (with different spins) at any energy level that is available. At $T \approx 0$ K, the electrons would take the lowest possible energy positions, meaning that the energy band would be filled up to the level with the index $n = N/2$. This energy level is called Fermi energy (E_{F0}). The example of a one-dimensional crystal with a square potential well illustrates that the Fermi energy does not depend on the size of the crystal:

$$E_{F0} = E_{N/2} = \left(\frac{N}{2}\right)^2 \frac{\pi^2\hbar^2}{2m(Na)^2} = \frac{\pi^2\hbar^2}{8ma^2} \tag{2.32}$$

By direct application of the definition of Fermi energy, we can say that the part of the energy band below E_{F0} is filled whereas the part above E_{F0} is empty at $T \approx 0$. If an electric field is applied to such a metal, some electrons will gain kinetic energy, moving to the empty part of the band as they conduct the current due to the applied electric field. The thermal energy at $T > 0$ K also increases the energy of some of the valence electrons in metals.

Using the analogy of electrons in a square well with a flat bottom at $E_1 = 0$, we are considering the valence electrons as *free* electrons inside the metal boundaries (inside the boundaries of the potential well). This means the energy of these electrons obeys the parabolic relationship to the momentum and the wave number [Eq. (2.24) and Fig. 2.5]. To summarize:

1. Approximating the E–k dependence for the valence electrons in metals by a *continuous parabola* is equivalent to adopting the *free-electron model.*
2. The E–x diagram for the valence electrons in metals is a half-filled energy band; the probability of finding any valence electron anywhere inside the metal is the same—there is no localization of these electrons.

2.2.3 Energy Gap and Energy Bands in Semiconductors and Insulators

The electrons creating the covalent bonds in *semiconductors* and *insulators* are neither free nor trapped in the potential wells of single atoms.[7] We cannot use calculations based on the square potential well (the free-electron model from the previous section) to draw any valid conclusions because that model assumes that the positive ions produce a uniform potential (the flat bottom of the well). In semiconductors and insulators, there are strong interactions between the valence electrons and the periodic potential due to the positive ions. These interactions alter the parabolic E–k dependence of free electrons in a way that leads to the appearance of *holes* as an additional type of current carriers.

Because we need to modify (or adapt) the free-electron model, it is best to analyze how imagined free electrons would behave in a periodic potential (so-called *nearly free electron* approach). As Eq. (2.23) shows, the general wave function of a free electron can be represented as the sum of two waves traveling in the opposite directions. Electrons with very low energies have small wave numbers ($E = \hbar^2 k^2/2m$) and long wavelengths ($\lambda = 2\pi/k$). As the electron energy is increased, the wavelength decreases and at some point the half-wavelength becomes equal to the crystal-lattice constant: $\lambda/2 = a$. This condition can also be expressed as $k = \pi/a$. An electron with this wavelength is reflected backward and forward by the periodic potential, which creates a standing wave. Importantly, two types of standing wave are possible at this situation. The first type has antinodes and the second type has nodes at the lattice sites. These two possibilities correspond to two sets of constants in the general solution [(Eq. (2.23)]: one for $A_+ = A_- = A$ and the other for $A_+ = -A_- = A$. With this, the two wave functions and their intensities can be expressed as

$$\begin{aligned} \psi_1(x) &= A\exp(j\pi x/a) + A\exp(-j\pi x/a) = 2A\cos(\pi x/a) \\ \psi_2(x) &= A\exp(j\pi x/a) - A\exp(-j\pi x/a) = 2jA\sin(\pi x/a) \end{aligned} \tag{2.33}$$

$$\begin{aligned} |\psi_1(x)|^2 &= 4A^2\cos^2(\pi x/a) \\ |\psi_2(x)|^2 &= 4A^2\sin^2(\pi x/a) \end{aligned} \tag{2.34}$$

[7] Refer to Sections 1.1.1 and 2.2.1 for descriptions of *covalent bonds*.

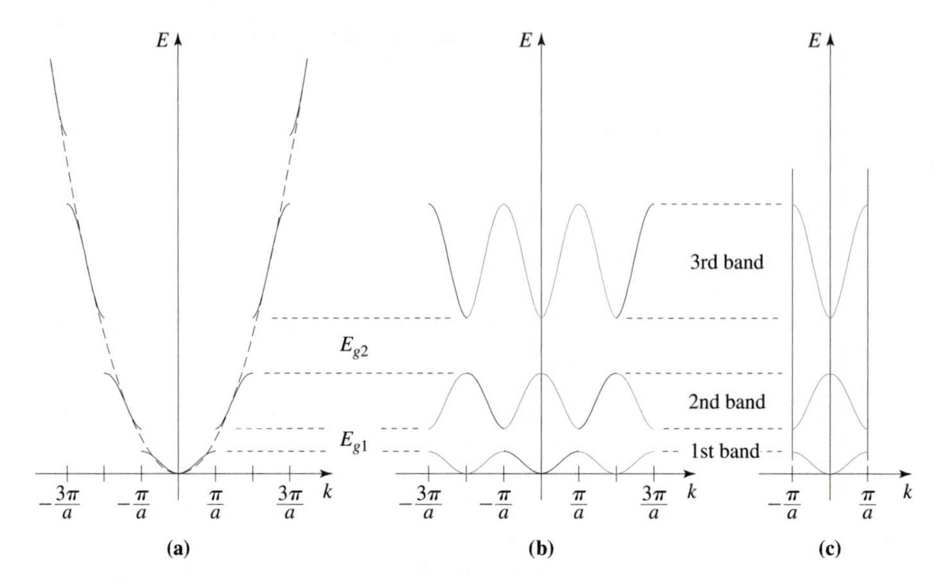

Figure 2.9 The dependence of energy on wave number when electrons interact with the periodic potential in a one-dimensional crystal. (a) Extended-zone presentation evolves from alterations of the free-electron (parabolic) $E(k)$. (b) Repeated-zone presentation illustrates that $E(k)$ becomes a multivalued periodic function. (c) Reduced-zone presentation is most commonly used.

For the case of $\psi_1(x)$, the maximum probability of finding an electron coincides with the crystal-lattice sites, as $x = ia$ $(i = 0, 1, 2, \ldots)$ leads to $\cos^2(i\pi) = 1$ in Eq. (2.34).[8] For the case of $\psi_2(x)$, the maximum probability of finding an electron appears halfway between the atoms, as $x = (2i-1)a/2$ $(i = 1, 2, 3, \ldots)$ leads to $\sin^2(i\pi) = 1$ in Eq. (2.34). The energy of an electron centered at the potential well $(x = ia)$ is lower than the energy of an electron centered at the potential barrier $(x = ia/2)$. The wave functions of both electrons $[\psi_1(x)$ and $\psi_2(x)]$ correspond to the same wave number $(k = \pi/a)$, yet they correspond to two different energies. This means that an electron with the wave number $k = \pi/a$ can have the lower or the higher energy but nothing between the two energy levels—there is an *energy gap* in the E–k dependence at $k = \pi/a$. A deviation from the parabolic E–k dependence to include the appearance of an energy gap is illustrated in Fig. 2.9a. Note that the E–k dependence for the negative values of k is just a reflection of the part for positive values of k; no additional information is implied by the extension of the graph, it is usually presented in this form for a mathematical convenience.

Standing waves, and accordingly energy gaps, appear for any $k_n = n\pi/a$ $(n = 1, 2, 3, \ldots)$, because the crystal-lattice constant is equal to a whole number of half-wavelengths: $n\lambda/2 = a$. This is also illustrated in Fig. 2.9a. The values of $k_n = n\pi/a$ are referred to as the boundaries between Brillouin zones, the first Brillouin zone being for $k < \pi/a$, the second for $\pi/a < k < 2\pi/a$, and so on.

[8]It is assumed that $x = 0$ coincides with a lattice site (the center of an atom).

Both wave functions [$\psi_1(x)$ and $\psi_2(x)$] are periodic in space, repeating themselves in each unit cell of the lattice. These wave functions are specific for $k_n = n\pi/a$, but the periodic property is general—any wave function of an electron in a periodic potential is periodic (Bloch theorem). As a result, the energy is a periodic function of k: $E(k+2\pi a) = E(k)$. The energy dependences from any Brillouin zone may be extended into other zones, as illustrated in Fig. 2.9b. This means that the energy is a multivalued function of k in general, not just for $k = n\pi/a$. In other words, the appearance of energy gaps creates multiple possible energy bands for any k. Given that $E(k)$ is periodic, it is quite sufficient to show it in the first Brillouin zone (it just repeats itself in the other zones). This most common way of presenting the E–k dependencies is called the *reduced-zone presentation* (Fig. 2.9c).

The E–k dependence shown in Fig. 2.9c may seem to be a significant modification of the E–k dependence for a free electron, but it evolved from the simplest considerations of how free electrons would behave when interacting with a periodic potential energy (the nearly free electron model). The valence electrons in real semiconductors and insulators are neither free nor localized within the potential wells of single atoms. Importantly, the alternative approach (so-called *tight binding* theory) leads to the same qualitative conclusions. In Section 2.3.1, new "features" will be added to the E–k dependence shown in Fig. 2.9c to more realistically represent E–k diagrams of real semiconductors. At this stage, it is important to consider the essential influence that the appearance of *energy gap* has on the electrons as current carriers.

The levels of the valence electrons appear in two energy bands with the *energy gap* between them in the case of *semiconductors* and *insulators,* as distinct from the case of *metals* (described in Section 2.2.2). The two bands of importance are illustrated in Fig. 2.10. The energy band below the gap is called *valence band,* whereas the energy band above the gap is called *conduction band.*

An electron at the bottom of the conduction band (E_C) is a standing wave because of the forward and backward reflections at the crystal-lattice sites. Because such an electron does not move through the crystal, it is convenient to define a kinetic energy that is zero for this electron (this means that the potential energy of this and all the other electrons

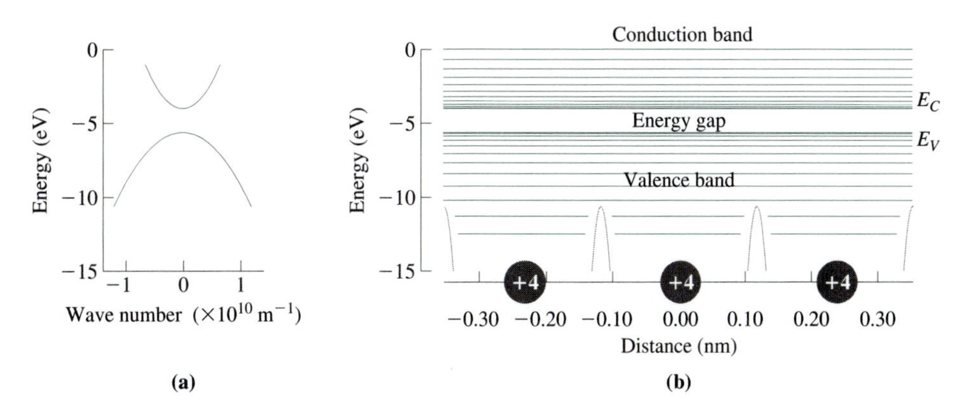

Figure 2.10 Energy bands in a semiconductor. (a) E–k diagram. (b) E–x diagram.

in the conduction band has to be equal to E_C). The electrons at higher energy levels in the conduction band will have kinetic energies according to the upper E–k branch, which can be approximated by a parabola. As shown in Example 2.1, electrons with parabolic E–k dependence appear as free electrons. Therefore, *the electrons in the conduction band are mobile particles*. This is not surprising, given that there are many free levels that the electrons can move on to when their energy is increased by electric field or temperature.

At $T \approx 0$ K, there would be no broken covalent bonds, and all the electrons would be in the valence band. Because half of the energy levels are in the valence band, this band would be full at $T \approx 0$ K (two electrons with different spins at each level). These electrons cannot move if an electric field is applied because there are no unoccupied energy levels for the electrons to move on to so that their energy can be increased. The lowest *free* levels are in the conduction band, but for electric fields lower than the breakdown field, the kinetic energy that the electrons could gain is lower than the energy gap. Therefore, the electrons in the valence band are immobile—they are tied in the covalent bonds that are not broken.

When a covalent bond is broken because sufficient thermal energy has been delivered to a valence electron, this electron jumps up into the conduction band, leaving behind a hole. If an electric field is applied now, an electron from a lower-energy position in the valence band can move to the upper empty position. Using the concept of *hole* as a current carrier (Section 1.2.1), the equivalent presentation is that the hole moves to a lower-energy position. Making an analogy with bubbles in liquid, we can see that the minimum energy for the holes in the valence band is the top of the valence band (E_V). When an electric field is applied, it can move the holes by pushing them deeper into the valence band. Figure 2.10 illustrates that the E–k dependence in the valence band looks like an inverted parabola, meaning that it can represent the kinetic energy of the holes as current carriers.

The existence of an energy gap distinguishes semiconductors and insulators from metals. Figure 2.11a shows again that there is no energy gap in metals (the conduction and the valence bands are merged). This makes all the electrons in the band conductive, as they all can move to a number of empty states therefore, metals are good current conductors even at temperatures very close to 0 K). The energy gap in semiconductors and insulators (Fig. 2.11b) separates the available energy levels into valence and conduction bands, with the valence electrons fully occupying the valence-band levels (assuming that no covalent bonds are broken). The electrons in the valence band are not mobile particles because they do not have empty states to move to; without electrons in the conduction band, intrinsic semiconductors and insulators would not conduct any current. Electrons may appear in the conduction band at high temperatures or at electric fields high enough to break the covalent

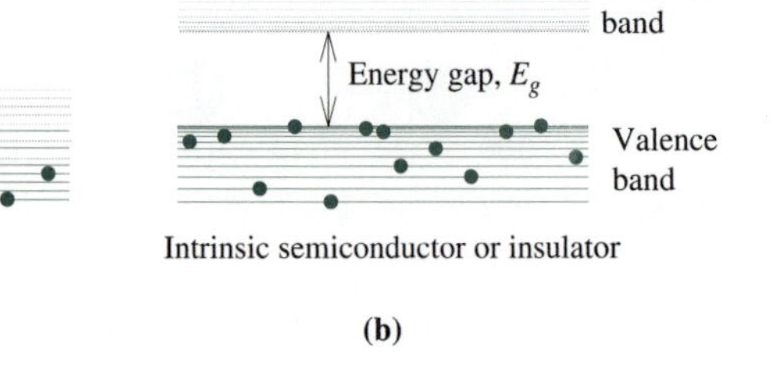

Figure 2.11 Energy-band diagrams for (a) a metal and (b) an intrinsic semiconductor or an insulator.

TABLE 2.1 **Energy-Gap Values for Different Materials at 300 K**

Material	E_g (eV)
Germanium	0.66
Silicon	1.12
Gallium arsenide	1.42
Silicon carbide (cubic)	2.4
Gallium nitride	3.4
Silicon nitride (Si_3N_4)	5
Diamond	5.5
Silicon dioxide (SiO_2)	9

bonds. A broken covalent bond will manifest itself by the appearance of an electron in the conduction band and a hole in the valence band; both these particles are mobile and will contribute to electric current. In practice, however, the most frequent reason for low conduction in insulators and intrinsic semiconductors is unintentional doping.

The energy-gap value is one of the most important parameters of semiconductors and insulators. Table 2.1 gives the values of the energy gap E_g at room temperature for different materials.

Semiconductors are distinguished from insulators by the possibility of increasing their conductivity to quite high levels by either N-type or P-type doping. The effects of doping are illustrated by the atomic-bond model in Fig. 1.19. An energy-band model presentation of the effects of doping is given in Fig. 2.12. For the case of a donor atom (N-type doping), four valence electrons play the role of the four valence electrons of the replaced silicon atom, so their energy levels replace the corresponding energy levels in the valence and the conduction bands. The wave function of the fifth electron in each donor atom does not interfere with any of the neighboring silicon atoms, and the neighboring donor atoms are too far from each other to permit any interference with the equivalent wave functions of their fifth electrons. Consequently, these electrons have energy levels that are localized to the parent atoms (they do not extend throughout the crystal). This is illustrated in Fig. 2.12, which also shows that the energy level of the fifth electron is inside the energy gap but very

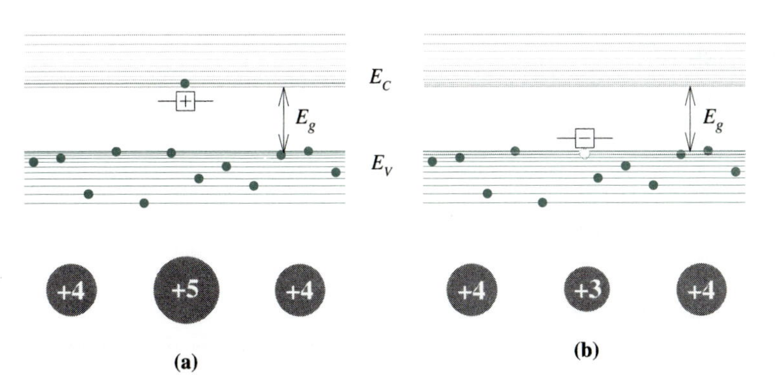

Figure 2.12 Effects of (a) N-type (b) and P-type doping of semiconductors in energy-band model presentation.

close to the bottom of the conduction band. This electron remains at its level in the energy gap at very low temperatures; at room temperature, however, it gets enough energy to jump into the conduction band, becoming a free electron. It leaves behind a positively charged donor atom, which is immobile.

The effect of P-type doping is similarly expressed in Fig. 2.12b. In this case, the acceptor atom introduces a *localized* energy level into the energy gap, which is close to the top of the valence band. As a consequence, an electron from the valence band jumps onto this level, leaving behind a mobile hole and creating a negatively charged and immobile acceptor atom.

†2.3 ELECTRONS AND HOLES AS PARTICLES

Considering electrons as *standing waves* (Sections 2.1 and 2.2) provides an important insight into the origin of energy bands and gap(s). However, the wave approach has its limitations—it deals with perfect waves, which do not exist in reality. The wave function of a free electron is perfectly periodic, from $-\infty$ to ∞ in both space and time (Fig. 2.2). This is a mathematical abstraction that does not perfectly match the fact that real electrons are localized. The Fourier transform [Eq. (2.12)] can be used to obtain a localized wave packet, but it employs another mathematical abstraction (integration of *infinitely* dense spectrum) to remove the abstraction of *perfectly* periodic functions (that is, to achieve a localization of the electron wave function).[9]

The unavoidable conclusion is that both the practical observations of particles and the mathematical wave abstractions have limitations. Our options are to either adapt the practical model of particles or adapt the mathematical model of waves. Kinetic phenomena in semiconductor devices are much easier to explain and understand if the already developed classical concepts, related to a *gas* of particles, are applied to the "free" electrons (the electrons in the conduction band) and the holes. We refer to the *free electrons* and the *holes* as the *electron gas* and the *hole gas,* or the *carrier gas* in general.

Electrons and holes are assumed and imagined as particles in this approach. However, the concepts of *mass* and *size* of the free electrons and holes have to be adapted to include the wave properties of electrons. This section addresses the questions of particle *mass* and *size* in the carrier-gas model.

2.3.1 Effective Mass and Real *E–k* Diagrams

As shown in Example 2.1, the kinetic (and the total) energy of a free electron is related to the velocity ($\boldsymbol{v}$) and the wave vector ($\boldsymbol{k}$) in the following way:

$$E = E_{kin} = \frac{m|\boldsymbol{v}|^2}{2} = \frac{\hbar^2|\boldsymbol{k}|^2}{2m} \tag{2.35}$$

[9]Note that the finite Fourier series [Eq. (2.11)] produces *periodic* wave packets—it does not remove the abstraction of a perfectly periodic function.

This is a parabolic E–k dependence (plotted in Fig. 2.5) with one parameter: the *mass* of the electron. Because of the parabolic dependence, the second derivative of the E–k dependence is a constant that is directly related to the mass:

$$\frac{d^2E}{dk^2} = \frac{\hbar^2}{m} \tag{2.36}$$

Therefore, the electron mass is inversely proportional to the second derivative of the E–k dependence:

$$m = \frac{\hbar^2}{d^2E/dk^2} \tag{2.37}$$

The E–k dependence for the electrons in the conduction band of a semiconductor can be approximated by a parabola (Fig. 2.10a). The second derivative of this parabola can be used to calculate the mass of electrons, according to Eq. (2.37). In general, the mass calculated in this way is different from the mass of a free electron in vacuum. To express this difference, the mass of a free electron in vacuum is labeled by m_0, whereas the mass obtained from Eq. (2.37) is labeled by m^* and called *effective mass.* The reason for the difference between m_0 and m^* is the same as the reason for the specific shape of an E–k dependence: interactions of the electron waves with the periodic potential of crystal atoms. Importantly, this means that the effective mass includes the influence of crystal atoms on the electrons in a conduction band. This enables us to model the electrons in a conduction band as classical particles in a gas; all we need to do is replace m_0 by m^*.

The E–k dependence for the *electrons in a valence band* can be approximated by an inverted parabola (Fig. 2.10a). This means that the mass calculated by Eq. (2.37) will have a negative value—a quantum-mechanical effect that is difficult to relate to classical particles. However, in selecting the *holes* as the valence-band carriers, this difficulty can be removed by applying the analogy with bubbles in a liquid (a *hole gas*). The minimum-energy position for the bubbles corresponds to the highest possible level; accordingly, the kinetic energy of holes is taken to be zero at the peak of the parabola. The kinetic energy of a hole has to be increased in order to "push" it to a lower total-energy level, analogous to the energy that is needed to push a bubble to a lower height in the liquid. With this model of a hole gas, the effective mass of the holes is positive and inversely proportional to the second derivative of the E–k dependence in a valence band. In general,

$$m^*_{e,h} = \pm\frac{\hbar^2}{d^2E/dk^2} \tag{2.38}$$

The actual effective-mass values for the electrons and holes in different semiconductors are different because the real E–k dependencies are different in different crystals. The most significant branches of the real E–k diagram of GaAs are illustrated in Fig. 2.13a. This E–k diagram represents the group of so-called *direct semiconductors,* because both the bottom of the conduction band and the top of the valence band are centered at $k = 0$ (the transitions between the valence and the conduction bands do not require a change in the wave number). For small and moderate values of k, the E–k diagram is

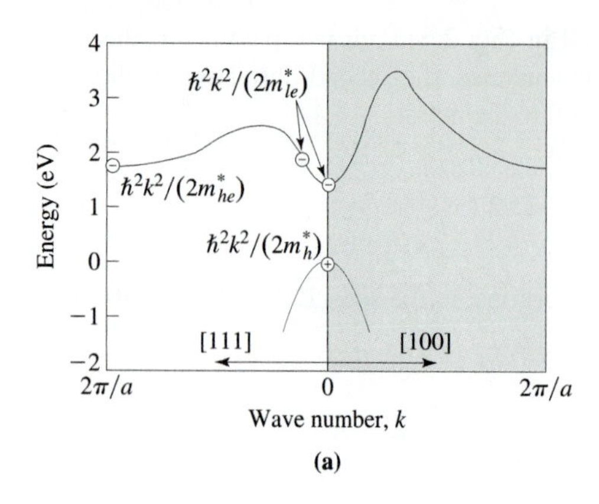

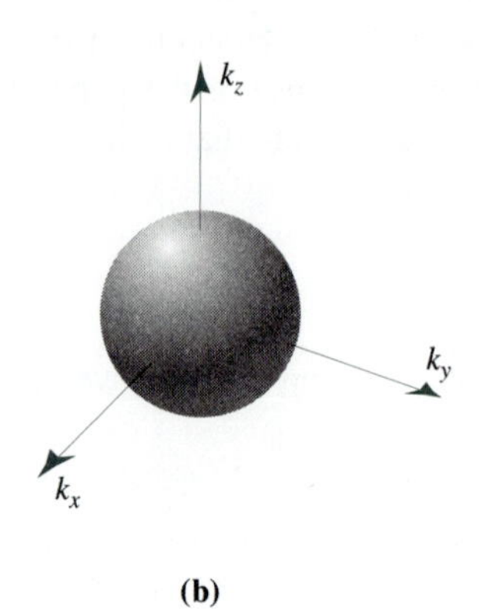

Figure 2.13 (a) E–k diagram and (b) spherical constant-energy surface for GaAs.

qualitatively the same as the ideal E–k diagram shown in Fig. 2.10a. This also means that the diagram is spherically symmetrical, which allows a straightforward transition from the one-dimensional presentation to three dimensions:

$$E = \pm \underbrace{\frac{\hbar^2}{2m^*_{e,h}} k^2}_{E_{kin}} + E_{C,V} \tag{2.39}$$

The kinetic-energy term in Eq. (2.39) is a sphere in the k_x–k_y–k_z coordinate system, as illustrated in Fig. 2.13b. For GaAs, the electron and hole effective masses are $m^*_e = 0.067m_0$ and $m_h = 0.45m_0$, respectively.

For high values of k, the minimum conduction-band energy is for $k = 2\pi/a$ when the electrons are moving along 111 direction (Fig. 2.13 a).[10] The parabola centered at $k = 0$ is "sharper" than the parabola centered at $k = 2\pi/a$, which means the second derivative is larger and the effective mass of the low-energy electrons (around $k = 0$) is smaller. When the energy of an electron is increased to values corresponding to the parabola centered at $k = 2\pi/a$, the effective mass of the electron increases as this parabola is "wider." The low-energy electrons in GaAs are referred to as the *light,* whereas the high-energy electrons are referred to as the *heavy* electrons.

The real E–k diagram of Si is more complicated (Fig. 2.14a): it represents the group of so-called *indirect semiconductors,* because the bottom of the conduction band and the top of the valence band appear for different values of k. Figure 2.14a shows the E–k diagram along two directions: [100] and [111]. It can be seen that the bottom of the conduction band is along the [100] direction. In three dimensions, the constant-energy surface is an ellipsoid:

$$E_{kin} = \frac{\hbar^2}{2m_l}\boldsymbol{k_x}^2 + \frac{\hbar^2}{2m_t}\boldsymbol{k_y}^2 + \frac{\hbar^2}{2m_t}\boldsymbol{k_z}^2 \tag{2.40}$$

---

[10]The E–k diagrams are different in different directions because the distance between atoms varies as the direction through the crystal changes.

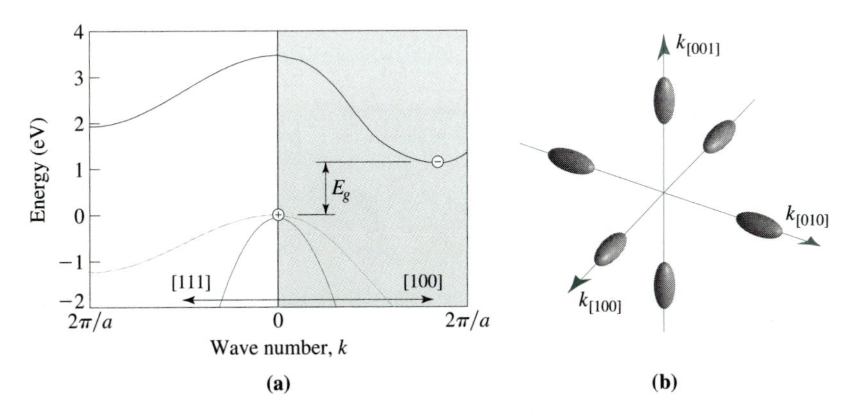

Figure 2.14 (a) E–k diagram of Si and (b) elliptical constant-energy surfaces in the conduction band.

where $\boldsymbol{k_x}$ coincides with a $< 100 >$ direction, so that m_l is the longitudinal effective mass and m_t is the transverse effective mass. Given that there are six equivalent $<100>$ directions ([100], $[\bar{1}00]$, [010], ...), there are six equivalent constant-energy ellipsoids, as illustrated in Fig. 2.14b. The values of the longitudinal and the transversal effective masses are $m_l = 0.98m_0$ and $m_t = 0.19m_0$, respectively.

The E–k diagram in the valence band is centered at k, however, it has two branches with approximately parabolic shapes and closely placed maxima. The sharper parabola corresponds to *light* holes, whereas the wider parabola corresponds to *heavy holes*. The effective masses of the light and the heavy holes are $m^* = 0.16m_0$ and $m^* = 0.49m_0$, respectively.

EXAMPLE 2.4 Conductivity Effective Mass

The kinetic energy of electrons in silicon is anisotropic with $<100>$ symmetry, which can be described as follows (refer to Fig. 2.14b): (1) one-third of the electrons move along the [100] and $[\bar{1}00]$ directions with effective mass m_l, but if scattered so that they start moving in a transverse direction—for example, [010] or $[00\bar{1}]$—their effective mass is m_t; (2) the second third of the electrons move along [010] and $[0\bar{1}0]$ direction with the longitudinal mass m_l and along the four transverse directions with the transverse mass m_t; (3) the longitudinal effective mass, m_l, of the final third of electrons appears for motion along [001] and $[00\bar{1}]$ directions. Knowing that longitudinal and transverse effective masses are $m_l = 0.98m_0$ and $m_t = 0.19m_0$, respectively, determine the average-conductivity effective mass of electrons in silicon so that the isotropic electron-gas model can be used.

SOLUTION

The dependence of kinetic energy on electron momentum for the case of isotropic electron gas is given by

$$E_{kin} = \frac{\hbar^2 \boldsymbol{k}^2}{2m^*} = \frac{\boldsymbol{p}^2}{2m^*} = \frac{p_x^2}{2m^*} + \frac{p_y^2}{2m^*} + \frac{p_z^2}{2m^*}$$

The dependence of kinetic energy on momentum in the case of silicon can be written as

$$\begin{aligned} E_{kin} =& \frac{1}{3}\left(p^2_{[100],[\bar{1}00]}/2m_l + p^2_{[010],[0\bar{1}0]}/2m_t + p^2_{[001],[00\bar{1}]}/2m_t\right) \\ &+ \frac{1}{3}\left(p^2_{[100],[\bar{1}00]}/2m_t + p^2_{[010],[0\bar{1}0]}/2m_l + p^2_{[001],[00\bar{1}]}/2m_t\right) \\ &+ \frac{1}{3}\left(p^2_{[100],[\bar{1}00]}/2m_t + p^2_{[010],[0\bar{1}0]}/2m_t + p^2_{[001],[00\bar{1}]}/2m_l\right) \end{aligned}$$

Taking into account the existing symmetry, we can compress the preceding equation as follows:

$$E_{kin} = \frac{p_l^2}{2m_l} + \frac{p_t^2}{2m_t} + \frac{p_t^2}{2m_t} = \frac{p_l^2}{2m_l} + 2\left(\frac{p_t^2}{2m_t}\right)$$

The kinetic energy for the case of isotropic model is equal to the kinetic energy given by the previous equation if

$$\frac{3}{m^*} = \frac{1}{m_l} + \frac{2}{m_t}$$

Therefore,

$$m^* = \frac{3}{1/m_l + 2/m_t} = \frac{3}{1/0.98 + 2/0.19} m_0 = 0.26 m_0$$

2.3.2 The Question of Electron Size: The Uncertainty Principle

The classical view of particles involves the concept of particle *size*. We used the concept of atom radius to calculate the number of atoms per unit volume (the concentration) or per unit area (the area density) in crystalline materials. The concentration and/or the area density of current carriers are essential components of the carrier-gas model. This raises the question of carrier size.

When considered as standing waves (superpositions of two plane waves traveling in the opposite directions), electrons lack any localization. Although this does not mean that an electron is everywhere in a given crystal, it certainly implies that it *can be* everywhere in that crystal at any instant of time. This model is adequate for metals, but it does not match observed phenomena in semiconductors. A good example is the Haynes–Shockley experiment. A flash of light illuminates a semiconductor bar through a narrow slit to generate free electrons and holes in a very narrow region (Δx) for a very short time interval (Δt); an electric field is applied to drive the electrons to one end of the semiconductor bar and through an ammeter, but no electric current is detected until $t \gg \Delta t$. The fact that no current is detected immediately after the generation of free electrons shows that the electrons were localized for at least a considerable time interval.

In the framework of the wave approach, the Fourier transform [Eq. (2.12)] is used to generate wave packets that provide some localization. By adjusting the spectral function $f(k)$, wave packets of different shapes can be achieved. This is nicely illustrated by the case of normal distribution for $f(k)$. The intensity of the wave packet obtained in this way is a normal distribution in x, as shown by Eq. (2.16). The relationship between the standard deviations of the spectral function (σ_k) and the wave packet intensity (σ_x) is

$$\sigma_x = \frac{1}{\sqrt{2}\sigma_k} \tag{2.41}$$

The wave packets obtained from Eq. (2.16) provide the smallest product $\sigma_x \sigma_k$. However, an electron represented by these wave packets has to be spread from $x = -\infty$ to $x = \infty$, according to the normal distribution. This means that the wave packets obtained from Eq. (2.16) can be used as approximative models at best. In other words, *the mathematical precision of the wave approach does not mean that this approach provides the perfect match to reality*. This problem remains even if a different spectral function $f(k)$ is used in the Fourier transform to properly limit the size of a wave packet—the wave number remains unlimited, extending from $k = -\infty$ to $k = \infty$. This leads to a problem when the wave number is related to the particle momentum ($p = \hbar k$) as the wave theory is applied to reality, because a real particle cannot exhibit infinite momentum. On the other hand, if the range for the wave number is limited in Eq. (2.12), the result is a wave packet that extends from $x = -\infty$ to $x = \infty$.

This wave–particle issue can be summarized as follows: neither the mathematical wave abstractions nor the classical observations of particles are precise enough to describe electrons. The wave approach leads to the result that either the size or the momentum of every electron can be infinitely large; in spite of the inherent mathematical precision, the abstraction of infinite size/momentum *cannot be applied without an alteration* of this mathematically precise approach. On the other hand, the model of classical particles cannot explain experimentally observed phenomena such as diffraction and interference of electrons. Perfectly defined particles do not exist: no particle has a precisely determined momentum and a precisely determined size. This general fact, applicable to any object, is known as the Heisenberg uncertainty principle. The minimum possible uncertainties in the size and the momentum of a particle are related to each other so that their product is equal to Planck's constant:[11]

$$\Delta x \Delta p_x \geq h \tag{2.42}$$

A wide range of considerations unavoidably lead to the uncertainty principle in general, and specifically to Eq. (2.42). Staying with the wave packets, we observe that a wave packet is made up by a superposition of plane sinusoidal waves with the wavelengths distributed around a central wavelength (λ_0). If the size (the width) of the wave packet is Δx, then it consists of n wavelengths, where

$$n = \frac{\Delta x}{\lambda_0} \tag{2.43}$$

Outside Δx, the superimposing sinusoids have to add up to zero, which is possible only if they have varying wavelengths. Then, there are some sinusoids with shorter wavelengths so that at least $n + 1$ wavelengths fit in Δx, and there are some sinusoids with longer wavelengths so that no more than $n - 1$ wavelengths fit in Δx:

$$\begin{aligned} \frac{\Delta x}{\lambda_0 - \Delta\lambda} &\geq n + 1 \\ \frac{\Delta x}{\lambda_0 + \Delta\lambda} &\leq n - 1 \end{aligned} \tag{2.44}$$

Combined with Eq. (2.43), this leads to

$$\begin{aligned} \frac{\Delta x \Delta\lambda}{\lambda_0^2} &\geq \frac{1}{1 + 1/n} \\ \frac{\Delta x \Delta\lambda}{\lambda_0^2} &\geq \frac{1}{1 - 1/n} \end{aligned} \tag{2.45}$$

For large n (so that $1/n \ll 1$), the considerations of both the longer and the shorter wavelengths lead to the following common result:

$$\frac{\Delta x \Delta\lambda}{\lambda_0^2} \geq 1 \tag{2.46}$$

We can use the relationship between the wavelength and the wave number ($k = 2\pi/\lambda$) to move from the uncertainty in the wavelength ($\Delta\lambda$) to uncertainty in the wave number (Δk).

---

[11] W. Heisenberg, *The Physical Principles of the Quantum Theory,* Dover, New York, 1949, p. 14. P. A. M. Dirac, *The Principles of Quantum Mechanics,* 4th ed., Oxford University Press, Oxford, 1958, p. 98. R. P. Feynman, R. B. Leighton, and M. Sands, *The Feynman Lectures on Physics: Quantum Mechanics,* Vol. III, Addison-Wesley, Reading, MA, 1965, p. 2–3.

As $dk/d\lambda = -2\pi/\lambda^2$, and given that the minus sign has no importance when considering uncertainties, the following relationship is established:

$$\Delta k_x = 2\pi \frac{\Delta\lambda}{\lambda_0^2} \tag{2.47}$$

With this, the following condition is obtained:

$$\Delta x \Delta k_x \geq 2\pi \tag{2.48}$$

This result is derived without any quantum-mechanical or particle-related considerations—it simply shows the relationship between the uncertainties in the width (Δx) and the wave number (Δk) of an abstract wave packet. To apply it to a particle, we relate the wave number to the momentum of the particle ($p_x = \hbar k_x$), which leads to the Heisenberg uncertainty relationship: $\Delta x \Delta p_x \geq h$.

There is no need to rely on the concept of wave packet (and the associated Fourier transform) to arrive at the result expressed by Eq. (2.48). In its simplest form, the standing-wave approach (the model of free electrons in a metal, used in Section 2.2.2) suggests uncertainty in the electron positions that is equal to the size of the crystal. However, this uncertainty can be significantly reduced if we take into account that the wave functions of electrons in a periodic potential are periodic (Section 2.2.3). One way of interpreting this result is to say that all the valence electrons, with their periodic wave functions, extend through the whole crystal. Another way is to say that an electron is localized within a cube $\Delta x \Delta y \Delta z$ in three dimensions, or just within an interval Δx for one-dimensional considerations. This would have to mean that a segment of a wave function that is repeated in the remaining Δx intervals corresponds to totally symmetrical electrons inside those intervals. The smallest Δx that we can take is equal to one period of the wave functions, which is equal to the crystal-lattice constant. For any valid Δx, a periodic wave function has to satisfy the following boundary condition:

$$\psi(x + \Delta x, t) = \psi(x, t) \tag{2.49}$$

It is not difficult to show that the wave function for a free electron [Eq. (2.6)] satisfies this condition if k_x is restricted to the following values:

$$k_x = n_x(2\pi/\Delta x) \qquad (n_x = 1, 2, 3, \ldots) \tag{2.50}$$

The difference between the closest k_x values is $\Delta k_x = (2\pi/\Delta x)$, which is the minimum uncertainty according to Eq. (2.48).

Another demonstration of the uncertainty principle relates to an attempt to arbitrarily limit the x size of electrons that have no uncertainty regarding the p_x momentum ($p_x = \Delta p_x = 0$), because they are moving in the y-direction. The condition of $p_x = 0$ is satisfied by a plane wave moving in the y-direction, and for as long as Δp_x remains zero, there is no localization in the x-direction: $\Delta x \to \infty$. To reduce Δx to a finite value, the electrons are made to pass through a slit of a small width, equal to Δx. As experimentally confirmed, electrons will experience diffraction, so that the beam of electrons beyond the slit has a finite angle of divergence (α). By the laws of optics,

$$\sin\alpha \approx \frac{\lambda}{\Delta x} \tag{2.51}$$

where λ is the wavelength of the electrons. The divergence of electrons in the x-direction leads to a spread of the p_x momentum,

$$\Delta p_x = p\sin\alpha \tag{2.52}$$

where p is the total momentum. Given that $p = h/\lambda$, Eqs. (2.51) and (2.52) lead to $\Delta x \Delta p_x \approx h$.

2.3.3 Density of Electron States

The uncertainty about the size of electrons opens the question of how to determine how many of them can fit into a unit of volume.

Given that the size uncertainty is linked to the momentum uncertainty, a solution is to define a six-dimensional x-y-z-p_x-p_y-p_z space. The "volume" of a minimum cell of this space is

$$\Delta V \Delta V_p = \Delta x \Delta y \Delta z \Delta p_x \Delta p_y \Delta p_z = \underbrace{\Delta x \Delta p_x}_{h}\,\underbrace{\Delta y \Delta p_y}_{h}\,\underbrace{\Delta z \Delta p_z}_{h} = h^3 \tag{2.53}$$

Because two electrons (with different spins) can fit into a cell like this, the maximum number of electrons per unit "volume" of the six-dimensional space is $2/(\Delta V \Delta V_p) = 2/h^3$. This means the maximum concentration of electrons (the maximum number of electrons per unit of space volume ΔV) is

$$C = \frac{2}{\Delta V} = \frac{2}{h^3}\Delta V_p \tag{2.54}$$

We note that C depends on $\Delta V_p = \Delta p_x \Delta p_y \Delta p_z$. This dependence can be converted into a dependence on ΔE_{kin}, using the relationship between the kinetic energy and the momentum ($E_{kin} = |p|^2/2m^*$ for an isotropic electron gas). Having this in mind, we can discuss the meaning of the *maximum* concentration C. The maximum electron concentration corresponds to the case of every *electron state* being occupied. At $T \approx 0$ K, all the states up to the Fermi energy level are occupied, and all the states above are empty. At $T > 0$ K, there is a temperature-dependent *distribution of state occupancy*. At this stage we do not know this distribution (it is the subject of Section 2.4). Nonetheless, if we think of C as the *concentration of states* that may or may not be occupied by electrons, rather than *maximum concentration of electrons,* then C is relevant for any occupancy distribution. Given the dependence of C on ΔV_p [Eq. (2.54)], its consequent dependence on kinetic energy will take the following form:

$$C = D(E_{kin})\Delta E_{kin} \approx D(E_{kin})\, dE_{kin} \tag{2.55}$$

The function $D(E_{kin})$ is referred to as the *density of electron states,* its meaning being the number of states per unit volume and unit energy. Note that ΔE_{kin} is the smallest possible difference between two discrete energy levels; in its usual meaning, dE_{kin} is an infinitesimal energy interval in the sense that there are no variations of $D(E_{kin})$ within dE_{kin}, but $dE_{kin} \gg \Delta E_{kin}$ so that the discrete energy levels can be approximated by the continuous function $D(E_{kin})$.

Analogous considerations apply for a hole gas, the analogous concept being the *density of hole states.*

EXAMPLE 2.5 Density of Electron States

Determine the density of states for a three-dimensional and isotropic electron gas, taking into account that the relationship between the kinetic energy and the momentum is given by

$$E_{kin} = \frac{\boldsymbol{p}^2}{2m^*}$$

SOLUTION

Given that

$$(\Delta x \Delta p_x)(\Delta y \Delta p_y)(\Delta z \Delta p_z) = h^3$$

we can define the volume of a double electron state ("double" to include the spin factor) as:

$$\Delta V = \Delta x \Delta y \Delta z = \frac{h^3}{\Delta p_x \Delta p_y \Delta p_z} = \frac{h^3}{\Delta V_p}$$

where $\Delta p_x \Delta p_y \Delta p_z = \Delta V_p$ is by analogy the *momentum volume* of a double electron state; we will refer to this concept as the unit of momentum volume. Given that there can be two electrons per volume ΔV, the number of electron states per unit volume is

$$C = \frac{2}{\Delta V} = \frac{2}{h^3} \Delta V_p$$

We need to determine the number of states per unit volume and unit energy, thus $D = C/\Delta E_{kin}$. This task requires us to express ΔV_p in terms of ΔE_{kin}. The unit of momentum volume (ΔV_p) can be imagined as a cube whose sides are Δp_x, Δp_y, and Δp_z. This model, however, does not include the special relationship that exists between the momentum as a vector and the energy as a scalar: $E_{kin} = \boldsymbol{p}^2/2m^* = |\boldsymbol{p}|^2/2m^*$. The important thing is that a single-energy value relates to all the combinations of p_x, p_y, and p_z that satisfy the condition $|\boldsymbol{p}|^2 = p_x^2 + p_y^2 + p_z^2$. This constant-energy situation can be visualized as a sphere with radius $|\boldsymbol{p}| = p$: as long as the position of a (p_x, p_y, p_z) point is on the sphere, the energy remains constant. The energy does change if the radius is changed by Δp. Therefore, the suitable unit of momentum volume can be visualized as a sphere with radius p and thickness Δp, meaning that ΔV_p is the difference of the volumes of spheres with radii $p + \Delta p$ and p, respectively: $\Delta V_p = V_{p+\Delta p} - V_p$. This enables V_p and ΔV_p to be related to E_{kin} and ΔE_{kin} in a straightforward way:

$$V_p = \frac{4}{3}\pi p^3$$

$$E_{kin} = \frac{p^2}{2m^*} \Rightarrow p = (2m^* E_{kin})^{1/2}$$

$$V_p = \frac{4}{3}\pi (2m^* E_{kin})^{3/2}$$

$$\frac{dV_p}{dE_{kin}} = 4\sqrt{2}\pi (m^*)^{3/2} E_{kin}^{1/2}$$

$$\Delta V_p = 4\sqrt{2}\pi (m^*)^{3/2} E_{kin}^{1/2} \Delta E_{kin}$$

Therefore, the number of electron states per unit volume is

$$C = \frac{2}{h^3}\Delta V_p = \frac{8\sqrt{2}\pi}{h^3}(m^*)^{3/2} E_{kin}^{1/2} \Delta E_{kin}$$

whereas the number of electron states per unit volume and unit energy (the density of electron states) is

$$D = \frac{C}{\Delta E_{kin}} = \frac{8\sqrt{2}\pi}{h^3}(m^*)^{3/2} E_{kin}^{1/2}$$

EXAMPLE 2.6 Density-of-States Effective Mass

As described in Example 2.4, the kinetic energy of electrons in silicon is anisotropic:

$$E_{kin} = \frac{p_x^2}{2m_l} + \frac{p_y^2}{2m_t} + \frac{p_z^2}{2m_t}$$

with $\langle 100 \rangle$ symmetry, meaning that the equation applies to all the six equivalent directions (x-axis aligned to [100], [$\bar{1}$00], [010], [0$\bar{1}$0], [001], and [00$\bar{1}$]). More descriptively, the constant-momentum sphere ($|\boldsymbol{p}|$) is converted into six constant-energy ellipsoids with the elongations aligned to each of the six equivalent $\langle 100 \rangle$ directions (refer to Fig. 2.14b). Knowing that longitudinal and transverse effective masses are $m_l = 0.98m_0$ and $m_t = 0.19m_0$, respectively, determine the density-of-state effective mass of electrons in silicon, so that the density-of-states equation derived in Example 2.5 can be used. Plot the density-of-state dependence on energy in the energy range 0 to 1 eV, taking the energy values in steps of 0.01 eV.

SOLUTION

The general equation of an ellipsoid in the (p_x, p_y, p_z) coordinate system is

$$\frac{p_x^2}{a^2} + \frac{p_y^2}{b^2} + \frac{p_z^2}{c^2} = 1$$

where a, b, and c are the semiaxes of the ellipsoid, and $(4\pi/3)abc$ is the volume of the ellipsoid. Dividing the kinetic-energy equation by E_{kin}:

$$\frac{p_x^2}{2m_l E_{kin}} + \frac{p_y^2}{2m_t E_{kin}} + \frac{p_z^2}{2m_t E_{kin}} = 1$$

we find that the semiaxes of the ellipsoid are: $a = \sqrt{2m_l E_{kin}}$ and $b = c = \sqrt{2m_t E_{kin}}$. Therefore, the unit of momentum volume, ΔV_p, can be determined as follows:

$$V_p = \frac{4}{3}\pi abc = \frac{8\sqrt{2}\pi}{3}(m_l m_t m_t)^{1/2} E_{kin}^{3/2}$$

$$\Delta V_p = 4\sqrt{2}\pi (m_l m_t m_t)^{1/2} E_{kin}^{1/2} \Delta E_{kin}$$

Given that there are $M = 6$ equivalent ellipsoids, the number of states per unit volume is

$$C = M\frac{2}{h^3}\Delta V_p = \frac{8\sqrt{2}\pi}{h^3} M (m_l m_t m_t)^{1/2} E_{kin}^{1/2} \Delta E_{kin}$$

and the number of states per unit volume and unit energy (the density of states) is

$$D = \frac{8\sqrt{2}\pi}{h^3} \underbrace{M (m_l m_t m_t)^{1/2}}_{(m^*)^{3/2}} E_{kin}^{1/2}$$

which is identical to the isotropic equation when the density-of-states effective mass is defined as

$$m^* = M^{2/3}(m_l m_t m_t)^{1/3}$$

For the case of silicon:

$$m^* = 6^{2/3} \underbrace{(0.98 \times 0.19 \times 0.19)^{1/3}}_{0.328} m_0 = 1.08 m_0$$

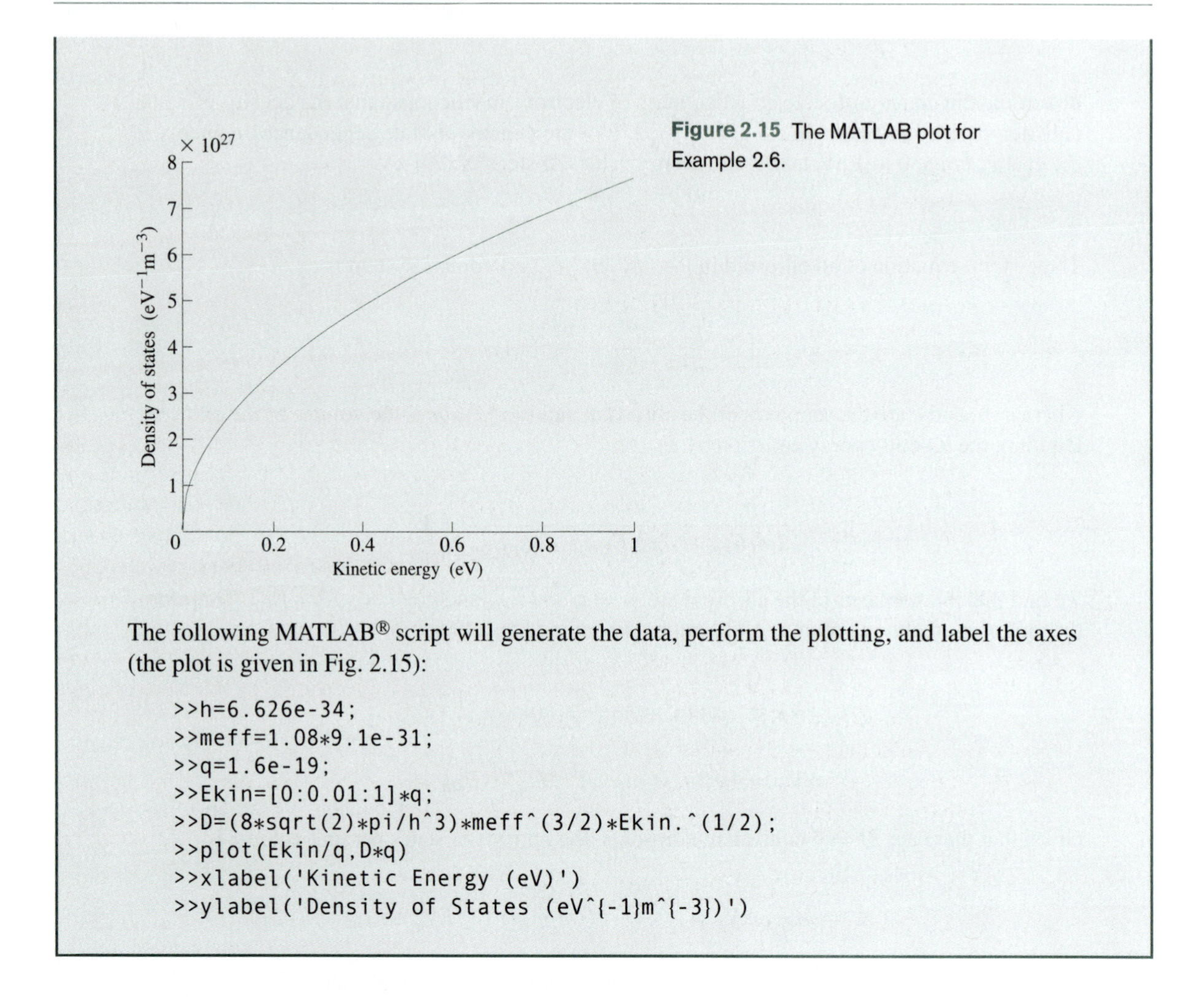

Figure 2.15 The MATLAB plot for Example 2.6.

The following MATLAB® script will generate the data, perform the plotting, and label the axes (the plot is given in Fig. 2.15):

```
>>h=6.626e-34;
>>meff=1.08*9.1e-31;
>>q=1.6e-19;
>>Ekin=[0:0.01:1]*q;
>>D=(8*sqrt(2)*pi/h^3)*meff^(3/2)*Ekin.^(1/2);
>>plot(Ekin/q,D*q)
>>xlabel('Kinetic Energy (eV)')
>>ylabel('Density of States (eV^{-1}m^{-3})')
```

2.4 POPULATION OF ELECTRON STATES: CONCENTRATIONS OF ELECTRONS AND HOLES

To convert the density of electron states (the number of *states* per unit volume and unit energy) into the number of *electrons* per unit volume and unit energy, we have to multiply the density of states by the *probability that each of the states is occupied by an electron,* $f(E)$. To obtain the total concentration of free electrons, regardless of what their kinetic energy may be, the product $f(E)D(E_{kin})$ has to be integrated across the entire range of kinetic energies:

$$n_0 = \int_0^\infty f(E)D(E_{kin})\, dE_{kin} \tag{2.56}$$

The probability that an electron state (corresponding to an energy level E) is occupied by an electron is an energy-dependent probability-density function, referred to as the *Fermi–Dirac distribution*. This function is general: it incorporates the minimum-energy principle[12] and the Pauli exclusion principle,[13] and is independent of the material containing the considered electrons. The properties of a specific material are taken into account by $D(E_{kin})$ when Eq. (2.56) is used to determine the concentration of free electrons.

2.4.1 Fermi–Dirac Distribution

The aim of this section is to demonstrate how the Fermi–Dirac distribution incorporates the minimum-energy and Pauli exclusion principles.

The minimum-energy principle has to be considered in the context of *thermal equilibrium*. This means that the electrons would occupy the states corresponding to the lowest possible energy levels at $T = 0$ K; but at higher temperatures, the free electrons (the electrons in the electron gas) share the thermal energy of the crystal. The state of thermal equilibrium is established and maintained by continuous exchanges of thermal energy between the electrons and the crystal lattice, and to some extent between the electrons themselves. This means that the free electrons continuously gain and give away thermal energy, jumping up and falling down in terms of energy levels corresponding to the states they occupy. Let us consider the transitions between two electron states, labeled by 1 and 2 so that $E_2 > E_1$ (E_1 and E_2 are the energy levels corresponding to the states 1 and 2, respectively). The probability for the transition of an electron from state 1 to state 2 is smaller than the probability for the transition from state 2 to state 1 ($P_{12} < P_{21}$). Moreover, P_{12} decays with an increase in the energy difference ($E_2 - E_1$), but increases as the thermal energy (kT) increases. Assuming exponential dependencies, P_{12} can be expressed as

$$P_{12} = P_{21} \exp\left(-\frac{E_2 - E_1}{kT}\right) \tag{2.57}$$

where k is the Boltzmann constant and T is the absolute temperature (temperature expressed in Kelvins).[14] This equation shows that it is increasingly less likely that an electron will make a transition from state 1 to state 2 when the temperature is decreased or the energy difference $E_2 - E_1$ is increased.

Importantly, the probabilities P_{12} and P_{21} are for specific cases. It is inherently assumed that state 1 is occupied and state 2 is empty when the transition from state 1

---

[12]The minimum-energy principle is analogous to the *maximum entropy* principle, used later in this section (Example 2.9) to derive the Fermi–Dirac distribution.

[13]Particles obeying the Pauli exclusion principle, and consequently the Fermi–Dirac distribution, are called *fermions*. Other types of particles, such as photons, follow different types of energy distribution.

[14]The exponential dependence for the decrease with $E_2 - E_1$ and the increase with kT appears the most logical. In Example 2.9, the Fermi–Dirac distribution is derived without this kind of intuitive assumption.

to state 2 is considered (P_{12}). Quite clearly, there can be no electron transition from state 1 if state 1 is empty. In addition, there can be no transition to state 2 if state 2 is occupied. This is because of the Pauli exclusion principle: if an electron state is occupied, no other electron can move to this state. Introducing the probabilities that state 1 is occupied (f_1) and that state 2 is empty ($1 - f_2$), the probability for the transition of an electron from state 1 to state 2 can be generalized as follows:

$$P'_{12} = P_{12} f_1 (1 - f_2) \tag{2.58}$$

Likewise, the probability that an electron will move from state 2 to state 1 is equal to the probability that state 2 is occupied (f_2), times the probability that state 1 is empty ($1 - f_1$), and times the probability for this transition when the conditions of occupied state 2 and empty state 1 are satisfied (P_{21}):

$$P'_{21} = P_{21} f_2 (1 - f_1) \tag{2.59}$$

In thermal equilibrium, the average occupancies of states 1 and 2 do not change over time. This is possible if the number of transitions from state 1 to 2 is equal to the number of transitions from state 2 to 1, $P'_{12} = P'_{21}$:

$$P_{12} f_1 (1 - f_2) = P_{21} f_2 (1 - f_1) \tag{2.60}$$

At this stage we see that $f_1(1-f_2)$ has to be larger than $f_2(1-f_1)$ to compensate the lower probability for transition from state 1 to state 2 ($P_{12} < P_{21}$). The condition $f_1(1 - f_2) > f_2(1 - f_1)$ means that $f_1 > f_2$—a state corresponding to a lower energy level is more likely to be occupied by an electron.

From Eqs. (2.57) and (2.60), we obtain

$$\frac{f_1}{1 - f_1} \exp(E_1/kT) = \frac{f_2}{1 - f_2} \exp(E_2/kT) \tag{2.61}$$

According to this result, the probability f that an arbitrary state, corresponding to energy level E, is occupied by an electron is given by

$$\frac{f}{1 - f} \exp(E/kT) = A \tag{2.62}$$

where A is a constant. Therefore,

$$f = \frac{1}{1 + (1/A)\exp(E/kT)} \tag{2.63}$$

This equation shows that the probability f decreases with energy. The decrease, however, is more complex than the simple $\exp(-E/kT)$ dependence [Eq. (2.57)] due to the involvement of the Pauli exclusion principle in Eqs. (2.58) and (2.59). The constant A can be determined so that $f = 0.5$ for $E = E_F$, where E_F is the maximum energy level

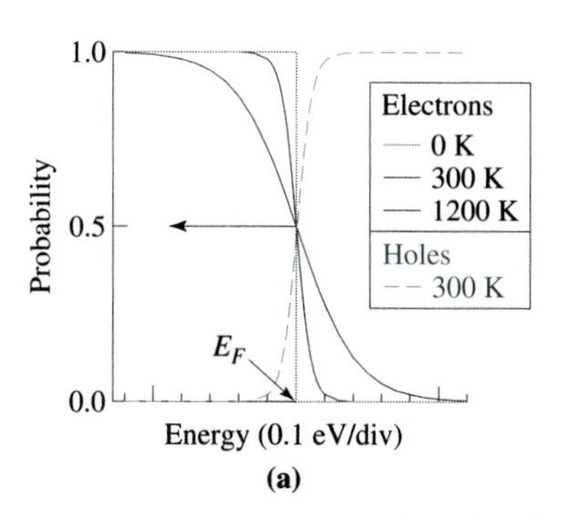

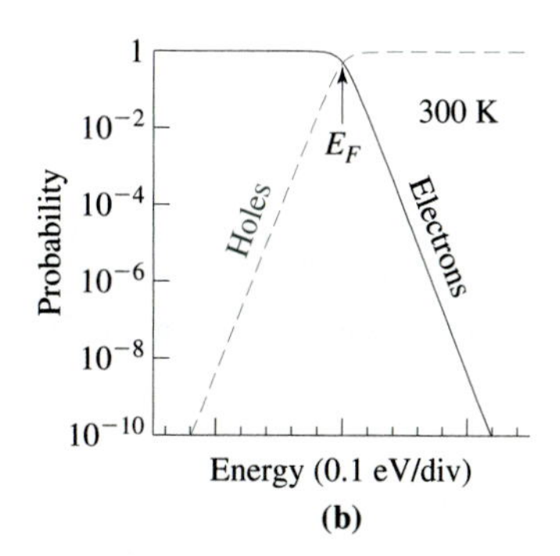

Figure 2.16 Fermi–Dirac distributions for electrons (solid lines) and holes (dashed lines), plotted on (a) linear and (b) logarithmic axes.

corresponding to filled electron states at $T = 0$ K. We will refer to E_F as the Fermi level.[15] It can easily be seen from Eq. (2.62) that $A = \exp(E_F/kT)$ for $f = 0.5$ and $E = E_F$. With this value for A, Eq. (2.63) takes the final form of the Fermi–Dirac distribution:

$$f = \frac{1}{1 + e^{(E-E_F)/kT}} \tag{2.64}$$

The Fermi–Dirac distribution for the holes, current carriers of the second type, can be obtained as the probability that an electron state is empty:

$$f_h = 1 - f = \frac{1}{1 + e^{(E_F-E)/kT}} \tag{2.65}$$

Equations (2.64) and (2.65) are plotted with solid and dashed lines, respectively, in Fig. 2.16. It can be seen that the electron and hole probabilities are completely symmetrical, always giving 1 when added to each other. The solid line for 0 K shows that f is equal to 1 for $E < E_F$ and is equal to 0 for $E > E_F$. Because f should express that at 0 K the probability of having an electron in the conduction band is equal to 0, while it is equal to 1 for the valence band, it is obvious that E_F has to be somewhere between the conduction and the valence bands, therefore in the energy gap. This is not in a contradiction with the fact that there are no allowed energy states in the energy gap, as E_F is a reference energy level only. In spite of the fact that $f = 0.5$ at $E = E_F$, there will be no electrons with $E = E_F$ because the electron concentration is obtained when the probability f is multiplied by the density of states [Eq. (2.56)], which is zero in the energy gap.

---

[15] Conceptually, the Fermi level E_F is not different from the Fermi energy E_{F0}, used in Eq. (2.32). The differences in the definitions and the values of E_F and E_{F0} are due to different reference levels. The Fermi level is defined with respect to the energy of a free electron in vacuum; therefore, $E_F < 0$ for electrons in solids, because these electrons need to gain energy to reach the zero level (free electrons in vacuum). The Fermi energy $E_{F0} > 0$ is defined with respect to the bottom of a potential well that models the trapping of free electrons in a solid material.

EXAMPLE 2.7 Probabilities of Finding Electrons and Holes at E_C and E_V

(a) Assuming that the Fermi level is at the midgap in the intrinsic silicon, calculate the probability of finding an electron at the bottom of the conduction band ($E = E_C$) for three different temperatures: 0 K, 20°C, and 100°C.

(b) How are these probabilities related to the probabilities of finding a hole at $E = E_V$, which is the top of the valence band?

SOLUTION

(a) To calculate these probabilities, the Fermi–Dirac distribution is used:

$$f(E_C) = 1/\left[1 + e^{(E_C - E_F)/kT}\right]$$

where $E_C - E_F = 0.56$ eV, which is half the value of the energy gap $E_g = 1.12$ eV. For the case of 0 K, $f(E_C) = 1/\left[1 + e^{\infty}\right] = 0$. For 20°C the following result is obtained:

$$f(E_C) = 1/\left[1 + e^{0.56/(8.62\times10^{-5}(20+273.15))}\right] = 2.3744692 \times 10^{-10}$$

In a similar way, $f(E_C)$ is determined for 100°C, the result being 2.7476309×10^{-8}.

(b) The probability of finding a hole at $E = E_V$ is equal to the probability of finding an electron at E_C in the case of an intrinsic semiconductor (based on the assumption from the text of the example that E_F is at the midgap).

EXAMPLE 2.8 Two-Dimensional Electron Gas (2DEG)

The density of states in two-dimensional systems (such as the channel in high-electron-mobility transistors) is given by (Problem 2.15):

$$D = \frac{4\pi}{h^2} m^*$$

Determine the dependence of the number of free electrons per unit area (n_{2D}) on the Fermi-level position with respect to the bottom of the two-dimensional subband (E_1).

SOLUTION

This example enables a straightforward application of Eq. (2.56), adapted for the areal density of electrons. The free electrons are those at the energy levels $E > E_1$, where $E - E_1$ is their kinetic energy. Therefore,

$$n_{2D} = \int_0^{\infty} Df(E)\, dE_{kin} = D \int_0^{\infty} \frac{dE_{kin}}{1 + e^{(E_{kin}+E_1-E_F)/kT}}$$

$$n_{2D} = DkT \ln\left[1 + e^{(E_F - E_1/kT)}\right] = \frac{4\pi m^* kT}{h^2} \ln\left[1 + e^{(E_F - E_1)/kT}\right]$$

*EXAMPLE 2.9 Derivation of Fermi–Dirac Distribution from the Maximum-Entropy Principle

Entropy is a quantity that characterizes the degree of uncertainty of random events, such as population of electron states. Labeling a specific electron state by i, the probability that the state is occupied by $p_{i,1} = f_i$, and the probability that the state is empty by $p_{i,2} = (1 - f_i)$, the entropy (S_i) of this set of events is given by

$$S_i = \sum_{j=1}^{2} p_{i,j} \ln p_{i,j} = f_i \ln f_i + (1 - f_i) \ln(1 - f_i) \tag{2.66}$$

If we consider a system of M electron states, the entropy of this system is the sum of the entropies of the individual states:

$$S = \sum_{i=1}^{M} S_i = \sum_{i=1}^{M} f_i \ln f_i + (1 - f_i) \ln(1 - f_i) \tag{2.67}$$

Spontaneous events occur in a way that maximizes the entropy of a system (this maximum-entropy principle is analogous to the minimum-energy principle). Occupancy of electron states is spontaneous with some restrictions (conditions): (1) no more than one electron can occupy a single state (this is the Pauli exclusion principle, where the factor of electron spin is taken into account in the count of the number of states), (2) the number of electrons (N) is fixed, and (3) the energy of the system of electrons (E_{system}) is fixed. These conditions can be expressed by the following two equations:

$$N = \sum_{i=1}^{M} f_i \tag{2.68}$$

$$E_{system} = \sum_{i=1}^{M} E_i f_i \tag{2.69}$$

where E_i is the energy level corresponding to the ith state.

Obtain the distribution of electron occupancy (f_i) so that the entropy S is maximum under the conditions given by Eqs. (2.68) and (2.69).

SOLUTION

The method of Lagrange multipliers can be used to find the maximum of a function under specified conditions. According to this method, the conditions are incorporated in a newly defined function, $H(f)$:

$$H(f) = \sum_{i=1}^{M} [f_i \ln f_i + (1 - f_i) \ln(1 - f_i)] - \lambda_1 \sum_{i=1}^{M} f_i - \lambda_2 \sum_{i=1}^{M} E_i f_i = \sum_{i=1}^{M} h_i(f_i)$$

where λ_1 and λ_2 are constants to be determined from the conditions given by Eqs. (2.68) and (2.69). With this, the problem is reduced to maximizing $H(f)$. It is known from the calculus of variations that the function f has to satisfy the Euler–Ostrogradski equation so that the first-order variation of $H(f)$ is zero [$H(f)$ has either maximum or minimum]. In this simple case, the Euler–Ostrogradski equation is simple: the first derivative of $h_i(f_i)$ equal to zero,

$$\frac{d}{df_i}[f_i \ln f_i + (1 - f_i)\ln(1 - f_i) - \lambda_1 f_i - \lambda_2 E_i f_i] = 0 \Rightarrow f_i = \frac{1}{1 + \exp(\lambda_1 + \lambda_2 E_i)}$$

The Lagrange constants λ_1 and λ_2 should be determined so that the function f_i satisfies the conditions given by Eqs. (2.68) and (2.69). Constants λ_1 and λ_2 can be transformed into a different set of mathematical constants that have physical meaning:

$$\lambda_1 = -\frac{E_F}{kT}$$

$$\lambda_2 = -\frac{1}{kT}$$

where E_F is the Fermi level, and kT is the thermal energy. With this, the function f_i becomes

$$f_i = \frac{1}{1 + \exp[(E_i - E_F)/kT]} \tag{2.70}$$

whereas the conditions [Eqs. (2.68) and (2.69)] take the following forms:

$$N = \sum_{i=1}^{M} \frac{1}{1 + \exp[(E_i - E_F)/kT]} \tag{2.71}$$

$$E_{system} = \sum_{i=1}^{M} \frac{E_i}{1 + \exp[(E_i - E_F)/kT]} \tag{2.72}$$

Equation (2.70) is the Fermi–Dirac distribution function. Equation (2.71) shows that the Fermi level is (mathematically) a Lagrange constant that is set by the number of electrons in the considered system. Equation (2.72) shows that the thermal energy is related to the total energy of the system of N electrons.

2.4.2 Maxwell–Boltzmann Approximation and Effective Density of States

The logarithmic plots of f and f_h, given in Fig. 2.16b by the solid and dashed lines, respectively, more clearly express the electron probabilities for $E > E_F$ (toward the conduction band) and the hole probabilities for $E < E_F$ (toward the valence band).

The linear segments in these semilogarithmic plots indicate the regions of practically exponential energy dependencies of f and f_h. For the case of electrons, the region of exponential energy dependence is for energies above the Fermi level. This is so because the condition $E - E_F \gg kT$ leads to $\exp[(E - E_F)/kT] \gg 1$, and by neglecting 1 in the denominator of Eq. (2.64) we obtain

$$f(E) \approx e^{-(E-E_F)/kT} \tag{2.73}$$

The removal of 1 from the denominator of the Fermi–Dirac distribution effectively means that the Pauli exclusion principle is removed. So, in the region where 1 can be neglected, the Pauli exclusion principle is not pronounced. This region is for $E - E_F \gg kT$, where most of the electron states are unoccupied, so there is almost no "competition" for electron states—the action of the Pauli principle is avoided. Because Eq. (2.73) does not involve the Pauli exclusion principle, it applies to classical particles and is known as the Maxwell–Boltzmann distribution.

Analogously, the Fermi–Dirac distribution for holes can be approximated by the Maxwell–Boltzmann distribution for $E_F - E \gg kT$. In this region, $\exp[(E_F - E)/kT \gg 1]$, so Eq. (2.65) can be approximated by

$$f_h(E) \approx e^{-(E_F-E)/kT} \tag{2.74}$$

The simplification achieved by the Maxwell–Boltzmann approximation may not seem significant, but its full importance becomes obvious when the distribution function is combined with the density of states to solve the integral in Eq. (2.56) for the electron concentration. For the case of three-dimensional density of states, there is no primitive function for the integral in Eq. (2.56) if the Fermi–Dirac distribution is used.[16] However, this integral can be solved with the Maxwell–Boltzmann distribution. Noting that $E = E_C + E_{kin}$,[17] and inserting $D(E_{kin})$ from Example 2.5 and $f(E) = f(E_C + E_{kin})$ as given by Eq. (2.73), we obtain

$$\begin{aligned} n_0 &= \int_0^\infty f(E_C + E_{kin}) D(E_{kin})\, dE_{kin} \\ &= \frac{8\sqrt{2}\pi (m^*)^{3/2}}{h^3} e^{-(E_C-E_F)/kT} \underbrace{\int_0^\infty E_{kin}^{1/2} e^{-E_{kin}/kT}\, dE_{kin}}_{\sqrt{\pi}(kT)^{3/2}/2} \end{aligned} \tag{2.75}$$

Therefore, the concentration of free electrons can be expressed in the following simple way:

$$n_0 = N_C e^{-(E_C-E_F)/kT} \tag{2.76}$$

[16]This is the reason for using the two-dimensional density of states for the straightforward illustration in Example 2.8.

[17]Refer to Section 2.2.3 for the relationship between the total energy (E) and the kinetic energy (E_{kin}) of electrons in the conduction band.

where N_C is a temperature-dependent material constant,

$$N_C = \frac{4\sqrt{2}(\pi m^* kT)^{3/2}}{h^3} \tag{2.77}$$

The term $\exp[-(E_C - E_F)/kT]$ in Eq. (2.76) represents the probability of finding an electron at the bottom of the conduction band (E_C) according to the Maxwell–Boltzmann distribution. If we assume that the density of states at (and close to) the bottom of the conduction band is N_C, then the concentration of electrons at (and close to) the bottom of the conduction band is equal to the density of states times the probability that each of these states is occupied:

$$n_0 = N_C f(E_C) = N_C e^{-(E_C - E_F)/kT} \tag{2.78}$$

which is identical result to the one given by Eq. (2.76). This gives a specific meaning to the constant N_C in Eq. (2.76). N_C is an *effective density of states,* allowing us to assume that all the electrons in the conduction band are at energy levels that are very close to the bottom of the conduction band.

Analogously, the effective density of states in the valence band is labeled by N_V. Given that the probability of finding a hole at $E = E_V$ is equal to $\exp[-(E_F - E_V)/kT]$, according to Eq. (2.74), the concentration of holes is obtained as

$$p_0 = N_V e^{-(E_F - E_V)/kT} \tag{2.79}$$

As can be seen from Eq. (2.77), the effective density of states N_C and N_V are temperature-dependent parameters for a given semiconductor. Their room-temperature values for the three most common semiconductors are given in Table 2.2. The room-temperature value for N_C in silicon is 2.86×10^{19} cm^{-3}, which means the probability of finding an electron at the bottom of the conduction band in the intrinsic silicon is $n_i/N_C = 1.02 \times 10^{10}/2.86 \times 10^{19} = 3.6 \times 10^{-10}$. In the case of a moderately doped N-type silicon (say, $n_0 = N_D = 2.86 \times 10^{16}$ cm^{-3}), the probability of finding an electron at the bottom of the conduction band is $n_0/N_C = 0.001$. One in a thousand may seem to be a small probability in everyday life, but it is very significant in terms of the electron-state occupancy in semiconductors.

It should be stressed again that Eqs. (2.76) and (2.79) are based on the Maxwell–Boltzmann *approximation* of the Fermi–Dirac distribution. In the case of electrons, the approximation can be used for $E - E_F \gg kT$. For the electrons at the bottom of the conduction band, this means $E_C - E_F \gg kT$. In other words, Eq. (2.76) can be used when the Fermi level (E_F) is well below the bottom of the conduction band. In the case of holes, Eq. (2.79) can be used when E_F is well above the top of the valence band. These conditions are not satisfied in heavily doped semiconductors when there are so many carriers in the bands that they start competing for empty states (the action of the Pauli exclusion principle becomes pronounced). Heavily doped semiconductors, where the Maxwell–Boltzmann approximation cannot be used, are also referred to as *degenerate semiconductors.* Suitable approximations for the concentration of electrons in a heavily doped semiconductor are derived in Example 2.11.

TABLE 2.2 Effective Density of States at 300 K

	Effective Density of States (cm^{-3})	
	Conduction Band	Valence Band
Semiconductor	$N_C = A_C T^{3/2}$	$N_V = A_V T^{3/2}$
Silicon	2.86×10^{19}	3.10×10^{19}
Gallium arsenide	4.7×10^{17}	7.0×10^{18}
Germanium	1.0×10^{19}	6.0×10^{18}

EXAMPLE 2.10 Energy Distribution of Electron Concentration

As shown in Example 2.5, the density of states in the conduction band of silicon is given by

$$D = \frac{8\sqrt{2}\pi}{h^3}(m^*)^{3/2} E_{kin}^{1/2}$$

where $E_{kin} = E - E_C$ and $m^* = 1.08m_0$ (refer to Example 2.6 for the value of the density-of-state effective mass). Plot the energy distributions of electron concentration for the following two positions of the Fermi level E_F: $E_C - E_F = 5kT$ and $E_C - E_F = 0$. Assume room temperature ($kT = 0.02585$ eV) and use both the Fermi–Dirac distribution and the Maxwell–Boltzmann approximation for the probability of state occupancy. Take the energy range between 0 and 200 meV and plot the energy distribution of electron concentration on a logarithmic axis.

SOLUTION

The energy distribution of electron concentration shows how the value of the fraction of electron concentration (dn_0), that is within a narrow energy range (dE), changes when the narrow energy range is centered at different energy levels E. We will label it by $g(E) = dn_0/dE$ because it expresses the concentration of electrons per unit energy. It is closely related to the density of states (D), so let us compare the definitions for $g = dn_0/dE$ and D:

1. D, the density (concentration) of electron *states* per unit energy,
2. $g = dn_0/dE$, the concentration of *electrons* (occupied states) per unit energy.

Therefore, D is multiplied by the occupancy probability (f) to obtain dn_0/dE:[18]

$$\frac{dn_0}{dE} = D(E)f(E)$$

[18]Mathematically, this result can be obtained by a direct differentiation of Eq. (2.56).

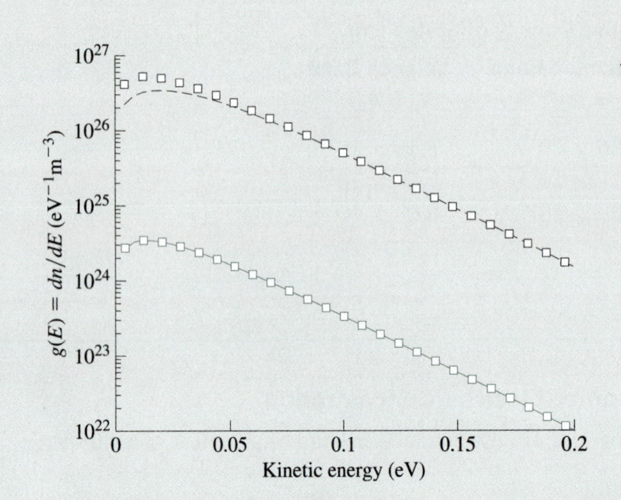

Figure 2.17 The MATLAB plot for Example 2.10.

where

$$f(E) = \begin{cases} 1/\left[1+e^{(E-E_F)/kT}\right] & \text{Fermi–Dirac distribution} \\ e^{-(E-E_F)/kT} & \text{Maxwell–Boltzmann approximation} \end{cases}$$

Taking into account the fact that the total energy E is the sum of the potential energy at the bottom of the conduction band (E_C) and the kinetic energy (E_{kin}), the occupancy-probability functions can be expressed as

$$f(E) = \begin{cases} 1/\left[1+e^{(E_{kin}+E_C-E_F)/kT}\right] & \text{Fermi–Dirac distribution} \\ e^{-(E_{kin}+E_C-E_F)/kT} & \text{Maxwell–Boltzmann approximation} \end{cases}$$

Now we can perform the calculations and the plotting (the plot is given in Fig. 2.17):

```
>>q=1.6e-19;
>>h=6.626e-34;
>>kT=0.02585;
>>meff=1.08*9.1e-31;
>>Ekin=[0:0.004:0.2]*q;
>>D=(8*sqrt(2)*pi/h^3)*meff^(3/2)*Ekin.^(1/2);
>>g1mb=D.*exp(-(Ekin/q+5*kT)/kT);
>>g1fd=D./(1+exp((Ekin/q+5*kT)/kT));
>>g2fd=D./(1+exp((Ekin/q+0)/kT));
>>g2mb=D.*exp(-(Ekin/q+0)/kT);
>>hold on;
>>plot(Ekin/q,g1fd*q,'-b')
>>plot(Ekin/q,g1mb*q,'ob')
>>plot(Ekin/q,g2fd*q,'--k')
```

```
>>plot(Ekin/q,g2mb*q,'squarek')
>>hold off;
>>set(gca,'YScale','log')
>>xlabel('Kinetic Energy (eV)')
>>ylabel('g(E)=dn/dE (eV^{-1}m^{-3})')
```

The lower curves in Fig. 2.17 are for the case of $E_C - E_F = 5kT$. We can see that there is no observable difference between the calculations based on the Fermi–Dirac and Maxwell–Boltzmann distributions. The upper curves are for $E_C - E_F = 0$, and we can see that the Maxwell–Boltzmann distribution (the square symbols) overestimates the concentration of electrons. In general, the distribution of electron concentration starts from zero at $E_{kin} = 0$ (because the density of states in a 3D system is zero at E_C) and then rises rapidly; after reaching its maximum value, it starts dropping exponentially (straight-line segments of the semilog plots) because of approximately exponential dependence of $f(E)$.

EXAMPLE 2.11 Heavily Doped Semiconductors

In heavily doped semiconductors, the Fermi level is very close to the energy levels in the bands and the state-occupancy approximation by the Maxwell–Boltzmann distribution is no longer accurate. For an N-type semiconductor, this occurs when the condition $\exp[(E - E_F)/kT] \gg 1$ is no longer satisfied because the bottom of the conduction band is too close to the Fermi level ($E_C - E_F$ is comparable to kT). In this case, the Fermi–Dirac distribution has to be used:

$$n_0 = \int_0^\infty D(E_{kin}) f(E)\, dE_{kin} = \frac{8\sqrt{2}\pi}{h^3}(m^*)^{3/2} \int_0^\infty \frac{E_{kin}^{1/2}\, dE_{kin}}{1 + \exp[(E_{kin} + E_C - E_F)/kT]}$$

(a) Express the electron concentration n_0 in terms of the effective density of states N_C and the Fermi integral, given by

$$F_{1/2}(\eta) = \int_0^\infty \frac{x^{1/2}\, dx}{1 + e^{x-\eta}}$$

(b) Derive the electron-concentration equation for heavily doped semiconductors using the following approximation for the Fermi integral $F_{1/2}$:

$$F_{1/2} \approx \begin{cases} \dfrac{\sqrt{\pi}}{2} e^{\eta} & \text{for } -\infty < \eta \le -1 \\[2ex] \dfrac{\sqrt{\pi}}{2} \dfrac{1}{1/4 + e^{-\eta}} & \text{for } -1 < \eta < 5 \\[2ex] \dfrac{2}{3}\eta^{3/2} & \text{for } 5 \le \eta < \infty \end{cases}$$

SOLUTION

(a) Taking $x = E_{kin}/kT$, $dx = dE_{kin}/kT$, and $\eta = (E_F - E_C)/kT$, we obtain

$$n_0 = \int_0^\infty D(E_{kin}) f(E)\, dE_{kin} = \frac{8\sqrt{2}\pi}{h^3}(m^* kT)^{3/2} \underbrace{\int_0^\infty \frac{x^{1/2}\, dx}{1 + e^{x-\eta}}}_{F_{1/2}}$$

N_C is given by Eq. (2.77). Therefore,

$$n_0 = \frac{2}{\sqrt{\pi}} N_C F_{1/2}$$

(b)

$$n_0 \approx \begin{cases} N_C e^{-(E_C - E_F)/kT} & \text{for } \dfrac{E_F - E_C}{kT} \le -1 \\[2ex] \dfrac{N_C}{1/4 + e^{-(E_F - E_C)/kT}} & \text{for } -1 < \dfrac{E_F - E_C}{kT} < 5 \\[2ex] \dfrac{4}{3\sqrt{\pi}} N_C \left(\dfrac{E_F - E_C}{kT}\right)^{3/2} & \text{for } \dfrac{E_F - E_C}{kT} \ge 5 \end{cases}$$

2.4.3 Fermi Potential and Doping

Equations (2.76) and (2.79) clearly account for the influence of temperature on the concentration of free electrons and holes. It is less obvious how these equations take into account the influence of doping level, which is even more important because doping is used to control the concentration of electrons and holes in semiconductors. As explained in Section 1.2.2, in an N-type semiconductor, the concentration of electrons is approximately equal to the donor-atom concentration ($n_0 \approx N_D$), and similarly $p_0 \approx N_A$ in P-type semiconductors. The fact that n_0 is much higher in an N-type semiconductor than in the intrinsic semiconductor means that the occupancy of the states in the conduction band has to be higher. Figure 2.18a and 2.18b shows that this is possible if E_C is closer to the Fermi level than in the case of intrinsic semiconductor. Indeed, if $E_C - E_F$ in Eq. (2.76) is smaller than in the intrinsic semiconductor, then the concentration n_0 would exceed the intrinsic concentration. Analogously, E_V moves closer to the Fermi level to express a higher occupancy of the valence band by holes in a P-type semiconductor (Fig. 2.18c). *The position of the energy bands (E_C and E_V) with respect to the Fermi level in the equations for electron and hole concentrations expresses the doping type and level.*

Using the fact that $n_0 \approx N_D$ in N-type semiconductors and $p_0 \approx N_A$ in P-type semiconductors, the position of E_C and E_V with respect to the Fermi level is obtained from

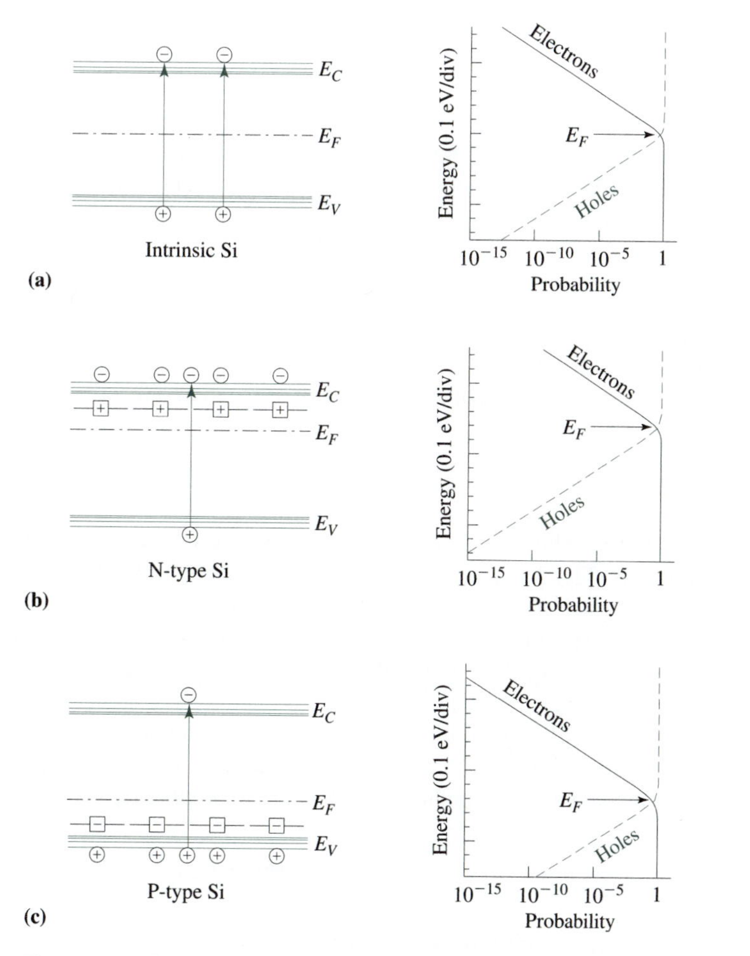

Figure 2.18 Position of the Fermi level expresses doping type and level. (a) Intrinsic semiconductor. (b) N-type semiconductor. (c) P-type semiconductor.

Eqs. (2.76) and (2.79) as

$$E_C - E_F = kT \ln \frac{N_C}{N_D} \qquad \text{N type} \tag{2.80}$$

$$E_V - E_F = -kT \ln \frac{N_V}{N_A} \qquad \text{P type} \tag{2.81}$$

Equations (2.80) and (2.81) are correct for doping levels that are lower than the effective density of states; that is, $N_D < N_C$ and $N_A < N_V$. When $N_D > N_C$, E_C approaches the Fermi level and the exponential approximation given by Eq. (2.73) is no longer good, so the complete Fermi–Dirac distribution has to be used; the electron gas becomes degenerate. Analogously, the hole gas becomes degenerate when N_A exceeds the density of states N_V.

The semilogarithm plots of Fig. 2.18b show that the increase in the occupancy of the conduction-band levels by electrons (E_C closer to E_F) is accompanied by a reduction in the occupancy of the valence-band levels by holes (E_V further from E_F). This is due to the effect of electron and hole recombination discussed in Section 1.2.4. Using Eqs. (2.76) and (2.79), we can now easily show that the product of electron and hole concentrations np is indeed a constant, thus independent of the doping level (position of the bands with respect to the Fermi level):

$$n_0 p_0 = N_C N_V e^{-(E_C - E_V)/kT} \tag{2.82}$$

Moreover, knowing that $n_0 p_0 = n_i^2$ [Eq. (1.6)], we can now express the intrinsic carrier concentration n_i in terms of the energy gap of the material $E_g = E_C - E_V$:

$$n_i = \sqrt{N_C N_V} e^{-E_g/2kT} \tag{2.83}$$

Equation (2.83) shows in its own way that the materials with large energy gaps E_g appear as insulators (Fig. 2.11), because the intrinsic concentration of the carriers is fairly small.

Equations (2.80) and (2.81) and Fig. 2.18 show how the doping level sets the position of E_C (with respect to the Fermi level) in the case of N-type doping and the position of E_V in the case of P-type doping. It will prove very useful to define the position of the energy bands by an energy level that is the same for both N-type and P-type doping. The position of the Fermi level in the intrinsic semiconductor is the best choice. Labeling the intrinsic Fermi-level position by E_i, we can express its position with respect to the Fermi level position in a doped semiconductor by the *Fermi potential* $q\phi_F$, defined as

$$q\phi_F = E_i - E_F \tag{2.84}$$

This way, $q\phi_F = 0$ for the intrinsic case, $q\phi_F < 0$ for N-type doping, and $q\phi_F > 0$ for P-type doping (Fig. 2.19). Note that $q\phi_F$ is in the units of energy (eV) whereas ϕ_F is in the units of voltage (V).

Equations (2.76) and (2.79) can be transformed to express the electron and hole concentrations in terms of the Fermi potential:

$$n_0 = N_C e^{-(E_C - E_F)/kT} = \underbrace{N_C e^{-(E_C - E_i)/kT}}_{n_0 = n_i} e^{-(E_i - E_F)/kT} = n_i e^{-q\phi_F/kT} \tag{2.85}$$

$$p_0 = N_V e^{-(E_F - E_V)/kT} = \underbrace{N_V e^{-(E_i - E_V)/kT}}_{p_0 = n_i} e^{(E_i - E_F)/kT} = n_i e^{q\phi_F/kT} \tag{2.86}$$

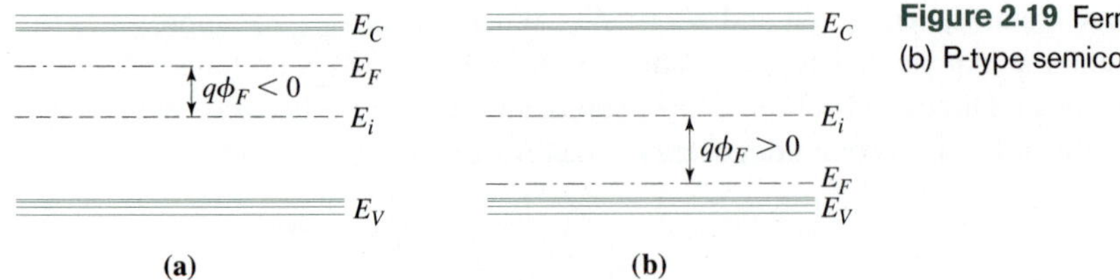

Figure 2.19 Fermi potential ($q\phi_F$) in (a) N-type and (b) P-type semiconductors.

When the Fermi-level position is set by N-type doping, the condition $n_0 \approx N_D$ and Eq. (2.85) can be used to determine the Fermi potential

$$q\phi_F = -kT \ln \frac{N_D}{n_i} \tag{2.87}$$

In the case of P-type doping, the Fermi potential can be calculated from the analogous equation:

$$q\phi_F = +kT \ln \frac{N_A}{n_i} \tag{2.88}$$

EXAMPLE 2.12 Intrinsic Fermi-Level Position

The energy gap of GaAs is $E_g = 1.42$ eV, and experimental values for the effective densities of states at room temperature are $N_C = 4.7 \times 10^{17}$ cm^{-3} and $N_V = 7.08 \times 10^{18}$ cm^{-3}. Determine the room-temperature position of the Fermi level with respect to E_V in intrinsic GaAs.

SOLUTION

The intrinsic concentrations of electrons and holes are equal,

$$n_0 = p_0 = n_i$$

where n_0 and p_0 are determined by the effective density of states in the conduction and valence bands, respectively, and the position of the Fermi level [Eqs. (2.76) and (2.79)]:

$$n_0 = N_C e^{-(E_C - E_F)/kT}$$

$$p_0 = N_V e^{-(E_F - E_V)/kT}$$

If N_C and N_V were equal, the midgap position of the Fermi level would provide the condition for $n = p$. However, $N_V > N_C$ in GaAs, so the intrinsic position of the Fermi level will be closer to the conduction band to compensate for the lower density of states in the conduction band:

$$N_C e^{-(E_C - E_F)/kT} = N_V e^{-(E_F - E_V)/kT}$$

$$\frac{N_C}{N_V} = e^{(E_C + E_V - 2E_F)/kT}$$

$$E_C + E_V - 2E_F = kT \ln \frac{N_C}{N_V}$$

$$E_g = E_C - E_V$$

$$E_g - 2(E_F - E_V) = kT \ln \frac{N_C}{N_V}$$

$$E_F - E_V = \frac{E_g - kT \ln(N_C/N_V)}{2}$$

$$E_F - E_V = 0.745 \text{ eV} > E_g/2$$

EXAMPLE 2.13 Doping and Fermi Level

If the donor concentration in N-type silicon is $N_D = 10^{15}$ cm^{-3}, calculate the probability of finding an electron at $E = E_C$ and determine the probability of finding a hole at $E = E_V$ at 20°C. Compare these results to the results of Example 2.7.

SOLUTION

As the doping shifts the bands, it is important to first determine the position of E_C with respect to E_F. Using Eq. (2.80), one obtains

$$E_C - E_F = kT \ln(N_C/N_D) = 8.62 \times 10^{-5}(20 + 273.15) \ln(2.86 \times 10^{19}/10^{15}) = 0.259 \text{ eV}$$

When this value is used in the Fermi–Dirac distribution, the following result is obtained: $f(E_C) = 3.5 \times 10^{-5}$. The probability of finding a hole at $E = E_V$ is

$$f_h(E_V) = 1 - 1/\left[1 + e^{(E_V - E_F)/kT}\right] = 1/\left[1 + e^{(E_F - E_V)/kT}\right]$$

$E_F - E_V$ is determined in the following way:

$$E_F - E_V = E_C - E_V - (E_C - E_F) = E_g - (E_C - E_F) = 1.12 - 0.259 = 0.861 \text{ eV}$$

Using this result, the probability of finding a hole at $E = E_V$ is calculated as $f_h(E_V) = 1.6 \times 10^{-15}$.

Comparing these results to the results obtained in Example 2.7 for the case of the intrinsic silicon, we see that the probability of having electrons in the conduction band is much higher in the N-type semiconductor ($3.5 \times 10^{-5}/2.4 \times 10^{-10} = 145,833$ times). The probability of having holes in the valence band, however, is much smaller.

EXAMPLE 2.14 Work Function and Doping

For metals, the *work function* is defined as the energy needed to liberate an average electron—in other words, the energy needed to bring an electron from the Fermi level to the level of a free electron in vacuum (vacuum level). Generalizing this definition for semiconductors, with no electrons at the Fermi level when it appears in the energy gap, the work function can be defined as the negative value of the Fermi-level position with respect to the vacuum level. Determine the work functions of N-type and P-type silicon samples with equal doping levels: $N_D = N_A = 10^{16}$ cm^{-3}. The difference between the vacuum level and the bottom of the conduction band (the electron affinity) in silicon is $q\chi_s = 4.05$ eV, the energy gap is $E_g = 1.12$ eV, and the effective densities of states in the conduction and valence bands are $N_C = 2.86 \times 10^{19}$ cm^{-3} and $N_V = 3.1 \times 10^{19}$ cm^{-3}.

SOLUTION

For the case of N-type silicon,

$$N_D = N_C e^{-(E_C - E_F)/kT}$$

$$E_C - E_F = kT \ln \frac{N_C}{N_D} = 0.21 \text{ eV}$$

$$q\phi_s = q\chi_s + (E_C - E_F) = 4.05 + 0.21 = 4.26 \text{ eV}$$

For the case of P-type silicon,

$$N_A = N_V e^{-(E_F - E_V)/kT}$$

$$E_F - E_V = kT \ln \frac{N_V}{N_A} = 0.21 \text{ eV}$$

$$q\phi_s = q\chi_s + E_g - (E_F - E_C) = 4.05 + 1.12 - 0.21 = 4.96 \text{ eV}$$

EXAMPLE 2.15 Occupancy of Doping States

When applied to the energy levels of donors and acceptors (E_D and E_A), the Fermi–Dirac distribution function involves a degeneracy factor g:

$$f = \frac{1}{1 + \dfrac{1}{g} e^{(E_{D,A} - E_F)/kT}}$$

A commonly used value for the degeneracy factor for electrons at the donor levels is $g = 2$, whereas the degeneracy factor for the electrons at the acceptor states is normally taken as $g = 1/4$. The reason for the appearance of the degeneracy factor in the Fermi–Dirac distribution is explained in Section 10.1.5.

(a) Derive equations for the fraction of donor states occupied by electrons and the fraction of acceptor states occupied by holes.

(b) Given that the doping level of phosphorus in silicon is 0.045 eV below the bottom of the conduction band, calculate the percentage of ionized phosphorus atoms at 300 K if the phosphorus concentration is $N_D = 10^{15}$ cm^{-3}. Based on the result, comment on the validity of the approximation that *practically all donors are ionized at room temperature*. Assume that $E_C - E_F \gg kT$ and $E_D - E_F \gg kT$. The effective density of states in the conduction band is $N_C = 2.86 \times 10^{19}$ cm^{-3}.

SOLUTION

(a) The donor atoms are assumed to be sufficiently apart from each other that they all have the same energy level (the electrons associated with the donor states are isolated from each other). The Fermi–Dirac distribution f applied to the level E_D gives the probability that a state with this energy level is occupied by an electron. This probability can also be expressed as n_D/N_D, where n_D is the concentration of donor states occupied by electrons and N_D is the total concentration of the donor states (equal to the concentration of donor atoms). Therefore,

$$\frac{n_D}{N_D} = \frac{1}{1 + \frac{1}{2}e^{(E_D - E_F)/kT}}$$

Analogous equation for the case of holes occupying acceptor levels is

$$\frac{p_A}{N_A} = f_h$$

where f_h is the probability that an acceptor state is occupied by a hole:

$$f_h = 1 - f = 1 - \frac{1}{1 + \frac{1}{g}e^{(E_A - E_F)/kT}} = \frac{1}{1 + ge^{(E_F - E_A)/kT}}$$

Therefore,

$$\frac{p_A}{N_A} = \frac{1}{1 + ge^{(E_F - E_A)/kT}}$$

(b) Neglecting the thermally generated electrons, we can write

$$N_D = n_0 + n_D$$

where n_0 is the concentration of electrons in the conduction band:

$$n_0 = N_C e^{-(E_C - E_F)/kT}$$

and n_D is the concentration of electrons remaining on the donor levels:

$$n_D = \frac{N_D}{1 + \frac{1}{2}e^{(E_D - E_F)/kT}} \approx 2N_D e^{-(E_D - E_F)/kT}$$

The percentage of the ionized donors is $100 \times n_0/N_D = 100 \times n_0/(n_0 + n_D)$:

$$\frac{n_0}{N_D} = \frac{N_C e^{-(E_C-E_F)/kT}}{N_C e^{-(E_C-E_F)/kT} + 2N_D e^{-(E_D-E_F)/kT}} = \frac{1}{1 + 2\frac{N_D}{N_C} e^{(E_C-E_D)/kT}} = 0.9996$$

$100 \times n_0/N_D = 99.96\%$. The error of the assumption that all the donor atoms are ionized is less than 0.05% in this case.

EXAMPLE 2.16 Partial Doping Ionization

(a) Express the fraction of the ionized donors (n_0/N_D) as a function of the donor depth ($E_C - E_D$) without the assumption $E_D - E_F \gg kT$ (this is a more general case that applies to both shallow and deep donors at different temperatures, which is not the case with the equation derived in Example 2.15), while maintaining the assumption of a nondegenerate semiconductor ($E_C - E_F \gg kT$).

(b) Based on the derived equation, find the condition that corresponds to 100% ionization.

(c) Simplify the derived equation for deep donor levels and/or low temperatures.

(d) Calculate the percentage of ionized phosphorus atoms in silicon at 300 K for the doping level of $N_D = 5 \times 10^{18}$ cm^{-3}. The donor level of phosphorus in silicon is 0.045 eV below the bottom of the conduction band.

SOLUTION

(a) Labeling the concentration of occupied (neutral) donors by n_D and noting that the concentration of ionized donors is equal to the electron concentration $n_0 = N_C \exp[-(E_C - E_F)/kT]$, we have

$$n_D = \frac{N_D}{1 + (1/g)e^{(E_D-E_F)/kT}}$$

$$\frac{n_0}{N_D} = \frac{n_0}{n_0 + n_D} = \frac{n_0}{n_0 + \frac{N_D}{1+(1/g)\exp[(E_D-E_F)/kT]}} = \frac{n_0 + (n_0/g)e^{(E_D-E_F)/kT}}{n_0 + (n_0/g)e^{(E_D-E_F)/kT} + N_D}$$

$$= \frac{ge^{-(E_D-E_F)/kT} + 1}{ge^{-(E_D-E_F)/kT} + 1 + g(N_D/n_0)e^{-(E_D-E_F)/kT}}$$

$$= \frac{1 + ge^{-(E_D-E_F)/kT}}{1 + g(1 + N_D/n_0)e^{-(E_D-E_F)/kT}}$$

$$= \frac{1 + g e^{(E_C - E_D)/kT} e^{-(E_C - E_F)/kT}}{1 + g(1 + N_D/n_0) e^{(E_C - E_D)/kT} e^{-(E_C - E_F)/kT}}$$

$$= \frac{1 + g(n_0/N_C) e^{(E_C - E_D)/kT}}{1 + g(1 + N_D/n_0)(n_0/N_C) e^{(E_C - E_D)/kT}}$$

$$= \frac{\frac{N_C}{N_D} + g\frac{n_0}{N_D} e^{(E_C - E_D)/kT}}{\frac{N_C}{N_D} + g\frac{n_0}{N_D} e^{(E_C - E_D)/kT} + g e^{(E_C - E_D)/kT}}$$

$$\frac{n_0}{N_D}\frac{N_C}{N_D} + g\left(\frac{n_0}{N_D}\right)^2 e^{(E_C - E_D)/kT} + g\frac{n_0}{N_D} e^{(E_C - E_D)/kT} - \frac{N_C}{N_D} - g\frac{n_0}{N_D} e^{(E_C - E_D)/kT} = 0$$

$$g e^{(E_C - E_D)/kT}\left(\frac{n_0}{N_D}\right)^2 + \frac{N_C}{N_D}\frac{n_0}{N_D} - \frac{N_C}{N_D} = 0$$

$$g\frac{N_D}{N_C} e^{(E_C - E_D)/kT}\left(\frac{n_0}{N_D}\right)^2 + \frac{n_0}{N_D} - 1 = 0$$

$$\frac{n_0}{N_D} = \frac{-1 + \sqrt{1 + 4g\frac{N_D}{N_C} e^{(E_C - E_D)/kT}}}{2g\frac{N_D}{N_C} e^{(E_C - E_D)/kT}}$$

(b) Consider the following condition:

$$x = 4g\frac{N_D}{N_C} e^{(E_C - E_D)/kT} \ll 1$$

Taking into account that $\sqrt{1 + x} \approx 1 + x/2$ for small x, we obtain

$$\frac{n_0}{N_D} \approx \frac{-1 + 1 + 2g\frac{N_D}{N_C} e^{(E_C - E_D)/kT}}{2g\frac{N_C}{N_D} e^{(E_C - E_D)/kT}} = 1$$

The ionization fraction is 100% for $x \ll 1$, which corresponds to shallow donor levels (small $E_C - E_D$), high temperatures, and low/moderate doping levels.

(c) For the case of deep levels or low temperatures,

$$x = 4g\frac{N_D}{N_C} e^{(E_C - E_D)/kT} \gg 1$$

$$\frac{n_0}{N_D} = \frac{\sqrt{4g\frac{N_D}{N_C} e^{(E_C - E_D)/kT}}}{2g\frac{N_D}{N_C} e^{(E_C - E_D)/kT}} \approx \sqrt{\frac{N_C}{g N_D}} e^{-(E_C - E_D)/2kT}$$

In this case, the ionization fraction increases exponentially with the temperature, the activation energy being $E_A = (E_C - E_D)/2$. This means that if the activation energy for the increase of the free-electron concentration is measured, the depth of the donor level can be calculated as $E_C - E_D = 2E_A$.

(d) In this case,

$$x = 4g\frac{N_D}{N_C}e^{(E_C-E_D)/kT} = 4 \times 2 \times \frac{5 \times 10^{18}}{2.86 \times 10^{19}}e^{0.045/0.02585} = 7.97$$

$$\frac{n_0}{N_D} = \frac{-1 + \sqrt{1+x}}{x/2} = 0.500$$

The ionization fraction is 50.0%.

2.4.4 Nonequilibrium Carrier Concentrations and Quasi-Fermi Levels

The Fermi–Dirac distribution [Eq. (2.64)] was derived for a system in thermal equilibrium. Accordingly, the considerations of electron-state occupancy and all the related equations are for a semiconductor in thermal equilibrium. This includes the Fermi level and its relationships to the electron and hole concentrations. The concept of Fermi level as a reference energy of the Fermi–Dirac distribution is meaningless for nonequilibrium cases. Yet, nonequilibrium concentrations of electrons and holes appear quite frequently in semiconductor devices. It will be very useful to establish relationships between nonequilibrium concentrations of electrons and holes and representative energy levels in the energy-band diagrams. This is possible by defining quasi-Fermi levels for electrons (E_{Fn}) and holes (E_{Fp}), to be used instead of the Fermi level (E_F) in Eqs. (2.76) and (2.79):

$$\begin{aligned} n &= n_0 + \delta n = N_C e^{-(E_C-E_{Fn})/kT} \\ p &= p_0 + \delta p = N_V e^{-(E_{Fp}-E_V)/kT} \end{aligned} \tag{2.89}$$

By extending the relationship between n_0 and E_F (the equilibrium case) to n and E_{Fn} (a nonequilibrium case), we enable the presentation of nonequilibrium cases by energy-band diagrams. As in the equilibrium cases, the position of the quasi-Fermi level for electrons indicates the electron concentration. The difference in nonequilibrium cases is that the concentration of holes is related to a different energy level, the quasi-Fermi level for holes. If the positions of the quasi-Fermi levels for electrons and holes are expressed with reference to the equilibrium Fermi level in the intrinsic semiconductor (E_i), Eqs. (2.89) take the following forms:

$$n = n_i e^{(E_{Fn} - E_i)/kT}$$
$$p = n_i e^{(E_i - E_{Fp})/kT} \qquad (2.90)$$

In nonequilibrium conditions, the product of electron and hole concentrations is

$$np = n_i^2 e^{(E_{Fn} - E_{Fp})/kT} \qquad (2.91)$$

This shows again that the product of electron and hole concentrations is equal to a constant (n_i^2) only in the case of thermal equilibrium. In nonequilibrium conditions, np can either be higher or lower than n_i^2, depending on whether $E_{Fn} - E_{Fp}$ is larger or smaller than zero. This is so because external conditions can lead to either excess or deficiency of current carriers in a considered region of a semiconductor device.

EXAMPLE 2.17 Quasi-Fermi Levels

Determine E_{Fn} and E_{Fp}, with respect to the bottom of the conduction band and the top of the valence band, respectively, if a slab of N-type silicon ($N_D = 10^{16}$ cm^{-3}) is illuminated so that the steady-state concentration of the additional electron–hole pairs is $\delta n = 2 \times 10^{16}$ cm^{-3}. What is the difference between the quasi-Fermi levels?

SOLUTION

Given that the nonequilibrium concentrations are $n \approx N_D + \delta n = 3 \times 10^{16}$ cm^{-3} and $p \approx \delta p = 2 \times 10^{16}$ cm^{-3}, we find that

$$E_C - E_{Fn} = kT \ln(N_C/n) = 0.02585 \times \ln(2.86 \times 10^{19}/3 \times 10^{16}) = 0.178 \text{ eV}$$

$$E_{Fp} - E_V = kT \ln(N_V/p) = 0.02585 \times \ln(3.1 \times 10^{19}/2 \times 10^{16}) = 0.190 \text{ eV}$$

The difference between the quasi-Fermi levels is

$$E_{Fn} - E_{Fp} = E_g - \left[(E_C - E_{Fn}) + (E_{Fp} - E_V)\right] = 1.12 - (0.178 + 0.190) = 0.752 \text{ eV}$$

SUMMARY

1. Quantum mechanics establishes a fundamental link between the wavelength (the main parameter of waves) and the momentum (the main parameter of particles):

$$\lambda p = h$$

where $h = 6.626 \times 10^{-34}$ J · s is the Planck constant.

2. A perfect sinusoidal wave, propagating along the x-direction, can be represented by the following complex *wave function*

$$\psi(x,t) = Ae^{-j(\omega t - kx)}$$

where $\omega = 2\pi/T$ is the *rate of change of phase with time* and $k = 2\pi/\lambda$ is the *rate of change of phase with distance*. The wave velocity (v) relates the wavelength (λ) to the period (T), and the angular frequency (ω) to the wave number (k): $\lambda = vT$, and $\omega = vk$. If the wave number is used as the wave parameter instead of the wavelength ($k = 2\pi/\lambda$), the fundamental $\lambda p = h$ relationship becomes $p = \hbar k$, where $\hbar = h/2\pi$.

3. There are no perfect waves in reality because no wave can extend from $-\infty$ to ∞ in terms of either time or space. The finite size of waves is modeled by *wave packets* obtained as the infinite sum (integral) of perfect sinusoids with different frequencies and wave numbers (the integral is the Fourier transform and the individual sinusoids are different harmonics). The wave packet, represented by its complex wave function $\psi(x,t)$, is an abstraction. What relates to reality is the *intensity of the wave function:* $\psi(x,t)\psi^*(x,t)$. The intensity of the wave function can be comprehended as the probability of finding the object modeled by the wave function at point x and time t.

4. The Schrödinger equation can be used to obtain the wave function that models a particular object in specific conditions. The time-independent form of the Schrödinger equation is given by

$$-\frac{\hbar^2}{2m}\frac{d^2\psi(x)}{dx^2} + E_{pot}(x)\psi(x) = E\psi(x)$$

The wave function and its first derivative have to be finite, continuous, and single-valued. In addition, a specific wave function has to satisfy specific boundary and initial conditions.

   - The solution of the Schrödinger equation for a totally free electron (an abstraction) is the wave function of the perfect sinusoid. As this electron wave propagates in space, it possesses a kinetic energy that is related to the wave number: $E = (\hbar^2/2m)k^2$.
   - The solution of the Schrödinger equation for an electron inside an energy well with infinite walls shows that the electron can exist only as a standing wave. The largest possible wavelength (when a single half-wavelength is equal to the width of the potential well) corresponds to the lowest possible energy: $E = (\hbar^2/2m)k^2 = (h^2/2m)/\lambda^2$. The next possible energy corresponds to the wavelength that is equal to the width of the potential well. Discretization in the possible wavelengths leads to *energy-level quantization*.
   - Solving the Schrödinger equation for an electron approaching a potential-energy barrier can establish the probability that the electron will pass through the barrier (tunneling coefficient) or be reflected by the barrier (reflection coefficient).

5. The potential-energy well, created by the attractive force of a positively charged atom core, confines the electrons belonging to that atom. In a crystal, the potential-energy barriers between neighboring atoms are reduced, so some of the electrons are shared by all the atoms (these are the valence electrons creating the atomic bonds). In metals, the valence electrons can be considered as standing waves in a potential well with space dimensions equal to the crystal size (free-electron model). The large crystal size enables the existence of "free electrons" with large wavelengths, and therefore small kinetic energies, as well as small differences between subsequent energy levels. The set of a large number of very close energy levels is modeled as *energy band.*
6. In semiconductors, there is a significant interaction between the valence electrons and the periodic potential of the crystal atoms. The valence electrons with the wave numbers $k_n = n(2\pi/a)$, where $n = 1, 2, 3, \ldots$ and a is the crystal-lattice constant, form standing waves, as they are reflected backward and forward by the energy barriers between neighboring atoms. There are two standing waves, with different energies, for each k_n: one with the nodes and the other with the antinodes at the crystal-lattice sites. The difference between these two energy levels is referred to as the *energy gap,* given that there are no electron–wave functions that would correspond to the energy values inside the energy gap. The continuous parabolic E–k dependence of "free" electrons in metals is split into many energy bands, separated by energy gaps. The two bands of interest in semiconductors are those that are neither completely filled nor completely empty. They are referred to as the *valence band* and the *conduction band.*
7. The E–k dependence in the conduction band can be approximated by a parabola, $E = (\hbar^2/2m^*)k^2$. The parameter of the parabola (m^*) can be considered as the *effective mass* (or apparent mass) of the electrons in the conduction band to allow to model these electrons as "free" electrons. The E–k dependence in the valence band is an inverted parabola, allowing us to use the model of "free" holes in the valence band. With this model, the kinetic energy of an electron at the bottom of the conduction band (E_C) is zero, and the kinetic energy of a hole at the top of the valence band (E_V) is zero. The kinetic energies of electrons increase above E_C, whereas the kinetic energies of holes increase below E_V (according to the inverter E–k parabola).
8. With the model of "free" electrons, along with the concept of effective mass, the electrons in the conduction band can be considered as a gas of negatively charged particles (the electron-gas model). Likewise, the holes in the valence band can be considered as a gas of positively charged particles. To complete the carrier-gas model, the *concentration of carriers* (the number of carriers/particles per unit volume) needs to be established. There are two steps toward this aim: (1) determining the maximum possible concentration of electrons/holes (the density of electron/hole states) and (2) determining the occupancy of the possible electron/hole states.
9. In the classical particle model, the "maximum possible particle concentration" can be determined from the particle size. The concept of *particle size,* however, has it limits: particles just do not have precisely defined sizes. According to the Heisenberg uncertainty principle, the minimum possible uncertainties in the particle size and momentum are related so that their product is equal to Planck's constant:

$$\Delta x \Delta p_x \geq h$$

10. Although the minimum size in x space cannot be determined, the Heisenberg uncertainty relation enables us to determine the minimum size in x–p_x space. The minimum volume of a cell in the six-dimensional x–y–z–p_x–p_y–p_z space is

$$\underbrace{\Delta x \Delta y \Delta z}_{\Delta V}\,\underbrace{\Delta p_x \Delta p_y \Delta p_z}_{\Delta V_p} = h^3$$

Given that two electrons (with different spins) can fit into the elementary volume $\Delta V \Delta V_p$, the *maximum* concentration of electrons can be defined, but remains dependent on ΔV_p:

$$C = \frac{2}{\Delta V} = \frac{2}{h^3}\Delta V_p$$

Given the kinetic-energy dependence on momentum [$E_{kin} = (\hbar^2/2m^*)k^2 = p^2/2m^*$], ΔV_p can be converted into ΔE_{kin} to express the maximum concentration of electrons as

$$C = D(E_{kin})\Delta E_{kin}$$

where

$$D(E_{kin}) = \frac{8\sqrt{2}\pi}{h^3}(m^*)^{3/2}E_{kin}^{1/2}$$

$D(E_{kin})$ is referred to as the *density of electron states*. The density of electron states gives the concentration of electron states per unit energy, so that the electron concentration can be determined as

$$n_0 = \int_0^\infty f(E)D(E_{kin})\,dE_{kin}$$

where $f(E)$ is the probability that the electron state corresponding to the energy level E is occupied by an electron.

11. The probability $f(E)$ is given by the Fermi–Dirac distribution:

$$f(E) = \frac{1}{1 + e^{(E-E_F)/kT}}$$

The probability that an electron state, corresponding to an energy level E, is empty is equal to the probability that this state is occupied by a hole:

$$f_h(E) = 1 - f(E) = \frac{1}{1 + e^{(E_F-E)/kT}}$$

The occupancy probability $f(E)$ is different from the classical Maxwell–Boltzmann distribution,

$$f(E) = e^{-(E-E_F)/kT}$$

because it includes the Pauli exclusion principle (in addition to the minimum-energy principle). Nonetheless, the Fermi–Dirac distribution can be approximated by the Maxwell–Boltzmann distribution when $E - E_F \gg kT$. This is the case when there are plenty of unoccupied states in the conduction band, so that the effects of the Pauli exclusion principle are not pronounced (nondegenerate electron gas).

12. With the Maxwell–Boltzmann approximation for $f(E)$, the integration of $f(E)D(E_{kin})$ leads to

$$n_0 = N_C e^{-(E_C-E_F)/kT}$$

where

$$N_C = \frac{4\sqrt{2}(\pi m^* kT)^{3/2}}{h^3}$$

is called the *effective density of states* for the conduction band. Analogously,

$$p_0 = N_V e^{-(E_F-E_V)/kT}$$

where N_V is the effective density of states for the valence band.

13. The reference energy E_F in the Fermi–Dirac distribution and in the equations for the concentrations of electrons and holes is called *Fermi level*. From the Fermi–Dirac distribution, it can be defined as the energy level of the state whose occupancy by an electron (or hole) is equal to $1/2$. From the equations for the electron and hole concentrations, it can be seen that the positions of E_C and E_V with respect to E_F correspond to the concentrations of electrons and holes. As E_C approaches E_F, n_0 increases and p_0 decreases (and vice versa). When the concentration of electrons or holes is set by donor or acceptor doping, we say that the position of the bands is set by the doping type and level. E_F above its intrinsic position (E_i) corresponds to N-type doping, whereas E_F below E_i corresponds to P-type doping. The difference between E_i and E_F (called Fermi potential) corresponds to the doping level:

$$\phi_F = \frac{E_i - E_F}{q} = \mp V_t \ln(N_{D,A}/n_i)$$

where $V_t = kT/q$ is the thermal voltage.

14. The product between the electron and hole concentrations does not depend on the Fermi level position, but it depends on the energy gap:

$$n_0 p_0 = n_i^2 = N_C N_V e^{-(E_C-E_V)/kT}$$

$$n_i = \sqrt{N_C N_V} e^{-E_g/2kT}$$

15. Quasi-Fermi levels for electrons (E_{Fn}) and holes (E_{Fp}) are introduced to express the concentrations of electrons and holes under nonequilibrium conditions:

$$n = N_C e^{-(E_C - E_{Fn})/kT} = n_i e^{(E_{Fn} - E_i)/kT}$$

$$p = N_V e^{-(E_{Fp} - E_V)/kT} = n_i e^{(E_i - E_{Fp})/kT}$$

Nonequilibrium conditions can be due to either excess ($np > n_i^2$) or deficiency ($np < n_i^2$) of current carriers.

PROBLEMS

2.1 The wavelength of a monochromatic red light is $\lambda = 440$ nm. Determine the velocity of an electron moving in free space, if the wavelength of the electron is the same: 440 nm. Are the electron and light (photon) momenta equal?

2.2 The wavelength of green light is $\lambda = 550$ nm. Determine the wave number, the angular frequency, and the period of this wave. The speed of light is 3×10^8 m/s.

2.3 Find the wave number, the wavelength, the angular frequency, and the period of an electron wave that corresponds to an electron moving in free space by velocity

(a) $v_1 = 5 \times 10^5$ m/s
(b) $v_2 = 5 \times 10^7$ m/s **A**

2.4 Electrons in solids have negative potential energies with respect to the energy of a free electron (this means that these electrons need to gain energy to liberate themselves from the solid). Determine and plot the E–k dependence for an electron in a field of constant potential energy $E_{pot} = -10$ eV. For plotting, use values for k between -10^{10} m^{-1} and 10^{10} m^{-1}.

2.5 **(a)** An electron leaves a heated cathode and enters the free space with kinetic energy $E_{kin} = 1$ eV. Determine the velocity, the wave number, the wavelength, and the period of this electron wave.

(b) Repeat the calculations for the case of an electron that enters the free space after being accelerated to the energy of 1000 eV. **A**

2.6 Show that the following relationship between the energy and the period is valid for a free electron:

$$ET = \frac{h}{2}$$

2.7 An electron is confined inside a potential well with infinite walls. The width of the well is $W = 5$ nm. What is the probability of finding the electron within 1 nm from either wall, if the electron is at

(a) the lowest energy level
(b) the second-lowest energy level **A**

2.8 The probability that an electron with energy E will tunnel through a rectangular energy barrier with height E_{pot} and width W can be approximated by

$$T \approx 16\left(\frac{E}{E_{pot}}\right)\left(1 - \frac{E}{E_{pot}}\right)\exp(-2\kappa W)$$

where $\kappa = \sqrt{\frac{2m}{\hbar^2}(E_{pot} - E)}$.

(a) Calculate the tunneling probability if $E = 0.1$ eV, $E_{pot} = 1$ eV, and $W = 1$ nm.
(b) If the tunneling current is 1 mA, how many electrons hit the barrier each second?

2.9 A rectangular barrier is narrowed by 0.2 nm. Determine how many times the tunneling probability is increased if the electron energy is 26 meV and the barrier height is 0.3 eV.

2.10 There is an approximation method, known as the Wentzel–Kramers–Brillouin (WKB) method, that can be used to determine the tunneling probability for the case of an arbitrary barrier. An arbitrary energy barrier $E_{pot}(x)$ is represented by

$$\kappa(x)=\sqrt{\frac{2m}{\hbar^2}\left[E_{pot}(x)-E\right]} \quad \text{for } E<E_{pot}, a\leq x\leq b$$

where a and b are the points at which the particle energy E intersect the tunneling barrier $E_{pot}(x)$. According to the WKB approximation, the tunneling probability is

$$T \approx \exp\left(-2\int_a^b \kappa(x)\,dx\right)$$

(a) Using the WKB approximation, determine the tunneling probability for the case considered in Problem 2.8a ($E = 0.1$ eV and a rectangular barrier with $E_{pot} = 1$ eV and $W = 1$ nm). Compare the results. **A**

(b) Determine the tunneling probability for a triangular barrier with the same parameters: the electron energy $E = 0.1$ eV, the barrier height $E_H = 1$ eV, and the barrier width at the energy level of the electron $W = 1$ nm.

2.11 The valence electrons in a 10-cm metal wire are modeled as free electrons in a square-well box. Determine the longest wavelength and the smallest velocity an electron can have in this wire.

2.12 Analogously to the one-dimensional illustration [Eq. (2.32)], the Fermi energy in real metals (3D) does not depend on the size of the crystal and is given by

$$E_{F0} = \frac{\hbar^2}{2m}(3\pi^2 n)^{2/3}$$

where n is the concentration of valence electrons.

(a) Determine the highest electron velocity in a copper crystal ($n = 8.45 \times 10^{22}$ cm^{-3}) that is cooled down to $T \approx 0$ K.

(b) What is the corresponding wavelength? **A**

2.13 There is no pronounced anisotropy associated with the kinetic energy of holes in silicon; however, the kinetic-energy dependence on the momentum is split, giving rise to the appearance of *light* and *heavy* holes. The kinetic energy of holes in silicon can be expressed in the following way:

$$E_{kin} = \frac{p^2}{2m_0}\left(A \pm \sqrt{B^2 + \frac{C_2}{5}}\right)$$

where A, B, and C are dimensionless constants approximately equal to 4.1, 1.6, and 3.3, respectively, whereas the plus/minus signs relate to the light and heavy holes. Determine the effective masses of the light and heavy holes. **A**

2.14 There is no pronounced anisotropy associated with the kinetic energy of electrons in GaAs, so that the effective mass is a scalar: $m^* = 0.067m_0$. Given that the average kinetic energy of the free electrons at room temperature is $E_{kin} = 38.8$ meV, determine the average electron momentum. If the electrons move, on average, for 0.1 ps between two scattering events, determine the average distance they travel without scattering.

2.15 Many semiconductor devices (such as MOSFET and HEMT) confine the motion of free electrons in two dimensions. Derive the dependence of the density of states on kinetic energy for a two-dimensional gas. For the case of GaAs (effective mass $m^* = 0.067m_0$), determine the number of states per unit area that are in the energy range between 0 and $2kT \approx 0.052$ eV.

2.16 Derive the equation for the density of states for a one-dimensional electron gas (a "quantum wire") and plot the dependence on kinetic energy for the case of GaAs (effective mass $m^* = 0.067m_0$). Take the energy range between 0 and 1 eV. *Note:* Make sure to include the fact that a single energy level E_{kin} corresponds to two momentum levels p: $E_{kin} = (\pm p)^2/2m^*$; in other words, the density of states is twice as high as the value obtained by ignoring this effect. **A**

2.17 There is no pronounced anisotropy associated with the kinetic energy of holes in silicon; however, there are *light* and *heavy* holes with effective masses as determined in Problem 2.13.

(a) Determine the density-of-state effective mass so that the density-of-state equation for a three-dimensional uniform-particle gas

$$D = \frac{8\sqrt{2}\pi}{h^3}(m^*)^{3/2}E_{kin}^{1/2}$$

can be applied to the holes in silicon.

(b) What are the density-of-state effective masses for two-dimensional and one-dimensional hole gases in silicon? **A**

2.18 The kinetic energy of electrons in germanium has longitudinal–transverse anisotropy and such $\langle 111 \rangle$ symmetry that the elongations of four constant-energy ellipsoids are aligned along the following directions: (1) $[111]$–$[\bar{1}\bar{1}\bar{1}]$, (2) $[\bar{1}11]$–$[1\bar{1}\bar{1}]$, (3) $[1\bar{1}1]$–$[\bar{1}1\bar{1}]$, and (4) $[11\bar{1}]$–$[\bar{1}\bar{1}1]$. Knowing that the longitudinal and transverse are $m_l = 1.64m_0$ and $m_t = 0.082m_0$, respectively, determine the density-of-state effective mass.

2.19 Select an answer for each of the following questions:

(a) What is the value of the Fermi–Dirac function at E_F?

(b) What is the probability of finding a hole at E_F?

(c) What is the probability of finding an electron at E_V at room temperature?

(d) What is the probability of finding a hole at E_V at room temperature?

(e) What is the probability of finding an electron at E_C at 0 K?

(f) What is the probability of finding a hole at E_V at 0 K?

The available answers are

(1) 0 **(2)** 0.25 **(3)** 0.5

(4) 0.56 **(5)** 1

(6) something between 0 and 1

2.20 The Fermi level in the channel of a GaAs HEMT is at the bottom of a two-dimensional subband. What is the areal density of electrons in the channel? At what position, with respect to the bottom of the subband, is the Fermi level when the electron density is increased 100 times? The effective mass of electrons is $m^* = 0.067m_0$.

2.21 For the case of a Fermi level deep inside a two-dimensional subband, the areal electron density (n_{2D}) can be obtained by assuming that all the energy levels below E_F are filled and all the energy levels above E_F are empty. Starting from the density of states given in Example 2.8, obtain the electron density (n_{2D}) in this approximation. Can you determine the mathematical condition that would convert the general solution from Example 2.8 into the form identical to the solution obtained in this problem?

2.22 The position of the Fermi level in a silicon sample is 0.3 eV above the bottom of the valence band. For a 3D hole gas, the concentration of holes first increases and then decreases as the total energy is reduced from the top of the valence band—there is a distribution of the hole concentration along energy in the valence band.

(a) Determine the values of the kinetic energy corresponding to the maxima of the distributions of hole concentrations at the following three temperatures: 280 K, 300 K, and 320 K.

(b) Plot the distributions for these three temperatures. Use the energy range between 0 and 200 meV. The density-of-state effective mass is $m^* = 0.54m_0$.

2.23 Assuming that the effective mass of electrons in the conduction band of GaAs is temperature independent and equal to $m^* = 0.067m_0$, determine the effective density of states N_C at $T_l = 77$ K, $T_r = 300$ K, and $T_h = 450$ K. Given that the density of state distribution (Example 2.5) is temperature-independent, explain the origin of the temperature dependence of the effective density of states.

2.24 The hole gas in many semiconductors consists of two types of holes: holes with light and heavy effective masses. Determine the room-temperature effective density of states in the valence band for

(a) Si: $m_l = 0.16m_0$, $m_h = 0.49m_0$

(b) GaAs: $m_l = 0.082m_0$, $m_h = 0.45m_0$ **A**

2.25 Express the hole concentration in a degenerate semiconductor in terms of the effective density of states N_V and the Fermi integral $F_{1/2}$:

$$F_{1/2}(\eta) = \int_0^\infty \frac{x^{1/2}\,dx}{1 + e^{x-\eta}}$$

2.26 The doping level of N-type semiconductor is $N_D = 5 \times 10^{19}$ cm^{-3}. Determine the room-temperature position of the Fermi level (with respect to the bottom of the conduction band) if the semiconductor is

(a) Si ($N_C = 2.86 \times 10^{19}$ cm^{-3})

(b) GaAs ($N_C = 7 \times 10^{18}$ cm^{-3}) **A**

What would be the relative errors (expressed in percent) if a Maxwell–Boltzmann distribution was assumed?

2.27 If $(E_F - E_C)/kT = -1$ is taken as the upper limit for the applicability of approximations based on the Maxwell–Boltzmann distribution, determine the corresponding upper-limit doping levels for N-type silicon. What is the corresponding doping level for P-type silicon ($N_C = 2.86 \times 10^{19}$ cm^{-3}, $N_V = 3.10 \times 10^{19}$ cm^{-3})?

2.28 A heavily doped N-type semiconductor, with the Fermi level deep in the conduction band ($E_F - E_C \geq 5kT$) resembles a metal. In that case, the electron concentration can be obtained by assuming that all the energy levels below E_F are filled and all the energy levels above E_F are empty (a step approximation for the Fermi–Dirac distribution). Starting from the equation for the density of states

$$D(E_{kin}) = \frac{8\sqrt{2}\pi}{h^3}(m^*)^{3/2}E_{kin}^{1/2}$$

express the electron concentration in terms of the effective density of states (N_C) and the Fermi-level position ($E_F - E_C$). Compare the result to the one obtained in Example 2.11.

2.29 The energy gap of GaAs is $E_g = 1.42$ eV and experimental values for the effective densities of states at room temperature are $N_C = 4.7 \times 10^{17}$ cm^{-3} and $N_V = 7.08 \times 10^{18}$ cm^{-3}. Assuming $T^{3/2}$ dependencies for N_C and N_V (Example 2.10 and Table 2.2) and temperature-independent energy gap, determine the intrinsic position of the Fermi level (with respect to the top of the valence band) at 200°C. What is the intrinsic carrier concentration at that temperature?

2.30 The probability that an energy state in the conduction band is occupied by an electron is 0.001. Is this N-type, P-type, or intrinsic silicon? **A**

2.31 Find the room-temperature position of the Fermi level with respect to the top of the valence band for N-type silicon doped with $N_D = 10^{16}$ cm^{-3} donor atoms. Is it closer to the top of the valence band, or to the bottom of the conduction band?

2.32 Calculate the electron concentration in a doped silicon if the Fermi level is as close to the top of the valence band as to the midgap. What is the hole concentration? Is this N-type or P-type silicon?

2.33 Determine the doping type and level if the Fermi potential in a silicon sample is
(a) $q\phi_F = -0.35$ eV **(b)** $q\phi_F = 0.30$ eV **A**

2.34 Two silicon samples are doped with $N_D = 10^{15}$ cm^{-3} and $N_A = 10^{17}$ cm^{-3}, respectively. What is the difference between the Fermi potentials in these samples at
(a) room temperature **(b)** $T = 200$°C

2.35 An aluminum layer having the work function $q\phi_m = 4.1$ eV is to be deposited onto 6H SiC substrate.

(a) Determine the doping type and level so that the work function of the SiC substrate matches the work function of the aluminum layer at room temperature. The electron affinity of 6H SiC is $q\chi_s = 3.9$ eV, the energy gap is $E_g = 3.0$ eV, and the effective densities of states in the conduction and valence bands are $N_C = N_V = 2.51 \times 10^{19}$ cm^{-3}.

(b) What is the work function at 150°C? **A**

2.36 N_C and N_V equations given in Table 2.2 show that the effective density of states depends on temperature. The energy gap is also slightly temperature-dependent; in the case of silicon, this dependence can be fitted by $E_g\,[\text{eV}] = 1.17 - 7.02 \times 10^{-4}T^2/(T\,[\text{K}] + 1108)$. Find the intrinsic carrier concentration of silicon at $T = 300$°C. Compare this value to the doping level of $N_D = 10^{16}$ cm^{-3}.

2.37 Repeat Problem 2.36 for 6H silicon carbide ($E_g = 3$ eV) to see the importance of the energy-gap value for high-temperature applications. The room-temperature densities of states are $N_C = N_V = 2.51 \times 10^{19}$ cm^{-3}, while the $E_g(T)$ dependence can be fitted by $E_g\,[\text{eV}] = 3.0 - 3.3 \times 10^{-4}(T\,[\text{K}] - 300)$. **A**

2.38 N-type silicon is doped with $N_D = 5 \times 10^{15}$ cm^{-3}. Calculate the concentration of holes at 300°C ($E_g\,[\text{eV}] = 1.17 - 7.02 \times 10^{-4}T^2/(T\,[\text{K}] + 1108)$).

2.39 Including the effect of partial ionization, determine the room-temperature electron concentration in silicon doped by 2×10^{18} cm^{-3} phosphorus atoms. Explain why the ionization is significantly below 100%. The doping level of phosphorus in silicon is 0.045 eV below the bottom of the conduction band, and the effective density of states in the conduction band is $N_C = 2.86 \times 10^{19}$ cm^{-3}.

2.40 For the case of $N_D = 2 \times 10^{18}$ cm^{-3}, determine $E_C - E_F$ for the following cases:

(a) total ionization is assumed **A**

(b) the effect of partial ionization is included **A**

2.41 Including the effects of partial ionization, calculate the concentration of electrons in silicon doped by 2×10^{18} cm^{-3} phosphorus atoms if the specified operating temperature of the device is 85°C. Determine the relative change in electron concentration if the operating temperature increases by 10%. Assume constant effective density of states of $N_C = 2.86 \times 10^{19}$ cm^{-3}, donor level of 0.045 eV below the conduction band, and degeneracy factor of $g = 2$.

2.42 At sufficiently low temperatures, the ionization fraction drops exponentially with temperature reduction (the semiconductor is under *carrier freezeout* conditions). Assuming constant effective density of states ($N_C = 2.86 \times 10^{19}$ cm^{-3}), determine the temperature that corresponds to ionization fraction of 10% in silicon doped with 10^{15} cm^{-3} phosphorus atoms.

2.43 Silicon is doped with 10^{16} cm^{-3} boron atoms. Determine the carrier freezeout temperature range if the freezeout boundary is defined as the temperature at which the ionization of the doping atoms is 90%. The doping level of boron in silicon is 0.045 eV, and the degeneracy factor is $g = 1/4$. Assume constant effective density of states in the valence band, $N_V = 3.10 \times 10^{19}$ cm^{-3}.

2.44 P-type SiC is doped with $N_A = 10^{18}$ cm^{-3} of aluminum atoms. The ionization fraction is 4.7% at room temperature and 12.2% at 125°C. Determine the aluminum energy level with respect to the top of the valence band ($E_A - E_V$).

2.45 It is possible to bring a low-doped germanium sample to the state of $n \approx p = 10^{17}$ cm^{-3} in two ways:

(a) by illumination at room temperature

(b) by heating

What are the quasi–Fermi level positions with respect to the midgap in each of these situations? Take the $T^{3/2}$ temperature dependency of the effective density of states in both the conduction and the valence bands, and take the following room-temperature values: $N_C = 1.0 \times 10^{19}$ cm^{-3} and $N_V = 6.0 \times 10^{18}$ cm^{-3}. Assume a constant energy gap of $E_g = 0.66$ eV.

2.46 The quasi-Fermi levels E_{Fn} and E_{Fp} in an illuminated Si sample are such that $E_C - E_{Fn} \approx E_C - E_F = 150$ meV and $E_{Fp} - E_V = 0.300$ meV, where E_C, E_V, and E_F are the bottom of the conduction band, the top of the valence band, and the equilibrium Fermi level, respectively.

(a) Calculate the excess carrier concentration in this sample.

(b) Determine $E_C - E_{Fn}$ if the light intensity is increased so that $E_{Fp} - E_V$ becomes equal to 150 meV.

REVIEW QUESTIONS

R-2.1 Does the relationship $\lambda p = h$ mean that a wave with wavelength λ has the momentum of $p = h/\lambda$ and that a particle with momentum p has the wavelength of $\lambda = p/h$? If so, what is the fundamental difference between a *wave* and a *particle*.

R-2.2 Is the relationship $\lambda p = h$ applicable only to small things?

R-2.3 Do large particles possess wavelike properties? Can they be observed (like "diffraction of a soccer ball")?

R-2.4 In what units is the wave number (wave vector) expressed? Is it meaningful to say that the wave number is related to the wavelength in the same way that the angular frequency is related to the period?

R-2.5 What complex wave function is used to express the perfect wave with angular frequency ω, propagating in the x-direction in space? What is the ratio of the angular frequency to the wave number?

R-2.6 The complex wave function is a mathematical abstraction. What is the link to reality when applying the wave-function concept to electrons?

R-2.7 Is the intensity of the wave function $\psi(x, t) = A \exp[-j(\omega t - kx)]$ constant for $\infty < x < \infty$ and $-\infty < t < \infty$? If so, can we make use of this wave function to describe real objects that are not uniformly spread in space and time?

R-2.8 What is a single harmonic in the Fourier transform, used as a wave packet model?

R-2.9 What is the physical interpretation of the wave packet intensity?

R-2.10 Solving the Schrödinger equation, one obtains the electron wave function. Is this the function that describes the "shape" of the wave packet?

R-2.11 What is the wave function good for in semiconductor-device electronics? Can it help determine the electron concentration?

R-2.12 Does solving the Schrödinger equation provide any other information about the electrons, such as the electron energy?

R-2.13 Can an electron, trapped inside a potential well, have arbitrary energy? Is the energy of this electron related to the possible wavelength values? Can we use the "perfect-wave" concept and wave function to describe an electron in a potential well with infinite walls?

R-2.14 The probability that an electron wave will be reflected by an energy barrier is 0.9. What is the remaining probability of 0.1?

R-2.15 Is the electron wave function related to the tunneling probability?

R-2.16 Are the wavelengths of the valence electrons related to the width of the potential well that confines them? If so, what are the effects of the much wider potential well of crystals, compared to single atoms, on the wavelength and the energy of valence electrons?

R-2.17 Is the shape of the E–k dependence of a valence electron in a metal different from the E–k dependence of a free electron in vacuum? How about the energy position of the E–k dependence?

R-2.18 Is there any space localization of the valence electrons in metals?

R-2.19 A valence electron in semiconductors can appear as a standing wave with either the nodes or the antinodes at the crystal-lattice sites. Does this mean that a single wavelength can correspond to two different energy levels? If so, can standing-wave electrons take energy values between the two different energy levels?

R-2.20 How does the appearance of energy gap in semiconductors influence the parabolic E–k dependence corresponding to free electrons in metals?

R-2.21 What are the two bands, important for the properties of semiconductors and insulators, called?

R-2.22 Where do the mobile electrons appear (in the conduction band or in the valence band)?

R-2.23 Are there electrons in the valence band? Where do the mobile holes appear?

R-2.24 Can the electron-gas model be applied to the electrons in the conduction band of a semiconductor? If so, what adjustment needs to be made?

R-2.25 In the carrier-gas model, what is the kinetic energy of an electron at the bottom of the conduction band? What is the kinetic energy of a hole at the top of the valence band?

R-2.26 The E–k dependence in the valence band of a semiconductor can be approximated as an "inverted" parabola. Does this mean that a higher kinetic energy of a hole corresponds to a lower energy position on the E–k diagram?

R-2.27 The bottom of the conduction band in silicon appears for $k \neq 0$. Does this mean that the conduction-band electrons are distributed over several equivalent conduction-band minima? If yes, how many and why?

R-2.28 A spectrum of sinusoidal wave functions, with distributed wave numbers, can be integrated to create a wave packet that is "localized" in space. Is the degree of space localization independent of the width of the wave number distribution. If not, describe the dependence.

R-2.29 One way of expressing the Heisenberg uncertainty principle is that we cannot measure the position of an electron with better accuracy than $\Delta x = h/\Delta p$ if the momentum is determined with precision Δp. Is this a fundamental limitation of our measurement techniques, or is it simply the way the small things are?

R-2.30 The number of electrons per unit "volume" of the six-dimensional x–y–z–p_x–p_y–p_z space is limited by the uncertainty principle to what value?

R-2.31 What is the difference between the *density of electron states* [$D(E_{kin})$] and the *number of electrons per unit volume and unit energy*?

R-2.32 How does the density of electron state relate to electron concentration?

R-2.33 What fundamental principle is included in the Fermi–Dirac distribution of the electron-state occupancy that is not involved by the classical Maxwell–Boltzmann (exponential) distribution?

R-2.34 What mathematical condition has to be satisfied to be able to approximate the electron-state occupancy by the Maxwell–Boltzmann distribution? Under this condition, why is there no problem with the fundamental principles that the Maxwell–Boltzmann distribution does not include?

R-2.35 What is the difference between the density of electron states $D(E_{kin})$ and the effective density of states in the conduction band N_C?

R-2.36 Does the electron concentration change with energy E, where $E \geq E_C$?

R-2.37 What is the Fermi level E_F?

R-2.38 How is the doping expressed in an energy-band diagram?

R-2.39 What is the Fermi potential ϕ_F?

3 Drift

Following the introduction of carrier properties in Chapters 1 and 2, this chapter begins the descriptions of carrier-related phenomena. The fundamental effect is that a current carrier in electric field $\boldsymbol{E}$ experiences force $\boldsymbol{F} = \pm q\boldsymbol{E}$, where the plus sign is for the positively charged holes and the minus sign is for the negatively charged electrons. This force leads to carrier motion and consequently to electric current. An electron in constant field and *free space* would move with a constant acceleration due to the electric-field force ($F = m_0 a$). In crystals, however, electrons collide with the crystal imperfections (including the thermal vibrations of crystal atoms, the doping atoms, and the mobile electrons themselves). These collisions lead to a constant average velocity, as distinct from the constant acceleration that would cause ever increasing velocity. This type of carrier motion is called *drift,* the average carrier velocity is called *drift velocity,* and the associated electric current is called *drift current.* It is important to mention at this stage that the *drift* (described in this chapter) is not the only way of creating electric current in semiconductors. Another important current mechanism, *diffusion,* is considered in Chapter 4.

3.1 ENERGY BANDS WITH APPLIED ELECTRIC FIELD

As described in Section 2.2.3 (Fig. 2.10), the bottom of the conduction band (E_C) and the top of the valence band (E_V) correspond to the potential energies of electrons and holes, respectively. Furthermore, the potential energy (E_{pot}) is linked to the electric potential (φ) by the following fundamental relation:

$$E_{pot} = -q\varphi \tag{3.1}$$

where q is the unit of charge ($q = +1.6 \times 10^{-19}$ C). If the potential energy is expressed in eV, then the electric potential and the potential energy have the same numerical values with different signs. This means that if a voltage of $V = \varphi_1 - \varphi_0 = -1$ V is applied across a slab of a semiconductor, the difference between the potential energies of electrons/holes at the ends of the slab will be 1 eV. If the electric potential changes linearly from φ_0 to φ_1, the potential energies of both electrons and holes (which means both E_C and E_V) also change linearly from $-q\varphi_0$ to $-q\varphi_1$. Given that the negative gradient of electric potential is equal to the electric field ($E = -d\varphi/dx$), the electric field in a semiconductor is expressed by the positive gradient of E_C or E_V:

$$E = \frac{1}{q}\frac{dE_{C,V}}{dx} \tag{3.2}$$

This means that the case of zero electric field can be visualized as flat bands (E_C and E_V do not change with x), whereas the sign and the strength of a nonzero electric field can be visualized as the slope of E_C and E_V. Of course, the slope and any spatial changes are equal for both E_C and E_V; the difference between E_C and E_V is equal to E_g everywhere in the semiconductor and is independent of the value of any applied electric field. Therefore, the E–x diagram shown in Fig. 2.10 is for the special case of zero electric field; in general, an E–x diagram shows *parallel* E_C and E_V lines that are changing to correspond to any changes in the electric potential.

3.1.1 Energy-Band Presentation of Drift Current

Any application of nonzero electric field causes motion of the electrons and holes in the semiconductor. The momentum that the electrons and holes gain due to the action of the electric-field force corresponds to their kinetic energy $E_{kin} = p^2/2m^*$. As explained in Section 2.2.3, the total-energy level of electrons (holes) with nonzero kinetic energy

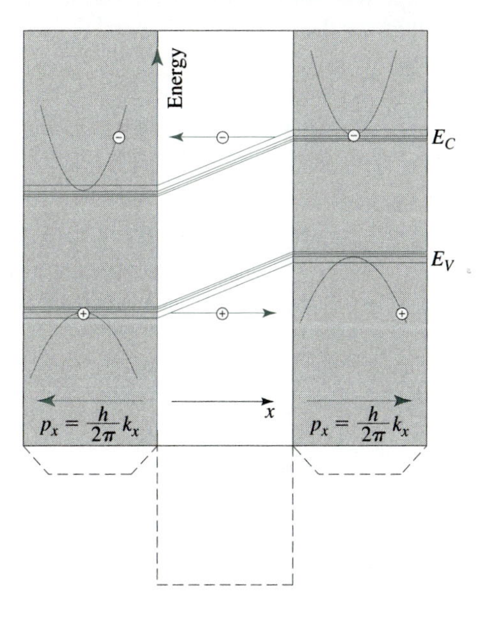

Figure 3.1 The relationship between E–k and E–x diagrams with applied electric field.

is above E_C (below E_V) for the value of the kinetic energy. Figure 3.1 shows how the parabolic E–k diagrams are related to the E–x diagrams for the case of nonzero electric field. It also illustrates the positions of electrons and holes with zero and nonzero kinetic energies. The arrows illustrate that the total energies of a moving electron or hole do not change. What happens is that an electron starting from the bottom of the conduction band (zero kinetic energy) gains kinetic energy as it moves in the direction of falling potential energy (falling level of E_C). Analogously, an increase in the potential energy sets a moving hole at a level below E_V, the difference being exactly equal to the gained kinetic energy. No energy loss or gain is illustrated in Fig. 3.1.

The effective carrier mass takes into account the periodic potential of the crystal lattice, so a carrier should move through the crystal as through a free space (just with a different mass). This would be the case in a perfect crystal, and Fig. 3.1 would be the complete presentation of the carrier motion. At temperatures $T > 0$ K, however, the lattice atoms vibrate around the lattice sites, causing disturbances in the periodic potential of the lattice. These disturbances (thermal vibrations of crystal atoms) can be modeled as particles, in which case they are called *phonons*. The phonons can scatter moving carriers. Carriers can also be scattered by other crystal imperfections, in particular by the charged doping atoms; this type of scattering is referred to as the Coulomb scattering. Similar to the scattering of molecules in a classical gas, carrier–carrier scattering is also possible, although it is much less pronounced than the phonon and Coulomb scattering.

Carrier scattering can be *elastic* (change in the carrier direction but no exchange of energy), or it can involve energy exchange. As the moving carriers gain kinetic energy from the electric field that is above the average thermal energy, they are likely to give the excess energy to the scattering centers. Figure 3.2 illustrates that an electron giving its kinetic energy as a result of a scattering event falls down toward the bottom of the conduction band. As this electron moves further in the direction of decreasing E_C, it gains kinetic energy again, only to lose it again in some of the forthcoming scattering events. The situation with holes is analogous: they bubble up, toward the top of the valence band, as they give their kinetic energy to the scattering centers.

The energy-band diagrams (E–x) with applied field enable the introduction of a very illustrative model for the drift current. *The conduction band can be considered as a vessel containing electrons that tend to roll down when the bottom of the conduction band is*

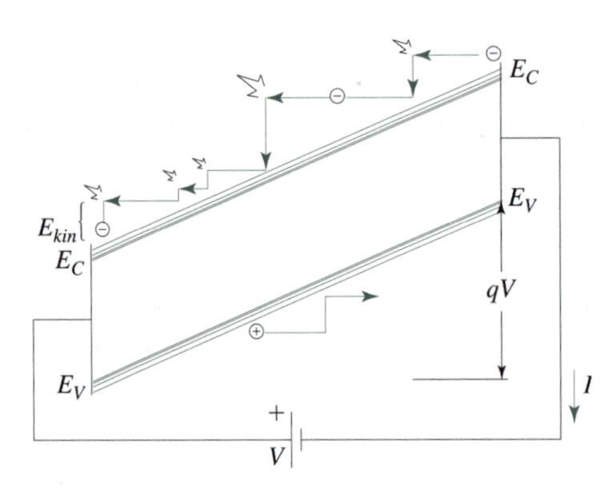

Figure 3.2 Energy-band (E–x) diagram with applied electric field, illustrating the *rolling down* of electrons and *bubbling up* of holes.

tilted. Analogously, the valence band can be considered as a vessel containing a liquid and bubbles (holes) in it. When the top of the valence band is tilted due to appearance of electric field, the holes tend to bubble up along the top of the valence band, moving effectively in the direction of the electric field applied. This model is very useful for illustration of the principle of operation of semiconductor devices.

3.1.2 Resistance and Power Dissipation Due to Carrier Scattering

The carrier scattering is manifested as *resistance* to the flow of current carriers, limiting the electric current. Because of the scattering, the energy that the carriers gain from an applied electric field is converted to thermal energy, increasing the crystal temperature (the vibration of the crystal atoms). The rate of this conversion to thermal energy is called *power dissipation*.

EXAMPLE 3.1 Resistance and Power Dissipation

The voltage across an N-type semiconductor slab, conducting 10 mA of current, is equal to 1 V. How much energy is delivered to the crystal by every electron that passes through the slab? Relate this energy to the dissipated power.

SOLUTION

If the potential energy at one end of the slab is taken as the reference energy (zero), the potential energy of the electrons at the other end is qV, where V is the voltage across the slab. Every electron flowing through this slab gains energy qV from the applied voltage and delivers it to the crystal lattice through the collisions that bring the energy (temperature) of the electrons to thermal equilibrium with the crystal lattice. Given that the voltage across the track is 1 V, *the energy delivered by every electron is* $qV = 1$ eV. If there are N electrons, the total energy will be qVN. The dissipated power is the energy delivered to the crystal per unit time; therefore,

$$P = qVN/t = \underbrace{\frac{qN}{t}}_{I} V = IV$$

where $I = qN/t$ is the electric current flowing through the track. The value of the dissipated power is $P = IV = 10$ mW.

3.2 OHM'S LAW, SHEET RESISTANCE, AND CONDUCTIVITY

If an electric-potential difference ($V = \varphi_1 - \varphi_0$) is established between two points in a semiconductor, a limited current (I) will flow through the semiconductor. The actual value of the current will depend on (1) how large the potential difference is and (2) how large the resistance (R) of the semiconductor is. An increase in the potential difference (applied voltage) increases the current; an increase in the resistance, however, reduces the current. *Assuming constant resistance,* the current dependencies on the voltage and resistance are expressed by the widely used Ohm's law:

$$I = \frac{V}{R} \tag{3.3}$$

The assumption of constant resistance holds very well in the cases of metals, carbon-based resistors, and the whole variety of thick- and thin-film resistors. It also works well for the case of semiconductor-based integrated-circuit resistors. In some important semiconductor devices, however, the current flowing through a resistive body does not depend linearly on the voltage applied; in other words, the resistance in Eq. (3.3) is not a constant but depends on the value of the current. This effect will be considered in Section 3.3. The aim of this section is to introduce the concepts related to design of integrated-circuit resistors and to relate the integral form of Ohm's law [Eq. (3.3)] to its differential form that is suitable for modeling of effects *inside* semiconductor devices.

3.2.1 Designing Integrated-Circuit Resistors

Resistors in integrated circuits are made of a resistive body surrounded by an insulating medium and contacted at the ends by conductive tracks, as shown in Fig. 3.3. The resistance of the resistor depends on the resistive property of the body, called resistivity (ρ), and its dimensions in the following way: (1) an increase in the length of the resistor, L, increases its resistance as the carriers making the current have to travel a longer distance; (2) an increase in the cross-sectional area, $A = x_j W$, decreases the resistance as more current carriers can flow in parallel. Therefore, the resistance can be expressed as

$$R = \rho \frac{L}{x_j W} \tag{3.4}$$

Note that the curved ends of the resistive body are neglected and the resistor shape is considered as a rectangular prism with dimensions L, W, and x_j.

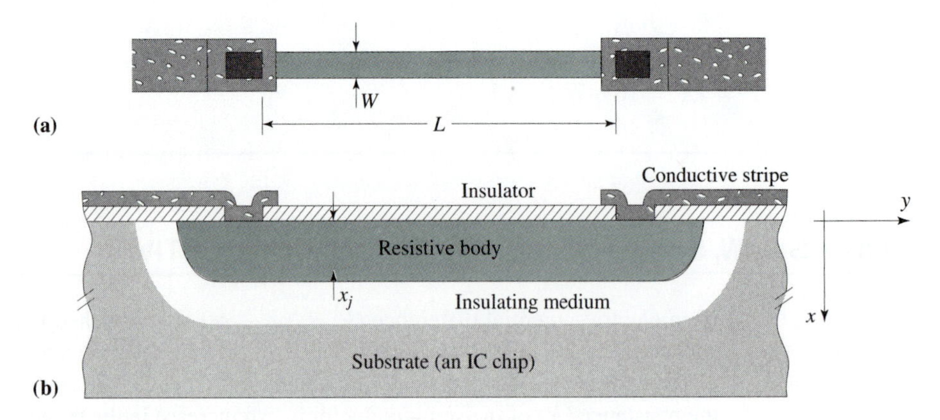

Figure 3.3 Integrated-circuit resistor: (a) top view and (b) cross section.

The reciprocal value of the resistivity, called conductivity, is more frequently used to characterize the resistive/conductive properties of semiconductor layers. The conductivity is denoted by σ; thus

$$\sigma = \frac{1}{\rho} \tag{3.5}$$

When designing an integrated-circuit layout, the designer sets L and W of each resistor individually. Taking into account that there is a minimum dimension achievable by the selected technology, the designer determines the ratio L/W to achieve the desired resistance, and it determines the actual L and W values to minimize the area of the resistor. The conductivity (σ) and the thickness of the resistive layer (x_j) are technological parameters, and they have the same values for all the resistors made with one type of resistive layer. There will be only a few different types of layers (with different conductivities and thicknesses) available for making resistors. The term $\rho/x_j = 1/(\sigma x_j)$ in Eq. (3.4), which brings together the technological parameters, is frequently expressed as one variable called the sheet resistance, R_S:

$$R_S = \frac{1}{\sigma x_j} = \frac{\rho}{x_j} \tag{3.6}$$

The sheet resistance is a quantity that can be measured more easily than the conductivity σ and the layer thickness x_j. If the resistance of a test resistor with known L and W is measured, the sheet resistance of the associated layer can easily be calculated from the following equation:

$$R = R_S \frac{L}{W} \tag{3.7}$$

Equation (3.7) is obviously obtained after replacement of ρ/x_j in Eq. (3.4) by R_S. The unit of the sheet resistance is essentially Ω; however, it is typically expressed as $\Omega/\square$, indicating that it represents the resistance of a squared resistor ($L = W$).

The difference and relationship between the resistance R and resistivity ρ needs a more careful consideration. We can assign a resistivity value to any point in the resistive body. If the resistive body is a homogeneous one, then the resistivity will be the same throughout the body; but if it is not homogeneous, the resistivity will have different values at different points. Typically, the resistivity is not uniform in integrated-circuit resistors, but is smallest at the surface, increasing inside the material. Equivalently, the conductivity is highest at the surface, reducing inside the material, as illustrated in Fig. 3.4. Because of this property, the conductivity and the resistivity are called local or *differential* quantities. Differential quantities are needed to describe the internal structure (sometimes called microstructure) of semiconductor devices. Information on the conductivity at different points of a resistive body, however, cannot be used to directly determine to what value will the resistor limit the terminal current (I) when a certain terminal voltage (V) is applied. That is why it is necessary to determine the overall resistance, or *integral* resistance R, after which Ohm's law (Eq. 3.3) can directly be applied. Therefore, the integral quantities are needed to describe the terminal, or integral, characteristics of semiconductor devices.

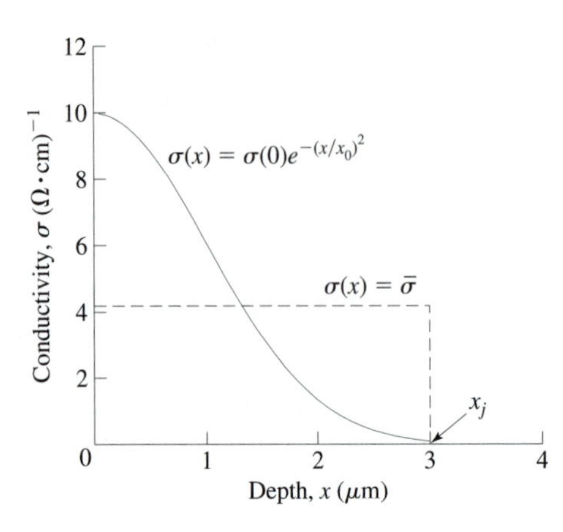

Figure 3.4 A typical variation of conductivity from the surface into the bulk of an integrated-circuit resistor (solid line) and the uniform-conductivity approximation (dashed line); in the example, σ (0)=10 $(\Omega \cdot \text{cm})^{-1}$, and $x_0 = \sqrt{2}\ \mu\text{m}$.

The relationships between the resistance R as an integral quantity and the conductivity σ as a differential quantity [Eqs. (3.4), (3.6), and (3.7)] appear very simple, but this is because they are simplified to account for only the simplest case of uniform (or homogeneous) resistive bodies. In the case of the resistive layer illustrated in Fig. 3.4, we see that Eq. (3.6) cannot be directly used to calculate the sheet resistance R_S. To be able to use Eq. (3.6), it is necessary to find a single value for the conductivity that will provide the most suitable uniform-conductivity approximation of the real conductivity dependence. Typically, the average value of the conductivity is used as the best uniform-conductivity approximation, as shown in Fig. 3.4 by the dashed lines. The average conductivity, $\overline{\sigma}$, can be found by equating the rectangular area $\overline{\sigma}x_j$ defined by the dashed lines in Fig. 3.4 with the area enclosed between the real conductivity curve and x- and y-axes, which is calculated as

$$\int_0^{x_j} \sigma(x)\,dx \approx \int_0^{\infty} \sigma(x)\,dx \tag{3.8}$$

Therefore,

$$\overline{\sigma} \approx \frac{1}{x_j}\int_0^{\infty} \sigma(x)\,dx \tag{3.9}$$

The average value of the conductivity determined in this way can now be used to calculate the sheet resistance by Eq. (3.6): $R_S = 1/(\overline{\sigma}x_j)$.

The preceding approach to calculating the sheet resistance enables a satisfactory estimation of the overall resistance, and its result can be used to determine the terminal voltage and current using Ohm's law. This agreement is due to the fact that the terminal current is an integral quantity, integrating any possible current flow through the resistive body in a similar way to what Eq. (3.9) does with the conductivity. It is not difficult to imagine that a larger portion of the terminal (integral) current I will flow close to the surface, due to the higher conductivity, than through the lower part of the resistor where the conductivity is low, and that all these larger and smaller current streams integrate into the terminal current I at the resistor terminals. However, the use of the approach of averaging the conductivity to find the terminal current does not give quantitative information on how the current flow is distributed inside the resistive body.

EXAMPLE 3.2 Resistance, Sheet Resistance, and Resistivity

The thickness of a copper layer, deposited to create the interconnecting tracks in an IC, is 100 nm. The copper resistivity is given in Table 3.1.

(a) What is the sheet resistance of the copper layer at 27°C (room temperature)?

(b) The minimum width of the interconnecting tracks is set at 0.5 μm. What is the maximum resistance per unit length?

(c) The length of a minimum-width track, connecting two components, is 300 μm. What is the resistance of this track?

(d) What is the resistance of the track at 75°C?

SOLUTION

(a) The sheet resistance incorporates the resistivity ρ and the layer thickness x_j [Eq. (3.6)]:

$$R_S = \rho / x_j = 0.17\ \Omega/\square$$

(b) According to Eq. (3.7), the resistance is $R = R_S L/W$, which means that the resistance per unit length is

$$\frac{R}{L} = \frac{R_S}{W} = 3.4 \times 10^5\ \Omega/\text{m} = 0.34\ \Omega/\mu\text{m}$$

TABLE 3.1 Resistivities and Temperature Coefficients for Selected Metals

Metal	ρ at 27°C ($\mu\Omega \cdot$ cm)	$\alpha = \frac{\Delta\rho/\rho}{\Delta T}$ (°C^{-1})
Aluminum	2.82	0.0039
Copper	1.7	0.0039
Gold	2.44	0.0045
Iron	9.7	0.0050
Lead	22	0.0039
Molybdenum	5.2	0.0040
Nichrome (Ni–Cr)	150	0.0004
Nickel	6.9	0.0038
Platinum	11	0.00392
Silver	1.59	0.0038
Tungsten	5.6	0.0045

(c) The total resistance is

$$R = R_S \frac{L}{W} = 102\ \Omega$$

(d) We can see from Table 3.1 that the temperature coefficient of resistivity α is given by

$$\alpha = \frac{\Delta\rho/\rho}{\Delta T}$$

The resistivity ρ is the only temperature-dependent quantity in Eq. (3.4). This means that the temperature coefficient α can also be expressed as

$$\alpha = \frac{\Delta R/R}{\Delta T}$$

A mathematical proof for this is as follows:

$$\frac{dR}{dT} = \frac{d\left[\rho L/(x_j W)\right]}{dT} = \frac{L}{x_j W}\frac{d\rho}{dT} = \underbrace{\frac{\rho L}{x_j W}}_{R}\frac{1}{\rho}\frac{d\rho}{dT}$$

$$\frac{1}{R}\frac{dR}{dT} = \frac{1}{\rho}\frac{d\rho}{dT}$$

ΔT and ΔR can be expressed as $T_1 - T_0$ and $R_1 - R_0$, where $T_0 = 27°\text{C}$, $T_1 = 75°\text{C}$, R_0 is the resistance at T_0, and R_1 is the resistance at T_1. Therefore, we can establish the following equation for the resistance R_1:

$$\alpha = \frac{(R_1 - R_0)/R_0}{T_1 - T_0}$$

$$R_1 - R_0 = R_0\alpha(T_1 - T_0)$$

$$R_1 = R_0\left[1 + \alpha(T_1 - T_0)\right] = 121\ \Omega$$

EXAMPLE 3.3 Design for a Minimum Integrated-Circuit Area

(a) Design a resistor of 3.5 kΩ using a layer with sheet resistance of 200 Ω/□. The minimum dimension achievable by the particular technology is 1 μm.

(b) An additional resistor with the resistance of 0.5 kΩ is needed. There are four layers available with the following sheet resistances: $R_{S1} = 5\ \Omega/\square$, $R_{S2} = 200\ \Omega/\square$, $R_{S3} = 1.5\ \text{k}\Omega/\square$, $R_{S4} = 4\ \text{k}\Omega/\square$. Which one of the four layers would you use?

SOLUTION

(a) From Eq. (3.7) we find

$$\frac{L}{W} = \frac{R}{R_S} = \frac{3500}{200} = 17.5$$

Any combination of L and W that gives the ratio of $L/W = 17.5$ will provide the needed resistor of 3.5 kΩ. However, the performance of integrated circuits is maximized if the integrated-circuit area is minimized. Because of that, the optimum solution is $W = 1\ \mu$m, and $L = 17.5\ \mu$m.

(b) The value of the resistor is closest to the sheet resistance of the second layer (200 Ω/□); therefore the smallest resistor (in terms of area $L \times W$) can be obtained with this layer. It is easy to find that $L/W = 2.5$.

EXAMPLE 3.4 Average Conductivity

(a) Find the average conductivity for the layer shown in Fig. 3.4.

(b) Design a 100-kΩ resistor using this layer, if it is known that the minimum dimension achievable by the particular technology is 1 μm.

SOLUTION

(a) Putting the conductivity function as given in Fig. 3.4 into the integral of Eq. (3.9), one obtains

$$\overline{\sigma} = \frac{10}{x_j}\int_0^{\infty} e^{-x^2/2}\, dx$$

where $x_j = 3.0\,\mu$m. It is known from mathematics that the solution of this integral (Laplace integral) is

$$\int_0^{\infty} e^{-x^2/2}\, dx = \sqrt{\pi/2}\ \mu\text{m}$$

The solution of the integral is expressed in μm as x and $x_0 = \sqrt{2}$ are given in μm. Therefore, the average conductivity is

$$\overline{\sigma} = \frac{10\ (\Omega \cdot \text{cm})^{-1}}{3\ \mu\text{m}} \sqrt{\pi/2}\ \mu\text{m} = 4.18\ (\Omega \cdot \text{cm})^{-1}$$

(b) Calculate first the sheet resistance:

$$R_S = \frac{1}{\overline{\sigma} x_j} = \frac{1}{4.18 \times 3 \times 10^{-4}} = 797.4\ \Omega/\square$$

Thus, the number of squares needed is

$$\frac{L}{W} = \frac{R}{R_S} = \frac{100{,}000}{797.4} = 125$$

As $W = 1\ \mu\text{m}$, the length of the resistor needs to be $L = 125\ \mu\text{m}$.

3.2.2 Differential Form of Ohm's Law

The approach of averaging the conductivity of semiconductor layers, described in the previous section, enables someone to design resistors without knowing much of what is happening inside the resistor. Can you imagine, however, the feeling of the designer if it happens that a proudly designed resistor does not show a linear current-to-voltage (I–V) characteristic but rather a saturating-type curve? There must have been a limitation the designer was not aware of! Was that a new effect? Could it be exploited for something useful? Was the effect reproducible? The answers to all these questions lie in the knowledge of what is happening inside the resistor. The *integral form* of Ohm's law links integral (terminal) voltages and currents, so it cannot be used to describe any current distribution or electric-potential variations inside a resistive body. What is needed is the *differential form* of Ohm's law that can be applied to a single point inside the resistive body.

To illustrate the link between the two forms of Ohm's law, consider the simplest (even though rarely realistic) case of a homogeneous resistive body with the length L and the cross-sectional area A (can be $A = W x_j$ as in Fig. 3.3). Let us transform the integral quantities from the integral form of Ohm's law [Eq. (3.3)] into their differential counterparts. We already know that for this simple case we can easily transform the resistance R into the conductivity σ using Eqs. (3.6) and (3.7). The differential counterpart for the terminal current I is the current density. In this chapter, we will use the symbol j_{dr} for the current density to specifically indicate that we are dealing with *drift* current. The current density in a homogeneous resistive body has the same value at any point of the resistor cross section, which is simply calculated as

$$j_{dr} = \frac{I}{A} \tag{3.10}$$

The differential counterpart of the terminal voltage is electric field, denoted by E. If the electric field has a constant value in the considered domain, as is the case for our example, then the electric field can be calculated as

$$E = \frac{V}{L} \tag{3.11}$$

Replacing I, V, and R in the integral form of Ohm's law (Eq. 3.3) by j_{dr}, E, and σ as given in Eqs. (3.6), (3.7), (3.10), and (3.11), the differential form of Ohm's law is obtained:

$$j_{dr} = \sigma E \tag{3.12}$$

The differential form of Ohm's law is said to be more general than the integral form because the former can also be applied in the case of nonhomogeneous resistive bodies when the current density is not uniformly distributed over the resistor cross section.

The differential form of Ohm's law as given by Eq. (3.12) is applicable in the case in which the current flows in one direction only, which is basically the direction of the electric field. In the example of our resistor (Fig. 3.4), this one-dimensional form of Ohm's law can be applied to the central part of the resistor where the current flow is parallel to the surface and an axis, say y-axis, can be placed along the current path. The one-dimensional form of Ohm's law, however, is not useful if we want to study the corner effects, because two dimensions are needed to express the circle-like current paths in these regions. That is why in the most general case the current density is considered to be a vector quantity, expressing not only the value of the current density at a certain point of the resistive body, but also the direction of the current flow. Note that the current density can in general take any direction in space, unlike the terminal current, whose direction is successfully expressed by making the current value positive or negative depending on the direction. The electric current takes no other path but the one traced by the electric-field lines, which means that the direction of the current-density vector and the electric-field vector are the same. Expressing the current density and the electric field as vectors, $\boldsymbol{j}_{dr}$ and $\boldsymbol{E}$, the most general form of Ohm's law is obtained:

$$\boldsymbol{j}_{dr} = \sigma \boldsymbol{E} \tag{3.13}$$

The electric field E is related to the electric potential φ, the electric field being equal to the negative value of the slope of electric-potential change along a direction in space (say the y-axis):

$$E = -d\varphi/dy \tag{3.14}$$

In the specific case of $E = \text{const}$, $d\varphi/dy = -(\varphi_1 - \varphi_0)/L = -V/L$, which reduces Eq. (3.14) to Eq. (3.11). In the three-dimensional case, Eq. (3.14) is generalized as

$$\boldsymbol{E} = -\frac{\partial\varphi}{\partial x}\boldsymbol{x}_u - \frac{\partial\varphi}{\partial y}\boldsymbol{y}_u - \frac{\partial\varphi}{\partial z}\boldsymbol{z}_u \tag{3.15}$$

where $\boldsymbol{x}_u$, $\boldsymbol{y}_u$, and $\boldsymbol{z}_u$ are the unit vectors in x-, y-, and z-directions, respectively, while ∂'s denote partial derivatives.[1] Using Eqs. (3.14) and (3.15), the one-dimensional (along the

[1] Partial derivatives are used to denote that a derivation is with respect to one variable, while others are treated as constants for that particular derivative; if there is one only variable, then the partial derivative reduces to the ordinary derivative, as in Eq. (3.14).

y-axis) and three-dimensional drift currents can be expressed as

$$j_{dr} = -\sigma \frac{d\varphi}{dy}, \qquad j = -\sigma \left(\frac{\partial}{\partial x} x_u + \frac{\partial}{\partial y} y_u + \frac{\partial}{\partial z} z_u \right) \varphi = -\sigma \nabla \varphi \tag{3.16}$$

Equation (3.13), or equivalently Eq. (3.16), is a mathematical model for the drift current. Equation (3.16) can directly be related to the following physical description of *drift* as a carrier-transport mechanism. If an electric potential difference is created across a medium having mobile charged particles, a nonequilibrium situation is created, and the particles will start flowing as a reaction aimed at diminishing the potential difference. This means that if the carriers are negatively charged—for example, electrons—they will flow toward the higher potential in order "to find" the positive charges creating the potential difference and neutralize them.

EXAMPLE 3.5 Current Density Versus Terminal Current

Calculate the maximum current density and the terminal current for the resistor designed in Example 3.4b if a voltage of 5 V is applied to the resistor terminals. Neglect the corner effects.

SOLUTION

Neglecting the corner effects, the one-dimensional form of Ohm's law [Eq. (3.12)] can be used to calculate the current density. In that case the electric field is uniform and equal to $E = V/L = 5/(125 \times 10^{-6}) = 40{,}000$ V/m $= 40$ V/mm. The maximum current is obtained for the maximum conductivity, which is $\sigma_{max} = \sigma(0) = 10\ (\Omega \cdot \text{cm})^{-1}$ as can be seen from Fig. 3.4. Therefore, $j_{dr-max} = \sigma_{max} E = 4 \times 10^7$ A/m^2.

The terminal current can be obtained by integrating the current density as distributed from the surface to x_j (or to ∞ which is the same as no significant current flows between x_j and ∞) and multiplying by the resistor width:

$$I = W \int_0^\infty j_{dr}(x)\, dx = W \frac{V}{L} \sigma(0) \int_0^\infty e^{-(x/x_0)^2} dx$$

The integral that appears in the foregoing equation was discussed in Example 3.4b. Using the values for all the quantities in SI units, one obtains

$$I = 10^{-6} \frac{5}{125 \times 10^{-6}} 1000 \sqrt{\pi/2} = 5 \times 10^{-5}\ \text{A} = 50\ \mu\text{A}$$

Note that the integral form of Ohm's law gives the same result for the terminal current ($I = V/R = 5\ \text{V}/100\ \text{k}\Omega = 50\ \mu\text{A}$), but it cannot be used to find the maximum current density inside the resistor body.

3.2.3 Conductivity Ingredients

Ohm's law [Eq. (3.13)] is a phenomenological relationship between the current density and the electric field, with the conductivity as the proportionality coefficient. Hence, the conductivity incorporates all the material-related factors. There are basically two things that influence the conductivity: (1) concentration of the carriers available to contribute toward the electrical current and (2) mobility of these carriers. With the concentration of the carriers, it is clear that a higher carrier concentration means a higher current for the same electric field, which further means that the conductivity is higher. The second factor, the mobility, accounts for an effect that different carriers, or the same carriers in different conditions, do not flow equally easily. To illustrate this effect, consider the ions in a metal lattice: these are charged particles with a finite concentration, but they would not make any current if an electric field is applied as they are completely immobile—their mobility is equal to zero. Therefore, *the conductivity is proportional to the carrier concentration and the carrier mobility*. Given that there are two types of current carriers in semiconductors, free electrons and holes, there are both electron and hole components in the conductivity equation:

$$\sigma = qn\mu_n + qp\mu_p \tag{3.17}$$

In Eq. (3.17), n and p are concentrations of the free electrons and holes, respectively, μ_n and μ_p are the free-electron and hole mobilities, respectively, and q is the electronic charge ($q = 1.6 \times 10^{-19}$ C). Obviously, $qn\mu_n$ is the electrons contribution to the conductivity, whereas $qp\mu_p$ is the contribution of the second type of current carriers, the holes. Usually, it is either $n \gg p$ or $p \gg n$, so one of the two terms in Eq. (3.17) is dominant and the other can be neglected.

Given that the unit of the conductivity is $(\Omega \cdot \text{m})^{-1}$, the mobility is in units of $\text{m}^2/\text{V} \cdot \text{s}$. The next section describes the mobility in more detail.

EXAMPLE 3.6 Conductivity and Carrier Concentration

P-type silicon has a resistivity of $0.5\,\Omega \cdot \text{cm}$. Find the following, assuming that $\mu_n = 1450\ \text{cm}^2/\text{V} \cdot \text{s}$ and $\mu_p = 500\ \text{cm}^2/\text{V} \cdot \text{s}$:

(a) the hole and electron concentrations

(b) the maximum change in resistivity caused by a flash of light, if the light creates 2×10^{16} additional electron–hole pairs/cm^3

SOLUTION

(a) In a P-type semiconductor the conductivity due to electrons can be neglected because $p \gg n$. Therefore, $\sigma \approx q\mu_p N_A$. The resistivity is

$$\rho \approx \frac{1}{q\mu_p N_A}$$

The concentration of acceptor ions, and therefore holes, is then $N_A \approx p = 1/(q\mu_p\rho) = 2.5 \times 10^{16}\ \text{cm}^{-3}$. The concentration of electrons is found as

$$n = \frac{n_i^2}{p} = 4.2 \times 10^3\ \text{cm}^{-3}$$

(b) The flash of light produces excess electrons and holes, reducing therefore the resistivity. When the light is removed, the excess electrons and holes will gradually recombine with each other, increasing the resistivity to its original value—that is, the equilibrium value. To find the maximum change in the resistivity, we need to determine the resistivity of the specimen when the light is on. In that case the concentration of holes is $p = 2.5 \times 10^{16} + 2 \times 10^{16} = 4.5 \times 10^{16}\ \text{cm}^{-3}$. The concentration of electrons is $n = 2 \times 10^{16}\ \text{cm}^{-3}$, as generated by the light, and in this specific case it cannot be neglected when compared to the concentration of holes. The conductivity is calculated as

$$\sigma = q\mu_p p + q\mu_n n = 8.24\ (\Omega \cdot \text{cm})^{-1}$$

The corresponding resistivity is $\rho = 1/\sigma = 0.12\ \Omega \cdot \text{cm}$. The maximum difference is, therefore, $\Delta\rho_{max} = 0.5 - 0.12 = 0.38\ \Omega \cdot \text{cm}$.

EXAMPLE 3.7 *TCR*

The temperature coefficient of a resistor, *TCR*, is defined as

$$TCR = \frac{1}{R}\frac{dR}{dT} \times 100\ [\%/^\circ\text{C}]$$

Find the *TCR* (at room temperature) of a resistor made of N-type silicon. If the resistance of the resistor is 1 kΩ at 27°C, estimate the resistance at 75°C. The temperature dependence of electron mobility is approximated by

$$\mu_n = \text{const}\ T^{-3/2}\ [T \text{ in K}]$$

SOLUTION

The resistance of a resistor depends on the resistivity ρ, resistor length L, and resistor cross-sectional area $x_j W$ [Eq. (3.4)], where the resistivity is the only temperature-dependent parameter. Replacing the resistance R in our definition of *TCR* by Eq. (3.4), the *TCR* is obtained as

$$TCR = \frac{1}{\rho}\frac{d\rho}{dT} \times 100$$

The resistivity of N-type silicon is given by

$$\rho = 1/(q\mu_n N_D)$$

where, obviously, the mobility is the only temperature-dependent parameter. Using the given temperature dependence of mobility, the resistivity is expressed by

$$\rho = \frac{1}{qN_D \text{ const}} T^{3/2}$$

Finding that

$$\frac{d\rho}{dT} = \frac{3}{2}\frac{1}{qN_D \text{ const}} T^{1/2}$$

the *TCR* is obtained as

$$TCR = \frac{3}{2T} \times 100$$

Therefore,

$$TCR = \frac{3}{2 \times 300} \times 100 = 0.5\%/\text{K} = 0.5\%/^\circ\text{C}$$

When the temperature changes from 27°C to 75°C, the resistance changes by

$$\Delta R \approx \frac{TCR}{100} R\Delta T = 0.005 \times 1\,\text{k}\Omega \times (75^\circ\text{C} - 27^\circ\text{C}) = 0.24\,\text{k}\Omega$$

The resistance at 75°C is, therefore, estimated at 1.24 kΩ.

3.3 CARRIER MOBILITY

3.3.1 Thermal and Drift Velocities

The carrier-gas model assumes that the kinetic energy of any single carrier is given by

$$E_{kin} = \begin{cases} m^*|\vec{v}|^2/2 = |\vec{p}|^2/2m^* & \text{general case} \\ m^* v_x^2/2 = p_x^2/2m^* & \text{one-dimensional case} \end{cases} \tag{3.18}$$

where m^* is the effective mass, introduced in Section 2.3.1. Like the molecules in the air, the current carriers in semiconductors possess kinetic energy even when no electric

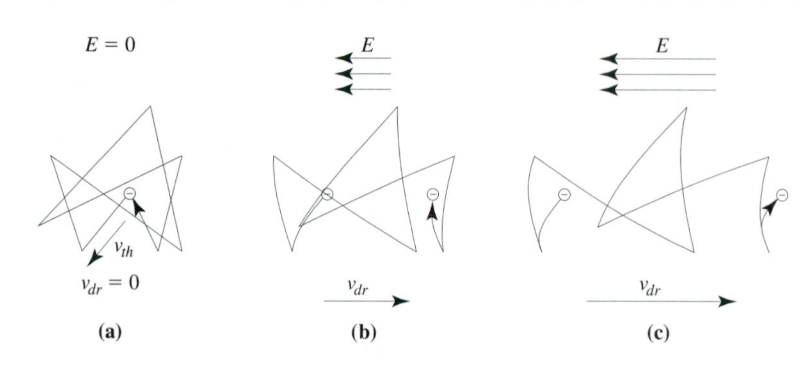

Figure 3.5 The concept of drift velocity. (a) No electric field is applied. (b) A small electric field is applied. (c) A larger electric field is applied.

field is applied. In that case, the kinetic energy of the carriers is related to the crystal temperature T:

$$E_{kin} = \frac{m^* v_{th}^2}{2} = \begin{cases} \frac{3}{2}kT & \text{three-dimensional case} \\ \frac{1}{2}kT & \text{one-dimensional case} \end{cases} \tag{3.19}$$

where k is the Boltzmann constant, T is the absolute temperature, and v_{th} is the *thermal velocity*. The thermal motion of carriers is essentially random due to the scattering from the imperfections of the crystal lattice (doping atoms, phonons, etc.). Carriers that move randomly do not make any drift current because they do not effectively move from one point in the semiconductor to another. Figure 3.5a illustrates an electron that returns to the initial position after a number of scattering events, performing therefore no effective motion in any particular direction in the crystal. It is said that the *drift velocity* of the carriers is equal to zero.

If an electric field is applied to the electron gas, the electric-field force will cause slight deviations of the electron paths between collisions, producing an effective shift of the electrons in the direction opposite to the direction of the electric field, as illustrated in Fig. 3.5b. The effective shift of carriers per unit time is the drift velocity. Figure 3.5c illustrates that an increase in the electric field increases the drift velocity.

The concept of *drift velocity* as the flow of a carrier is quite fundamentally linked to the concept of *electric current,* which is the flow of many carriers. To establish this fundamental relationship, let us consider a semiconductor bar that has a cross-sectional area A and good contacts at the ends. A voltage applied at the ends of the sample establishes an electric field within the sample that forces the carriers, say electrons, to move at an average drift velocity of v_{dr}. If we count how many electrons per unit time t are collected at the positive contact, we will find that this number is equal to the number of electrons that are not further than $v_{dr}t$ from the contact at the beginning of the counting ($v_{dr}t$ is the average distance that electrons travel in time t when moving with the average velocity v_{dr}). In other words, this number is equal to the number of electrons confined within the volume $v_{dr}tA$. If there are n electrons per unit volume—that is, the concentration of electrons is n—there will be $nv_{dr}tA$ electrons that reach the positive contact during time t. The number of electrons that are collected per unit time t, which is $nv_{dr}A$, multiplied by the charge that every electron carries ($-q$), is the electric current flowing through the terminal, I.

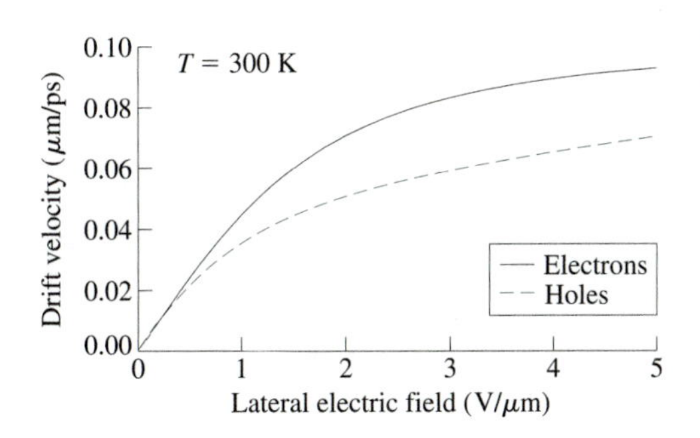

Figure 3.6 Drift velocity versus electric field in Si.

Therefore,

$$I = -qnv_{dr}A \tag{3.20}$$

The current I (expressed in A, which is C/s) turns into current density j_{dr} (expressed in A/m^2) when divided by the area A:

$$j_{dr} = \begin{cases} -qnv_{dr} & \text{electrons} \\ qpv_{dr} & \text{holes} \end{cases} \tag{3.21}$$

Equation (3.21) provides the relationship between the electric-current density and the drift velocity of carriers. It is not difficult to understand this equation: If there are more carriers per unit volume (n higher) that move faster (v_{dr} larger), the current density j will be proportionally larger. The factor q is there to convert particle current density expressed in $s^{-1}\,m^{-2}$ into electric current density expressed in $C \cdot s^{-1}\,m^{-2}$—that is, $A \cdot m^{-2}$.

As illustrated in Fig. 3.5, the drift velocity depends on the electric field applied. Figure 3.6 shows measured values of the drift velocities of electrons and holes versus electric field in silicon. The expected linear relationship between the drift velocity and the electric field is observed only for small electric fields—that is, small drift velocities. As the electric field is increased, the drift velocity tends to saturate at about 0.1 μm/ps in the case of Si crystal. Let us think of what is going to happen with the electron trajectory in Fig. 3.5 if we continue to increase the electric field. We can see that there is a limit to the increase in drift velocity because there is a limit to how much the scatter-like electron trajectory can be stretched in the direction of the electric field. This provides an insight into the effect of drift-velocity saturation.

EXAMPLE 3.8 Drift Velocity

A uniformly doped semiconductor resistor (doping level $N_D = 10^{16}$ cm^{-3} and cross-sectional area $A = 20$ μm^2) conducts 2 mA of current. The resistor is connected by copper wires with cross section of 0.1 mm^2. Determine and compare the drift velocities of the electrons in the semiconductor and copper regions. The concentration of free electrons in copper is $n_{Cu} = 8.1 \times 10^{22}$ cm^{-3}.

SOLUTION

The relationship between the drift velocity and the current density is given by Eq. (3.21). Therefore, the absolute value of the drift velocity is

$$v_{dr} = \frac{I}{Aqn}$$

where $n = N_D$ and $n = n_{Cu}$ in the semiconductor and the copper wires, respectively.

The drift velocities of electrons in the semiconductor and the copper wires are 62.5 km/s and 1.54 μm/s, respectively. There are so many more electrons in the copper wires that they can move $6.25 \times 10^4/1.54 \times 10^{-6} = 4 \times 10^{10}$ times slower and still supply the necessary current to the resistor.

3.3.2 Mobility Definition

Let us consider the semiconductor bar having good contacts at the ends again. If electric field E is applied and the conductivity of the semiconductor is σ, then the current density is obtained from the differential form of Ohm's law [Eq. (3.12)]. If the semiconductor is N type, the conductivity is given by $\sigma = q\mu_n n$. Therefore, Ohm's law expresses the current density in the following way:

$$j_{dr} = \begin{cases} q\mu_n nE & \text{electrons} \\ q\mu_p pE & \text{holes} \end{cases} \tag{3.22}$$

Ohm's law is not derived from principles of solid-state physics; it simply expresses an observation that the current density depends linearly on the electric field applied. Lacking a detailed physical background, it has to involve a proportionality constant, which is the mobility μ_n in Eq. (3.22), that needs to be determined experimentally. In the previous section, a current equation is derived [Eq. (3.21)] from a microscopic consideration of the flow of current carriers. Given that Eqs. (3.21) and (3.22) have to give the same current densities, the elimination of j_{dr} from these equations leads to the following relationship between v_{dr} and E:

$$v_{dr} = \begin{cases} -\mu_n E & \text{electrons} \\ \mu_p E & \text{holes} \end{cases} \tag{3.23}$$

Equation (3.23) reveals that Ohm's law implicitly assumes a linear relationship between the drift velocity and the electric field. The experimental results given in Fig. 3.6 show that this assumption is valid only at small electric fields. This clearly shows an important limitation of Ohm's law when applied to semiconductor devices. The mobility has to be adjusted to preserve the validity of Eq. (3.22). This complicates the concept of mobility due to the fact that it can be considered as a constant only at low electric fields, whereas it becomes an electric-field-dependent parameter at high electric fields.

Equation (3.23) is basically the definition of mobility. *The carrier mobility is the proportionality coefficient in the dependence of drift velocity on the applied field.* The unit for mobility is given by the velocity unit over the electric field unit, which is $(\text{m/s})/(V/m) = \text{m}^2/\text{V} \cdot \text{s}$.

3.3.3 Scattering Time and Scattering Cross Section

Electrons appearing in electric field E experience force $F = -qE$, which accelerates them: $F = m^* dv/dt$. For small electric fields, the acceleration can be expressed as v_{dr}/τ_{sc}, where v_{dr} is the average velocity (the drift velocity) and τ_{sc} is the average scattering time. With this, $-qE = m^* v_{dr}/\tau_{sc}$, which means that the linear relationship between v_{dr} and E can be expressed as

$$v_{dr} = -\frac{q\tau_{sc}}{m^*}E \tag{3.24}$$

Given that the mobility is the proportionality coefficient in Eq. (3.24) ($v_{dr} = -\mu_n E$), the mobility is directly proportional to the scattering time and inversely proportional to the effective mass:

$$\mu_n = \frac{q\tau_{sc}}{m^*} \tag{3.25}$$

Equation (3.25) shows that carriers with a smaller effective mass have a higher mobility. The concept of effective mass was introduced in Section 2.3.1. Let us now gain an insight into the scattering time. When a carrier moves toward a scattering center (phonon, ion, etc.), a scattering event occurs if the carrier hits the effective area of the scattering center, called scattering cross section (σ_{sc}). If there are N_{sc} scattering centers per unit volume, the product $N_{sc}\sigma_{sc}$ has the meaning of scattering probability per unit length of the carrier path. The reciprocal value, $1/N_{sc}\sigma_{sc}$, is the *scattering length*—that is, the average distance that a carrier travels between two collisions with the scattering centers. The average time between two collisions—the scattering time, τ_{sc}—is obtained as the ratio between the scattering length and the average thermal velocity of the carrier, v_{th}:

$$\tau_{sc} = \frac{1}{v_{th}\sigma_{sc}N_{sc}} \tag{3.26}$$

The reciprocal value of τ_{sc} is the probability that a carrier will be scattered per unit time. The probability that a carrier will be scattered by a phonon depends on the phonon size that can be related to the radius of crystal-atom vibrations. The radius of the atom vibrations increases with an increase in the temperature, thereby increasing the probability that carriers will collide with vibrating atoms. Therefore, the phonon-limited mobility decreases with temperature.

The doping atoms are another important source of electron and hole scattering. The doping atoms are ionized particles (donors positively and acceptors negatively); therefore they repel or attract electrons or holes that appear in their vicinity, changing consequently the direction of their motion. This is referred to as *Coulomb scattering*. Coulomb scattering is more pronounced at lower temperatures because the thermal velocity of the carriers is smaller and the carriers stay for a longer time in electric contact with the ionized doping atoms, making the scattering more efficient. A deeper insight into the temperature dependences of phonon and Coulomb scattering is provided by Examples 3.9 and 3.10, respectively.

EXAMPLE 3.9 Scattering Cross Section of Phonons

(a) If the phonon-limited scattering time in silicon is $\tau_{sc-ph} = 0.2$ ps, determine the scattering cross section of phonons at 300 K. The effective mass of electrons in silicon is $m^* = 0.26 m_0$.

(b) Assuming that the scattering cross section of phonons is proportional to the temperature, determine the temperature dependence of the phonon-limited mobility.

SOLUTION

(a) From Eq. (3.26),

$$\sigma_{sc-ph} = \frac{1}{\tau_{sc-ph} v_{th} N_{sc-phonons}}$$

where N_{sc-ph} is equal to the concentration of silicon atoms (5×10^{22} cm^{-3} according to Example 1.3), and v_{th} can be obtained from the energy-balance equation $m^* v_{th}^2/2 = 3kT/2$:

$$v_{th} = \sqrt{3kT/m^*} = 2.29 \times 10^5 \text{ m/s}$$

With this value for v_{th} and $\tau_{sc-ph} = 0.2$ ps, we obtain $\sigma_{sc-ph} = 4.36 \times 10^{-22}$ m^2 = 4.36×10^{-18} cm^2.

(b) The dependence of phonon-limited mobility on temperature is due to σ_{sc-ph} and v_{th}:

$$\mu_{ph} = \frac{q\tau_{sc-ph}}{m^*} = \frac{q}{m^* v_{th} \sigma_{sc-ph} N_{sc-ph}} \propto \frac{1}{\sigma_{sc-ph} v_{th}}$$

$$\mu_{ph} \propto \frac{1}{T\sqrt{3kT/m^*}} \propto \frac{1}{T^{3/2}}$$

$$\mu_{ph} = A_p T^{-3/2}$$

EXAMPLE 3.10 Cross Section of Coulomb Scattering Centers

The scattering cross section of a donor ion can be related to the spherical region where the thermal energy of a carrier is smaller than the energy associated with the Coulomb attraction.

(a) Estimate the cross section of Coulomb scattering centers in silicon. The dielectric constant of silicon is $\varepsilon_s/\varepsilon_0 = 11.8$.

(b) Determine the temperature dependence of the Coulomb-limited mobility.

SOLUTION

(a) The energy of Coulomb attraction/repulsion is $q^2/(4\pi\varepsilon_s r)$, where r is the distance between the center of the ion and the carrier, whereas the kinetic energy of the carrier is $3kT/2$. Therefore, the radius of the spherical region can be found from the following condition:

$$\frac{q^2}{4\pi\varepsilon_s r} = \frac{3}{2}kT$$

$$r = \frac{q^2}{6\pi\varepsilon_s kT}$$

The cross section of the spherical region (the scattering cross section) is then $\sigma_{sc-C} = \pi r^2 = 3.1 \times 10^{-17}\ \mathrm{m}^{-2} = 3.1 \times 10^{-13}\ \mathrm{cm}^{-2}$.

(b) The temperature-dependent factors are σ_{sc-C} and v_{th}, where $\sigma_{sc-C} = \pi r^2 \propto 1/T^2$:

$$\tau_{sc-C} \propto \frac{1}{\sigma_{sc-C} v_{th}} \propto \frac{1}{(1/T^2)\sqrt{T}} = T^{3/2}$$

$$\mu_C = A_c T^{3/2}$$

3.3.4 Mathieson's Rule

The concept of scattering probability per unit time $(1/\tau_{sc})$ can be used to combine the effects of independent scattering mechanisms. For example, if the probabilities that a carrier is scattered by lattice vibrations (phonon scattering) and doping ions (Coulomb scattering) are $1/\tau_{sc-ph}$ and $1/\tau_{sc-C}$, respectively, the total probability of scattering per unit time is the sum of the two:

$$\frac{1}{\tau_{sc}} = \frac{1}{\tau_{sc-ph}} + \frac{1}{\tau_{sc-C}} \tag{3.27}$$

Combining Eqs. (3.25) and (3.27), we can see that the reciprocal values of the phonon-limited (μ_{ph}) and the Coulomb-limited (μ_C) mobilities are added to obtain the reciprocal value of the total mobility: $1/\mu = 1/\mu_{ph} + 1/\mu_C$. This is known as Mathieson's rule. In general,

$$\frac{1}{\mu} = \sum_{i=1}^{N} \frac{1}{\mu_i} \tag{3.28}$$

where μ_i is the mobility limited by the action of the ith scattering mechanism, and μ is the total mobility.

Because temperature has opposite effects in the cases of phonon and Coulomb scattering, the dependence of mobility on temperature is not a straightforward function. If the doping concentration is high enough, causing Coulomb scattering to dominate over phonon scattering, mobility continuously increases with the temperature as Coulomb scattering is weakened. Such a situation appears in silicon for a doping level of 10^{19} cm^{-3} and temperatures up to 500 K (Fig. 3.7). As the doping concentration is reduced, the phonon scattering dominates at higher temperatures, which means that the increase in mobility at lower temperatures (due to weakened Coulomb scattering) is followed by a decrease in mobility at higher temperatures (due to strengthened phonon scattering). Figure 3.7 shows such behavior by the mobility, observed for low and medium doping levels (10^{15} and 10^{17} cm^{-3}, respectively).

Figures 3.8 and 3.9 provide dependencies of electron and hole mobilities on doping concentration at different temperatures for silicon and gallium arsenide, respectively.

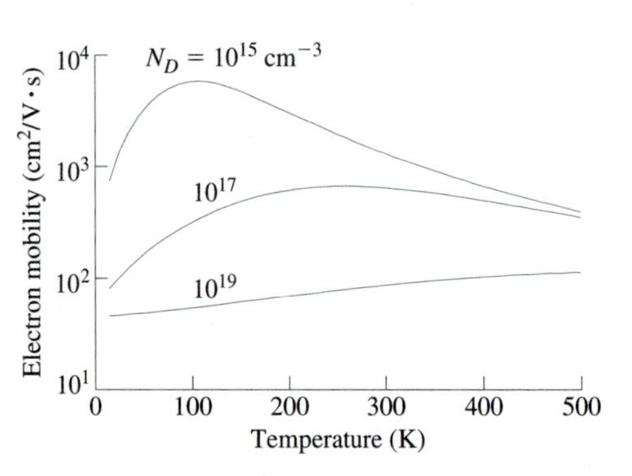

Figure 3.7 Temperature dependence of mobility for three different doping levels in Si.

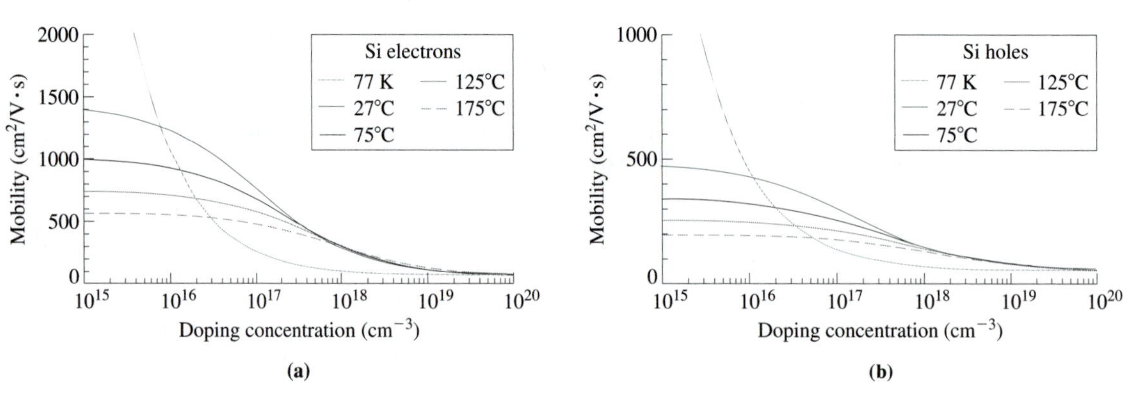

Figure 3.8 (a) Low-field electron and (b) low-field hole mobilities in Si.

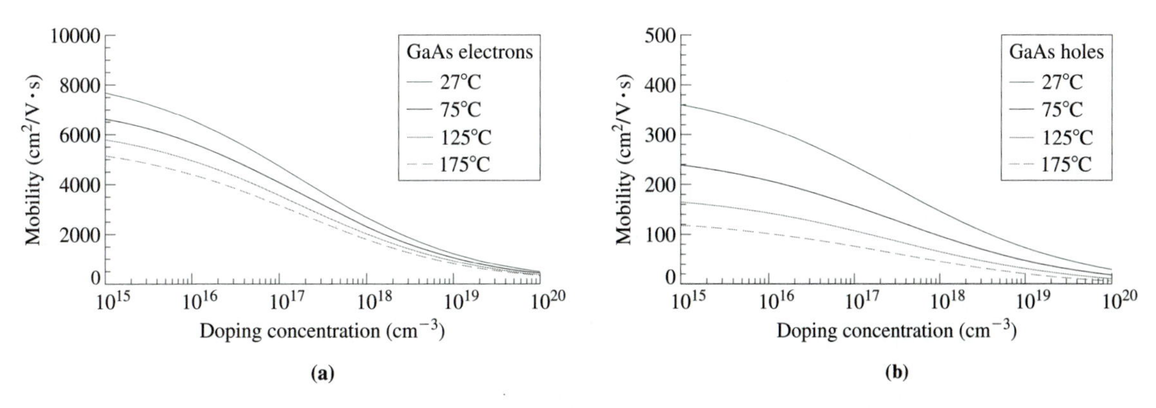

Figure 3.9 (a) Low-field electron and (b) low-field hole mobilities in GaAs.

EXAMPLE 3.11 Mobility Plots

Based on the results from Exercises 3.9 and 3.10, plot the following mobility dependencies:

(a) Electron mobility versus temperature for $N_D = 10^{16}$ cm^{-3} in the temperature range 30 K to 300 K (use 10-K steps to calculate the data).

(b) Electron mobility versus donor concentration for 300 K in the doping concentration range 10^{15} cm^{-3} to 10^{20} cm^{-3} (use 10 concentration points per decade).

SOLUTION

(a) Given that the phonon-limited scattering time at 300 K is $\tau_{sc-ph} = 0.2$ ps, the room-temperature phonon-limited mobility is $\mu_{ph}(300) = A_p \times 300^{-3/2}$, and the coefficient A_p can be determined from

$$A_p = \mu_{ph}(300) \times 300^{3/2}$$

The room-temperature Coulomb-limited mobility is

$$\mu_C(300) = \frac{q}{m^* \, N_D \, \sigma_{sc-C}(300) \, v_{th}(300)}$$

where $\sigma_{sc-C}(300) = 3.1 \times 10^{-17}$ m^{-2} (according to Example 3.10a) and $v_{th}(300) = 3.92 \times 10^5$ m/s (according to Example 3.9a). Knowing $\mu_C(300)$, the coefficient A_c can be determined from

$$A_c = \mu_C(300)/300^{3/2}$$

The following MATLAB® script can be used to calculate the numerical values:

```
>>q=1.6e-19;
>>meff=0.26*9.1e-31;
>>tauph=0.2e-12;
>>muph300=q*tauph/meff
muph300 =
    0.1352
>>Ap=muph300*300^1.5
Ap =
    702.7763
>>vth300=3.92e5;
>>sigma300=3.1e-17;
>>Nd=1e22;
>>muC300=q/(meff*Nd*sigma300*vth300);
>>Ac=muC300/300^1.5
Ac =
    0.0011
```

$A_p = 702.78\ \text{m}^2 \cdot \text{K}^{3/2}/\text{V} \cdot \text{s}$, $A_c = 0.0011\ \text{m}^2/\text{V} \cdot \text{s} \cdot \text{K}^{3/2}$.

The total mobility is determined according to the Mathieson's rule [Eq. (3.28)]:

$$\frac{1}{\mu} = \frac{1}{\mu_{ph}} + \frac{1}{\mu_C}$$

$$\mu = \frac{\mu_{ph}\mu_C}{\mu_{ph} + \mu_C}$$

The following MATLAB script will generate the data, perform the plotting, and label the axes (the plot is shown in Fig. 3.10):

```
>>T=[30:10:300];
>>muph=Ap*T.^(-1.5);
```

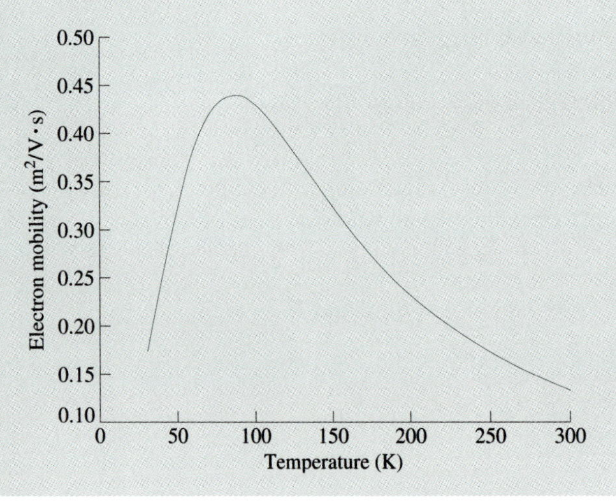

Figure 3.10 The MATLAB plot for Example 3.11a.

```
>>muC=Ac*T.^1.5;
>>mu=muph.*muC./(muph+muC);
>>plot(T,mu)
>>xlabel('Temperature (K)')
>>ylabel('Electron Mobility (m^2/Vs)')
```

(b) In this case, the phonon mobility is constant $\mu_{ph} = 0.1352$ m²/V · s, and the Coulomb mobility changes with the doping concentration as

$$\mu_C = \underbrace{q/m^* \, \sigma_{sc-C}(300) \, v_{th}(300)}_{A_{CN}} \frac{1}{N_D}$$

where $A_{CN} = 5.5649 \times 10^{22}$ (mV · s)$^{-1}$. The following is the MATLAB script for the calculations and the plot is shown in Fig. 3.11.

```
>>Acn=5.5649e22;
>>Ndlog=[21:0.1:26];
>>Nd=10.^Ndlog;
>>muC=Acn./Nd;
>>mu=muph300*muC./(muph300+muC);
>>semilogx(Nd,mu)
>>xlabel('Doping Concentration (m^-3)')
>>ylabel('Electron Mobility (m^2/Vs)')
```

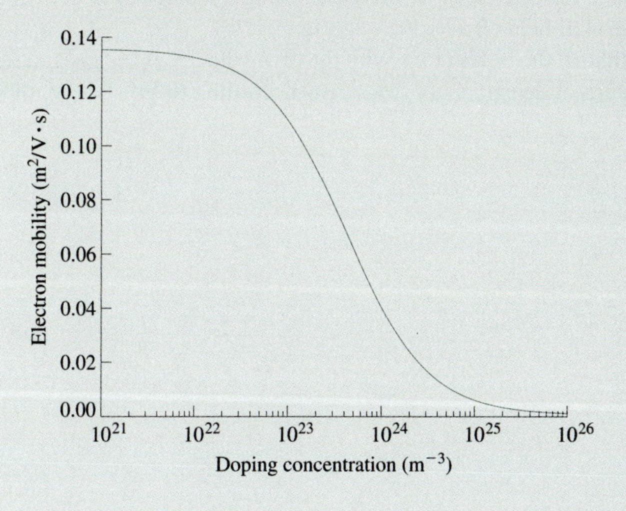

Figure 3.11 The MATLAB plot for Example 3.11b.

*3.3.5 Hall Effect

The Hall effect is related to the force that acts on a charged particle that moves in a magnetic field. In the general case of arbitrary velocity and magnetic field directions, the force is expressed as

$$\boldsymbol{F} = qv \times \boldsymbol{B} \tag{3.29}$$

where v is the particle velocity, $\boldsymbol{B}$ is the magnetic flux density, and the vector cross product is $v \times \boldsymbol{B} = |v||\boldsymbol{B}| \sin [\angle(v, \boldsymbol{B})]$. Figure 3.12 illustrates the Hall effect. Let us assume that a current of holes (I) is established through a P-type semiconductor bar so that the holes move in the y-direction with the velocity equal to drift velocity $v_{dr} = v_y$. At time $t = 0$, a magnetic field perpendicular to the current flow, B_z, is established. As a hole enters the semiconductor bar with velocity v_y, the force $F_x = -qv_y B_z$ causes its trajectory to deviate and to hit the side of the semiconductor bar (Fig. 3.12a). The holes accumulating at one side of the bar create an electric field $-E_x$, which counteracts the magnetic-field force:

$$qE_x = qv_y B_z \sin \underbrace{\left[\angle(v_y, B_z)\right]}_{-90^\circ} \Rightarrow -E_x = v_y B_z \tag{3.30}$$

This establishes a steady state where the holes continue to flow through the semiconductor bar in the y-direction (Fig. 3.12b). If the width of the semiconductor bar (the dimension in the x-direction) is W, a voltage V_H equal to $-E_x W$ can be measured between the opposite sides. This voltage is referred to as the Hall voltage, whereas the corresponding field is referred to as the Hall field ($E_H = -E_x$ in Fig. 3.12b).

Assuming that all the holes move with the drift velocity $v_y = v_{dr}$ and relating the drift velocity to the current density j_y [$j_y = qpv_{dr}$ according to Eq. (3.21)], the Hall field can

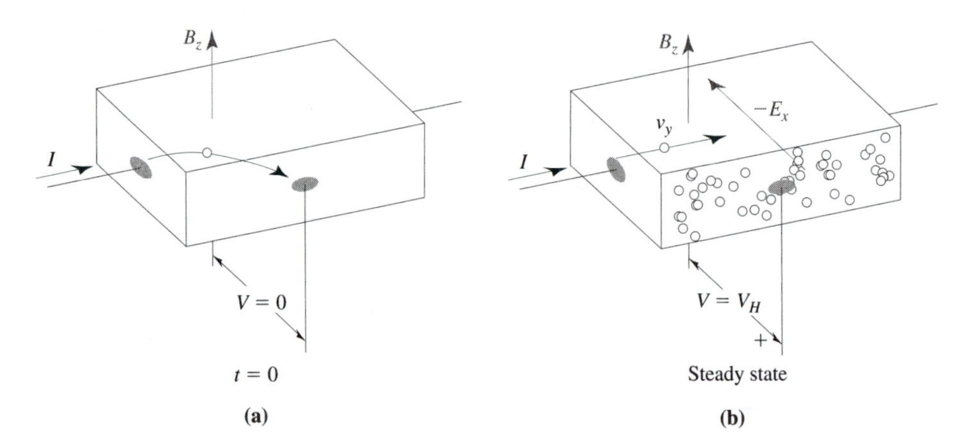

Figure 3.12 Illustration of the Hall effect. (a) The force due to the magnetic field B_z causes the hole trajectory to deviate. (b) Accumulated holes create a Hall field $E_H = -E_x$ that counteracts the force from the magnetic field B_z.

be expressed as

$$E_H = \frac{j_y}{qp} B_z \tag{3.31}$$

which further leads to the following equation for the Hall voltage:

$$V_H = \frac{1}{qp} \frac{I}{t_s} B_z \tag{3.32}$$

where t_s is the sample thickness. When the effect was discovered, Hall observed that the field E_H is directly proportional to the magnetic flux density and the current density and is inversely proportional to the sample thickness,

$$V_H = R_H \frac{I B_z}{t_s} \tag{3.33}$$

where the proportionality constant R_H is now known as the Hall coefficient. Comparing the empirical Eq. (3.33) to Eq. (3.32), the Hall coefficient is obtained as

$$R_H = \frac{1}{qp} \tag{3.34}$$

Alternatively, R_H can be expressed as

$$R_H = \frac{\mu_p}{\sigma_p} \tag{3.35}$$

given that $\sigma_p = q\mu_p p$.

In reality, the drift velocity v_{dr} represents only the average carrier velocity, and it cannot be said that $v_y = v_{dr}$ for any carrier. This means that this theory can strictly be applied only to the carriers moving with average velocity. Consequently, the mobility in Eq. (3.35) is called the Hall mobility, to indicate that it generally has different value from the drift mobility. This inaccuracy can be compensated by introducing a factor r in the equation relating the empirical Hall coefficient to the semiconductor properties:

$$R_H = \begin{cases} r/qp & \text{for holes} \\ -r/qn & \text{for electrons} \end{cases} \tag{3.36}$$

The factor r is typically between 1 and 2.

Using Eq. (3.33), the Hall coefficient can experimentally be determined by measuring the Hall voltage V_H for a given current I and magnetic flux density B_z and knowing the sample thickness t_s. The measured value of the Hall coefficient can be used to calculate the carrier concentration, p or n, using Eq. (3.36).

It is interesting to note that the polarity of the Hall voltage V_H depends on the type of semiconductor used. If an N-type semiconductor is used in the example of Fig. 3.12, with the same current direction, the electrons will move in the negative y-direction; however, the force due to the magnetic field B_z is in the same direction because of the negative charge of the electrons: $(-q)(-v_y)B_z = qv_yB_z$. This means that the electrons will accumulate at the same side as the holes in Fig. 3.12b, leading to an E_x field in the opposite direction and, therefore, a V_H voltage with the opposite polarity. Therefore, the polarity of the Hall voltage can be used to determine the semiconductor type.

Practical applications of the Hall effect include techniques for characterization of semiconductor materials and magnetic sensors.

SUMMARY

1. A slope in an energy–space diagram indicates an electric field: an energy difference due to band bending in space, expressed in eV, is numerically equal to the negative electric-potential difference between the two points ($E = -q\varphi$). To visualize the drift current, the electrons in the conduction band can be thought of as balls on a solid surface; an analogy for the holes in the valence band is bubbles in water.
2. *Sheet resistance* $R_S = 1/(\bar{\sigma}x_j)$ involves all the technological parameters that influence the resistance of a semiconductor layer. It is expressed in $\Omega/\square$ to indicate that it means resistance per square, so that the total resistance is

$$R = R_S \frac{L}{W}$$

where the geometric parameters determine the number of squares L/W.
3. An electric field E_y causes *drift current* of charged particles, as expressed by the differential form of Ohm's law:

$$j_y = \sigma E_y = -\sigma \frac{d\varphi}{dy}$$

4. The conductivity of a semiconductor is given by

$$\sigma = qn\mu_n + qp\mu_p$$

where n and p are the *concentrations,* and μ_n and μ_p are the *mobilities* of electrons and holes, respectively.
5. The concept of *current density* relates to the *velocity* of a certain *concentration* of particles. In the case of drift current of electrons, $j_{dr} = -qnv_{dr}$, where v_{dr} is the *drift velocity.* The drift velocity is directly related to the electric field causing the drift:

$$v_{dr} = \begin{cases} -\mu_n E & \text{for electrons} \\ \mu_p E & \text{for holes} \end{cases}$$

where the proportionality coefficient $\mu_{n,p}$ is the *mobility.* By introducing the concept of mobility, the fundamental current equation ($j_{dr} = -qnv_{dr}$) is converted into the differential Ohm's law ($j_{dr} = qn\mu_n E = \sigma E$). The mobility is constant at low electric fields, but it drops at high fields due to the *drift-velocity saturation* effect.

6. The mobility value is basically determined by the average time between two scattering events (τ_{sc}) and the effective mass (m^*):

$$\mu = \frac{q\tau_{sc}}{m^*}$$

The reciprocal value of the scattering time, $1/\tau_{sc}$, has the meaning of the probability that a carrier is scattered per unit time. This probability depends on the concentration of scattering centers (N_{sc}), their scattering cross section σ_{sc}, and the thermal velocity of the carriers (v_{th}):

$$\frac{1}{\tau_{sc}} = v_{th}\sigma_{sc}N_{sc}$$

7. Two dominant scattering mechanisms are phonon scattering ($\mu_{ph} \propto T^{-3/2}$) and Coulomb scattering by the ionized doping atoms ($\mu_C \propto T^{3/2}$).
8. Given that the total scattering probability is equal to the sum of the scattering probabilities due to different and independent scattering mechanisms, the total mobility can be expressed by Mathieson's rule:

$$\frac{1}{\mu} = \sum_{i=1}^{N} \frac{1}{\mu_i}$$

where μ_i is the mobility limited by the action of ith scattering mechanism.

PROBLEMS

3.1 The bottom of the conduction band at one end of a silicon resistor is at the same energy level as the top of the valence band at the other end. What current flows through the resistor if the resistance is 1.12 kΩ?

3.2 A test resistor with length $L = 50$ μm and width $W = 5$ μm is used to measure the sheet resistance of a resistive layer. What is the sheet resistance if a test voltage of 1 V produces current of $I = 0.50$ mA?

(a) 50Ω/□
(b) 100Ω/□
(c) 200Ω/□
(d) 300Ω/□
(e) 400Ω/□
(f) 450Ω/□

3.3 A resistor of 50 Ω is to be designed for fabrication by a standard bipolar integrated-circuit technology incorporating base- and emitter-type diffusion layers. The sheet resistance of the base-type diffusion layer is 200 Ω/□, whereas the sheet resistance of the emitter-type diffusion layer is 5 Ω/□. The minimum width of the diffusion lines is 5 μm. Which diffusion layer (emitter type or base type) would you use for this resistor, and why? Determine the dimensions (length and width) of the resistor.

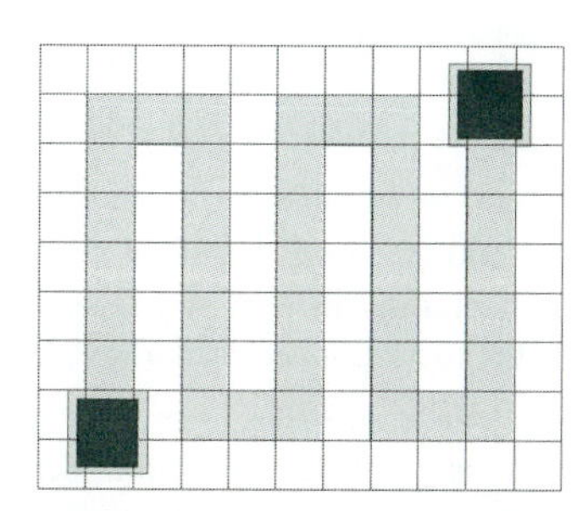

Figure 3.13 Top view of an IC resistor.

3.4 A diffusion layer with sheet resistance R_S = 200 $\Omega/\square$ is used for the resistor of Fig. 3.13. If the total resistance of this resistor is $R = 6.5$ kΩ, what is the resistance of each corner square?

3.5 Estimated operating temperature of an IC chip is 70°C. Design a molybdenum resistor so that its resistance is 75 Ω at the estimated operating temperature. The thickness of the molybdenum film is 100 nm and the minimum track width is limited to 2 μm. What will the resistance be if the actual operating temperature is 75°C? **A**

3.6 A copper track is to connect two device terminals that are 100 μm apart. If the thickness of the copper film is 200 nm and the operating temperature is 80°C, design the width of the track so that the maximum dissipated power is 1 mW. The maximum current flowing through the track is 200 mA.

3.7 An N-type diffusion layer creates a semiconductor resistor. The length and the width of the resistor are $L = 100$ μm and $W = 5$ μm, respectively. The applied voltage across the resistor is 1 V.

(a) What is the electric field inside the resistor?

(b) If the doping level at the surface is $N_D(0) = 10^{16}$ cm^{-3} and the doping level at 0.5 μm below the surface is $N_D(0.5\ \mu\text{m}) = 10^{15}$ cm^{-3}, determine the current density at the semiconductor surface and at $x = 0.5$ μm below the surface.

Assume constant electron mobility of μ_n = 500 cm^2/V · s.

3.8 The conductivity of a 1-μm-deep semiconductor layer changes as

$$\sigma(x) = \sigma(0)\exp(-x/x_0)$$

where $\sigma(0) = 15(\Omega \cdot \text{cm})^{-1}$ and $x_0 = 0.2$ μm. What is the average conductivity of this layer?

3.9 The conductivity of a 3-μm-deep semiconductor layer changes as

$$\sigma(x) = \sigma(0)\exp(-x/x_0)$$

where $\sigma(0) = 100\ (\Omega \cdot \text{cm})^{-1}$ and $x_0 = 0.5$ μm. What is the sheet resistance of the layer that is obtained after the top 1 μm is etched away? **A**

3.10 The voltage across a 200-Ω resistor in a bipolar integrated circuit is $\leq$5 V. Reliability considerations limit the average current density to $\overline{j_{dr-max}} = 10^9\,\text{A}/m^2$. Design this resistor so that it can be implemented as a base diffusion resistor, with the sheet resistance and the junction depth of the base diffusion layer being $R_S = 100\,\Omega/\square$ and $x_j = 2$ μm, respectively.

3.11 Repeat Problem 3.10 for the case in which the reliability constraint means that the current density should not exceed $j_{dr-max} = 10^9$ A/m^2 at any point, and the nonuniform conductivity can be expressed as $\sigma(x) = \sigma(0)\exp-(x/x_0)^2$, where $x_0 = 1$ μm. Assume constant electric field $E = V/L$. It is also known that $\int_0^\infty \exp(-u^2/2)\,du = \sqrt{\pi/2}$. **A**

3.12 To measure the sheet resistance of a resistive layer, taking into account the parasitic series contact resistance, a test structure consisting of resistors with the same width and different lengths is provided. Measuring the resistances of the resistors with lengths $L_1 = 10$ μm and $L_2 = 30$ μm, the following values are obtained: $R_1 = 365$ Ω, and $R_2 = 1085$ Ω, respectively. If the width of the resistors is 5 μm, determine the sheet resistance and the contact resistance values. **A**

3.13 The concentration of donor atoms in N-type silicon is $N_D = 10^{16}$ cm^{-3}. Determine the conductivity of this material, assuming that the electron mobility is $\mu_n = 1450$ cm^2/V · s. **A**

3.14 P-type doped semiconductor layer has approximately uniform acceptor concentration $N_A = 5 \times 10^{16}$ cm^{-3} and thickness $x_j = 4$ μm. Calculate the sheet resistance, if the hole mobility is μ_p = 450 cm^2/V · s.

3.15 In a silicon crystal, $N_D = 10^{17}$ cm^{-3} and $N_A = 10^{16}$ cm^{-3}. Find the resistivity of the crystal if the electron mobility is $\mu_n = 770$ cm^2/V · s.

3.16 An N-type silicon substrate with $N_D = 10^{15}$ cm^{-3} is to be converted into P type by boron diffusion, so that the resistivity at $T_o = 75$°C is $\rho = 5\ \Omega \cdot$ cm. What should be the doping level? What is the resistivity at room temperature? The hole mobilities at 75°C and room temperature are 341 cm^2/V · s and 467 cm^2/V · s, respectively. The intrinsic carrier concentration at 75°C is $n_i = 2.9 \times 10^{11}$ cm^{-3}. **A**

3.17 What is the tolerance $\Delta R/R$ of an N-type diffusion resistor if the tolerances of the resistor dimensions (including junction depth) are ± 0.3 μm and the tolerance of the doping density is 5%?

3.18 A 2-cm-long silicon piece, with cross-sectional area of 0.1 cm^2, is used to measure electron mobility. What is the electron mobility if 90 Ω of resistance is measured and the doping level is known to be $N_D = 10^{15}$ cm^{-3}? Neglect any contact resistance.

3.19 The resistance of a P-type semiconductor layer is 1 kΩ. In what range will the resistance change if the operating temperature is 75°C ± 10°C? The temperature dependence of hole mobility is given by $\mu_p = \text{const}\ T^{-3/2}$.

3.20 A bar of N-type semiconductor ($N_D = 5 \times 10^{17}$ cm^{-3}, $L = 1$ cm, $W \times T = 3$ mm $\times$ 0.5 mm) is used as a resistor. Calculate and compare the sheet resistances at 27°C and 700°C if the semiconductor material is

(a) silicon
(b) 6H silicon carbide **A**

Assume constant mobilities $\mu_n = 400$ cm^2/V · s and $\mu_p = 200$ cm^2/V · s, and use the energy-gap and density-of-states data from Problems 2.36 and 2.37.

3.21 An electric field $E = 1$ V/μm produces current density $j = 0.8 \times 10^9$ A/m^2 through an N-type doped semiconductor ($N_D = 10^{17}$ cm^{-3}). What will the current density be if the electric field is increased five times, so that the electrons reach the velocity saturation $v_{sat} = 0.1$ μm/ps?

(a) 0
(b) 0.8×10^9 A/m^2
(c) 1.6×10^9 A/m^2
(d) 4.0×10^9 A/m^2
(e) 3.2×10^9 A/m^2
(f) 8.0×10^9 A/m^2

3.22 A bar of silicon 1 cm long, 0.5 cm wide, and 0.5 mm thick has a resistance of 190 Ω. The silicon has a uniform N-type doping concentration of 10^{15} cm^{-3}.

(a) Calculate the electron mobility.
(b) Find the drift velocity of the electrons, when 10 V is applied to the ends of the bar of silicon.
(c) Find the corresponding electric field, and relate it to the drift velocity and the electron mobility.

3.23 **(a)** The channel of a 0.1-μm MOSFET can be considered as a resistor with length $L = 0.1$ μm. Because the channel is short, the electrons drift through the channel with the saturation velocity ($v_{sat} = 0.1$ μm/ps) when nominal voltage is applied across the channel. How long does it take an average electron to drift across the whole channel?

(b) The MOSFET channel is connected by a 5-mm-long copper track to another device. How long does it take an average electron to drift the distance between the two devices connected by the copper track if the current that the electrons conduct is 10 mA? The cross section of the stripe is 0.1 μm^2, and the number of free electrons in copper is $n_{Cu} = 8.1 \times 10^{22}$ cm^{-3}. **A**

(c) Based on the speed of light in vacuum ($c = 3 \times 10^8$ m/s), estimate the time that it takes an electromagnetic wave to propagate along the MOSFET channel and along the 5-mm copper track. Based on the results, answer the following question: Does drift velocity impose limitation to the speed of signal/energy propagation? If not, what is the meaning of the drift velocity?

3.24 **(a)** Design a 1-kΩ diffused N-type resistor. The technology parameters are as follows: the concentration of donor impurities $N_D = 10^{17}$ cm^{-3}, the junction depth $x_j = 2$ μm, and the minimum diffusion width as constrained by the particular photolithography process and lateral diffusion is 4 μm.

(b) How long should the resistor be if P-type silicon with the same doping density is used instead of the N-type silicon?

(c) What will the resistance be if the concentration of donors is increased to $N_D = 10^{19}$ cm^{-3}? **A**

3.25 For the resistor designed in Problem 3.24a, find the resistance at 125°C?

3.26 What is the scattering time of N-type silicon ($N_D = 10^{15}$ cm^{-3}) whose measured conductivity is $\sigma = 0.224$ $(\Omega \cdot \text{cm})^{-1}$? The effective mass of electrons in silicon is $m^* = 0.26 m_0$.

3.27 Determine the scattering lengths of electrons and holes in an N-type silicon sample with the following values for the electron and hole mobilities: $\mu_n = 1000$ cm^2/V · s and $\mu_p = 350$ cm^2/V · s. The sample is at room temperature. Assume that the effective masses of electrons and holes are $0.26 m_0$ and $0.50 m_0$, respectively. (**A** for holes)

3.28 Determine the scattering time due to the Coulomb scattering of silicon doped at $N_D = 10^{17}$ cm^{-3} if the mobility of the doped silicon is $\mu_{total} =$

$800\ \text{cm}^2/\text{V}\cdot\text{s}$ and the mobility of pure silicon is $\mu_0 = 1500\ \text{cm}^2/\text{V}\cdot\text{s}$. The effective mass of silicon is $m^* = 0.26m_0$.

3.29 When the doping level of N-type GaAs is increased from $N_{D1} = 10^{15}\ \text{cm}^{-3}$ to $N_{D2} = 10^{17}\ \text{cm}^{-3}$, the conductivity is increased from $\sigma_1 = 1.36\ (\Omega\cdot\text{cm})^{-1}$ to $\sigma_2 = 800\ (\Omega\cdot\text{cm})^{-1}$. What is the scattering time of the higher-doped sample if the scattering time of the lower-doped sample is $\tau_{sc} = 0.324$ ps? **A**

3.30 The scattering times in a GaAs sample due to phonon scattering, Coulomb scattering, and all the other scattering mechanisms are $\tau_{sc-ph} = 0.35$ ps, $\tau_{sc-C} = 0.20$ ps, and $\tau_{sc-other} = 43$ ps, respectively. Determine the carrier mobility if the effective mass is $m^* = 0.067m_0$.

3.31 For silicon doped with $N_D \approx 10^{16}\ \text{cm}^{-3}$, the electron mobility at the liquid-nitrogen temperature ($T = 77$ K) is approximately the same as that at room temperature ($T = 27°\text{C}$): $\mu_n = 1250\ \text{cm}^2/\text{V}\cdot\text{s}$. Assuming that the mobility is determined by Coulomb ($\mu_C = A_C T^{3/2}$) and phonon ($\mu_{ph} = A_p T^{-3/2}$) scattering mechanisms, determine the maximum mobility and the temperature at which the maximum mobility is observed. Which scattering mechanism dominates at this temperature?

3.32 The phonon (μ_{ph}) and Coulomb (μ_C) limited mobilities in N-type silicon can be expressed as

$$\mu_{ph} = A_p T^{-3/2}$$

$$\mu_C = A_{cn} N_D^{-1} T^{3/2}$$

where $A_p = 700\ \text{m}^2\text{V}^{-1}\text{s}^{-1}\text{K}^{3/2}$ and $A_{cn} = 1.1 \times 10^{19}\ \text{V}^{-1}\text{s}^{-1}\text{K}^{-3/2}\text{m}^{-1}$.

(a) What doping is needed to achieve temperature-independent resistivity for temperatures around 300 K?

(b) What would be the temperature coefficient of this resistive film (TCR) if the operating temperature is 85°C?

3.33 Derive the equation that expresses the scattering length in terms of carrier mobility and the necessary physical constants. Calculate the scattering length for carriers with effective mass $m^* = 0.26m_0$ and mobility $\mu_n = 500\ \text{cm}^2/\text{V}\cdot\text{s}$ at room temperature.

3.34 The phonon-limited mobility of electrons in silicon at 75°C is $\mu_n = 1000\ \text{cm}^2/\text{V}\cdot\text{s}$. Determine the scattering cross section of phonons and the effective phonon radius. The effective mass of electrons in Si is $m^* = 0.26m_0$.

3.35 A slab of N-type silicon, doped at $N_D = 10^{15}\ \text{cm}^{-3}$, is biased so that $I = 200$ mA of current flows through its cross-sectional area, $A = 100\ \mu\text{m} \times 100\ \mu\text{m}$. The electron mobility is $\mu_n = 1500\ \text{cm}^2/\text{V}\cdot\text{s}$. Consider a scattering event that drops the kinetic energy of the electron to $E_{kin} = 3kT/2 = 38.8$ meV and scatters the electron exactly in the direction opposite to the electric-field direction. The electron is now accelerated by the electric field until the next scattering event that occurs after $l_{sc} = 20$ nm.

(a) Apply the energy-band presentation of current flow, illustrated in Fig. 3.2, to determine the kinetic energy of the electron just before the second scattering event.

(b) How much faster is this electron just before the second scattering event in comparison to its average thermal velocity at room temperature? The effective mass of the electron is $m^* = 0.26m_0$.

REVIEW QUESTIONS

R-3.1 What does a nonzero slope of a E–x band diagram mean?

R-3.2 Can E_C and E_V have different slopes at a single point in space?

R-3.3 How do the electrons and holes behave in the region of tilted energy bands?

R-3.4 What is the sheet resistance? In what units is it expressed?

R-3.5 What is the difference between the integral and the differential forms of Ohm's law?

R-3.6 Do the positive donor ions in an N-type semiconductor contribute to the conductivity σ? If not, why not?

R-3.7 Is the linear relationship between the current density and the electric field (Ohm's law) always valid? What about the linear relationship between the current density and the drift velocity?

R-3.8 What is the proportionality coefficient in the linear relationship between the drift velocity and the applied electric field called?

R-3.9 How is the effect of drift-velocity saturation included in Ohm's law?

R-3.10 Is the ratio between the scattering length and the scattering time equal to the thermal velocity or the drift velocity of the carriers?

R-3.11 Does v_{dr}/τ_{sc} depend on the applied electric field? What about v_{th}/τ_{sc}?

R-3.12 The product of the concentration of scattering centers and their scattering cross section ($N_{sc}\sigma_{sc}$) has the unit of length. What is the physical meaning of $N_{sc}\sigma_{sc}$? What does it give when multiplied by the thermal velocity (v_{th}) of a carrier?

R-3.13 What are the two dominant scattering mechanisms in semiconductors?

R-3.14 How are the contributions of different scattering mechanisms combined to obtain the total carrier mobility?

R-3.15 How does the mobility of carriers change with temperature?

4 Diffusion

This chapter deals with carrier *diffusion* as a current mechanism that is additional to, and in general independent from, the carrier *drift*. Following the introduction of the diffusion-current equation and the diffusion coefficient as its parameter, the diffusion coefficient and its relationship to the mobility (the parameter of the drift-current equation) are considered in more detail. Because the diffusion current is closely related to time and spatial variations of carrier concentrations, the chapter concludes with the basic form of the equation that links these variations to both the diffusion and the drift currents—the basic form of the continuity equation.

4.1 DIFFUSION-CURRENT EQUATION

Diffusion appears due to random thermal motion of the diffusing particles (these can be electrons, holes, doping atoms such as boron or phosphorus, molecules in the air, or smoke particles). Imagine that thousands of smoke particles are produced near the wall of a house (Fig. 4.1). If there is an open window in the wall, some particles will pass through the window. In their random motion, half the outside particles move toward the window, as well as half the inside particles, with the same chance of passing through it. However, if there are more particles outside the window, more particles will be entering the house, compared to those going in the opposite direction.

Diffusion of particles creates an effective particle current toward the points of lower particle concentration. There is no effective particle current when the concentration of particles is uniform, because in this case an equal number of particles move either way. The current of particles (charged or uncharged) produced by a difference in the particle concentration is called *diffusion current*. The force behind this current has nothing to do with gravity or with electric-field forces; these are separate forces that can produce a current of particles on their own. The force behind the diffusion current is the random thermal motion.

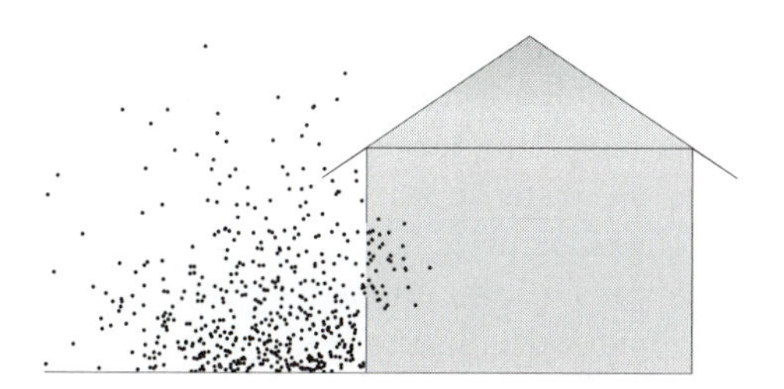

Figure 4.1 The concept of diffusion.

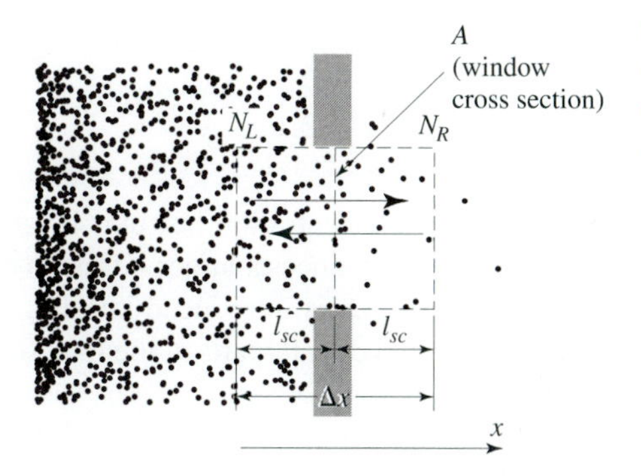

Figure 4.2 Diffusion current: particles flow through the window in both directions, but more from left to right because there are more particles on the left-hand side.

Diffusion is not limited to gases; it happens in liquids and solids as well, although a very high temperature is typically needed for the diffusion of atoms to be clearly observed in solids. If a semiconductor crystal is heated, the diffusion of doping atoms can be achieved in a way similar to the diffusion of the smoke particles into the house. Figure 1.20 is analogous to Fig. 4.1 except that it shows the doping atoms diffusing into a piece of semiconductor through a window in a diffusion-protective "wall."

Figure 4.2 illustrates in more detail that a difference in concentration of particles, performing random thermal motion, is the only driving force behind the diffusion current. It is useful to express this understanding of the diffusion current by a mathematical equation. It should be noted that the current considered here is not necessarily electric current—that is, current of charged particles. Uncharged particles can create diffusion current as well. The unit for the current of uncharged particles is not the ampere (A = C/s), but simply 1/s, expressing the number of particles per unit time. The current density (which is current per unit area) as a differential quantity is more general and more convenient to work with than the overall current (i.e., integral current), as discussed in Section 3.2.2. The diffusion current density will be denoted by J_{diff} (when meant as particle current and expressed in units of $s^{-1}m^{-2}$) and by j_{diff} (when meant as electric current and expressed in units of A/m^2).

As already mentioned, the current density of the particles flowing from left to right ($J_{\rightarrow}$) is proportional to the concentration of particles to the left of the window (N_L). The current density of the particles flowing in the opposite direction ($J_{\leftarrow}$) is proportional to the concentration of particles to the right of the window (N_R). The effective current density is equal to the difference between those two currents:

$$J_{diff} = J_{\rightarrow} - J_{\leftarrow} \propto \underbrace{N_L - N_R}_{\equiv -\Delta N} \tag{4.1}$$

Note that the minus sign in front of ΔN expresses that the concentration N is decreasing along the x-axis; $\Delta N \equiv N(x + \Delta x) - N(x)$.

The difference in concentration is not the only thing that is important for the diffusion current; equally important is the distance at which this difference appears. The same difference in the concentrations of particles appearing on each side of the window, ΔN, would lead to a smaller current density if the difference were across a larger Δx. Therefore, the current density is proportional to the difference in concentration ΔN appearing across a distance Δx:

$$J_{diff} \propto -\frac{\Delta N}{\Delta x} \tag{4.2}$$

The ratio $\Delta N/\Delta x$ is the change in the concentration per unit length, or concentration gradient. A steeper concentration gradient produces a larger diffusion current.

If a proportionality factor, denoted by D, is introduced into Eq. (4.2), and the finite differences Δ replaced by their infinitesimal counterparts d, the final form of the diffusion-current equation is obtained:

$$J_{diff} = -D\frac{dN}{dx} \tag{4.3}$$

The proportionality factor D is called *diffusion coefficient*. Equation (4.3) is the one-dimensional form of the diffusion-current equation. It can be expanded to include all the three space dimensions, in which case the current density is expressed as a vector:

$$\mathbf{J}_{diff} = -D\left(\frac{\partial N}{\partial x}\mathbf{x}_u + \frac{\partial N}{\partial y}\mathbf{y}_u + \frac{\partial N}{\partial z}\mathbf{z}_u\right) = -D\nabla N \tag{4.4}$$

To convert Eq. (4.3) for the particle current density (expressed in $s^{-1}\,m^{-2}$) into electric current of electrons or holes (expressed in A/m^2), the particle concentration N can be replaced by p or n and multiplied by $\pm q$:

$$j_{diff} = \begin{cases} qD_n dn/dx & \text{electrons} \\ -qD_p dp/dx & \text{holes} \end{cases} \tag{4.5}$$

The random thermal motion of particles is more pronounced at higher temperatures; therefore, the diffusion current is expected to be larger. It appears that the diffusion-current equation does not take into account the influence of the temperature on diffusion. The dependence of the random thermal motion of particles on temperature, however, is different in different materials and for different particles (different in gases and solids for that matter), so it is not possible to establish a general temperature dependence of the diffusion current. To deal with this problem, the temperature dependence of the diffusion current is taken into account by the diffusion coefficient, which has to be determined for any individual material and any kind of diffusing particles.

4.2 DIFFUSION COEFFICIENT

The diffusion coefficient of electrons and holes can be related to the electron and hole mobilities. This relationship, known as the Einstein relationship, is introduced in this section. In spite of this relationship, the diffusion and drift mechanisms act independently, as is nicely illustrated by the Haynes–Shockley experiment. The Haynes–Shockley experiment and its use for measurements of mobility and the diffusion coefficient are also described in this section. The doping atoms are neutral particles, so there is no mobility parameter and no Einstein relationship for the diffusion coefficient of these particles. The diffusion coefficient of the doping atoms follows the Arrhenius-type temperature dependence, as described in Section 4.2.3.

4.2.1 Einstein Relationship

Equation (4.5) expresses a phenomenological relationship between the diffusion current and the concentration gradient in a similar way to the expression by Ohm's law of a phenomenological relationship between the drift current and the gradient of electric potential [Eq. (3.16)]. As opposed to this, Eq. (3.21) is the fundamental link between the drift current and the average drift velocity of the current carriers. Analogously, the thermal velocity v_{th} can be linked to a thermal current. Referring to Fig. 4.2, let us estimate the current of particles moving through the window from left to right, as a result of the random thermal motion. The particles that have a chance to pass through the window during the time interval equal to the average scattering time τ_{sc} are confined within the volume $l_{sc}A$, where A is the area of the window, $l_{sc} = v_{th}\tau_{sc}$ is the average scattering length, and v_{th} is the thermal velocity. Labeling the concentration of particles by N_L, the number of particles within the volume $l_{sc}A$ is $N_L l_{sc}A = N_L v_{th}\tau_{sc}A$. We can assume that one-third of these particles move in the x-direction and that the other two-thirds move in the y- and z-directions. Of those moving in the x-direction (assumed to be normal to the window), one-half would be moving toward the window and the other half away from it. Therefore, the number of particles passing through the window is estimated as $N_L v_{th}\tau_{sc}A/6$. Because these are the particles moving from the left-hand side to the right during the time interval equal to τ_{sc}, the current density of particles moving to the right is

$$J_{\rightarrow} = \frac{N_L v_{th}}{6} \tag{4.6}$$

Analogously, the current density of particles passing through the window from the right-hand side to the left is

$$J_{\leftarrow} = \frac{N_R v_{th}}{6} \tag{4.7}$$

The difference between $J_{\rightarrow}$ and $J_{\leftarrow}$ is the effective current through the window—that is, the diffusion-current density:

$$J_{diff} = J_{\rightarrow} - J_{\leftarrow} = (N_L - N_R)\frac{v_{th}}{6} \tag{4.8}$$

Obviously, the diffusion current depends on the difference between the particle concentrations on the left and on the right (there is no effective current when the concentrations are equal). Presenting the concentration difference as concentration gradient (dN/dx),

$$\frac{N_R - N_L}{2l_{sc}} = \frac{dN}{dx} \tag{4.9}$$

we obtain the diffusion-current density as

$$J_{diff} = -\underbrace{\frac{v_{th}l_{sc}}{3}}_{D}\frac{dN}{dx} \tag{4.10}$$

Comparing Eqs. (4.10) and (4.3), we find that the diffusion coefficient is given by

$$D = \frac{v_{th}l_{sc}}{3} = \frac{\tau_{sc}v_{th}^2}{3} \tag{4.11}$$

Therefore, we conclude that the diffusion constant is determined by temperature (the thermal velocity) and by the particle scattering (l_{sc} or τ_{sc}). The temperature can be shown explicitly, given the relationship between the kinetic ($m^*v_{th}^2/2$) and the thermal energy ($3kT/2$) [Eq. (3.19)]. Using this relationship to eliminate v_{th} from Eq. (4.11), we obtain

$$D = kT\frac{\tau_{sc}}{m^*} \tag{4.12}$$

The derivation presented has involved some coarse approximations, such as the assumption that one-sixth of particles move toward the window and Eq. (4.9) for the concentration gradient. Nonetheless, rigorous analyses also lead to the result given by Eq. (4.12). The derivation has not involved any assumptions regarding the particle charge, so Eq. (4.12) is valid for both neutral particles (such as gas molecules) and charged particles (such as electrons and holes). Importantly, it has been assumed that particle scattering alone limits thermal motion. This means that Eq. (4.12) cannot be used when other limiting factors are pronounced. For example, the diffusion of doping atoms is not limited by scattering but by availability of empty sites for the doping atoms to move to. Consequently, a different equation is needed for the diffusion coefficient of doping atoms (Section 4.2.3). Similarly, the availability of empty states influences the electron/hole transport in heavily doped (degenerate) semiconductors. Again, Eq. (4.12) cannot be used for degenerate semiconductors.

The fact that the diffusion constant depends on scattering indicates that there is a link between the diffusion constant and the mobility. The relationship between τ_{sc} and mobility is given by Eq. (3.25). Using D_n and D_p as symbols for the diffusion coefficients of electrons and holes, respectively, and eliminating τ_{sc} from Eqs. (4.12) and (3.25) lead to the following relationship between $D_{n,p}$ and $\mu_{n,p}$:

$$D_{n,p} = (kT/q)\mu_{n,p} = V_t\mu_{n,p} \tag{4.13}$$

Equation (4.13) is known as the Einstein relationship.

The fact that *drift* and *diffusion* are independent current mechanisms, yet there is a link between their parameters ($\mu_{n,p}$ and $D_{n,p}$), should not be confusing. The independence relates to the independent forces causing the current: gradient of electric potential (that is, electric field) in the case of drift and gradient of concentration in the case of diffusion. The link between $\mu_{n,p}$ and $D_{n,p}$ exists because the properties of the current carriers are the same in both cases. As an illustration, take particles with a low mobility due to their heavy effective mass. If there is a concentration gradient, a diffusion current would exist. Because particles would move very slowly in this case, it would take a significant time for the particles to be transported from a higher to a lower concentration region, which means that the diffusion current is small. This example illustrates that the mobility $\mu_{n,p}$ is proportional to the diffusion coefficient $D_{n,p}$.

Consider the effect of temperature because both the mobility and the diffusion coefficient are found to be temperature-dependent. The influence of the temperature on mobility is basically through scattering, more pronounced scattering leading to reduced scattering length, and consequently reduced mobility. Reduction in the scattering length would proportionally reduce the diffusion coefficient, implying again that the mobility is proportional to the diffusion coefficient. The temperature, however, is essential for the process of diffusion. Imagine an intrinsic semiconductor (Coulomb scattering negligible) at a very low temperature. The thermal velocity of the carriers would be very small; but this would not stop the electric field from driving the carriers very efficiently, producing a significant drift current, which means the mobility is quite high. When the thermal velocity of the carriers is small, however, the diffusion process goes very slowly because the only driving force behind the diffusion is the random thermal motion. If the temperature is increased, the random thermal motion as well as the diffusion process become more pronounced, which means that the diffusion coefficient is increased.

These considerations illustrate that the diffusion coefficient $D_{n,p}$ of a carrier gas is proportional to the mobility $\mu_{n,p}$ of the carriers as well as to the thermal energy kT (or thermal voltage $V_t = kT/q$), just as given by Eq. (4.13).

Given that Eq. (4.12) does not hold for degenerate gases, when the carriers start competing for vacant positions, the Einstein relationship [Eq. (4.13)] cannot be used for heavily doped semiconductors when the electron–hole gas cannot be approximated by the Maxwell–Boltzmann distribution.

EXAMPLE 4.1 Alternative Derivation of the Einstein Relationship

A constant electric field E is established inside a semiconductor. The direction of the field is normal to the surface of the semiconductor, so that the flow of any current is prevented. Using the fact that the net currents of both electrons and holes have to be zero, determine the relationship between the diffusion constant $D_{n,p}$ and the mobility $\mu_{n,p}$.

SOLUTION

Labeling the direction normal to the semiconductor surface by x, we write the total current density of electrons:

$$j_n = \underbrace{q\mu_n n(x)E}_{\text{drift}} + \underbrace{qD_n\frac{dn(x)}{dx}}_{\text{diffusion}}$$

The drift and the diffusion components of the total current must balance each other so that the total current is zero:

$$\mu_n n(x)E = -D_n\frac{dn(x)}{dx} \tag{4.14}$$

The concentration of electrons (n), and therefore the concentration gradient (dn/dx), can be related to the electric field by expressing both n and E in terms of E_C (the bottom of the conduction band). Taking into account that E_F = const for a system in equilibrium, and according to Eq. (2.76),

$$n(x) = N_C e^{-[E_C(x)-E_F]/kT} \tag{4.15}$$

$$\frac{dn(x)}{dx} = -\frac{1}{kT}\underbrace{N_C e^{-[E_C(x)-E_F]/kT}}_{n(x)}\frac{dE_C(x)}{dx}$$

As the gradient of E_C is related to the electric field [Eq. (3.2)],

$$\frac{1}{q}\frac{dE_C(x)}{dx} = E$$

we obtain

$$\frac{dn(x)}{dx} = -\frac{q}{kT}n(x)E \tag{4.16}$$

From Eqs. (4.14) and (4.16),

$$\mu_n = \frac{q}{kT}D_n$$

which is the Einstein relationship for electrons. The Einstein relationship for holes can be obtained by analogous derivation. Note that involving Eq. (4.15) in the derivation means that the Maxwell–Boltzmann distribution is assumed. This is the case of nondegenerate semiconductors. When the Fermi level is too close to E_C (degenerate semiconductors), the Maxwell–Boltzmann approximation and the Einstein relationship cannot be used.

*4.2.2 Haynes–Shockley Experiment

As mentioned before, drift and diffusion are two independent current mechanisms. This is nicely demonstrated by the Haynes–Shockley experiment, illustrated in Fig. 4.3. To monitor the process of drift and diffusion, a pulse of minority carriers is generated in a very narrow region ($\Delta x \to 0$) of a semiconductor bar for a very short time interval ($\Delta t \to 0$). In the case shown in Fig. 4.3, this is achieved by a flash of light that illuminates a narrow region at the end of an N-type silicon bar at time $t = 0$. The light generates both electrons and holes; however, the direction of the electric field applied is such that the electrons are quickly collected by the positive contact, while the holes have to travel through the silicon bar to be collected by the negatively biased contact. The drift velocity of holes is $v_d = \mu_p E$, and if the diffusion did not exist, a short current pulse would be detected by the ammeter as the holes are collected by the negative contact after the time interval equal to L/v_d. However, as the holes diffuse in either direction, the hole distribution widens with time, as illustrated in Fig. 4.3. The hole distribution can be expressed by the following form of the Gauss distribution:

$$p = p_{max} e^{-(x-x_{max})^2/4D_p t} \tag{4.17}$$

The position of the distribution peak ($x = x_{max}$) shifts as the holes drift along the electric field: $x_{max} = v_d t$.

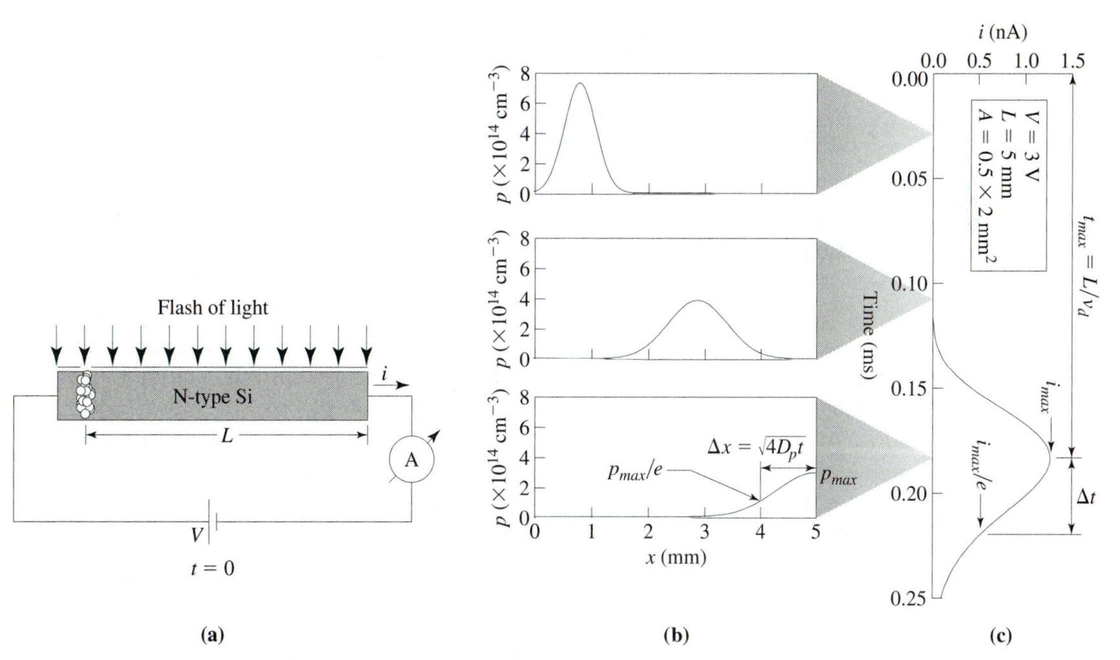

Figure 4.3 Haynes–Shockley experiment. (a) A flash of light is used to generate a narrow pulse of minority carriers (holes). (b) The motion of the hole peak illustrates the hole drift, while the widening of the hole distribution illustrates the hole diffusion. (c) The measured current at the end of the semiconductor can be used to calculate μ_p and D_p.

Both the minority-carrier mobility and diffusion coefficient can be extracted from the measured time dependence of the current. Figure 4.3c illustrates that the maximum of the current coincides with the arrival of the peak of the hole distribution at the negative contact. The drift velocity can be calculated from the measured time between the current maximum and the flash of light as $v_d = L/t_{max}$. The mobility is then calculated using Eq. (3.23):

$$\mu_p = \frac{v_d}{E} = \frac{L^2}{t_{max}V} \tag{4.18}$$

The diffusion coefficient can be determined from the width of the current pulse. Equation (4.17) shows that $p = p_{max}/e$ for $(x - x_{max})^2 = 4D_p t$; that is, $\Delta x = x_{max} - x = \sqrt{4D_p t}$. The current that corresponds to this point of the hole distribution is i_{max}/e. Because it takes the time of $\Delta t = \Delta x/v_d$ for the holes to travel the distance of $\Delta x = \sqrt{4D_p t} = \sqrt{4D_p(t_{max} + \Delta t)}$, the following equation can be written:

$$\Delta t = \frac{\Delta x}{v_d} = \frac{\sqrt{4D_p(t_{max} + \Delta t)}}{v_d} \tag{4.19}$$

By using $v_d = L/t_{max}$, the diffusion coefficient can be expressed in terms of the measured values of t_{max} and Δt:

$$D_p = \frac{\Delta t^2 \left(\frac{L}{t_{max}}\right)^2}{4(t_{max} + \Delta t)} \tag{4.20}$$

4.2.3 Arrhenius Equation

Equation (4.12) was derived with an assumption that the particle motion is determined by particle scattering. In the case of the diffusion of doping atoms into a semiconductor, the motion of the doping atoms is limited by the availability of empty sites that the doping atoms can move to. Therefore, the diffusion coefficient of the doping atoms does not follow the temperature dependence predicted by Eq. (4.12). Doping atoms can diffuse through vacancies in the crystal lattice. In this case, the dominating role of temperature is related to the generation of vacancies. This is a process that follows the common exponential dependence on the needed *activation energy* (E_A), normalized by the thermal energy (kT): $D \propto \exp(-E_A/kT)$. Doping atoms can also diffuse as interstitial atoms. In this case, they need to overcome the energy barriers between the interstitial positions. If the barrier height is E_A, the probability that a doping atom will gain this energy is again proportional to $\exp(-E_A/kT)$. Introducing the proportionality constant D_0, the diffusion coefficient can be expressed by

$$D = D_0 e^{-E_A/kT} \tag{4.21}$$

The diffusion constant for doping atoms follows the so-called Arrhenius type of temperature dependence. The parameters E_A (the activation energy) and D_0 (the frequency factor) have to be determined for any individual semiconductor material and for individual doping species.

How strongly the diffusion coefficient depends on the temperature can be illustrated by considering the ratio of the diffusion coefficient at 1000°C and at room temperature. For the case of $E_A = 3.5$ eV, which is a typical value, this ratio is $D(T = 1273 \text{ K})/D(T = 300 \text{ K}) = 8.5 \times 10^{44}$. This means that if an hour is needed to obtain a doped layer in the silicon at $T = 1000°\text{C}$, the same process would take 8.5×10^{44} hours at room temperature. This time expressed in years is 10^{41}, or 10^{38} millennia. It is certainly correct to say that there is no diffusion in the silicon at room temperature.

EXAMPLE 4.2 Diffusion Coefficient for Boron

Calculate the diffusion coefficient for boron at 1000°C and 1100°C, using the following values for the frequency factor and the activation energy: $D_0 = 0.76$ cm²/s, $E_A = 3.46$ eV. Comment on the results. The Boltzmann constant is $k = 8.62 \times 10^{-5}$ eV/K.

SOLUTION

The diffusion coefficient can be calculated using Eq. (4.21). Note that the temperature should be expressed in K; therefore, $T_1 = 1000 + 273.15 = 1273.15$ K and $T_2 = 1100 + 273.15 = 1373.15$ K, respectively. The results are $D_1 = 1.54 \times 10^{-14}$ cm²/s and $D_2 = 1.53 \times 10^{-13}$ cm²/s for T_1 and T_2, respectively. An increase in the temperature from 1000°C to 1100°C (10%) increases the diffusion coefficient 10 times.

4.3 BASIC CONTINUITY EQUATION

The concentration profile $[N(x)]$ of a set of particles changes with time as the diffusion process transports particles from higher- to lower-concentration regions. The concentration of diffusing particles is, therefore, a function of two variables: $N(x,t)$. Let us consider how the particle concentration at the window in Fig. 4.4 changes in time. To do so, consider the change in the number of particles confined in the little box at the window during a small time interval Δt. This change is equal to the difference between the number of particles that enter the box and the number of particles that come out of the box in the time interval Δt. If ΔA is the area of the side of the box, the number of particles entering the box per unit time is $J(x)\Delta A$, where $J(x)$ is the density of particle current at x. Analogously, the number of particles coming out of the box per unit time is $J(x+\Delta x)\Delta A$, where $J(x+\Delta x)$ is the density of particle current at $x+\Delta x$. Therefore, the change in the number of particles inside the box per unit time can be expressed as

$$\frac{\Delta NUM}{\Delta t} = \underbrace{[J(x) - J(x+\Delta x)]}_{-\Delta J(x)} \Delta A = -\Delta J(x)\Delta A \tag{4.22}$$

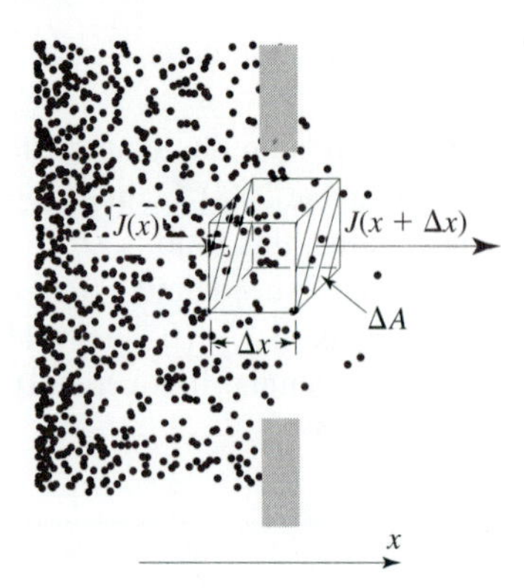

Figure 4.4 If the particle current flowing out of the box is different from the particle current flowing into the box, the number of particles in the box changes in time: $\partial N/\partial t \neq 0$.

Because our aim is to express the change of *concentration* in time, Eq. (4.22) can be divided by the volume of the box $(\Delta A \Delta x)$ to convert the *number* of particles *NUM* into the *concentration* of particles $N(x, t)$:

$$\frac{\overbrace{\Delta NUM/\Delta A \Delta x}^{\Delta N/\Delta t}}{\Delta t} = -\frac{\Delta J(x)}{\Delta x} \tag{4.23}$$

Finally, by taking infinitesimal values, we obtain the following differential equation:

$$\frac{\partial N(x,t)}{\partial t} = -\frac{\partial J(x)}{\partial x} \tag{4.24}$$

Partial derivatives are used to express that the derivative of $N(x, t)$ is only with respect to time, whereas the derivative of $J(x)$ is with respect to space. Equation (4.24) is called the continuity equation. It represents a general conservation principle. A nonzero gradient of particle current $[\partial J(x)/\partial x \neq 0]$ means that the numbers of particles flowing into and out of a specified point in space are different. For particles *that are not being generated or annihilated,* this means that the particle concentration at that point must change in time $[\partial N(x,t)/\partial t \neq 0]$. It also means that the rate of concentration change $[\partial N(x,t)/\partial t]$ is determined by the current gradient.

The density of particle current in Eq. (4.24) is in the units of $s^{-1} m^{-2}$. It can be converted into the electric-current density of electrons, in the units of $C \cdot s^{-1}\ m^{-2} =$ A/m^2, by taking into account that each electron carries the charge of $-q$: $j_n = -qJ(x)$. Analogously, the electric-current density of holes is $j_p = qJ(x)$. With this, and using the usual labels for the concentrations of electrons and holes (n and p), the continuity equations for electrons and holes take the following forms:

$$\begin{aligned} \partial n(x,t)/\partial t &= (1/q)\partial j_n(x)/\partial x \qquad &\text{for electrons} \\ \partial p(x,t)/\partial t &= -(1/q)\partial j_p(x)/\partial x \qquad &\text{for holes} \end{aligned} \tag{4.25}$$

The current densities $j_{n,p}$ in Eqs. (4.25) are the total current densities of electrons and holes. When both diffusion and drift currents exist, they are equal to $j_n = j_{n\text{-}dr} + j_{n\text{-}diff}$

and $j_p = j_{p\text{-}dr} + j_{p\text{-}diff}$. It should be mentioned again that Eq. (4.25) is for the case of no effective generation or recombination of the carriers. We will refer to Eq. (4.25) as the *basic continuity equation*. The effects of generation and recombination due to nonequilibrium concentrations of electrons and holes are considered in Chapter 5, where the complete form of the continuity equation is presented.

EXAMPLE 4.3 Gradient of Current and Change of Particle Concentration in Time

After 1 h of phosphorus diffusion into silicon, the concentration of phosphorus at and around 1 μm from the surface can be expressed as shown in Table 4.1.

(a) Knowing that the diffusion coefficient at the diffusion temperature is $D = 3.43 \times 10^{-13}$ cm²/s, estimate the phosphorus current density at $x = 0.975$ μm and $x = 1.025$ μm.

(b) If the diffusion is performed through 100 μm × 10 μm window, estimate how many phosphorus atoms arrive at 1 ± 0.025 μm during a time interval of 1 s, and how many phosphorus atoms leave this segment during a second.

(c) The results of parts (a) and (b) show that the number of phosphorus atoms leaving the segment 1 ± 0.025 μm is lower than the number of arriving atoms. As a result, the phosphorus concentration at 1 ± 0.025 μm increases with time (*the continuity equation*). Estimate how long it will take for the concentration to rise by 1%.

SOLUTION

(a)

$$J = -D\frac{dN}{dx} \approx -D\frac{\Delta N}{\Delta x}$$

$$J(0.975\ \mu\text{m}) = -3.43 \times 10^{-17} \times \frac{(1.32 - 1.61) \times 10^{25}}{0.05 \times 10^{-6}} = 1.989 \times 10^{15}\ \text{s}^{-1}\text{m}^{-2}$$

$$J(1.025\ \mu\text{m}) = -3.43 \times 10^{-17} \times \frac{(1.07 - 1.32) \times 10^{25}}{0.05 \times 10^{-6}} = 1.715 \times 10^{15}\ \text{s}^{-1}\text{m}^{-2}$$

(b)

$$NUM = I\Delta t = JA\Delta t$$

$$NUM_{IN} = J(0.975\ \mu\text{m})A\Delta t = 1.989 \times 10^{15} \times 100 \times 10^{-6} \times 10 \times 10^{-6} \times 1 = 1{,}989{,}000$$

$$NUM_{OUT} = J(1.025\ \mu\text{m})A\Delta t = 1.715 \times 10^{15} \times 100 \times 10^{-6} \times 10 \times 10^{-6} \times 1 = 1{,}715{,}000$$

(c) One way to estimate this time is to utilize the result of part (b), which is that the number of phosphorus atoms in the volume $A\Delta x$ increases by $NUM_{IN} - NUM_{OUT} = 1{,}989{,}000 - 1{,}715{,}000 = 274{,}000$ every second. To be able to find how many atoms correspond to 1% of concentration reduction, we need to determine the number of

TABLE 4.1

x (μm)	N (cm^{-3})
0.95	1.61×10^{19}
1.00	1.32×10^{19}
1.05	1.07×10^{19}

phosphorus atoms in the volume $A\Delta x$. The total number of phosphorus atoms in $A\Delta x$ is

$$NUM = NA\Delta x = 1.32 \times 10^{25} \times 1000 \times 10^{-6} \times 10^{-6} \times 0.05 \times 10^{-6} = 6.6 \times 10^{8}$$

therefore, a 1% increase corresponds to an increase by 6.6×10^{6} atoms. Given that the increase is 274,000 atoms per second, it will take $6.6 \times 10^{6}/274{,}000 = 24.1$ s for the 1% increase. Another way to estimate this time is to work with the concentration directly: that is, to use the continuity equation in its usual form:

$$\frac{\partial N}{\partial t} = -\frac{\partial J}{\partial x}$$

$$\frac{\Delta N}{\Delta t} \approx -\frac{\Delta J}{\Delta x}$$

$$\Delta t \approx -\frac{\Delta N \Delta x}{\Delta J}$$

ΔN is equal to 1% of 1.32×10^{25} m^{-3} (with a minus sign because the concentration is smaller for a larger x), $\Delta J = (1.989 - 1.715) \times 10^{15}$ s^{-1}m^{-2}, and $\Delta x = 0.05 \times 10^{-6}$ m. Therefore,

$$\Delta t \approx \frac{0.01 \times 1.32 \times 10^{25} \times 0.05 \times 10^{-6}}{(1.989 - 1.715) \times 10^{15}} = 24.1 \text{ s}$$

EXAMPLE 4.4 Nonuniform Electron Concentration in Equilibrium

Determine the equilibrium distribution of electron concentration, $n(x)$, for the sample described in Example 4.1. Set $x = 0$ at the surface of the semiconductor and assume that the semiconductor extends to $x \to \infty$.

SOLUTION

There is no time change of $n(x)$ when the equilibrium condition is reached. Consequently, $\partial n/\partial t = 0$. According to the continuity equation [Eq. (4.25)], the first derivative of the total current has to be equal to zero (in addition to the fact that $j_n = 0$). Given that the total current is equal to the sum of the drift and the diffusion currents,

$$j_n = q\mu_n n(x)E + qD_n \frac{dn(x)}{dx}$$

and that $E =$ const, we obtain

$$\frac{dj_n}{dx} = q\mu_n E \frac{dn(x)}{dx} + qD_n \frac{d^2n(x)}{dx^2}$$

The condition $dj_n/dx = 0$ leads to the following differential equation:

$$\frac{d^2n(x)}{dx^2} = -\underbrace{(\mu_n/D_n)}_{q/kT} E \frac{dn(x)}{dx}$$

The variables can be separated by introducing $u(x) = dn(x)/dx$:

$$\frac{du(x)}{dx} = -\frac{qE}{kT}u(x)$$

$$\frac{du(x)}{u(x)} = -\frac{qE}{kT}dx$$

$$\ln u(x) \underbrace{- \ln A}_{\text{integ. const.}} = -\frac{qE}{kT}x$$

$$u(x) = \frac{dn(x)}{dx} = Ae^{-qEx/kT}$$

$$n(x) = -\frac{kT}{qE}Ae^{-qEx/kT} + B$$

The integration constants A and B can be determined from the boundary conditions. The term $-kTA/qE$ is an integration constant (A) multiplied by another constant $(-qE/kT)$, so it can be replaced by a single constant C:

$$n(x) = Ce^{-qEx/kT} + B$$

The two boundary conditions are for $x = 0$ and $x \to \infty$: $n(0)$ and $n(\infty)$. Therefore,

$$n(0) = C + B$$

$$n(\infty) = B$$

Given that $C = n(0) - B = n(0) - n(\infty)$, the distribution of electron concentration can be expressed as

$$n(x) = [n(0) - n(\infty)]\, e^{-qEx/kT} + n(\infty)$$

The concentration at $x \to \infty$ is equal to the equilibrium concentration without the electric field E. Upon labeling it by $n_0 = n(\infty)$, for convenience, we can express the concentration distribution in the following way:

$$n(x) - n_0 = [n(0) - n_0]\, e^{-qEx/kT}$$

If the direction of the electric field is such that electrons are accumulated at the surface $[n(0) - n_0 > 0]$, the difference in electron concentration inside the semiconductor $[n(x) - n_0]$ will drop exponentially toward zero. If the field is such that the surface is depleted of electrons $[n(0) - n_0 < 0]$, the difference in electron concentration will rise exponentially toward zero.

SUMMARY

1. Random thermal motion of either charged or neutral particles results in an effective diffusion current when there is a gradient of the particle concentration. The density of particle diffusion current, in units of $s^{-1}m^{-2}$, is

$$J_{diff} = -D\frac{\partial N}{\partial x}$$

The density of electric diffusion current, in units of A/m^2, is

$$j_{diff} = \begin{cases} qD_n\, dn/dx & \text{for electrons} \\ -qD_p\, dp/dx & \text{for holes} \end{cases}$$

2. The diffusion coefficient of electrons and holes depends on the scattering time, the effective mass, and the thermal energy:

$$D_{n,p} = kT\frac{\tau_{sc}}{m^*_{n,p}}$$

Because the carrier mobility also depends on the scattering time and the effective mass, there is also a relationship between the coefficients of diffusion and drift (two otherwise independent current mechanisms):

$$D_{n,p} = (kT/q)\mu_{n,p} = V_t\mu_{n,p}$$

3. The diffusion coefficient of doping atoms is determined by the thermal energy and the activation energy of the diffusion process:

$$D = D_0 \exp\left(-\frac{E_A}{kT}\right)$$

4. A current-density gradient (a difference in the number of particles flowing *in* and *out* per unit time) leads to a change in the particle concentration in time, as expressed by the basic continuity equation:

$$\partial N(x,t)/\partial t = -\partial J(x)/\partial x \qquad \text{for neutral particles}$$
$$\partial n(x,t)/\partial t = (1/q)\partial j_n(x)/\partial x \qquad \text{for electrons}$$
$$\partial p(x,t)/\partial t = -(1/q)\partial j_p(x)/\partial x \qquad \text{for holes}$$

PROBLEMS

4.1 Identify

(a) drift-current equation
(b) diffusion-current equation
(c) continuity equation

applied to electrons. The list of options is

(1) $j = qD_n(dn/dx)$
(2) $j = qn\mu_n E$
(3) $j = \sigma(dn/dx)$
(4) $\partial j/\partial x = q(\partial n/\partial t)$
(5) $j = -qD_n(d\varphi/dx)$
(6) $j = qD_n nE$

4.2 Certain carrier-transport conditions result with a steady-state distribution of electrons $n(x)$. The distribution can be approximated by a linear function of x. Knowing that the maximum electron concentration is $n(0) = 10^{16}$ cm^{-3} and that it drops to one-third of this value at $x = 100$ μm, determine the diffusion current density. The diffusion coefficient of electrons is $D_n = 25$ cm^2/s.

4.3 A constant diffusion current of electrons is established through a semiconductor material. The value of the current density is $j_n = -0.5$ A/cm^2 (the minus sign indicates that the current direction is opposite to the chosen direction of the x-axis). The electron concentration at $x = 0$ is $n(0) = 2 \times 10^{15}$ cm^{-3}. Determine the electron concentration at $x = 25$ μm if the material is

(a) Si ($D_n = 35$ cm^2/s),
(b) GaAs ($D_n = 220$ cm^2/s). **A**

4.4 The concentration of holes in a silicon sample can be approximated by

$$p(x) = p_{max} e^{-x/L_p}$$

where $p_{max} = 5 \times 10^{15}$ cm^{-3} and $L_p = 50$ μm. Determine the hole diffusion current density at $x_1 = 0$ and $x_2 = L_p = 50$ μm. The diffusion coefficient for holes is $D_p = 10$ cm^2/s.

4.5 Metal film with area $A_C = 2.25 \times 10^{-4}$ cm^2 provides ohmic contact to P-type silicon. If the concentration of holes is $p_0 = 10^{16}$ cm^{-3}, calculate how many holes hit the contact with metal per unit time. Express this number as hole current. If this contact is in thermal equilibrium, what mechanism provides the balancing current so that the total diffusion current is equal to zero? Assume that the thermal velocity of holes is $v_{th} = 10^7$ cm/s.

4.6 Consider a bar of N-type silicon with cross-sectional area $A = 100$ μm $\times$ 100 μm and concentration of free electrons $n_0 = 10^{15}$ cm^{-3}.

(a) Estimate the number of electrons that pass through a cross-sectional plane in only one direction each second. The temperature is 300 K, and the effective mass of electrons in silicon is $m^* = 0.26m_0$.
(b) Express the flow of electrons, calculated in part (a), as electric current in amperes.

(c) Based on the diffusion-current equation, determine the concentration gradient dn_0/dx that corresponds to the diffusion current of $I_{diff} = 100$ mA. Assuming linear concentration change, determine the distance Δx that is needed for the concentration to change from n_0 to 0. The diffusion constant for the electrons is $D_n = 38$ cm²/s.

(d) Based on the results obtained in parts (b) and (c), answer the following question: Can the diffusion current exceed the current value that corresponds to the flow of electrons in one direction only, which is determined in part (b), to reach the value of 100 mA at 300 K that is assumed in part (c)?

4.7 The diffusion constant of electrons in a low-doped silicon is dominated by phonon scattering and is equal to $D_n = 36$ cm²/s at room temperature. What is the diffusion constant at 125°C?

4.8 Room-temperature values of hole mobilities in P-type silicon are 470 cm²/V · s and 150 cm²/V · s for $N_A = 10^{15}$ cm^{-3} and $N_A = 10^{18}$ cm^{-3}, respectively. Calculate the diffusion constants of holes at 125°C for these two doping levels.

4.9 For silicon doped with $N_D \approx 10^{16}$ cm^{-3}, the electron mobility at liquid-nitrogen temperature ($T = 77$ K) is approximately the same as at room temperature ($T = 27°$C): $\mu_n = 1250$ cm²/V · s. Calculate the diffusion coefficient at these temperatures.

4.10 The total hole-current density is constant and equal to $j_p = 1$ A/cm² in the region of a silicon sample extending from $x = 0$ to $x \to \infty$. The distribution of hole concentration is given by

$$p(x) = p(0)e^{-x/L_p} + p_0$$

where $L_p = 100$ μm and $p_0 = 10^{15}$ cm^{-3}. At $x = 0$, the drift current is equal to zero so that $j_{p-diff}(0) = j_p$. Determine the electric field as a function of x and calculate the values of the electric field at the following points: $x = 0$, $x \to \infty$, and $x = L_p$. The hole mobility is $\mu_p = 350$ cm²/V · s, the diffusion coefficient is $D_p = V_t\mu_p$, and $V_t = kT/q = 0.02585$ V (given that the sample is at room temperature). [**A** for $E(L_p)$]

4.11 The donor distribution of a nonuniformly doped silicon sample can be approximated by $N(x) = N_s \exp(-x/x_0)$, where $x_0 = 1$ μm. As no current flows under the open-circuit condition, a built-in electric field is established so that the drift current exactly compensates the diffusion current. Calculate the built-in electric field.

4.12 Internal power dissipation increases the temperature of a silicon device to $T_{oper} = 75°$C. This reduces the scattering length of electrons from $l_{sc-room} = 45.7$ nm at room temperature to $l_{sc-oper} = 36.5$ nm at T_{oper}. Determine the relative change in the diffusion coefficient. The effective mass of electrons is $m^* = 0.26m_0$. **A**

4.13 The drift velocities of holes in a silicon sample are $v_{dr} = 0.02$ μm/ps, $v_{dr} = 0.04$ μm/ps, and $v_{dr} = 0.07$ μm/ps at $E = 0.5$ V/μm, $E = 1.0$ V/μm, and $E = 5$ V/μm, respectively. Determine the corresponding diffusion coefficients and comment on the results. The temperature is $T = 300$ K.

4.14 The electron mobility in a silicon sample is determined by the Coulomb ($\mu_C = A_c T^{3/2}$) and the phonon ($\mu_{ph} = A_p T^{-3/2}$) scattering mechanisms, where $A_c = 1.85 \times 10^{-4}$ m²/K$^{3/2}$V · s and $A_p = 650$ m²K$^{3/2}$/V · s.

(a) Determine the room-temperature value of the diffusion coefficient.

(b) Determine the value of the diffusion coefficient at the liquid-nitrogen temperature ($T = 77$ K). **A**

(c) Determine the maximum value of the diffusion coefficient and the temperature at which this value is observed.

4.15 The concentration profile of ion-implanted boron atoms in silicon can be approximated by the Gaussian

$$N(x) = N_{max}e^{-(x-x_0)^2/2\sigma^2}$$

where $x_0 = 0.3$ μm, $\sigma = 0.07$ μm, and $N_{max} = 10^{18}$ cm^{-3}. After the ion implantation, the sample is heated to a high temperature to activate and diffuse the implanted boron atoms. At the selected diffusion temperature, the diffusion coefficient of boron in silicon is $D = 1.54 \times 10^{-14}$ cm²/s. As the diffusion process begins, determine

(a) the current density of boron atoms (in s^{-1} cm^{-2}) at $x = x_0 - \sigma$, $x = x_0$, and $x = x_0 + \sigma$

(b) the rate of concentration change ($\partial N/\partial t$) at $x = x_0$, $x = x_0 + \sigma$, and $x = x_0 + 2\sigma$

Explain why the obtained values are negative, positive, or zero.

4.16 A P-type silicon is covered by a metal that totally reflects any incident light. There is a narrow slit in the metal film, allowing a planelike generation of electron–hole pairs when the sample is illuminated. The distribution of hole concentration in the sample, following the exposure to a short light pulse, is given by

$$p(x,t) = \frac{\Phi}{\sqrt{2\pi}\sigma} e^{-x^2/2\sigma^2} + N_A$$

where x is the normal distance from the slit, $\sigma = \sqrt{2D_p t}$, t is the time after the light pulse, $D_p = 10$ cm²/s, and $\Phi = 5 \times 10^{13}$ cm^{-2}. Determine the changes of hole concentrations at $x = 10$ μm during the following time intervals after the light pulse:

(a) 10 ns $\pm\Delta t/2$

(b) 50 ns $\pm\Delta t/2$

(c) 100 ns $\pm\Delta t/2$ **A**

where $\Delta t = 1$ ns. Explain why the obtained values are negative, positive, or zero.

4.17 A constant electric field, normal to the surface of a semiconductor, establishes the following equilibrium distribution of electron concentration:

$$n(x) = 2N_D e^{-qEx/kT} + N_D \qquad (\text{for } x > 0)$$

where $x = 0$ is at the surface of the semiconductor, the electric field is $E = 100$ V/m, and $N_D = 10^{15}$ cm^{-3}. At time $t = 0$, the electric field is set to zero.

(a) Determine the current density at $x = 50$ μm, appearing immediately after the removal of the electric field.

(b) Determine the initial rate of concentration change at $x = 50$ μm. **A**

(c) Can the diffusion-current and continuity equations be used to determine the current density and the rate of concentration change at the surface of the semiconductor ($x = 0$)? If not, why not?

The diffusion coefficient is $D_n = 35$ cm²/s.

REVIEW QUESTIONS

R-4.1 What force causes the diffusion current?

R-4.2 Can charge-neutral particles make diffusion current? If so, what is the unit for the density of the particle-diffusion current?

R-4.3 If *drift* and *diffusion* are independent current mechanisms, why are their coefficients related?

R-4.4 Is the Einstein relationship, $D = V_t\mu$, applicable to any electron–hole gas in semiconductors?

R-4.5 Can the total current density of electrons be zero if there is a nonzero gradient of electron concentration?

R-4.6 If the concentration of particles in a considered point does not change in time, does this mean that the diffusion-current density is equal to zero? Does it mean that the total current density is equal to zero?

R-4.7 Can the currents of particles flowing into and out of a given elementary volume be different? If so, how can we account for the difference in the number of particles that flow into and out of the volume per unit time?

R-4.8 An electric field is applied to the surface of a semiconductor, causing a nonuniform distribution of electrons by attracting electrons to the surface. A steady-state concentration profile is established as the diffusion current balances the drift current. Then, liquid nitrogen is poured over the sample. Will this cause a current transient? If so, will the electrons move toward the surface or away from the surface?

5 Generation and Recombination

Electron–hole generation is a process that creates free electrons and holes. The opposite process, which results with annihilation of free electrons and holes, is called *recombination*. The balance of generation and recombination rates maintains the constant electron–hole concentrations in thermal equilibrium, as explained in Section 1.2.4. When the carrier concentrations are increased above or reduced below the thermal-equilibrium levels by external factors, the recombination and generation processes act to bring the system back into equilibrium. This action of the recombination and generation processes could be accompanied by a current flow (typically due to diffusion) to simultaneously cause changes in the carrier concentrations ($\partial n/\partial t \neq 0$, $\partial p/\partial t \neq 0$). Therefore, recombination and generation terms have to be added to the basic continuity equation of Section 4.3 to obtain its general form.

Following an introduction to the generation and recombination mechanisms, this chapter describes the generation and recombination rates as they appear in the general form of the continuity equation. This is followed by an in-depth generation–recombination physics and models for the effective recombination rate. Finally, a section is devoted to surface recombination, due to its practical importance.

5.1 GENERATION AND RECOMBINATION MECHANISMS

The generation and recombination processes involve absorption and release of energy. The generation of carriers due to thermal energy is called *spontaneous generation,* whereas generation caused by absorption of light is called *external generation*. In the case of *spontaneous recombination,* the released energy can be in the form of either heat or light. Nonspontaneous recombination, also known as *stimulated recombination,* may occur

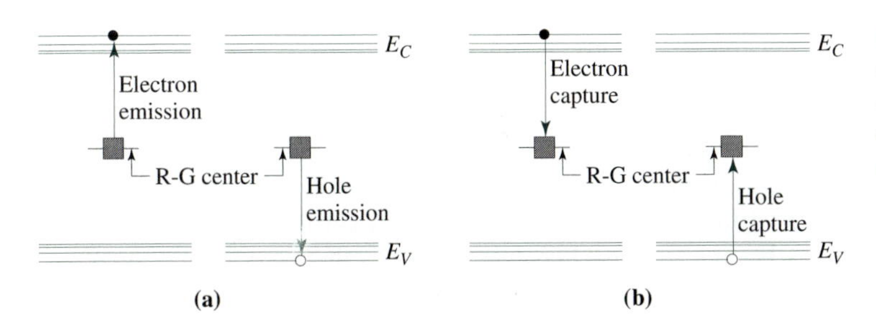

Figure 5.1 (a) Electron and hole emissions from an R–G center and (b) electron and hole captures by an R–G center lead to indirect generation and recombination, respectively.

when light of appropriate wavelength initiates recombination events to cause emission of additional photons with the same wavelength (the laser effect).[1]

Direct or *band-to-band* generation and recombination are conceptually the simplest mechanisms. In the generation process, electrons move from the valence band to the conduction band, which means that each generation event creates a pair of a free electron and a hole. In the recombination process, electrons drop from the conduction band to the valence band, so each recombination event annihilates a pair of a free electron and a hole. In direct semiconductors (E_C and E_V appear for the same wave vector $\boldsymbol{k}$), the momentum of an electron recombining with a hole does not change, so the electron energy is typically given away as a photon (radiative recombination). In the case of indirect semiconductors (E_C and E_V appear for different wave vectors $\boldsymbol{k}$), a recombination event necessitates a change in both electron energy and electron momentum. Accordingly, phonons are typically involved in the recombination process, and the energy is typically released to the phonons (nonradiative recombination).

It was explained in Section 2.2.3 that the doping atoms introduce energy levels into the energy gap. Analogously, other impurity atoms and crystal defects introduce energy levels into the energy gap as well. Figure. 5.1 illustrates that an impurity atom or a defect with the energy level in the energy gap can emit and capture electrons and holes. Two consecutive steps of electron and hole emissions, in either order, lead to generation of an electron–hole pair. Two consecutive steps of electron and hole captures lead to recombination of an electron–hole pair. This is *indirect* recombination–generation, also called R–G center recombination–generation, where the R–G centers are impurity atoms or defects with energy levels in the energy gap. The energy needed for indirect generation is typically supplied by the phonons (thermal energy) and also phonons are released during indirect recombination (nonradiative recombination).

Two consecutive steps of electron and hole emissions can also be described as two consecutive steps of electron jumps—for example, (1) an electron jump from the valence band onto the R–G center (this is equivalent to hole emission) and (2) an electron jump from the R–G center into the conduction band (the electron emission). An analogous description can be developed for the recombination process: (1) a drop of an electron from the conduction band to the level of the R–G center and (2) a drop of an electron from the R–G center to the valence band. With this view, the R–G centers are sometimes described as "stepping-stones" in analogy with the use of stepping-stones to jump across

[1]The conditions needed for the stimulated recombination are considered in Section 12.3.

large *space* distances in smaller steps. This analogy, however, can be misleading.[2] Based on this analogy, it may seem that generation–recombination by multiple energy levels in the energy gap (the use of more than one "stepping-stone" in the analogy) is much easier and therefore more likely than recombination–generation through a single R–G center (one "stepping-stone"). In reality, however, electron transitions from an R–G center to another R–G center that may have a different energy level in the energy gap (two "stepping-stones") are very unlikely. This is because the R–G centers are states that are not only *localized* in space but also too far apart—in terms of space, not energy—to allow any transition from one R–G center to another. The concept of *isolated* centers is valid if the distance between them is such that any overlap between the electron wave functions, associated with neighboring centers, is insignificant. Assuming that 10 nm is a large enough distance, we can easily work out that R–G centers placed at distance of 10 nm in a cubic-like crystal define a unit cell with volume of 10^{-24} m^3. This means that there can be as many as $1/10^{-24} = 10^{24}$ R–G centers per m^3 and yet, they are still electronically isolated from one another. In other words, the concentration of defects or impurity atoms can be as high as 10^{18} cm^{-3}, and electron transitions between them are still insignificant. Therefore, the transitions that are of practical importance are the transitions between the energy bands and the energy level of the R–G centers (a single "stepping-stone").

A related question here is the case of a very high concentration of impurity atoms or defects—for example, $>10^{20}$ cm^{-3}. In that case, the overlap of wave functions of electrons at neighboring impurity atoms or defects is strong, and by the force of the Pauli exclusion principle it causes a split in the energy levels to create an energy band. Therefore, electron transitions from either the conduction or the valence band to a band associated with the impurity atoms or defects is better described by the model of direct recombination–generation. There can be isolated R–G centers helping the transitions from either the valence or the conduction band to the band of high-concentration defects or impurity atoms, but this is no different from the model of a single "stepping-stone."

A third type of generation–recombination mechanism is illustrated in Fig. 5.2. The carrier generation in this case is due to high kinetic energy that the carriers can gain when accelerated by an electric field. Figure 5.2a illustrates an electron that gains kinetic energy $E_{kin} = -qEl_{sc}$ between two scattering events, where l_{sc} is the scattering length and E is the electric field (the slope of the energy bands). When the electron hits an atom, it may break a covalent bond to generate an electron–hole pair, if the kinetic energy is larger than the energy needed to generate the pair. This mechanism of electron–hole generation is called *impact ionization*. Typically, the process continues with the newly generated electrons, leading to *avalanche generation* of electrons and holes (Fig. 5.2a). The opposite process of impact ionization is *Auger recombination* (pronounced "oh-zhay"). In semiconductors with high concentration of carriers (for example, in highly doped regions), the carrier–carrier scattering becomes pronounced. During the collision of two carriers, one of the carriers may give its energy to the other carrier (Fig. 5.2b). The carrier that loses the energy is either trapped by an R–G center or recombined by a minority carrier. The carrier that gains the energy, appearing with a high kinetic energy initially, will typically lose it as heat (thermal energy) in subsequent phonon-scattering events.

[2]The stepping-stone analogy is about transitions in *space*. Recombination and generation transitions are in *energy*, but importantly, *space* cannot be ignored to enable the use of a simple space-to-energy analogy.

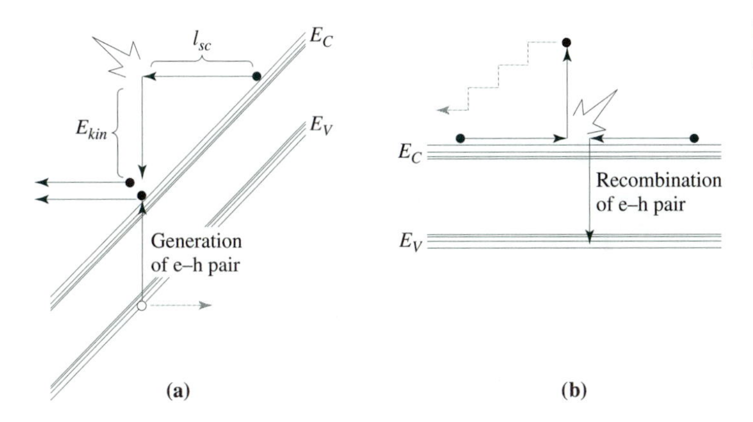

Figure 5.2 (a) Avalanche generation and (b) Auger recombination.

5.2 GENERAL FORM OF THE CONTINUITY EQUATION

The continuity equations for electrons and holes express the changes of electron and hole concentrations in time. These changes can be due to carrier motion (drift and diffusion), but can also be due to carrier generation and recombination. Accordingly, recombination and generation *rates* (the concentrations of carriers that are recombined and generated per unit time) have to be included in the continuity equation. In this section, the effective thermal (or spontaneous) generation–recombination rate and the external-generation rate are defined. After that, the typically used form of the effective thermal generation–recombination term is introduced to define the *minority-carrier lifetime* and the *diffusion length.*

5.2.1 Recombination and Generation Rates

The basic continuity equation, introduced in Section 4.3, is for the case of no carrier recombination or generation. Figure 5.3a illustrates this case: any change of the electron concentration ($\partial n/\partial t \neq 0$) in the space element indicated by the dashed rectangle is due to the difference in electron current densities flowing into and out of the space element ($\partial j_n/\partial x \neq 0$). To include the effects of recombination, consider Fig. 5.3b. In this case the electron concentration changes not only because more electrons enter into than come out of the considered space element, but also because some of the electrons are recombined inside the element. The recombination contribution to the change of electron concentration is exactly equal to the recombination rate, in units of concentration by time: $\text{m}^{-3}\,\text{s}^{-1}$.

Before we include the recombination rate in the continuity equation, let us consider the following "idea." Assume a P-type semiconductor with some concentration of electrons (minority carriers) and some (does not matter how small) recombination rate. As electron by electron is recombined, after some time (does not matter how long) all the electrons would disappear if electron generation did not occur. This thinking shows that recombination is inseparable from the opposite process, thermal generation. As emphasized in Section 1.2.4, the recombination and thermal-generation rates are equal in thermal equilibrium, so that the concentration of minority carriers remains constant. If *excess* carrier concentration is created, then the recombination rate exceeds the thermal generation rate so that the excess carrier concentration is reduced (as in Fig. 5.3b). Likewise, if *deficiency* of the carriers is created, the thermal-generation rate will exceed

(a)

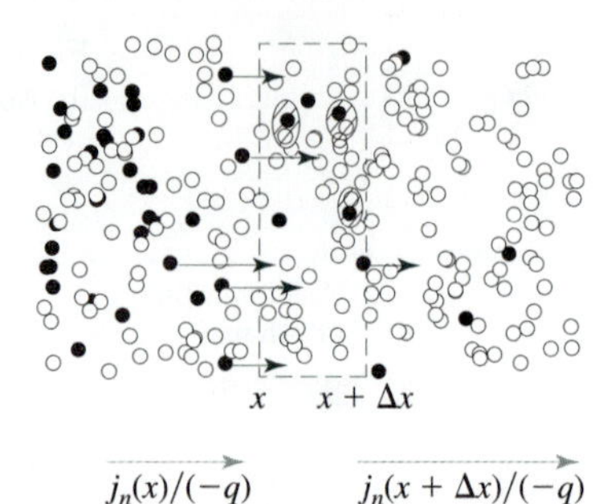

(b)

Figure 5.3 Illustration of the continuity equation (a) without effective recombination and (b) with the effective recombination term included.

the recombination rate, increasing the carrier concentration toward the equilibrium level. It is quite convenient to define *the difference between the recombination and the thermal generation rates* as *effective thermal generation–recombination rate*. If U is this effective rate, $U = 0$ represents the thermal equilibrium case of no effective change in the carrier concentrations. A positive effective thermal recombination rate ($U > 0$) indicates that recombination mechanisms prevail, reducing the excess carrier concentration toward the equilibrium level. Likewise, $U < 0$ indicates that thermal generation mechanisms prevail, increasing the carrier concentration toward the equilibrium level.

The need to bring the system into equilibrium arises because external factors can take it out of equilibrium. For example, application of an electric field may cause a sudden depletion of the carriers. As opposed to this, external carrier generation (due to mechanisms such as light absorption and impact ionization) may cause excess concentrations of the carriers. External generation usually extends over time and is usually nonuniform, which means that it has to be expressed as a time- and space-dependent function. If the *external generation rate* (G_{ext}) is defined to represent the overall effect of the external generation mechanisms, it can simply be added as an additional term in the continuity equation.

With the addition of the effective thermal generation–recombination rate (U) and the external generation rate (G_{ext}), the general form of the continuity equation is obtained:

$$\begin{aligned}\frac{\partial n}{\partial t} &= \frac{1}{q}\frac{\partial j_n}{\partial x} - U + G_{ext}\\ \frac{\partial p}{\partial t} &= -\frac{1}{q}\frac{\partial j_p}{\partial x} - U + G_{ext}\end{aligned} \tag{5.1}$$

The general three-dimensional continuity equations for the electrons and holes are

$$\frac{\partial n}{\partial t} = \frac{1}{q}\left(\frac{\partial \boldsymbol{j_n}}{\partial x}\boldsymbol{x_u} + \frac{\partial \boldsymbol{j_n}}{\partial y}\boldsymbol{y_u} + \frac{\partial \boldsymbol{j_n}}{\partial z}\boldsymbol{z_u}\right) - U + G_{ext} = \frac{1}{q}\nabla \boldsymbol{j_n} - U + G_{ext} \tag{5.2}$$

$$\frac{\partial p}{\partial t} = -\frac{1}{q}\left(\frac{\partial \boldsymbol{j_p}}{\partial x}\boldsymbol{x_u} + \frac{\partial \boldsymbol{j_p}}{\partial y}\boldsymbol{y_u} + \frac{\partial \boldsymbol{j_p}}{\partial z}\boldsymbol{z_u}\right) - U + G_{ext} = -\frac{1}{q}\nabla \boldsymbol{j_p} - U + G_{ext} \tag{5.3}$$

When one is using the continuity equations, G_{ext} appears as a time- and space-dependent function.[3] U can also vary in time and space, but additionally, it depends on the instantaneous values of the electron and hole concentrations. It is necessary to establish the dependence of U on the electron and hole concentrations to be able to use the continuity equation. The next section describes the simplest and the most commonly used form of this dependence. A more general model is considered in Section 5.3.

5.2.2 Minority-Carrier Lifetime

Because both electrons and holes are needed for recombination, the recombination rate is directly proportional to the concentrations of available electrons and available holes. Given that the *effective* recombination rate is zero in thermal equilibrium, when the product of the electron and hole concentrations is equal to n_i^2, it can be expressed as

$$U = \alpha_r \left(np - n_i^2\right) \tag{5.4}$$

where α_r is the proportionality constant. U is positive for $np > n_i^2$ because the recombination mechanisms prevail (as they act to reduce np to n_i^2), and U is negative for $np < n_i^2$ because the generation mechanisms prevail (as they act to increase np to n_i^2). The electron and hole concentrations can be expressed as sums of the equilibrium concentrations (n_0 and p_0) and *excess* concentrations, labeled by δn and δp:

$$\begin{aligned} n &= n_0 + \delta n \\ p &= p_0 + \delta p \end{aligned} \tag{5.5}$$

With this, the equation for U becomes

$$U = \alpha_r (n_0 \delta p + p_0 \delta n + \delta n \delta p) \tag{5.6}$$

If we assume an N-type semiconductor ($p_0 \ll n_0$) and a *small deviation from equilibrium*, so that $|\delta n| \ll n_0$, the effective recombination rate can be approximated by

$$U = \alpha_r n_0 \delta p \tag{5.7}$$

Given that U has to be expressed in the units of concentration by time, Eq. (5.7) can be written in the following form

$$U = \frac{\delta p}{\tau_p} \tag{5.8}$$

where $\tau_p = 1/\alpha_r n_0$ is a time constant that replaces the constant α_r.

To develop a physical meaning for the constant τ_p, let us consider the example of external generation of minority carriers (for example, due to absorption of light) that increases the excess minority-carrier concentration to $|\delta p(0)| \ll n_0$ and then suddenly

[3] Obviously, this function has to be determined so as to properly represent the carrier-generation rate due to the external causes (light absorption, impact ionization, etc.).

stops at time $t = 0$. The continuity equation for holes can be used to determine how the excess hole concentration $\delta p(t)$ decays to zero for $t > 0$, as the effective recombination acts to bring the system back to equilibrium. Assuming no space variations of the electron and hole concentrations, the continuity equation for holes can be simplified to

$$\frac{dp(t)}{dt} = -\frac{\delta p(t)}{\tau_p} \tag{5.9}$$

This equation is equivalent to

$$\frac{d\delta p(t)}{dt} = -\frac{\delta p(t)}{\tau_p} \tag{5.10}$$

because $dp/dt = d(p_0 + \delta p)/dt = d\delta p/dt$. The solution of Eq. (5.10) is

$$\delta p(t) = \delta p(0)e^{-t/\tau_p} \tag{5.11}$$

Therefore, the excess minority-carrier concentration decays exponentially, with the time constant τ_p. Accordingly, the time constant τ_p is referred to as the *excess carrier lifetime*. If we attempted to apply the concept of τ_p to a single minority carrier, we would come to the conclusion that it represents the average lifetime of the minority carrier. Therefore, τ_p can be considered as the minority-hole lifetime, but only under the considered condition that $|\delta p| \ll n_0$, which in other words means that the minority carriers are surrounded by abundant majority carriers.

Analogous conclusions can be derived for the case of a P-type semiconductor with a small deviation from equilibrium ($|\delta n| \ll p_0$). In that case, the minority-electron lifetime (τ_n) shows how quickly the system will spring back to equilibrium when any external disturbance is removed:

$$\delta n(t) = \delta n(0)e^{-t/\tau_n} \tag{5.12}$$

EXAMPLE 5.1 Effective Thermal Generation

The minority-carrier lifetime and the doping level of an N-type region in a silicon device are $\tau_p = 10\ \mu s$ and $N_D = 10^{16}\ cm^{-3}$, respectively. An applied electric field causes depletion of both the majority and the minority carriers. At time $t = 0$, the electric field is removed and the thermal generation acts to bring the silicon region back into thermal equilibrium. If direct generation dominates, it can be assumed that the concentration of electron–hole pairs generated per unit time does not change if majority and minority carriers are suddenly depleted.[4] Under these conditions, estimate the effective generation rates for the cases of

[4]This is because the generation mechanism does not depend on the availability of free electrons and holes. In the case of Shockley–Read–Hall generation (Section 5.3), this assumption cannot be used because the associated depletion of the G–R centers changes the generation conditions.

(a) lightly depleted region: $\delta p/p_0 = 0.1$ and $\delta n/n_0 = 0.1$
(b) fully depleted region: $\delta p = p - p_0 \approx -p_0$ and $\delta n = n - n_0 \approx -n_0$

SOLUTION

Replacing the proportionality constant α_r by the minority-carrier lifetime ($\alpha_r = 1/\tau_p n_0$), Eq. (5.6) takes the following form:

$$U = \frac{n_0 \delta p + p_0 \delta n + \delta n \delta p}{n_0 \tau_p}$$

(a) *Light Depletion.* The equilibrium concentrations of the majority and minority carriers are $n_0 = N_D = 10^{16}$ cm^{-3} and $p_0 = n_i^2/n_0 = 10^4$ cm^{-3}. The concentrations of the depleted carriers are then $\delta n = -0.1 n_0 = -10^{15}$ cm^{-3} and $\delta p = -0.1 p_0 = -10^3$ cm^{-3}. Therefore,

$$U = \frac{-10^{16} \times 10^3 - 10^4 \times 10^{15} + 10^{15} \times 10^3}{10^{16} \times 10^{-5}} = -1.9 \times 10^8 \text{ cm}^{-3}\text{ s}^{-1}$$

(b) *Full Depletion.* In this case,

$$U = \frac{-n_0 p_0 - p_0 n_0 + n_0 p_0}{n_0 \tau_p} \approx -\frac{p_0}{\tau_p} = -10^9 \text{ cm}^{-3}\text{ s}^{-1}$$

EXAMPLE 5.2 Balance of External Generation and Recombination

(a) A slab of N-type silicon with the doping level $N_D = 10^{16}$ cm^{-3} and minority-carrier lifetime $\tau_p = 10$ μs is exposed to a light source. Assuming a uniform external generation rate of $G_{ext} = 5 \times 10^{18}$ cm^{-3} s^{-1}, determine the steady-state concentration of minority carriers.
(b) At time $t = 0$, the light source is switched off. Determine the effective recombination rate. Noting that the values of the doping level and the minority-carrier lifetime are the same as in Example 5.1, compare the rates of effective recombination and generation and comment on the results.

SOLUTION

(a) Mathematically, steady-state concentration of holes means that $\partial p/\partial t = 0$. Given that there is no current flow in the sample, we obtain from the continuity equation for the holes [Eq. (5.1)] that

$$G_{ext} = U$$

As expected, the external generation rate has to be balanced by the effective recombination rate to reach the steady-state condition. Assuming that the term $n_0 \delta p$ dominates in Eq. (5.6), the effective recombination rate is given by Eq. (5.8), so

$$G_{ext} = \frac{\delta p}{\tau_p}$$

From here, $\delta p = 5 \times 10^{18} \times 10^{-5} = 5 \times 10^{13}$ cm^{-3}. Therefore the total concentration of holes is

$$p = p_0 + \delta p \approx \delta p = 5 \times 10^{13} \text{ cm}^{-3}$$

Now we can verify the assumption that $n_0 \delta p$ term dominates in Eq. (5.6): $n_0 \delta p = 5 \times 10^{29}$ cm^{-6}, $p_0 \delta n = p_0 \delta p = 5.2 \times 10^{17}$ cm^{-6}, and $\delta n \delta p = \delta p^2 = 2.5 \times 10^{27}$ cm^{-6}.

(b) The effective recombination rate is

$$U = \frac{\delta p}{\tau_p} = 5 \times 10^{18} \text{ cm}^{-3}\text{ s}^{-1}$$

Comparing to the effective generation rates in Example 5.1, we see that the change of minority-carrier concentration is many orders of magnitude slower when the generation dominates ($|U| \leq 10^9$ cm^{-3} s^{-1} in Example 5.1) compared to a typical case when the recombination dominates ($U = 5 \times 10^{18}$ cm^{-3} s^{-1} in this example).

5.2.3 Diffusion Length

The concept of *diffusion length* characterizes the steady-state nonuniform distribution of excess minority carriers. Take the following example: electrons are injected into a P-type region at a constant rate so that a steady-state current of electrons is established. The concentration of electrons at the edge of the P-type region (taken as $x = 0$) is $n(0)$, but it decays inside the P-type region ($x > 0$) as the electrons are recombined by the majority holes. The task is to determine the profile of electron concentration, $n(x)$.

To begin with, note that we are considering a steady-state case, which means the electron concentration $n(x)$ does not change in time: $\partial n(x)/\partial t = 0$. Next, note that the current of the minority electrons j_n is the diffusion current given by Eq. (4.5). With $\partial n(x)/\partial t = 0$, $U = \delta n/\tau_n$, and j_n given by Eq. (4.5), the continuity equation takes the following form:

$$0 = D_n \frac{d^2 n(x)}{dx^2} - \frac{\delta n}{\tau_n} \tag{5.13}$$

This equation can be transformed into the following form:

$$\frac{d^2\delta n}{dx^2} - \frac{\delta n}{L_n^2} = 0 \tag{5.14}$$

where

$$L_n = \sqrt{D_n \tau_n} \tag{5.15}$$

is a constant with the unit of length. To solve Eq. (5.14), its characteristic equation is written as

$$s^2 - \frac{1}{L_n^2} = 0 \tag{5.16}$$

and solved, which gives the following two roots: $s_{1,2} = \pm(1/L_n)$. The general solution of Eq. (5.14) is then expressed as

$$\delta n(x) = A_1 e^{s_1 x} + A_2 e^{s_2 x} = A_1 e^{x/L_n} + A_2 e^{-x/L_n} \tag{5.17}$$

where A_1 and A_2 are integration constants to be determined from the boundary conditions. One boundary condition is $\delta n(\infty) = 0$, which turns the constant A_1 to zero [if A_1 was not zero, $\delta n(\infty)$ would be infinitely large, which is physically impossible]. Using the other boundary condition, which is $\delta n(0)$, shows that the constant A_2 has to be equal to $\delta n(0)$. With this, the solution of Eq. (5.14) is obtained as

$$\delta n(x) = \delta n(0) e^{-x/L_n} \tag{5.18}$$

Obviously, the excess electron concentration drops exponentially with the length constant L_n. The constant L_n is called the *diffusion length of minority electrons*. It indicates how deeply electrons can penetrate when injected into a P-type region. As Eq. (5.15) shows, the diffusion length depends on the diffusion coefficient and the excess carrier lifetime. For a shorter τ_n (because of a stronger recombination), the electrons are recombined closer to the injection point, so the diffusion length is shorter. For a smaller D_n, a larger gradient of electron concentration is needed so that the diffusion current can balance the injection current, and a larger concentration gradient corresponds to a shorter diffusion length.

Analogously, the diffusion length of the minority holes is given by

$$L_p = \sqrt{D_p \tau_p} \tag{5.19}$$

†5.3 GENERATION AND RECOMBINATION PHYSICS AND SHOCKLEY–READ–HALL (SRH) THEORY

Equation (5.8) for the effective thermal generation–recombination rate was written intuitively. In a sense, it is a phenomenological equation. It does enable generation–recombination-related changes of electron and hole concentrations to be incorporated into the continuity equation, but it involves an empirical constant (either α_r or $\tau_{p,n}$) that is not related to known technological parameters and physical constants. An approach that is based on a more fundamental physics is needed to link the generation–recombination-related changes of n and p to specific technological parameters and physical constants. A theory that addresses this need was developed by Shockley and Read[5] and, independently of them, by Hall.[6] It has been widely accepted as the model for indirect recombination/generation and is known as the SRH theory. This theory and its results will be presented in this section.

5.3.1 Capture and Emission Rates in Thermal Equilibrium

To develop a physical model for the changes in n and p due to the generation and recombination mechanisms illustrated in Fig. 5.1, the four mechanisms (electron and hole emissions and captures) should be represented by their *rates*—how many electrons or holes are emitted or captured *per unit volume and per unit time*. The symbols that will be used for the four rates are as follows: $r_{c,n}$ for the rate of electron capture, $r_{c,p}$ for the rate of hole capture, $r_{e,n}$ for the rate of electron emission, and $r_{e,p}$ for the rate of hole emission. All these rates are in units of $\text{m}^{-3}\,\text{s}^{-1}$. The first steps in developing the theory are to establish physically based equations for these rates.

To develop the equation for $r_{c,n}$, let us begin the consideration with a single electron in the conduction band. As this electron moves with the thermal velocity, it has a chance to hit some of the R–G centers that exist in the material. This chance is determined by the thermal velocity of the electron (v_{th}), the capture cross section of the R–G centers (σ_n), and the concentration of the R–G centers (N_t).[7] In analogy with the description of Eq. (3.26) for the scattering time, the product of the capture cross section and the concentration of R–G centers, $\sigma_n N_t$, is the probability that the moving electron will hit an R–G center per unit length. Given that the length traveled per unit time is the thermal velocity, the probability that a carrier will come in contact with an R–G center *per unit time* is

$$p_1 = v_{th}\sigma_n N_t \tag{5.20}$$

Being in contact with an R–G center, an event characterized by the probability p_1, is not equivalent to electron capture. If the R–G center is already filled by an electron, then it will not capture the considered electron. To obtain the probability per unit time that the electron will be captured, p_1 has to be multiplied by the probability that the trap is empty. This probability is $1 - f_t$, where f_t is the value of the Fermi–Dirac distribution at the energy of the R–G center. Therefore,

$$p_2 = p_1(1 - f_t) = v_{th}\sigma_n N_t(1 - f_t) \tag{5.21}$$

---

[5]W. Shockley and W. T. Read, Statistics of the recombination of holes and electrons, *Phys. Rev.*, vol. 87, p. 835 (1952).

[6]R. N. Hall, Electron–hole recombination in germanium, *Phys. Rev.*, vol. 87, p. 387 (1952).

[7]The index t comes from the term *trap,* which is used for R–G centers that capture carriers.

The step that remains is to convert the probability per unit time that a single electron will be captured (p_2) to the concentration of electrons captured per unit time ($r_{c,n}$). To this end, the probability p_2 is multiplied by the concentration of electrons that have a chance of being captured:

$$r_{c,n} = v_{th}\sigma_n N_t(1 - f_t)n_0 \tag{5.22}$$

Obviously, n_0 is the concentration of electrons in the conduction band.

The analogous equation for holes is

$$r_{c,p} = v_{th}\sigma_p N_t f_t p_0 \tag{5.23}$$

where σ_p is the capture cross section for holes.

Turning to the emission rate, the SRH theory does not develop an equation for the probability per unit time that an electron will be emitted from a considered R–G center. Thus, there is no equation that is analogous to the equation for p_2 in the case of capturing process. The SRH theory introduces this probability as a parameter e_n that is to be determined so that the emission rate is equal to the capture rate in thermal equilibrium, $r_{e,n} = r_{c,n}$.

The emission probability e_n is converted into the concentration of emitted electrons per unit time in a way that is fully analogous to the case of the capturing process. Specifically, e_n is multiplied by the concentration of electrons that have a chance of being emitted into the conduction band. This concentration of electrons is equal to the concentration of filled R–G centers, $N_t f_t$. Therefore,

$$r_{e,n} = e_n N_t f_t \tag{5.24}$$

The analogous equation for holes is

$$r_{e,p} = e_p N_t(1 - f_t) \tag{5.25}$$

where e_p is the probability that a hole will be emitted from an R–G center into the valence band per unit time.

To satisfy the thermal equilibrium condition, the emission and capture rates have to be equal for both electrons and holes. This is frequently referred to as the *detailed balance principle*. From the condition $r_{e,n} = r_{c,n}$, the emission coefficient e_n is determined as

$$e_n = v_{th}\sigma_n n_0 \frac{1 - f_t}{f_t} \tag{5.26}$$

Given that

$$f_t = \frac{1}{1 + e^{(E_t - E_F)/kT}} \tag{5.27}$$

the equation for the emission coefficient can be expressed in the following way:

$$e_n = v_{th}\sigma_n n_0 e^{(E_t - E_F)/kT} \tag{5.28}$$

We can see from Eq. (5.28) that the emission probability increases exponentially with the increase in the energy level of the R–G centers (E_t).

The analogous equation for the hole emission rate is

$$e_p = v_{th}\sigma_p p_0 e^{-(E_t - E_F)/kT} \tag{5.29}$$

EXAMPLE 5.3 Emission and Capture Rates in Thermal Equilibrium

Determine the capture rates for electrons and holes in a P-type silicon ($N_A = 10^{18}$ cm^{-3}), having midgap R–G centers with concentration of $N_t = 10^{15}$ cm^{-3}. Assume that $\sigma_n = \sigma_p = 10^{-15}$ cm^2 and that the room-temperature thermal velocity of both electrons and holes is equal to $v_{th} = 10^7$ cm/s. Compare and discuss the results. What are the electron and hole emission rates?

SOLUTION

The concentration of electrons as minority carriers is $n_0 = n_i^2/N_A = 104$ cm^{-3}. Because this is a P-type silicon, the Fermi level is well below the midgap and $f_t \ll 1$. With this, the electron capture rate is calculated from Eq. (5.22) as

$$r_{c,n} \approx v_{th}\sigma_n N_t n_0 = 10^9 \text{ cm}^{-3}\,\text{s}^{-1}$$

To determine the hole capture rate using Eq. (5.23), we need to determine the value of f_t. The position of the midgap with respect to the Fermi level is

$$q\phi_F = kT \ln(N_A/n_i) = 0.476 \text{ eV}$$

Therefore,

$$f_t = \frac{1}{1 + e^{(E_t - E_F)/kT}} = \frac{1}{1 + e^{\phi_F/V_t}} = 1.0 \times 10^{-8}$$

$$r_{c,p} = v_{th}\sigma_p N_t f_t N_A = 1.0 \times 10^{17} \text{ cm}^{-3}\,\text{s}^{-1}$$

Comparing $r_{c,p}$ and $r_{c,n}$, we see that the capture rate of holes is eight orders of magnitude higher. The high capture rate of holes is balanced by the hole emission rate: $r_{e,p} = r_{c,p} = 1.0 \times 10^{17}$ cm^{-3} s^{-1}. Analogously, the electron capture rate is balanced by the electron emission rate: $r_{e,n} = r_{c,n} = 10^9$ cm^{-3} s^{-1}. The interpretation of the high capture rate for holes can be related to the fact that the concentration of holes is so much higher than the concentration of electrons.

The interpretation of much higher emission rate for holes can be related to the fact tht the concentration of holes at the R–G centers is much higher than the concentration of electrons, $(1 - f_t)N_t \gg f_t N_t$.

EXAMPLE 5.4 Capture Cross Section

Donor-type R–G centers are due to impurity atoms or defects that can be either positively charged or neutral. As in Example 3.10 for the case of scattering, the capture cross section of a donor ion can be related to the spherical region where the thermal energy of an electron is smaller than the energy associated with the Coulomb attraction. This is relevant for the capture cross section of electrons (σ_n), given that positive donor ions capture electrons. In the case of capture cross section of holes (σ_p), it is the cross section of the neutral donor atom that is relevant, given that neutral donor atoms capture holes to become positive. Estimate and compare the values for σ_n and σ_p if the atom radius is 0.175 nm. Draw a conclusion for the relationship between σ_n and σ_p for the case of acceptor-type R–G centers.

SOLUTION

For the case of σ_n, when a positively charged donor captures an electron by Coulomb attraction, the solution is the same as in Example 3.10a. Therefore, $\sigma_n = 3.1 \times 10^{-13}$ cm^2. For the case of σ_p, it is the cross section of the neutral atom that matters:

$$\sigma_p = \pi r^2 = \pi \times (0.175 \times 10^{-9})^2 = 9.6 \times 10^{-20}\ \text{m}^2 = 9.6 \times 10^{-16}\ \text{cm}^2$$

We can see that σ_n is $3.1 \times 10^{-13}/9.6 \times 10^{-16} = 323$ times larger than σ_p.

Acceptor-type R–G centers are negative when filled with electrons and are neutral when empty. Therefore, neutral centers capture electrons to become negative, which means that σ_n corresponds to the cross section of the atoms/defects that cause the R–G centers. As opposed to this, negative acceptor ions capture holes, which means that σ_p corresponds to the area of effective Coulomb attraction. This means $\sigma_p \gg \sigma_n$ for the case of acceptor-type R–G centers.

5.3.2 Steady-State Equation for the Effective Thermal Generation–Recombination Rate

The next step in the SRH theory is to assume that the equilibrium equations for the capture and emission rates retain their mathematical forms for nonequilibrium cases. This approach is frequently utilized to model nonequilibrium processes by equations developed for equilibrium, and it is generally shown to be valid when the deviations from equilibrium are sufficiently small. Because this specific case is about the change of nonequilibrium electron and hole concentrations due to recombination–generation, the equilibrium concentrations

n_0 and p_0 have to be replaced by their nonequilibrium counterparts, $n = n_0 + \delta n$ and $p = p_0 + \delta p$. Importantly, the SRH theory does this for the case of capture rates only. Therefore, Eqs. (5.22) and (5.23) become

$$r_{c,n} = v_{th}\sigma_n N_t (1 - f_t) n \tag{5.30}$$

$$r_{c,p} = v_{th}\sigma_p N_t f_t p \tag{5.31}$$

No replacement of the equilibrium variables is performed for the emission rates. It will be pointed out later that the theory collapses if nonequilibrium values are consistently used in the equations for both emission and capture rates. Therefore, Eqs. (5.30) and (5.31) for the nonequilibrium capture rates and Eqs. (5.24), (5.25), (5.28), and (5.29) for the equilibrium emission rates will be used in the further derivation. In the original derivation, n_0 and p_0 in Eqs. (5.28) and (5.29) are replaced by the Maxwell–Boltzmann *approximations* given by Eqs. (2.85) and (2.86). At this stage we will retain the symbols n_0 and p_0 for the more general case to show that the Maxwell–Boltzmann approximation is not necessary.

In addition to n and p, another variable that changes its value under nonequilibrium conditions is the probability that an R–G center is occupied by an electron, f_t. In the case of equilibrium, this value can be obtained from the Fermi–Dirac distribution, which was used to convert the emission rate from the form given by Eq. (5.26). As already explained, the SRH theory uses the equilibrium equations for the emission rates; thus it uses the Fermi–Dirac distribution for f_t and the equilibrium electron and hole concentrations. As opposed to this, nonequilibrium values are used in the equations for the capture rates. Then it become necessary to determine the nonequilibrium value for f_t. The solution used in the SRH theory is to obtain it from a condition that can be established for a steady-state case.

Steady-state conditions are special nonequilibrium cases. Under steady-state conditions, the nonequilibrium electron and hole concentrations, n and p, do not change in time: $dn/dt = dp/dt = 0$. This is possible when the effective thermal recombination rate is perfectly balanced by a constant external generation rate. For example, the effective recombination rate for electrons is given by the difference between the electron capture and emission rates:

$$U = r_{c,n} - r_{e,n} \tag{5.32}$$

In thermal equilibrium $U = 0$, so $U > 0$ is a nonequilibrium case. For $U > 0$, the rate of electron capture exceeds the rate of electron emission, which means that the thermal processes reduce the concentration of electrons in the conduction band. This reduction can be perfectly balanced by an increase in electron concentration due to external generation—for example, by light. A specific generation mechanism, which is practically most important, is selected for the external generation. The mechanism is direct, or band-to-band, generation, where the rate of external electron generation is directly linked to the rate of external hole generation. Given that external generation increases the hole concentration, there has to be a difference between the rates of hole capture and thermal emission that balances the increase due to external generation. Thus, the effective recombination rate for holes is also equal to U:

$$U = r_{c,p} - r_{e,p} \tag{5.33}$$

Equations (5.32) and (5.33) establish the following steady-state condition:

$$r_{c,n} - r_{e,n} = r_{c,p} - r_{e,p} \tag{5.34}$$

This is the condition used by the SRH theory to determine the nonequilibrium f_t. Inserting the emission and capture rates from Eqs. (5.28), (5.29), (5.30), and (5.31) into Eq. (5.34), we obtain the following equation for the nonequilibrium f_t:

$$f_t = \frac{\sigma_n n + \sigma_p p_0 e^{-(E_t - E_F)/kT}}{\sigma_n \left[n + n_0 e^{(E_t - E_F)/kT}\right] + \sigma_p \left[p + p_0 e^{-(E_t - E_F)/kT}\right]} \tag{5.35}$$

Inserting f_t back into Eq. (5.30), and inserting the obtained electron-capture rate with the emission rate given by Eqs. (5.24) and (5.28) into Eq. (5.32), we obtain the following result for the effective recombination rate:

$$U = \frac{\sigma_n \sigma_p v_{th} N_t (np - n_0 p_0)}{\sigma_n \left[n + n_0 e^{(E_t - E_F)/kT}\right] + \sigma_p \left[p + p_0 e^{-(E_t - E_F)/kT}\right]} \tag{5.36}$$

This equation is for a steady-state nonequilibrium case, but it does contain equilibrium variables: n_0, p_0, and E_F. It is obvious from the derivation that these equilibrium variables originate from the use of equilibrium equations for the emission probabilities [Eqs. (5.28) and (5.29)]. It is quite clear that if the equilibrium concentrations n_0 and p_0 were to be replaced by the nonequilibrium values n and p in the equations for the emission rates, the term $np - n_0 p_0$ in Eq. (5.36) would become $np - np$. Therefore, the asymmetrical use of nonequilibrium and equilibrium values in the equations for capture and emission rates is necessary to avoid this collapse of the theory, where $U = 0$ for the nonequilibrium steady-state case.

By using the Maxwell–Boltzmann approximations for n_0 and p_0 [Eqs. (2.85) and (2.86)], all the equilibrium values, n_0, p_0, and E_F, can be converted into material constants:

$$U = \frac{\sigma_n \sigma_p v_{th} N_t \left(np - n_i^2\right)}{\sigma_n \left[n + n_i e^{(E_t - E_i)/kT}\right] + \sigma_p \left[p + n_i e^{-(E_t - E_i)/kT}\right]} \tag{5.37}$$

This is the general equation for steady-state nonequilibrium rate of effective thermal generation–recombination in the SRH theory. In this equation, the term $np - n_i^2$ is significant, given that it provides a model for the dependence of the effective thermal generation–recombination rate U on the degree of deviation from the equilibrium. $U = 0$ for equilibrium ($np = n_i^2$), but then it increases with increasing difference between the product of nonequilibrium n and p and the equilibrium level n_i^2. For negative $np - n_i^2$, we have the case of electron–hole deficiency and negative U, indicating that thermal generation prevails over the recombination as the thermal processes work to bring the system back into equilibrium.

The term $np - n_i^2$ also appears in the empirical equation (5.4) used in Section 5.2. Comparing Eqs. (5.37) and (5.4), we conclude that the empirical proportionality coefficient α_r involves the following technological and physical parameters: N_t, E_t, $\sigma_{n,p}$, v_{th}, and n_i. This comparison also shows that the coefficient α_r depends on the nonequilibrium concentrations of electrons and holes, n and p, which means that α_r is not a constant when n and p change over a wide range.

EXAMPLE 5.5 Dependence of the Effective Thermal Generation and Recombination Rates on the Energy Position of the R–G Centers

For N-type silicon with $N_D = 10^{15}$ cm^{-3} and $N_t = 10^{15}$ cm^{-3}, plot the dependencies of U on $E_t - E_i$ for the following cases:

(a) Low-level injection of excess carriers: $\delta n = \delta p = 10^{12}$ cm^{-3} (this is 0.1% of n_0)
(b) Slight depletion: $\delta n = -10^{12}$ cm^{-3}, $p = 0$
(c) Full depletion: $n = p = 0$

Use the capture cross section values obtained in Example 5.4, the thermal velocity $v_{th} = 10^7$ cm/s, and $E_t - E_i$ in the range of -0.56 eV to 0.56 eV with the energy step of 0.01 eV. Simplify Eq. (5.37) for each of these cases to perform an analysis of the obtained graphs and comment on the results.

SOLUTION

The following are the corresponding MATLAB® scripts followed by the comments. The plots are shown in Figs. 5.4, 5.5, and 5.6 for low-level injection, slight depletion, and full depletion, respectively.

(a) *Low-level injection of excess carriers.*

```
>>kT=8.62e-5*300;
>>ni=1.02e16;
>>vth=1e5;
>>sigman=3.1e-17;
>>sigmap=9.6e-20;
>>Nd=1e21;
>>Nt=1e21;
>>n=Nd+1e18;
>>p=Nd/ni^2+1e18;
>>EtEi=[-0.56:0.01:0.56];
>>U=sigman*sigmap*vth*Nt*(n*p-ni^2)./
        (sigman*(n+ni*exp(Eti./kT))+sigmap*(p+ni*exp(-Eti./kT)));
>>plot(EtEi,U*1e-6)
>>xlabel('E_t-E_i (eV)')
>>ylabel('U(cm^{-3}s^{-1})')
```

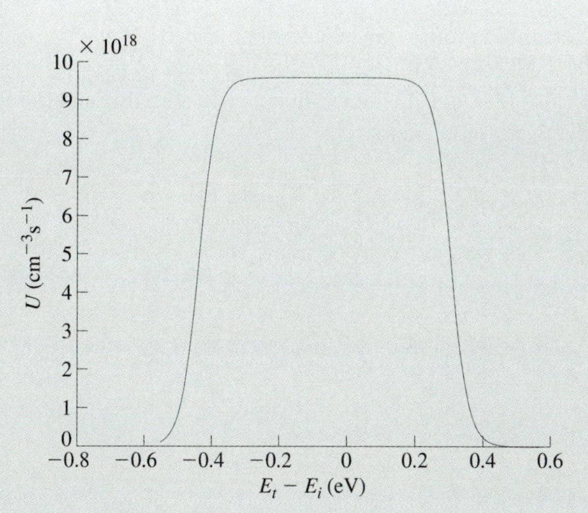

Figure 5.4 The MATLAB plot for Example 5.5a.

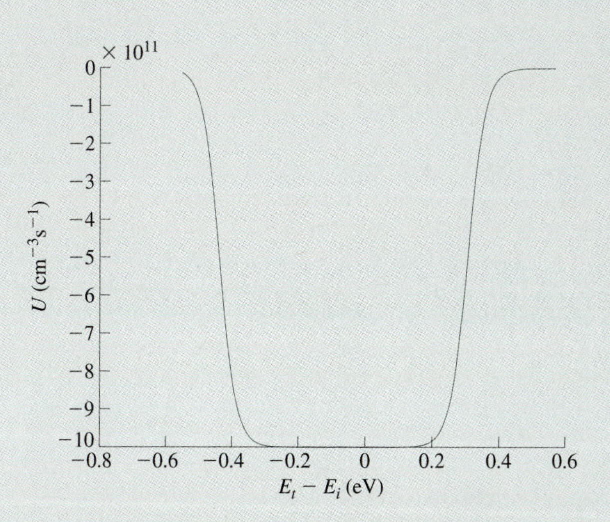

Figure 5.5 The MATLAB plot for Example 5.5b.

The plot in Fig. 5.4 shows that the effective recombination rate is constant for R–G centers with energy levels in a wide energy range around E_i. This range is not precisely centered at E_i because $\sigma_n \gg \sigma_p$ in this specific example. The effective recombination rate drops rapidly for R–G centers with energy levels close to either the top of the valence band or the bottom of the conduction band. Mathematically, this is because either the term $\exp[(E_t - E_i)/kT]$ or the term $\exp[-(E_t - E_i)/kT]$ in Eq. (5.37) becomes large for E_t levels away from E_i. Because we know that the origin of these terms is in the emission probabilities e_n and e_p [Eqs. (5.28) and (5.29)], we conclude that the emission rates of electrons/holes are so high in these cases that any captured electron/hole is quickly emitted back into the conduction/valence bands. As opposed to this, the emission terms are much smaller than the capture terms for the energies closer to E_i. As a consequence, an electron capture is likely to be followed by a hole capture by the same R–G center, which results in recombination of an electron–hole pair. Mathematically, the emission-related terms are small enough to allow $\sigma_n n$ to dominate in the denominator of Eq. (5.37) (this conclusion also includes the fact that $n \gg p$). With this and the fact that $np \gg n_i^2$, Eq. (5.37) is effectively reduced to

$$U \approx \sigma_p v_{th} N_t p = 9.6 \times 10^{18}\ \text{cm}^{-3}\,\text{s}^{-1}$$

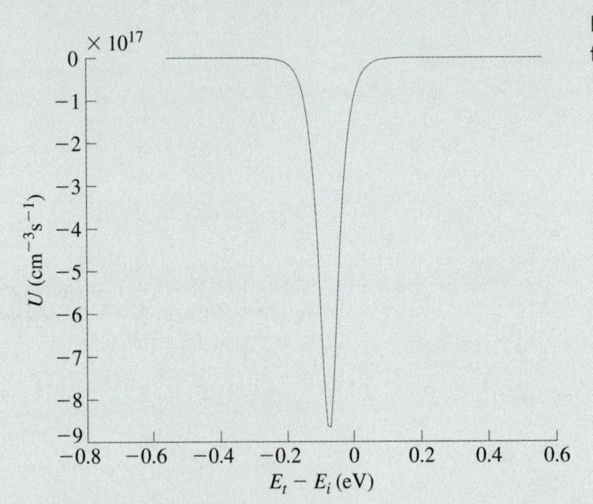

Figure 5.6 The MATLAB plot for Example 5.5c.

This is the value of U in the flat part of the U-versus-$(E_t - E_i)$ graph. This analysis shows that the effective recombination rate is limited by the minority carriers (p and σ_p).

(b) *Slight depletion.*

```
>>n=Nd-1e18;
>>p=0;
>>U=sigman*sigmap*vth*Nt*(n*p-ni^2)./
        (sigman*(n+ni*exp(Eti./kT))+sigmap*(p+ni*exp(-Eti./kT)));
>>plot(EtEi,U*1e-6)
>>xlabel('E_t-E_i (eV)')
>>ylabel('U(cm^{-3}s^{-1})')
```

In this case $U < 0$, corresponding to the case of effective thermal generation. Even though the generation mechanisms dominate, the generation rate is still very small when energy level of the R–G centers is away from E_i and close to the band edges. This is again because the carriers are exchanged (captured and emitted) between the R–G centers and the conduction/valence band. The generation rate becomes significant for energy levels that are closer to E_i. In this case the terms $n_i \exp[(E_t - E_i)/kT]$ and $n_i \exp[-(E_t - E_i)/kT]$ are much smaller than n, so that

$$U \approx -\frac{\sigma_p v_{th} N_t n_i^2}{n} = 10^{12}\ \text{cm}^{-3}\,\text{s}^{-1}$$

We can see that the maximum effective generation rate is much smaller than the maximum effective recombination rate, even though the deviation of electron concentration is the same in both cases ($\pm 0.1\%$ of n).

(c) *Full depletion.*

```
>>n=0;
>>U=sigman*sigmap*vth*Nt*(n*p-ni^2)./
      (sigman*(n+ni*exp(Eti./kT))+sigmap*(p+ni*exp(-Eti./kT)));
>>plot(EtEi,U*1e-6)
>>xlabel('E_t-E_i (eV)')
>>ylabel('U(cm^{-3}s^{-1})')
```

In this case,

$$U = -\frac{\sigma_n \sigma_p v_{th} N_t n_i}{\sigma_n e^{(E_t - E_i)/kT} + \sigma_p e^{-(E_t - E_i)/kT}}$$

If σ_n were equal to σ_p, the maximum generation rate would be for $E_t = E_i$ when both exponential terms in the denominator are equal to 1. Given that $\sigma_n \gg \sigma_p$, the maximum is shifted toward the top of the valence band. Importantly, one of the two exponential terms becomes very large on either side of the peak, dropping the effective generation rate to very small values.

5.3.3 Special Cases

As demonstrated by Example 5.5, Eq. (5.37) can be simplified for some specific cases. Two cases are of special interest: (1) low-level injection of carriers so that recombination dominates and (2) depletion of carriers so that generation dominates.

Minority-Carrier Lifetime at Low-Level Injection

Example 5.5a and Fig. 5.4 show that the effective recombination rate is high and constant for a relatively wide energy range of the R–G centers for the case of low-level injection. It is very likely that there will be high enough concentration of R–G centers with E_t in this range to allow them to dominate the effective recombination rate. Therefore, we will focus on this range, which is defined by the conditions $n \gg n_i \exp[(E_t - E_i)/kT]$ and $p \gg n_i \exp[-(E_t - E_i)/kT]$ for the case of an N-type semiconductor. With these conditions, and given that $n \gg p$ and $np \gg n_i^2$, Eq. (5.37) is simplified to

$$U = \sigma_p v_{th} N_t p \tag{5.38}$$

The low-level injection for an N-type semiconductor is defined in Section 5.2.2 by the condition $\delta n \ll n_0$. The derivation of Eq. (5.37) assumed that the steady state is maintained by a constant external generation of electron–hole pairs, so that $\delta n = \delta p$. Even though we discuss the case of low-level injection ($\delta n \ll n_0$), the concentration of minority carriers in an N-type semiconductor is so low that $\delta p = \delta n$ is likely to be much higher than p_0 for most cases of practical interest. This means that $p = p_0 + \delta p \approx \delta p = \delta n$. With this, Eq. (5.38) becomes

$$U = \sigma_p v_{th} N_t \delta p \tag{5.39}$$

Comparing this simplified equation for U with Eq. (5.8) obtained in Section 5.2.2, we find that the minority-carrier lifetime is given by

$$\tau_p = 1/\sigma_p v_{th} N_t \tag{5.40}$$

The physical meaning of the term $\sigma_p v_{th} N_t$, labeled by p_1 in Section 5.3.1, was found to be *the probability that a carrier would come in contact with an R–G center per unit time*. The reciprocal value of p_1 is the average time that it takes a carrier to come in contact with an R–G center. This *average time* is τ_p, according to Eq. (5.40).

The following consideration provides the link between *the average time that it takes a carrier to come in contact with an R–G center* and *the minority-carrier lifetime*. The carriers we are considering are the holes as the minority carriers in an N-type semiconductor. Most of the R–G centers in an N-type semiconductor are likely to be filled by electrons. In equilibrium, the Fermi level is close to the bottom of the conduction band, which means well above E_t for the R–G center(s) of interest. According to the Fermi–Dirac distribution, the probability of finding electrons at E_t is close to 1, and there is no reason for this high occupancy of the R–G centers to be any lower when the external electron–hole generation takes the semiconductor out of equilibrium. Therefore, practically any event of a hole hitting an R–G center results in capture of the hole. Furthermore, this hole capture is equivalent to a removal of an electron from the R–G center whose occupancy by an electron should remain close to 1. Consequently, the R–G center practically instantaneously captures an electron, which means that the captured hole is recombined. Therefore, the average time that it takes a hole to hit an R–G is practically equal to the average lifetime of the hole.

Analogously, the lifetime of electrons as minority carriers is

$$\tau_n = 1/\sigma_n v_{th} N_t \tag{5.41}$$

The considerations in this section show that the minority-carrier lifetime is just that—the average lifetime of a minority carrier in the environment of abundant majority carriers. Mathematically, Eqs. (5.40) and (5.41) are obtained by simplifications of the SRH equation for the effective generation rate that is derived for a steady-state condition. Nonetheless, the physical considerations provided in this section show that Eqs. (5.40) and (5.41) can be used even when carrier concentrations change in time.

EXAMPLE 5.6 Minority-Carrier Lifetime for R–G Centers with Multiple Energy Levels

Two types of R–G center have energy levels close enough to E_i that both are effective recombination centers. One of them is donor type ($\sigma_{n1} \approx 10^{-13}$ cm^2) with $N_{t1} = 10^{13}$ cm^{-3}, and the other is acceptor type ($\sigma_{n2} \approx 10^{-15}$ cm^2) with $N_{t2} = 10^{15}$ cm^{-3}. Determine the lifetime of electrons as minority carriers. The thermal velocity is $v_{th} = 10^7$ cm/s.

SOLUTION

The probability that a minority electron will hit an R–G center (and will soon after recombine with a hole) should be calculated as the sum of probabilities that the electron will hit an R–G center of the first and the second types:

$$\frac{1}{\tau_n} = \frac{1}{\tau_{n1}} + \frac{1}{\tau_2} = v_{th}\sigma_{n1}N_{t1} + v_{th}\sigma_{n2}N_{t2} = 2 \times 10^7 \text{ s}^{-1}$$

Therefore, $\tau_n = 50$ ns.

Generation Time Constant in Depletion

The second special case of interest is a semiconductor depleted of the free carriers so that the generation mechanisms prevail. In practice, a semiconductor region can be depleted of carriers by application of external field that drifts electrons and holes away from the depleted region. Usually, the concentrations of both electrons and holes are very small under these conditions, so it is of particular interest to consider the case of $n \approx 0$ and $p \approx 0$. For this, Eq. (5.37) becomes

$$U = -\frac{\sigma_n \sigma_p v_{th} N_t n_i}{\sigma_n e^{(E_t - E_i)/kT} + \sigma_p e^{-(E_t - E_i)/kT}} \tag{5.42}$$

This equation can be written in the following simple form

$$U = -n_i/\tau_g \tag{5.43}$$

where

$$\tau_g = \frac{\sigma_n e^{(E_t - E_i)/kT} + \sigma_p e^{-(E_t - E_i)/kT}}{\sigma_n \sigma_p v_{th} N_t} = \tau_p e^{(E_t - E_i)/kT} + \tau_n e^{-(E_t - E_i)/kT} \tag{5.44}$$

The unit of τ_g is s, so it is a time constant. In analogy with the minority-carrier lifetime in Eq. (5.8), τ_g is frequently referred to as the generation lifetime. However, this name and the definition of τ_g as "the time to generate one electron–hole pair by thermal emission processes"[8] are physically ambiguous (as opposed to the clear concept of minority-carrier lifetime).

To analyze this issue, we can take specific examples. As far as the minority-carrier lifetime in Eq. (5.8) is concerned, a value of $\tau_p = 1\ \mu s$ means that the average lifetime of a minority hole, when surrounded by majority electrons, is 1 μs. This is true for silicon and for any other material. For *generation lifetime* of $\tau_g = 1\ \mu s$, the generation rates are different in different materials because of the factor n_i in Eq. (5.43). As specific examples, 10^{10} electron–hole pairs would be generated per μs and per cm^3 in Si, but only 10^{-7} electron–hole pairs would be generated per μs and per cm^3 in 4H SiC.[9] To have one electron–hole pair generated in SiC per μs, we need to have a volume of 10^7 cm^3 $= 10\ m^3$! A statement that $\tau_g = 1\ \mu s$ is the time that it takes to generate one electron–hole pair is incomplete because the statement does not specify the volume in which the generation processes act. *To avoid confusion, we will refer to τ_g given by Eq. (5.44) as the generation time constant.*

The physically meaningful generation counterpart of the minority-carrier lifetime is the average time that it takes a single R–G center to generate an electron–hole pair. We will label this time by τ_t. The reciprocal value $1/\tau_t$ represents the number of electron–hole pairs generated by a single R–G center per unit time. Multiplying $1/\tau_t$ by the concentration of R–G centers N_t, we obtain the effective generation rate:

$$U = -\frac{N_t}{\tau_t} \qquad (5.45)$$

The minus sign is needed because of the convention that $U < 0$ when the generation mechanisms prevail. Equation (5.45) is the generation counterpart of Eq. (5.8) and τ_t is the generation counterpart of the minority-carrier lifetimes $\tau_{p,n}$. The dependence of τ_t on physical and material constants is derived in Example 5.7.

Equation (5.44) for the generation time constant has two terms because an electron–hole generation event is a two-step process: (1) emission of an electron and (2) emission of a hole from the R–G center (in either order). Depending on the energy position of the R–G centers, E_t, one of the two steps may dominate the time constant (being the limiting step in the generation process) or both steps may have significant impact on the generation process. An inspection of Eq. (5.44) shows that the first term in τ_g increases exponentially for E_t values above E_i whereas the second term increases exponentially for E_t values below E_i. Accordingly, τ_g has a sharp minimum that corresponds to the sharp peak in U shown in Fig. 5.6 (the solution of Example 5.5c). It is useful to determine the E_t position of this peak as a reference point. Moreover, the peak position is of practical importance when there are R–G centers with almost continuous distribution of E_t levels. In that case, it is the R–G centers with E_t corresponding to the peak in U (the minimum τ_g) that provide the fastest effective generation rate.

---

[8] D. K. Schroder, *Semiconductor Material and Device Characterization,* 2nd ed., Wiley, New York, 1998, p. 428.

[9] This is because n_i in 4H SiC is about 17 orders of magnitude lower than that in Si: $n_i \approx 10^{-7}$ cm^{-3}.

The E_t level corresponding to minimum τ_g can be determined by the following straightforward procedure:

$$\frac{d\tau_g}{dE_t} = 0 \Rightarrow \tau_p e^{(E_t - E_i)/kT} = \tau_n e^{-(E_t - E_i)/kT} \tag{5.46}$$

$$E_t - E_i = \frac{kT}{2} \ln \frac{\tau_n}{\tau_p} \tag{5.47}$$

This equation shows that the R–G centers with E_t at the intrinsic Fermi level (E_i) are the most active for a very specific case of $\tau_n = \tau_p$. Equations (5.40) and (5.41) show that this case means that the R–G centers have equal capture cross sections for both electrons and holes ($\sigma_n = \sigma_p$). In practice, this would mean that the R–G centers with $E_t = E_i$ are amphoteric defects: that is, they can act as both donors and acceptors. If no such R–G centers are present, then the most active R–G centers would have E_t either above or below E_i. This depends on whether the R–G centers are donor type ($\sigma_n > \sigma_p$ and $\tau_n < \tau_p$) or acceptor type ($\sigma_n < \sigma_p$ and $\tau_n > \tau_p$).[10]

The minimum generation time constant for R–G centers with spread out E_t can be obtained by inserting Eq. (5.47) into Eq. (5.44):

$$\tau_g = 2\sqrt{\tau_n \tau_p} \tag{5.48}$$

If there is insignificant concentration of R–G centers with an energy level that is close to E_t given by Eq. (5.47), then one of the two terms in Eq. (5.44) will dominate. For example, if $(E_t - E_i)/kT \gg \frac{1}{2}\ln(\tau_n/\tau_p)$, then the first term in Eq. (5.44) dominates:

$$\tau_g \approx \tau_p e^{(E_t - E_i)/kT} \tag{5.49}$$

This is the case when the emission of holes limits the generation process because E_t is closer to the conduction band and further from the valence band or σ_p is smaller than σ_n. Analogously,

$$\tau_g \approx \tau_n e^{-(E_t - E_i)/kT} \tag{5.50}$$

for the case when $(E_t - E_i)/kT \ll \frac{1}{2}\ln(\tau_n/\tau_p)$.

[10] Refer to Example 5.4 for estimates of σ_n and σ_p for donor- and acceptor-type R–G centers.

EXAMPLE 5.7 The Average Time a Single R–G Center Takes to Generate an Electron–Hole Pair, τ_t

Derive the dependence of τ_t on the relevant physical and material parameters. Simplify the obtained equation for the R–G centers with the energy level that minimizes τ_t. Calculate the minimum τ_t for Si, GaAs, and 3C SiC.

SOLUTION

From Eqs. (5.45) and (5.43), we obtain

$$\tau_t = \frac{N_t}{n_i}\tau_g$$

Replacing τ_g from Eq. (5.44), the following equation for τ_t is obtained:

$$\tau_t = \frac{e^{(E_t-E_i)/kT}}{\sigma_p v_{th} n_i} + \frac{e^{-(E_t-E_i)/kT}}{\sigma_n v_{th} n_i}$$

The minimum τ_t corresponds to the condition for minimum τ_g, given by Eq. (5.46). With this condition,

$$\tau_{t-min} = \frac{1}{\sigma_g v_{th} n_i} \tag{5.51}$$

where

$$\sigma_g = \frac{\sqrt{\sigma_n \sigma_p}}{2}$$

Comparing Eq. (5.51) to Eqs. (5.40) and (5.41), it can be seen that τ_{t-min} is the generation counterpart of the minority-carrier lifetimes τ_p and τ_n. The capture cross section σ_g appears as an effective capture cross section when the generation dominates and it is approximately equal for both donor-type ($\sigma_n > \sigma_p$) and acceptor-type ($\sigma_p > \sigma_n$) R–G centers. Using the values for the capture cross -sections of neutral and charged atoms, estimated in Example 5.4, the value for σ_g is

$$\sigma_g = \frac{\sqrt{3.1 \times 10^{-13} \times 10^{-15}}}{2} = 8.8 \times 10^{-15}\ \text{cm}^2$$

Assuming $v_{th} \approx 1.5 \times 10^7$ cm/s in Si, GaAs, and 3C SiC, we obtain

$$\tau_{t-min} = \frac{1}{8.8 \times 10^{-15} \times 1.5 \times 10^7 \times 10^{10}} = 0.76\ \text{ms}$$

$$\tau_{t-min} = \frac{1}{8.8 \times 10^{-15} \times 1.5 \times 10^7 \times 2.1 \times 10^6} = 3.6\ \text{s}$$

$$\tau_{t-min} = \frac{1}{8.8 \times 10^{-15} \times 1.5 \times 10^{7} \times 1} = 7.6 \times 10^{6} \text{ s}$$

respectively. Therefore, the average time it takes a single R–G center to generate an electron–hole pair in a fully depleted Si sample is 0.76 ms; in a fully depleted GaAs sample, 3.6 s; and in a fully depleted 3C SiC sample, 7.6×10^{6} s.

5.3.4 Surface Generation and Recombination

The surface of a semiconductor is of significant practical importance. Conceptually, the generation and recombination mechanisms illustrated in Fig. 5.1 are not different for the case of a semiconductor surface. There are, however, important practical and theoretical differences that have to be taken into account in the modeling of surface generation and recombination.

Given that the crystal lattice of a semiconductor is terminated at the surface, the terminating bonds are crystal defects. Many of these bonds will be tied to *passivating* atoms at the semiconductor surface, so that they are electronically passive defects. The passivating atoms can be individual atoms, typically hydrogen, or the atoms from a passivating material that is in contact with the semiconductor surface. The best passivating material in practice has proved to be SiO_2 as the native oxide of Si and SiC. The Si–SiO_2 interface is by far the best interface in terms of the density of interface defects that remain unpassivated either by oxygen from the SiO_2 or by individual hydrogen or nitrogen atoms. The *unpassivated* interface defects are said to be those that introduce energy levels in the energy gap of the semiconductor. Thus, these interface defects act as R–G centers, similar to the R–G centers in the bulk of the semiconductor. They also have a significant impact on other device parameters, in particular in the case of surface-based devices. A commonly used term for these defects is *interface traps,* so this term will be used in this section. The density of interface traps per unit area, in units of m^{-2}, will be labeled N_{it}. Importantly, the physical structure of the interface defects is not precisely defined; as a consequence, the energy levels they introduce into the energy gap tend to be continuously distributed. The energy distribution of the interface traps is labeled D_{it}, and because it expresses the number of interface traps per unit of energy and unit of area, it is given in units of $eV^{-1}\,m^{-2}$.

The SRH theory can be adapted for surface generation–recombination in two different ways: (1) to convert the concentration of R–G centers (N_t) into density of interface traps (N_{it}) and (2) to introduce the effect of continuous distribution of the interface traps (D_{it}).

In mathematical terms, the first approach is a simple replacement of N_t by N_{it}. Also, the label for the energy level will be changed from E_t to E_{it} to ensure that it is clearly associated with the interface traps. With this, and emphasizing that the electron and hole concentrations are to be considered at the semiconductor surface (n_s and p_s instead of n

and p), Eq. (5.37) can be rewritten as

$$U_s = \frac{\sigma_n \sigma_p v_{th} N_{it} \left(n_s p_s - n_i^2\right)}{\sigma_n \left[n_s + n_i e^{(E_{it}-E_i)/kT}\right] + \sigma_p \left[p_s + n_i e^{-(E_{it}-E_i)/kT}\right]} \tag{5.52}$$

This equation is the SRH equivalent for the effective surface recombination–generation rate. Because of the change of units from m^{-3} for N_t to m^{-2} for N_{it}, the unit for U_s is $m^{-2}s^{-1}$—the number of generated/recombined carriers per unit time and per unit *area* (rather than per unit volume as in the case of U).

Clearly, Eq. (5.52) gives U_s due to interface traps having a single energy level E_{it} and density N_{it}. As already mentioned, real interface traps have energy levels that are continuously distributed in the energy gap. Therefore, the second approach—the introduction of the continuous distribution $D_{it}(E)$—is more general. Replacing N_t in Eq. (5.37) by D_{it} results in the following equation for the differential surface recombination rate:

$$u_s = \frac{\sigma_n \sigma_p v_{th} D_{it}(E) \left(n_s p_s - n_i^2\right)}{\sigma_n \left[n_s + n_i e^{(E_{it}-E_i)/kT}\right] + \sigma_p \left[p_s + n_i e^{-(E_{it}-E_i)/kT}\right]} \tag{5.53}$$

The unit of u_s is $eV^{-1}m^{-2}\,s^{-1}$, meaning that u_s is the effective thermal generation–recombination rate per unit area and unit energy. Integrating u_s over the entire energy gap leads to the total rate, U_s:

$$U_s = \int_{E_V}^{E_C} u_s \, dE \tag{5.54}$$

Surface Recombination Velocity at Low-Level Injection

In the first approach—replacing N_t by N_{it}—the simplified equation for the special case of low-level injection can be obtained directly from Eqs. (5.38) and (5.39):

$$U_s \approx v_{th} \sigma_p N_{it} \delta p_s \approx \sigma_p v_{th} N_{it} p_s \tag{5.55}$$

In analogy with Eqs. (5.38) and (5.39), $\delta p_s \approx p_s$ is the excess concentration of the minority holes at the semiconductor surface. These holes are surrounded by abundant majority electrons, so it is the holes that limit the surface recombination rate. The analogy, however, cannot be extended to Eq. (5.40), which links $v_{th}\sigma_p N_t$ to the minority-carrier lifetime. Because the unit of the factor $v_{th}\sigma_p N_{it}$ is m/s, its reciprocal value does not relate to the minority-carrier lifetime in the way the reciprocal value of $v_{th}\sigma_p N_t$ does in Eq. (5.40). The product $\sigma_p N_{it}$ has a different unit and a different meaning from the product $\sigma_p N_t$. It no longer represents the probability that a considered hole will hit an R–G center *per unit length* of its path. For the case of interface traps, $\sigma_p N_{it}$ is simply the fraction of the semiconductor surface that is covered by the capture cross sections of the interface traps. Thus, when a hole hits the semiconductor surface, $\sigma_p N_{it}$ shows the probability that the hole will be in contact with an interface trap. This leads to the question about the physical meaning of $v_{th}\sigma_p N_{it}$.

To derive the physical meaning of $v_{th}\sigma_p N_{it}$, we note that we are again considering the specific case when the majority electrons are in abundance as the minority holes hit the surface. Because of the abundance of electrons at the surface, it can be assumed that nearly all interface traps have captured electrons. For a recombination event to occur, an interface trap also must capture a hole. Thus, nearly every hit of a minority hole into the capture cross section of an interface trap results in a recombination event. The recombination of the holes that hit the surface results in effective flow of these holes to the surface. This picture is analogous to the case of drifting carriers collected by the biased contact of the semiconductor bar considered during the derivation of Eq. (3.21) for the drift-current density. In the analogy with Eq. (3.21), the drift velocity v_{dr} has to be replaced by the thermal velocity of the holes that not only hit the surface but also hit it in the recombination-active area. The fraction of the surface that is covered by the capture cross sections of the interface traps is $\sigma_p N_{it}$. Therefore, the thermal velocity has to be multiplied by $\sigma_p N_{it}$ to obtain the surface recombination counterpart of the drift velocity:

$$s_p = v_{th}\sigma_p N_{it} \tag{5.56}$$

This velocity is referred to as the *surface recombination velocity*. For electrons, the surface recombination velocity is

$$s_n = v_{th}\sigma_n N_{it} \tag{5.57}$$

In summary, the minority-carrier lifetime is replaced by the surface recombination velocity when N_t is replaced by N_{it} to consider surface recombination:

$$U_s = \begin{cases} s_p \delta p_s & \text{for } n_s \gg p_s \\ s_n \delta n_s & \text{for } p_s \gg n_s \end{cases} \tag{5.58}$$

The more general approach is to introduce the continuous energy distribution of the interface traps, D_{it}. For low-level *injection,* the condition $n_s p_s \gg n_i^2$ is satisfied for any energy level E_{it}. With this condition and for the case of $n_s \gg p_s$ (N-type semiconductor), Eq. (5.53) can be simplified to the following form:

$$u_s = \frac{\sigma_p v_{th} D_{it}(E) n_s p_s}{n_s + n_i e^{(E_{it}-E_i)/kT} + (\sigma_p/\sigma_n) n_i e^{-(E_{it}-E_i)/kT}} \tag{5.59}$$

The exponential terms in the denominator can be neglected for energy levels E_{it} that are not too far from E_i, leading to the result that can be obtained by direct replacement of N_t in Eq. (5.39) with D_{it}:

$$u_s \approx \sigma_p v_{th} D_{it}(E) p_s \approx \sigma_p v_{th} D_{it}(E) \delta p_s \tag{5.60}$$

However, for E_{it} levels that are well above E_i, the term $\exp[(E_{it} - E_i)/kT]$ begins to dominate, causing an exponential drop of u_s with increasing E_{it}. Analogously, the term $(\sigma_p/\sigma_n) n_i \exp[-(E_{it} - E_i)/kT]$ causes an exponential drop of u_s for E_{it} levels that are well below E_i. This behavior of u_s is illustrated in Fig. 5.7 for $D_{it} = \text{const}$. Representing the interface traps by a constant D_{it} is a convenient assumption that is far more realistic than the assumption of interface traps with a single E_{it} level (the first approach).

The two threshold energies shown in Fig. 5.7 are obtained as follows. The upper threshold energy corresponds to the condition $n_s = n_i \exp[(E_{it} - E_i)/kT]$. For low-level injection, n_s is close to the equilibrium electron concentration in the N-type semiconductor that is given by Eq. (2.85). Therefore, this condition is equivalent to $E_{it} = E_F$, which is the upper threshold for E_{it}. The lower threshold energy corresponds to the condition

$$\frac{\sigma_p n_i}{\sigma_n n_s} e^{-(E_N - E_i)/kT} = 1 \tag{5.61}$$

For the specific case of $\sigma_n = \sigma_p$, $E_i - E_N = E_F - E_i$, which means that the flat region of u_s is centered at E_i and that its width is $2(E_F - E_i)$.

Approximating the differential surface rate u_s by its maximum value given by Eq. (5.60) for $E_N \le E_{it} \le E_F$ and neglecting it outside this energy range, the total surface rate U_s can be estimated from

$$U_s \approx s_p p_s \approx s_p \delta p_s \tag{5.62}$$

where the surface-recombination velocity is

$$s_p = v_{th} \sigma_p D_{it}(E_F - E_N) \tag{5.63}$$

Comparing Eqs. (5.63) and (5.56), we find that $D_{it}(E_F - E_N)$ corresponds to N_{it}. For electrons, the surface-recombination velocity is

$$s_n = v_{th} \sigma_n D_{it}(E_P - E_F) \tag{5.64}$$

where E_P is obtained from the following condition:

$$\frac{\sigma_n n_i}{\sigma_p p_s} e^{(E_P - E_i)/kT} = 1 \tag{5.65}$$

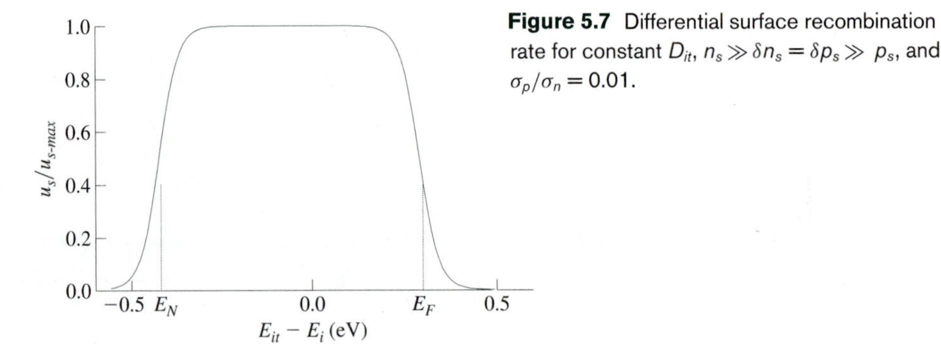

Figure 5.7 Differential surface recombination rate for constant D_{it}, $n_s \gg \delta n_s = \delta p_s \gg p_s$, and $\sigma_p/\sigma_n = 0.01$.

Surface Generation Constant in Depletion

Upon defining surface depletion by $n_s \approx 0$ and $p_s \approx 0$, Eq. (5.53) becomes

$$u_s = -\frac{\sigma_n \sigma_p v_{th} D_{it} n_i}{\sigma_n e^{(E_{it}-E_i)/kT} + \sigma_p e^{-(E_{it}-E_i)/kT}} \tag{5.66}$$

Figure 5.8 illustrates this dependence for a constant D_{it}. The peak of u_s is at

$$E_{it} - E_i = \frac{kT}{2} \ln \frac{\sigma_p}{\sigma_n} \tag{5.67}$$

and is equal to

$$u_{s-peak} = -\frac{1}{2}\sqrt{\sigma_n \sigma_p} v_{th} D_{it} n_i \tag{5.68}$$

To obtain the total rate U_s, the differential rate u_s, given by Eq. (5.66), should be integrated from E_V to E_C. Alternatively, we can approximate u_s by its peak for the interface traps in the active energy region and neglect it outside this region. An analysis of Eq. (5.66) shows that one of the two terms in the denominator drops e times for $E_{it} - E_i = \pm kT$. Thus, the energy width of the active interface traps can be approximated by $2kT$. Therefore, $U_s = 2kTu_{s-peak}$ leads to

$$U_s = -s_g n_i \tag{5.69}$$

where

$$s_g = \sqrt{\sigma_n \sigma_p} v_{th} kT D_{it} \tag{5.70}$$

Different names are used for the constant s_g in the literature, including *surface generation velocity*. This term has problems analogous to those associated with *generation lifetime,* described in Section 5.3.3. Accordingly, a much less confusing name for s_g is *surface generation constant.*

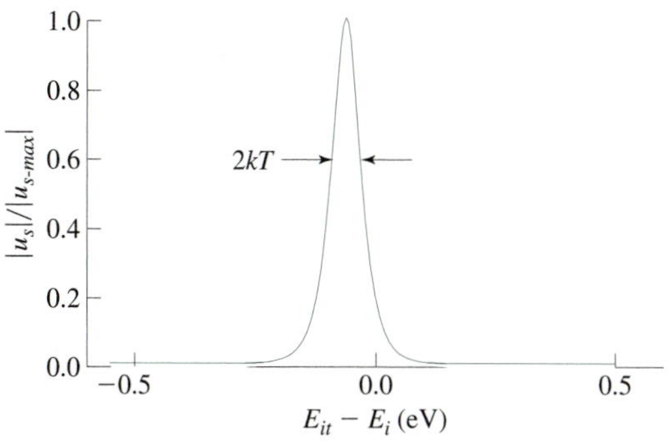

Figure 5.8 Differential surface generation rate for constant D_{it}, $n_s = p_s = 0$, and $\sigma_p/\sigma_n = 0.01$.

SUMMARY

1. A direct or band-to-band *generation* event occurs when an electron–hole pair is created because an electron moves from the valence into the conduction band. The opposite process, when an electron drops from the conduction into the valence band, is a direct or band-to-band *recombination* event. A photon can be absorbed to provide the energy for a direct generation event, and a photon can be emitted during a direct recombination event (radiative generation and recombination).
2. Usually more frequent generation and recombination events occur through R–G centers—defects with energy levels in the energy gap. These are two-step processes: (1) an electron and a hole emission from an R–G center (in either order) is a generation event and (2) an electron and a hole capture by an R–G center (in either order) is a recombination event. These mechanisms are known as SRH generation and recombination.
3. Another pair of generation–recombination mechanisms consists of avalanche generation and Auger recombination. The energy provided for avalanche generation is due to the kinetic energy when carriers are accelerated by electric field; the energy released by Auger recombination is in the form of kinetic energy that the accelerated carrier loses as heat in subsequent phonon-scattering events.
4. The rates of spontaneous generation and recombination are combined into a single quantity, called the effective generation–recombination rate U. The effective generation–recombination rate (U) and any external generation rate (G_{ext}) directly influence the change of carrier concentrations per unit time. The continuity equations for electrons and holes that include these rates are as follows:

$$\frac{\partial n}{\partial t} = \frac{1}{q}\frac{\partial j_n}{\partial x} - U + G_{ext}$$

$$\frac{\partial p}{\partial t} = -\frac{1}{q}\frac{\partial j_p}{\partial x} - U + G_{ext}$$

This form of the continuity equations does not include the possible rate of stimulated recombination that occurs in lasers.
5. The effective generation–recombination rate is proportional to $(np - n_i^2)$. In thermal equilibrium, $U = 0$. When recombination prevails, $U > 0$, and when generation prevails, $U < 0$. For a small deviation from equilibrium in an N-type semiconductor, the effective generation–recombination rate can be expressed as

$$U = \frac{\delta p}{\tau_p}$$

where τ_p is the lifetime of minority holes.
6. When minority carriers are injected into a neutral semiconductor, their steady-state concentration decays exponentially from the injecting plane. For electrons as minority carriers, $\delta n(x) = \delta n(0)\exp(-x/L_n)$, where L_n is the diffusion length. The diffusion

length is determined by the minority-carrier lifetime and the diffusion constant:

$$L_{n,p} = \sqrt{D_{n,p}\tau_{n,p}}$$

7. According to the Shockley–Read–Hall (SRH) theory, the effective generation–recombination rate for a steady-state nonequilibrium case is given by

$$U = \frac{\sigma_n \sigma_p v_{th} N_t \left(np - n_i^2\right)}{\sigma_n\left[n + n_i e^{(E_t - E_i)/kT}\right] + \sigma_p\left[p + n_i e^{-(E_t - E_i)/kT}\right]}$$

where n and p are the nonequilibrium electron and hole concentrations, n_i is the intrinsic-carrier concentration, $\sigma_{n,p}$ are electron–hole capture cross sections, v_{th} is the thermal velocity, N_t is the concentration of R–G centers, and $E_t - E_i$ is the energy position of the R–G centers with respect to E_i.

- For the case of a low-level injection of minority carriers,

$$U \approx \frac{\delta p}{\tau_p}$$

where

$$\tau_p = \frac{1}{v_{th}\sigma_p N_t}$$

is the minority-carrier lifetime. This result shows that the reciprocal value of the minority-carrier lifetime is equal to the probability that a minority hole will hit an R–G center per unit time ($v_{th}\sigma_p N_t$).

- For the case of depletion,

$$U = -\frac{n_i}{\tau_g}$$

where τ_g is the generation time constant. The dominant generation time constant is for R–G centers with energy levels $E_t - E_i = (kT/2)\ln(\tau_n/\tau_p)$, its value being $\tau_g = 2\sqrt{\tau_n \tau_p}$.

8. Analogously to the equation for U, the effective surface generation/recombination rate is given by

$$U_s = \frac{\sigma_n \sigma_p v_{th} N_{it} \left(n_s p_s - n_i^2\right)}{\sigma_n\left[n_s + n_i e^{(E_{it} - E_i)/kT}\right] + \sigma_p\left[p_s + n_i e^{-(E_{it} - E_i)/kT}\right]}$$

where N_{it} is the density of *effective* interface traps (in m^{-2}) and n_s and p_s are the surface nonequilibrium concentrations of electrons and holes. The unit of U_s is $m^{-2}s^{-1}$.

- For low-level injection of minority carriers,

$$U_s = \begin{cases} s_p \delta p_s & \text{for } n_s \gg p_s \\ s_n \delta n_s & \text{for } p_s \gg n_s \end{cases}$$

where

$$s_{p,n} = v_{th} \sigma_{p,n} N_{it}$$

is the surface-recombination velocity (in m/s). For a constant energy distribution of interface traps, $D_{it} = \text{const}$, the total density of *effective* interface traps in the equation for $s_{p,n}$ is $N_{it} = 2|E_F - E_i|D_{it}$.

- For depletion,

$$U_s = -s_g n_i$$

where

$$s_g = \frac{1}{2}\sqrt{\sigma_n \sigma_p} v_{th} N_{it}$$

is the surface generation constant. For a constant energy distribution of interface traps, $D_{it} = \text{const}$, the total density of *effective* interface traps in the equation for s_g is $N_{it} = 2kT D_{it}$.

PROBLEMS

5.1 Two N-type silicon samples ($N_D = 10^{15}$ cm^{-3}) are exposed to light, which in both samples generates electron–hole pairs at the rate of $G_{ext} = 10^{19}$ cm^{-3} s^{-1}. The minority-carrier lifetime in the first sample is $\tau_p = 100$ μs, whereas the minority-carrier lifetime in the second sample is reduced to $\tau_p = 100$ ns by intentionally introduced R–G centers. For each sample, determine

(a) the steady-state excess concentration of holes, δp

(b) how long it takes for δp to drop by 10% when the light is switched off

(c) how long it takes for δp to drop to the value that is 10% higher than the thermal equilibrium level

5.2 A region of N-type semiconductor, doped by $N_D = 10^{15}$ cm^{-3} is fully depleted of both majority and minority carriers by an external electric field. Assuming that the concentration of electron–hole pairs that is generated per unit time does not change when the semiconductor is depleted, determine the effective generation rate if the semiconductor is

(a) Si ($n_i = 1.02 \times 10^{10}$ cm^{-3})

(b) GaAs ($n_i = 2.1 \times 10^6$ cm^{-3})

The minority-carrier lifetime is the same in both cases: $\tau_p = 1$ μs.

5.3 Consider silicon in thermal equilibrium, with the equilibrium concentration of electrons $n_0 = 10^{16}$ cm^{-3}. The minority-carrier lifetime is $\tau_p = 0.4$ μs.

(a) What is the effective recombination rate?

(b) Assuming that the concentration of electron–hole pairs that are generated per unit time does not change if the semiconductor is suddenly fully depleted of both the electrons and holes, determine the concentration of holes generated per unit time in thermal equilibrium.

(c) What is the concentration of holes recombined per unit time in thermal equilibrium? **A**

5.4 Consider GaAs in thermal equilibrium, with the equilibrium concentration of electrons $n_0 = 10^{16}$ cm^{-3} and the minority-carrier lifetime $\tau_p = 0.4\ \mu$s. If the direct generation–recombination mechanism is dominant, what is the concentration of holes recombined per unit time in thermal equilibrium? Compare the result with the case of silicon (Problem 5.3c) and explain the difference.

5.5 A light source uniformly generates carriers in a slab of P-type silicon doped at the level of $N_A = 10^{17}$ cm^{-3}. The minority-carrier lifetime is $\tau_n = 1\ \mu$s. At time $t = 0$, the light source is removed. Determine how long it takes for the minority-carrier concentration to

(a) drop by 10% from the steady-state level
(b) drop to the level that is 10% higher than the thermal equilibrium value

for the following two values of the external generation rate: (1) $G_{ext} = 10^{12}$ cm^{-3} s^{-1} and (2) $G_{ext} = 10^{18}$ cm^{-3} s^{-1}.

5.6 A photoresistor is made of a P-type semiconductor doped at the level of $N_A = 10^{15}$ cm^{-3}. Determine the change in resistivity caused by an exposure to light that uniformly generates carriers at the rate of $G_{ext} = 5 \times 10^{19}$ cm^{-3} s^{-1}. The electron and hole mobilities are $\mu_n = 250$ cm^2/V · s and $\mu_p = 50$ cm^2/V · s, and the minority-carrier lifetime is $\tau_n = 20\ \mu$s.

5.7 **(a)** Design a photoresistor, based on a P-type semiconductor doped with $N_A = 10^{15}$ cm^{-3}, so that the resistance at the external generation rate of 10^{22} cm^{-3} s^{-1} is 1 kΩ. The following technological parameters are known: the thickness of the semiconductor film is 5 μm, the minimum width of the semiconductor strip is 500 μm, the electron mobility is $\mu_n = 250$ cm^2/V · s, the hole mobility is $\mu_p = 50$ cm^2/V · s, and the minority-carrier lifetime is $\tau_n = 20\ \mu$s.
(b) What is the resistance in the dark?

5.8 The equilibrium concentration of minority holes in a semiconductor is p_0 and their lifetime is τ_p. At time $t = 0$, light begins to uniformly generate carriers at the rate of G_{ext}. Derive the function that describes the increase of minority-carrier concentration in time, $p(t)$. **A**

5.9 The equilibrium concentration of minority electrons in a semiconductor is $n_0 = 5 \times 10^4$ cm^{-3} and their lifetime is $\tau_n = 1\ \mu$s. At time $t = 0$, light begins to uniformly generate carriers at the rate of $G_{ext} = 10^{20}$ cm^{-3} s^{-1}.

(a) Determine the steady-state electron concentration under the light illumination.
(b) How long does it take for the electron concentration to increase from 10% to 90% of the steady-state level?

5.10 Holes are uniformly injected into the neutral N-type region of a semiconductor device at the rate of 5×10^{15} holes per second. The area of uniform hole injection is $A = 0.01$ mm^2. Initially, the holes accumulate close to the injection plane ($x = 0$), increasing the concentration difference from the equilibrium level in the N-type region away from the injection plane. When the gradient of the hole concentration is large enough to allow the diffusion current to take holes away at the same rate as the injection rate, the steady state is established.

(a) What is the steady-state diffusion current per unit area? **A**
(b) If the diffusion constant of the minority holes is $D_p = 10$ cm^2/s and their lifetime is 1 μs, determine the steady-state value of the excess hole concentration at the injection plane.

5.11 A sample with different concentrations of G–R centers but all other parameters identical to the sample of Problem 5.10 is exposed to the same injection conditions. If the steady-state value of the excess hole concentration at the injection plane is 10 times smaller than the sample of Problem 5.10, what is the minority-carrier lifetime?

5.12 The concentration of R–G centers in an N-type silicon sample is $N_t = 10^{12}$ cm^{-3}. Based on the estimates of the capture cross sections from Example 5.4, determine the minority-carrier lifetime if the R–G centers are

(a) donor type
(b) acceptor type **A**

Assume that the thermal velocity of holes is $v_{th} = 10^7$ cm/s.

5.13 Determine the minority-carrier lifetimes in a P-type silicon for the same other conditions and assumptions as in Problem 5.12, including the value of thermal velocity ($v_{th} = 10^7$ cm/s).

5.14 Assuming $\tau_n = \tau_p = 1\ \mu s$ and the dominance of SRH generation, determine the maximum effective generation rate in a fully depleted semiconductor if the semiconductor is

(a) Si

(b) GaAs **A**

5.15 The temperature dependence of the effective SRH generation rate in a fully depleted silicon is to be expressed by the Arrhenius equation,

$$|U| = Ae^{-E_A/kT}$$

Determine the activation energy E_A, neglecting any temperature dependence of the generation time constant.

5.16 A silicon sample has $N_t = 10^{11}$ cm^{-3} donor-type R–G centers with the energy level that corresponds to the maximum generation rate. The following values can be assumed for the capture cross sections of the donor-type R–G centers: $\sigma_n = 3 \times 10^{-13}$ cm^2 for the electron capture and $\sigma_p = 10^{-15}$ cm^2 for the hole capture.

(a) Determine the generation time constant.

(b) Determine the effective generation rate at room temperature if the sample is fully depleted.

(c) Determine the effective generation rate in a fully depleted sample if the temperature is increased to 100°C ($n_i \approx 10^{12}$ cm^{-3} at 100°C; the generation time constant remains the same). **A**

Assume that the thermal velocity is $v_{th} = 10^7$ cm/s in all cases.

5.17 An N-type silicon sample with passivated surface has interface traps with uniform distribution across the entire energy gap, $D_{it} = 5 \times 10^{10}$ cm^{-2} eV^{-1}. The doping level is $N_D = 10^{16}$ cm^{-3}. Determine the number of effective interface traps per unit area (N_{it}) for the following two cases:

(a) the recombination rate dominates to eliminate excess carriers

(b) the generation rate dominates because the surface is fully depleted

Assume that the capture cross sections for electrons and holes are equal.

5.18 For $\sigma_n = \sigma_p = 10^{-14}$ cm^2, $v_{th} = 10^7$ cm/s, and the densities of effective interface traps determined in Problem 5.17, calculate

(a) the effective surface recombination rate if there are excess holes at the surface, $\delta p_s = 10^{12}$ cm^{-3}

(b) the effective surface generation rate if the surface is fully depleted of carriers

REVIEW QUESTIONS

R-5.1 List the three most frequent pairs of generation and recombination mechanisms.

R-5.2 Can the effective generation–recombination rate (U) be equal to zero? Are there any generation or recombination events when $U = 0$?

R-5.3 Is the thermal equilibrium condition when the effective generation–recombination rate is equal to the external generation rate, $U = G_{ext}$? If not, what is the condition $U = G_{ext}$ called?

R-5.4 For $G_{ext} = 0$, does $U \neq 0$ show how many electrons/holes are recombined per unit volume and unit time?

R-5.5 What is the physical meaning of *minority-carrier lifetime*? Does it represent the average lifetime of a minority carrier? If so, is it similar to the average lifetime of a majority carrier?

R-5.6 Does diffusion length depend on carrier recombination? If so, how?

R-5.7 In thermal equilibrium, can the rate of direct generation be balanced by the rate of SRH recombination?

R-5.8 Can the rates of SRH generation and recombination balance each other?

R-5.9 In thermal equilibrium, is it necessary that the emission and capture rates be balanced separately for the electrons and the holes?

R-5.10 At low-level injection, how is the effective recombination rate related to the minority-carrier lifetime?

R-5.11 In depletion, how is the effective generation rate related to the generation time constant? Is it related to the minority-carrier lifetime?

R-5.12 What is the physical meaning of the surface recombination velocity? What are the similarities and what are the differences between the drift current and the surface current due to the surface recombination?

R-5.13 What is the effective energy range in the energy gap in terms of recombination-active interface traps?

R-5.14 In depletion, how is the effective surface generation rate related to the surface generation constant?

R-5.15 What is the effective energy range in the energy gap in terms of generation-active interface traps?

6 P–N Junction

The P–N junction—a contact between P-type and N-type semiconductor regions—is a fundamental structure. A P–N junction on its own performs the function of a *rectifying diode,* as illustrated in Fig. 6.1. It can also be operated in its breakdown mode as a *reference diode*. The capacitance of a P–N junction is voltage-dependent, so it can be used as a *variable capacitor*. Transistor structures, including *MOSFET* and *BJT,* involve combinations of P–N junctions.

This chapter introduces the P–N junction. The principles are described in the first section using energy-band and concentration diagrams. The second section deals with the DC model (current–voltage equations) in a way that links the theory to the practice of related SPICE parameters. The capacitance of a reverse-biased P–N junction is analyzed in the third section to introduce the concept of a voltage-dependent capacitance and the equations for a very important concept: the depletion-layer width. The fourth section describes stored-charge effects and their influence on the switching characteristics.

6.1 P–N JUNCTION PRINCIPLES

6.1.1 P–N Junction in Thermal Equilibrium

As the Fermi level depends on doping (Section 2.4.3), the average electron energy is different in differently doped semiconductors. When an electrical contact is made between two materials having different Fermi levels, the higher-energy electrons move into the region of the lower-energy electrons to establish thermal equilibrium—the state where the average electron energy is the same across the entire electrical system. With this, we can establish the following important fact: *for a system in thermal equilibrium, the Fermi level is constant throughout the system.*

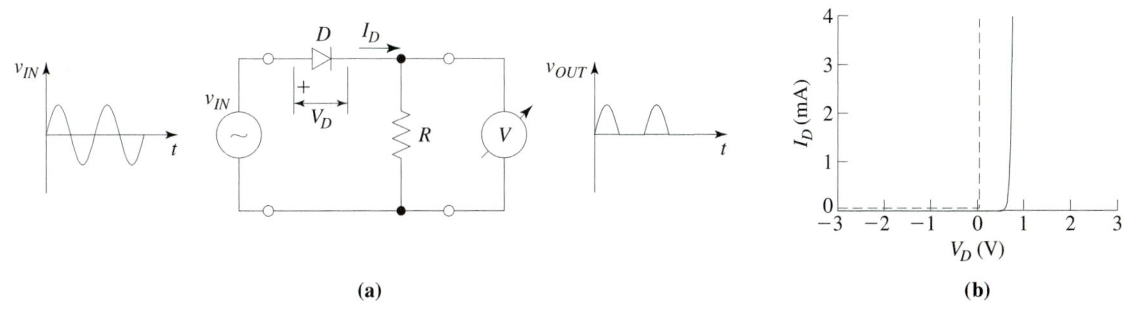

Figure 6.1 (a) Basic rectifying circuit. (b) *I–V* characteristics of the ideal rectifying device (dashed line) and the P–N junction diode (solid line).

Energy-Band Diagram of a P–N Junction

The energy-band diagrams of an N-type and P-type semiconductor, shown in Fig. 2.19, are drawn so that the E_C and E_V levels are aligned.[1] The energy-band diagrams in Fig. 2.19 show that the difference between the Fermi levels in an N-type and a P-type semiconductor is

$$qV_{bi} = |q\phi_{Fn}| + |q\phi_{Fp}| = kT \ln \frac{N_D}{n_i} + kT \ln \frac{N_A}{n_i} = kT \ln \frac{N_D N_A}{n_i^2} \tag{6.1}$$

If the energy-band diagrams are redrawn so that the Fermi levels are aligned, as in Fig. 6.2, we see that there is a difference between E_C levels (as well as E_V levels) in the N-type and the P-type regions. Clearly, this difference is also equal to qV_{bi} given by Eq. (6.1). The energy offset between the E_C (E_V) levels in the N-type and the P-type regions is such that aligned energy levels in the bands have equal occupancy probability throughout the system. This is because E_F is constant so that the distance of aligned energy levels from E_F is constant, and the value of the Fermi–Dirac distribution (f) at that energy level is equal throughout the system. To visualize this effect, the diagrams given in Fig. 6.2 indicate electrons in the conduction band (filled circles) and holes in the valence band (open circles).

When the energy-band diagrams of an N-type and a P-type region are drawn with aligned Fermi levels, as in Fig. 6.2, they correspond to a P–N junction in equilibrium. To complete the energy-band diagram in the missing part at and around the junction itself, the E_C and E_V lines from the two regions are joined as in Fig. 6.3a to present both the bottom of the conduction band and the top of the valence band by continuous E_C and E_V lines, respectively. Clearly, there has to be a nonzero slope of the E_C and E_V lines at and around the P–N junction. This slope expresses the existence of an electric field.[2]

---

[1]The position of E_C with respect to the energy level of a free electron in space (vacuum level) is a material constant, $q\chi$, where χ is the electron affinity. Given that $q\chi$ does not change with doping, the vacuum levels in the energy-band diagrams in Fig. 2.19 are also aligned (although not shown).

[2]The electric field is equal to the negative gradient of the electric potential ($-d\varphi/dx$), which means it is proportional to the positive gradient of the potential energy $[(1/q)\,dE_{pot}/dx]$.

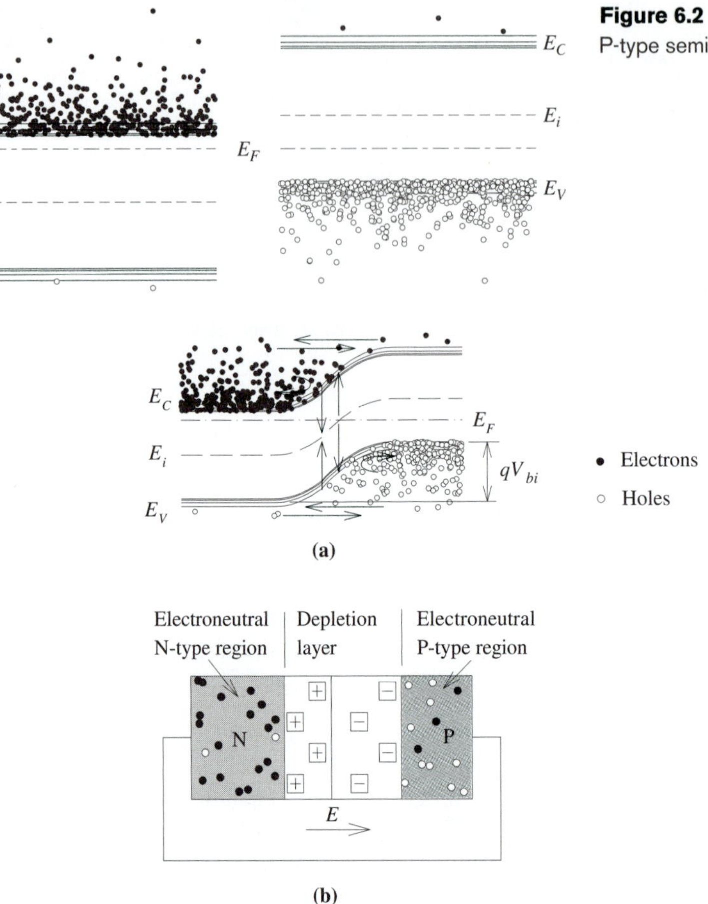

Figure 6.2 Energy-band diagrams of N-type and P-type semiconductors with aligned Fermi levels.

Figure 6.3 (a) Energy-band diagram and (b) cross section of a P–N junction in thermal equilibrium.

Depletion Layer

The cross-sectional diagram shown in Fig. 6.3b helps explain the source of the electric field at a P–N junction in equilibrium. The source of this electric field is uncompensated donor and acceptor ions, illustrated by the square symbols in Fig. 6.3b to distinguish them from the circles used as symbols for mobile charge. These ions are uncompensated because some electrons from the N-type region and some holes from the P-type region are removed in the process of setting the thermal equilibrium condition at the P–N junction. We can think of electrons moving from the N-type region to the P-type region (both across the junction and through the short-circuiting wire) to recombine with holes in the P-type region. Also, holes can move to the N-type region to get recombined with electrons. This process of electron–hole recombination leads to the creation of a *depletion layer* at the P–N junction. Although commonly used, the term *depletion* can be confusing, and it can even be wrongly extended beyond its intended meaning. The use of *depletion* for P–N junctions in equilibrium does not mean the same thing as *depletion* as used to describe nonequilibrium deficiency of electrons and holes in Chapter 5. The term *depletion layer* is used for a region that has fewer majority carriers than there are in the *electroneutral* regions, irrespective of whether the P–N junction is in equilibrium or not.[3] As a result, a depletion layer has an uncompensated charge of fixed acceptor and/or donor ions. This uncompensated charge creates an electric field that is referred to as the *built-in field*. As distinct from the depletion layer, there is no built-in field in the electroneutral regions, which is indicated by the flat E_C and E_V lines.

The built-in electric field results in a built-in voltage across the depletion layer. This voltage corresponds to the difference between E_C (or E_V) at the edges of the depletion layer, and it appears as the energy barrier qV_{bi} in Fig. 6.3a. Given that qV_{bi} is in units of energy, the built-in voltage, V_{bi} is in units of voltage. From Eq. 6.1, the built-in voltage at a P–N junction is given by

$$V_{bi} = V_t \ln \frac{N_D N_A}{n_i^2} \tag{6.2}$$

where V_t is the thermal voltage ($V_t = kT/q$).

It is important to clarify that the appearance of the built-in voltage at the P–N junction by no means violates Kirchhoff's second law, which states that the sum of voltage drops along any closed loop is equal to zero. Imagine that the semiconductor is a circle, half of which is doped as P type and half as N type. There will be two P–N junctions with two equivalent built-in voltages that cancel each other when added up to make the sum of the voltages along the closed loop. If the P–N junction is contacted by metal tracks, built-in voltages will appear at the contacts,[4] so that the sum of built-in voltages at the contacts and at the P–N junction is equal to zero.

Balance of Currents

The net current through the circuit in Fig. 6.3b is equal to zero because there is no external voltage source (a battery) to cause a current by attracting electrons toward its positive pole and holes toward its negative pole. Although the net current is equal to zero, there are individual electrons and holes that do pass through the P–N junction. Figure 6.3a shows that most of the electrons in the N-type region are unable to move onto the P-type side because the energy they possess is smaller than the energy barrier at the P–N junction. These electrons hit the "wall" (bottom of the conduction band) when they move toward the P-type side. Only the electrons possessing energy larger than the energy barrier can make the transition to the P-type side, as shown by the arrow in the conduction band pointing toward the P-type region (Fig. 6.3a). However, the concentration of electrons in the N-type region that possess higher energy than the barrier is the same as the concentration of electrons in the P-type region (the minority carriers). This is so because all these electrons appear at the same energy levels with respect to the constant Fermi level and, according to the Fermi–Dirac distribution, the population of the leveled density of states is the same. Accordingly, an equal number of electrons move from the P-type to the N-type side, as

---

[3]The term *depletion* in this meaning is retained even for the case of excess carriers injected in the region around the P–N junction.

[4]In general, built-in voltage appears at any junction of materials that have different positions of the Fermi level with respect to the vacuum level.

indicated by the conduction-band arrow pointing from the N-type toward the P-type region. This is the current of minority electrons that balances the current of the fraction of majority electrons possessing energies larger than the energy barrier.

The situation is analogous with the holes. The difference is that holes in the energy-band diagram are like bubbles in water, which tend to bubble up. The top of the valence band at the P–N junction appears as a barrier for the holes from the P-type side (the majority carriers). The arrows in the valence band (Fig. 6.3a) indicate the balance between the current of minority holes and the current of the fraction of majority holes that can get through the junction without hitting the energy barrier.

In addition to the currents of electrons and holes moving from the N-type neutral region to the P-type neutral region and vice versa, there are balanced currents due to electron–hole recombination and generation inside the depletion layer (the vertical arrows in Fig. 6.3a). The recombination current is due to the flow of electrons from the N-type and holes from the P-type regions that hit each other and get recombined in the depletion layer. The balancing generation current is due to generated electrons and holes that are pushed by the electric field in the depletion layer to return to the N-type and P-type regions, respectively.

Concentration Diagrams

The concentrations of carriers in the electroneutral regions are the same as in the case of separate N-type and P-type semiconductors. As already explained, the concentrations of majority carriers in the depletion layer are smaller. This can be deduced from the energy-band diagram (Fig. 6.3a), recalling the dependencies of electron and hole concentrations on the position of the Fermi level with respect to the bands. According to Eq. (2.76), the concentration of electrons drops as $E_C - E_F$ is increased in the depletion layer to reach its minority-carrier level in the neutral P-type region. Likewise, $E_F - E_V$ increases in the direction from the P-type toward the N-type region, as the concentration of holes drops from the majority-carrier level in the neutral P-type region to the minority-carrier level in the N-type region [Eq. (2.79)]. The "proximity" of E_F to E_C (E_V) can be used to visualize the concentrations of electrons (holes), but keep in mind that we are dealing with exponential dependencies [Eqs. (2.76) and (2.79)]. Figure 6.4a shows the linear plots of electron and hole concentrations in a P–N junction doped with $N_D = 10^{16}$ cm^{-3} (the N-type region) and $N_A = 3 \times 10^{15}$ cm^{-3} (the P-type region). It can be seen that the concentrations of electrons and holes drop rapidly at the edges of the depletion layer. Linear concentration diagrams are not usually used, however, because beyond illustrating the "sharpness" of the depletion-layer edges, they provide no useful information. Instead, the concentration is commonly plotted on a logarithmic axis, as in Fig. 6.4b. This diagram shows complete information on the electron–hole concentrations (throughout the P–N junction system), but again, it should be kept in mind that these are logarithmic plots and that the concentrations drop by orders of magnitude at very short distances.

6.1.2 Reverse-Biased P–N Junction

When a reverse-bias voltage is applied to a P–N junction (Fig. 6.5), the energy-band diagram is changed to express the applied voltage. Because of the applied voltage, the P–N junction is no longer in thermal equilibrium, and the Fermi level is not constant throughout

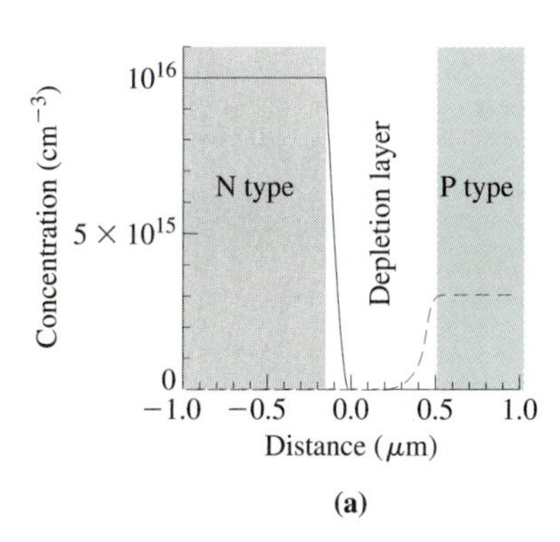

(a)

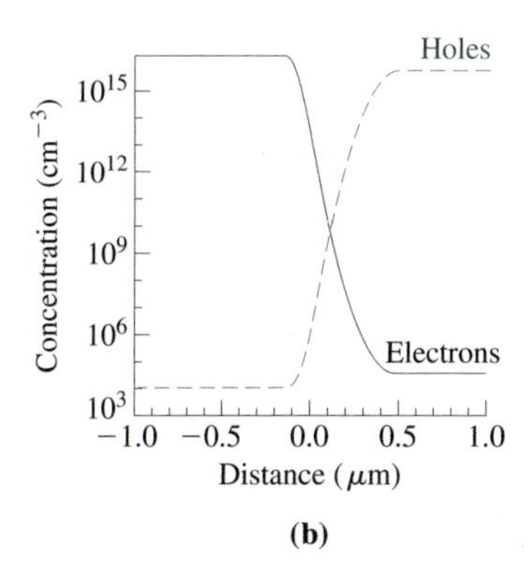

(b)

Figure 6.4 (a) Linear and (b) logarithmic concentration diagrams for a P–N junction ($N_D = 10^{16}$ cm^{-3} and $N_A = 3 \times 10^{15}$ cm^{-3}).

the system. The voltage applied to the P–N junction appears mainly across the depletion layer, expanding the depletion-layer width and increasing the electric field in the depletion layer (there cannot be a significant increase in the field in the neutral regions because it would move the mobile electrons and holes, producing an unsustainable current flow). A voltage drop $-V_D$ is expressed in the band diagram as energy difference of qV_D. Figure 6.5 shows that the initially leveled Fermi levels in the N-type and P-type side are now separated by qV_D to express the voltage across the depletion layer. This increases the barrier height between the N-type and P-type region by qV_D.

The energy-band diagram of Fig. 6.5a is therefore constructed in the following way:

1. Fermi-level lines (the dashed–dotted line in the figure) separated by qV_D are drawn first. The *lower* Fermi-level line is for the N-type region, indicating that the N-type region is at *higher* electric potential ($E_{pot} = -q\varphi$). The Fermi-level lines are drawn inside the electroneutral regions but not in the depletion layer. The depletion layer is not in equilibrium and the Fermi level is not defined, although the electron and hole concentrations in the depletion layer can be related to quasi-Fermi levels.
2. The conduction and valence bands are drawn for the N-type and P-type neutral regions (outside the depletion layer). The bands are placed appropriately with respect to the Fermi level, indicating the band diagrams of N-type and P-type semiconductors in equilibrium, as in Fig. 6.2.
3. The conduction- and valence-band levels are joined by curved lines to complete the diagram.

Since the P–N junction is not in thermal equilibrium, there must be a current flowing through the junction that works to bring the system back in equilibrium. A comparison between the energy-band diagrams in Figs. 6.3a and 6.5a shows that the increased energy barrier prevents the flow of majority carriers through the junction, which means that the current of minority carriers is no longer balanced. This current is usually labeled by I_S and consists of electron (I_{Sn}) and hole (I_{Sp}) components: $I_S = I_{Sn} + I_{Sp}$. Furthermore, a comparison of the electron and hole concentrations in the widened depletion layer in the reverse-biased P–N junction (Fig. 6.5a) and the depletion layer in the equilibrium P–N junction (Fig. 6.3a) illustrates that the probability of electrons and holes meeting each other in the depletion layer is significantly reduced by the reverse bias. As a consequence, the

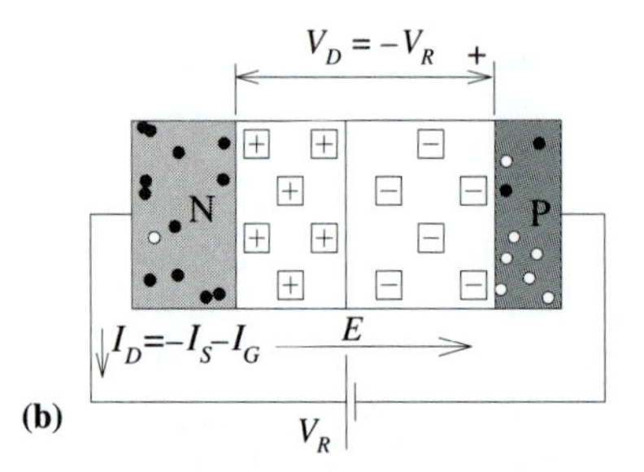

Figure 6.5 (a) Energy-band diagram and (b) cross section of a reverse-biased P–N junction.

generation current I_G is no longer balanced by the recombination current and it contributes to the total current through the P–N junction.

The current of minority carriers (I_S) does not increase with an increase in the reverse-bias voltage. An increase in the reverse-bias voltage would increase the split between the Fermi levels and consequently the slope of the energy bands (increased electric field in the depletion layer). It may become easier for the electrons to roll down and the holes to bubble up through the depletion layer; however, it is not the slope of the bands in the depletion layer that limits the minority-carrier current—it is always favorable enough for any electron appearing at the top to roll down, and for any hole appearing at the bottom to bubble up to the opposite side. The current of the minority carriers is limited by the number of minority electrons and holes appearing at the edges of the depletion layer. A waterfall provides a good analogy for this effect: the water current is not limited by the height or steepness of the fall but by the amount of water that reaches the fall.

At this point, it is important to recall that average number of minority carriers in a neutral region of a semiconductor device is almost always smaller than one carrier (Example 1.8). The consequences of this fact will be considered in Section 10.1. At this stage, it is important to note that current of minority carriers is negligibly small. *In practical diodes, it is the generation current I_G that determines the total current through a reverse-biased P–N junction.* If we label the average time a single R–G center takes to generate an electron–hole pair by τ_t, the generation current due to this center is q/τ_t. If there are M_t R–G centers of the same type in the depletion layer, the total generation current is $I_G = M_t q/\tau_t$. The total number of the R–G centers can be expressed by their concentration N_t multiplied by the volume of the depletion layer, $A_J w_d$, where A_J is the junction area and w_d is the depletion-layer width. Therefore, the generation current can be expressed as

$$I_G = q \underbrace{\frac{N_t}{\tau_t}}_{|U|} w_d A_J = q \underbrace{\frac{n_i}{\tau_g}}_{|U|} w_d A_J \tag{6.3}$$

where $|U|$ is the effective generation rate and $\tau_g = \tau_t n_i / N_t$ is the generation constant that can be related to the minority-carrier lifetimes τ_n and τ_p (Section 5.3.3).

Figure 6.5a illustrates that a generated electron in the depletion layer rolls down into the N-type region, and it can be imagined that the electron is further attracted to the positive pole of the battery V_R, where it is neutralized by a positive charge unit from the battery. This is half the circuit for the generation current. The other half is completed by the generated hole that is attracted to the negative pole of the battery. It can be said that the aim of the generation current is to neutralize the battery charge as the cause for this net current and to bring the P–N junction back to equilibrium, where $qV_D = 0$ and the Fermi levels are leveled off again. The situation in which the battery voltage is kept, or assumed, constant is referred to as *steady state*.

The generation current is very small (usually more than 10 orders of magnitude smaller than the current of the majority carriers); if it is neglected as a small leakage current, we can say that a reverse-biased P–N junction (diode) acts as an open circuit.

6.1.3 Forward-Biased P–N Junction

Figure 6.6a illustrates that the energy barrier height is reduced for the case of *forward-biased P–N junction*. The forward-bias voltage V_D appearing across the depletion layer splits the Fermi levels in the opposite direction, to reduce the barrier height from qV_{bi} to $q(V_{bi} - V_D)$. One effect is a significant increase in the concentration of electrons in the N-type and holes in the P-type regions with energies higher than the reduced barrier height. This results in a significant increase of the number of majority carriers passing through the depletion layer and appearing as minority carriers on the other side. The concentration of the minority carriers (electrons in the P-type and holes in the N-type region) is, consequently, increased along the depletion-layer boundary, as shown in Fig. 6.6c.

Holes from the P-type region are pushed toward the N-type region by the positive pole of the battery; likewise, electrons are pushed by the negative pole of the battery. This results in increased chances of electrons and holes meeting each other and recombining in the depletion layer or either the P-type or the N-type region, as illustrated in Fig. 6.6 by the vertical arrows. The process of recombination reduces the minority-carrier concentration in the neutral regions from the highest value at the boundary of the depletion layer to the equilibrium level, as shown in the concentration diagram of Fig. 6.6c. The concentration of the majority carriers is not significantly altered, since any recombined carrier is quickly replaced from the power supply.

Similar to the case of reverse bias and the generation current, a hole travels the part of the circuit between the positive pole of the battery and the recombination point, whereas an

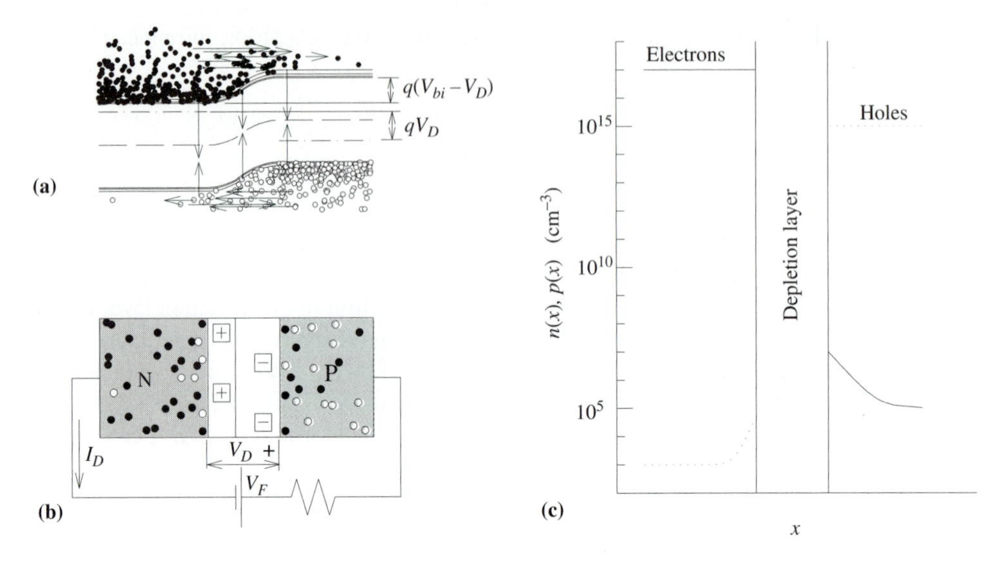

Figure 6.6 (a) Energy-band diagram, (b) cross section, and (c) concentration diagram of a forward-biased P–N junction.

electron travels the other part (from the negative pole of the battery to the recombination point). Again, it can be said that the current flowing through the junction is aimed at removing the battery charge that is causing the current, except in this case the current flows in the opposite direction from the reverse-bias current.

The number of majority carriers able to go over the barrier in the depletion layer increases as the barrier height $q(V_{bi} - V_D)$ is lowered by an increase in the forward-bias voltage V_D. This means that the current of the majority carriers flowing through the depletion layer is increased with an increase in the forward-bias voltage V_D. A P–N junction diode is said to be forward-biased when the voltage V_D is large enough to produce a significant flow of the majority carriers through the P–N junction—that is, when it produces a significant effective diode current (in this case we can neglect the minority-carrier current, which remains constant and small).

To gain insight into the type of current–voltage dependence of the forward-biased diode, it is necessary to refer to the type of electron/hole energy distribution in the conduction/valence band. The energy distribution of the electrons/holes is according to the Fermi–Dirac distribution, the tails of which are very close to an exponential distribution, as explained in Section 2.4 (Fig. 2.18). Therefore, as the energy barrier height in the depletion layer $q(V_{bi} - V_D)$ (Fig. 6.6a) is lowered by the increasing V_D voltage, the number of majority electrons able to go over the barrier increases exponentially. The same occurs with the majority holes. As a result, the current of a forward-biased diode increases exponentially with the voltage. When the applied voltage is normalized by the thermal voltage $V_t = kT/q$, the current–voltage dependence of a forward-biased diode can be expressed as

$$I_D \propto e^{V_D/V_t} \tag{6.4}$$

6.1.4 Breakdown Phenomena

It has been explained that almost the whole reverse-bias voltage applied to a diode drops across the P–N junction depletion layer, giving rise to the electric field E in the depletion layer. The reverse-bias current is small and independent of the reverse-bias voltage, but only to a certain value of the reverse bias, which is referred to as *breakdown voltage*. When the breakdown voltage is reached, the reverse-bias current increases sharply, and the diode is said to be in breakdown mode. This electrical breakdown does not cause permanent damage by itself, provided the breakdown current is limited by a resistance connected in series with the diode. In fact, the breakdown mode can be utilized, in particular as a voltage reference. The simplest reference-voltage circuit is shown in Fig. 6.7a. As Fig. 6.7b illustrates, the steep (nearly vertical) portion of the I_D–V_D characteristic in the breakdown is used to provide the reference voltage. Diodes that are operated in their breakdown mode are called Zener diodes, and they are presented with a modified symbol, also illustrated in Fig. 6.7a.

There are two entirely different physical mechanisms that can lead to electrical breakdown in reverse bias, and they are described in the following text.

Avalanche Breakdown

The direction of the electric field in the depletion layer of a reverse-biased P–N junction is such that it takes the minority carriers through the P–N junction, causing the reverse-bias current of the diode (I_S). Let us consider an electron attracted by the electric field E from the P-type region (Fig. 6.8a). This electron is accelerated by the electric field in the depletion layer, which means that the electron gains kinetic energy $E_{kin} = -qEx$ as the distance traveled by the electron (x) increases. As the electron moves through the depletion layer, there is a high probability that it will be scattered by a phonon (vibrating crystal atom) or an impurity atom. In the process of this collision, the electron may deliver its kinetic energy to the crystal. If the reverse-bias voltage V_R (and consequently the electric field E) is increased so much that the electron gains enough kinetic energy to break the covalent bonds when the electron collides, an electron–hole pair is generated during this collision. This process is illustrated in Fig. 6.8a. Now we have two electrons

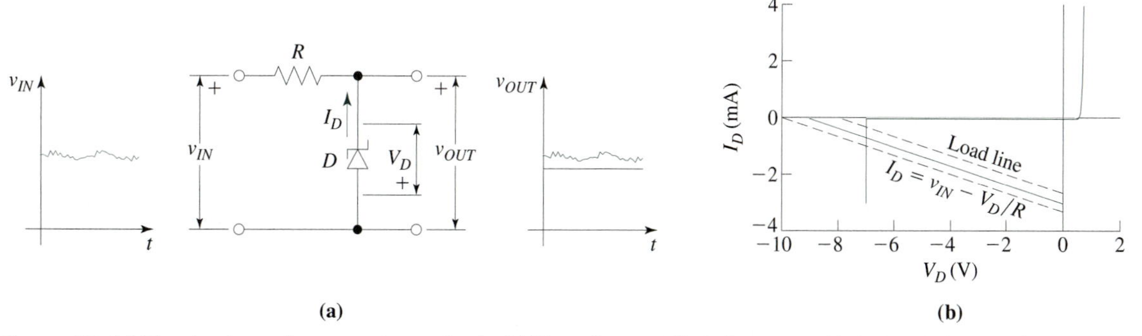

Figure 6.7 (a) The simplest reference-voltage circuit. (b) The reference diode is operated in its breakdown region. Variations of the input voltage (indicated by the dashed load lines) do not change the voltage across the diode; it is only the current that is influenced.

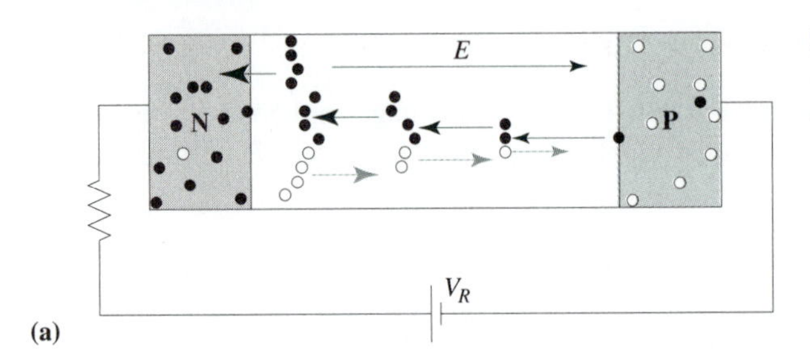

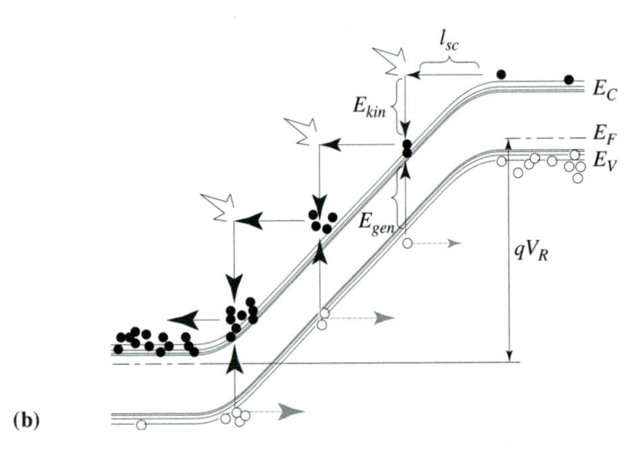

Figure 6.8 Illustration of the avalanche breakdown using (a) a diode cross section and (b) the energy bands.

that can generate two additional electron–hole pairs. The four new electrons can generate an additional four to become eight, then sixteen, and so on, in this avalanche process.

The electron–hole generation process is repeated many times in the depletion layer, because the average distance between two collisions (or also called scattering length) l_{sc} is much smaller than the depletion-layer width. Therefore, once the electric field is strong enough that the kinetic energy gained between two collisions ($E_{kin} = -qEl_{sc}$) is larger than the threshold needed for electron–hole generation, the avalanche process is triggered and an enormous number of free carriers is generated in the depletion layer. This produces the sudden diode current increase in the breakdown region. Appropriately, this type of breakdown is called *avalanche breakdown*.

The energy-band diagram, shown in Fig. 6.8b, provides a deeper insight into the avalanche mechanism. We have frequently said that electrons roll down along the bottom of the conduction band, while the holes bubble up along the top of the valence band. This rolling and bubbling, however, is not quite as smooth as this simplified model may suggest. At this point we should remember the details of the electron transport along tilted energy bands, explained in Section 3.1 (Fig. 3.2). This process is again illustrated in the band diagram of Fig. 6.8b. The horizontal motion of the electrons between two collisions expresses the fact that the total electron energy is preserved. However, the potential energy of the electrons is transformed into kinetic energy when electrons move through the depletion layer as the energy difference from the bottom of the conduction band increases. When the electrons collide and deliver the kinetic energy E_{kin} to the crystal, they fall down on the energy-band diagram. Therefore, the rolling down of electrons and bubbling up of the holes occurs in staircase fashion.

As mentioned, the kinetic energy E_{kin} gained between two collisions can be delivered to the crystal as thermal energy. In fact, this is the only possibility when $E_{kin} < E_g$, where E_g is the energy gap. However, as the slope of the bands is increased by increase in the reverse-bias voltage, there will be a voltage V_R at which E_{kin} becomes large enough to move an electron from the valence band into the conduction band. This is the process of electron–hole pair generation. As mentioned, this process is repeated many times in the depletion layer, where the newly generated electrons also gain enough energy to generate additional electron–hole pairs. This is the avalanche mechanism. Note that, theoretically, the holes can also generate electron–hole pairs; however, this process is rarely observed because the electron-induced avalanche requires smaller energy and happens at lower voltages.

Understanding the mechanism of the avalanche process gives us a basis for understanding the temperature behavior of the avalanche breakdown. If the temperature is increased above room temperature, the phonon scattering (Section 3.3.3) is enhanced and the average distance between two collisions (the scattering length l_{sc}) is reduced. This means that a larger electric field E is needed to achieve the threshold kinetic energy ($E_{kin} = -qEl_{sc}$). *An increase in the temperature increases the breakdown voltage. The avalanche breakdown is said to have a positive temperature coefficient.* This temperature dependence of avalanche breakdown is generally undesirable because we do not want the reference voltage to be temperature-dependent.

We can also understand now that the avalanche breakdown voltage is concentration-dependent. A diode with higher doping levels in the P-type and N-type regions has a larger built-in electric field (electric field due to the ionized doping atoms in the depletion layer). Therefore, less external voltage V_R is needed to achieve the critical breakdown field. Adjusting the doping levels enables us to make diodes with avalanche breakdown voltage as low as ≈ 6 V and as high as thousands of volts.

Tunneling Breakdown

The tunneling effect is related to the wave properties of electrons, as explained in Section 2.1.3. Figure 6.9a illustrates the bound electrons (electrons in the valence band) in the P-type region of the diode. The energy gap at the P–N junction separates the valence electrons of the P-type region from the N-type side of the diode. This, however, does not cause the electron wave function to abruptly drop to zero at the energy barrier (the top of the valence band in the depletion layer). This would mean that the electrons could be confined in a strictly defined space, denying them the wave properties they in fact exhibit. The wave function does have a tail expanding beyond the position of the top of the valence band.

Figure 6.9b illustrates the case in which the width of the energy barrier d is reduced by the increased reverse-bias voltage. This reduced energy barrier width is smaller than the tail of the electron wave function, which means that there is some probability that a valence electron from the P-type region is found on the N-type side of the P–N junction. This is the *tunneling* phenomenon. The probability of finding an electron (or hole) on the other side of energy barrier is called the *tunneling probability*.

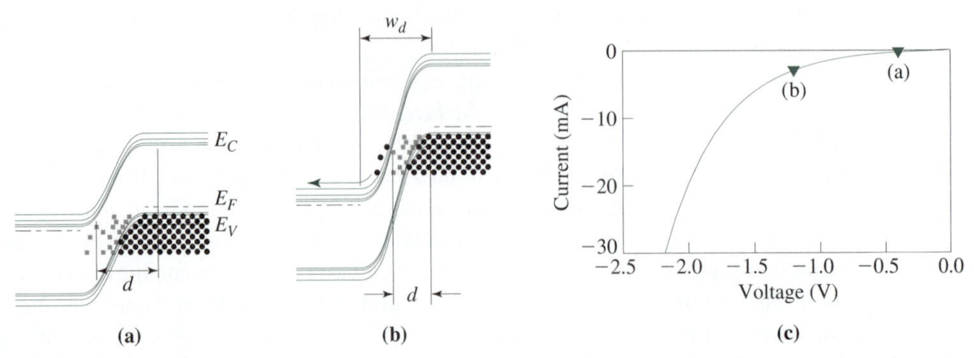

Figure 6.9 Illustration of electron tunneling through a P–N junction. (a) Small reverse-bias voltage, no tunneling is observed. (b) The tail of the wave function for valence electrons in the P-type region expands onto the N-type side—this is the phenomenon of electron tunneling through the P–N junction energy barrier. (c) Typical tunneling breakdown I_D–V_D characteristic.

The tunneling effect cannot be neglected even though the tunneling probability is typically very small. To explain this, assume a tunneling probability of 10^{-7}. Because there are more than 10^{22} electrons per cm^3 in the valence band, the concentration of electrons tunneling through the energy barrier of a P–N junction is higher than $10^{22} \times 10^{-7} = 10^{15}\ cm^{-3}$. This concentration of electrons is high enough to make a significant current when the tunneling electrons are attracted to the positive battery terminal connected to the N-type side. This current is called *tunneling current.*

If the reverse-bias voltage is further increased, the associated energy-band bending will further reduce the energy barrier width d, increasing the tunneling probability. Because the tunneling probability depends exponentially on the energy barrier width, the tunneling current depends exponentially on the reverse-bias voltage. This is illustrated in Fig. 6.9c.

Tunneling breakdown cannot occur unless the depletion layer, and consequently the energy barrier, is very narrow. A narrow depletion layer appears when the doping levels of both the N-type and the P-type region of the diode are very high. In addition, an increase in the reverse-bias voltage V_R does not significantly expand the depletion-layer width, w_d. This makes the energy barrier width (d) reduction due to the band bending quite significant.

We mentioned before that the doping concentration influences the avalanche breakdown voltage as well. In fact, it is the level of doping concentration that determines whether the avalanche-breakdown voltage or the tunneling-breakdown voltage will be smaller, thereby determining which type of breakdown will actually occur. If the doping concentration is such that the breakdown occurs at more than about 6 V, then the breakdown is generally avalanche-type breakdown. At higher doping levels, the tunneling occurs at lower voltages than those needed to trigger the avalanche mechanism; therefore, the tunneling breakdown is actually observed.

The I_D–V_D characteristic associated with the tunneling breakdown is not as abrupt as the I_D–V_D characteristic of the avalanche breakdown. This can be seen by comparing the I_D–V_D characteristics of Fig. 6.13 (Section 6.2.2) and Fig. 6.9c. Consequently, the stability of the reference voltage provided by the diodes above 6-V diodes is a lot better than what can be achieved by the sub-5-V diodes.

Another difference is the temperature coefficient. Whereas for avalanche breakdown the temperature coefficient is positive, it is negative in the case of tunneling breakdown. An increase in the temperature reduces the tunneling breakdown voltage. This enables us to reduce the temperature dependence of the reference voltage by designing diodes that operate in the mixed-breakdown mode. The breakdown voltage of these diodes is generally in the 5- to 6-V range.

6.2 DC MODEL

The term device *model* is used for a set of mathematical equations that can be used to calculate the electrical characteristics of the device. A *DC model* is a set of mathematical equations that can be used to calculate the time-independent current–voltage (I–V) characteristics. These equations are mainly needed for the design and analysis of circuits involving the considered device(s). A widely used computer program for this purpose is SPICE. Even with the help of SPICE, a good understanding of device models is necessary, in particular in terms of the adjustable parameters that are included in these models.

In this section, the basic I–V equation is derived from the first principles of diode operation and is upgraded to its typical SPICE form. In this way, a link is made between the physics of diode operation and the pragmatic SPICE parameters. The influence of temperature on the I–V characteristic is also considered.

6.2.1 Basic Current–Voltage (I–V) Equation

To derive the basic I–V equation, let us assume the bias arrangement as shown in Fig. 6.10a and consider the corresponding concentration diagram shown in Fig. 6.10b. The characteristic points labeled in the concentration diagram of Fig. 6.10b are w_p, the depletion-layer edge on the P-type side; w_n, the depletion-layer edge on the N-type side (appearing with a minus sign because the origin of the x-axis is placed at the P–N junction); n_{pe} and p_{ne}, the equilibrium concentrations of the minority carriers; and $n_p(w_p)$ and $p_n(-w_n)$, the minority-carrier concentrations at the edges of the depletion layer.

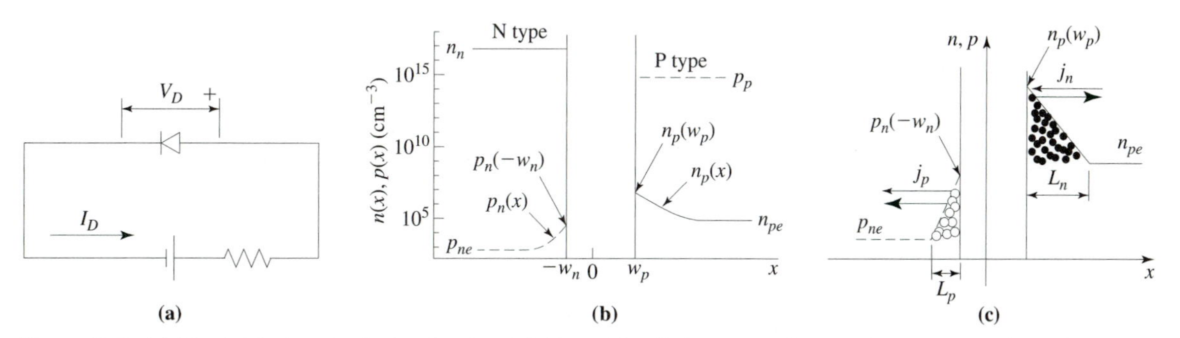

Figure 6.10 (a) Simple biasing circuit showing the definition of the diode current direction I_D and the polarity of the diode voltage V_D. (b) Carrier-concentration diagram illustrating the characteristic points and symbols used. (c) Carrier-concentration diagram illustrating the diffusion current of the minority carriers.

Following again the path of electrons, starting at the negative battery terminal, we come to the neutral N-type region of the diode. Electrons can move through the neutral N-type region because the electric field from the negative battery terminal pushes them toward the depletion layer (this is the *drift current*). It is very important to clarify the question of how strong this field is. The concentration of electrons in the N-type region (the majority carriers) is huge, $n_n = 10^{17}$ cm^{-3} in the example of Fig. 6.10b. Because the drift electron current density is given by $j_n = \sigma_n E = q\mu_n n_n E$ (the differential form of Ohm's law), only a small electric field E is needed to produce a moderate $n_n E$ value, thus a moderate current density. Assuming $\mu_n = 625$ cm^2/V $\cdot$ s, we find that the electric field needed to produce 1 μA/μm^2 of drift-current density is only $E = 1$ V/mm. If the width of the N-type region is 10 μm, the voltage across the neutral region is only 10 mV. It is rightly assumed that almost the whole diode voltage V_D appears across the depletion layer.

The electrons that reach the depletion layer with large enough energies can go through the depletion layer to appear at the P-type side (the point $x = w_p$ on the concentration diagram). In the P-type region, the electrons are the minority carriers. The equilibrium concentration of the minority electrons in the example shown in Fig. 6.10b is about 10^5 cm^{-3}. If the same electric field and electron mobility are assumed in the P-type region ($E = 1$ V/mm and $\mu_n = 625$ cm^2/V $\cdot$ s), the drift current of the minority electrons in the P-type region is $q\mu_n n_{pe} E = 10^{-12}$ μA/μm^2. This means that the drift current taking minority electrons away from the depletion-layer edge (the point $x = w_p$ on the concentration diagram) is negligible compared to the drift current that brings electrons to that point. In the absence of any other current, minority electrons accumulate at the depletion-layer edge. As Fig. 6.10b illustrates, this causes an increased concentration of minority electrons $[n_p(w_p) > n_{pe}]$. The increasing concentration gradient increases the diffusion current of the minority electrons, and at some point the diffusion current is strong enough to balance the incoming flow of electrons (the drift current of electrons from the N-type region). Analogous considerations apply to the holes. With the balance of incoming and outgoing currents, the steady state is achieved with steady but nonuniform profiles of the minority-carrier concentrations.

Therefore, *the currents of minority carriers are basically diffusion currents.* The minority carriers move due to the diffusion until recombined. Once recombination has occurred, the electric circuit is closed—the other part is completed by the recombining majority carriers as explained previously.

To derive the I–V equation of the diode, it is sufficient to find the diffusion currents of the minority carriers at the edges of the depletion layer.

To be able to use the diffusion-current equation [Eq. (4.5)], we need the profiles of minority carriers, $n_p(x)$ and $p_n(x)$. Based on the considerations presented in Section 5.2.3, we know that these profiles follow exponential dependencies, with the concentrations dropping e times at distances equal to the diffusion lengths, L_n and L_p. Noting that $\delta n(0)$ in Eq. (5.18) is equal to $n_p(w_p) - n_{pe}$ in Fig. 6.10, and using this concentration profile and the analogous profile for the minority holes in the diffusion-current equations [Eq. (4.5)], the following result is obtained:

$$
\begin{aligned}
j_n &= qD_n\left[n_{pe} - n_p(w_p)\right]/L_n \\
j_p &= -qD_p\left[p_n(-w_n) - p_{ne}\right]/L_p
\end{aligned}
\qquad (6.5)
$$

In Eq. (6.5), the diffusion currents of minority electrons and holes are labeled by j_n and j_p, respectively, in accordance with the labeling in Fig. 6.10c. The result presented by Eq. (6.5) would be obtained if the minority-carrier concentrations $n_p(x)$ and $p_n(x)$ were approximated by linear functions, changing from the maximum values at the edges of the depletion layer to the equilibrium values at distances L_n and L_p, respectively. The linear-profile approximations are illustrated in Fig. 6.10c.

In Eq. (6.5), the concentrations of the minority carriers at the depletion-layer edges, $n_p(w_p)$ and $p_n(-w_n)$, are the only parameters that significantly depend on the applied voltage. Remember that $n_p(w_p)$ and $p_n(-w_n)$ relate to the electrons and holes in the N-type and P-type regions, respectively, that can overcome the energy barrier in the depletion layer. The energy-barrier height is $q(V_{bi} - V_D)$, as shown previously in Fig. 6.6c, which means that it depends on the applied voltage V_D. To find out the concentration of electrons having kinetic energies larger than the barrier height $q(V_{bi} - V_D)$, we should count and add all the electrons appearing at the energy levels higher than $q(V_{bi} - V_D)$. We know that the energy distribution of the electrons is given by the Fermi–Dirac distribution, which can be approximated by the exponential function when the Fermi level is inside the energy gap (Section 2.4.2):

$$f = \frac{1}{1 + e^{(E-E_F)/kT}} \approx e^{-(E-E_F)/kT} \tag{6.6}$$

Given that the function f represents the probability of having an electron with energy E, the total number of electrons having energies larger than $q(V_{bi} - V_D)$ can be found by adding up the probabilities of finding an electron at any single energy level above $q(V_{bi} - V_D)$. Because the energy levels in the conduction band are assumed to constitute a continuous energy band, the *adding up* is performed by way of the following *integration:*

$$n_p(w_p) \propto \int_{q(V_{bi}-V_D)}^{\infty} f\, dE = \int_{q(V_{bi}-V_D)}^{\infty} e^{(E-E_F)/kT}\, dE \tag{6.7}$$

which gives the following result:

$$n_p(w_p) \propto e^{qV_D/kT} = e^{V_D/V_t} \quad \Rightarrow \quad n_p(w_p) = C_A e^{V_D/V_t} \tag{6.8}$$

In Eq. (6.8), $V_t = kT/q$ is the thermal voltage ($V_t \approx 25.85$ mV at room temperature), and C_A is a constant that needs to be determined. To determine the constant C_A, we can consider the case of zero bias ($V_D = 0$), because we know that the electron concentration is equal to the equilibrium value n_{pe} throughout the P-type region in that case. Therefore, the constant C_A must be chosen so to express the fact that $n_p(w_p) = n_{pe}$ for $V_D = 0$. It is not difficult to see that the constant C_A must be equal to the equilibrium value of the electron concentration in the P-type region n_{pe}; thus,

$$n_p(w_p) = n_{pe} e^{V_D/V_t} \tag{6.9}$$

In a similar way it can be shown that

$$p_n(-w_n) = p_{ne}e^{V_D/V_t} \tag{6.10}$$

Replacing $n_p(w_p)$ and $p_n(-w_n)$ in Eq. (6.5) by Eqs. (6.9) and (6.10), the current densities j_n and j_p are directly related to the voltage applied, V_D:

$$\begin{aligned} j_n &= -\frac{qD_n}{L_n} n_{pe}\left(e^{V_D/V_t} - 1\right) \\ j_p &= -\frac{qD_p}{L_p} p_{ne}\left(e^{V_D/V_t} - 1\right) \end{aligned} \tag{6.11}$$

The total current density is obtained as the sum of the electron and hole current densities:

$$j = j_n + j_p = -\left(\frac{qD_n}{L_n} n_{pe} + \frac{qD_p}{L_p} p_{ne}\right)\left(e^{V_D/V_t} - 1\right) \tag{6.12}$$

Using the fact that the product of the minority- and the majority-carrier concentrations is constant in the thermal equilibrium [Eq. (1.6)] and that the majority-carrier concentrations are approximately equal to the doping levels N_D and N_A, the equilibrium minority-carrier concentrations are determined from the doping levels as

$$\begin{aligned} n_{pe} &= \frac{n_i^2}{N_D} \\ p_{ne} &= \frac{n_i^2}{N_A} \end{aligned} \tag{6.13}$$

in which case the current density becomes

$$j = -qn_i^2\left(\frac{D_n}{L_n N_A} + \frac{D_p}{L_p N_D}\right)\left(e^{V_D/V_t} - 1\right) \tag{6.14}$$

Finally to convert the current density j (A/m^2) into the terminal current I_D (in A), the current density is multiplied by the area of the P–N junction A_J. In addition, the sign of the terminal current (positive or negative) needs to be properly used. The convention is that the current is expressed as positive if it flows in an indicated direction (as in Fig. 6.10a, for example), and it is expressed as negative if it flows in the opposite direction. On the other hand, the minus sign in Eq. (6.14) for the current density means that the current density vector and the x-axis are in the opposite directions; thus the current flows from the P-type toward the N-type side. The current arrow in Fig. 6.10a is consistent with this but is not consistent with the x-axis direction; therefore, the minus sign in Eq. (6.14) should be dropped. With this the diode current can be expressed as

$$I_D = I_S\left(e^{V_D/V_t} - 1\right) \tag{6.15}$$

where I_S is used as a single parameter to replace the following:

$$I_S = A_J q n_i^2 \left(\frac{D_n}{L_n N_A} + \frac{D_p}{L_p N_D} \right) \tag{6.16}$$

In practice, the widths of the effective neutral regions are much smaller than the electron and hole diffusion lengths: $W_{anode} \ll L_n$ and $W_{cathode} \ll L_p$. As a consequence, the minority-carrier concentrations will drop to the equilibrium levels at the ends of the neutral regions. The concepts shown in the concentration diagrams of Fig. 6.10 and the corresponding form of Eq. (6.16) are still valid, so the diffusion lengths can be replaced by the actual widths of the anode and cathode neutral regions:

$$I_S = A_J q n_i^2 \left(\frac{D_n}{W_{anode} N_A} + \frac{D_p}{W_{cathode} N_D} \right) \tag{6.17}$$

The thermal voltage V_t appearing in Eq. (6.15) is approximately equal to 0.026 V at room temperature. Normally, either the voltage V_D is much larger than 0.026 V (forward bias) or its absolute value is much larger than 0.026 V (reverse bias). This means that $\exp(V_D/V_t)$ is either much larger than 1 (forward bias) or much smaller than 1 (reverse bias). In the case of the reverse bias $[\exp(V_D/V_t) \ll 1]$, the diode current due to minority carriers is $I_D \approx -I_S$. This expresses the fact that the current of minority carriers does not depend on the voltage applied as discussed in the preceding section. Consequently, the parameter I_S is called *saturation current*. In the case of the forward bias $[\exp(V_D/V_t) \gg 1]$, the diode current is $I_D \approx I_S \exp(V_D/V_t)$. This expresses the fact that the forward-bias current increases exponentially with V_D, again as discussed earlier [Eq. (6.4)].

EXAMPLE 6.1 Dominance of the Generation Current (I_G) Over the Saturation Current (I_S) in a Reverse-Biased P–N Junction

The following are typical parameters of a P–N junction diode: $N_A \gg N_D = 10^{15}$ cm^{-3}, $w_d \approx w_n = 1.2$ μm, $L_p \gg W_{cathode} = 10$ μm, $D_p = 12$ cm^2/s, $\tau_g = 1$ μs, and $A_J = 100$ μm $\times$ 100 μm. Calculate and compare the contributions of the saturation and the generation currents to the reverse-bias current in this diode.

SOLUTION

The saturation current is given by Eq. (6.17):

$$I_S = A_J q n_i^2 \left(\frac{D_n}{W_{anode} N_A} + \frac{D_p}{W_{cathode} N_D} \right) \approx A_J q n_i^2 \frac{D_p}{W_{cathode} N_D}$$

$$= 10^{-8} \times 1.6 \times 10^{-19} \times (10^{16})^2 \frac{12 \times 10^{-4}}{10 \times 10^{-6} \times 10^{21}} = 1.92 \times 10^{-14} \text{ A}$$

The generation current is given by Eq. (6.3):

$$I_G = q\frac{n_i}{\tau_g}w_d A_J = 1.6 \times 10^{-19}\frac{10^{16}}{10^{-6}}1.2 \times 10^{-6} \times 10^{-8} = 1.92 \times 10^{-11}\ \text{A}$$

Obviously, $I_G \gg I_S$, which means that the reverse-bias current of the diode is $I_R = -I_D \approx I_G$. The ratio between I_G and I_S is even higher for higher doping levels (an increase in N_D will reduce I_S) and for real diodes where surface generation adds to I_G.

EXAMPLE 6.2 A Diode Circuit

For the circuit of Fig. 6.1a, find the output voltage v_{OUT} for $v_{IN} = 5$ V. The value of the resistor is $R = 1$ kΩ.

SOLUTION

To begin with, it is obvious that

$$v_{OUT} = v_{IN} - V_D$$

The voltage across the diode can be expressed in terms of the diode current,

$$I_D = I_S e^{V_D/V_t} \quad \Rightarrow \quad V_D = V_t \ln(I_D/I_S)$$

and the diode current I_D is related to the output voltage as $v_{OUT} = I_D R$. Therefore,

$$v_{OUT} = v_{IN} - V_t \ln v_{OUT}/(I_S R)$$

The voltage v_{OUT} cannot be explicitly expressed from this equation. However, the equation can be solved by an iterative method: assume v_{OUT} value, calculate the value of the right-hand side (RHS), and compare to v_{OUT}, which is the left-hand side (LHS). Table 6.1 gives an example, where 5 V is used as the initial guess, and the obtained RHS value is used as the guess for the next iteration. It can be seen in Table 6.1 that after three iterations, the LHS and RHS become equal, which is the solution: $v_{OUT} = 4.308$ V.

TABLE 6.1

LHS	RHS
5.000 V	4.304 V
4.304 V	4.308 V
4.308 V	4.308 V

EXAMPLE 6.3 Layout Design and Current Rating of an IC Diode

The doping levels and the neutral-region widths of N- and P-type layers to be utilized for designing an IC diode are as follows: $N_D = 10^{21}$ cm^{-3}, $N_A = 10^{18}$ cm^{-3}, and $W_N = W_P = 1$ μm. Determine the area of the P–N junction so that the turn-on voltage of the designed diode will be 0.7 V if the current rating of the diode is to be specified as 1 mA at room temperature.

SOLUTION

The current rating of a diode, for a specified voltage, is determined by the value of the saturation current:

$$I_D = I_S e^{V_D/V_t}$$

$$I_S = I_D e^{-V_D/V_t} = 10^{-3} e^{-0.7/0.02585} = 1.74 \times 10^{-15} \text{A}$$

Given that $N_D \gg N_A$, Eq. (6.17) for the saturation current can be simplified:

$$I_S = A_J q n_i^2 \frac{D_n}{W_P N_A}$$

To determine the junction area A_J, which is the layout design parameter, the values of the technological parameters (N_A and W_P) and the physical constants (q, n_i, and D_n) are needed. The diffusion constant for minority electrons in a P-type silicon doped at $N_A = 10^{18}$ cm^{-3} can be determined from the Einstein relationship ($D_n = \mu_n V_t$), where the mobility μ_n can be estimated from the mobility plots shown in Fig. 3.8a. For the doping level of 10^{18} cm^{-3}, the electron mobility from Fig. 3.8a is estimated as $\mu_n = 350$ cm^2/V · s. This means that $D_n = 0.035 \times 0.02585 = 9 \times 10^{-4}$ m^2/s $= 9$ cm^2/s. Therefore,

$$A_J = \frac{I_S W_P N_A}{q n_i^2 D_n} = \frac{1.74 \times 10^{-15} \times 10^{-6} \times 10^{24}}{1.6 \times 10^{-19} \times (1.02 \times 10^{16})^2 \times 9 \times 10^{-4}} = 1.15 \times 10^{-7} \text{ m}^2$$

If the diode is to have a square shape, the side of the square will be $\sqrt{A_J} = \sqrt{1.15 \times 10^{-7}} = 3.4 \times 10^{-4}$ m $= 340$ μm.

***EXAMPLE 6.4 Derivation of the Electron-Diffusion Current for a Diode with a Short Anode**

The boundary condition for minority electrons in the infinitely long anode, $\delta n_p(\infty) = 0$, cannot be used in the case of a diode with an anode that is not much longer than the diffusion length of electrons (in other words, $W_{anode} \gg L_n$ is not satisfied). In this case, the boundary conditions for the excess electron concentrations at the beginning ($x = w_p$) and the end ($x = W_{anode} + w_p$) of the neutral region are given by $\delta n_p(w_p) = n_p(w_p) - n_{pe}$ and $\delta n_p(W_{anode} + w_p) = 0$, respectively.

(a) Apply the method of solving the continuity equation presented in Section 5.2.3 to derive the concentration distribution of minority carriers in the general case (neither $W_{anode} \gg L_n$ nor $W_{anode} \ll L_n$ can be used as a good approximation).

(b) Using the concentration distribution from part (a), derive the equation for the diffusion-current density of minority electrons at the edge of the depletion layer ($x = w_p$ in Fig. 6.10b).

(c) Simplify the equation obtained in part (b) for the case of a short diode ($W_{anode} \ll L_n$) using the following approximation: $\exp(\pm W_{anode}/L_n) \approx 1 \pm W_{anode}/L_n$.

SOLUTION

(a) Using the notation from Fig. 6.10 (the symbol n_p for the concentration of minority electrons), Eq. (5.14) and its general solution [Eq. (5.17)] can be rewritten as follows:

$$\frac{d^2 \delta n_p}{dx^2} - \frac{\delta n_p}{L_n^2} = 0$$

$$\delta n_p(x) = A_1 e^{s_1 x} + A_2 e^{s_2 x} = A_1 e^{x/L_n} + A_2 e^{-x/L_n} \tag{6.18}$$

where $\delta n_p(x) = n_p(x) - n_{pe}$. The constants A_1 and A_2 should be determined from the following boundary conditions:

$$\delta n_p(x) = \delta n_p(w_p) \qquad \text{for } x = w_p$$
$$\delta n_p(x) = 0 \qquad \text{for } x = w_p + W_{anode}$$

Applying these boundary conditions, the following system of linear equations is obtained:

$$A_1 e^{w_p/L_n} + A_2 e^{-w_p/L_n} = \delta n_p(w_p)$$
$$A_1 e^{(w_p+W_{anode})/L_n} + A_2 e^{-(w_p+W_{anode})/L_n} = 0$$

The solutions of this system are

$$A_1 = \delta n_p(w_p) \frac{e^{-(w_p+W_{anode})/L_n}}{e^{-W_{anode}/L_n} - e^{W_{anode}/L_n}}$$

$$A_2 = -\delta n_p(w_p) \frac{e^{(w_p+W_{anode})/L_n}}{e^{-W_{anode}/L_n} - e^{W_{anode}/L_n}}$$

Putting these constants into Eq. (6.18), one obtains

$$\delta n_p(x) = \delta n_p(w_p)\,\frac{e^{[W_{anode}-(x-w_p)]/L_n} - e^{-[W_{anode}-(x-w_p)]/L_n}}{e^{(W_{anode}/L_n)} - e^{-(W_{anode}/L_n)}} \tag{6.19}$$

(b) The diffusion current of the electrons [Eq. (4.5)] is

$$j_n(x) = qD_n\frac{dn_p(x)}{dx} = qD_n\frac{d\delta n_p(x)}{dx}$$

Replacing $\delta n_p(x)$ by the concentration profile given by Eq. (6.19), we obtain

$$j_n(x) = -qD_n\frac{\delta n_p(w_p)}{L_n}\,\frac{e^{[W_{anode}-(x-w_p)]/L_n} + e^{-[W_{anode}-(x-w_p)]/L_n}}{e^{(W_{anode}/L_n)} - e^{-(W_{anode}/L_n)}}$$

At $x = w_p$, this equation becomes

$$j_n(w_p) = -qD_n\frac{\delta n_p(w_p)}{L_n}\,\frac{e^{W_{anode}/L_n} + e^{-W_{anode}/L_n}}{e^{W_{anode}/L_n} - e^{-W_{anode}/L_n}} = -qD_n\frac{\delta n_p(w_p)}{L_n}\,\frac{\cosh(W_{anode}/L_n)}{\sinh(W_{anode}/L_n)}$$

Using the result for $n_p(w_p)$ given by Eq. (6.9), we find that $\delta n_p(w_p)$ is related to the applied voltage V_D:

$$\delta n_p(w_p) = n_p(w_p) - n_{pe} = n_{pe}e^{V_D/V_t} - n_{pe} = n_{pe}\left(e^{V_D/V_t} - 1\right)$$

Therefore, the current density can be expressed as

$$\begin{aligned} j_n(w_p) &= -qD_n\frac{n_{pe}}{L_n}\,\frac{\cosh(W_{anode}/L_n)}{\sinh(W_{anode}/L_n)}\left(e^{V_D/V_t} - 1\right) \\ &= -q\frac{D_n}{L_n}\frac{n_i^2}{N_A}\,\frac{\cosh(W_{anode}/L_n)}{\sinh(W_{anode}/L_n)}\left(e^{V_D/V_t} - 1\right) \end{aligned}$$

(c) Given that

$$\frac{\cosh(W_{anode}/L_n)}{\sinh(W_{anode}/L_n)} = \frac{e^{W_{anode}/L_n} + e^{-W_{anode}/L_n}}{e^{W_{anode}/L_n} - e^{-W_{anode}/L_n}} \approx \frac{1 + W_{anode}/L_n + 1 - W_{anode}/L_n}{1 + W_{anode}/L_n - (1 - W_{anode}/L_n)} = \frac{L_n}{W_{anode}}$$

the current density becomes

$$j_n(w_p) = -qD_n\frac{n_{pe}}{W_{anode}}\left(e^{V_D/V_t} - 1\right) = -q\frac{D_n}{W_{anode}}\frac{n_i^2}{N_A}\left(e^{V_D/V_t} - 1\right)$$

Comparing this result to Eq. (6.11) for the case of a long anode ($L_n \ll W_{anode}$), we see that L_n is basically replaced by W_{anode}. This confirms the approach used to write Eq. (6.17).

6.2.2 Important Second-Order Effects

The diode theory, presented in the previous sections, introduced the rudimentary diode model [Eq. (6.15)]. This model takes into account only the effects that are of principal importance for the diode operation. However, there are a number of second-order effects that significantly influence the characteristics of real diodes. Consequently, the rudimentary model very frequently fails to accurately fit the experimental data. This section introduces the most important second-order effects: carrier recombination and high-level injection effects, parasitic resistance effects, and breakdown.

Recombination Current and High-Level Injection

The following two assumptions were made while developing the rudimentary model in the previous section:

1. No electron–hole recombination occurs in the depletion layer.
2. The electric field in the neutral regions is negligibly small, meaning that there is no voltage drop in the neutral regions.

These assumptions did simplify the modeling; however, most frequently they are not justified in terms of model accuracy.

As for the electron–hole recombination in the depletion layer, it is obvious that some of the electrons and holes will inevitably meet in the depletion layer, recombining with each other there. This recombination process enables electrons and holes with kinetic energies smaller than the barrier height $q(V_{bi} - V_D)$ to contribute to the current flow, increasing the total current. Figure 6.6 helps explain this phenomenon. Obviously, an electron and a hole cannot come to a common point if their kinetic energies (distances from the bottom and the top of the conduction and valence bands, respectively) do not add up to $q(V_{bi} - V_D)$, which means that such an electron–hole pair does not get recombined. An important point is, however, that they do not need independently to have energies larger than $q(V_{bi} - V_D)$, as is necessary for the diffusion current described in the previous sections. In fact, an electron and a hole with kinetic energies as low as $q(V_{bi} - V_D)/2$ can meet each other in the middle and get recombined. If the recombination current was exclusively due to recombination of the electrons and holes satisfying this minimum energy condition, it would be proportional to $\exp[-(V_{bi} - V_D)/2V_t]$, thus proportional to $\exp[V_D/2V_t]$. *The maximum recombination current is proportional to* $\exp[V_D/2V_t]$, *as distinct from the diffusion current, which is proportional to* $\exp[V_D/V_t]$.

The voltage dependencies of the diffusion and the maximum recombination currents are similar, for they are both exponential-type dependencies. Indeed, they can be expressed in the following common way:

$$I_D \propto e^{V_D/nV_t} \tag{6.20}$$

where $n = 1$ for the diffusion current and $n = 2$ for the maximum recombination current. More importantly, varying the coefficient n between 1 and 2 would enable us to fit any combination of diffusion and recombination currents.

Furthermore, the introduction of the variable coefficient n eliminates the need for the second assumption used to develop the rudimentary model, namely the assumption of zero voltage drop in the neutral regions. The existence of excess minority electrons and holes (illustrated in Fig. 6.10b and 6.10c) means that there must be an electric field associated with this charge. In the case of high injection levels, the voltage drop associated with this field cannot be neglected.

The electric field in the neutral regions leads to voltage drops v_n between the depletion-layer edge and the N-type bulk and v_p between the P-type bulk and the corresponding edge of the depletion layer. Therefore, the voltage applied V_D is distributed as $v_n + v_{depl} + v_p$, where v_{depl} is the voltage across the depletion layer. It is the voltage across the depletion layer that alters the height of the potential barrier in the depletion layer: $q(V_{bi} - v_{depl})$. The assumption used to derive the rudimentary model is that $v_n, v_p \ll v_{depl}$, thus $v_{depl} \approx V_D$.

In the case of high-level injection, the voltage that drops in the neutral regions becomes pronounced, which means that the rudimentary model overestimates the current. This is because the barrier height is reduced by only a fraction of qV_D, not by the whole value of qV_D, as assumed in that model. However, the coefficient n introduced in Eq. (6.20) enables us to involve this effect because the term qV_D/nV_t implies that the barrier is reduced by qV_D/n $(n > 1)$, which is a fraction of qV_D.

With the variable coefficient n, the current–voltage equation of a forward-biased diode becomes

$$I_D = I_S e^{V_D/nV_t} \tag{6.21}$$

The coefficient n is called the *emission coefficient*.

To illustrate the importance of the emission coefficient n, the experimental data of a real diode are fitted with the rudimentary model ($n = 1$) and the model that allows n to be adjusted [Eq. (6.21)] in Fig. 6.11. It is obvious that the rudimentary model (the colored

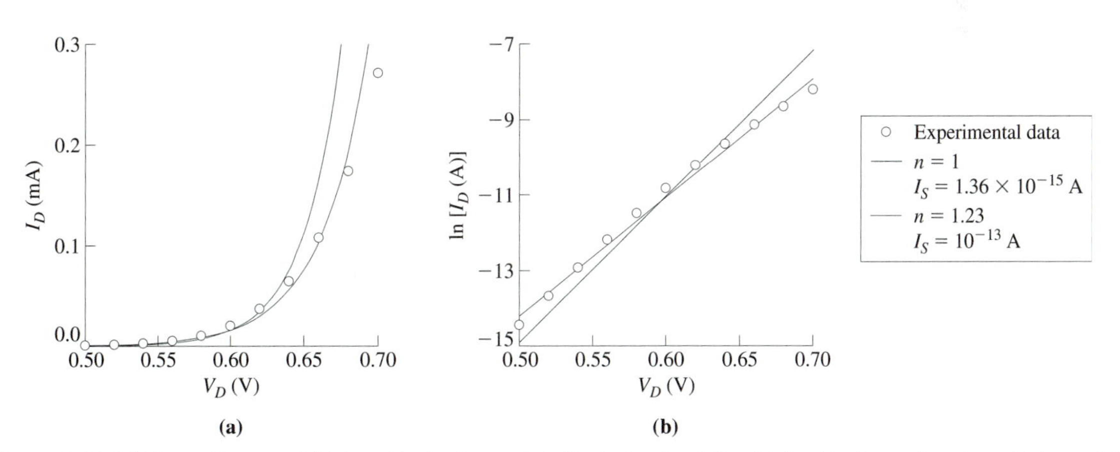

Figure 6.11 (a) Linear–linear and (b) logarithmic–linear plots illustrating best fits obtained by the rudimentary diode model ($n = 1$, colored lines) and the model with variable emission coefficient n (black lines). The experimental data (symbols) are for an LM3086 base-to-emitter P–N junction diode.

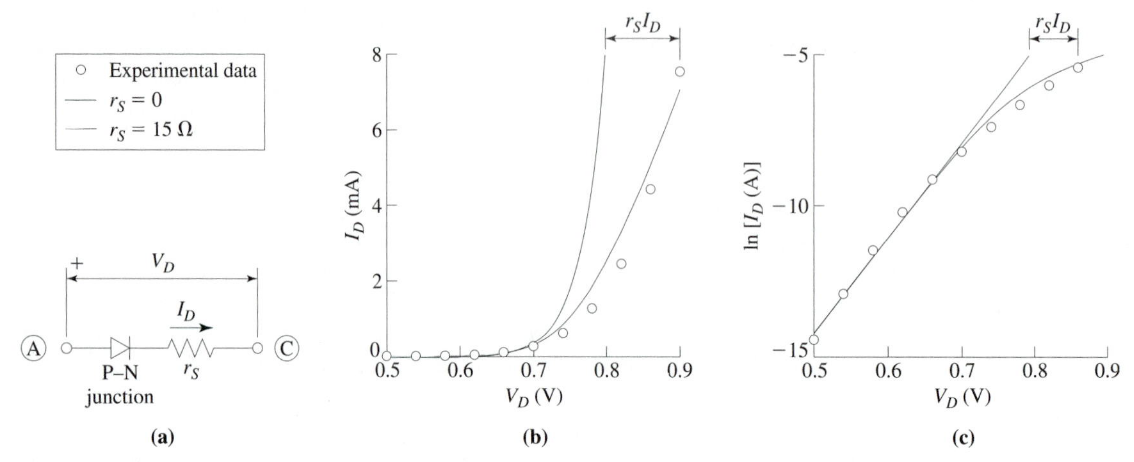

Figure 6.12 Illustration of the influence of the parasitic resistance on P–N junction diode characteristic. (a) Electric-circuit model. (b) Linear–linear and (c) logarithmic–linear plots of the diode forward characteristic (the symbols show the experimental data for an LM3086 base-to-emitter P–N junction diode; the difference between the black lines and the colored lines shows the effect of the parasitic resistance r_S).

lines, $n = 1$) does not properly fit the data, while quite satisfactory fitting can be achieved when the emission coefficient is set to $n = 1.23$ (the black lines).

The emission coefficient n can be introduced into the more general diode equation [Eq. (6.15)], covering both the forward-bias and the reverse-bias regions. The emission coefficient accounts for the effect in the forward-bias region while not altering the model predictions in the reverse-bias region:

$$I_D = I_S \left[\exp(V_D/nV_t) - 1\right] \tag{6.22}$$

Equation (6.22) is called the Shockley equation. It is also the basic SPICE model for the I_D–V_D characteristic of a diode. Clearly, I_S and n are the parameters of this model, so they will be referred to as SPICE parameters. The thermal voltage $V_t = kT/q$ is calculated in SPICE from the set value for the temperature T (the default value is 27°C).

Parasitic Resistance

Parasitic resistances exist in the structure of any real P–N junction diode. The parasitic resistances are especially pronounced at the metal–semiconductor contacts (both the cathode and the anode contacts) and in the neutral regions of the P-type and/or N-type bodies. The effects of all these resistances can be expressed by a single parasitic resistor r_S connected in series with the P–N junction diode itself, as illustrated in Fig. 6.12a. The series resistance r_S is a direct SPICE parameter.

Figure 6.12b and 6.12c illustrates that the diode model of Eq. (6.22) (the colored lines, $r_S = 0$, in Fig. 6.12) fits the experimental data properly only in the region of relatively small diode currents (<1 mA in this example). Clearly, the voltage across the parasitic resistance ($r_S I_D$) is not significant compared to the voltage applied (V_D) when the current I_D is small. However, as the current I_D is increased, the voltage drop $r_S I_D$

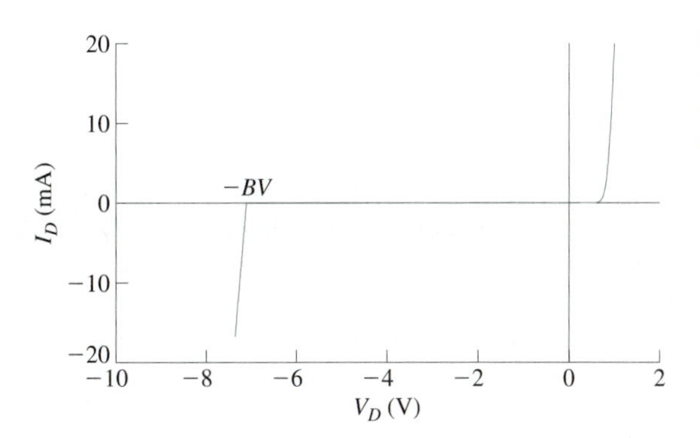

Figure 6.13 Experimental diode characteristic (LM3086 base-to-emitter P–N junction diode) illustrating the breakdown at $V_D = -BV$.

becomes pronounced, which means that the voltage across the P–N junction itself is smaller than V_D; therefore the current is smaller. Equation (6.22) can be used to model the diode characteristic in this case if $V_{D0} = V_D - r_S I_D$ is used instead of V_D. In other words, Eq. (6.22) is applied to the P–N junction diode in the equivalent circuit of Fig. 6.12a, and the resistance r_S is added between the cathode and the anode terminals.

Breakdown

The experimental characteristic of the P–N junction diode given in Fig. 6.13 shows that the reverse-bias current is small ($I_D \approx I_S$) only in the region $-BV < V_D \leq 0$. When the reverse-bias voltage becomes $V_D = -BV$, the current sharply increases due to the electrical breakdown. The P–N junction breakdown mechanisms are described in more detail in Section 6.1.4. Here, the appearance of the breakdown is introduced as a second-order effect that limits the voltage range in which a diode can be used as a rectifier. Although the breakdown voltage of $BV \approx 7$ V (as shown in Fig. 6.13) is typical for the base-to-emitter type diodes, it is by no means representative for all possible P–N junction diodes. The breakdown voltage can be smaller than that; however, more frequently it is much larger than 7 V and can be as large as hundreds or even thousands of volts in specifically designed high-voltage diodes. Obviously, the breakdown voltage is a diode parameter that has to be specified for any particular type of diode.

Generation Current as the Dominant Reverse-Bias Current

The Shockley equation predicts $I_D = -I_S$ as the reverse-bias current of a P–N junction: for $V_D < 0$ and $\exp(V_D/nV_t) \ll 1$, the current in Eq. (6.22) becomes $I_D = -I_S$. As already discussed in Section 6.1.2 and shown in Example 6.1, I_S is much smaller in real diodes than the generation current I_G. Furthermore, I_S is a voltage-independent current, which is not the case with the generation current. The dependence of I_G on the applied reverse-bias voltage ($V_R = -V_D$) is due to the depletion layer (w_d), which directly determines I_G value, as can be seen from Eq. (6.3). It has already been mentioned that w_d increases with V_R because the negative terminal of V_R pulls holes away from the junction and the positive terminal pulls electrons away from the junction to increase the electric field in the depletion layer. Therefore, there is an increase in I_G with V_R. The actual dependence of w_d on V_R is derived in Section 6.3.2, where it is shown that w_d is proportional to $(V_R + V_{bi})^m$, where V_{bi} is the built-in voltage and m is a parameter that takes a value between 1/3 and 1/2.

SPICE does not include a model for the generation current. There is a parallel conductance G_{MIN} set in parallel with the diode, predominantly as a way to manage numerical problems when I_S becomes too small. Although this conductance sets a linear relationship between the current and the voltage ($I_D = G_{MIN} V_D$), it can be used as an approximation of the generation current when it is really necessary to include it in the simulation. In most cases, the reverse-bias current of a diode can be neglected anyway.

6.2.3 Temperature Effects

The exponential term $\exp(V_D/nV_t) = \exp(qV_D/nkT)$ of the static diode model shows the explicit temperature dependence of the diode characteristic. More importantly, there is a very strong implicit dependence through the saturation current I_S. Equation (6.16) shows that I_S is proportional to n_i^2, where the intrinsic carrier concentration n_i strongly depends on temperature. The temperature dependence of the intrinsic carrier concentration is explicitly shown by Eq. (2.83). E_g in Eq. (2.83) is the energy gap, which is different for different materials, meaning that diodes based on different materials exhibit different temperature behavior. $E_g = 1.12$ eV for Si, which is the default value taken in SPICE; however, E_g is considered as a parameter, and it can be set to a different value to express a different material. It should be noted that E_g itself depends on the temperature, but this dependence does not significantly influence the diode characteristics.

To illustrate the temperature effects, the diode characteristics (originally shown in Fig. 6.11 for $T = 27°$C) are calculated by the SPICE model for temperatures $T = 100°C$ and $T = -50°$C, and they are shown in Fig. 6.14. It can be seen that the temperature basically shifts the I_D–V_D characteristic along the V_D axis by approximately -2 mV/°C. This temperature coefficient is negative because an increase in the temperature reduces the voltage across the diode needed to produce the same current.

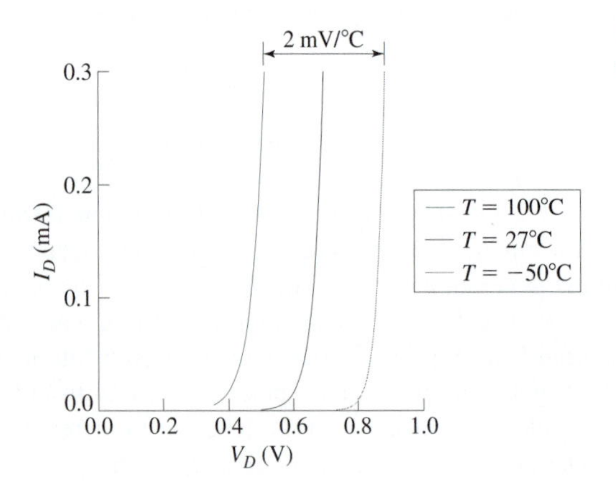

Figure 6.14 Temperature dependence of the diode characteristic.

EXAMPLE 6.5 Temperature Dependence of V_D

The anode of a P–N junction diode is connected to the positive terminal of a 12-V voltage source through a 1.2-kΩ resistor. This forward-biased diode is used as a temperature sensor. The room-temperature ($T = 27°C$) parameters of the diode are $I_S = 10^{-12}$ A, $n = 1.4$, and $r_S = 10\ \Omega$. What temperature is measured by the diode if the voltage across the diode is $V_D = 0.807$ V?

SOLUTION

The voltage across the diode changes by −2 mV/°C. We need to find V_D at room temperature to be able to determine the voltage shift, and consequently the temperature shift. The voltage across the diode is

$$V_D = nV_t \ln(I_D/I_S) + r_S I_D$$

As the current through the diode is

$$I_D = \frac{V_{supply} - V_D}{R}$$

the voltage can be expressed as

$$V_D = nV_t \ln\left(\frac{V_{supply} - V_D}{RI_S}\right) + r_S \frac{V_{supply} - V_D}{R}$$

This equation has to be solved numerically. If we assume $V_D = 0.7$ V, the value of the right-hand side (RHS) of the equation is 0.925 V. Assuming now $V_D = 0.925$ V, the RHS is calculated as 0.923 V. Finally, assuming $V_D = 0.923$ V gives 0.923 V for the RHS as well, and this is the voltage across the diode at room temperature. Therefore, the voltage shift is $0.807 - 0.923 = -0.116$ V, which gives a temperature shift of -0.116 V/(-0.002 V/°C) = 58°C. The temperature that is measured by the diode is $27 + 58 = 85$°C.

6.3 CAPACITANCE OF REVERSE-BIASED P–N JUNCTION

The capacitance effect is achieved when two conductive "plates" are placed close to each other, but are still electrically isolated by a dielectric. The dielectric suppresses any direct current (DC) flow through the capacitor, causing accumulation of positive and negative charge at the capacitor plates. If the voltage applied across the capacitor is varied, the charge stored at the capacitor plates changes, which appears as current flow through the circuit. A reverse-biased P–N junction can act as a capacitor. The two conductive plates are the N-type region on one side and the P-type region on the other side. These two conductive plates are separated by the depletion layer that acts as the capacitor dielectric.

As in any capacitor, the strength of the capacitance effect depends on three factors: (1) the area of the P–N junction, as larger conductive plates accommodate more charge, (2) the depletion-layer width, as thinner dielectrics provide more effective penetration of the electric field, and (3) the degree to which the internal structure of the dielectric material permits the penetration of the electric field, which is called *dielectric permittivity*. Given that the P–N junction area is a geometric design parameter, it is useful to introduce the concept of capacitance per unit area—a differential quantity that is basically a technological parameter. For the case of the capacitance per unit area due to the depletion layer—the depletion-layer capacitance—the capacitance per unit area is

$$C_d = \frac{\varepsilon_S}{w_d} \tag{6.23}$$

where ε_s is the silicon permittivity ($\varepsilon_s = 11.8 \times \varepsilon_0 = 11.8 \times 8.85 \times 10^{-12}$ F/m) and w_d is the depletion-layer width. Given that ε_s is a material constant, the depletion-layer capacitance per unit area is basically determined by the depletion-layer width.

Following a brief introduction to the effect of voltage dependence of the depletion-layer capacitance, the bulk of this section is devoted to the determination of the depletion-layer width and the associated effects. The importance of the depletion-layer width is not limited to calculation of the depletion-layer capacitance but also for general understanding and design of devices based on or utilizing P–N junctions.

6.3.1 *C–V* Dependence

Figure 6.15 shows a reverse-biased P–N junction used as a capacitor in a simple R–C circuit. It can be seen from Fig. 6.15b that the increase in the voltage applied increases the depletion-layer width as this voltage attracts electrons and holes toward the contacts. As a result, more uncompensated donor ions on the N-type side, along with uncompensated

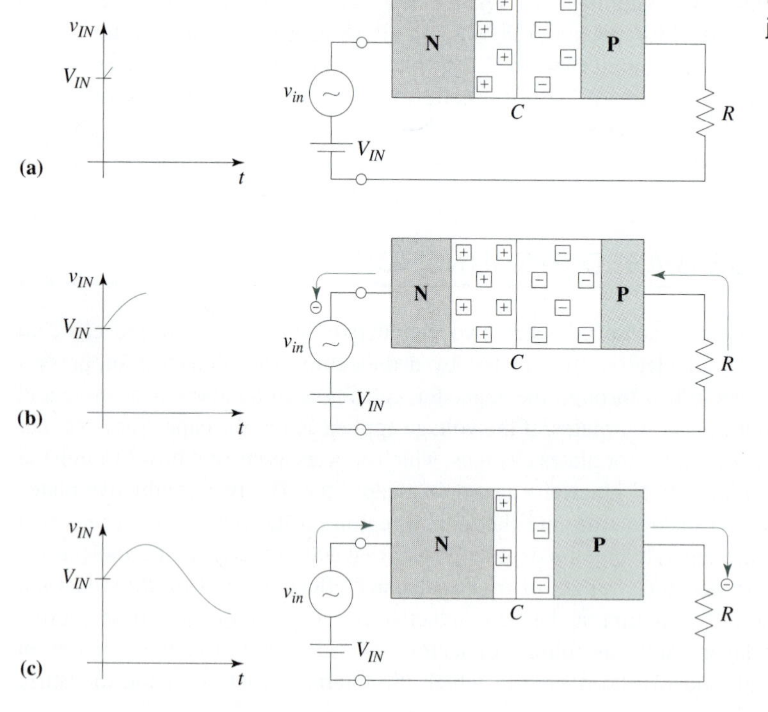

Figure 6.15 Reverse-biased P–N junction used as a capacitor.

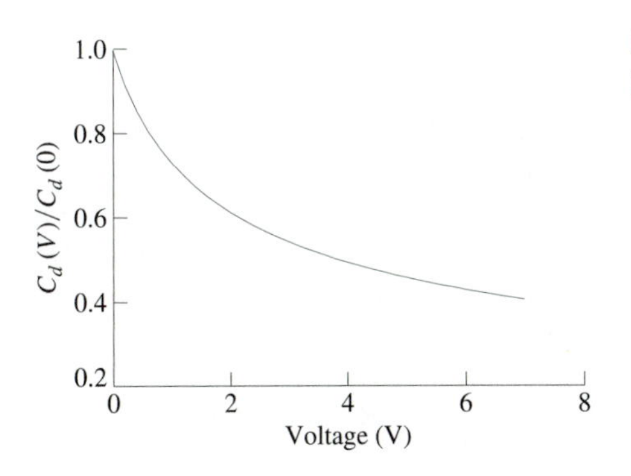

Figure 6.16 C–V curve for a typical P–N junction.

acceptor ions on the P-type side, appear in the depletion layer. Figure 6.15c illustrates that the decrease in the voltage causes a reduction of the depletion-layer width, and consequently a reduction of the depletion-layer charge. These changes in the capacitor charge, induced by the changes of the applied voltage, represent exactly the capacitance effect ($C = dQ_C/dV_C$). The current that charges and discharges the capacitor is also indicated in Fig. 6.15b and 6.15c.

The fact that the externally applied voltage changes the width of the depletion layer makes the P–N junction capacitor voltage-dependent. If the DC component of the voltage, V_{IN}, in the circuit of Fig. 6.15 is increased, the average value of the depletion layer is also increased. In effect, the capacitance is reduced [Eq. (6.23)]. The capacitance–voltage (C–V) dependence of a typical P–N junction is shown in Fig. 6.16. This voltage dependence of the capacitance is a feature of the P–N junction capacitor that does not exist in discrete ceramic capacitors. The voltage dependence of the capacitance is very useful as it enables variable capacitors that are electrically controlled.

6.3.2 Depletion-Layer Width: Solving the Poisson Equation

The relationship between the electric potential and the charge concentration at any single point in the depletion layer is given by the Poisson equation. By solving the Poisson equation, we can find the relationship between the depletion-layer width w_d and the voltage applied; and then by using Eq. (6.23), we can express the depletion-layer capacitance C_d in terms of the voltage applied.

Neglecting any edge effects at the perimeter of the considered P–N junction (mathematically, this is equivalent to the assumption of infinite junction along y- and z-axes), the problem becomes one-dimensional and the Poisson equation can be expressed as

$$\frac{d^2\varphi}{dx^2} = -\frac{\rho_{charge}(x)}{\varepsilon_s} \tag{6.24}$$

In Eq. 6.24, $\rho_{charge}(x)$ is the net charge concentration[5] at point x in the units of C/m^3. The net charge concentration in the neutral regions is equal to zero; however, it has nonzero values in the depletion layer due to the uncompensated donor and acceptor ions.

The mathematical equations for $\rho_{charge}(x)$ that provide good fit to the real doping profiles, such as the diffusion doping profile shown previously in Fig. 1.14, turn out to be unsuitable in terms of obtaining analytical solutions of Eq. (6.24). Nonetheless, it will prove very useful to solve Eq. (6.24) for two simple and extreme approximations: (1) *abrupt* junction and (2) *linear* junction. The solution for abrupt junctions is provided in the following text, whereas the linear junction is consider shortly (Example 6.8). Then Section 6.3.3 will show that the SPICE model for the depletion-layer capacitance is actually a generalization of the analytical solutions for the two extreme approximations.

A P–N junction is considered as abrupt if there are only donors on one side of the junction and only acceptors on the other side, as shown in Fig. 6.17a. Four regions can be identified:

1. The electroneutral N^+-type region in which the donor ions are compensated for by the electrons. The charge concentration in this region is $\rho_{charge} = 0$.
2. The depletion layer on the N^+-type side in which the donor ions appear as the only charge centers. There are N_D donor ions per unit volume, and every donor ion carries one unit (q) of positive charge. Therefore, the charge concentration in this region is $\rho_{charge} = qN_D$.
3. The depletion layer on the P-type side in which the negative acceptor ions appear as the only charge centers. Because there are N_A negative charge centers per unit volume, the charge concentration in this region is given as $\rho_{charge} = -qN_A$.
4. The electroneutral P-type region in which the acceptor ions are compensated for by the holes. The charge concentration in this region is $\rho_{charge} = 0$.

The charge concentration at and around the P–N junction is graphically presented in Fig. 6.17b. To solve the Poisson equation, it is useful to split it into four parts, corresponding to the four regions:

$$\frac{d^2\varphi}{dx^2} = \begin{cases} 0 & \text{for } x \leq -w_n \\ -\frac{qN_D}{\varepsilon_s} & \text{for } -w_n \leq x \leq 0 \\ \frac{qN_A}{\varepsilon_s} & \text{for } 0 \leq x \leq w_p \\ 0 & \text{for } x \geq w_p \end{cases} \tag{6.25}$$

[5]In electromagnetics literature, ρ_{charge} is frequently referred to as the "charge density." Here the term "charge density" is reserved for charge per unit area, expressed in C/m^2.

(a)

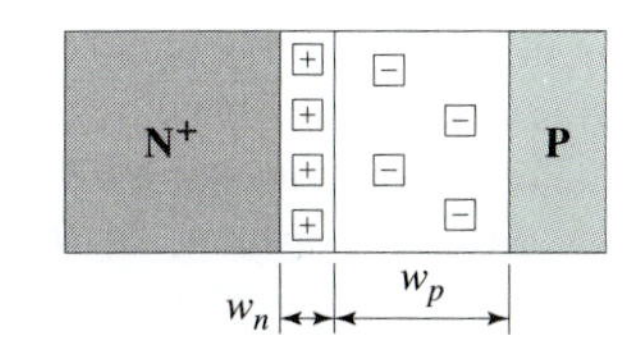

(b)

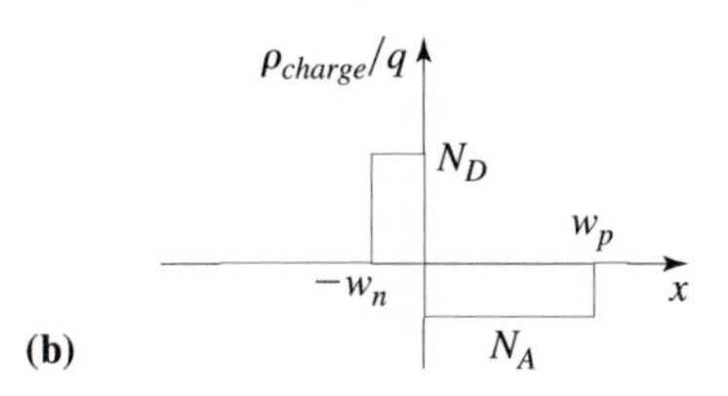

(c)

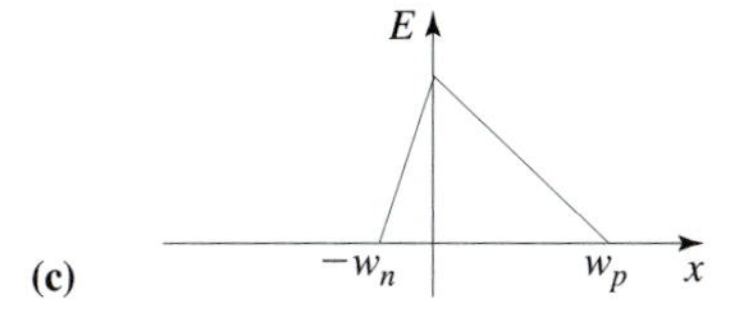

(d)

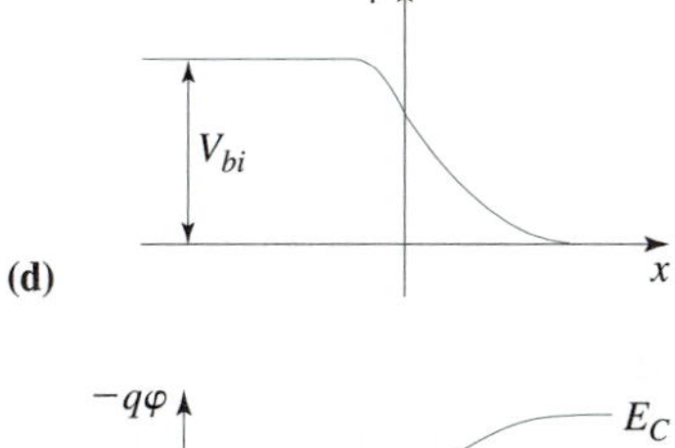

(e)

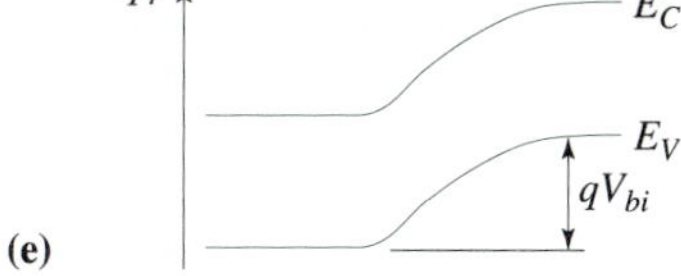

Figure 6.17 An abrupt P–N junction. (a) Illustration. (b) Distribution of charge concentration. (c) Electric-field distribution. (d) Electric-potential distribution. (e) Energy bands.

If the left- and right-hand sides of Eq. (6.25) are integrated, the following is obtained

$$\frac{d\varphi}{dx} = \begin{cases} C_1 & \text{for } x \leq -w_n \\ -\frac{qN_D}{\varepsilon_s}x + C_2 & \text{for } -w_n \leq x \leq 0 \\ \frac{qN_A}{\varepsilon_s}x + C_3 & \text{for } 0 \leq x \leq w_p \\ C_4 & \text{for } x \geq w_p \end{cases} \tag{6.26}$$

where C_1, C_2, C_3, and C_4 are integration constants. Remembering that the electric field is $E = -d\varphi/dx$, it can be seen that Eq. (6.26) gives the electric field multiplied by a minus sign. The electric field in the electroneutral regions ($x \leq -w_n$ and $x \geq w_p$) must be zero. If the field was not zero, the existing free electrons and holes would be moved by the field producing a current flow, which means that the P–N junction would not be in equilibrium.

We wish to consider the P–N junction in equilibrium here. Thus, the constants $C_1 = 0$ and $C_4 = 0$. To obtain the other two integration constants, the boundary conditions at $x = -w_n$ and $x = w_p$ should be used. We concluded earlier that the field at $x = -w_n$ and $x = w_p$ (edges of the electroneutral regions) is zero. By applying Eq. (6.26) to the boundaries

$$0 = \begin{cases} \frac{qN_D}{\varepsilon_s} w_n + C_2 \\ \frac{qN_A}{\varepsilon_s} w_p + C_3 \end{cases} \tag{6.27}$$

we obtain the constants

$$\begin{aligned} C_2 &= -\frac{qN_D}{\varepsilon_s} w_n \\ C_3 &= -\frac{qN_A}{\varepsilon_s} w_p \end{aligned} \tag{6.28}$$

After putting these constants into Eq. (6.26), we have

$$\frac{d\varphi}{dx} = \begin{cases} 0 & \text{for } x \le -w_n \\ -\frac{qN_D}{\varepsilon_s}(x + w_n) & \text{for } -w_n \le x \le 0 \\ \frac{qN_A}{\varepsilon_s}(x - w_p) & \text{for } 0 \le x \le w_p \\ 0 & \text{for } x \ge w_p \end{cases} \tag{6.29}$$

The integration of the Poisson equation practically provides equations for the electric field at and around the P–N junction because it is necessary only to multiply Eq. (6.29) by -1 to convert $d\varphi/dx$ into E:

$$E(x) = \begin{cases} 0 & \text{for } x \le -w_n \\ \frac{qN_D}{\varepsilon_s}(x + w_n) & \text{for } -w_n \le x \le 0 \\ -\frac{qN_A}{\varepsilon_s}(x - w_p) & \text{for } 0 \le x \le w_p \\ 0 & \text{for } x \ge w_p \end{cases} \tag{6.30}$$

The electric field, as given by Eq. (6.30), is plotted in Fig. 6.17c. It can be seen that the maximum of the electric field appears right at the P–N junction ($x = 0$). Moreover, the maximum electric field $E_{max} = E(0)$ is expressed in two ways: (1) through N_D and w_n [the second line in Eq. (6.30)] and (2) through N_A and w_p [the third line in Eq. (6.30)]. This indicates that there is a relationship between (a) N_D and w_n on the N^+-type side of the junction and (b) N_A and w_p on the P-type side of the junction. This relationship can be found using the two equations for E_{max}:

$$E_{max} = E(0) = \frac{qN_D}{\varepsilon_s} w_n = \frac{qN_A}{\varepsilon_s} w_p \tag{6.31}$$

which gives

$$N_D w_n = N_A w_p \tag{6.32}$$

It is not hard to understand the meaning of Eq. (6.32). The number of donors (acceptors) N_D (N_A) per unit volume, multiplied by the depletion-layer width w_n (w_p) gives the number of donors (acceptors) per unit of junction area. As the junction area is the same for both donors and acceptors, Eq. (6.32) means that there is equal number of donor and acceptor ions in the depletion layer of the P–N junction. We could not expect any other result, because every electric-field line that originates at a positive donor ion should terminate at a negative acceptor ion.

Equation (6.32) means that if N_D is larger than N_A (as in the case shown in Fig. 6.17), the depletion-layer width w_n is smaller than w_p. *The depletion layer expands more on the side of the lower-doped material.*

The electric-potential distribution at and around the P–N junction can be obtained if Eq. (6.29) is integrated (the second integration of the Poisson equation; the first integration provided the electric field). The second integration gives

$$\varphi = \begin{cases} C_5 & \text{for } x \le -w_n \\ -\frac{qN_D}{\varepsilon_s}\left(\frac{x^2}{2} + w_n x\right) + C_6 & \text{for } -w_n \le x \le 0 \\ \frac{qN_A}{\varepsilon_s}\left(\frac{x^2}{2} - w_p x\right) + C_7 & \text{for } 0 \le x \le w_p \\ C_8 & \text{for } x \ge w_p \end{cases} \tag{6.33}$$

As the field in the electroneutral regions ($x \le -w_n$ and $x \ge w_p$) is zero, the electric potential is constant. Let us take the electric potential in the P-type region as the reference potential (P-type side of the P–N junction is grounded). This means that $C_8 = 0$. As explained in Section 6.1.1, the built-in electric field results in a potential difference appearing across the P–N junction, referred to as the built-in voltage and denoted by V_{bi}. This means that $C_5 = V_{bi}$. To find the other two constants, the following boundary conditions should be used: $\varphi(-w_n) = V_{bi}$, $\varphi(w_p) = 0$. Applying Eq. (6.33) to the boundaries $x = -w_n$ and $x = w_p$,

$$\begin{aligned} V_{bi} &= -\frac{qN_D}{\varepsilon_s}\frac{w_n^2}{2} + C_6 \\ 0 &= \frac{qN_A}{\varepsilon_s}\frac{w_p^2}{2} + C_7 \end{aligned} \tag{6.34}$$

the constants are obtained as

$$\begin{aligned} C_6 &= V_{bi} - \frac{qN_D}{\varepsilon_s}\frac{w_n^2}{2} \\ C_7 &= \frac{qN_A}{\varepsilon_s}\frac{w_p^2}{2} \end{aligned} \tag{6.35}$$

By putting the obtained constants into Eq. (6.33), we find the electric-potential distribution as follows:

$$\varphi = \begin{cases} V_{bi} & \text{for } x \le -w_n \\ V_{bi} - \frac{qN_D}{2\varepsilon_s}(x + w_n)^2 & \text{for } -w_n \le x \le 0 \\ \frac{qN_A}{2\varepsilon_s}(x - w_p)^2 & \text{for } 0 \le x \le w_p \\ 0 & \text{for } x \ge w_p \end{cases} \tag{6.36}$$

The electric potential, as given by Eq. (6.36), is plotted in Fig. 6.17d. The electric-potential distribution at and around the P–N junction can be used to plot the energy bands as well. The electric potential is related to the potential energy: $E_{pot}(x) = -q\varphi(x)$. Using this equation and knowing that the bottom of the conduction band is separated from the top of the valence band by the energy-gap value (E_g), we can construct the energy bands as illustrated in Fig. 6.17e.

The earlier-derived Eq. (6.32) provides a relationship between the depletion-layer width components (w_n and w_p), expressed in terms of doping concentrations N_D and N_A. The electric-potential distribution, given by Eq. (6.36), can be used to provide the second relationship between w_n and w_p. With two equations, the two unknown components of the depletion layer, w_n and w_p, can be found. To obtain the second relationship between w_n and w_p, observe that the electric potential at $x = 0$ is expressed in two ways: (1) through the parameters of the N^+-type side [the second line in Eq. (6.36)] and (2) through the parameters of the P-type side [the third line in Eq. (6.36)]. Therefore, by using the fact that the electric potential at $x = 0$ is unique, we can establish a relationship between the parameters of the N^+-type and P-type sides:

$$\varphi(0) = V_{bi} - \frac{qN_D}{2\varepsilon_s}w_n^2 = \frac{qN_A}{2\varepsilon_s}w_p^2 \tag{6.37}$$

Solving the system of two equations, (6.32) and (6.37), we obtain w_n and w_p as

$$w_n = \sqrt{\frac{2\varepsilon_s V_{bi}}{qN_D\left(1 + \dfrac{N_D}{N_A}\right)}} \tag{6.38}$$

$$w_p = \sqrt{\frac{2\varepsilon_s V_{bi}}{qN_A\left(1 + \dfrac{N_A}{N_D}\right)}} \tag{6.39}$$

The total depletion-layer width is, obviously, given as

$$w_d = w_n + w_p \tag{6.40}$$

Equations (6.38), (6.39), and (6.40) can be used to calculate the components of the depletion-layer width, as well as the total depletion-layer width, for an abrupt P–N junction in thermal equilibrium (zero-bias applied). Our final aim, however, is to determine the dependence of the depletion-layer width (and thus depletion-layer capacitance) on the reverse-bias voltage applied to the terminals of the P–N junction capacitor. If the current of the reverse-biased P–N junction is neglected (the leakage current), the Poisson equation can be solved in a similar way as for the zero-bias case. An important difference would be the boundary condition for the electric potential at $x = -w_n$. If a reverse-bias voltage V_R is applied to the N^+-type region, it should be added to the existing built-in voltage V_{bi} to obtain the correct boundary condition in this case. This is analogous to the transition from Fig. 6.3 to Fig. 6.5. Therefore, in the final equations for w_n and w_p, $V_{bi} + V_R$ will appear instead of V_{bi} alone:

$$w_n = \sqrt{\frac{2\varepsilon_s(V_{bi} + V_R)}{qN_D\left(1 + \dfrac{N_D}{N_A}\right)}} \tag{6.41}$$

$$w_p = \sqrt{\frac{2\varepsilon_s(V_{bi} + V_R)}{qN_A\left(1 + \dfrac{N_A}{N_D}\right)}} \tag{6.42}$$

The depletion-layer capacitance, given by Eq. (6.23), can now be related to the applied reverse-bias voltage V_R through w_n and w_p:

$$C_d = A\frac{\varepsilon_s}{w_d} = A\frac{\varepsilon_s}{w_n + w_p} \tag{6.43}$$

It is important to note that in most practical cases it is either $N_D \gg N_A$ (the case in Fig. 6.17) or $N_A \gg N_D$. This is because either P-type substrate has to be converted into N-type layer at the surface by diffusing much higher concentration of donors at the surface, or vice versa, to create a P–N junction. Accordingly, the equations for the depletion-layer widths can be simplified as follows:

$$\left.\begin{aligned} w_p &\approx \sqrt{\tfrac{2\varepsilon_s(V_{bi}+V_R)}{qN_A}} \\ w_n &\ll w_p \\ w_d &\approx w_p \end{aligned}\right\} \quad \text{for } N_D \gg N_A \tag{6.44}$$

$$\left.\begin{aligned} w_n &\approx \sqrt{\tfrac{2\varepsilon_s(V_{bi}+V_R)}{qN_D}} \\ w_p &\ll w_n \\ w_d &\approx w_n \end{aligned}\right\} \quad \text{for } N_A \gg N_D \tag{6.45}$$

By using Eqs. (6.43), (6.44), and (6.45), we can express the depletion-layer capacitance in the following compact form

$$C_d = C_d(0)\left(1 + \frac{V_R}{V_{bi}}\right)^{-1/2} \tag{6.46}$$

where $C_d(0)$ is

$$C_d(0) = \begin{cases} \frac{A}{2}\sqrt{\frac{2\varepsilon_s q N_A}{V_{bi}}} & \text{for } N_D \gg N_A \\ \frac{A}{2}\sqrt{\frac{2\varepsilon_s q N_D}{V_{bi}}} & \text{for } N_A \gg N_D \end{cases} \tag{6.47}$$

$C_d(0)$ can be considered as a parameter in the equation for the capacitance versus applied voltage [Eq. (6.46)]. Its physical meaning is obvious: it represents the P–N junction capacitance at zero bias ($V_R = 0$). It is referred to as *zero-bias junction capacitance*.

In conclusion, the depletion-layer capacitance of the abrupt P–N junction is shown to be proportional to $(1 + V_R/V_{bi})^{-1/2}$, where V_R is the reverse-bias voltage.

EXAMPLE 6.6 Layout Design of a P–N Junction as a Varactor

A P–N junction is to be used as a varactor (a voltage-dependent capacitor used as a tuning element in microwave circuits). The doping concentrations in the P-type and N^+-type regions are known: $N_A = 10^{18}$ cm^{-3} and $N_D = 10^{20}$ cm^{-3}, respectively.

(a) Design the varactor (determine the needed junction area) so that the maximum capacitance is $C_{max} = 30$ pF.

(b) Calculate the varactor sensitivity (dC/dV_R) at $V_R = 5$ V.

SOLUTION

(a) The depletion-layer capacitance of a reverse-biased P–N junction is maximum when the reverse-bias voltage is zero. Any increase in the reverse-bias voltage increases the depletion-layer width, reducing the capacitance. Therefore, Eq. (6.47) can be used to calculate the maximum capacitance. We need to calculate V_{bi} first, which can be done using Eq. (6.2):

$$V_{bi} = \frac{kT}{q}\ln\frac{N_A N_D}{n_i^2} = 0.02585 \ln\frac{10^{18} \times 10^{20}}{(1.02 \times 10^{10})^2} = 1.07 \text{ V}$$

By using Eq. (6.47), the maximum capacitance per unit area is determined:

$$\frac{C_{max}}{A} = \frac{1}{2}\sqrt{\frac{2\varepsilon_s q N_A}{V_{bi}}} = 0.5\sqrt{\frac{2 \times 11.8 \times 8.85 \times 10^{-12} \times 1.6 \times 10^{-19} \times 10^{24}}{1.07}} = 2.8 \text{ mF/m}^2$$

The needed junction area is then

$$A = \frac{30 \text{ pF}}{2.8 \text{ mF/m}^2} = 1.07 \times 10^{-4} \text{ cm}^2 = 10{,}700\ \mu\text{m}^2$$

(b) The capacitance versus applied reverse-bias voltage is given by Eq. (6.46), which can be rewritten as

$$C = C_{max}(1 + V_R/V_{bi})^{-1/2}$$

The first derivative of the capacitance with respect to V_R will provide the equation for the capacitor sensitivity:

$$dC/dV_R = -\frac{1}{2}\frac{C_{max}}{V_{bi}}(1 + V_R/V_{bi})^{-3/2}$$

At $V_R = 5$ V, the sensitivity is $dC/dV_R = -1.0$ pF/V.

EXAMPLE 6.7 Design for Specified Reverse-Voltage Rating and Minimum Parasitic Resistance

For stable and reliable operation, the electric field inside a P–N junction diode should not exceed 2×10^5 V/cm. Design the P–N junction so that it can safely operate up to 12 V of reverse bias. The selection of design parameters should minimize the parasitic resistance.

SOLUTION

The maximum electric field is right at the P–N junction and is given by Eq. (6.31):

$$E_{max} = \frac{qN_D}{\varepsilon_s}w_n = \frac{qN_A}{\varepsilon_s}w_p$$

The depletion-layer widths w_n and w_p depend on the reverse bias, as given by Eqs. (6.44) and (6.45). Obviously, the N-type and P-type regions should be longer than w_n and w_p at the maximum reverse bias ($V_R = 12$ V). When the diode operates in forward-bias mode, the depletion-layer widths w_n and w_p shrink. This leaves neutral regions that act as resistors and contribute to the parasitic resistance. The change of the depletion layer is especially pronounced on the lower-doped side. For example, if $N_D \ll N_A$, w_p is negligible compared to w_n. For $N_A \ll N_D$, it is w_p that dominates. To minimize the parasitic resistance, it is better to have a longer N-type region than P-type region because the electron mobility is higher. This means that it is better to use the N–P$^+$ junction, where w_p is negligible and w_n is given by

$$w_n = \sqrt{\frac{2\varepsilon_s(V_{bi} + V_R)}{qN_D}}$$

Inserting the equation for w_n into the equation for the maximum electric field, we obtain

$$E_{max} = \sqrt{\frac{2qN_D}{\varepsilon_s}(V_{bi} + V_R)}$$

From here, the doping level on the N-type side can be expressed as

$$N_D = \frac{\varepsilon_s E_{max}^2}{2q(V_{bi} + V_R)} \tag{6.48}$$

The built-in voltage V_{bi} also depends on the doping levels:

$$V_{bi} = V_t \ln \frac{N_D N_A}{n_i^2}$$

As far as N_A is concerned, the highest practical doping should be selected; for example, $N_A = 10^{20}$ cm^{-3}. However, this still does not enable to explicitly express N_D if V_{bi} is to be eliminated from Eq. (6.48). To solve Eq. (6.48), we can begin by neglecting V_{bi}:

$$N_D = \frac{11.8 \times 8.85 \times 10^{-12}(2 \times 10^7)^2}{2 \times 1.6 \times 10^{-19} \times 12} = 1.0878 \times 10^{16}\ \text{cm}^{-3}$$

With this value for N_D, the built-in voltage is $V_{bi} = 0.026 \times \ln(10^{20} \times 10^{16}/10^{20}) = 0.96$ V. Recalculating N_D with this value for V_{bi},

$$N_D = \frac{11.8 \times 8.85 \times 10^{-12}(2 \times 10^7)^2}{2 \times 1.6 \times 10^{-19} \times (0.96 + 12)} = 1.0072 \times 10^{16}\ \text{cm}^{-3}$$

we can see that the change in V_{bi} does not cause a significant change in the calculated value for N_D. Therefore, the final design value for N_D is 10^{16} cm^{-3}.

EXAMPLE 6.8 Solving the Poisson Equation for the Linear P–N Junction

The linear P–N junction is the extreme approximation that is opposite to the abrupt-junction approximation: the charge concentration changes from the most positive value (the donors on the N-type side) to the most negative value (the acceptors on the P-type side) in the smoothest possible way (Fig 6.18a and 6.18b). The linear P–N junction is the one where the charge concentration in the depletion layer changes linearly,

$$\rho_{charge} = -ax$$

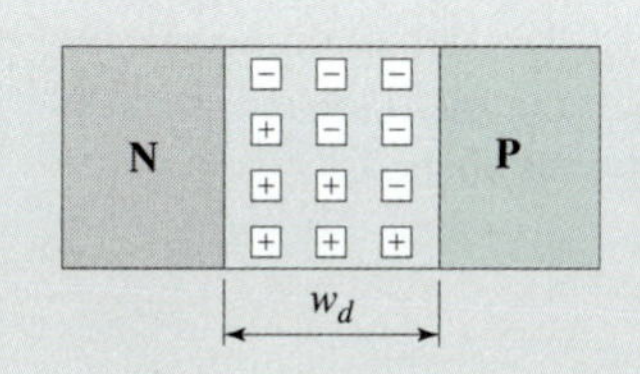

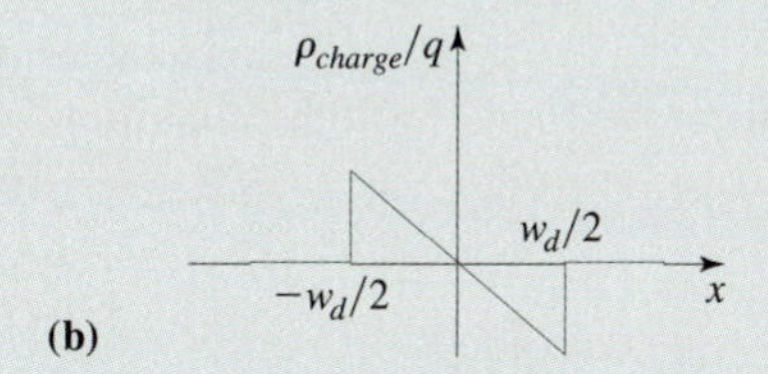

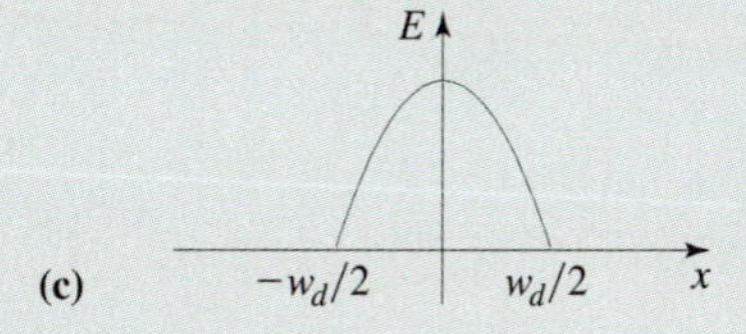

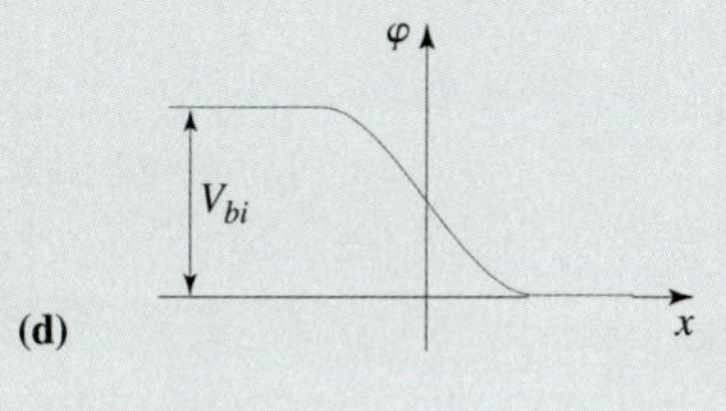

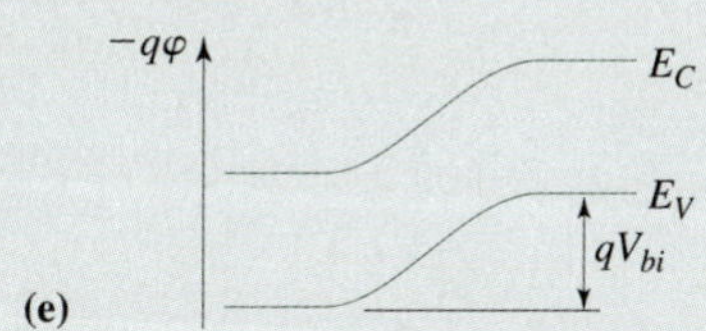

Figure 6.18 Linear P–N junction. (a) Illustration. (b) Distribution of charge concentration. (c) Electric-field distribution. (d) Electric-potential distribution. (e) Energy bands.

where a is the slope of the linear dependence. Solve the Poisson equation for this case to determine the dependence of the depletion-layer width and the depletion-layer capacitance on reverse-bias voltage.

SOLUTION

In this case, the Poisson equation (6.24) becomes

$$\frac{d^2\varphi}{dx^2} = \frac{a}{\varepsilon_s}x \tag{6.49}$$

Of course, Eq. (6.49) is correct only for $-w_d/2 \leq x \leq w_d/2$ (the depletion layer); however, solving the Poisson equation in the depletion layer will be quite enough to find the depletion-layer width, provided that the boundary conditions are properly established.

The first integration of Eq. (6.49) leads to

$$\frac{d\varphi}{dx} = \frac{a}{\varepsilon_s}\frac{x^2}{2} + C_a$$

where C_a is the integration constant. The integration constant can be found from the condition that the electric field, and therefore $d\varphi/dx$, is equal to zero at $x = -w_d/2$:

$$C_a = -\frac{a}{\varepsilon_s}\frac{w_d^2}{8}$$

When this value is used for the constant C_a, $d\varphi/dx$ becomes

$$\frac{d\varphi}{dx} = \frac{a}{2\varepsilon_s}\left[x^2 - \left(\frac{w_d}{2}\right)^2\right] \tag{6.50}$$

whereas the electric field ($E = -d\varphi/dx$) is

$$E(x) = \frac{a}{2\varepsilon_s}\left[\left(\frac{w_d}{2}\right)^2 - x^2\right] \tag{6.51}$$

The electric field, as given by Eq. (6.51), is plotted in Fig. 6.18c.

Integrating the left-hand and right-hand sides of Eq. (6.50) (the second integration of the Poisson equation) leads to an equation for the electric potential φ:

$$\varphi(x) = \frac{a}{2\varepsilon_s}\left[\frac{x^3}{3} - \left(\frac{w_d}{2}\right)^2 x\right] + C_b$$

The integration constant C_b can be found from the boundary condition $\varphi(w_d/2) = 0$ (it is again assumed that the potential of the P-type side is the reference potential). Using this boundary condition, the constant is found as

$$C_b = \frac{a}{\varepsilon_s}\frac{w_d^3}{24}$$

which means the electric-potential distribution in the depletion layer is given by

$$\varphi(x) = \frac{a}{2\varepsilon_s}\left[\frac{x^3}{3} - \left(\frac{w_d}{2}\right)^2 x + \frac{w_d^3}{12}\right] \tag{6.52}$$

The electric potential, as given by Eq. (6.52), is plotted in Fig. 6.18d. The electric-potential function $\varphi(x)$ can be used to plot the energy bands as well. Using the fact that the energy bands follow the potential-energy function $E_{pot}(x) = -\varphi(x)$ and that the bottom of the conduction band and the top of the valence band are separated by the energy gap E_g, the energy bands are constructed as in Fig. 6.18e.

The electric-potential distribution $\varphi(x)$ is used to determine the depletion-layer width. This can be achieved if the second boundary condition for the electric potential is employed, which is the electric potential at $x = -w_d/2$. Taking the electric potential at $x = w_d/2$ as the reference potential $[\varphi(w_d/2) = 0]$, the electric potential at the other side of the depletion layer ($x = -w_d/2$) must be equal to the built-in voltage V_{bi}. Therefore,

$$V_{bi} = \frac{a}{2\varepsilon_s}\left(-\frac{w_d^3}{24} + \frac{w_d^3}{8} + \frac{w_d^3}{12}\right) = \frac{a w_d^3}{12\varepsilon_s} \tag{6.53}$$

The depletion-layer width w_d is obtained from Eq. (6.53) as

$$w_d = \left(\frac{12\varepsilon_s V_{bi}}{a}\right)^{1/3}$$

This is the depletion-layer width of the linear P–N junction in thermal equilibrium (zero bias). The Poisson equation can similarly be solved for a reverse-biased linear P–N junction. An important difference will be the boundary condition at $x = -w_d/2$. In that case the reverse-bias voltage applied V_R has to be added to the existing built-in voltage V_{bi}, which means that the boundary condition becomes $\varphi(-w_d/2) = V_{bi} + V_R$. Accordingly, $V_{bi} + V_R$ will appear in the final equation for w_d instead of V_{bi} alone:

$$w_d = \left[\frac{12\varepsilon_s(V_{bi} + V_R)}{a}\right]^{1/3} \tag{6.54}$$

The depletion-layer capacitance of the linear P–N junction can now be related to the applied reverse-bias voltage V_R through w_d:

$$C_d = A\frac{\varepsilon_s}{w_d} = A\left[\frac{a\varepsilon_s^2}{12(V_{bi} + V_R)}\right]^{1/3}$$

The depletion-layer capacitance C_d can be written in a more suitable form:

$$C_d = C_d(0)\left(1 + \frac{V_R}{V_{bi}}\right)^{-1/3} \tag{6.55}$$

where

$$C_d(0) = A\left(\frac{a\varepsilon_s^2}{12V_{bi}}\right)^{1/3} \tag{6.56}$$

$C_d(0)$ in Eq. (6.55) can be considered as a parameter. This parameter represents the zero-bias junction capacitance, analogously to the case of the abrupt P–N junction.

In conclusion, the depletion-layer capacitance of the linear P–N junction is shown to be proportional to $(1 + V_R/V_{bi})^{-1/3}$, where V_R is the reverse-bias voltage.

EXAMPLE 6.9 Minimum P–N Junction Capacitance

Calculate the minimum capacitance that can be achieved by a linear P–N junction capacitor when the reverse-bias voltage varies between 0 and 5 V and the maximum capacitance is 2.5 pF. Assume that the built-in voltage is $V_{bi} = 0.8$ V.

SOLUTION

The capacitance of the linear P–N junction is given by Eq. (6.55). The capacitance is maximum for $V_R = 0$ V, which means that the parameter $C_d(0) = 2.5$ pF. The reverse-bias voltage V_R reduces the capacitance, which means that it is minimum for the largest V_R, which is $V_{R-max} =$ 5 V in this example. Therefore, the minimum capacitance is calculated as:

$$C_{min} = C_d(0)(1 + V_{R-max}/V_{bi})^{-1/3} = 2.5 \times (1 + 5/0.8)^{-1/3} = 1.3 \text{ pF}$$

6.3.3 SPICE Model for the Depletion-Layer Capacitance

Section 6.3.2 and Example 6.8 provide solutions of the Poisson equation for the two extreme approximations of the P–N junction, the abrupt and the linear P–N junction, respectively. The main result is that the capacitance is proportional to $(1 + V_R/V_{bi})^{-1/2}$ for the abrupt P–N junction and is proportional to $(1 + V_R/V_{bi})^{-1/3}$ for the linear P–N junction. It can be seen that the only difference between these two extreme cases is in the power coefficient, which is $-1/2$ for the abrupt and $-1/3$ for the linear P–N junctions. The dependence of the depletion-layer capacitance C_d on the reverse-bias voltage V_R can, therefore, be expressed in the following compact form:

$$C_d = C_d(0)\left(1 + \frac{V_R}{V_{bi}}\right)^{-m} \tag{6.57}$$

where $m = 1/2$ for the abrupt P–N junction and $m = 1/3$ for the linear P–N junction. It is obvious that Eq. (6.57) becomes applicable to any P–N junction when m is allowed to take any value between the two extremes $(1/2 \geq m \geq 1/3)$. Thus, considering m as a parameter whose value has to be determined experimentally, Eq. (6.57) can be used for real P–N junctions that are neither abrupt nor linear. The parameter m is referred to as *grading coefficient*. The other two SPICE parameters in Eq. (6.57) are $C_d(0)$, the zero-bias capacitance, and V_{bi}, the built-in voltage.

6.4 STORED-CHARGE EFFECTS

The excess minority carriers, accumulated at the edges of the depletion layer due to forward-bias current through a P–N junction, cannot instantly disappear when the forward bias is reduced, set to zero, or changed to reverse bias. The charge of the excess minority carriers at the depletion-layer edges is referred to as the *stored charge,* and the associated time effects as the *stored-charge effects*. In SPICE, the strength of stored-charge effects is set by a parameter called *transit time*. It will be shown in this section that the transit time is directly related to minority-carrier lifetime. The main effect of the stored charge relates to extended *reverse-recovery time*—the time needed for the reverse-bias current to drop to its small equilibrium level when the diode is suddenly switched off. Therefore, the transit time will be related to the reverse-recovery time.

6.4.1 Stored Charge and Transit Time

Figure 6.10 (Section 6.2) illustrates the appearance of excess minority carriers at the edges of the depletion layer when a forward-bias current flows through the diode. The charge associated with the excess minority carriers—the stored charge, Q_s—is directly proportional to the value of the current flowing through the diode:

$$Q_s = \tau_T I_D \tag{6.58}$$

The proportionality coefficient τ_T in Eq. (6.58) is called *transit time*.

Figure 6.19 shows that an increase in the voltage V_D by ΔV_D causes a corresponding increase in the minority-carrier charge stored at the sides of the depletion layer. The charge change ΔQ_s caused by the voltage change ΔV_D represents a capacitance effect. Therefore, stored-charge capacitance can be defined as $C_s = \Delta Q_s / \Delta V_D$. By using the relationship between Q_s and I_D [Eq. (6.58)], we can express the stored-charge capacitance as

$$C_s = \frac{dQ_s}{dV_D} = \tau_T \frac{dI_D}{dV_D} \tag{6.59}$$

which is the equation used in SPICE to model the stored-charge effects. The transit time, τ_T, is the SPICE parameter for this effect.

6.4.2 Relationship Between the Transit Time and the Minority-Carrier Lifetime

To gain insight into the dependence of the transit time on technological and physical parameters, we can use a familiar equation for the diode current (I_D), develop a similar equation for the stored charge (Q_s), and insert them into the equation defining the transit

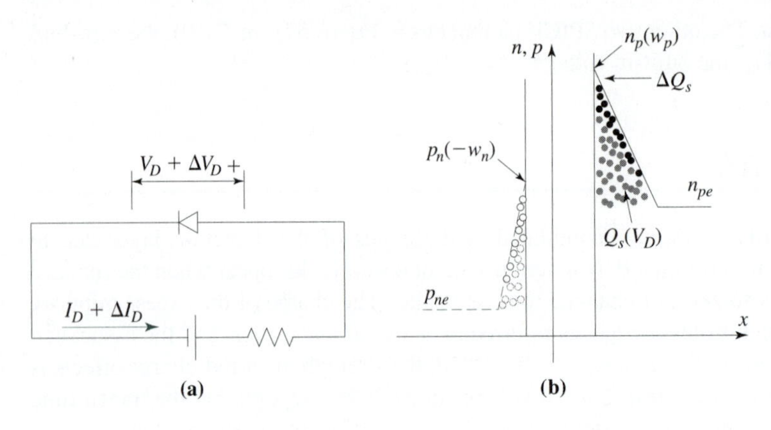

Figure 6.19 Illustration of the stored charge Q_s and the associated capacitance. (a) Voltage change ΔV_D leads to (b) change in the stored-charge $\Delta Q_s = C_s \Delta V_D$.

time:

$$\tau_T = \frac{Q_s}{I_D} \tag{6.60}$$

Let us assume the N^+–P junction, meaning that the doping level of the N-type is much higher than the doping level of the P-type region. With this assumption, we can neglect the current due to the minority holes [$p_{ne} \ll n_{pe}$ in Eqs. (6.11) and (6.12)]. In this case, the diode current density is $j \approx j_n$, which means that the current is

$$I_D = A_J j_n \tag{6.61}$$

where A_J is the area of the diode junction. Replacing j_n by Eq. (6.5), the diode current becomes

$$I_D = A_J q D_n \frac{n_{pe} - n_p(w_p)}{L_n} \tag{6.62}$$

Let us now express Q_s in terms of $n_p(w_p)$, n_{pe}, and L_n, as well, to replace it together with the relation for I_D in Eq. (6.60). Figures 6.10c and 6.19b illustrate Q_s. To obtain Q_s, the average excess minority-carrier concentration $\left[n_p(w_p) - n_{pe}\right]/2$ is multiplied by the charge that every electron carries (this gives the average charge in C/m^3), and is also multiplied by the volume $A_J L_n$, to obtain the charge in C:

$$Q_s = -q A_J L_n \frac{n_p(w_p) - n_{pe}}{2} \tag{6.63}$$

Replacing I_D and Q_s in Eq. (6.60) from Eqs. (6.62) and (6.63), respectively, the transit time is obtained as

$$\tau_T = \frac{L_n^2}{2D_n} \tag{6.64}$$

This equation is frequently used to estimate the transit time value, given that approximate values of the diffusion coefficient D_n and the diffusion length L_n are usually known.

If L_n in Eq. (6.64) is replaced by Eq. (5.15), the transit time becomes

$$\tau_T = \frac{\tau_n}{2} \tag{6.65}$$

This result shows that there is a direct relationship between the transit time and the minority-carrier lifetime (τ_n). For a shorter τ_n (because of a stronger recombination), the electrons are recombined closer to the injection point, so the diffusion length is shorter, in accordance with the lower level of stored charge.

Equation (6.65) is for an N^+–P junction. It can be similarly shown that in the case of a P^+–N junction, the transit time is approximately equal to $\tau_p/2$. In diodes where the foregoing approximations cannot be applied, the transit time is a combination of the minority hole and electron lifetimes. In either case, the important conclusion is that the transit time is determined by the recombination rate. This means that the transit time can be reduced by increasing the concentration of the recombination centers.

6.4.3 Switching Characteristics: Reverse-Recovery Time

Figure 6.20 illustrates the effect of the stored charge on the switching characteristics of a P–N junction diode and the importance of properly setting the value of the associated

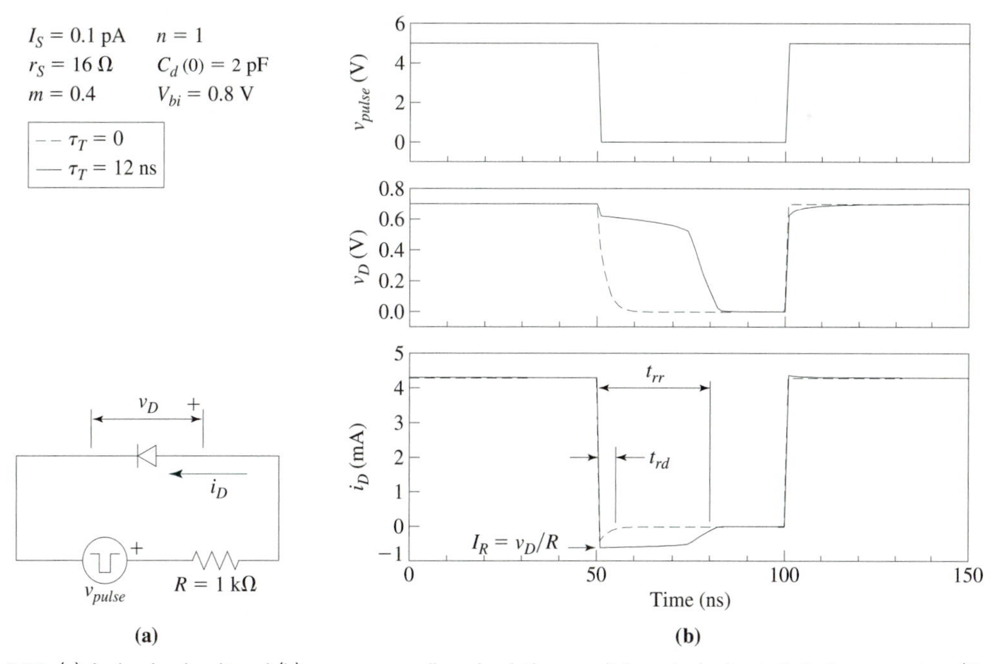

Figure 6.20 (a) A simple circuit and (b) a corresponding simulation result for a typical set of diode parameters (the solid lines) and for $\tau_T = 0$ (the dashed lines) to illustrate the effect of the stored-charge capacitance.

parameter τ_T when simulating switching circuits. To explain the effect of the stored charge, consider the behavior of the diode when the forward-bias voltage is suddenly switched off ($t = 50$ ns in Fig. 6.20b). With $\tau_T = 0$ (the dashed lines), the circuit simulation predicts quick exponential decay of the current i_D and the voltage v_D, as determined by the relatively small RC_d time constant. In reality, however, it takes much longer for the diode to turn off (i_D and v_D to drop to zero values). The thermal equilibrium ($i_D = 0$) is not achieved for as long as there is excess minority-carrier charge—that is, stored charge—at the P–N junction. Referring to Fig. 6.19b, the minority-carrier concentrations at the edges of the depletion layer, $n_p(w_p)$ and $p_n(w_n)$, should fall down to the equilibrium levels, n_{pe} and p_{ne}, respectively. The process of stored-charge removal determines the time-response and high-frequency behavior of the diode.

The process of stored-charge removal should be considered in more detail. When the forward-bias voltage is turned off ($t = 50$ ns in Fig. 6.20b), the electric field that injects the minority electrons and holes into the neutral regions is removed. The minority electrons and holes caught at the edges of the depletion layer (the stored charge) find it now easier to flow back through the depletion layer, returning to their respective original neutral regions. In the spirit of the band diagram of Fig. 6.6a, the excess electrons at the P-type side find it easier to roll down along the bottom of the conduction band back to the N-type region, and the holes at the N-type side find it easier to bubble up along the top of the valence band back to the P-type region. This produces the sudden change in the current i_D from positive to negative, shown in Fig. 6.20b.

The voltage drop v_D is no longer due to the externally applied forward-bias voltage, but due to the stored charge at the depletion-layer edges: $v_D = Q_s/C_s$. In this period, the draining current $i_D = I_R$ is limited by the external resistor R and the relatively small value of v_D. Consequently, the current $I_R = v_D/R$ appears as nearly constant (Fig. 6.20b). Assuming that the constant current I_R alone has to remove the stored charge, the discharge time can be calculated as

$$t_{rs} = \frac{Q_s}{I_R} \tag{6.66}$$

In Eq. (6.66), any contribution from the recombination mechanism is, clearly, neglected. Therefore, the time given by Eq. (6.66) is the worst-case estimate. The contribution of recombination can become significant if (1) the recombination mechanism is enhanced by special manufacturing techniques (some of them are discussed later) and (2) the resistor in the discharging circuit is too high, causing the discharging current to be too small. In general, if estimated t_{rs} is not significantly smaller than $2\tau_T$, then there is enough time for the recombination mechanism to remove a significant portion of the stored charge. Therefore, if t_{rs} calculated by Eq. (6.66) is larger than the minority-carrier lifetime, then it is significantly overestimated.

It should be noted here that the discharge current I_R will be higher in circuits in which the diode is not switched off from forward bias to zero but to a reverse bias. If the reverse-bias value V_R is much larger than $v_D \approx 0.65$ V, then the discharge current is $I_R \approx V_R/R$.

After the removal of the stored charge, the charge of the depletion-layer capacitance C_d (Section 6.3) has to be set at the level that corresponds to the reverse-bias voltage V_R ($V_R = 0$ in Fig. 6.20). This period starts at about 73 ns in the example of Fig. 6.20. During this process, the diode behaves more like an ordinary capacitor, with the current and voltage setting at a nearly exponential rate with time constant $t_{rd} = C_d R_{on}$. The times t_{rs} and t_{rd} add up to the total *reverse recovery time* t_{rr}.

The main problem with t_{rr}, especially t_{rs}, is that it limits the switching speed. Because Q_s and τ_T are related to the recombination rate and the minority-carrier lifetime, a recombination-rate increase would reduce these times. The recombination rate can be increased by increasing the concentration of recombination centers. Suitable elements as recombination centers in silicon are gold and platinum. Consequently, diffusion of gold or platinum is used to improve the switching performance of P–N junction diodes. Another technique is high-energy electron irradiation, which introduces recombination energy levels in the energy gap by creating damage in the silicon crystalline structure. When the concentration of recombination centers is too high, however, there is a sharp increase in the forward voltage drop, which limits the reduction of transit time by this approach.

SUMMARY

1. A *built-in voltage* V_{bi} appears at any junction of materials having mobile electrons with different potential energies—that is, different positions of the Fermi level with respect to the reference vacuum level—or expressed in yet another way, different *work functions*. The built-in voltage corresponds to energy-band bending of qV_{bi} that is needed to bring the Fermi levels to a constant level throughout the system, indicating that the system is in thermal equilibrium. For the case of a P–N junction,

$$qV_{bi} = |q\phi_{Fn}| + |q\phi_{Fp}| = kT \ln \frac{N_D}{n_i} + kT \ln \frac{N_D}{n_i} = kT \ln \frac{N_D N_A}{n_i^2}$$

   Energy-band bending is associated with the existence of built-in electric field, which is caused by removal of electrons from the N-type side and holes from the P-type side. The electric field appears in the *depletion layer*—the region that is depleted of majority carriers to have uncompensated donor and acceptor ions.
2. A reverse-biased P–N junction is not in thermal equilibrium: the Fermi level is split into electron and hole quasi-Fermi levels, so that the barrier height at the P–N junction is increased ($qV_{bi} + qV_R$). This is a barrier for the majority carriers; minority carriers flow easily through the junction, resulting in a small "leakage" current.
3. When a forward-bias voltage $V_D = V_F$ is applied across a P–N junction, the barrier height (due to the built-in voltage) is reduced, and a number of majority electrons and holes are able to go over the barrier, appearing as minority carriers on the other side. Because of their exponential distribution with energy, the number of electrons/holes with energies higher than the reduced barrier height $V_{bi} - V_D$ depends exponentially on V_D. Consequently, the concentration of minority carriers at the edges of the depletion layer depends exponentially on V_D:

$$n_p(|w_p|) = n_{pe} e^{V_D/V_t}, \qquad p_n(|w_n|) = p_{ne} e^{V_D/V_t}$$

where n_{pe} and p_{ne} are the equilibrium minority-carrier concentrations ($p_{ne} = n_i^2/N_D$, $n_{pe} = n_i^2/N_A$). The injected minority carriers move through the neutral regions by diffusion, so that the diffusion coefficients of minority carriers (D_n and D_p), in addition to the applied voltage, the geometrical parameters, and the doping levels, determine the overall diode current:

$$I_S = A_J q n_i^2 \left(\frac{D_n}{W_{anode} N_A} + \frac{D_p}{W_{cathode} N_D} \right)$$

$$I_D = I_S \left(e^{V_D/V_t} - 1 \right)$$

4. In the absence of second-order effects, the slope of the $\ln I_D$–V_D plot is $1/V_t$. However, electron–hole recombination increases the low-level current, and observable voltage drops in the neutral region decrease the high-level current. These effectively reduce the $\ln I_D$–V_D slope to $1/nV_t$, where $1 \leq n \leq 2$. Another important second-order effect is due to parasitic resistance r_S, appearing in series with the P–N junction. As a result, the P–N junction voltage V_{D0} is different from the voltage applied across the terminals of the diode: $V_D = V_{D0} + r_S I_D$. The static diode model in SPICE is a resistance r_S in series with a voltage-controlled current source:

$$I_D = I_S \left(e^{V_{D0}/nV_t} - 1 \right)$$

where the saturation current I_S and the emission coefficient n are model parameters, whereas V_t is the thermal voltage ($V_t \approx 26$ mV at room temperature).

5. The electric field in the depletion layer, especially due to a reverse bias, accelerates the minority electrons and holes, which lose the gained kinetic energy as they collide with the crystal atoms. If the critical electric field is reached, this energy is sufficient to generate electron–hole pairs: the new electrons are accelerated to further generate electron–hole pairs, leading to *avalanche breakdown*. Diodes operating in the avalanche mode can be used as voltage references, because large current variations are supported by a very small voltage change. Another type of breakdown is due to *tunneling:* heavily doped P- and N-type regions create a very narrow depletion layer (barrier), which is further narrowed by the reverse bias to enable the electrons/holes to tunnel through the barrier. The avalanche breakdown voltage has a positive temperature coefficient (V_{BR} increases with the temperature), whereas the tunneling breakdown voltage has a negative temperature coefficient. The forward diode voltage also has a negative temperature coefficient ($\Delta V_D/\Delta T \approx -2$ mV/°C).

6. The width of the depletion layer, w_d, defines the depletion-layer capacitance of a P–N junction:

$$C_d = \frac{\varepsilon_s}{w_d} \qquad (\text{F/m}^2)$$

The *Poisson equation,*

$$\frac{d^2\varphi}{dx^2} = -\frac{\rho_{charge}}{\varepsilon_s}, \qquad \text{where} \qquad \rho_{charge} = q(N_D - N_A + p - n)$$

has to be solved to establish the relationship between the depletion-layer width and the voltage across the depletion layer. In the case of abrupt and asymmetrical P–N junctions, the result is

$$\underbrace{w_d \approx w_p = \sqrt{\frac{2\varepsilon_s(V_{bi}+V_R)}{qN_A}}}_{\text{for } N_A \ll N_D}; \qquad \underbrace{w_d \approx w_n = \sqrt{\frac{2\varepsilon_s(V_{bi}+V_R)}{qN_D}}}_{\text{for } N_D \ll N_A}$$

The maximum electric field, appearing right at the P–N junction, is also found:

$$E_{max} = \frac{q}{\varepsilon_s}N_D w_n = \frac{q}{\varepsilon_s}N_A w_p$$

7. The SPICE equation for $C_d(V_R)$ is

$$C_d = C_d(0)\left(1 + \frac{V_R}{V_{bi}}\right)^{-m}$$

which generalizes the analytical solution for the two extreme cases ($m = 1/2$ for an abrupt and $m = 1/3$ for a linear P–N junction). The parameters of this equation are the zero-bias capacitance [$C_d(0)$], the grading coefficient (m), and the built-in voltage (V_{bi}).

8. The excess minority carriers, due to forward diode current, create *stored charge* at the edges of the depletion layer. The stored charge is directly proportional to the current

$$Q_s = \tau_T I_D$$

and it changes as the applied voltage changes: $C_s = dQ_s/dV_D = \tau_T dI_D/dV_D$. C_s is effectively a capacitance that appears in parallel with the depletion-layer capacitance. The proportionality coefficient, τ_T, is called *transit time* (a SPICE parameter). The transit time is directly related to the minority-carrier lifetimes $\tau_{n,p}$:

$$\tau_T = \begin{cases} \tau_n/2 & \text{for N}^+\text{–P junctions} \\ \tau_p/2 & \text{for P}^+\text{–N junctions} \end{cases}$$

The forward diode voltage cannot be instantly changed to zero- or reverse-bias voltage (the diode cannot be instantly turned off) before the stored charge is removed. The stored charge is removed by (1) approximately constant discharging current ($\approx \frac{V_R+0.7V}{R}$), flowing through the series-equivalent resistance R and the reverse-bias voltage source, in the direction opposite to that of the forward current, and (2) electron–hole recombination.

PROBLEMS

6.1 Assign each of the band diagrams from Fig. 6.21 to a statement describing the biasing condition.

6.2 Assign each of the concentration diagrams from Fig. 6.22 to a statement describing the biasing condition. Electron concentrations are presented with solid lines, and the hole concentrations are presented with dashed lines.

6.3 Which of the following statements, related to a P–N junction, are correct?

(a) The direction of the built-in electric field in the depletion layer is from the N-type toward the P-type region.
(b) At reverse bias, the Fermi level is constant for the entire P–N junction system.
(c) The net (effective) charge density at each point of the depletion layer is zero.
(d) A nonzero current of minority carriers flows through the junction at zero bias.
(e) The current of minority carriers depends on the voltage applied.
(f) At reverse bias, the depletion-layer width is saturated (does not depend on the voltage applied).
(g) The net charge outside the depletion layer of a reverse-biased P–N junction is zero.
(h) At reverse bias, any electron current is fully compensated by corresponding hole current.
(i) The voltage drop in the neutral regions can be neglected in comparison to the voltage drop in the depletion layer.
(j) The electric field in the neutral regions is not large enough to produce a significant drift current of the majority carriers.
(k) The diffusion current of the minority carriers is significant.
(l) The number of electrons able to overcome the energy barrier at the P–N junction increases exponentially with the forward-bias voltage.
(m) The drift current of the minority carriers is significant.
(n) Forward bias does not influence the barrier height at the P–N junction, because it does not affect the built-in voltage.

6.4 The N-type and P-type doping levels of a silicon P–N junction are $N_D = N_A = 5 \times 10^{16}$ cm^{-3}. Calculate the built-in voltage V_{bi}

(a) at room temperature
(b) at 300°C (n_i at 300°C is calculated in Problem 1.28) **A**
(c) at 700°C ($n_i = 1.10 \times 10^{18}$ cm^{-3}). Does a negative value for V_{bi} mean anything?

6.5 Calculate V_{bi} when the P–N junction of Problem 6.4 is implemented in GaAs ($n_i = 2.1 \times 10^6$ cm^{-3}) and Ge ($n_i = 2.5 \times 10^{13}$ cm^{-3}).

6.6 **(a)** Using the results of Problems 6.4a and 6.5, plot V_{bi} versus E_g for the cases of Ge (E_g = 0.66 eV), Si (E_g = 1.12 eV), and GaAs (E_g = 1.42 eV), respectively.
(b) Derive $V_{bi}(E_g)$ equation to explain the obtained linear correlation between V_{bi} and E_g.

6.7 A very-low-doped N$^-$ layer ($N_D = 5 \times 10^{14}$ cm^{-3}) is deposited onto a heavily doped N$^+$ substrate ($N_D = 10^{20}$ cm^{-3}). Calculate the built-in voltage at this N$^-$–N$^+$ junction.

6.8 Identify the mode of diode operation for each of the I_D–V_D characteristics shown in Fig. 6.23.

6.9 The donor and acceptor concentrations of a diode are $N_D = 10^{20}$ cm^{-3} and $N_A = 10^{16}$ cm^{-3}, respectively. Find the minority-carrier concentration at the edges of the depletion layer for

(a) forward bias $V_F = V_D = 0.65$ V
(b) reverse bias $V_R = |V_D| = 0.65$ V **A**

6.10 The neutral region of the anode and cathode of the diode from Problem 6.9 are much smaller than the diffusion lengths: W_{anode} = 4 μm, $W_{cathode}$ = 2 μm. Assuming linear distribution of minority carriers in the neutral regions, find the diffusion-current density of the minority electrons and holes at $V_F = 0.65$ V if the diode is implemented in

(a) silicon (μ_n=1450 cm^2/Vs, μ_p=500 cm^2/Vs, $n_i = 1.02 \times 10^{10}$ cm^{-3})
(b) GaAs (μ_n = 8500 cm^2/Vs, μ_p = 400 cm^2/Vs, $n_i = 2.1 \times 10^6$ cm^{-3}) **A**

6.11 For the diode considered in Problems 6.9 and 6.10, calculate the reverse-bias ($V_R = |V_D|$ = 0.65 V) and the forward-bias ($V_F = V_D$ = 0.65 V) current densities. Is the forward-bias current equal to the sum of electron and hole diffusion currents calculated in Problem 6.10? Perform the calculations for both Si and GaAs.

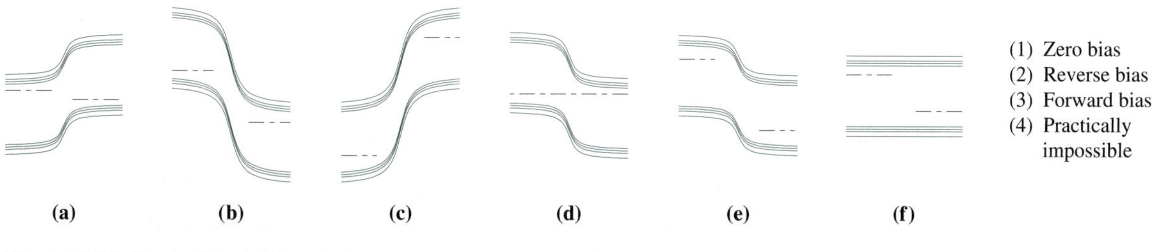

Figure 6.21 Energy-band diagrams.

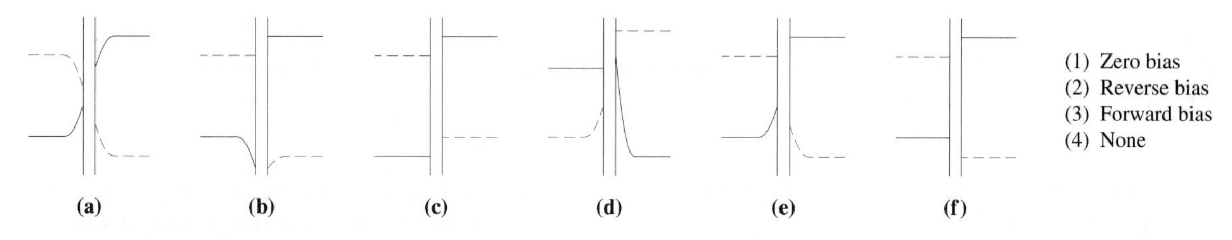

Figure 6.22 Concentration diagrams.

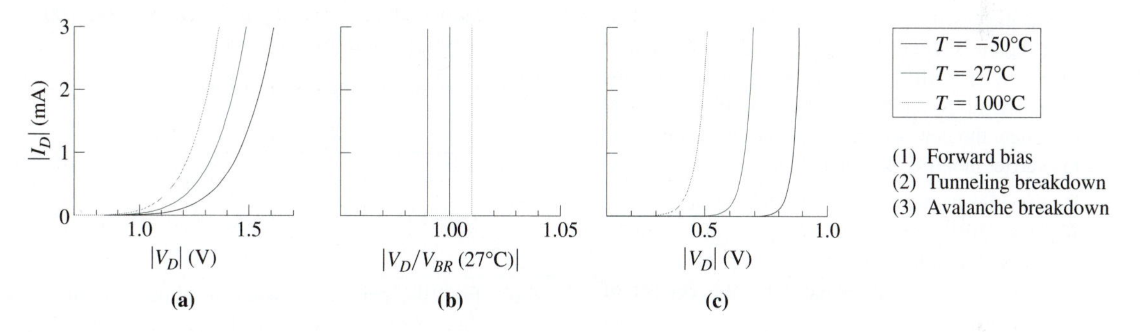

Figure 6.23 $|I_D|$–$|V_D|$ characteristics of P–N junction diodes in three different modes of operation.

6.12 If the widths of the anode and the cathode of a P–N junction diode are similar ($W_{anode} \approx W_{cathode} \ll L_{n,p}$), whereas the doping level in the cathode is 1000 times higher, estimate how much the saturation current will change if

(a) the doping of the anode is increased 10 times
(b) the doping of the cathode is increased 10 times

6.13 The doping and geometric parameters of a P–N junction diode are $N_D = 10^{20}$ cm^{-3}, $W_{cathode} = 1\ \mu$m $\ll L_p$, $N_A = 5 \times 10^{15}$ cm^{-3}, $W_{anode} = 10\ \mu$m $\ll L_n$, and the junction area $A_J = 500 \times 500\ \mu$m^2. Calculate the saturation current I_S, and then use this result to obtain the forward voltage that corresponds to a current of 100 mA, if the semiconductor material is

(a) Si ($\mu_n = 1450$ cm^2/V · s, $\mu_p = 500$ cm^2/V · s, $n_i = 1.02 \times 10^{10}$ cm^{-3})
(b) SiC ($\mu_n = 380$ cm^2/V · s, $\mu_p = 70$ cm^2/V · s, $n_i = 1.6 \times 10^{-6}$ cm^{-3}) **A**

6.14 As a layout designer, determine the radius of a P–N junction diode with a circular shape so that the current rating of the diode is 1 mA, specified at 0.6 V and room temperature. The needed technological parameters are as follows: the doping level of the N type is much higher ($N_D \gg N_A$), the doping level of the P type is 10^{17} cm^{-3}, the diffusion constant of minority electrons is 20 cm^2/s, and the width of

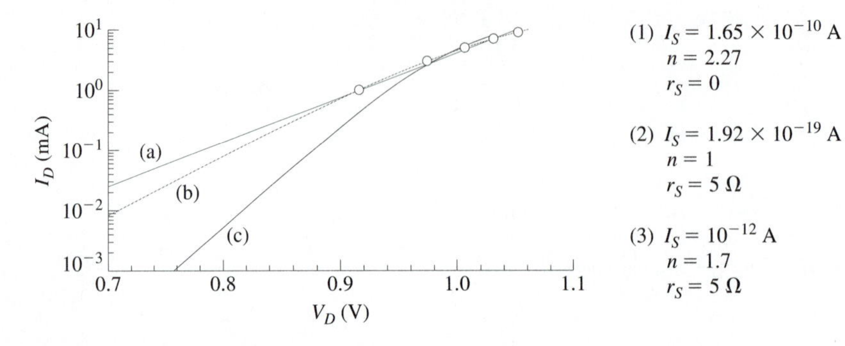

Figure 6.24 A set of measured I_D–V_D points (symbols) is fitted with three different sets of parameters (lines).

the neutral P-type region is 1 μm. Assume that the emission coefficient is $n = 1.2$.

6.15 The saturation current of a P–N junction diode is $I_S = 10^{-11}$ A. The breakdown voltage of this diode can be increased if the doping of the lower-doped N-type region is reduced from $N_D = 5 \times 10^{16}$ cm^{-3} to $N_D = 5 \times 10^{14}$ cm$^{-3} \ll N_A$. How would this reduction of N_D influence the saturation current? Calculate the new saturation current, assuming that the carrier mobility does not change.

6.16 Doping levels of a silicon P–N junction are $N_A = 10^{20}$ cm^{-3} and $N_D = 10^{17}$ cm^{-3}. In thermal equilibrium, the current of minority electrons ($I_{n,p\to n}$) is balanced by the current of majority electrons ($I_{n,n\to p}$). Likewise, the current of minority holes ($I_{p,n\to p}$) is balanced by the current of majority holes ($I_{p,p\to n}$). Calculate the currents of minority electrons and minority holes. Assume that the thermal velocity of both minority electrons and minority holes is the same: $v_{th} = 2 \times 10^5$ cm/s. The area of the P–N junction is $A_J = 2.25 \times 10^{-4}$ cm^2. **A**

6.17 The current of majority holes ($I_{p,p\to n}$), which is equal to the current of minority holes in thermal equilibrium ($I_{p,n\to p}$), reduces exponentially with the applied reverse-bias voltage V_R. Assuming that $I_{p,n\to p} = 10^{-13}$ A and that it remains constant, calculate the total reverse-bias current for $V_R = 0.01$ V, 0.05 V, 0.10 V, 0.20 V, and 0.50 V at room temperature.

6.18 The doping levels of a silicon P–N junction diode are $N_D = 10^{20}$ cm^{-3} and $N_A = 10^{18}$ cm^{-3}. The width of both neutral regions is 1 μm, and the diffusion constant of electrons is $D_n = 9$ cm^2/s. Assuming that the reverse-bias current is equal to the saturation current I_S, calculate the number of electrons that pass through the P–N junction each second if the area of the junction is (a) 100 μm $\times$100 μm and (b) 1 μm $\times$1 μm.

6.19 Find the missing result in Table 6.2.

TABLE 6.2

V_D (V)	0.70	0.72	0.74
I_D (mA)	0.6	?	2.3

6.20 Find the parasitic resistance, if the current of the diode from Problem 6.15 is $I_D = 17$ mA at $V_D = 0.88$ V. **A**

6.21 A set of experimental I_D–V_D data, shown by the symbols in Fig. 6.24, is fitted with three different sets of parameters (the lines). Identify the set of parameters correspondsing to each of the lines.

6.22 The SPICE parameters of the diode used in the circuit of Fig. 6.10 are $I_S = 10^{-12}$ A, $n = 1.4$, and $r_S = 10\ \Omega$. The current flowing through the circuit is found to be $I_D = 3.5$ mA. Knowing that the thermal voltage is $V_t = 26$ mV, determine the voltage between the diode terminals, V_D. **A**

6.23 The operating junction temperature of a P–N junction diode is estimated to be $T = 75$°C. To obtain the SPICE parameters for simulations with 75°C as the nominal temperature, two I_D–V_D points are measured at 75°C: $I_{D1} = 0.57$ mA, $V_{D1} = 0.67$ V; $I_{D2} = 1.28$ mA, $V_{D2} = 0.70$ V.

(a) Calculate I_S and n.

(b) What would I_S and n be if these measurements were performed at 0°C? **A**

6.24 A forward-biased diode is used as a temperature sensor. To calibrate the sensor, the diode is placed in a melting ice and boiling water, and the following forward voltage drops are measured: 0.680 V and 0.480 V, respectively. What is the temperature when the forward voltage drop of the diode is 0.618 V?

6.25 The avalanche breakdown of a silicon N^+–P junction, with the doping level of $N_A = 10^{16}$ cm^{-3}, occurs when the electric field at the junction reaches $E_{BR} = 4 \times 10^5$ V/cm.

(a) Calculate the kinetic energy of the minority electrons that trigger the avalanche process. Compare this energy to the energy gap of silicon. The electron mobility in the P-type region is $\mu_n = 1200$ cm^2/V · s and the electron effective mass is $m^* = 0.26m_0$.

(b) When the temperature of the P–N junction is increased to 125°C, the electron mobility drops to $\mu_n = 600$ cm^2/V · s. Calculate the average kinetic energy of the minority electrons for the same electric field ($E = 4 \times 10^5$ V/cm) at this temperature. Is this electric field sufficient to cause avalanche breakdown at this temperature?

6.26 Derive the equation for the temperature coefficient of the avalanche breakdown field,

$$TC_{av} = \frac{100}{E_{BR}} \frac{dE_{BR}}{dT} \left(\frac{\%}{°\mathrm{C}}\right)$$

assuming that the mobility is determined by the phonon scattering: $\mu_n = A_p T^{-3/2}$. Calculate TC_{av} at 300 K.

6.27 Estimate the ratio of generation currents through a reverse-biased silicon diode at $T_{oper} = 85$°C and $T_r = 27$°C if it is assumed that the generation process is dominated by midgap R–G centers.

6.28 **(a)** A P–N junction has zero-bias ($V_R = 0$ V) capacitance per unit area $C_d = 0.722$ mF/m^2. What is the depletion-layer width?

(b) Assuming equal and uniform doping on either side of the P–N junction, $N_A = N_D = 5 \times 10^{16}$ cm^{-3}, calculate the number of uncompensated donor ions per unit area that exist on either side of the P–N junction.

(c) Given that a unit of charge ($q = 1.6 \times 10^{-19}$ C) is associated with every uncompensated donor, calculate the density of the capacitor's positive charge. **A**

(d) Because $V_R = 0$ V, $Q_C = C_d V_R = 0$ C/m^2 gives a completely different result from the one obtained in part (c). What is the explanation?

(1) $Q_C = C_d V_R$ cannot be used in this case ($Q = CV$ relationship is valid only in special cases).

(2) There is equal density of negative charge, due to the acceptors in the depletion layer, so that the total charge is $Q_C = 0$ C/m^2.

(3) The built-in voltage V_{bi} is not included, i.e., $Q_C = C_d(V_R + V_{bi})$.

6.29 The C_d–V_R characteristic of a P–N junction capacitor is represented by Table 6.3. Assuming $N_D = N_A = 5 \times 10^{16}$ cm^{-3}, calculate the density of positive charge at $V_R = 5$ V. **A**

6.30 **(a)** If a sinusoidal signal voltage with zero-to-peak amplitude of 500 mV is superimposed to V_R, calculate the maximum density of positive charge for the P–N junction capacitor of Problem 6.29. Using the result for $V_R = 5$ V, obtained in Problem 6.29, calculate the charge density increase due to the peak signal voltage.

(b) By using the capacitance definition, the charge density increase can be approximated as $\Delta Q_C \approx C_d \Delta v_C$, where C_d is assumed to be constant and equal to the zero-signal value $C_d(V_R)$. Calculate ΔQ_C with this method, and compare it to the result from part (a). **A**

6.31 Assume the signal voltage of Problem 6.30 is increased to 5 V (zero-to-peak) so that the instantaneous voltage v_C oscillates between 0 and 10 V.

(a) Using the results of Problems 6.29 and 6.28c, calculate the charge density decrease due to the negative signal peak.

(b) Calculate the charge density decrease from $\Delta Q_C \approx C_d \Delta v_C$.

(c) The results from parts (a) and (b) of Problem 6.30 are very close; however, this time the $\Delta Q_C \approx C_d \Delta v_C$ method gives different result. Why?

(1) The method in part (a) ignores the negative charge due to the acceptors in the depletion layer.

(2) The method in part (b) ignores V_{bi}.

TABLE 6.3

V_R (V)	0.0	1.0	2.0	3.0	4.0	4.5	5.0	5.5
C_d (mF/m^2)	0.722	0.386	0.295	0.248	0.218	0.206	0.197	0.188

(3) The method in part (b) is meaningless for $\Delta v_C < 0$.

(4) C_d changes with V_R.

6.32 The doping levels of an abrupt P–N junction are such that $N_D \gg N_A$. At what point is the maximum of the electric field when a reverse bias is applied?

(1) $x = -w_n$ (the edge of the depletion layer in the N-type region)

(2) $x = -w_n/2$

(3) $x = 0$

(4) $x = w_p/2$

(5) $x = w_p$ (the edge of the depletion layer in the P-type region)

6.33 The N-type and P-type doping levels of an abrupt P–N junction are $N_D = 10^{17}$ cm^{-3}, and $N_A = 10^{15}$ cm^{-3}.

(a) Calculate and compare the zero-bias depletion layer widths in the N- and P-type regions.

(b) Calculate the maximum electric field in the depletion layer.

(c) Calculate the zero-bias capacitance per unit area. **A**

6.34 The range of operating voltages of an abrupt P–N junction capacitor is 0–10 V. What is the minimum capacitance that can be achieved, if the maximum capacitance is 2.5 pF? The built-in voltage is $V_{bi} = 0.71$ V.

6.35 The possibility to vary the capacitance by changing the reverse-bias voltage V_R enables the capacitor of Problem 6.15 to be used as a tuning element in microwave circuits. If the operating point of the capacitor is $V_R = 5$ V, determine the capacitor sensitivity dC/dV_R.

6.36 The donor and acceptor concentrations at each side of a P–N junction are given in Table 6.4. Determine $\rho_{charge}(x)$ in the depletion layer to demonstrate that this is a linear P–N junction.

(a) Calculate the built-in voltage V_{bi}. **A**

(b) Calculate the zero-bias capacitance per unit area and the capacitance at $V_R = 20$ V.

(c) Calculate N_D of an abrupt P–N junction ($N_A = 10^{20}$ cm^{-3}), so that it has the same zero-bias capacitance as the P–N junction in part (b). Calculate the capacitance at $V_R = 20$ V and compare it to the result in part (b). Comment on the result.

(d) Calculate and compare the maximum zero-bias electric fields at the P–N junctions of parts (b) and (c). **A**

TABLE 6.4

	x (μm)		
	−5.0	0.0	5.0
N_D (cm^{-3})	6×10^{15}	3.5×10^{15}	10^{15}
N_A (cm^{-3})	10^{15}	3.5×10^{15}	6×10^{15}

6.37 **(a)** The charge density at a linear P–N junction can be expressed by $\rho_{charge} = -5 \times 10^8 x$, where x is in m and ρ_{charge} in C/m^3. Design the donor concentration of an abrupt P–N junction ($N_A \gg N_D$), so that the zero-bias maximum field is the same as in the case of the linear junction. Assume $V_{bi} = 0.6$ V for the linear, and $V_{bi} = 0.9$ V for the abrupt P–N junction.

(b) Calculate and compare the maximum electric fields at these junctions at $V_R = 20$ V.

6.38 The capacitance of a P–N junction is $C_d(0) = 3.0$ pF and $C_d(5V) = 1.33$ pF at $V_R = 0$ V and $V_R = 5$ V, respectively. If $V_{bi} = 0.76$ V, determine the grading coefficient m (SPICE parameter).

6.39 Derive the equation for the dependence of the generation current (I_G) on the applied reverse-bias voltage (V_R) at an abrupt P$^+$–N junction. Use the derived equation to calculate the generation current at $V_R = 10$ V if it is known that the generation current for very small reverse-bias voltages is $I_G(V_R \approx 0) = 100$ pA and the built-in voltage at the P$^+$–N junction is $V_{bi} = 1.0$ V.

6.40 The generation current of a silicon diode is 112.5 pA at $V_R = 1$ V and 238.7 pA at $V_R = 10$ V. What is the grading coefficient of this diode if the built-in voltage is $V_{bi} = 0.80$ V?

6.41 Identify the correct statement:

(a) If the forward bias across a P–N junction is suddenly switched off, the excess minority carriers continue to flow in the same direction until their concentration drops to the equilibrium level.

(b) If the forward bias across a P–N junction is suddenly switched off, the current changes its direction but continues to flow until the minority-carrier concentrations reach the equilibrium levels.

6.42 A current of 1 mA flows through a P–N junction diode. Calculate the associated stored-charge capacitance, if the transit time is $\tau_T = 10$ ns, the emission coefficient of the diode is $n = 1.2$, and the thermal voltage is $V_t = 26$ mV.

6.43 A P–N junction diode with $\tau_T = 100$ ns conducts a forward-bias current $I_D = 4.3$ mA. At time $t = 0$, the circuit conditions change, so that a reverse-bias voltage $V_E = 20$ V is connected in series with the diode and a resistor $R = 1$ kΩ.

(a) Because of the strong discharging current ($\approx$ 20.7 V/1 kΩ = 20.7 mA), the contribution of the recombination to the stored charge removal can be neglected. How long will it take for the stored charge to be removed?

(b) If $V_E = 0$, the diode is discharged with a much smaller current ($\approx$ 0.7 V/1 kΩ = 0.7 mA), so that the contribution from the recombination cannot be neglected. If it is found that it takes $t_{rs} = 200$ ns for the stored charge to be removed, how much stored charge is removed by recombination? **A**

6.44 The SPICE parameters of a P–N junction diode are $I_S = 10^{-11}$ A, $n = 1.4$, $r_S = 0$, and $\tau_T = 10$ ns. What is the value of the stored charge if the forward voltage across the diode is $V_D = 0.7$ V? **A**

6.45 A diode circuit is simulated by SPICE with four different sets of parameters. The results for the current flowing through the diode are shown in Fig. 6.25. Identify which set of parameters corresponds to each of the simulation results.

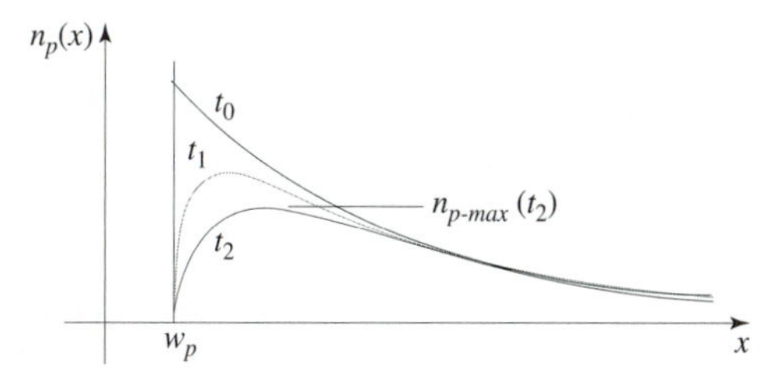

Figure 6.26 Concentration diagrams of minority electrons during diode turnoff (stored charge removal).

6.46 The voltage v_{pulse} in the circuit of Fig. 6.20a rapidly changes from 5 V to 0 V at $t = t_0$. As a result, the concentration of minority electrons starts changing as in Fig. 6.26. Solve the continuity equation to find the change of the maximum electron concentration with time: $n_{p-max}(t)$. If $\tau_n = 10$ ns, how long does it take for the maximum of the excess electron concentration to drop to half of its original value?

6.47 The technological parameters of a P–N$^+$ GaAs diode are as follows: $N_A = 5 \times 10^{16}$ cm^{-3}, $D_n = 30$ cm^2/V·s, $n_i = 2.1 \times 10^6$ cm^{-3}, and $L_n = 5$ μm. Calculate the transit time τ_T (SPICE parameter).

6.48 Derive the dependence of τ_T on D_n for the case of a short diode ($W_{anode} \ll L_n$), and calculate the

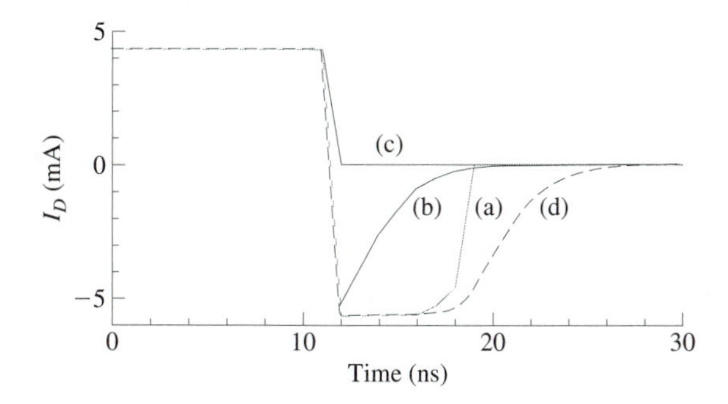

(1) $\tau_T = 0$, $C_d(0) = 0$

(2) $\tau_T = 12$ ns, $C_d(0) = 0$

(3) $\tau_T = 0$, $C_d(0) = 4$ pF

(4) $\tau_T = 12$ ns, $C_d(0) = 4$ pF

Figure 6.25 SPICE transient analysis of a diode circuit. The diode current is obtained with four different sets of diode parameters.

transit time for $W_{anode} = 1\ \mu$m. Use the same technological parameters as in Problem 6.47.

6.49 The doping levels in an abrupt P–N junction are $N_A = 10^{15}$ cm^{-3} and $N_D = 10^{17}$ cm^{-3}. Determine the generation current per unit area at reverse-bias voltage of $V_R = 5$ V and compare it to the diffusion-current density for

(a) Si ($\varepsilon_{Si} = 11.8\varepsilon_0$, $n_i = 1.02 \times 10^{10}$ cm^{-3})
(b) 3C SiC ($\varepsilon_{SiC} = 10\varepsilon_0$, $n_i = 1$ cm^{-3}) **A**

Assume midgap R–G centers and equal minority-carrier lifetimes for electrons and holes in both materials: $\tau_n = \tau_p = 0.5\ \mu$s. For simplicity, assume equal diffusion constants for both electrons and holes in both materials: $D_{n,p} = 10$ cm^2/s.

REVIEW QUESTIONS

R-6.1 The difference between the Fermi levels in an N-type and a P-type semiconductor is 0.8 eV. What is the built-in voltage if these materials create a P–N junction? What is the barrier height for the electrons and the holes at the P–N junction?

R-6.2 In thermal equilibrium, the current of majority electrons able to go over the barrier to appear in the P-type region is balanced by the current of minority electrons that easily roll down from the P-type to the N-type region. How is this balance maintained in materials with large barrier heights? For example, the barrier height at a P–N junction on SiC can be as high as 3 eV, meaning that practically no electron can move from the N-type to the P-type region. If the total current is to be zero, there should be no electrons rolling down from the P-type to the N-type region. Why?

R-6.3 If the temperature is increased, there are more electrons at higher energy levels (refer to Fermi–Dirac and Maxwell–Boltzmann distributions). This means the number of majority electrons able to go over the barrier is increased. If the balance with the current of the minority carriers is to be maintained, what does increase the minority-carrier current?

R-6.4 Does the minority-carrier current of a reverse-biased P–N junction depend significantly on the value of the reverse-bias voltage?

R-6.5 If the temperature is increased, will the minority-carrier current be increased as well? If so, is the temperature dependence linear, quadratic, exponential, or logarithmic?

R-6.6 If the junction is exposed to light so that the light generates electron–hole pairs, will the reverse-bias current (leakage current) be increased?

R-6.7 Consider a forward-biased P–N junction. Is the current of electrons in the N-type region (the majority carriers) due to the drift or diffusion?

R-6.8 Is the electric field in the neutral regions large or small? Why can the voltage drop in the neutral regions be neglected compared to the voltage drop in the depletion layer? Is the electric field in the neutral regions large enough to produce a significant drift current of the majority carriers? Why?

R-6.9 What is the effect of the forward bias on the energy barrier at the P–N junction?

R-6.10 Why is the increase in the number of electrons able to overcome the energy barrier at the P–N junction exponentially dependent on the applied forward-bias voltage?

R-6.11 What happens to the electrons that overcome the energy barrier at the P–N junction and appear in the P-type region? What are these electrons called?

R-6.12 Is the drift current of the minority carriers significant? The diffusion current?

R-6.13 How do the excess minority-carrier concentrations at the edges of the depletion layer depend on the forward-bias voltage? Are these the maximum concentrations of the minority electrons and holes, respectively?

R-6.14 How does the avalanche breakdown voltage depend on temperature? Why?

R-6.15 We have seen that energy barriers of sufficient height can block electron motion. Is the barrier width important as well? Can the energy barrier be so narrow that it lets some electrons through? If so, what is this effect called?

R-6.16 Can a P–N junction diode be used as a temperature sensor?

R-6.17 Why does the capacitance of a reverse-biased P–N junction depend on the reverse-bias voltage? What reverse-bias voltage provides maximum capacitance?

R-6.18 Abrupt and linear P–N junctions represent two extreme cases. What are the similarities and what are the differences in the derived equations for C_d–V_R dependence?

R-6.19 If the forward-bias voltage is suddenly switched off, what happens to the excess minority carriers? In which direction does the current flow immediately after the voltage has been switched off?

R-6.20 What is the SPICE parameter that accounts for the stored-charge effect?

R-6.21 Is the transit time related to the minority-carrier lifetime? If so, how?

R-6.22 Is the transit time related to the reverse-recovery time? If so, how?

7 Metal–Semiconductor Contact and MOS Capacitor

The P–N junction is a fundamental device structure. Many concepts from Chapter 6 on P–N junctions can be applied to junctions/contacts of other types, such as semiconductor–metal and semiconductor–dielectric contacts. There are, however, important differences. The essential similarity is that a built-in voltage appears at any junction of materials with different work functions; this is because the mobile electrons in such materials have different potential energies, so they are redistributed around the junction to bring the system into thermal equilibrium. A very important difference is that band offsets appear at the junction of two materials with different electron affinities; this is a new concept that did not appear in the P–N junction chapter because the electron affinities of both P-type and N-type silicon are equal. Both the built-in voltage and the band offset are directly related to potential-energy barriers, but there is a fundamental difference: the barrier due to the work-function difference can be altered by a voltage applied across the junction, whereas the barrier due to the band offset is a material constant.

Based on this difference, junctions between two materials can be classified into *homojunctions* (these are the P–N junctions of identical semiconductors) and *heterojunctions*. With this classification, it can be stated that this chapter extends the homojunction concepts from the previous chapter to the more general case of a heterojunction. The term *heterojunction,* however, tends to be used specifically for junctions of two different types of semiconductors.[1] Nonetheless, junctions between semiconductors and metals or between semiconductors and dielectrics are "heterojunctions" in qualitative terms. For example, dielectrics may be modeled by energy-band diagrams that are not qualitatively different from the energy-band diagram of semiconductors, which means that

[1]Specific heterojunctions of different semiconductors will be considered in the sections on heterojunction bipolar transistors (Section 9.4), lasers (Section 12.3), and high-electron-mobility transistors (HEMT) (Section 10.2).

a semiconductor–dielectric junction is not qualitatively different from the heterojunction of two different semiconductors.

The first part of this chapter is devoted to a specific type of "heterojunction"—the metal–semiconductor contact. The second part of the chapter is devoted to the metal–oxide–semiconductor (MOS) capacitor (obviously, the MOS structure can be considered as a system of two "heterojunctions").

7.1 METAL–SEMICONDUCTOR CONTACT

In practical terms, there are two different types of metal–semiconductor contact: (1) rectifying contacts that are widely known as Schottky diodes and (2) ohmic contacts.

Schottky diodes have current–voltage characteristics very similar to those of the P–N junctions, but there is also an important difference: Schottky diodes operate with a single type of carrier (the majority carriers). The irrelevance of the minority carriers means that the stored-charge effect does not exist, which makes the Schottky diodes suitable for fast switching applications.

Ohmic contacts have linear current–voltage characteristics, so they enable the semiconductor devices to be contacted by metal conductors.

7.1.1 Schottky Diode: Rectifying Metal–Semiconductor Contact

This section describes the principles of metal–semiconductor contacts on the example of rectifying contacts (Schottky diodes). Energy-band diagrams are used in the explanations because they are by far the best tool for comprehending junction phenomena.

Figure 7.1a shows the energy-band diagrams of an N-type semiconductor and a metal that are separated from each other. The energy-band diagrams of these two materials are

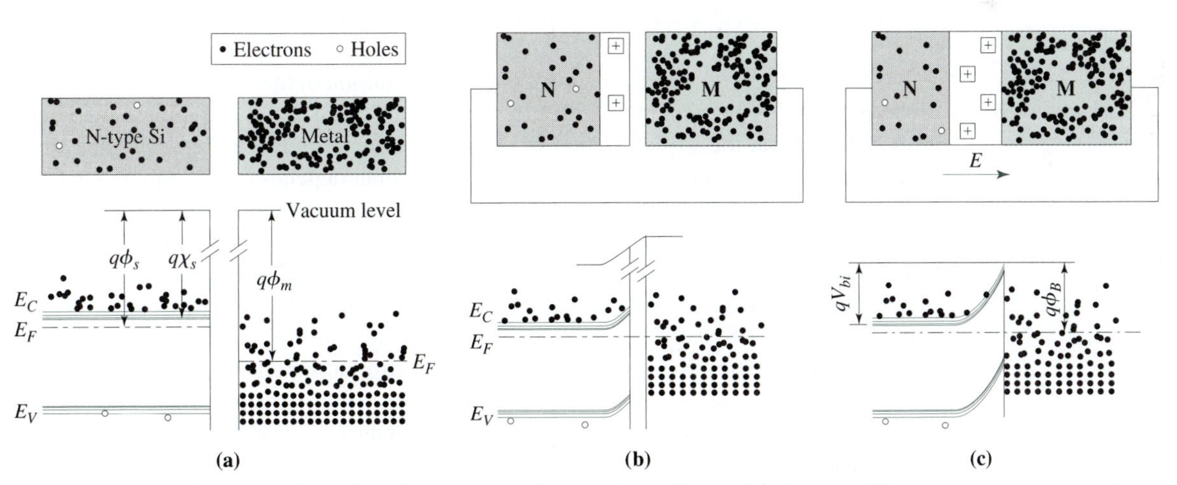

Figure 7.1 Illustration of metal–semiconductor contact in thermal equilibrium. (a) Separated N-type semiconductor and metal. (b) N-type semiconductor and metal in electrical contact. (c) Ideal physical contact between N-type semiconductor and metal.

TABLE 7.1 The Work Functions and Electron Affinities of the Most Frequently Used Materials at 300 K

Material	Work Function $q\phi_m$ (eV)	Electron Affinity $q\chi_s$ (eV)
Al	4.1	
Cr	4.5	
Ni	5.15	
Pt	5.7	
PtSi	5.4	
W	4.6	
WSi_2	4.7	
Si		4.05
GaAs		4.07
Ge		4.0
SiO_2		1
N^+ Si	4.05	4.05
P^+ Si	5.17	4.05

drawn so that the vacuum levels are matched. Remember, the vacuum level is the energy level of a completely free electron (an isolated electron in vacuum). The electrons in solids have negative energies with respect to the vacuum level: they need some energy to be able to liberate themselves from the attracting forces in the crystal and become free electrons. The energy (work) that is needed to remove an *average electron* (an electron at the Fermi level) from the metal is called the *work function* and is labeled by $q\phi_m$ in Fig. 7.1a. In other words, the work function expresses the position of the Fermi level with respect to the energy of a free electron in vacuum (vacuum level). The work functions for different metals are material constants, and some of them are given in Table 7.1.

When dealing with semiconductors, the same definition of the work function has to be used, regardless of the fact that, typically, there are no electrons at and around the Fermi level. In Fig. 7.1a, the work function of the semiconductor is labeled by $q\phi_s$. The work function of semiconductors is not a material constant but depends on the doping type and level because the position of the Fermi level is changed with the doping. However, the doping does not influence the position of the bands with respect to the vacuum level. The position of the bottom of the conduction band with respect to the vacuum level is referred to as the *electron affinity*, labeled $q\chi_s$ in Fig. 7.1a. The electron affinity is equal to the energy needed to remove a free electron with zero kinetic energy (an electron at the bottom of the conduction band) from the semiconductor. The electron affinities are material constants and for some important semiconductors and insulators are also given in Table 7.1.

In Fig. 7.1a, the Fermi level of the N-type semiconductor is higher than the Fermi level of the metal, which means that the mobile electrons in the semiconductor are at higher energy levels than the electrons in the metal.

Thermal Equilibrium

Consider now what happens when an electric contact is made between the two materials—for example, by the wire (short circuit) illustrated in Fig. 7.1b. In this case, the two materials represent a single system. The electrons from the higher energy levels in the semiconductor move through the wire into the metal, creating a depletion layer at the semiconductor surface. The uncompensated positive donor ions in the depletion layer create an electric field that is associated with the appearance of a potential difference between the surface of the metal and the bulk of the semiconductor. When thermal equilibrium is established, this potential difference is exactly equal to the initial difference between the Fermi levels, which means that the Fermi levels in the semiconductor and the metal are now perfectly aligned to each other. We already know that the Fermi level of a single system in thermal equilibrium is constant.

There is a gap between the metal and the semiconductor in Fig. 7.1b. In reality, it is very likely that there will be a minute gap of the order of atomic distances at the interface. Although this does influence the properties of the metal–semiconductor contact, it is not of essential importance. For easier understanding, it is helpful to assume ideal contact. The ideal metal–semiconductor contact is illustrated in Fig. 7.1c. The energy-band diagram is constructed in the following way:

1. The Fermi level (dash–dot line in Fig. 7.1c) is drawn first.
2. The conduction and valence bands are drawn in the neutral region of the semiconductor. The bands are placed appropriately with respect to the Fermi level, to express the doping level of the semiconductor (the energy-band diagram in the neutral region should be the same as in Fig. 7.1a).
3. Inside the metal, no change is practically observed because of the abundance of electrons and the related fact that any electric field penetrates only to atomic distances.
4. At the semiconductor–metal interface, the energy-band offset between the semiconductor and the metal is preserved. Accordingly, the interface positions of E_C and E_V are determined so that they remain at the same energy positions with respect to E_F in the metal. Hence, the difference between E_F and E_C at the interface is

$$q\phi_B = q\phi_m - q\chi_s \tag{7.1}$$

This is the energy barrier for the electrons in the metal. It is independent of either semiconductor doping or bias applied because both $q\phi_m$ and $q\chi_s$ are material constants.

5. The energy bands are bent in the depletion layer of the semiconductor to join the bulk and interface levels. This bending is equal to the original difference between the Fermi levels of the metal and the semiconductor:

$$qV_{bi} = q\phi_m - q\phi_s = q\phi_{ms} \tag{7.2}$$

This is the energy barrier for the electrons in the semiconductor. It is directly related to the built-in voltage V_{bi} in the depletion layer of the semiconductor.

Reverse Bias

Now, we can consider the effects of applied voltage between the metal and the semiconductor. Figure 7.2a illustrates the case of reverse bias. Applying *negative voltage* $V_D = -V_R$ to the metal with respect to the N-type semiconductor means that its Fermi level is set above the Fermi level of the semiconductor by the amount qV_R. Remember that *potential energy* $= -q$*(electric potential)*.

The reverse bias $V_D = -V_R$ increases the energy barrier height for the electrons in the semiconductor to $q(V_{bi} + V_R)$. The increased energy barrier in the depletion layer of the semiconductor prevents the electrons from the semiconductor from moving through the contact.

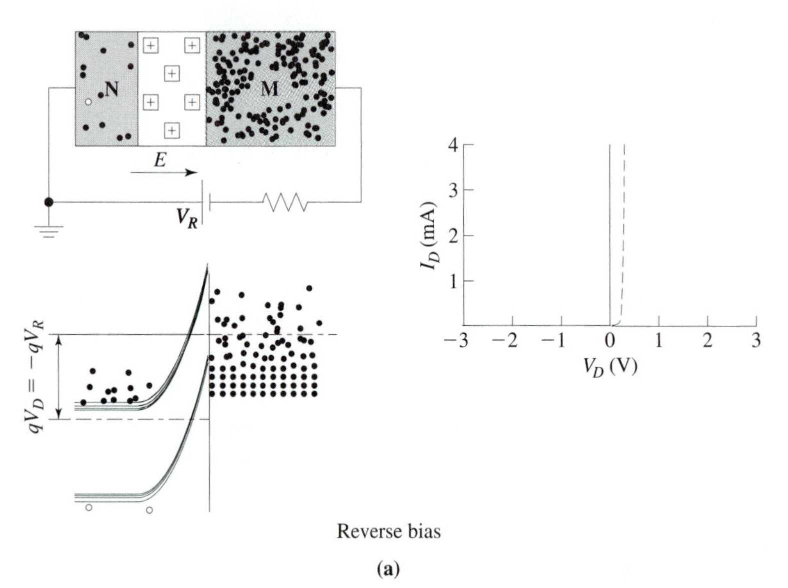

Reverse bias

(a)

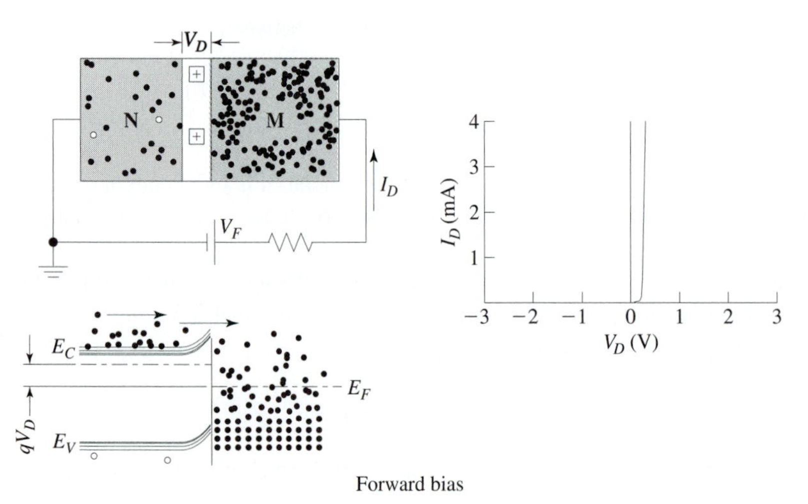

Forward bias

(b)

Figure 7.2 Illustration of (a) reverse-biased and (b) forward-biased Schottky diodes.

The barrier for the electrons in the metal ($q\phi_B$) does not change. The minor number of electrons in the metal that are able to overcome the potential barrier and appear in the N-type semiconductor make the reverse-bias current of the Schottky diode. Given that $q\phi_B$ is bias-independent, the reverse-bias current does not depend on the reverse-bias voltage:

$$I_D = -I_S \tag{7.3}$$

If I_S is considered as a parameter, as is the case in SPICE, the reverse-bias current of the Schottky diode is expressed in the same way as for P–N junction diodes. The difference between Schottky diodes and P–N junctions is in the ingredients of I_S. In the case of the Schottky diode, I_S is due to thermal emission of electrons over the barrier $q\phi_B$. This is different from the I_S current due to the minority carriers in P–N junctions. It is similar, however, to the injection of carriers over the reduced barrier in forward-biased P–N junctions. Following reasoning analogous to that used for carrier injection over a P–N junction barrier [Eq. (6.7)], we can conclude that the number of electrons able to go over $q\phi_B$ is proportional to $\exp(-q\phi_B/kT)$. More precisely, the current due to thermal emission is given by

$$I_S = A_J A^* T^2 e^{-q\phi_B/kT} \tag{7.4}$$

where A_J is the diode area, T is absolute temperature, $q\phi_B$ is the barrier height, and A^* is the so-called effective Richardson constant. For N-type Si, $A^* \approx 120\ \text{A cm}^{-2}\,\text{K}^{-2}$; and for N-type GaAs, $A^* \approx 140\ \text{A cm}^{-2}\,\text{K}^{-2}$.

There is a second-order effect that significantly influences the reverse-bias current of a Schottky diode. This effect is due to the appearance of an image force between the electrons in the semiconductor and the nearby highly conductive metal. It is shown that the barrier height reduction is

$$\Delta\phi_B = \sqrt{\frac{q E_{max}}{4\pi\varepsilon_s}} \tag{7.5}$$

where E_{max} is the maximum electric field.[2] This field is the same as for the abrupt P–N junction [Eqs. (6.31) and (6.41)] and is therefore proportional to $\sqrt{V_R}$. Clearly, this effect causes a significant increase in the reverse-bias current I_S of the Schottky diode with increasing reverse-bias voltage.

Forward Bias

Analogously to the case of the P–N junction, forward-bias voltage V_D reduces the energy barrier in the depletion layer to $q(V_{bi} - V_D)$ (Fig. 7.2b). This enables a number of electrons from the semiconductor to overcome the barrier and appear in the metal. The number of electrons that move in the opposite direction, from the metal into the semiconductor, is the same as in the case of reverse bias. Importantly, the current of the electrons from the

---

[2]E. H. Rhoderic, *Metal–Semiconductor Contacts,* Clarendon Press, Oxford, 1978; S. M. Sze, *Physics of Semiconductor Devices,* 2nd ed., Wiley, New York, 1981.

semiconductor dominates. Moreover, this current increases exponentially with the forward-bias voltage. Again, this is due to the fact that the electrons are exponentially distributed along the energy in the conduction band. As the energy barrier height is reduced by qV_D, the number of electrons able to go over the barrier increases exponentially with V_D.

The different physical background of I_S notwithstanding, the forward-bias current of the Schottky diode can be expressed in the same way as the forward-bias current of the P–N junction:

$$I_D = I_S e^{V_D/nV_t} \tag{7.6}$$

Of course, the values of the parameters I_S and n should be adjusted to fit the characteristic of a particular diode. In that sense, there is no different SPICE model for the Schottky diode. The values of the diode parameters are adjusted appropriately to reflect the characteristics of Schottky diodes.

Although the same mathematical equation can be used to model the I_D–V_D characteristic of Schottky diodes, there are important practical differences. Typically, the built-in voltage (V_{bi}) is smaller in the case of Schottky diodes.[3] The current of silicon P–N junctions becomes significant at about 0.7 V, whereas the same current can be achieved by a voltage as low as 0.2 V in the case of the Schottky diodes. This represents an advantage of the Schottky diodes in applications where the power loss $V_D I_D$ across the diode is critical. This improvement, however, is achieved at the expense of increased reverse-bias (saturation) current I_S. Mathematically, it can be seen that I_S in Eq. (7.6) has to be significantly increased to account for the significant increase in I_D. This can also be concluded from the energy bands. Using a metal with smaller $q\phi_m$ to reduce the built-in voltage (Eq. 7.2) leads to a related reduction of $q\phi_B$ [Eq. (7.1)], and hence an increase in I_S [Eq. (7.4)]. In the extreme case of $q\phi_B = 0$, electrons move freely from the metal into the semiconductor and vice versa—this is a perfect ohmic contact.

Another important difference between P–N junction and Schottky diodes is in the dynamic characteristics. In the case of a forward-biased P–N junction diode, the electrons from the N-type semiconductor appear as minority carriers in the P-type region after they pass through the junction. This leads to accumulation of the minority carriers at the sides of the depletion layer—the stored charge. When the forward-bias voltage is switched off, the steady-state reverse-bias current cannot be established before the stored charge is removed.[4] In Schottky diodes (Fig. 7.2b), the electrons from the silicon appear in the *metal* after they passed through the contact. These electrons appear among the huge number of electrons already existing in the metal. They are not minority carriers and do not make any difference. *Due to the absence of the stored-charge effects, the Schottky diode responds much more quickly to voltage changes than the P–N junction diode.*

---

[3]Note from Eq. (7.2) that V_{bi} of Schottky diodes can be technologically altered by using metals with different work functions (ϕ_m).

[4]The effects of the stored charge on the dynamic characteristics of the P–N junction diode are explained in Section 6.4.

EXAMPLE 7.1 Designing and Comparing PIN and Schottky Diodes for Power Applications

Both P–N junction and Schottky diodes are frequently used for power applications. To increase the breakdown voltage and reduce the parasitic resistance, a very-low-doped region, labeled I, is sandwiched between N^+ and P^+ regions to create so-called PIN diode, or between the metal and the highly doped semiconductor body in the case of the Schottky diode. This example compares Si PIN and Schottky power diodes, with the same area $A_J = 0.1$ cm^2, and the same doping level in the low-doped region of $N_D = 5 \times 10^{14}$ cm^{-3}. The Schottky diode is created by depositing tungsten onto the low-doped Si.

(a) Assuming maximum allowable field of $E_{max} = 20$ V/μ m, calculate the needed width of the low-doped (drift) layer so that the depletion layer does not extend into the N^+ region. What reverse-bias voltage (maximum operating voltage) corresponds to E_{max}?

(b) Calculate the saturation currents of the PIN and the Schottky diodes and compare them. Assume the following value for the diffusion coefficient of minority holes: $D_p = 50$ cm^2/s.

(c) Neglecting the parasitic resistances, calculate and compare the forward voltages for an operating current $I_F = 5$ A. Discuss the results in terms of power loss. (Assume an ideal emission coefficient $n = 1$.)

(d) Can the area of the PIN diode be designed to have the same forward voltage at $I_F = 5$ A as the Schottky diode?

(e) Assuming mobility of $\mu_n = 1400$ cm^2/V · s, calculate the resistance of the drift region. What are the forward voltage drops when this resistance is included?

(f) Including the barrier-lowering effect, calculate the reverse current of the Schottky diode at maximum reverse voltage. Discuss the result in terms of power dissipation.

SOLUTION

(a) The theory of asymmetrical abrupt P–N junction, presented in Section 6.3.2, can also be applied to the case of Schottky diode. This is because a significant depletion layer appears only in the N-drift region in both cases. From Eq. (6.31), the maximum depletion-layer width is obtained as

$$w_{n-max} = \varepsilon_s E_{max}/(qN_D) = \frac{11.8 \times 8.85 \times 10^{-12} 20 \times 10^6}{1.6 \times 10^{-19} \times 5 \times 10^{20}} = 26\ \mu\text{m}$$

Therefore, the width of the drift layer should be $w_{n-epi} \approx 26\ \mu$m. Neglecting the small V_{bi} in comparison to V_R, the maximum operating voltage is calculated from Eq. (6.45) as

$$V_{R-max} = w_{n-max}^2 qN_D/(2\varepsilon_s) = (26 \times 10^{-6})^2 \frac{1.6 \times 10^{-19} \times 5 \times 10^{20}}{2 \times 11.8 \times 8.85 \times 10^{-12}} = 260\ \text{V}$$

(b) Noting that $N_A \gg N_D$, Eq. (6.17) can be simplified to calculate the saturation current of the PIN diode:

$$I_S = A_J q n_i^2 \frac{D_p}{w_{n-epi} N_D}$$

$$= 10^{-5} \times 1.6 \times 10^{-19} \times (1.02 \times 10^{16})^2 \frac{0.005}{26 \times 10^{-6} \times 5 \times 10^{20}}$$

$$= 6.4 \times 10^{-11}\ \text{A}$$

The barrier height $q\phi_B$ is needed to be able to calculate the saturation current of the Schottky diode. From Eq. (7.1), we find

$$q\phi_B = q\phi_m - q\chi_s = 4.6 - 4.05 = 0.55\ \text{eV}$$

where the value for $q\phi_m$ is obtained from Table 7.1. From Eq. (7.4),

$$I_S = A_J A^* T^2 e^{-q\phi_B/kT} = 10^{-5} \times 120 \times 10^4 \times 300^2 \times e^{-0.55/0.2585} = 6.2 \times 10^{-4}\ \text{A}$$

The saturation current of the Schottky diode is $6.2 \times 10^{-4}/6.4 \times 10^{-11} \approx 10^7$ times higher.

(c) Neglecting the parasitic resistances, the forward voltage can be obtained from the equation

$$I_F = I_S e^{V_F/V_t}$$

for both PIN and Schottky diodes. For the PIN diode, it is

$$V_F = V_t \ln(I_F/I_S) = 0.02585 \times \ln(5/6.4 \times 10^{-11}) = 0.65\ \text{V}$$

For the Schottky diode,

$$V_F = 0.02585 \times \ln(5/6.2 \times 10^{-4}) = 0.23\ \text{V}$$

The series contact resistances, the resistance of the drift region, and the substrate resistance will add voltage to V_F, so the actual forward voltage will be significantly higher—part (d) of this example. Nonetheless, the difference between the PIN and Schottky diodes of about 0.4 V will remain. At 5 A and an assumed 50% duty cycle, this means a difference of $0.5 \times 0.4 \times 5 = 1$ W of power dissipation.

(d) The equation $I_F = I_S \exp(V_F/V_t)$, which is applicable to both devices, shows that I_S of the PIN diode will have to be increased to the same value as in the case of the Schottky diode so that $I_F = 5$ A corresponds to the same V_F in either case. Based on the results from part (b), the needed increase in I_S is $6.2\times10^{-4}/6.4\times10^{-11} \approx 10^7$ times. This increase can theoretically be achieved if the junction area is increased 10^7 times. Practically, this is impossible because the needed area would be $A_J = 0.1 \times 10^7\ \text{cm}^2$ (for the square shape, this is a square with the side equal to 10 m).

(e) The conductivity of the drift region is

$$\sigma = q\mu_n N_D = 1.6 \times 10^{-19} \times 0.14 \times 5 \times 10^{20} = 11.2\,(\Omega \cdot \text{m})^{-1}$$

The resistance is then

$$R = w_{n-epi}/(\sigma A_J) = \frac{26 \times 10^{-6}}{11.2 \times 10^{-5}} = 0.232\,\Omega$$

With $I_F = 5$ A, the forward voltages of the PIN and Schottky diodes are $V_F = 0.65 + 0.232 \times 5 = 1.81$ V and $V_F = 0.23 + 0.232 \times 10 = 1.39$ V, respectively.

(f) The barrier height lowering is given by Eq. (7.5):

$$\Delta\phi_B = \sqrt{qE_{max}/(4\pi\varepsilon_s)} = \sqrt{\frac{1.6 \times 10^{-19} \times 20 \times 10^6}{4\pi \times 11.8 \times 8.85 \times 10^{-12}}} = 0.05\ \text{V}$$

The reverse-bias current with the lower barrier $q\phi_B - q\Delta\phi_B = 0.55 - 0.05 = 0.50$ eV is

$$I_R = I_S(V_{R-max}) = 10^{-5} \times 120 \times 10^4 \times 300^2 \times e^{-0.50/0.2585} = 4.3\ \text{mA}$$

The power loss due to this current is not insignificant at $V_{R-max} = 260$ V. Assuming again a 50% duty cycle, it is $0.5 \times 260 \times 4.3 \times 10^{-3} = 0.56$ W.

7.1.2 Ohmic Metal–Semiconductor Contacts

The width of the energy barrier at a Schottky contact depends on the doping level in the semiconductor. This is due to the dependence of the depletion-layer width on doping (Section 6.3.2). Equation (6.45), which gives the depletion-layer width in an N-type semiconductor as a part of one-sided ($N_A \gg N_D$) abrupt P–N junction, would also be obtained for the case of a metal–N-type semiconductor contact if the Poisson equation was solved. The heavily doped P-type semiconductor in the one-sided P–N junction behaves similarly to the metal in the case of the Schottky diode, as far as the depletion layer is concerned.

Figure 7.3 illustrates the contact between metal and heavily doped N-type semiconductor. To express that the doping level is high, the semiconductor is called an N^+ type. Due to the high doping level, the energy barrier at the surface of the semiconductor is very narrow. As a consequence, the electrons from the metal can tunnel through the barrier when negative voltage is applied to the metal (Fig. 7.3a). As the applied voltage is increased, the associated splitting of the Fermi levels increases the band bending, further narrowing the energy barrier. This leads to a significant increase in the tunneling current.

Figure 7.3b illustrates the case of positive voltage applied to the metal. In this case the electrons from the semiconductor not only go over the reduced barrier (as in the case of the

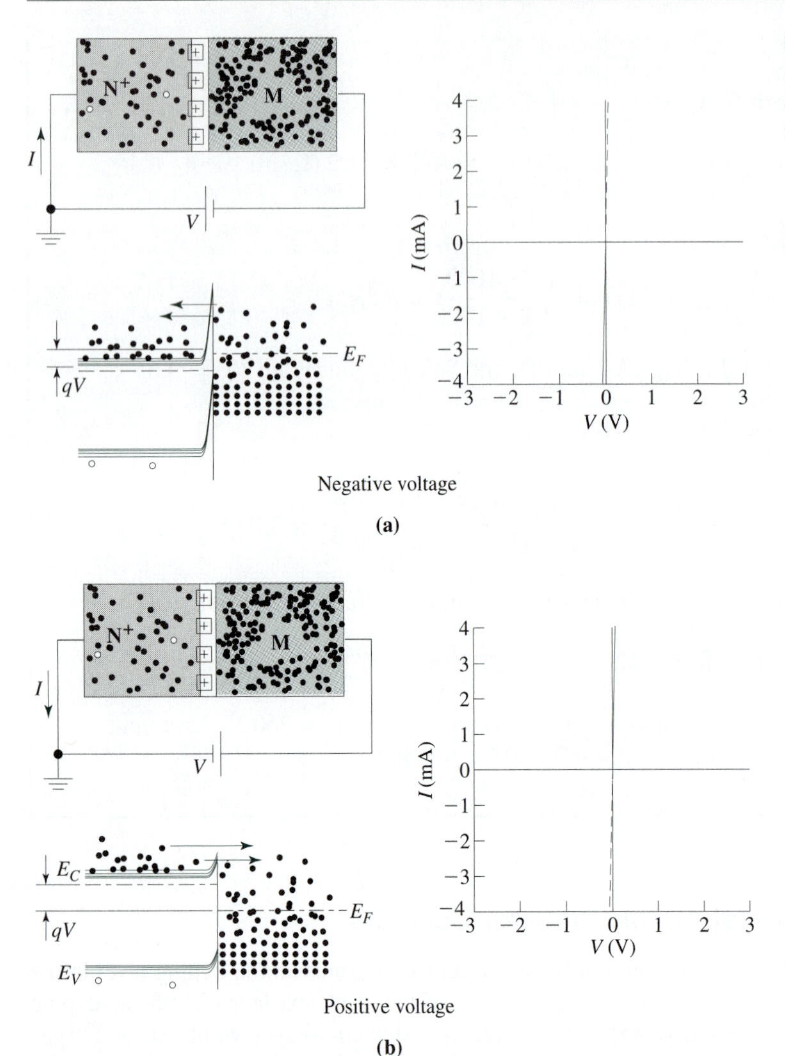

Figure 7.3 Illustration of ohmic metal–semiconductor contact with (a) negative and (b) positive voltage at the metal.

Schottky diode), but also tunnel through the barrier, significantly increasing the current. Therefore, positive voltage at the metal also produces a current flow through the contact that rapidly increases with the voltage increase. *The electric characteristic of the contact between a metal and a heavily doped semiconductor is equivalent to the characteristic of a small resistance.*

7.2 MOS CAPACITOR

Metal–oxide–semiconductor (MOS) capacitors exhibit sophisticated characteristics when compared to the ordinary metal–dielectric–metal capacitors. It is the replacement of one metal electrode by a semiconductor that leads to some unique effects. The dielectric in the MOS capacitor has almost always been the silicon dioxide, or *oxide,* for short, so the standard term is MOS (metal–oxide–semiconductor). The MOS capacitor can be seen as a structure consisting of two heterojunctions: (1) metal–dielectric and (2) dielectric–semiconductor, where the dielectric is the silicon dioxide.

It is the high quality of oxide–semiconductor interface that enables practical applications of this device structure. For decades, a device-quality oxide–semiconductor interface has been limited to one semiconductor only—silicon. It is this fact that makes silicon by far the dominant semiconductor, in spite of the fact that many other semiconductors have better bulk properties. Recently, device-quality oxide–semiconductor interfaces have been developed on silicon carbide—a wide-energy-gap semiconductor with excellent bulk properties.

The problems with the interface are due to the dangling atomic bonds at the semiconductor surface that have to be electronically passivated to enable existence of mobile charge at the semiconductor surface. This will be explained in Section 7.2.1. There are no special effects at the metal–oxide interface. In fact, heavily doped polysilicon has been typically used in the place of metal electrode, the reason being an important technological effect that relates to the MOSFET structure. Accordingly, the energy-band diagrams in this chapter will be shown for polysilicon gates.

MOS capacitors have been used in linear circuits and as the storage elements in random-access memories (RAMs) and charge-coupled devices (CCDs). The real importance of this structure, however, is that it is the central part of the most used device in electronics—the metal–oxide–semiconductor field-effect transistor (MOSFET). The metal (or heavily doped polysilicon) electrode of the MOS capacitor in a MOSFET is called the *gate*. Accordingly, the metal/polysilicon electrode of the MOS capacitor will be referred to as the gate, and the oxide will frequently be referred to as the gate oxide to distinguish it from oxide layers that play other roles in integrated circuits.

7.2.1 Properties of the Gate Oxide and the Oxide–Semiconductor Interface

The properties of thermally grown silicon dioxide films on silicon are summarized in Table 7.2. A two-dimensional chemical-bond model of the oxide–semiconductor interface is given in Fig. 7.4a. Although the oxide is not a crystal, the silicon and oxygen atoms are packed in an orderly manner: each silicon atom is bonded to four oxygen atoms, and each oxygen atom is bonded to two silicon atoms. Cells formed by one silicon atom and the

TABLE 7.2 Properties of Thermally Grown SiO_2

Structure	Amorphous silica in which Si atoms are surrounded tetrahedrally by four O atoms: Si–O distances vary from 0.152 to 0.169 nm, Si–O–Si angles vary from 120° to 180°, O–Si–O angle is about 109.5°.
Dielectric constant	3.9
Dielectric strength	$\approx 10^7$ V/cm
Energy gap	≈ 9 eV
Resistivity	10^{12}–10^{16} $\Omega \cdot$cm

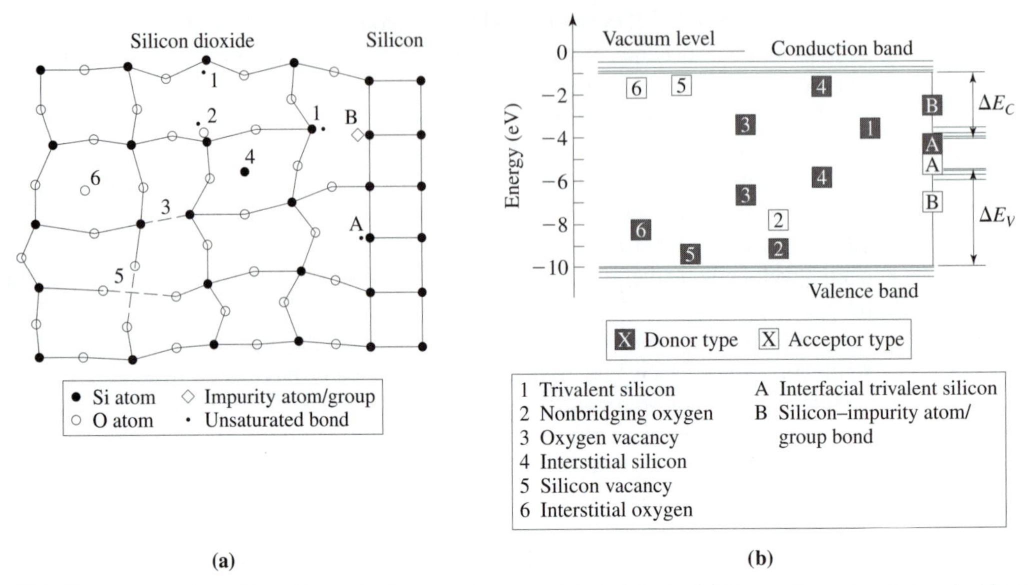

Figure 7.4 Illustration of the oxide–silicon interface and the associated defects. (a) A two-dimensional chemical-bond model. (b) The energy-band model at the flat-band condition.

four surrounding oxygen atoms have tetrahedral shapes in the reality (three dimensions) as explained in Table 7.2. The energy-band model of the oxide–silicon interface is shown in Fig. 7.4b. The energy gap of the oxide is about 9 eV. This value places the oxide among the very good insulators—its energy gap is more than eight times larger than that of silicon (1.12 eV). The large difference in the energy gaps of the oxide and the silicon means that there must be discontinuities in the energy bands at the oxide–silicon interface; these band discontinuities are called band offsets. The conduction-band offset is $\Delta E_C = 3.2$ eV. This is the barrier that the electrons from the silicon face when they move toward the oxide. This barrier is high enough to prevent any flow of electrons from the silicon into the oxide under normal conditions. The valence-band offset is even larger: $\Delta E_V = 9-3.2-1.12 \approx 4.7$ eV. Of course, this is the barrier that stops the flow of holes from the silicon into the oxide.

Interface Traps and Oxide Charge

The average distance between the oxygen atoms in the oxide is larger than the distance between the silicon atoms in the silicon. This means that some of the interface atoms from the silicon cannot create Si–O bonds because they are missing oxygen atoms. The atoms from the silicon that remain bonded only to three silicon atoms with the fourth bond unsaturated (trivalent interfacial silicon atoms) represent interface defects. The energy levels associated with the fourth unsaturated bond of the trivalent silicon atoms do not appear in the conduction or the valence band, but rather in the silicon energy gap. It is believed that every trivalent silicon atom introduces a pair of energy levels; one can be occupied by an electron (acceptor type), and the other can be occupied by a hole (donor type). Electrons and holes that appear on these levels cannot move freely because there is a relatively large distance between the neighboring interfacial trivalent silicon atoms (these levels are localized and isolated from each other). Because these levels can effectively trap mobile electrons and holes, they are called *interface traps*. Impurity atoms and groups (such as H, OH, and N) can be bonded to the unsaturated bonds of the interfacial trivalent silicon atoms, which results in a shift of the corresponding energy levels into the conduction and the valence bands (defect B in Fig. 7.4). Although this process effectively neutralizes the interface traps, it is not possible to enforce such a saturation of all the interfacial trivalent silicon atoms, which means that the density of the interface traps can never be reduced to zero. The interface trap density will be denoted by N_{it} to express the number of interface trap per unit area (in m^{-2}), or by qN_{it} to express the associated charge per unit area (in $C\ m^{-2}$).

Trivalent silicon atoms can also appear in the oxide; these are silicon atoms bonded to three neighboring oxygen atoms with the fourth bond unsaturated (defect 1 in Fig. 7.4). There are also a number of other possible defects in the oxide: nonbridging oxygen, oxygen vacancy, interstitial silicon, silicon vacancy, and interstitial oxygen. All these defects are illustrated in Fig. 7.4 as well. The oxide defects introduce energy levels in the oxide energy gap, which can trap electrons and holes. The charge due to the trapped electrons and holes onto the oxide defects is referred to as the *oxide charge*. Although the oxide traps do not continuously exchange electrons and holes with the silicon, the oxide charge does affect the electrons and holes in the silicon by its electric field. In general the oxide charge is usually positive and is mostly located close to the oxide–silicon interface. The density of oxide charge will be labeled N_{oc} to express the number of charge centers per unit area (in m^{-2}), or qN_{oc} to express it in $C\ m^{-2}$.

Oxide Growth

It can be imagined that the density of the interface traps and oxide charge is largely dependent on the processing conditions. Although it is possible to deposit oxide film onto the silicon surface, such a process does not provide an oxide–silicon interface good enough to be used in MOS capacitors. The density of interface traps in this case may exceed the density of electrons/holes that can ever be attracted to the surface. This means that the interface traps would make the appearance of any significant density of free carriers at the silicon surface impossible.

A high-quality oxide–silicon interface can be achieved if the oxide is thermally grown on the silicon surface. When the silicon is exposed to oxygen or water vapor at high temperature (around 1000°C), silicon dioxide is created through the following reactions:

$$\begin{aligned} &\mathrm{Si} + \mathrm{O_2} \Rightarrow \mathrm{SiO_2} && \text{(dry oxidation)} \\ &\mathrm{Si} + 2\mathrm{H_2O} \Rightarrow \mathrm{SiO_2} + 2\mathrm{H_2} && \text{(wet oxidation)} \end{aligned} \tag{7.7}$$

This process of thermal oxidation, when conducted in an ultrapure atmosphere and after a sophisticated cleaning of the silicon surface, produces a high-quality oxide–silicon interface. Importantly, hydrogen is always present at the oxide–semiconductor interface so it plays an important role in passivating dangling silicon bonds at the interface (refer to defect B in Fig. 7.4). A specific post-metalization annealing is also performed to enhance the effects of hydrogen-based passivation. With this, the density of the interface traps is reduced to the order of $10^{10}\ cm^{-2}$. This has proved sufficient for the integrated circuits that use gate oxides thicker than 5 nm. However, Si–H and S–OH bonds are rather weak and can be dissociated during device operation, especially when the use of ultrathin oxides in modern devices results in a relatively high electric field in the oxide. It has been established that Si≡N are much stronger and that they also provide interface passivation. Accordingly, the gate oxides in modern devices are subject to *nitridation* conditions to improve interface reliability. There are a number of different nitridation processes, the typical being high-temperature annealing or direct oxide growth in N_2O or NO. The percentage of nitrogen that accumulates at the interface is rather low (several percents), but it has a significant impact on the quality of the oxide–silicon interface.

The oxidation reaction takes place at the interface, which means that after a layer of the oxide has been created, the oxygen or the water molecules must diffuse through this already created layer to interact with the silicon. As a consequence, the growth rate is slowed down as the oxide thickness is increased. The growth rate is very dependent on the oxidation temperature, and it is also different for the dry (O_2) and the wet (H_2O) processes. An increase in the oxidation temperature causes a significant increase in the growth rate, which at any temperature is higher for wet processing.

The process of thermal oxidation cannot be used to create gate oxides of sufficient quality with GaAs substrates. This is because the quality of the native oxide of GaAs is not good enough to be used as a gate dielectric. Deposition of silicon dioxide onto GaAs substrate creates a high density of interface traps. These facts practically prevent implementation of a MOS capacitor with GaAs substrates.

EXAMPLE 7.2 Oxide Growth Kinetics

The dependence of the thermal oxide thickness (t_{ox}) on the oxidation time and temperature is frequently modeled by the following equation:

$$\frac{t_{ox}}{A/2} = \sqrt{1 + \frac{t + \tau}{A^2/4B}} - 1$$

where A, B, and τ are temperature-dependent coefficients. The values of the coefficients A, B, and τ are given in Tables 7.3 and 7.4, respectively (Source: L. E. Katz, Oxidation, in *VLSI Technology,* S. M. Sze, ed., McGraw-Hill, New York, 1983, pp. 131–167). How long would it take to grow 0.5 μm of SiO_2 at 920°C in a wet atmosphere? If dry oxidation is applied, what would be the oxide thickness? Comment on the difference. Repeat the calculations for 1000°C and comment on the results.

TABLE 7.3 Rate Constants for Wet Oxidation of Silicon

Oxidation Temperature (°C)	A (μm)	Parabolic Rate Constant B (μm^2/h)	Linear Rate Constant B/A (μm/h)	τ (h)
1200	0.05	0.720	14.40	0
1100	0.11	0.510	4.64	0
1000	0.226	0.287	1.27	0
920	0.50	0.203	0.406	0

TABLE 7.4 Rate Constants for Dry Oxidation of Silicon

Oxidation Temperature (°C)	A (μm)	Parabolic Rate Constant B (μm²/h)	Linear Rate Constant B/A (μm/h)	τ (h)
1200	0.040	0.045	1.12	0.027
1100	0.090	0.027	0.30	0.076
1000	0.165	0.0117	0.071	0.37
920	0.235	0.0049	0.0208	1.40
800	0.370	0.0011	0.0030	9.0
700	—	—	0.00026	81.0

SOLUTION

The oxidation time t can be expressed from the model given in the text of the example as

$$t = \left[\left(\frac{t_{ox}}{A/2} + 1\right)^2 - 1\right]\frac{A^2}{4B} - \tau$$

The values of the parameters are found in Table 7.3 in the row corresponding to $T = 920°\text{C}$. The calculated time is $t = 2.46$ h. If the dry oxidation process is used at the same temperature, the appropriate parameters are found in Table 7.4 in the row for $T = 920°\text{C}$. Putting these parameters and $t = 2.46$ h into the equation given in the text of the example, the oxide thickness is calculated to be $t_{ox} = 0.063\ \mu\text{m} = 63$ nm. The oxide grows much faster in the wet than in the dry ambient.

For the case of $T = 1000°\text{C}$, the time of the wet oxidation needed to grow 0.5 μm of oxide is found to be $t = 1.26$ h. The same time and temperature, used in the process of dry oxidation, would grow $t_{ox} = 78$ nm. When the temperature is increased from 920°C to 1000°C, the time required to grow 0.5 μm of "wet" oxide is approximately halved. The "dry" oxide grown for the same time is again much thinner, but somewhat thicker than that grown for twice as long at 920°C.

7.2.2 C–V Curve and the Surface-Potential Dependence on Gate Voltage

There are three characteristic modes of a MOS capacitor: accumulation, depletion, and strong inversion. For a MOS capacitor on a P-type semiconductor, these modes can briefly be described as follows: (1) in accumulation mode, a negative effective voltage between the gate and the substrate attracts holes to the semiconductor surface creating an accumulation layer at the semiconductor surface, (2) in depletion mode, a positive effective voltage repels the holes from the surface region, creating a depletion layer, and (3) in strong inversion, a strong positive voltage attracts electrons (the minority carriers) to the surface, creating an inversion layer. The capacitance–voltage dependence, illustrated in Fig. 7.5, reflects

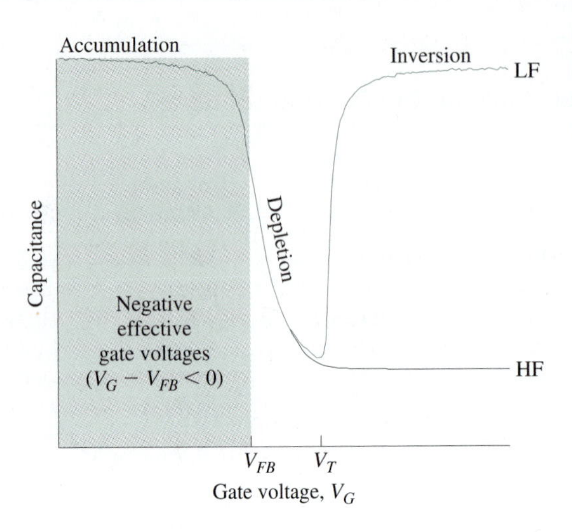

Figure 7.5 Typical C–V curves of a MOS capacitor on a P-type substrate: LF, low frequency (quasistatic); HF, high frequency.

these modes of operation. Accordingly, the capacitance–voltage dependences (C–V curves) are widely used for characterization of MOS capacitors and for monitoring and analyzing numerous phenomena related to the MOS structure. The C–V curve will be systematically described in this section. A very important related quantity in terms of understanding the modes of operation and the C–V curve is the *surface potential*. The surface potential is the electric potential between the semiconductor surface and the semiconductor bulk, which is assumed to be at the reference potential (ground). This means that the surface potential (φ_s) is just a fraction of the voltage applied between the gate and the substrate (V_G). The rest of the gate voltage, $V_G - \varphi_s$, appears across the gate oxide. The definition of the three characteristic modes (accumulation, depletion, and strong inversion) and the two boundary voltages (flat-band voltage and threshold voltage) will be directly linked to specific approximations of the surface-potential values and its dependence on the gate voltage.

Accumulation

There is a net charge of equal density and opposite signs at each plate of a charged capacitor. In accumulation, the net charge at the semiconductor surface is due to excess holes (assuming P-type semiconductor). An incremental change in the applied gate voltage (ΔV_G) causes a corresponding change in the density of the accumulation charge (ΔQ_A). This situation is equivalent to an ordinary metal–dielectric–metal capacitor. Therefore, the capacitance per unit area in the accumulation mode is determined by the gate-oxide thickness and is voltage-independent:

$$C_{ox} = \frac{\varepsilon_{ox}}{t_{ox}} \tag{7.8}$$

where C_{ox} is in F/m^2. This capacitance relates ΔQ_A to ΔV_G: $\Delta Q_A = C_{ox}\Delta V_G$.

Assuming an abundance of holes (the majority carriers), there is no significant penetration of the electric field into the semiconductor. The electric field lines originate from the excess holes at the semiconductor surface and terminate at the excess electrons at the metal–dielectric interface. The absence of electric field below the surface means that

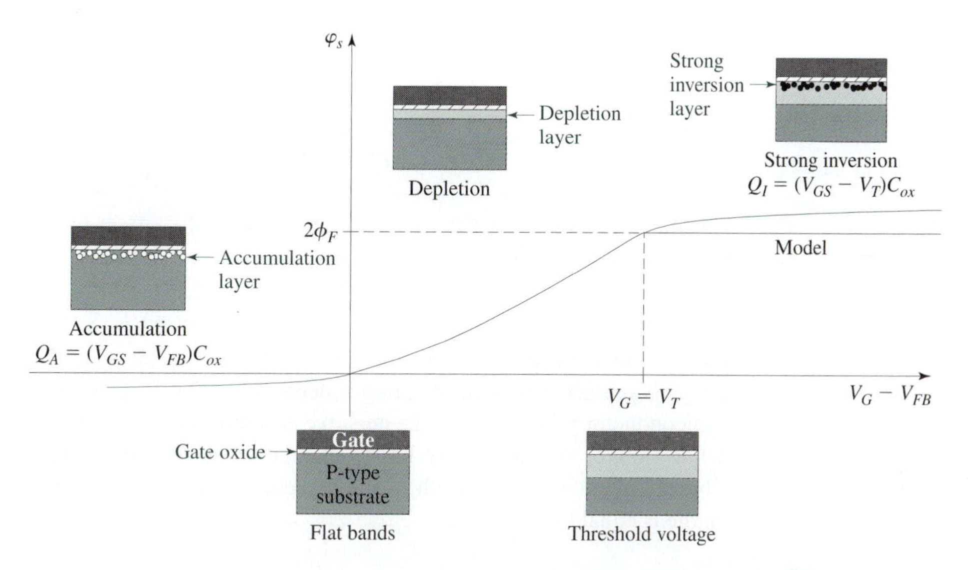

Figure 7.6 Approximations of the surface-potential dependence on gate voltage define the three MOS capacitor modes: (1) accumulation ($\varphi_s = 0$), (2) depletion (φ_s increases with the gate voltage and the depletion-layer widening), and (3) strong inversion ($\varphi_s = 2\phi_F$).

there is no potential difference between the surface and the bulk of the semiconductor—the surface potential is equal to zero. This approximation, based on the assumption of abundance of holes, is illustrated in Fig. 7.6.

The real φ_s–V_G curve (the colored line in Fig. 7.6) shows that the actual surface potential drops below zero in accumulation. This corresponds to some penetration of the electric field below the semiconductor surface that is needed to create the high density of holes in the accumulation layer. This effect is considered in more detail in Section 1.2.4.

Flat Bands

The point where the actual surface potential is equal to zero is of special interest. This is the point where the energy bands in the semiconductor are flat (no band bending) because there is no change in the electric potential in the surface region: the electric potential is equal to zero everywhere in the semiconductor. In the ideal case, this condition would occur for zero applied voltage at the gate. However, the difference in the work functions of the metal and the semiconductor creates a built-in voltage and a built-in electric field, analogously to the P–N junction and the metal–semiconductor contact. In addition, there is a "built-in" electric field from the oxide charge. As a result, the electric field in the oxide is typically not zero and the capacitor is not discharged at $V_G = 0$. The gate voltage that neutralizes the built-in electric field and sets the surface of the semiconductor at zero (the flat-band condition) is called flat-band voltage (V_{FB}). The equation that links the flat-band voltage to the technological parameters, the work-function difference and the oxide-charge density, is introduced in Section 7.2.3 when the energy bands of the MOS structure are drawn. At this stage, it is quite sufficient to know that V_{FB} appears as a gate-voltage offset. To draw a voltage dependence with $\varphi_s = 0$ as the reference point, either V_{FB} can be assumed to be equal to zero (the *ideal* MOS capacitor approach) or the gate voltage can be expressed as $V_G - V_{FB}$ (the *effective*-gate-voltage approach).

When one considers the total charge density at the capacitor plates, the effective gate voltage should be used in the capacitance–voltage equation ($Q = CV$). In accumulation, the net charge at the semiconductor plate is due to accumulated holes; therefore,

$$Q_A = (V_G - V_{FB})C_{ox} \tag{7.9}$$

This equation shows that the gate voltage is basically offset by V_{FB}. It also shows that $Q_A = 0$ at $V_G = V_{FB}$.

Depletion

When a small positive effective voltage is applied to the gate, holes are repelled from the semiconductor surface, creating a depletion layer. In this case, the net charge at the semiconductor plate is due to the negative acceptor ions in the depletion layer. This charge is not mobile, yet its density has to change when the gate voltage is changed. This is achieved by altering the width of the depletion layer (w_d), given that the depletion layer charge is equal to

$$Q_d = qN_A w_d \tag{7.10}$$

For example, an increase in $V_G - V_{FB}$ expands the depletion layer to increase Q_d. This situation is fully analogous to the variable depletion-layer capacitance in the case of a P–N junction. In fact, the equations for w_d derived in Section 6.3.2 for asymmetrical abrupt P–N junctions can be used in this case, provided the voltage across the depletion layer is properly specified.[5] In the case of a reverse-biased P–N junction, this voltage is $V_{bi} + V_R$. In the case of the surface-depletion layer in a MOS capacitor, this voltage is equal to the surface potential (φ_s). Therefore Eq. (6.44) for an N^+–P junction can be used for the depletion layer in a MOS capacitor on a P-type semiconductor if $V_{bi} + V_R$ is replaced by φ_s:

$$w_d = \sqrt{\frac{2\varepsilon_s \varphi_s}{qN_A}} \tag{7.11}$$

Analogously to the depletion-layer capacitance in P–N junctions, the MOS capacitance in the depletion region is reduced when the applied voltage is increased, as can be seen in Fig. 7.5.

In the depletion region, the MOS capacitance can be represented by a series connection of two capacitors: the gate-oxide capacitance ($C_{ox} = \varepsilon_{ox}/t_{ox}$) and the depletion-layer capacitance ($C_d = \varepsilon_s/w_d$). Therefore, the total MOS capacitance is

$$C = \frac{C_{ox}C_d}{C_{ox} + C_d} = \frac{C_{ox}}{1 + C_{ox}/C_d} = \frac{\varepsilon_{ox}}{t_{ox} + (\varepsilon_{ox}/\varepsilon_s)w_d} \tag{7.12}$$

Equation (7.12) shows that C is reduced as w_d is increased; however, Eq. (7.11) gives w_d in terms of the surface potential and not in terms of the applied gate voltage itself. To be able to use Eqs. (7.11) and (7.12) to calculate the C–V dependence, an equation linking

---

[5] The Poisson equation can be solved as in Section 6.3.2, the only difference being the boundary condition for the voltage across the depletion layer.

the gate voltage and the surface potential is needed. The difference between the effective gate voltage ($V_G - V_{FB}$) and the surface potential (φ_s) is the voltage drop across the oxide. When divided by the oxide thickness, this difference is related to the gate-oxide field:

$$E_{ox} = \frac{V_G - V_{FB} - \varphi_s}{t_{ox}} \tag{7.13}$$

The gate-oxide field is related to the electric field at the semiconductor surface (E_s). From Gauss's law, we have

$$\varepsilon_{ox} E_{ox} = \varepsilon_s E_s \tag{7.14}$$

The surface electric field can be found by solving the Poisson equation. This procedure is analogous to the solution given in Section 6.3.2, so we can simply adjust Eq. (6.31) for the maximum electric field at the P–N junction:

$$E_s = \frac{qN_A}{\varepsilon_s} w_d \tag{7.15}$$

From Eqs. (7.11), (7.13), (7.14), and (7.15), the relationship between the gate voltage and the surface potential can be expressed in the following form

$$V_G - V_{FB} = \varphi_s + \gamma\sqrt{\varphi_s} \tag{7.16}$$

where

$$\gamma = \sqrt{2\varepsilon_s q N_A}/C_{ox} \tag{7.17}$$

is called the body factor and incorporates all the technological parameters. The unit for the body factor is $V^{1/2}$. Figure 7.6 illustrates the dependence of φ_s on V_G in the depletion region.

Strong Inversion and Threshold Voltage

The validity of Eq. (7.16) is limited to the depletion region. For simplicity, the concentrations of both electrons and holes are neglected in Eq. (7.16). However, the surface potential increase can lead to a pileup of a significant concentration of electrons (minority carriers) at the semiconductor surface. A more general equation, which includes the minority carriers, is derived in Example 7.3. A deeper insight into the conditions for appearance of electrons at the semiconductor surface can be gained from the energy-band diagrams that will be considered in Section 7.2.3. At this stage, it is sufficient to realize that the energy-band bending brings the bottom of the conduction band closer to the Fermi level, and when $E_C - E_F$ becomes smaller than $E_F - E_V$, the semiconductor surface is inverted. Thermally generated electrons are collected at the surface, creating an inversion layer. The surface-potential increase slows down as the MOS capacitor enters the inversion region. The gate-voltage increase does increase the gate-oxide field, but this field is screened by increasing density of electrons in the inversion layer. The screening becomes more effective as the concentration of electrons in the inversion layer is increased, and when it becomes so strong that any further increase in the surface potential can be neglected, it is said that the MOS capacitor is in *strong inversion*. The threshold value of the surface potential, at the onset of strong inversion, is defined as

$$\varphi_s = 2\phi_F \tag{7.18}$$

where ϕ_F is the Fermi potential. This is an empirically based and convenient definition. Figure 7.6 illustrates that the surface potential in strong inversion is assumed to be constant and equal to $2\phi_F$. The gate voltage corresponding to the onset of strong inversion, when φ_s reaches $2\phi_F$, is called the *threshold voltage*. The threshold voltage will be labeled by V_T and should be distinguished from V_t used for the thermal voltage.[6] Applying Eq. (7.16) to the defined threshold condition, $V_G = V_T$ for $\varphi_s = 2\phi_F$, the following threshold-voltage equation is obtained:

$$V_T = V_{FB} + 2\phi_F + \gamma\sqrt{2\phi_F} \tag{7.19}$$

The strong-inversion mode of MOS capacitor operation is by far the most important, given that it is the inversion layer that is utilized for conduction in the most frequent electronics device—the MOSFET. Accordingly, the threshold voltage is a very important MOS parameter because it defines the voltage boundary of the strong-inversion region.

Two C–V lines are shown in Fig. 7.5 in the inversion region, one labeled by HF and the other by LF. The low-frequency (or more precisely, the quasistatic C–V line) shows that the capacitance increases as the concentration of electrons in the inversion layer increases to approach the C_{ox} level in the strong inversion. In the simple model, any increase in the gate-oxide field due to an increase in the gate voltage beyond the threshold voltage is perfectly screened by a corresponding increase of electrons in the inversion layer. Accordingly, the inversion-layer charge is modeled by

$$Q_I = (V_G - V_T)C_{ox} \qquad (V_G \geq V_T) \tag{7.20}$$

According to this equation, $Q_I = 0$ at the onset of strong inversion ($V_G = V_T$), whereas in the strong-inversion region, any increase in the gate voltage leads to corresponding increase in the inversion-layer charge: $\Delta Q_I = C_{ox}\Delta V_G$. This is consistent with the assumption that the strong-inversion capacitance is constant and equal to C_{ox}.

The described behavior of the MOS capacitor in strong inversion can be observed under the condition that there is a supply of electrons (minority carriers) to respond to the gate-oxide field changes. This supply of electrons is provided by specific electrodes in the case of a complete MOSFET. Therefore, the MOS capacitance in strong inversion is according to the LF (or quasistatic) model when measured by a MOSFET with a specific supply (source) of electrons. In the case of a simple MOS capacitor, the supply of electrons is limited to thermal generation. Thermal generation is a slow process and is not able to respond to fast oscillation of a measurement signal. It may take as long as seconds for thermal generation to provide electrons in the inversion layer of silicon, and it may take many years for this process to be completed in wide-energy-gap semiconductors. Therefore, if the capacitance is measured with a signal whose oscillations are faster than

[6] The two variables share a similar subscript in their labels, but they are otherwise completely separate and independent variables.

these times, oscillations in the inversion-layer charge are not possible. In this case, the oscillating field of the measuring signal has to penetrate through both the gate oxide and the depletion layer to oscillate the negative charge due to the acceptors in the depletion layer. This means the model of C_{ox} and C_d connected in series has to be applied in this case. This model is used for the depletion region, the difference this time being that the surface potential is approximately pinned at the strong-inversion value of $2\phi_F$. As a result, the strong-inversion capacitance remains constant at its minimum level that corresponds to the maximum value of $\varphi_s = 2\phi_F$ in Eqs. (7.11) and (7.12). This behavior of the capacitance in strong inversion is shown by the line labeled by HF in Fig. 7.5. Of course, this behavior will be observed under the condition that the inversion layer is fully formed at the considered gate bias V_G when the high-frequency measurement is performed. If the capacitor is created on a semiconductor with a much wider energy gap, such as silicon carbide, the formation of the inversion layer by thermal generation would take much longer than any practical time for room-temperature measurements. As a result, no inversion layer is formed, and the C–V curve behavior from the depletion region continues into what is referred to as *deep depletion*. Analogous behavior can be observed with silicon at very low temperatures when the thermal generation in silicon is practically inhibited.

For the case of capacitors made with N-type rather than P-type silicon substrate, the depletion-layer charge is positive because it originates from uncompensated positive donor ions. Accordingly, the concentration of donors (N_D) should replace the concentration of acceptors in the equations for both the body factor γ and the Fermi potential ϕ_F. The Fermi potential [as given by Eq. (2.8)] appears as negative; in the equation for the threshold voltage, it cannot be used as negative under the square root. Careful consideration of the boundary conditions for the electric potential in the process of solving the Poisson equation would indicate that the only thing that matters for w_d is the absolute value of $2\phi_F$. These considerations show that the threshold voltage in the case of N-type silicon substrate has to be modified in the following way:

$$V_T = V_{FB} - 2|\phi_F| - \gamma\sqrt{2|\phi_F|} \tag{7.21}$$

where the body factor γ is given in terms of the donor concentration:

$$\gamma = \sqrt{2\varepsilon_s q N_D}/C_{ox} \tag{7.22}$$

EXAMPLE 7.3 Surface Potential Versus Gate Voltage in Depletion and Strong Inversion

In the model that uses the concept of the threshold voltage, the surface potential in strong inversion is assumed to be constant, specifically, $\varphi_s = 2\phi_F$. In reality, there is a slight dependence of the surface potential on the gate voltage (the colored line in Fig. 7.6). Solve the Poisson equation, including both the depletion-layer charge and the minority electrons, to obtain a more precise model for the surface-potential dependence on the gate voltage.

SOLUTION

When both the acceptors in the depletion layer and the electrons in the inversion layer are included, the charge density used in the Poisson equation becomes

$$\rho(x) = -qN_A + n(x) = -q\left[N_A + n_0 e^{q\varphi(x)/kT}\right] = -q\left[N_A + n_0 e^{\varphi(x)/V_t}\right] \tag{7.23}$$

The exponential dependence of $n(x)$ on $\varphi(x)$ is again due the the Maxwell–Boltzmann distribution and the fact that $\varphi(x)$ corresponds to the bending of the conduction band. The constant n_0, which is the equilibrium concentration of electrons, is determined so that $n(x) = n_0$ for $\varphi(x) = 0$. Given that $n_0 p_0 = n_i^2$ and that $p_0 \approx N_A$, the charge density can be expressed as

$$\rho(x) = -qN_A\left[1 + \left(\frac{n_i}{N_A}\right)^2 e^{\varphi(x)/V_t}\right] = -qN_A\left[1 + \left(\frac{n_i}{p_0}\right)^2 e^{\varphi(x)/V_t}\right] \tag{7.24}$$

Using the relationship between p_0 and n_i [Eq. (2.86)], the term $(n_i/p_0)^2$ can be expressed through the Fermi potential: $(n_i/p_0)^2 = \exp[-q(2\phi_F)/kT] = \exp(-2\phi_F/V_t)$. With this, the charge density can be expressed in the following convenient form:

$$\rho(x) = -qN_A\left\{1 + e^{\varphi(x)-2\phi_F/V_t}\right\} \tag{7.25}$$

The Poisson equation is then

$$\frac{d^2\varphi}{dx^2} = \frac{qN_A}{\varepsilon_s}\left\{1 + e^{\varphi(x)-2\phi_F/V_t}\right\} \tag{7.26}$$

This equation has to be integrated once to obtain the electric field $E = -d\varphi/dx$ and, specifically, to obtain the equation for the electric field at the surface of the semiconductor that is valid in both the depletion region and the inversion region. This equation is to replace Eq. (7.15), which is limited to the depletion region. All other steps in the previously shown derivation of the relationship between the surface potential and the gate voltage remain the same.

To enable integration of Eq. (7.26), the following identity is utilized:

$$\frac{d^2\varphi}{dx^2} = \frac{1}{2}\frac{d}{d\varphi}\left(\frac{d\varphi}{dx}\right)^2 \tag{7.27}$$

With this, Eq. (7.26) becomes

$$\frac{1}{2}d\underbrace{\left[\left(\frac{d\varphi}{dx}\right)^2\right]}_{E^2} = \frac{qN_A}{\varepsilon_s}\left\{1 + e^{\varphi(x)-2\phi_F/V_t}\right\}d\varphi \tag{7.28}$$

and the integration is performed as follows:

$$\frac{1}{2}\int_{0}^{E_s^2} d(E^2) = \frac{qN_A}{\varepsilon_s}\int_{0}^{\varphi_s}\{1 + e^{\varphi(x)-2\phi_F/V_t}\}d\varphi \tag{7.29}$$

$$\frac{1}{2}E_s^2 = \frac{qN_A}{\varepsilon_s}\left[\varphi_s + V_t e^{(\varphi_s-2\phi_F)/V_t}\right] \tag{7.30}$$

$$E_s = \sqrt{\frac{2qN_A}{\varepsilon_s}}\sqrt{\varphi_s + V_t e^{(\phi_s-2\phi_F)/V_t}} \tag{7.31}$$

We can see that for small values of φ_s, when the exponential term can be neglected, this equation becomes identical to the surface electric field defined by Eqs. (7.15) and (7.11). However, this is the more general equation, and it will expand the $\gamma\sqrt{\varphi_s}$ term in Eq. (7.16) as follows:

$$V_G - V_{FB} = \varphi_s + \gamma\sqrt{\varphi_s + V_t e^{(\varphi_s-2\phi_F)/V_t}} \tag{7.32}$$

It is the exponential term that causes the apparent saturation of φ_s with V_G in the strong-inversion region. Because of this exponential dependence, very small increases in φ_s correspond to large increases in V_G. This exponential term is due to the exponential increase of electron concentration with φ_s; hence, it is said that the increase in electron concentration screens the penetration of the electric field.

7.2.3 Energy-Band Diagrams

The previous section defined the principal modes of MOS–capacitor operation using the idealized model for surface-potential dependence on the applied gate voltage (Fig. 7.6). As a summary, these modes are (1) *accumulation* for $V_G < V_{FB}$ ($\varphi_s \approx 0$), (2) *depletion* for $V_{FB} < V_G < V_T$, and (3) *strong inversion* for $V_G > V_T$ ($\varphi_s \approx 2\phi_F$). In this section, energy-band diagrams are used to provide a deeper insight into the physics of MOS capacitors.

Zero Bias Versus Flat Bands: Definition of the Fundamental Terms

Figure 7.7a illustrates a MOS capacitor with P-type substrate and N^+-type polysilicon gate when all the charges are compensated—that is, no net charge appears at the capacitor plates (the flat-band condition). The holes in the P-type substrate are compensated by the negative acceptor ions and the minority electrons, whereas the electrons in the N^+-type gate are compensated by the positive donor ions and the minority holes. In general, the flat-band condition does not appear for zero applied voltage. The fundamental source of the nonzero flat-band voltage can be illustrated by the corresponding energy-band diagrams.

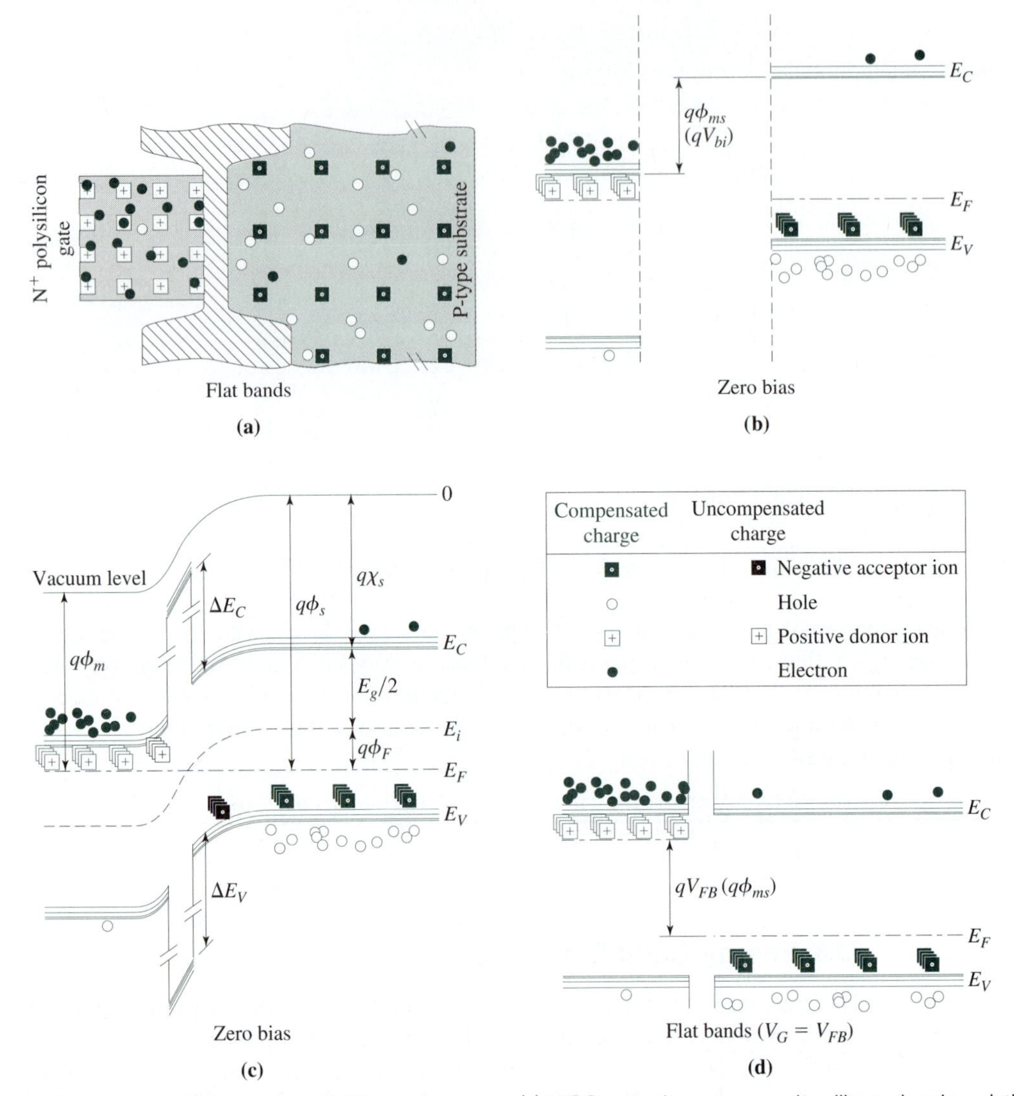

Figure 7.7 Illustration of the fundamental MOS-related terms. (a) MOS capacitor cross section, illustrating the existing types of charge. (b) The starting point in construction of the MOS energy-band diagram. (c) The energy-band diagram at zero-bias condition. (d) The energy-band diagram at flat-band condition.

To construct the energy-band diagram of a MOS capacitor, the procedure used for the case of P–N junctions (Section 6.1) has to be developed further to account for the existence of the oxide appearing between the P- and N^+-type silicon regions:

1. As before, the Fermi level lines are drawn first (dashed–dotted lines in Fig. 7.7):
   1.1 If the system is in thermal equilibrium (zero bias applied), the Fermi level is constant throughout the system. The Fermi level lines in the substrate and the gate have to be matched, as in Fig. 7.7b.

1.2 If a voltage is applied between the gate and the substrate, the Fermi level (or more precisely, the quasi-Fermi level) lines should be split to express this fact, as in Fig. 7.7d.

2. Analogously to the case of P–N junctions, the conduction and valence bands are drawn for the P-type and N-type neutral regions (away from the oxide–silicon interfaces). The bands are placed appropriately with respect to the Fermi level, so as to express the band diagrams of the P-type and N-type silicon, respectively. This is illustrated in Fig. 7.7b.

3. In the case of a P–N junction, the conduction- and the valence-band levels would simply be joined by sloped lines to complete the diagram. The oxide that appears between the P- and N^+-type regions in the case of the MOS capacitor, has a much larger energy gap ($E_g = E_C - E_V$) than the silicon regions. Figure 7.4b illustrates the existence of conduction-band and valence-band discontinuities at the oxide silicon interface. These discontinuities have to be expressed in the band diagram. The band diagram around the oxide is constructed in the following way (refer to Fig. 7.7c):

3.1 The bands are bent in the P- and N^+-type silicon regions toward each other, but they do not come to the same level. There is a difference between the bands (say the intrinsic Fermi level E_i) at the N^+–gate-oxide interface and at the oxide–P-type-substrate interface. This difference is due to the voltage across the oxide.

3.2 Lines expressing the discontinuities of the conduction and the valence bands at the oxide–silicon interfaces are drawn. The conduction-band discontinuity is $\Delta E_C = 3.2$ eV and the valence-band discontinuity is about $\Delta E_V = 4.7$ eV (this makes an oxide energy gap of about $3.2 + 4.7 + 1.12 \approx 9$ eV).

3.3 The conduction and valence bands in the oxide are drawn with straight lines, to express that the electric field in the oxide is constant.[7] It is assumed that there is no built-in charge in the bulk of the oxide, which would enforce change in the electric field in the oxide (any oxide charge is modeled as a sheet charge appearing along the oxide–silicon interface).

Figure 7.7b and 7.7c illustrates that there is a potential difference between the N^+-type gate and the P-type substrate at zero bias. This is analogous to the built-in voltage in P–N junctions. In the case of the MOS capacitor, the built-in potential difference is referred to as a *work-function difference* ($q\phi_{ms}$). The electric field associated with the nonzero work-function difference is due to uncompensated positive donor ions on the N^+-type gate side (Fig. 7.7c).

Considering the energy-band diagram, as shown in Fig. 7.7c, the work function of a semiconductor ($q\phi_s$) can be related to the electron affinity $q\chi_s$:

$$q\phi_s = q\chi_s + \frac{E_g}{2} + q\phi_F \tag{7.33}$$

[7]Constant electric field E corresponds to a linear electric potential, thus potential energy ($E = -d\varphi/dx \propto dE_{pot}/dx$).

where $E_g/2$ is the half-value of the energy gap and $q\phi_F$ is the Fermi potential (the difference between E_i and E_F).[8] The Fermi potential depends on the doping type and level, as expressed by Eqs. (2.87) and (2.88).

Using Eq. (7.33), the work-function difference can be expressed as

$$q\phi_{ms} = q\phi_m - q\phi_s = q\phi_m - q\left(\chi_s + \frac{E_g}{2q} + \phi_F\right) \tag{7.34}$$

To calculate the work-function difference, the Fermi potential is calculated first for the given doping level [using Eq. (2.87) or Eq. (2.88)] and, according to Eq. (7.34), combined with the values of $q\chi_s$, E_g, and $q\phi_m$ corresponding to the materials used (Table 7.1). In the case of metal gates, the value of $q\phi_m$ is a material constant. When a silicon gate is used, it may be necessary to calculate $q\phi_m$ from Eq. (7.33) using an appropriate doping level. Very frequently, however, the gates are very heavily doped (to provide as close an emulation of the metal properties in terms of conductivity as possible), which means the Fermi level is either very close to the bottom of the conduction band (N^+-type gate) or very close to the top of the valence band (P^+-type gate). Therefore, as given in Table 7.1, the work function of a heavily doped polysilicon gate can be approximated by $q\chi_s$ in the case of N^+-type doping and by $q\chi_s + E_g$ in the case of P^+-type doping.

Because of the work-function difference, creating the built-in field, the capacitor plates are not discharged at the zero bias. To achieve the zero-charge condition—that is, to flatten the bands—it is necessary to split the Fermi levels in the gate and the substrate by a value that will compensate for the work-function difference (Fig. 7.7d). This value is equal to the work-function difference. Because the Fermi levels are split by applying a voltage between the gate and the substrate, we conclude that the flat-band voltage is equal to the work-function difference ($V_{FB} = q\phi_{ms}$).

It is important to note, however, that the effects of a nonzero oxide charge are not shown in Fig. 7.7d. As described Section 7.2.1, the oxide charge is typically located close to the oxide–silicon interfaces. The charge that appears close to the oxide–gate interface does not influence the MOS capacitor properties significantly, because it is easily compensated by the charge from the heavily doped silicon gate. The charge sheet appearing close to the oxide–substrate interface, however, influences the mobile carriers in the substrate by its electric field. The electric field of this oxide charge is able to produce significant band bending in the surface area of the silicon substrate. To bring the bands in flat condition, this field should be compensated by an appropriate gate voltage, as well. In other words, it is necessary to apply a gate voltage to remove any charge attracted to the substrate surface by the oxide charge. If the gate-oxide capacitance is C_{ox} and the density of the charge is qN_{oc}, the needed voltage is qN_{oc}/C_{ox}. A negative voltage is needed in the case of positive oxide charge (to repel the electrons attracted to the surface by the positive oxide charge); analogously, a positive voltage is needed in the case of negative oxide charge.

---

[8]The concepts of *work function* and *electron affinity* are introduced in Section 7.1.1.

Combining the effects of the work-function difference (ϕ_{ms}) and the oxide charge ($-qN_{oc}/C_{ox}$), the flat-band voltage is expressed as

$$V_{FB} = \phi_{ms} - qN_{oc}/C_{ox} \tag{7.35}$$

Accumulation

A negative effective gate bias, $V_G - V_{FB} < 0$, produces an electric field that attracts holes to the surface of the silicon substrate. This is the accumulation mode. The density of holes in the surface layer of the silicon substrate is exactly matched by the density of electrons at the gate, induced by the negative gate bias applied. This is illustrated in Fig. 7.8a. The appearance of extra holes in the surface region of the silicon substrate means that the Fermi level in the surface region is closer to the top of the valence band than in the bulk. The energy bands are, therefore, bent upward going from the silicon substrate toward the gate,

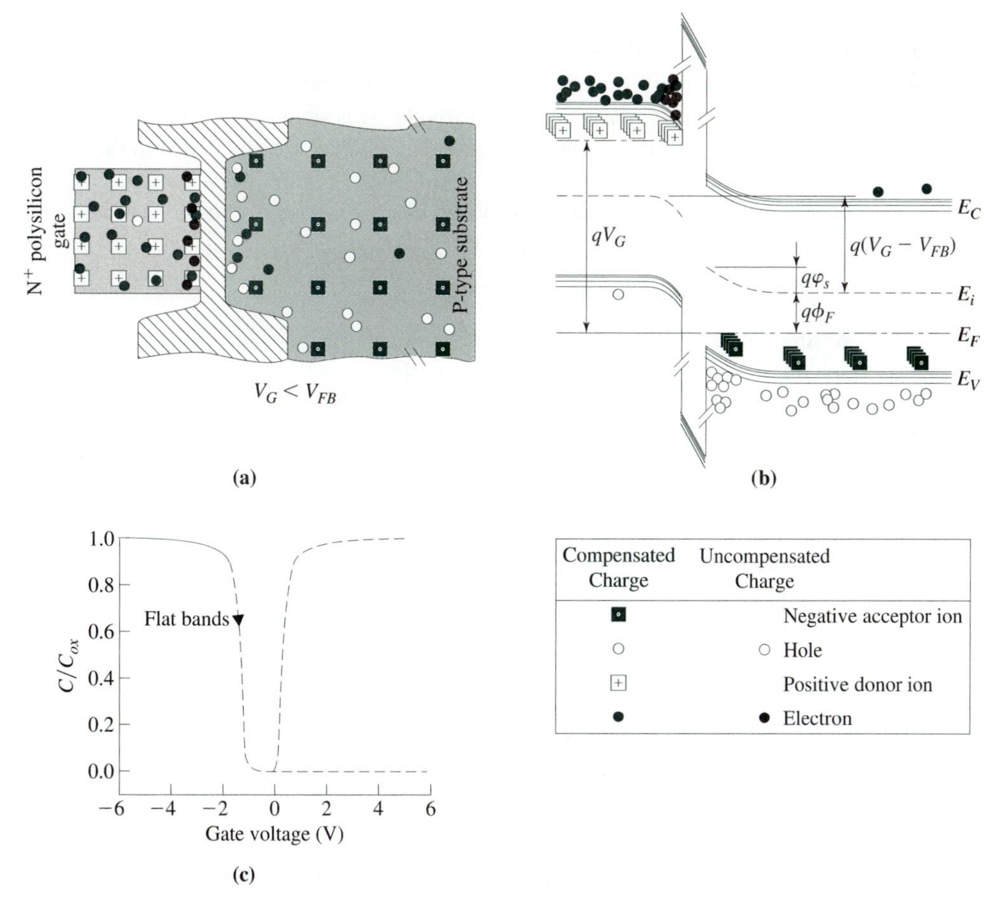

Figure 7.8 MOS capacitor in accumulation. (a) Cross section illustrating the type of charge at the capacitor plates. (b) The energy-band diagram. (c) C–V dependence.

as shown in Fig. 7.8b. This band bending is due to the difference between the energy bands in the bulk of the silicon and the gate, which is directly related to the effective bias applied ($qV_G - qV_{FB}$). The level of the band bending in the silicon substrate is directly related to the value of the surface potential (φ_s); the band bending is equal to $q\varphi_s$.

The gradient in the energy bands around the oxide–silicon interfaces and in the oxide expresses the existence of an electric field in the direction from the substrate toward the gate ($E = -d\varphi/dx \propto dE_{pot}/dx$). This is the field that keeps the excess holes in the substrate and the excess electrons in the gate close to the oxide–silicon interfaces.

Any change in the gate voltage ΔV_G will inevitably produce a change in the band bending. For example, if the voltage is decreased by ΔV_G, the bending is further increased, which means that the Fermi level at the surface of the silicon substrate is moved a bit closer to the top of the valence band. Thus, the density of the holes at the substrate surface is increased. Because the Fermi level is very close to the top of the valence band, only a slight shift in the top of the valence band toward the Fermi level (a slight reduction in the surface potential) is needed to significantly increase the density of the populated energy levels in the valence band by new holes. This is due to the fact that the probability of occupancy by holes of the levels in the valence band increases exponentially as the valence band moves toward the Fermi level (Fig. 2.18 in Section 2.4.3). Because the change in the surface potential is much smaller than the gate voltage change ΔV_G, almost the entire gate voltage change appears across the oxide. This situation is very much like the situation of a capacitor with metal plates separated by a dielectric equivalent to the gate oxide. The capacitance per unit area is, therefore, equal to the gate-oxide capacitance, C_{ox}. Figure 7.8c illustrates that the capacitance of the MOS capacitor in accumulation, being equal to the gate-oxide capacitance, is independent on the gate voltage.

Depletion and Weak Inversion

When relatively small positive effective gate bias ($V_G - V_{FB} > 0$) is applied, the electric field produced repels the holes from the surface, creating a depletion layer at the surface of the silicon substrate. The charge that appears at the substrate plate of the capacitor in this case is due to uncompensated negative acceptor ions, as illustrated in Fig. 7.9a and 7.9b. Although these ions are immobile, gate voltage changes must produce related changes in the density of the negative-ion charge. For this to happen, the electric-field lines from the gate have to penetrate through the depletion layer to either further repel the holes (an increase in the gate voltage) or attract the holes back toward the surface (a decrease in the gate voltage). The appearance of the depletion layer reduces the capacitance in the depletion mode.

Capacitance reduction is illustrated in Fig. 7.9c, which also illustrates that the capacitance behavior changes as the voltage is further increased. There is a characteristic point (V_{MG}), which can be explained using the energy-band diagram of Fig. 7.9b. The gate voltage increase above the flat-band level causes band bending in the direction that increases the difference between the top of the valence band and the Fermi level. This is the condition that is associated with the reduction (and eventual elimination) of the holes from the surface. As a consequence of this band bending, E_i approaches E_F. The characteristic point mentioned is the midgap line exactly at the Fermi level at the silicon surface (the situation illustrated in Fig. 7.9b). At this point, the surface of the silicon substrate is in the intrinsic silicon condition. For a larger band bending than this (due to a larger gate voltage

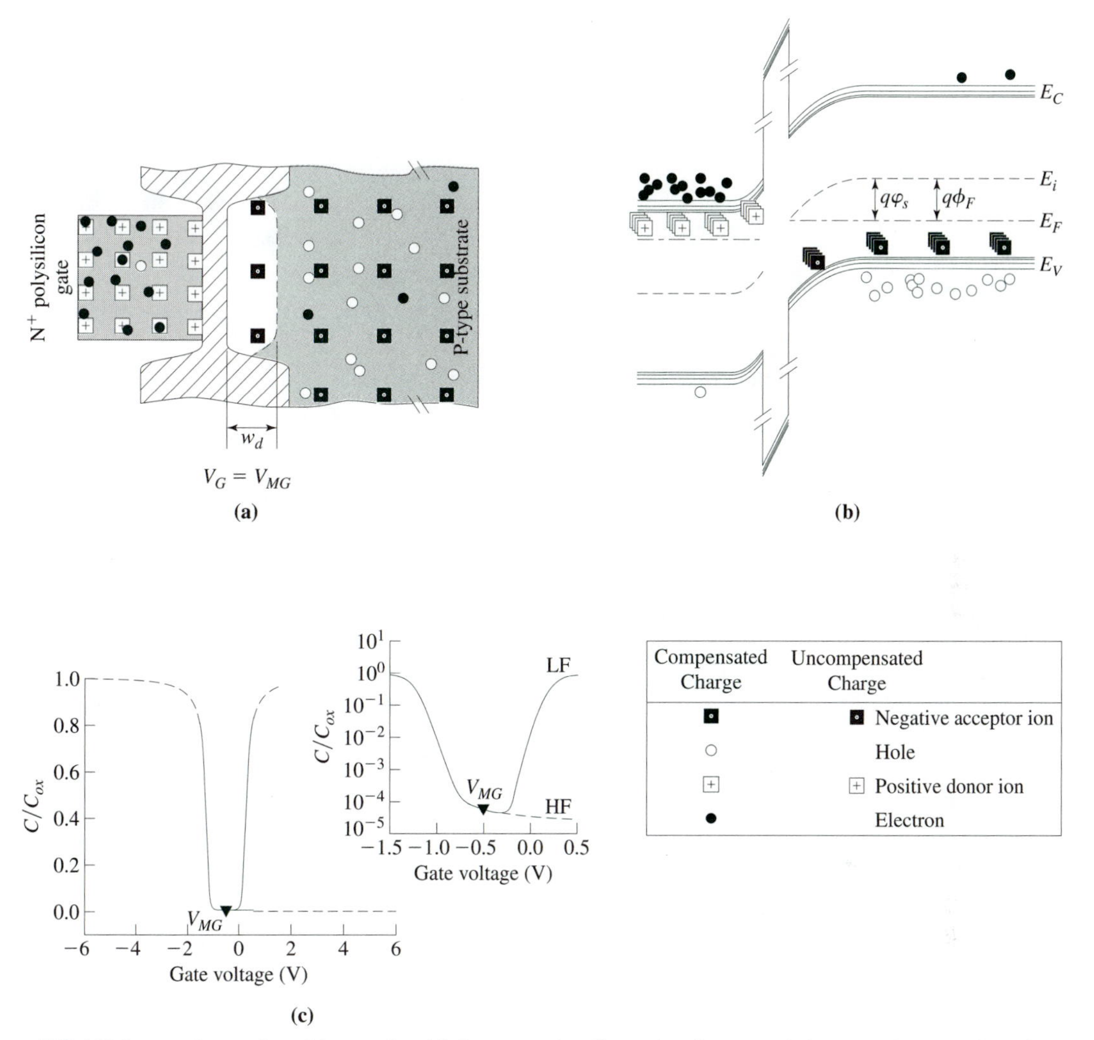

Figure 7.9 MOS capacitor at the midgap point. (a) Cross section illustrating the type of charge at the capacitor plates. (b) The energy-band diagram. (c) C–V dependence.

than V_{MG}), E_i crosses E_F at some distance from the surface. This means that the Fermi level is closer to the bottom of the conduction band than to the top of the valence band at the silicon surface. This further means that the concentration of electrons (minority carriers in the P-type substrate) is larger than the concentration of holes at the silicon surface—an *inversion layer* is created at the surface.

The appearance of some mobile charge at the silicon surface means that some of the electric field lines do not need to penetrate through the depletion layer to change the charge as a response to a gate voltage variation. The low-frequency (or quasistatic) capacitance increases as the density of the mobile charge (electrons in this case) is increased by the gate voltage. This mode is referred to as *weak inversion*.

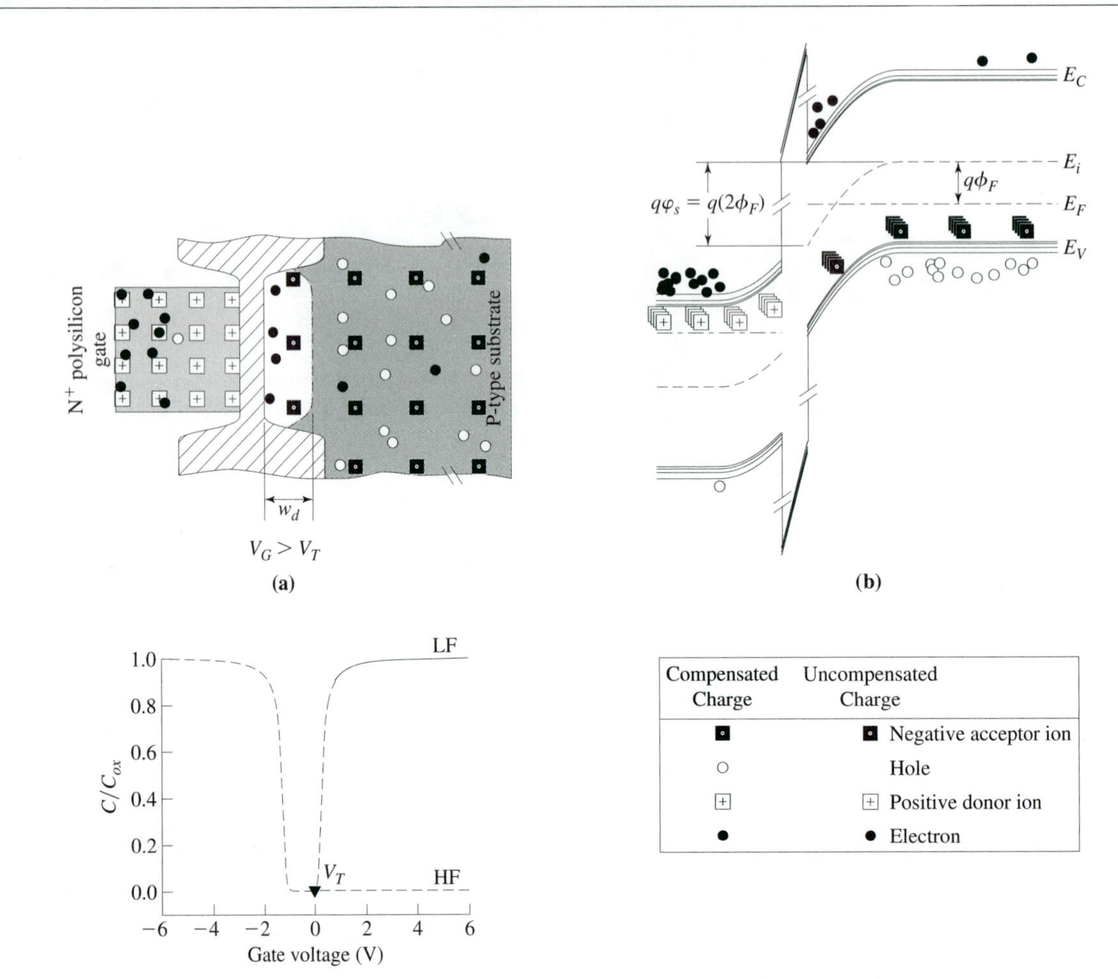

Figure 7.10 MOS capacitor in strong inversion. (a) Cross section illustrating the types of charge at the capacitor plates. (b) The energy-band diagram. (c) LF and HF C–V dependence.

Strong Inversion

The behavior of the MOS capacitor in the *strong-inversion mode* is of great importance for MOSFET operation. To understand the difference between weak and strong inversion, it is necessary to refer to the fact that the tail of the Fermi–Dirac distribution shows an exponential increase in the probability of electrons appearing at the conduction-band levels as the difference between the bottom of the conduction band E_C and the Fermi level E_F is reduced.[9] It is also useful to keep in mind that any increase in the gate voltage (ΔV_G)

[9] This effect is described in Section 2.4.2.

has to be accompanied by a corresponding increase in electron density in the inversion layer (ΔQ_I).

When the Fermi level E_F is not very close to the bottom of the conduction band (weak-inversion mode), the occupancy probability of the electron levels in the conduction band is relatively small. This means that the electron density increase ΔQ_I, necessary as a response to a gate voltage increase ΔV_G, can be achieved only by a significant band bending. As the $E_C - E_F$ difference is reduced, the probability of the electron-level occupancy in the conduction band rises exponentially. This means that a significant increase in ΔQ_I can now be achieved only by a slight reduction of the $E_C - E_F$ difference. The related slight change in the surface potential $\Delta\varphi_s$ is much smaller than the gate voltage change ΔV_G, which means that the increased gate voltage (ΔV_G) appears mostly across the gate oxide. With this, it can be assumed that the surface potential is pinned at $\varphi_s \approx 2\phi_F$ [Eq. (7.18)].

Figure 7.10a and 7.10b illustrates the appearance of electrons at the silicon surface as a response to a gate voltage increase beyond the threshold voltage. The situation in which applied voltage variations produce related variations in the charge located along the oxide interfaces is like the situation in an ordinary metal-plate capacitor with a dielectric equivalent to the gate oxide. The overall MOS capacitance is equal to the gate-oxide capacitance and is therefore voltage-independent. This is illustrated in Fig. 7.10c by the solid line labeled LF. As mentioned in Section 7.2.2, the thermally generated electrons in the inversion layer are unable to respond to the variations of signals with frequencies higher than the generation–recombination rates. In that case, the density of the electrons in the inversion layer appears "frozen" and the varying electric field penetrates through the depletion layer to repel/attract the majority carriers. The total capacitance also appears "frozen" at the minimum level, labeled HF.

EXAMPLE 7.4 Calculating the Threshold Voltage

Technological parameters of a MOS capacitor are given in Table 7.5, together with the values of the relevant physical parameters.

(a) Determine the value of the flat-band voltage.

(b) Calculate the charge density at the onset of strong inversion. Identify the type and origin of this charge.

TABLE 7.5 MOS Technological Parameters

Parameter	Symbol	Value
Substrate doping concentration	N_A	7×10^{16} cm^{-3}
Gate-oxide thickness	t_{ox}	30 nm
Oxide charge density	N_{oc}	10^{10} cm^{-2}
Type of the gate		N^+-polysilicon
Intrinsic carrier concentration	n_i	1.02×10^{10} cm^{-3}
Energy gap	E_g	1.12 eV
Thermal voltage at room temperature	$V_t = kT/q$	0.026 V
Oxide permittivity	ε_{ox}	3.45×10^{-11} F/m
Silicon permittivity	ε_s	1.04×10^{-10} F/m

(c) Calculate the value of the body factor.
(d) Calculate the value of the threshold voltage.
(e) Calculate the charge density in the inversion layer at $V_G = 5$ V.

SOLUTION

(a) To calculate the flat-band voltage V_{FB} using Eq. (7.35), the work-function difference ϕ_{ms} is needed; to obtain ϕ_{ms} using Eq. (7.34), the Fermi potential ϕ_F has to be determined first. Using Eq. (2.88), we write

$$\phi_F = V_t \ln \frac{N_A}{n_i} = 0.41 \text{ V}$$

Using this value of ϕ_F, and reading the values of ϕ_m and χ_s from Table 7.1, we can calculate the work-function difference by using Eq. (7.34):

$$\phi_{ms} = \phi_m - \left(\chi_s + \frac{E_g}{2q} + \phi_F\right) = 4.05 - \left(4.05 + \frac{1.12}{2} + 0.41\right) = -0.97 \text{ V}$$

After finding C_{ox} as $C_{ox} = \varepsilon_{ox}/t_{ox} = 3.45 \times 10^{-11}/30 \times 10^{-9} = 1.15 \times 10^{-3}$ F/m^2, we calculate the flat-band voltage by using Eq. (7.35):

$$V_{FB} = \phi_{ms} - \frac{qN_{oc}}{C_{ox}} = -0.96 - \frac{1.6 \times 10^{-19} \times 10^{14}}{1.15 \times 10^{-3}} = -0.974 \text{ V}$$

(b) At the onset of strong inversion ($V_G = V_T$), the inversion-layer charge is assumed to be zero. Consequently, the depletion-layer charge Q_d appears as the only uncompensated charge in the silicon substrate. Originating from the uncompensated negative acceptor ions (concentration N_A) in the depletion layer of width w_d, the depletion-layer charge density Q_d (C/m^2) can be expressed as

$$Q_d = qN_A w_d$$

Equation (7.11) gives w_d in terms of N_A and the Fermi potential $\varphi_s = 2\phi_F$, which after substitution into this equation leads to

$$Q_d = \sqrt{2\varepsilon_s q N_A (2\phi_F)}$$

Thus,

$$Q_d = \sqrt{2 \times 1.04 \times 10^{-10} \times 1.6 \times 10^{-19} \times 7 \times 10^{22} \times (2 \times 0.41)} = 1.38 \times 10^{-3} \text{ C/m}^2$$

(c) The body factor is defined by Eq. (7.17):

$$\gamma = \frac{\sqrt{2\varepsilon_s q N_A}}{C_{ox}} = \frac{\sqrt{2 \times 1.04 \times 10^{-10} \times 1.6 \times 10^{-19} \times 7 \times 10^{22}}}{1.15 \times 10^{-3}} = 1.327 \text{ V}^{1/2}$$

(d) The threshold voltage is given by Eq. (7.19):

$$V_T = V_{FB} + 2\phi_F + \gamma\sqrt{2\phi_F} = -0.974 + 2 \times 0.41 + 1.327\sqrt{2 \times 0.41} = 1.05 \text{ V}$$

(e) The assumptions are that the inversion-layer charge density is zero at $V_G = V_T$ and that the whole gate voltage increase beyond the threshold voltage is spent on creating the inversion-layer charge. Thus:

$$Q_I = (V_G - V_T)C_{ox} = (5 - 1.05) \times 1.15 \times 10^{-3} = 4.54 \times 10^{-3} \text{ C/m}^2$$

EXAMPLE 7.5 Designing the Threshold Voltage

In CMOS (complementary MOS) integrated circuits, it is required to provide equal absolute values of the threshold voltages of MOS structures on P-type and N-type substrates. Determine the value of the donor concentration N_D, necessary to provide an N-substrate MOS structure with the absolute value of the threshold voltage equal to the threshold voltage of the P-substrate MOS structure considered in Example 7.4. The values of all other parameters should remain the same.

SOLUTION

Equation (7.21) gives the threshold voltage for the case of N-type silicon substrate:

$$V_T = V_{FB} - 2|\phi_F| - \gamma\sqrt{2|\phi_F|}$$

Because the Fermi potential $2\phi_F$, the body factor γ, and the flat-band voltage V_{FB} depend on the donor concentration, it is necessary to express them in terms of N_D. Using Eq. (2.87) for $2\phi_F$, Eqs. (7.35) and (7.34) for V_{FB}, and Eq. (7.22) for γ, the threshold-voltage equation is developed as

$$V_T = \phi_m - \chi_s - \frac{E_g}{2q} + V_t \ln\frac{N_D}{n_i} - 2V_t \ln\frac{N_D}{n_i} - \frac{\sqrt{2\varepsilon_s q N_D}}{C_{ox}}\sqrt{2V_t \ln\frac{N_D}{n_i}}$$

where $V_t = kT/q$. Using the values of the known parameters, we simplify this equation to

$$V_T = -0.56 - 0.026 \ln\frac{N_D}{n_i} - 5.02 \times 10^{-12}\sqrt{N_D}\sqrt{0.052 \ln\frac{N_D}{n_i}}$$

Because $\ln N_D/n_i > 0$ (this is because $N_D > n_i$), all three terms are negative, which means that the threshold voltage is negative. To use the absolute value of the threshold voltage, all the minus

TABLE 7.6 Iterative Solutions for Example 7.5

N_D (m^{-3})	LHS
7×10^{22}	2.569
10^{21}	1.380
2×10^{21}	1.454
1.8×10^{21}	1.441

signs should be changed to pluses. Given that the absolute value of the threshold voltage should be 1.05 V (as obtained in Example 7.4), and developing $\ln(N_D/n_i)$ as $\ln N_D - \ln 1.02 \times 10^{16} = \ln N_D - 36.86$ (where N_D is in m^{-3}), the following equation is obtained:

$$1.05 = 0.56 + 0.026 \ln N_D - 0.026 \times 36.86 + 5.02 \times 10^{-12}\sqrt{N_D}\sqrt{0.052 \ln N_D - 0.052 \times 36.86}$$

Grouping the terms with the unknown N_D on the left-hand side leads to

$$0.026 \ln N_D + 5.02 \times 10^{-12}\sqrt{N_D}\sqrt{0.052 \ln N_D - 1.917} = 1.448$$

This equation can be solved iteratively: a guess is made for the value of N_D, and the left-hand side (LHS) is calculated and compared to the value on the right-hand side (1.448). This comparison provides an indication of whether the value of N_D should be increased or decreased before the next iteration is performed. When the difference between the LHS and RHS is acceptable, the value of N_D is taken as the solution. Perhaps it makes sense to take the value of the acceptor concentration from Example 7.4 as the initial guess: $N_D = 7 \times 10^{16}$ cm^{-3} = 7×10^{22} m^{-3}. Table 7.6 illustrates that the LHS value is 2.569, which is higher than the RHS value of 1.448; the concentration N_D should be reduced. Table 7.6 also illustrates that the next guess of 10^{21} m^{-3} is smaller than it should be, whereas the guess of 2×10^{21} m^{-3} is slightly larger than the proper concentration. Finally, $N_D = 1.8 \times 10^{21}$ m^{-3} is found to give a quite acceptable value of the LHS (1.441 as compared to the wanted 1.448). Therefore, the solution is taken to be $N_D = 1.8 \times 10^{21}$ m^{-3} = 1.8×10^{15} cm^{-3}.

*7.2.4 Flat-Band Capacitance and Debye Length

In the case of metal–dielectric–metal capacitors, any field penetration beyond the metal surfaces is limited to atomic distances, so it is practically negligible. In semiconductors, the field penetration may be quite significant. An obvious example is the case of a MOS capacitor in depletion mode. This section provides an analysis of the surface field penetration in MOS capacitors biased in the vicinity of the flat-band voltage (the reference voltage).

The surface potential is equal to zero when the effective gate voltage $V_G - V_{FB} = 0$. This also means that the electric field is equal to zero throughout the semiconductor.

Assume now that a small negative effective voltage is applied at the gate—for example, in the process of measuring the capacitance at the flat-band condition. The electric field created by this voltage will attract holes toward the surface. The question is how deeply the electric field enters into the semiconductor, or, in other words, how thick the accumulation layer is. The Poisson equation [Eq. (6.24)] can be used to obtain the quantitative estimate of this effect. In the general case, the charge density to be used in the Poisson equation is $\rho = q(p - n - N_A + N_D)$. In this specific case, $N_D = 0$ because we assumed P-type semiconductor and $p \gg n$ because we are focusing on the accumulation mode. Therefore,

$$\frac{d^2\varphi(x)}{dx^2} = -\frac{q}{\varepsilon_s}[p(x) - N_A] \tag{7.36}$$

Assuming Maxwell–Boltzmann distribution, we write

$$p(x) = p_0 e^{-q\varphi(x)/kT} \approx N_A e^{-q\varphi(x)/kT} \tag{7.37}$$

where $\varphi(x) \leq 0$ is the electric potential that varies from the surface potential φ_s at the semiconductor surface ($x = 0$) to 0 for $x \to \infty$. The exponential dependence on $\varphi(x)$ is due to the Maxwell–Boltzmann distribution and the fact that E_V (and E_C) bending follows $\varphi(x)$. The constant p_0 is determined from the condition that the concentration of holes is equal to the equilibrium level for $\varphi(x) = 0$. For a small $\varphi(x)$, the exponential dependence can be approximated by the following linear dependence:

$$p(x) = N_A e^{-q\varphi(x)/kT} \approx N_A\left[1 - \frac{q\varphi(x)}{kT}\right] \tag{7.38}$$

With this, Eq. (7.36) can be written in the following form:

$$\frac{d^2\varphi(x)}{dx^2} = \frac{\varphi(x)}{L_D^2} \tag{7.39}$$

where

$$L_D = \sqrt{\frac{\varepsilon_s kT}{q^2 N_A}} \tag{7.40}$$

The general solution of Eq. (7.39) is

$$\varphi(x) = A_1 e^{-x/L_D} + A_2 e^{x/L_D} \tag{7.41}$$

The constants A_1 and A_2 for this specific case are $A_2 = 0$ [because $\varphi(\infty) = 0$] and $A_1 = \varphi_s$ [because $\varphi(0) = \varphi_s$]. Therefore,

$$\varphi(x) = \varphi_s e^{-x/L_D} \tag{7.42}$$

This result shows that the electric potential drops exponentially from the surface value φ_s toward zero. The parameter of this exponential dependence, L_D, corresponds to the distance $x = L_D$ at which the electric potential drops e times. This type of exponential dependence is maintained for the electric field

$$E = -\frac{d\varphi}{dx} = \frac{\varphi_s}{L_D} e^{-x/L_D} \tag{7.43}$$

as well as for the hole distribution. From the linear approximation in Eq. (7.38) and Eq. (7.42) for $\varphi(x)$, the excess hole concentration is obtained as

$$\delta p(x) = p(x) - N_A = \frac{q\varphi_s}{kT} N_A e^{-x/L_D} \tag{7.44}$$

Therefore, the parameter L_D also characterizes the field penetration and the width of the accumulation layer. This parameter is called the *Debye length*. Equation (7.40) shows that the Debye length is inversely proportional to the doping level. Because of the electroneutrality equation, $p_0 = N_A$ in this case, the Debye length is in principle inversely proportional to the equilibrium concentration of the mobile charge. The physical meaning here is that a higher concentration of mobile charge can provide much better screening, so the penetration of the electric field is shallower. In the case of metals, the concentration of the mobile charge is so high that the Debye length drops to subatomic levels.

As a direct consequence of the field penetration in semiconductors, the accumulation-mode capacitance is actually smaller than C_{ox}. The accumulation capacitance approaches the C_{ox} level with increasing negative voltage because the increasing concentration of holes in the accumulation layer provides more efficient field screening. Figure 7.5 shows that the actual accumulation capacitance increases as the negative voltage is increased. Nonetheless, the model of accumulation capacitance that is constant and equal to C_{ox} throughout the accumulation region is still useful. For example, it enables a straightforward calculation (estimate) of the accumulation-layer charge: $Q_A = C_{ox}(V_G - V_{FB})$.

Apart from the model of constant accumulation capacitance, it is sometimes quite useful to determine the actual capacitance at flat bands. In particular, this can be used for experimental determination of the value of the flat-band voltage from a measured C–V curve. At flat bands, the MOS capacitance can be represented by a series connection of two capacitors: the gate-oxide capacitance and the semiconductor capacitance that is due to the penetration of the electric field into the semiconductor. The gate-oxide capacitance per unit area is $C_{ox} = \varepsilon_{ox}/t_{ox}$. Analogously, the semiconductor capacitance per unit area is $C_s = \varepsilon_s/L_D$. Therefore

$$\frac{1}{C_{FB}} = \frac{1}{C_{ox}} + \frac{1}{C_s} = \frac{t_{ox}}{\varepsilon_{ox}} + \frac{L_D}{\varepsilon_s} = \frac{t_{ox} + (\varepsilon_{ox}/\varepsilon_s)L_D}{\varepsilon_{ox}} \tag{7.45}$$

$$C_{FB} = \frac{\varepsilon_{ox}}{t_{ox} + \left(\dfrac{\varepsilon_{ox}}{\varepsilon_s}\right)\sqrt{\varepsilon_s kT/(q^2 N_A)}} \tag{7.46}$$

The voltage that corresponds to this capacitance on the accumulation side of the C–V curve is the flat-band voltage.

SUMMARY

1. A difference in the work functions of a metal ($q\phi_m$) and a semiconductor ($q\phi_s$) creates a built-in voltage at the metal–semiconductor interface:

$$V_{bi} = \phi_m - \phi_s$$

This built-in voltage is associated with a depletion layer at the semiconductor surface, and it represents a barrier for the electrons in the semiconductor. The electrons in the metal also face a barrier, which is

$$q\phi_B = q\phi_m - q\chi_s$$

$q\phi_B$ is voltage-independent, whereas the barrier from the semiconductor side can be changed by applied voltage ($V_{bi} - V_F$; $V_{bi} + V_R$), enabling the metal–semiconductor contact to operate as a rectifying diode—a *Schottky diode*.

2. The current–voltage characteristic of the Schottky diode has the same form as for the P–N junction diode:

$$I_D = I_S\left(e^{V_D/nV_t} - 1\right)$$

By proper selection of the metal electrode, Schottky diodes with smaller built-in voltages can be created, resulting in a smaller forward voltage for the same current. In the model, this is accounted for by a larger I_S, which shows that there is a direct link to an increase in the reverse-bias current.

3. Schottky diodes are single-carrier devices (e.g., the minority holes in metal–N-type Schottky diodes do not play a significant role in the current flow). There is no stored charge of minority carriers, so there is no stored-charge capacitance and the associated switch-off delay.
4. A contact between a metal and a *heavily doped* semiconductor leads to a very narrow barrier (narrow depletion layer), enabling the carriers to tunnel in either direction. This type of metal–semiconductor contact acts as a small-resistance (ohmic) contact.
5. Oxidizing silicon in strictly controlled conditions creates a dielectric–semiconductor interface of unique quality in terms of electronic properties. When a metal or heavily doped polysilicon is deposited on the thermal oxide, a metal–oxide–semiconductor (MOS) capacitor is created. The metal (polysilicon) electrode is referred to as *gate*.
6. Depending on the gate voltage applied, a MOS capacitor is said to be in one of the three modes defined in Table 7.7.
7. At flat-band conditions ($V_G = V_{FB}$), the net charge at the capacitor plates, the electric field, and the semiconductor *surface potential* (φ_s) are all zero. A nonzero *flat-band voltage* appears due to the metal–semiconductor *work-function difference* ($q\phi_{ms}$) and the effects of the oxide charge (N_{oc}):

$$V_{FB} = \underbrace{\phi_{ms}}_{\phi_m - \phi_s} - \frac{qN_{oc}}{C_{ox}}$$

TABLE 7.7 Potential–Capacitance–Charge Equations for a MOS Capacitor

	Accumulation	Depletion	Strong Inversion
P type	$V_G < V_{FB}$	$V_{FB} \le V_G \le V_T$	$V_T < V_G$
N type	$V_{FB} < V_G$	$V_T \le V_G \le V_{FB}$	$V_G < V_T$
φ_s	Small	$\lvert\varphi_s\rvert \le \lvert 2\phi_F\rvert$	$\varphi_s \approx 2\phi_F$
C	$C = C_{ox} = \varepsilon_{ox}/t_{ox}$	$\frac{1}{C} = t_{ox}/\varepsilon_{ox} + w_d/\varepsilon_s$	LF: $C = C_{ox} = \varepsilon_{ox}/t_{ox}$
		$w_d = \sqrt{2\varepsilon_s\lvert\varphi_s\rvert/(qN_{A,D})}$	HF: $\frac{1}{C} = t_{ox}/\varepsilon_{ox} + w_{d-inv}/\varepsilon_s$
			$w_{d-inv} = \sqrt{2\varepsilon_s\lvert 2\phi_F\rvert/(qN_{A,D})}$
$\lvert Q\rvert$	$Q_A = \lvert V_{FB} - V_G\rvert C_{ox}$	$Q_d = qN_{A,D}w_d$	$Q_I = \lvert V_G - V_T\rvert C_{ox}$
		$= \sqrt{2\varepsilon_s qN_{A,D}\lvert\varphi_s\rvert}$	$Q_d = \sqrt{2\varepsilon_s qN_{A,D}\lvert 2\phi_F\rvert}$
			$= \gamma C_{ox}\sqrt{\lvert 2\phi_F\rvert}$

8. The Fermi level of a heavily doped N^+ silicon (polysilicon) is approximately at the bottom of the conduction band, so $q\phi_m = q\chi_s$, where the electron affinity $q\chi_s$ is a material constant. The Fermi level of a heavily doped P^+ silicon (polysilicon) is approximately at the top of the valence band; therefore, $q\phi_m = q\chi_s + E_g$. The Fermi level in the moderately doped semiconductor substrate depends on the doping level, the Fermi potential $q\phi_F$ expressing the difference between the midgap, and the position of the Fermi level:

$$q\phi_s = q\left(\chi_s + \frac{E_g}{2q} + \phi_F\right)$$

$$\phi_F = \begin{cases} +\frac{kT}{q}\ln\frac{N_A}{n_i} & \text{for P-type} \\ -\frac{kT}{q}\ln\frac{N_D}{n_i} & \text{for N-type} \end{cases}$$

9. At the onset of strong inversion ($V_G = V_T$), the surface potential φ_s is approximately at its maximum value of $|2\phi_F|$. In strong inversion, the depletion-layer width and the depletion-layer charge do not change significantly, because the voltage across the depletion layer (φ_s) remains approximately at the constant level of $2\phi_F$. Any gate-voltage change in the strong inversion results in a proportional change of the inversion-layer charge: $Q_I = C_{ox}(V_G - V_T)$. The threshold voltage is given by

$$V_T = V_{FB} \pm 2|\phi_F| \pm \gamma\sqrt{2|\phi_F|}$$

where the upper and lower signs are for P-type and N-type substrates, respectively, and the body factor γ is given by

$$\gamma = \frac{\sqrt{2\varepsilon_s qN_{A,D}}}{C_{ox}}$$

PROBLEMS

7.1 Each of the graphs in Fig. 7.11 shows a pair of P–N junction and Schottky diode characteristics. Identify the combination that properly groups three characteristics that all belong to one of the two diodes.

7.2 The graphs of Fig. 7.12 show the energy-band diagrams of three different pairs of metal–semiconductor materials. State whether each of the systems would create a Schottky or an ohmic contact if placed in contact.

7.3 A Schottky diode with N-type silicon ($N_D = 10^{15}$ cm^{-3}) is designed to have built-in voltage $V_{bi} = 0.40$ V and barrier height $\phi_B = 0.65$ V. If the actual donor concentration is 4.66×10^{15} cm^3, what are the built-in voltage and the barrier height?

(a) 0.40 V; 0.65 V
(b) 0.36 V; 0.65 V
(c) 0.44 V; 0.65 V
(d) 0.40 V; 0.61 V
(e) 0.40 V; 0.69 V
(f) 0.36 V; 0.69 V

7.4 The energy barrier and the built-in voltage of a Schottky diode are $q\phi_B = 0.65$ eV and $V_{bi} = 0.40$ V, respectively. What are the barrier heights for the electrons in semiconductor and metal, respectively, at

(a) forward bias $V_F = 0.2$ V
(b) reverse bias $V_R = 5$ V **A**

7.5 Chromium (Cr) and tungsten silicide (WSi_2) are available as the metal electrodes for a Schottky diode. Design the minimum-area Schottky diode so that $V_F = 0.2$ V at $I_F = 100$ mA (this current is low enough so that the emission coefficient is $n = 1$). Neglect the image-force effect and use $T = 300$ K.

7.6 For the diode designed in Problem 7.5, determine the forward voltage for the same forward current at

(a) 85°C **A**
(b) 100°C

7.7 A Schottky diode is created by depositing tungsten ($q\phi_m = 4.6$ eV) onto N-type silicon. By appropriate heating, tungsten silicide ($q\phi_m = 4.7$ eV) can be created, becoming effectively the metal electrode of the Schottky diode. Would this heating process increase or decrease the saturation current? How many times?

7.8 A P–N junction diode and a Schottky diode have the same emission coefficient $n = 1.25$. The forward voltage (measured at the same current) of the Schottky diode is 0.5 V smaller. How many times is the reverse-bias saturation current of the Schottky diode larger? The thermal voltage is $V_t = 0.026$ V.

7.9 A Schottky diode, created on N-type silicon ($N_D = 10^{15}$ cm^{-3}) has $V_{bi} = 0.4$ V. Determine the depletion-layer capacitance per unit area at

(a) zero bias **A**
(b) reverse bias $V_R = 25$ V

7.10 To reduce the series resistance, the width of the low-doped region of the PIN and Schottky diode considered in Example 7.1 is cut down to $w_{n-epi} = 10$ μm.

(a) Calculate the achieved reduction in forward voltage ΔV_F.
(b) Calculate the associated reduction in maximum reverse-bias voltage ΔV_{R-max}.

7.11 Obtain the SPICE parameters I_S and n of a Schottky diode, using the data given in Table 7.8. Assume room temperature ($V_t = 0.02585$ V).

TABLE 7.8 I_D–V_D **Data for a Schottky Diode**

Current (mA)	Voltage (V)
60.4	0.40
101.2	0.42

7.12 Design thermal oxidation conditions to grow 500-nm SiO_2 that is required in an IC technology process. The temperature limits are 900°C and 1200°C, and both dry and wet processes are available. To maximize throughput, the design should minimize the growth time.

7.13 Which of the following statements, related to a MOS capacitor, is correct?

(a) The condition of zero net charge at the MOS capacitor plates is referred to as zero-bias condition.
(b) There is no field at the semiconductor surface at $V_{GS} = V_{FB}$.

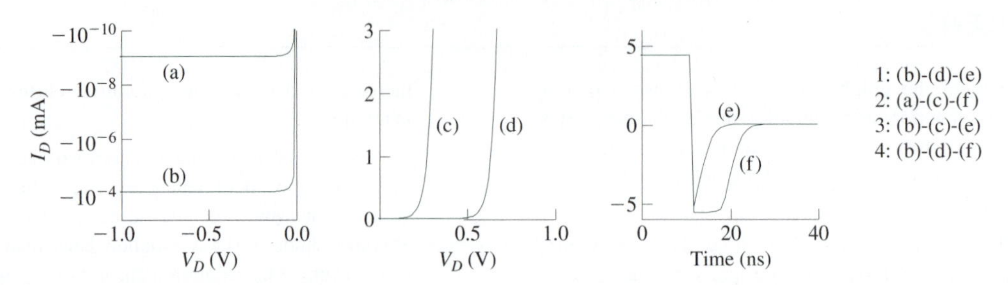

Figure 7.11 Characteristics of a P–N junction and a Schottky diode.

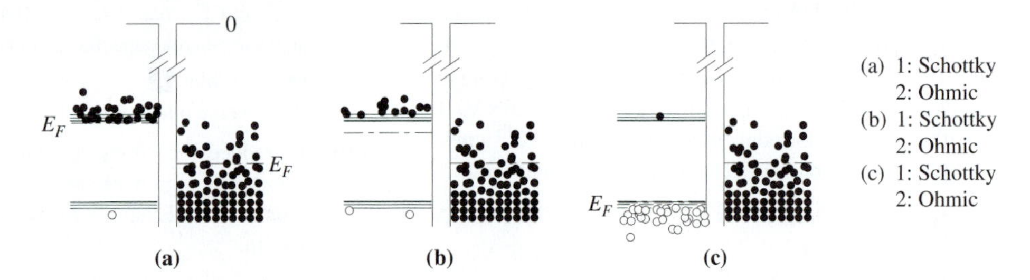

Figure 7.12 Semiconductor–metal energy-band diagrams.

(c) The net charge at the MOS capacitor plates has to be zero at $V_{GS} = 0$.

(d) The density of inversion-layer charge is expressed as $Q_I = C_{ox}(V_{GS} - V_{FB})$.

(e) The surface potential φ_s increases exponentially with gate voltage in strong inversion.

(f) The density of inversion-layer charge at the onset of strong inversion is $Q_I = \gamma\sqrt{2\phi_f}$.

(g) The capacitance in depletion mode does not depend on the gate voltage applied.

(h) The inversion-layer capacitance is proportional to $(V_{GS} - V_T)$.

7.14 Which mode of MOS capacitor operation (accumulation, depletion, or strong inversion) is expressed by the energy-band diagram of Fig. 7.13? Knowing that the energy gap of silicon is 1.12 eV, estimate the voltage applied between the gate and the silicon substrate.

7.15 How is the corresponding mode referred to and what type of mobile and/or fixed charge appears at the semiconductor surface of a MOS capacitor on a P-type substrate when

(a) a negative effective voltage ($V_G - V_{FB} < 0$) is applied to the gate?

(b) a small positive effective gate voltage is applied to the gate ($V_{FB} < V_G < V_T$)?

(c) a large positive effective gate voltage ($V_G > V_T$) is applied to the gate?

Assume $N_{oc} = 0$.

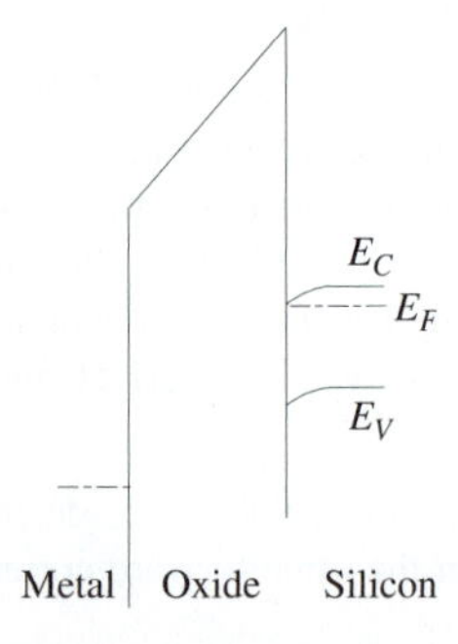

Figure 7.13 MOS energy-band diagram.

7.16 The flat-band voltage of a MOS capacitor on N-type substrate is $V_{FB} = -1$ V. If the gate-oxide thickness is 5 nm, calculate the value and determine the direction(s) of the electric fields in the gate oxide and at the surface of the semiconductor at zero gate bias ($V_G = 0$). **A**

7.17 Two different MOS capacitors, with different gate-oxide thicknesses (3 nm and 15 nm), have the same density of positive oxide charge ($N_{oc} = 5 \times 10^{10}$ cm^{-2}) close to the oxide–semiconductor interface. Find the flat-band voltage shifts due to this positive oxide charge for these two MOS capacitors. What are the threshold voltage shifts?

7.18 The C–V curves of a MOS capacitor before and after a gate-oxide stressing are shown in Fig. 7.14. Find the density of gate-oxide charge, N_{oc}, created by this stress.

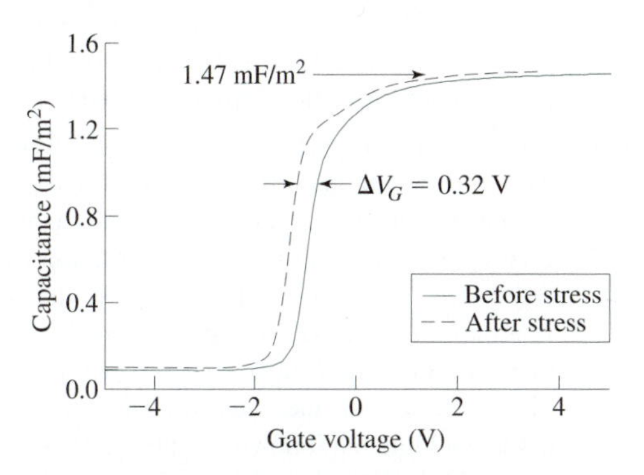

Figure 7.14 C–V curves of a MOS capacitor before and after gate-oxide stressing.

7.19 What is the work-function difference between heavily doped polysilicon and P-type silicon that is doped with 10^{16} cm^{-3} acceptor atoms if the heavily doped polysilicon is

(a) P^+ type **A**
(b) N^+ type

7.20 **(a)** The flat-band voltage of a MOS capacitor with N^+ polysilicon gate is $V_{FB} = -0.25$ V. Assuming zero oxide charge, determine the type and level of the substrate doping.
(b) What would be the flat-band voltage if P^+ polysilicon gate is used with the same type and level of substrate doping? **A**

7.21 The threshold voltage of a MOS capacitor on a P-type silicon substrate is $V_T = 1.0$ V. Give the type and density of the mobile charge at the silicon surface if 5 V is applied between the gate and the substrate. The gate-oxide thickness is 15 nm. **A**

7.22 The threshold voltage of a MOS capacitor on a P-type substrate is $V_T = 0.25$ V, and the gate-oxide capacitance is $C_{ox} = 6.5$ mF/m^2. How many electrons can be found in 0.1 μm × 0.1 μm of capacitor area if the gate voltage is $V_G = 0.5$ V?

7.23 The threshold and the flat-band voltages of a MOS capacitor are $V_T = -1.0$ V and $V_{FB} = -0.5$ V, respectively. Is this capacitor created on an N-type or P-type semiconductor? What is the density of minority carriers (in C/m^2) at the semiconductor surface when the voltage applied between the metal and semiconductor electrodes is $V_G = -0.75$ V? **A**

7.24 The flat-band voltage and the threshold voltage of a MOS capacitor are $V_{FB} = -3.0$ V and $V_T = -1.0$ V, respectively. The gate-oxide capacitance is $C_{ox} = \varepsilon_{ox}/t_{ox} = 3.45 \times 10^{-3}$ F/m^2.

(a) Is this capacitor created on an N-type or P-type semiconductor? Explain your answer.
(b) What is the density of minority carriers (in C/m^2) at the semiconductor surface when the voltage applied between the metal and semiconductor electrodes is $V_G = -2.0$ V?
(c) What is the density of minority carriers (in C/m^2) at the semiconductor surface when no voltage is applied across the capacitor ($V_G = 0$)?

7.25 Calculate the threshold voltage of a P^+ polysilicon-gate MOS capacitor on an N-type substrate ($N_D = 5 \times 10^{16}$ cm^{-3}) if the gate oxide thickness is 7 nm. Neglect the gate-oxide charge.

7.26 The oxide thickness of a MOS capacitor is 4 nm. The silicon substrate is N-type (doping level $N_D = 7 \times 10^{16}$ cm^{-3}), and the gate is heavily doped N^+-type polysilicon. Assuming zero oxide charge density, determine the density of mobile charge at the semiconductor surface if the voltage applied between the gate and the substrate is −1.5 V. **A**

7.27 Calculate high- and low-frequency strong-inversion capacitances per unit area of a MOS capacitor having 50-nm-thick oxide as the dielectric and P-type substrate doped with $N_A = 10^{15}$ boron atoms per cm^3.

7.28 The accumulation and strong-inversion capacitances of a MOS capacitor, measured by a high-frequency signal, are 9 mF/m^2 and 3 mF/m^2, respectively. These capacitances are measured at $V_G = -2$ V and $V_G = 2$ V, respectively.

(a) Is the substrate N or P type?

(b) Determine the gate-oxide thickness.

(c) Assuming a uniform substrate doping, calculate the substrate doping concentration. **A**

7.29 A MOS capacitor with N^+ polysilicon gate is biased in strong inversion. What is the surface potential φ_s? Determine the voltage across the gate oxide if the gate to substrate voltage is 5 V. The doping level of the silicon substrate is $N_A = 5 \times 10^{16}$ cm^{-3}. **A**

7.30 The oxide breakdown electric field is 1 V/nm. Design the oxide thickness of a MOS capacitor so that the breakdown voltage in strong inversion is 5 V. N^+ polysilicon is to be used for the gate, and the substrate doping is $N_A = 7.5 \times 10^{16}$ cm^{-3}. What is the breakdown voltage in accumulation if the surface potential is neglected in comparison to the breakdown voltage?

7.31 The following relationship among the oxide field, the semiconductor field, and the density of oxide charge close to the oxide–semiconductor interface can be derived from the integral form of Gauss's law:

$$\varepsilon_s E_s - \varepsilon_{ox} E_{ox} = q N_{oc}$$

where both the semiconductor field at the surface (E_s) and the oxide field (E_{ox}) are in the direction toward the substrate. Calculate E_{ox} and E_s for $V_G = V_{FB}$ and

(a) $N_{oc} = 0$ **A**

(b) $N_{oc} = 5 \times 10^{10}$ cm^{-2}

7.32 A MOS capacitor has a P^+ polysilicon gate, substrate doping of $N_D = 10^{16}$ cm^{-3} and gate-oxide thickness of $t_{ox} = 80$ nm. The breakdown field of the oxide is 1 V/nm. Calculate the breakdown voltage in strong inversion if

(a) the oxide charge can be neglected.

(b) as a result of exposure to high electric field, a positive charge with density of $N_{oc} = 5 \times 10^{11}$ cm^{-2} is created close to the silicon–oxide interface. **A**

7.33 A MOS capacitor on a P-type substrate with $\phi_{ms} = 1.0$ V is biased by a constant gate voltage $V_G = -7$ V, in order to test the integrity of its 10-nm gate oxide. If this stress creates $N_{oc} = 10^{10}$ cm^{-2} of positive charge close to the silicon–oxide interface every hour, how long will it take before the oxide field reaches the critical level of 1 V/nm?

7.34 Five polysilicon-gate N-channel MOSFETs, each with a different gate-oxide thickness t_{ox} (45 nm, 47 nm, 50 nm, 53 nm, and 55 nm), are made on P-type silicon substrates having the same doping level $N_A = 5 \times 10^{16}$ cm^{-3}. The gate-oxide charge Q_{oc} is assumed to be equal for all the MOSFETs because they are processed in a single batch of wafers. Measurements of the threshold voltages V_T are made, yielding the following values: 1.10 V, 1.16 V, 1.25 V, 1.33 V, and 1.39 V. Discuss the shape and the slope of $V_T(t_{ox})$ dependence, and explain the meaning of the intercept $V_T(t_{ox} = 0)$. Determine the value of the oxide charge Q_{oc} and the metal–semiconductor work-function difference ϕ_{ms}.

REVIEW QUESTIONS

R-7.1 Make an analogy between a P–N junction and a Schottky diode. What is the origin of the built-in voltages in either case? How do typical values compare?

R-7.2 What is the origin and meaning of the barrier potential ϕ_B? Does it depend on the bias applied?

R-7.3 Why do the electrons from an N-type semiconductor not appear as minority carriers after they pass through metal–semiconductor contact?

R-7.4 How does the absence of stored charge influence the Schottky diode characteristics?

R-7.5 Is there a net charge at MOS capacitor plates at $V_G = 0$ V? If there is, there must be an electric field at the semiconductor surface to keep that charge at the capacitor plates. With no gate voltage applied, where can this electric field originate from?

R-7.6 How is the condition of zero charge at MOS capacitor plates referred to? Is there any field in the oxide or the substrate? Is there any potential difference between the surface and the bulk of the silicon substrate?

R-7.7 How is the flat-band voltage expressed in terms of the work-function difference and the oxide charge?

R-7.8 What type of charge appears at MOS capacitor plates when negative effective gate voltage ($V_G - V_{FB} < 0$) is applied? How is this mode referred to? Assume a P-type semiconductor.

R-7.9 What determines the MOS capacitance in accumulation mode? Does it depend on the voltage applied?

R-7.10 What type of charge appears at MOS capacitor plates when small positive effective gate voltage ($V_G - V_{FB} > 0$) is applied? How is this mode referred to? Assume a P-type semiconductor.

R-7.11 What determines the MOS capacitance in depletion mode? Does it depend on the voltage applied? Does the surface potential change as the gate voltage is changed?

R-7.12 What type of charge appears at MOS capacitor plates when the gate voltage applied is larger than the threshold voltage ($V_G > V_T$)? How is this mode referred to? Assume a P-type semiconductor.

R-7.13 Write the capacitance–voltage–charge equation ($Q = CV$) for a MOS capacitor in strong inversion to show the relationship of different types of charge in the silicon substrate and the effective gate voltage across the gate oxide.

R-7.14 Does surface potential depend significantly on the gate voltage in strong inversion?

R-7.15 What determines the surface potential in strong inversion? How is it expressed?

R-7.16 What is the density of the inversion-layer charge (electrons in the case of a P-type substrate) at the onset of strong inversion?

R-7.17 What is the voltage at the gate that sets the MOS structure at the onset of strong inversion called?

R-7.18 Obtain the threshold-voltage equation from the equation written in Question 13.

8 MOSFET

The first MOSFET (metal–oxide–semiconductor field-effect transistor) was fabricated in 1960, only a year after the beginning of the integrated-circuit era in 1959. The MOSFET became the basic building block of very-large-scale integrated (VLSI) circuits, therefore becoming the most important microelectronic device. Huge investments have been made in what is known as CMOS technology, a technology used to manufacture circuits consisting of complementary pairs of MOSFETs. Those investments, having been quite favorable, consequently led to the rapid progress in computer and communication integrated circuits that we have seen in the past decades.

However, the application of MOSFETs is not limited to VLSI circuits. MOSFETs play an important role in power-electronic circuits, and they are becoming increasingly popular and suitable for microwave applications.

This chapter explains MOSFET principles and characteristics, and it describes MOSFET models and parameters used in circuit simulation. The metal–oxide–semiconductor (MOS) capacitor, dealt with in Section 7.2, represents the basis of a MOSFET. Therefore, a good grasp of the effects explained in the MOS capacitor section is necessary for effective understanding of this chapter. Also, the MOSFET involves two P–N junctions, which means that the P–N junction concepts introduced in Chapter 6 need to be understood as well.

8.1 MOSFET PRINCIPLES

8.1.1 MOSFET Structure

As indicated, the MOSFET is developed from the MOS capacitor. The voltage applied to the gate of the MOS capacitor controls the state of the silicon surface underneath. Negative gate voltages attract the holes from the P-type silicon to the surface (accumulation),

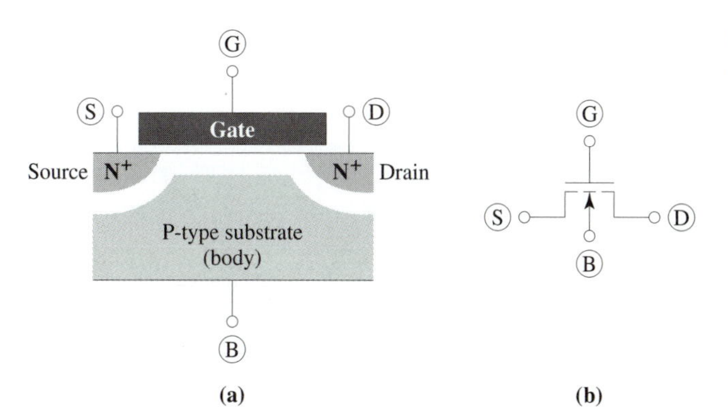

Figure 8.1 (a) Schematic cross section and (b) the circuit symbol of an N-channel MOSFET.

whereas positive voltages larger than the threshold voltage create a layer of electrons at the surface (inversion).

These two states of the MOS capacitor can be used to make a voltage-controlled switch. To achieve this, the layer of electrons at the surface is contacted at the ends by N^+ regions referred to as *source* and *drain,* as illustrated in Fig. 8.1a. The existence of the electron layer, also referred to as *channel,* corresponds to the *on* state of the switch as the electron channel virtually short circuits the source and the drain regions, which are used as the switch terminals. When the gate voltage is below the threshold voltage, the electron layer (the channel) disappears from the surface, and the source and drain N^+ regions are isolated by the P-type substrate. This is the *off* state of the switch.

The same structure, shown in Fig. 8.1, can be used to create a voltage-controlled current source. This is possible because at higher drain-to-source voltages the current flowing through the channel (*on* mode) does not increase linearly with the drain-to-source voltage but saturates. The mechanisms of current saturation will be explained in detail in Section 8.1.4. Because the saturation current is independent of the voltage between the source and drain, the device behaves as a current source. In addition, it is possible to alter the value of this current between its maximum value and zero. Therefore, the MOSFET appears as a voltage-controlled current source at higher drain-to-source voltages.

It is obvious from Fig. 8.1 that the MOSFET is essentially a four-terminal device. The four terminals are as follows: the silicon substrate (body) (B), the gate (G), the source (S), and the drain (D). Very frequently, the body and the source are connected together, so that the controlling voltage applied to the gate, as well as the driving voltage applied to the drain, can be expressed with respect to the common reference potential of the short-circuited source and body. In some integrated circuits, the body and the source cannot be short-circuited, or a voltage is deliberately applied between the body and the source. The effect of the body-to-source voltage is called *body effect* and is explained in Section 8.1.3.

Figure 8.1 illustrates one type of MOSFET, which uses a P-type substrate (body), an N^+-type source, and drain layers and also needs positive voltage at the gate to turn the MOSFET *on* by creating channel of electrons between the source and the drain. Because this type of MOSFET operates with N-type channel (electrons), it is referred to as an *N-channel* MOSFET. It is possible to make a complementary MOSFET using N-type substrate (body) and P^+-type source and drain layers. In this type of MOSFET, the channel connecting the source and the drain in *on* mode has to be created of holes (P-type carriers),

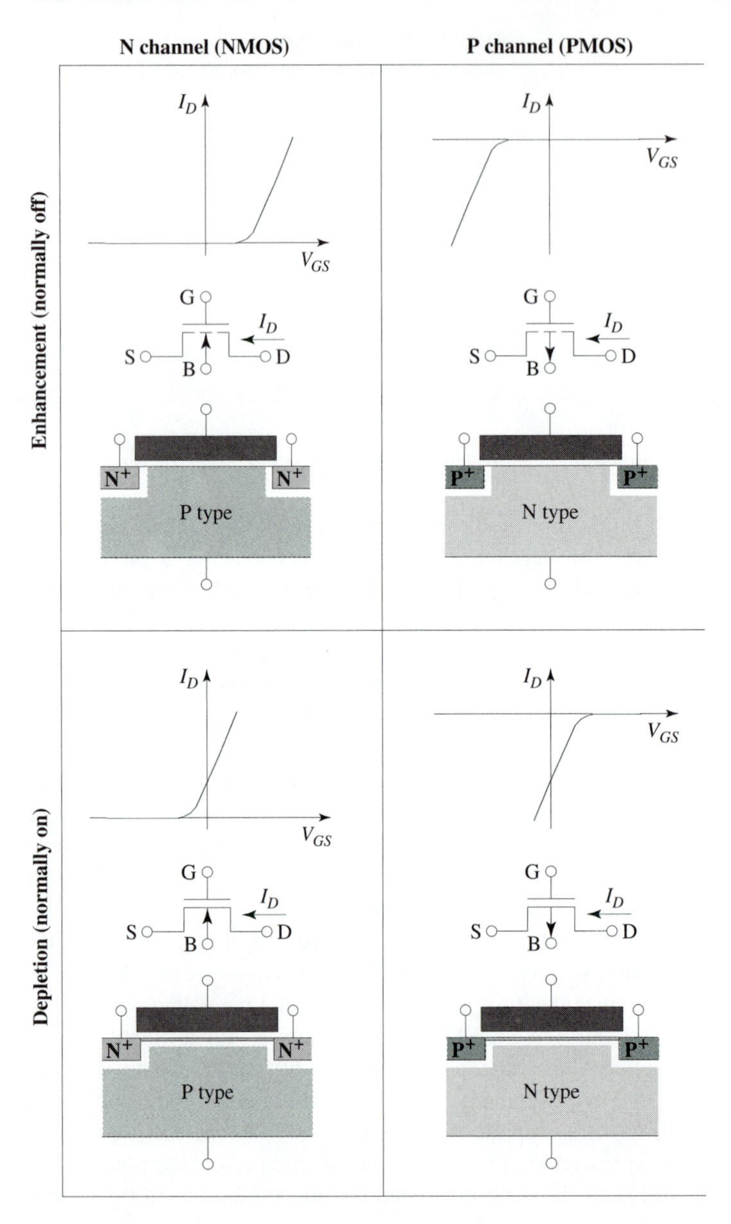

Figure 8.2 Types of MOSFETs.

because of which it is called a *P-channel* MOSFET. The terms N-channel and P-channel MOSFETs are frequently replaced by the shorter terms *NMOS* and *PMOS*.

A common characteristic of these N-channel and P-channel MOSFETs is that they are in the *off* mode when no gate bias is applied. This is because there is no channel between the source and the drain, and therefore the drain current is zero. Consequently, these MOSFETs are classified as *normally off* MOSFETs. The transfer characteristics (I_D–V_{GS}) in Fig. 8.2 show that the drain current appears for (a) sufficiently large positive gate voltages in the case of an N-channel and (b) negative gate voltages in the case of a P-channel MOSFET.

This is because appropriate gate voltages are needed to create the channel of electrons and holes in N-channel and P-channel MOSFETs, respectively. As a consequence, MOSFETs of this type are also referred to as *enhancement* MOSFETs.

MOSFETs can be created with technologically built-in channels. Because no gate voltage is needed to set these MOSFETs in the *on* state, they are called *normally on* MOSFETs. To turn this type of MOSFETs *off*, the channels have to be depleted of electrons or holes, so they are also referred to as *depletion-type* MOSFETs. The transfer characteristics in Fig. 8.2 illustrate that negative voltage is needed to stop the drain current in the case of an N-channel MOSFET, and similarly positive voltage is needed to set a P-channel MOSFET in the *off* state.

Defining the threshold voltage as the gate voltage at which the channel is just formed (or depleted), we can say that the threshold voltage of the enhancement-type N-channel MOSFETs is positive, whereas it is negative in the case of the depletion-type N-channel MOSFETs. The situation is opposite with the P-channel MOSFETs: negative threshold voltage in the case of enhancement type, and positive threshold voltage in the case of the depletion type.

The main MOSFET type is the N-channel enhancement type. The P-channel enhancement-type MOSFET is used as a complementary transistor in circuits known as CMOS (complementary MOS) technology. The N-channel depletion-type MOSFET is used as a kind of complementary transistor in circuits using only N-channel MOSFETs (NMOS technology).

8.1.2 MOSFET as a Voltage-Controlled Switch

Cross-Sectional Illustration and I_D–V_{DS} Characteristics

Figure 8.3 illustrates the two modes of a MOSFET used as a voltage-controlled switch. The switch is between the source and drain terminals, whereas the gate is the controlling electrode. In the *off* mode, the structure between the drain and the source terminals is equivalent to two back-to-back P–N junction diodes. A positive voltage between the drain and the source ($V_{DS} > 0$) sets the drain-to-body diode in *off* mode (reverse bias); hence there is no significant current through the switch (refer to the I_D–V_D characteristic in Fig. 8.3c). The reverse bias expands the depletion layer at the drain-to-body junction, as illustrated in Fig. 8.3a.

The switch remains in *off* mode for as long as the gate voltage is below the threshold voltage ($V_{GS} < V_T$). When the gate voltage is higher than the threshold voltage, the channel between the drain and source is formed, thereby enabling a current flow through the switch. The channel under the gate electrode appears as a resistor between the drain and the source electrodes, so the channel current increases linearly with the drain-to-source voltage (Fig. 8.3c). The channel resistance can be considered as a parasitic resistance of a switch in *on* mode. In other words, the switch is not ideal because it does not provide the perfect short circuit in *on* mode.

Voltage-controlled switches are typically used in circuits that operate with two voltage levels: low ($V_L \approx 0$) and high ($V_H \approx V_+$, where V_+ is the positive power-supply voltage in the circuit). For enhancement-type MOSFETs, $V_L < V_T$ (*off* mode) and $V_+ > V_T$ (*on* mode). The two colored lines in the I_D–V_{DS} characteristic of Fig. 8.3c correspond to these two cases.

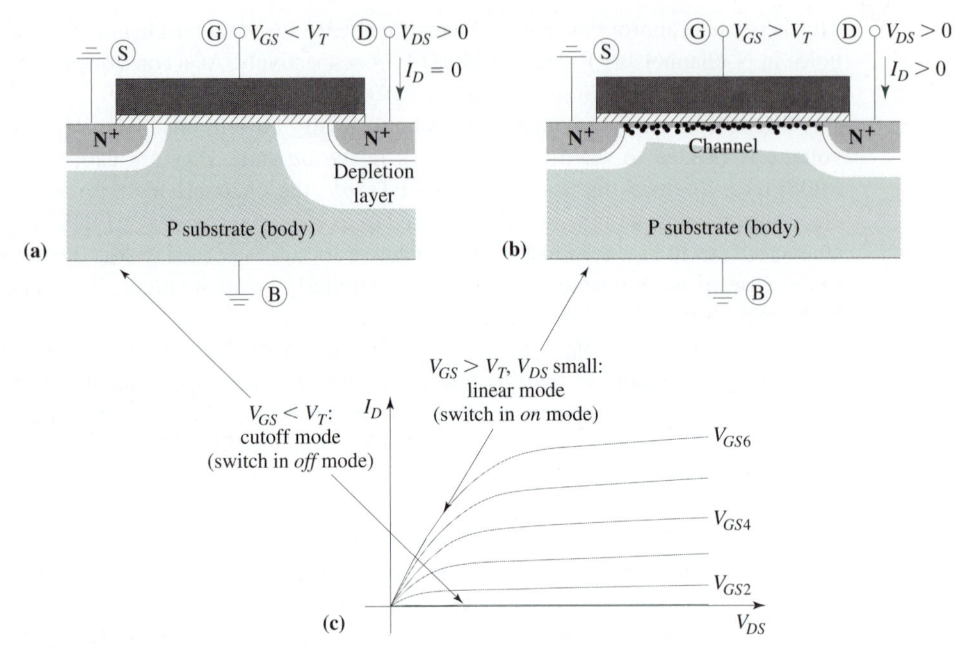

Figure 8.3 Cross-sectional illustrations of a MOSFET in (a) *off* and (b) *on* modes, along with (c) the corresponding current–voltage characteristics.

The region of the I_D–V_{DS} characteristics corresponding to the negligible I_D current of a MOSFET used as a switch in *off* mode ($V_{GS} < V_T$) is referred to as the *cutoff* region. The region of the I_D–V_{DS} characteristics corresponding to the use of a MOSFET as a switch in *on* mode is referred to as the *linear* region. The black lines in Fig. 8.3c illustrate the I_D–V_{DS} characteristics for gate voltages between $V_{GS1} = V_L$ and $V_{GS6} = V_H$. In the linear region, the slope of the I_D–V_{DS} characteristic is smaller for lower gate voltages than $V_{GS6} = V_H$. This is because there are fewer electrons in the channel for a lower gate voltage, meaning that the channel resistance is higher. Therefore, the MOSFET acts as a voltage-controlled resistor when biased in the linear region. As Fig. 8.3c illustrates, the linear region is limited to small drain-to-source voltages; the deviation from the linear dependence and the current saturation at higher drain-to-source voltages will be considered in Section 8.1.4.

Equations (3.4) and (3.17) can be used to relate the channel resistance to the technological and the geometric parameters:

$$R = \rho \frac{L}{W x_{ch}} = \frac{1}{q n \mu_n} \frac{L}{W x_{ch}} \tag{8.1}$$

where W is the channel width, L is the channel length, x_{ch} is the channel thickness, n is the average electron concentration in the channel, and μ_n is the average electron mobility in the channel. The average electron concentration depends on the gate voltage, and it can be related to the charge density per unit area in the inversion layer of a MOS capacitor in strong inversion (Q_I). The dependence of Q_I on the gate voltage is given by Eq. (7.20).

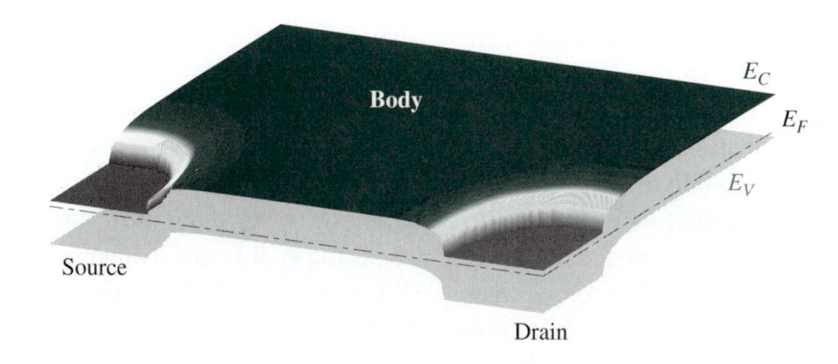

Figure 8.4 Two-dimensional energy-band diagram for the semiconductor part of an N-channel MOSFET in the flat-band condition. The two different colors in the conduction band indicate the two different types of doping: N type in the source and drain regions, and P type in the body. Darker colors indicate higher carrier concentration, and the nearly white areas indicate the depletion region.

Given that Q_I/q is the number of electrons per unit of channel area, Q_I/qx_{ch} is the number of electrons per unit volume, which is the electron concentration. Replacing the electron concentration n in Eq. (8.1) by $n = Q_I/qx_{ch}$ and using Eq. (7.20) for Q_I, the following equation for the resistance of the channel is obtained:

$$R = \frac{1}{\mu_n C_{ox}(V_{GS} - V_T)} \frac{L}{W} \tag{8.2}$$

The current through the channel—that is, the drain current—is then

$$I_D = \frac{V_{DS}}{R} = \frac{\mu_n C_{ox} W}{L}(V_{GS} - V_T)V_{DS} \tag{8.3}$$

This equation models both the linear dependence of I_D on V_{DS} (the output characteristics) and the linear dependence of I_D on $V_{GS} - V_T$ (the transfer characteristic). Of course, these linear models are applicable only to the linear region of MOSFET operation.

Energy Bands

As with the other devices, the energy-band model can be used to provide a deeper insight into the MOSFET operation.[1] The MOSFET effects have to be considered in two dimensions. The first dimension is needed to express the gate-voltage-related effects—that is, to follow the direction from the silicon surface into the silicon body. The second dimension is needed to express the drain-to-source voltage-related effects—that is, along the silicon surface. Because of that, two-dimensional energy-band diagrams will be used when we explain MOSFET operation.

A two-dimensional energy-band diagram for an N-channel MOSFET in the flat-band condition is shown in Fig. 8.4. Looking at the right-hand side—the cross section of the diagram that goes through the drain region and the body—we can identify the familiar energy-band diagram of the N–P junction formed by the N-type drain and the P-type body. Just as in Fig. 6.3, showing the energy-band diagram of a P–N junction, E_F is closer to E_C in the N-type drain region, whereas it is closer to E_V in the P-type body.

[1]This may not be very important for the basic operation of a MOSFET in the cutoff and linear regions, but the energy bands will appear as an irreplaceable tool for the forthcoming explanations of more advanced effects, such as the body effect, the current saturation due to channel pinch-off, and the drain-induced barrier lowering.

The front cross section of the two-dimensional energy-band diagram of Fig. 8.4 goes along the semiconductor surface. This is the energy-band diagram of the N–P–N structure formed by the N-type source, P-type body, and N-type drain at the semiconductor surface. This one-dimensional energy-band diagram can easily be deduced from the energy-band diagram of a P–N junction, given that the N–P–N structure can be split into N–P and P–N structures with the P regions merged with one another. The energy-band diagram of the N-type source and P-type body is constructed in the same way as the energy-band diagram of the N–P junction shown in Fig. 6.3. The remaining part (P-type body and N-type drain) is a mirror image of the source–body part, so the energy-band diagram is completed by the mirror image of the band diagram for the N–P part.

If a cross section is made in the central part in the direction from the surface into the semiconductor body, the energy bands would be flat, just as in Fig. 7.7d showing the flat-band condition of a MOS capacitor.

The energy barriers at the P–N junctions, which are equal to qV_{bi}, create potential wells that confine the source and the drain electrons in the N-type source and drain regions, respectively. A helpful analogy for the electrons in the source and the drain regions is to consider the source and the drain as "lakes" with water (or "electron fluid" if that seems closer to the reality). The electrons in the source and drain regions are separated by the potential barrier created by the P-type body (refer to Fig. 8.4).

When a positive voltage is applied between the drain and the grounded source and body contacts, the drain-to-body N–P junction is set at reverse bias. The quasi-Fermi level in the N-type drain is lowered by qV_{DS} with respect to the Fermi-level position in the grounded body and source regions, as shown in Fig. 8.5a. This lowering of the energy bands in the drain causes any electrons appearing in the surrounding depletion layer to roll down into the drain. This is certainly applicable to minority electrons in the P-type body that contribute to the reverse-bias current of the drain-to-body N–P junction. The electrons in the source are at a higher energetic position compared to the electrons in the drain, but they do not roll down into the drain because they are separated by the energy barrier of the P-type body. There is no significant current between the source and the drain, even for relatively high drain-to-source voltages—the MOSFET acts as a switch in *off* mode.

With a gate voltage higher than the threshold voltage, the surface potential is set to approximately $2\phi_F$ [Fig. 7.6 and Eq. (7.18)]. The surface potential change from 0 (the flat bands) to $2\phi_F$ corresponds to lowering of the energy bands at the surface by $2q\phi_F$, as illustrated in Fig. 8.5b. In other words, the barrier height between the source and the drain is lowered by $2q\phi_F$ to $qV_{bi} - 2q\phi_F$ at the surface of the P-type body. As Fig. 8.5b shows, the electrons from the source can now flow through the channel into the drain. In fact, the energy bands in the channel region correspond to the energy bands of a biased resistor, discussed in Chapter 3 and shown in Fig. 3.2. Therefore, the analogy of electron fluid together with the energy bands in Fig. 8.5b provide a clear illustration of the previously discussed linear dependence of I_D on V_{DS}: an increase in V_{DS} corresponds to an increase in the slope of the bands in the channel (qV_{DS}/L), which causes a proportional increase in the current flow.

CMOS Inverter

A typical application of MOSFETs as controlled switches is in complementary MOS (CMOS) digital circuits. The CMOS inverter, shown in Fig. 8.6a, is the representative

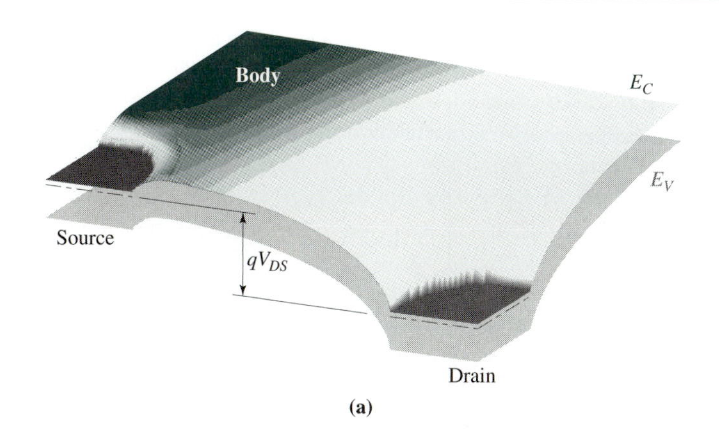

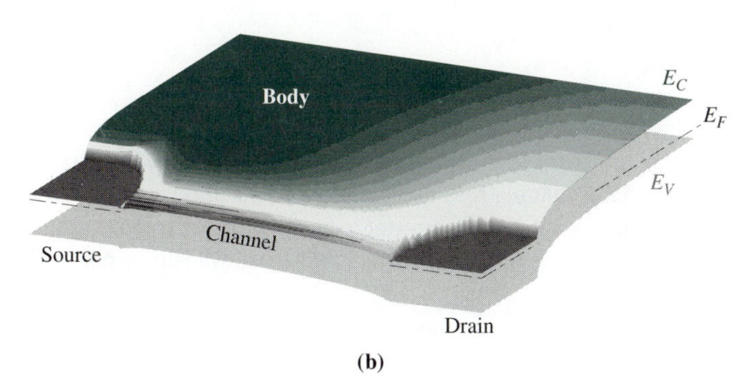

Figure 8.5 Two-dimensional energy-band diagrams for an N-channel MOSFET acting as a switch in (a) *off* mode and (b) *on* mode. The two colors in the conduction band indicate the concentrations of electrons and holes: darker colors correspond to higher carrier concentrations, whereas the nearly white areas indicate depleted regions. Note that the depleted regions correspond to the areas with sloped energy bands and therefore with existence of built-in and externally applied electric field.

CMOS circuit. CMOS digital circuits operate with two voltage levels, the low level being $V_L \approx 0$ and the high level being $V_H = V_+$. Digital circuits can be built using complementary pairs of N-channel and P-channel MOSFETs, also referred to as NMOS and PMOS transistors, respectively. This technology is known as the complementary MOS (CMOS) technology.

At high input voltage ($V_H \approx V_+$), the NMOS is in *on* mode because its gate-to-source voltage is higher than its threshold voltage. However, the PMOS is in *off* mode because its gate-to-source voltage ($v_H - V_+$) is close to zero (in absolute terms, it is smaller than the threshold voltage). As Fig. 8.6b shows, no DC current flows through the inverter in this logic state. The other logic state is for a low input voltage ($V_L \approx 0$). In this case, the PMOS is in *on* mode because its gate-to-source voltage is close to the negative value of V_+ ($V_{GS} = V_L - V_+ \approx -V_+$), which means that its absolute value is larger than the PMOS threshold voltage. In this case, however, the NMOS is in *off* mode because its gate-to-source voltage is at $V_L \approx 0$, thus below the threshold voltage. Again, no DC current flows through the inverter (Fig. 8.6b). Consequently, no static power dissipation is needed to maintain any logic state by a CMOS digital circuit.

With this conclusion, it is important to emphasize that CMOS circuits do dissipate power when the logic states are changed (dynamic power dissipation). This is because the outputs of CMOS logic cells, such as the inverter in Fig. 8.6, are loaded by the parasitic input capacitances of the connected logic cells (represented by C_L in Fig. 8.6a). To change

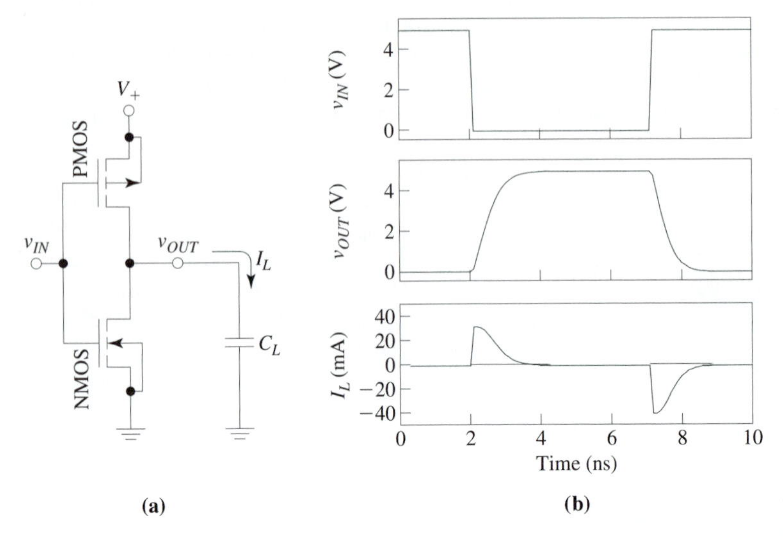

Figure 8.6 (a) The circuit of the CMOS inverter. (b) Typical input/output signals.

the output of the inverter of Fig. 8.6 from low to high level, the capacitor C_L has to be charged by current flowing through the channel resistance of the PMOS in *on* mode. When the output is changed from high to low level, the capacitor C_L has to be discharged through the channel resistance of the NMOS in *on* mode. Obviously, in these transition periods, some power is dissipated by the inverter.

The low-power-dissipation characteristic of the CMOS circuits has expanded the applications of digital circuits enormously, ranging from battery-supplied portable and entertainment electronics to computer applications as we now know them. The switching speed (maximum operating frequency) was initially a disadvantage of the CMOS circuits, but aggressive MOSFET dimension reduction has led to a dramatic increase in the speed. The dimension reduction has also enabled increased levels of integration, leading to powerful digital ICs. Dimension reduction, or so-called MOSFET downscaling, is described in more detail in Section 8.4. The CMOS technology has become the dominant electronics technology today.

8.1.3 The Threshold Voltage and the Body Effect

It has been assumed so far that the source and the body of the MOSFET are short-circuited ($V_{BS} = 0$). Although MOSFETs are very frequently used in this way, there are some applications where the body and the source cannot be short-circuited, or nonzero voltage is deliberately applied between the body and the source. In the case of N-channel MOSFETs, therefore P-type body and N-type source, the voltage applied between the body and the source should not be positive, because it would bias the body-to-source P–N junction in the forward mode, opening a current path between the source and the body contacts. Negative body-to-source voltages ($V_{BS} < 0$), or equivalently positive source-to-body voltages ($V_{SB} > 0$), set the body-to-source P–N junction in reverse-bias mode. This reverse bias increases the threshold voltage of the MOSFET, which is the effect referred to as the *body effect*.

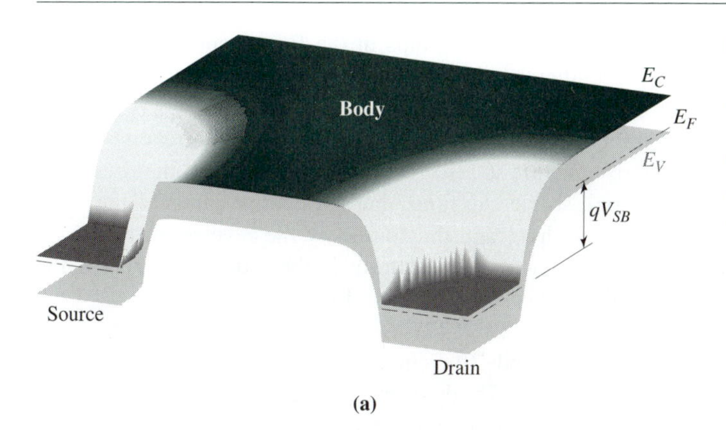

(a)

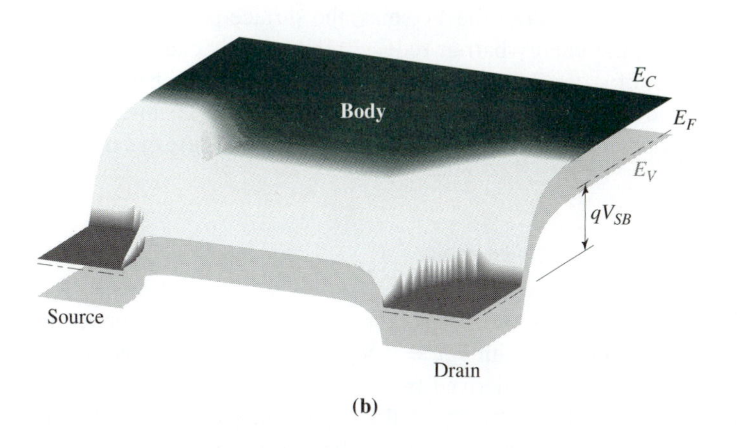

(b)

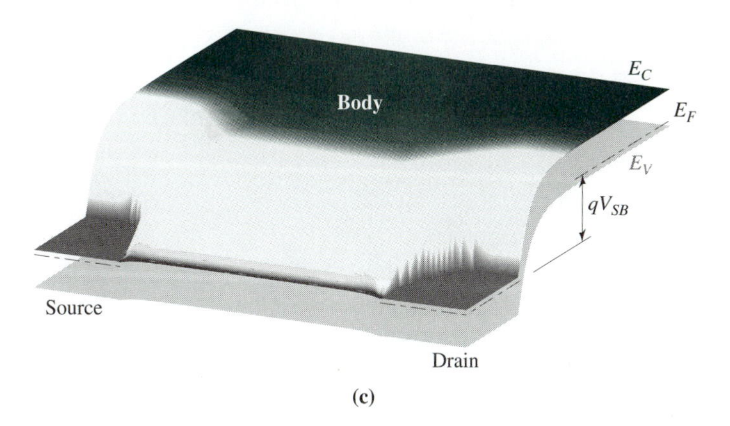

(c)

Figure 8.7 Illustration of the body effect. (a) V_{SB} voltage increases the barrier between the electrons in the source and the drain. (b) The surface potential of $2\phi_F$ does not reduce the barrier sufficiently for the electrons to be able to move into the channel. (c) The surface potential needed to form the channel is $2\phi_F + V_{SB}$.

The body effect may be hard to comprehend with cross-sectional diagrams in mind, but the two-dimensional energy-band diagrams illustrate this effect clearly. Figure 8.7 shows the two-dimensional band diagrams for three different values of the surface potential φ_s. In Fig. 8.7a, the surface potential in the P-type region is zero with respect to the electric potential in the neutral region of the body—flat bands in the P-type substrate along an imagined cross section that is perpendicular to the surface. The effect of the applied V_{SB} voltage can be seen at the cross section that goes through the drain and is visible at at the right-hand side of the diagram. This is the energy-band diagram of a reverse-biased N–P junction. The source-to-body junction is also reverse-biased. In fact, the drain and the source are at the same potential, so we can think that the energy bands of the body are lifted up with respect to both the drain and the source by the reverse bias $-qV_{BS}$. As a consequence, the energy barrier between the electrons in the source and the drain is increased by qV_{SB} when compared to the case of $V_{SB} = 0$ (Fig. 8.5a).

Figure 8.7b illustrates the case of $\varphi_s = 2\phi_F$. This is the value of the surface potential that corresponds to the strong-inversion mode (the channel of electrons is formed) in the case $V_{SB} = 0$. As can be seen from Fig. 8.7b, the surface potential of $\varphi_s = 2\phi_F$ does not reduce the energy barrier between the source and the drain sufficiently in the case of $V_{SB} > 0$. To compensate for the effect of the V_{SB} bias, the surface potential needs to be further increased (therefore, the energy barrier reduced), and that is exactly by V_{SB}. Figure 8.7c illustrates the case of $\varphi_s = 2\phi_F + V_{SB}$ when the energy barrier between the source and the drain is reduced sufficiently to enable electrons to form the channel.

In conclusion, the surface potential in strong inversion for the general case of a nonzero source-to-bulk voltage is

$$\varphi_s = 2\phi_F + V_{SB} \tag{8.4}$$

The threshold-voltage equation for a MOS capacitor, which is Eq. (7.19) derived in Section 7.2.2, can be used for a MOSFET with $V_{SB} = 0$ V. For the case of a nonzero V_{SB}, the threshold-voltage equation can also be derived from the capacitance–voltage–charge relationship in the depletion mode. If the density of the depletion-layer charge (Q_d) is divided by the gate-oxide capacitance per unit area (C_{ox}), the *effective* voltage across the capacitor dielectric is obtained:

$$\underbrace{\underbrace{V_{GS} + V_{SB} - V_{FB}}_{\text{effective gate-to-body voltage}} \quad -\varphi_s}_{\text{voltage across the gate oxide}} = \frac{Q_d}{C_{ox}} \qquad (\text{for } V_{FB} \leq V_{GS} \leq V_T) \tag{8.5}$$

Using $\varphi_s = 2\phi_F + V_{SB}$ and $V_{GS} = V_T$ as the values of the surface potential and the gate voltage at the onset of strong inversion, Eq. (8.5) becomes

$$V_T + V_{SB} - V_{FB} - (2\phi_F + V_{SB}) = \frac{Q_d}{C_{ox}} \tag{8.6}$$

which leads to the following equation for the threshold voltage:

$$V_T = V_{FB} + 2\phi_F + \frac{Q_d}{C_{ox}} \tag{8.7}$$

It is the depletion-layer charge (Q_d) that depends on V_{SB} in this equation. Given that the voltage across the depletion layer is $\varphi_s = 2\phi_F + V_{SB}$, it is this value of φ_s that should be used in Eq. (7.11). With this change, Eqs. (7.10) and (7.11) lead to the following equation for Q_d:

$$Q_d = \sqrt{2\varepsilon_s q N_A (2\phi_F + V_{SB})} \tag{8.8}$$

and the following equation for the threshold voltage:

$$V_T = V_{FB} + 2\phi_F + \gamma\sqrt{2\phi_F + V_{SB}} \tag{8.9}$$

The threshold-voltage increase ΔV_T, caused by the voltage V_{SB} is

$$\Delta V_T = V_T(V_{SB}) - V_T(V_{SB} = 0) = \gamma\left(\sqrt{2\phi_F + V_{SB}} - \sqrt{2\phi_F}\right) \tag{8.10}$$

For a MOSFET on an N-type substrate (a P-channel MOSFET), Eq. (7.21) for the case of $V_{BS} = 0$ can be extended in a similar way to obtain the threshold-voltage equation that includes the effects of body bias ($V_{BS} > 0$ in this case):

$$V_T = V_{FB} - 2|\phi_F| - \gamma\sqrt{2|\phi_F| + V_{BS}} \tag{8.11}$$

The body factor γ is given by Eqs. (7.17) and (7.22) for P-type and N-type substrates, respectively.

EXAMPLE 8.1 Threshold Voltage with $V_{SB} = 0$ and $V_{SB} \neq 0$ (Body Effect)

An N-channel MOSFET with N^+-type polysilicon gate has oxide thickness of 10 nm and substrate doping $N_A = 5 \times 10^{16}$ cm^{-3}. The oxide charge density is $N_{oc} = 5 \times 10^{10}$ cm^{-2}. Find the threshold voltage if the body is biased at 0 V and −5 V, respectively. The following constants are known: the thermal voltage $V_t = 0.026$ V, the oxide permittivity $\varepsilon_{ox} = 3.9 \times 8.85 \times 10^{-12}$ F/m, the silicon permittivity $\varepsilon_s = 11.8 \times 8.85 \times 10^{-12}$ F/m, the intrinsic carrier concentration $n_i = 1.02 \times 10^{10}$ cm^{-3}, and the silicon energy gap $E_g = 1.12$ eV.

SOLUTION

To use Eq. (8.9) to calculate the threshold voltage, the Fermi potential ϕ_F, the flat-band voltage V_{FB}, and the body factor γ should be obtained first. According to Eq. (2.88), the Fermi potential is

$$\phi_F = +V_t \ln\frac{N_A}{n_i} = 0.401 \text{ V}$$

The flat-band voltage equation [Eq. (7.35)] shows that the work-function difference $q\phi_{ms}$ and the gate-oxide capacitance per unit area C_{ox} are needed as well. The work-function difference is given by Eq. (7.34). In the N^+ type gate, the Fermi level is very close to the bottom of the conduction band, which means that the work function of the gate $q\phi_m$ is approximately equal to the electron affinity $q\chi_s$ (Section 7.2.3); therefore,

$$\phi_{ms} = -\frac{E_g}{2q} - \phi_F = -0.961 \text{ V}$$

Because the gate-oxide capacitance per unit area is

$$C_{ox} = \varepsilon_{ox}/t_{ox} = 3.45 \times 10^{-3} \text{ F/m}^2$$

the flat-band voltage is obtained as

$$V_{FB} = \phi_{ms} - \frac{qN_{oc}}{C_{ox}} = -0.961 - \frac{1.6 \times 10^{-19} \times 5 \times 10^{14}}{3.45 \times 10^{-3}} = -0.984 \text{ V}$$

The body factor is given by Eq. (7.17):

$$\begin{aligned}\gamma &= \sqrt{2\varepsilon_s q N_A}/C_{ox} \\ &= \sqrt{2 \times 11.8 \times 8.85 \times 10^{-12} \times 1.6 \times 10^{-19} \times 5 \times 10^{22}}/3.45 \times 10^{-3} \\ &= 0.375 \text{ V}^{1/2}\end{aligned}$$

The threshold voltage at $V_{SB} = 0$ V is calculated as

$$V_T(0) = V_{FB} + 2\phi_F + \gamma\sqrt{2\phi_F + V_{SB}} = -0.984 + 2 \times 0.401 + 0.375\sqrt{2 \times 0.401 + 0} = 0.15 \text{ V}$$

To calculate the threshold voltage at $V_{SB} = 5$ V, find the threshold-voltage difference ΔV_T using Eq. (8.10):

$$\Delta V_T = 0.375\left(\sqrt{2 \times 0.401 + 5} - \sqrt{2 \times 0.401}\right) = 0.57 \text{ V}$$

Therefore, $V_T(5V) = V_T(0) + \Delta V_T = 0.72$ V.

8.1.4 MOSFET as a Voltage-Controlled Current Source: Mechanisms of Current Saturation

When a MOSFET is operated as a switch in *on* mode ($V_{GS} > V_T$ and $V_{DS} < V_{DSsat}$ for an N-channel MOSFET), the normal electric field from the gate voltage V_{GS} holds the electrons in the inversion-layer channel while the lateral electrical field due to the drain-to-source voltage V_{DS} rolls them into the drain. The channel of electrons extends all the way from the source to the drain, the resistance of which determines the slope of the linear I_D–V_{DS} characteristic.

To use the MOSFET as a constant-current source, I_D should become independent of V_{DS}. This happens at larger drain-to-source voltages ($V_{DS} > V_{DSsat}$), an effect referred to as current saturation. There are two different mechanisms that can cause drain-current saturation in MOSFETs. These two mechanisms are considered in the following text.

Channel Pinch-off

As the drain-to-source voltage V_{DS} is increased, the lateral electric field in the channel is increased as well and may become stronger than the vertical electric field due to the gate voltage. This would first happen at the drain end of the channel. In this situation, the vertical field is unable to keep the electrons at the drain end of the channel as the stronger lateral field sweeps them into the drain. The channel is pinched off at the drain end. The drain-to-source voltage at which this happens is called the *saturation voltage,* V_{DSsat}.

An increase of V_{DS} beyond V_{DSsat} expands the region in which the lateral field is stronger than the vertical field, effectively moving the pinch-off point closer to the source. A MOSFET with the pinch-off point between the source and the drain is illustrated in Fig. 8.8a. The region created between the pinch-off point and the drain is basically the depletion layer at the reverse-biased drain–substrate junction. Note that we are considering the surface area of the junction, which is influenced by the gate field. Consequently, the surface region of the P–N junction is not in the reverse-bias mode until the drain voltage reaches V_{DSsat}. This is different from the bulk region of the junction, which is in the reverse-bias mode for any positive V_{DS} voltage.

The voltage across the depletion region at the surface (the reverse bias) is $V_{DS} - V_{DSsat}$, which is the voltage increase beyond V_{DSsat}. The remaining part of the drain-to-source voltage, which is V_{DSsat}, drops between the pinch-off point and the source. In this region the vertical field is stronger than the lateral field, and the inversion layer (channel of electrons) still exists. It is in fact this part of the source-to-drain region that determines the value of the drain current. *Given that the voltage across the channel of electrons is fixed to V_{DSsat} for $V_{DS} > V_{DSsat}$, the drain current remains fixed to the value corresponding to V_{DSsat}.* This effect is called drain-current saturation, and the $V_{DS} > V_{DSsat}$ region of the MOSFET operation is referred to as the *saturation region.*

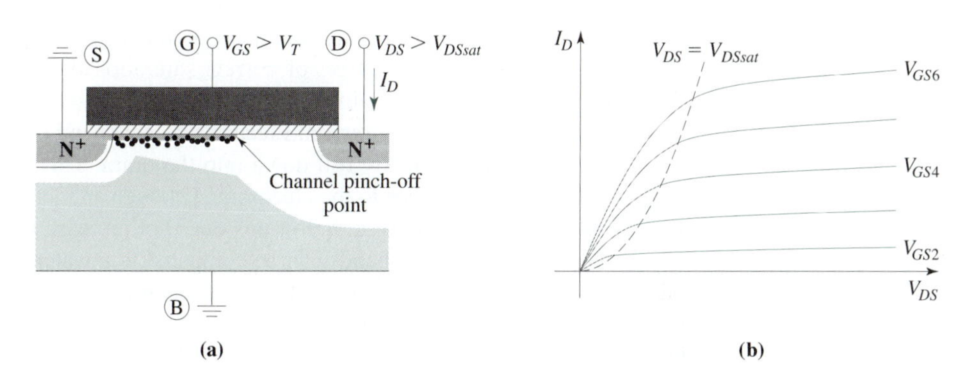

Figure 8.8 (a) Cross section of a MOSFET in the saturation region. (b) The corresponding I_D–V_{DS} characteristics.

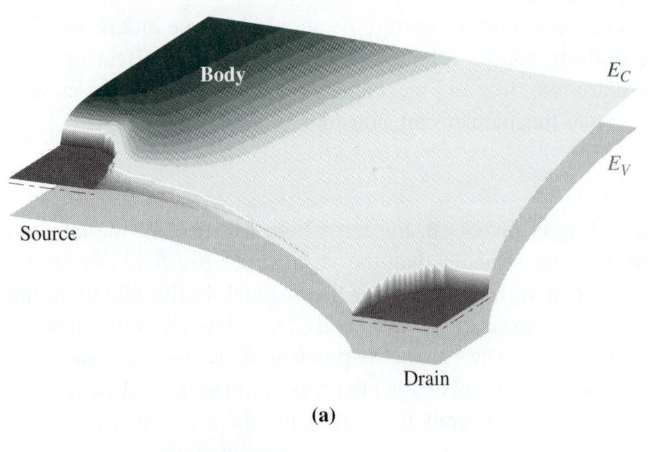

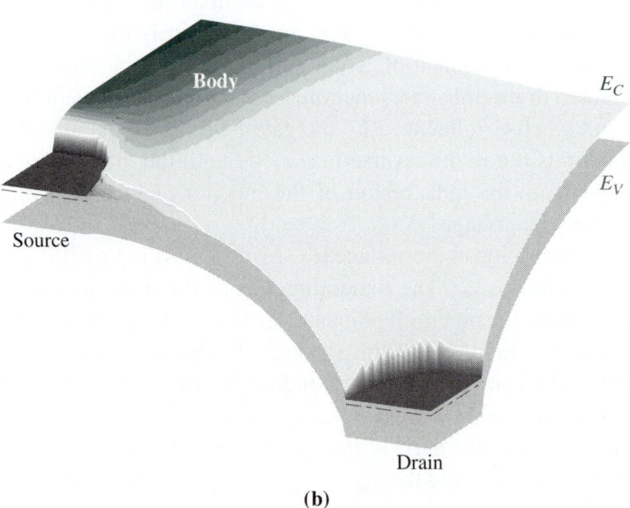

Figure 8.9 Two-dimensional energy-band diagrams for an N-channel MOSFET in saturation due to channel pinch-off. A comparison of the smaller V_{DS} bias in (*a*) to the larger V_{DS} value in (*b*) shows that the channel is shortened by the increased drain-to-source bias, but the concentration of electrons in the channel is not changed. As in the waterfall analogy, the drain current is limited by the concentration of the electrons in the channel and not by the height of the fall.

The two-dimensional energy-band diagrams of the MOSFET, shown in Fig. 8.9, provide a clearer insight into the effect of current saturation due to the channel pinch-off. The energy bands are very steep in the depletion region, which represents the situation of a very strong lateral field in this region. Electrons do not spend much time on this very steep part of E_C; they very quickly roll down into the drain. This part of the source-to-drain region offers little resistance to the electrons. Although an increase of V_{DS} continues to lower E_C in the drain region (Fig. 8.9b), this does not increase the drain current. The electrons in this shape of energy bands can be compared to a waterfall: the water current depends on the quantity of water before the fall (the channel) and not on the height of the waterfall ($qV_{DS} - qV_{DSsat}$).

In summary,

1. The depletion region has little influence on the drain current.
2. The value of the drain current is limited by the number of electrons that appear at the edge of the depletion region (the pinch-off point) per unit time. In the waterfall

analogy, the current of water is determined by the water flow before the fall and cannot be increased by increasing the height of the fall.

3. The number of the electrons in the channel, and therefore the number of electrons that hit the pinch-off point per unit time, is controlled by the gate voltage and not the drain voltage.
4. As a consequence, the drain current is controlled by the gate voltage and is independent of the drain voltage—the MOSFET acts as a voltage-controlled current source.
5. As it should be for a current source providing a current that does not depend on the voltage across the current source, the I_D–V_{DS} characteristics of a MOSFET in saturation ($V_{DS} > V_{DSsat}$) are horizontal. Figure 8.8b shows a set of horizontal lines because the saturation current I_D depends on the gate voltage—these are the current–voltage characteristics of a *voltage-controlled* current source.

Drift Velocity Saturation

Minimum lateral and vertical dimensions of MOSFETs are continuously being reduced in order to increase the density and the speed of modern ICs. These MOSFETs are referred to as *short-channel MOSFETs*. The operating V_{DS} voltages cannot be proportionally reduced because the maximum operating voltage has to be kept well above the MOSFET threshold voltage. As a consequence, short-channel MOSFETs operate with significantly increased lateral and vertical electric fields in the channel. Although the relative relationship between the lateral and the vertical electric fields is roughly maintained, these MOSFETs typically exhibit a different type of drain-current saturation. It happens that the current saturates at a drain-to-source voltage smaller than the voltage that would cause channel pinch-off at the drain end.

To explain this effect, refer to Eq. (3.21) in Section 3.3, which shows that the current is directly proportional to the drift velocity of the carriers. As explained in Section 3.3, the drift velocity follows a linear dependence on the lateral electric field in the channel up to a certain level, and saturates if the field is increased beyond that level (Fig.3.6). The lateral electric field in short-channel MOSFETs is stronger than the critical velocity saturation value, while not being stronger than the vertical electric field at the drain end of the channel (the channel is not pinched off). Nonetheless, the drain current saturates due to the carrier velocity saturation in the channel.

Although this is a different mechanism of current saturation than the pinch-off, the MOSFET can equally well be used as a voltage-controlled current source.

Related to modeling of this saturation mechanism, it is worth considering the following issue. If the channel is not pinched off, it can be modeled as a resistor between the source and the drain. If that is so, it could be expected that the current should not saturate but increase linearly with V_{DS}, according to Ohm's law! However, the velocity saturation and the related current saturation are real effects. They do happen in semiconductor devices in practice, although they cannot be observed in metals (too-high carrier concentration makes it impossible to reach high electric fields). If we wish to continue using Ohm's law, we have to alter appropriately the value of the mobility in Eq. (3.22) so that it models this effect properly. The next section, on MOSFET modeling, will describe the mobility models used in SPICE to account for this effect.

8.2 PRINCIPAL CURRENT–VOLTAGE CHARACTERISTICS AND EQUATIONS

Mathematical equations are needed to calculate the drain current (I_D) for a set of applied voltages, V_{GS}, V_{DS}, and V_{SB}. In general, $I_D = f(V_{GS}, V_{DS}, V_{SB})$, where the function f depends on a number of geometrical and technological parameters. For convenience, a set of mathematical equations is typically used to present the function f. This set of mathematical equations is referred to as the MOSFET model. There are a large number of models that differ in terms of their accuracy and complexity. Additionally, the models of the principal effects (first-order models) are modified in virtually countless ways to include observed second-order effects. The consideration of MOSFET models, presented in this section, is limited to a selection of SPICE equations (even in SPICE, there are three basic model options, referred to as LEVEL 1, LEVEL 2, and LEVEL 3, plus additional more sophisticated or specific models).

In the most important practical models, the function f is different for the triode, saturation, and subthreshold regions of MOSFET operation. The output characteristics shown in Fig. 8.10a illustrate that I_D increases with V_{DS} in the triode region, whereas it is almost independent of V_{DS} in the saturation region. The transfer characteristic shown in Fig. 8.10b shows that $I_D \approx 0$ in the subthreshold region ($V_{GS} < V_T$). Accordingly, the general form of these first-order models is as follows:

$$I_D = \begin{cases} 0 & \text{subthreshold or cutoff } (V_{GS} < V_T) \\ f(V_{GS}, V_{SB}, V_{DS}) & \text{triode region } (0 \le V_{DS} \le V_{DSsat}) \\ f(V_{GS}, V_{SB}, V_{DSsat}) & \text{saturation } (V_{DS} \ge V_{DSsat}) \end{cases} \quad (8.12)$$

Basically, the function f is derived for the triode region, and its prediction for the current at the onset of saturation ($V_{DS} = V_{DSsat}$) is used for the saturation region (the current in the subthreshold region is assumed to be zero).

Different functions $f(V_{GS}, V_{SB}, V_{DS})$ are used in LEVELS 1, 2, and 3 of the principal SPICE model. The SPICE LEVEL 1 model is the simplest MOSFET model that is typically used in circuit design books. The SPICE LEVEL 2 model is a physically based model presented in a number of semiconductor books as the MOSFET model. The LEVEL 2 model frequently appears as unnecessarily complex, whereas the LEVEL 1 model is rarely

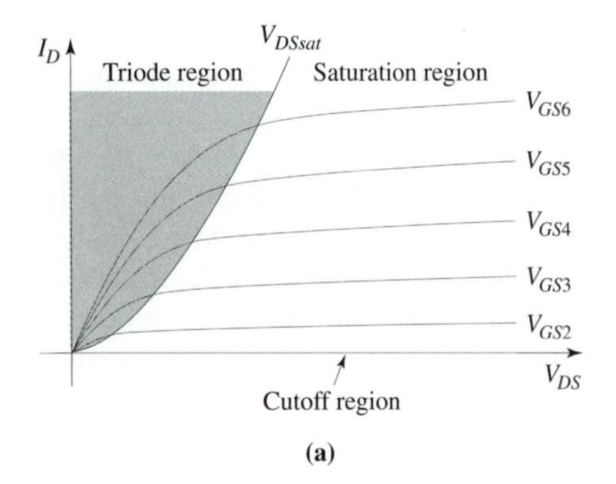

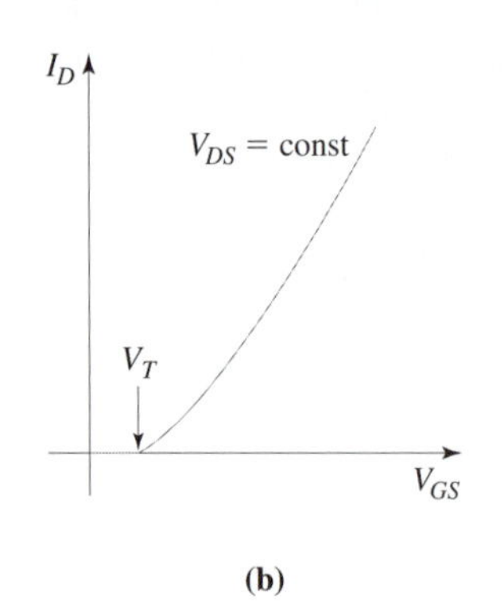

Figure 8.10 (a) Output and (b) transfer characteristics of a MOSFET.

accurate enough. The SPICE LEVEL 3 model is almost as simple as the LEVEL 1 model (the equations resemble the LEVEL 1 equations) and is almost as accurate as the LEVEL 2 model. Technically, the LEVEL 3 model is the best choice. Moreover, the equation of the LEVEL 3 model can be obtained by simplifying the equation of the LEVEL 2 model, and the LEVEL 3 equation can be reduced to the equation of the LEVEL 1 model. To show these relationships between different models, all three first-order equations are presented in this section. The second-order effects included in SPICE are presented in Section 8.3 for the LEVEL 3 model.

8.2.1 SPICE LEVEL 1 Model

It is assumed in the SPICE models that all the majority carriers flowing through the channel terminate at the drain. In the case of N-channel MOSFETs, this means that the terminal current I_D is equal to the current of the electrons in the channel. The electron flow in the channel is caused by the electric field due to the drain-to-source bias V_{DS}. This current mechanism, the *drift current,* is modeled by Ohm's law [Eq. (3.12)]:

$$j = \sigma E \tag{8.13}$$

where j is the current density in A/m^2, E is the electric field, and σ is the conductivity. The conductivity can be expressed in terms of the electron concentration n and electron mobility μ_0 [refer to Eq. (3.17)], which leads to

$$j = q\mu_0 n E \tag{8.14}$$

It appears that the current density of the electrons in the channel j, given in units of A/m^2, can simply be multiplied by the channel cross-sectional area to convert it into the drain current I_D, which is expressed in A. As discussed in Section 3.2.2, this would implicitly assume a uniform current density. If the current density changes, it means that the average value of the current density is actually taken. If this approach is taken, all the other differential quantities in Eq. (8.14) should be represented by their average values. Denoting the channel cross section by $x_{ch}W$ and expressing the average value of the electric field by V_{DS}/L_{eff},

$$\underbrace{j\ x_{ch}W}_{I_D\ [A]} = \mu_0\ \underbrace{qnx_{ch}}_{\overline{Q}_I\ [C/m^2]}\ W\ \underbrace{E}_{V_{DS}/L_{eff}} \tag{8.15}$$

the following equation is obtained:

$$I_D = \frac{\mu_0 W}{L_{eff}}\overline{Q}_I V_{DS} \tag{8.16}$$

$\overline{Q}_I$ in Eqs. (8.15) and (8.16) is the average value of the inversion-layer charge density, expressed in C/m^2. Small drain-to-source voltages ($V_{DS} \ll V_{GS}$) do not disturb the channel, in which case $\overline{Q}_I$ is very close to the inversion-layer charge density of the MOS structure [Eq. (7.20)],

$$\overline{Q}_I \approx Q_I = (V_{GS} - V_T)C_{ox} \qquad \text{for } (V_{GS} \geq V_T) \tag{8.17}$$

Using this equation for $\overline{Q}_I$ in Eq. (8.16) leads to the drain-current equation for the linear region [Eq. (8.3)].

To model the drain current in the whole triode region (not its linear part only), the influence of V_{DS} voltage on the inversion-layer charge density cannot be neglected. As V_{DS} voltage is increased, the drain end of the inversion layer is being gradually depleted, until it is completely pinched off at the drain end, which is the point of drain-current saturation ($V_{DS} = V_{DSsat}$). Therefore, the V_{DS} voltage causes a nonuniform distribution of the inversion-layer charge density along the channel. To include this effect, the MOS capacitor equation for Q_I [Eq. (7.20)] can be modified as follows:

$$Q_I(v) = (V_{GS} - V_T - v)C_{ox} \tag{8.18}$$

where v is electric potential whose value changes from 0 at the source end of the channel to V_{DS} at the drain end of the channel. Therefore, Q_I changes from $(V_{GS} - V_T)C_{ox}$ at the source end of the channel (no influence from the drain bias) to $(V_{GS} - V_T - V_{DS})C_{ox}$ at the drain end of the channel. The average value of the inversion-layer charge density is then

$$\begin{aligned}\overline{Q}_I &= \frac{1}{V_{DS}}\int_0^{V_{DS}} Q_I(v)dv \\ &= \frac{C_{ox}}{V_{DS}}\int_0^{V_{DS}} (V_{GS} - V_T - v)dv \\ &= \left(V_{GS} - V_T - \frac{V_{DS}}{2}\right) C_{ox}\end{aligned} \tag{8.19}$$

With this equation for $\overline{Q}_I$, Eq. (8.16) becomes

$$I_D = \beta\left[(V_{GS} - V_T)V_{DS} - \frac{V_{DS}^2}{2}\right] \tag{8.20}$$

where β is called the gain factor and is defined as

$$\beta = \mu_0 C_{ox}\frac{W}{L_{eff}} \tag{8.21}$$

The gain factor involves the two geometric or layout-design variables (the ratio of channel width to channel length W/L_{eff}) and two technological parameters, μ_0 and C_{ox}. In SPICE, the technological parameters are frequently grouped together to specify them by a single so-called transconductance parameter KP:

$$\beta = \mu_0 C_{ox}(W/L_{eff}) = KP(W/L_{eff}) \tag{8.22}$$

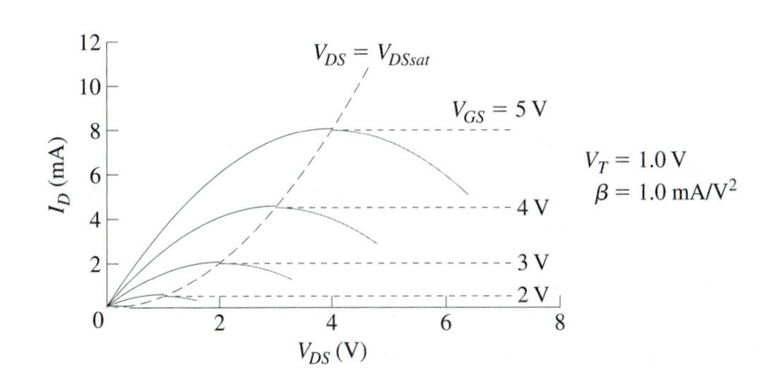

Figure 8.11 Output characteristics corresponding to SPICE LEVEL 1 model. Solid lines, Eq. (8.20); dashed lines, saturation current.

The unit for both the gain factor and the transconductance parameter is A/V^2.

Equation (8.20) is the principal SPICE LEVEL 1 model in the triode region. Importantly, this equation cannot be used in the saturation region. The plots of Eq. (8.21), given in Fig. 8.11, show that it predicts a current reduction with V_{DS} increase in the saturation region. This result is due to the integration of negative values of $(V_{GS} - V_T - v)C_{ox}$, appearing between the channel pinch-off point and the drain ($v > V_{GS} - V_T$ in this region). However, the negative values of $(V_{GS} - V_T - v)C_{ox}$ should not be integrated when one is deriving the drain current because there is no current-reduction mechanism in the depletion layer between the pinch-off point and the drain. Accordingly, the derivation leading to Eq. (8.20) for the drain current is limited to nonnegative values of $Q_I(v) = (V_{GS} - V_T - v)C_{ox}$.

The condition of zero inversion-layer charge at the drain end of the channel, $(V_{GS} - V_T - V_{DS})C_{ox} = 0$, corresponds to the case of the drain end of the channel being pinched off. This is the onset of saturation region. From this condition, the drain-to-source voltage at which the MOSFET enters saturation is obtained:

$$V_{DSsat} = V_{GS} - V_T \tag{8.23}$$

Another way of obtaining this saturation voltage equation is by determining the V_{DS} voltage that correspond to the maximum of I_D given by Eq. (8.20):

$$\frac{\partial I_D}{\partial V_{DS}} = 0 \Rightarrow V_{DSsat} = V_{GS} - V_T \tag{8.24}$$

Thus, the triode-region current I_D reaches maximum for $V_{DS} = V_{DSsat}$. This maximum current is taken as the MOSFET current in saturation ($V_{DS} \geq V_{DSsat}$). Putting Eq. (8.23) for V_{DSsat} in place of V_{DS} in the current equation (8.20), the saturation current is obtained as

$$I_{Dsat} = I_D(V_{DSsat}) = \frac{\beta}{2}(V_{GS} - V_T)^2 \tag{8.25}$$

Therefore, the saturation current in the principal model is independent of V_{DS}, and it depends parabolically on $V_{GS} - V_T$.

The SPICE LEVEL 1 MOSFET model in the above-threshold region can be summarized as follows:

$$I_D = \begin{cases} \beta\left[(V_{GS} - V_T)V_{DS} - V_{DS}^2/2\right] & \text{if } 0 \leq V_{DS} < V_{DSsat} \\ \frac{\beta}{2}(V_{GS} - V_T)^2 & \text{if } V_{DS} \geq V_{DSsat} \end{cases} \tag{8.26}$$

where the gain factor β is given by Eq. (8.21) and the drain saturation voltage V_{DSsat} is given by Eq. (8.23). The difference between the triode-region equation and the simplest model for the linear region [Eq. (8.3)] is due to the term $V_{DS}^2/2$. This equation is the simplest model that accounts for the reduction of the inversion-layer charge when V_{DS} is increased toward V_{DSsat}. This simple approach does not fully incorporate the effects of the reverse-biased drain-to-body junction, and as a result the simple LEVEL 1 model overestimates the drain current.

8.2.2 SPICE LEVEL 2 Model

The drain-current equation used as the basic SPICE LEVEL 2 model is quite different from the LEVEL 1 model. The difference emerges from a different approach used to incorporate the effects of the reverse-biased drain-to-body junction on Q_I reduction toward the drain end of the channel. This approach is based on the model for the threshold-voltage increase caused by the body effect. As described in Section 8.1.3, the increase of strong-inversion surface potential due to the reverse bias of the body-to-source and body-to-drain junctions leads to a threshold-voltage increase, which is known as the *body effect*. As shown by Eq. (8.4), the surface potential is increased from the usual $2\phi_F$ to $2\phi_F + V_{SB}$ to include the body bias. In Fig. 8.12, this is the situation at the source end of the channel. At the drain end of the channel, the surface potential in strong inversion is increased further to $2\phi_F + V_{SB} + V_{DS}$, because the complete reverse-bias voltage across the drain–body P–N junction is $V_{DS} + V_{SB}$. Consequently, the threshold voltage at the drain end of the channel is larger than at the source end (stronger body effect due to the larger body bias, $V_{DS} + V_{SB}$).

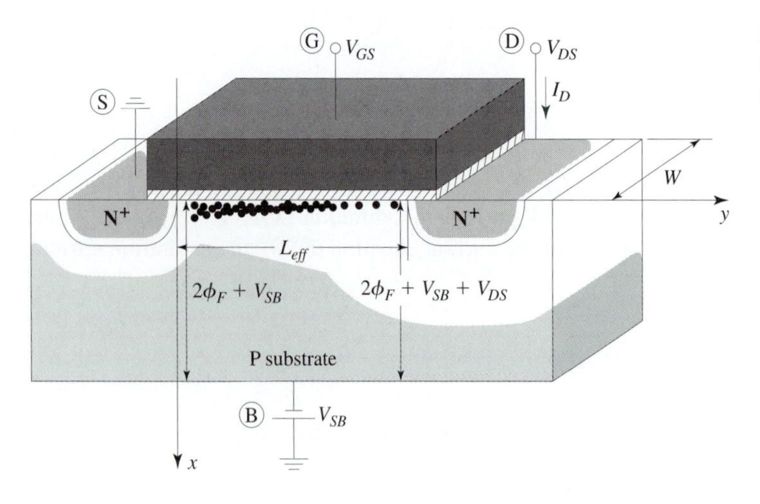

Figure 8.12 N-channel MOSFET diagram, indicating the surface potential at the source and drain ends of the channel.

According to Eq. (8.17), the larger threshold voltage causes smaller inversion-layer charge density Q_I at the drain end of the channel.

The varying threshold voltage along the channel can uniquely be expressed in terms of the surface potential by generalizing Eq. (8.9) in the following way:

$$V_T = V_{FB} - V_{SB} + \underbrace{2\phi_F + V_{SB}}_{\varphi_s} + \gamma\sqrt{\underbrace{2\phi_F + V_{SB}}_{\varphi_s}} \tag{8.27}$$

As long as φ_s is fixed to $2\phi_F + V_{SB}$, this threshold-voltage equation is equivalent to Eq. (8.9). However, if φ_s is allowed to take any value between $2\phi_F + V_{SB}$ (the surface potential at the source end) and $2\phi_F + V_{SB} + V_{DS}$ (the surface potential at the drain end of the channel), the threshold voltage becomes a differential quantity that varies along the channel due to the surface-potential variation:

$$v_T(\varphi_s) = V_{FB} - V_{SB} + \varphi_s + \gamma\sqrt{\varphi_s} \tag{8.28}$$

In principle, the average inversion-layer charge density, to be used in Eq. (8.16), should be obtained through the following averaging formula:

$$\overline{Q}_I = \frac{1}{L_{eff}} \int_0^{L_{eff}} Q_I(y)\, dy = \frac{1}{L_{eff}} \int_0^{L_{eff}} [V_{GS} - v_T(\varphi_s)]\, C_{ox}\, dy \tag{8.29}$$

However, the integration with respect to y (the space coordinate along the channel) is not possible as the surface-potential dependence, and for that matter the threshold-voltage dependence on y is not established. Instead, the averaging is performed as follows:

$$\begin{aligned}\overline{Q}_I &= \frac{1}{(2\phi_F + V_{SB} + V_{DS}) - (2\phi_F + V_{SB})} \int_{2\phi_F+V_{SB}}^{2\phi_F+V_{SB}+V_{DS}} Q_I(\varphi_s)\, d\varphi_s \\ &= \frac{1}{V_{DS}} \int_{2\phi_F+V_{SB}}^{2\phi_F+V_{SB}+V_{DS}} [V_{GS} - v_T(\varphi_s)]\, C_{ox}\, d\varphi_s\end{aligned} \tag{8.30}$$

Replacing $v_T(\varphi_s)$ from Eq. (8.28) and solving the integral in Eq. (8.30), the following result is obtained:

$$\begin{aligned}\overline{Q}_I =& \frac{C_{ox}}{V_{DS}} \left\{ \left(V_{GS} - V_{FB} - 2\phi_F - \frac{V_{DS}}{2} \right) V_{DS} \right. \\ & \left. - \frac{2}{3}\gamma \left[(2\phi_F + V_{SB} + V_{DS})^{3/2} - (2\phi_F + V_{SB})^{3/2} \right] \right\}\end{aligned} \tag{8.31}$$

Inserting the obtained equation for the average inversion-layer charge density $\overline{Q}_I$ into Eq. (8.16), the drain current is obtained as

$$I_D = \beta \left\{ \left(V_{GS} - V_{FB} - 2\phi_F - \frac{V_{DS}}{2} \right) V_{DS} - \frac{2}{3}\gamma \left[(2\phi_F + V_{SB} + V_{DS})^{3/2} - (2\phi_F + V_{SB})^{3/2} \right] \right\} \tag{8.32}$$

Equation (8.32) is the SPICE LEVEL 2 model in the triode region. Analogously to the LEVEL 1 equation (Fig. 8.11), the $I_D(V_{DS})$ dependence reaches maximum at the saturation voltage V_{DSsat}. Using the condition that the first derivative of $I_{DS}(V_{DS})$ is zero at $V_{DS} = V_{DSsat}$ (the maximum of the current I_D), the saturation voltage is obtained as

$$\frac{\partial I_D}{\partial V_{DS}} = 0 \Rightarrow V_{DSsat} = V_{GS} - V_{FB} - 2\phi_F - \frac{\gamma^2}{2}\left[\sqrt{1 + \frac{4}{\gamma^2}(V_{GS} - V_{FB} + V_{SB})} - 1 \right] \tag{8.33}$$

The SPICE LEVEL 2 model works in the following way: (1) the saturation drain voltage V_{DSsat} is calculated first, using Eq. (8.33); (2) if $V_{DS} < V_{DSsat}$, V_{DS} itself is used in Eq. (8.32) to calculate the current; (3) if $V_{DS} \geq V_{DSsat}$, V_{DSsat} is used in Eq. (8.32) to calculate the current.

As can be seen from Eq. (8.32), the threshold voltage does not appear as a parameter of the LEVEL 2 model. This is not helpful, given that the threshold voltage is the most important parameter of a MOSFET used as a voltage-controlled switch.

8.2.3 SPICE LEVEL 3 Model: Principal Effects

The LEVEL 3 model can be obtained by simplifying the equation of the LEVEL 2 model. Equation (8.32) is approximated by the first three terms of the Taylor series:

$$I_D \approx I_D(0) + I'_D(0)V_{DS} + I''_D(0)\frac{V_{DS}^2}{2} \tag{8.34}$$

The first three terms of the Taylor series are taken to achieve a good compromise between accuracy and simplicity. Taking only the first two terms would be even simpler; however, this would lead to the linear $I_{DS}(V_{DS})$ dependence, which is obviously not good enough as a MOSFET model in the complete triode region. $I_D(0)$, $I'_D(0)$, and $I''_D(0)$ are obtained as follows:

$$I_D(0) = 0$$

$$I_D' = \frac{\partial I_D}{\partial V_{DS}} = \beta(V_{GS} - V_{FB} - 2\phi_F - V_{DS}) - \beta\gamma(2\phi_F + V_{SB} + V_{DS})^{1/2}$$

$$I_D'(0) = \beta(V_{GS} - V_{FB} - 2\phi_F) - \beta\gamma(2\phi_F + V_{SB})^{1/2} \tag{8.35}$$

$$I_D'' = \frac{\partial^2 I_D}{\partial V_{DS}^2} = \frac{\partial}{\partial V_{DS}}\left(\frac{\partial I_D}{\partial V_{DS}}\right) = -\beta - \frac{1}{2}\beta\gamma(2\phi_F + V_{SB} + V_{DS})^{-1/2}$$

$$I_D''(0) = -\beta - \frac{1}{2}\beta\gamma(2\phi_F + V_{SB})^{-1/2}$$

Putting the obtained $I_D(0)$, $I_D'(0)$, and $I_D''(0)$ into Eq. (8.34), the following drain-current equation is obtained:

$$I_D = \beta\left[V_{GS} - \underbrace{(V_{FB} + 2\phi_F + \gamma\sqrt{2\phi_F + V_{SB}})}_{V_T} - \frac{1}{2}\left(1 + \overbrace{\frac{\gamma}{2\sqrt{2\phi_F + V_{SB}}}}^{F_B}\right)V_{DS}\right]V_{DS} \tag{8.36}$$

It can be seen that the threshold voltage, as defined by Eq. (3.36), appears in the drain-current equation. Also, a new factor is introduced to additionally simplify this equation. This factor is F_B and is defined by

$$F_B = \frac{\gamma}{2\sqrt{2\phi_F + V_{SB}}} \tag{8.37}$$

Because the original LEVEL 2 Eq. (8.36) is valid only in the triode region ($0 \le V_{DS} \le V_{DSsat}$), the simplified LEVEL 3 Eq. (3.36) is also valid only in the triode region. The saturation drain voltage V_{DSsat} can be determined in analogous way:

$$\frac{\partial I_D}{\partial V_{DS}} = 0 \;\Rightarrow\; V_{DSsat} = \frac{V_{GS} - V_T}{1 + F_B} \tag{8.38}$$

The triode-region current I_D reaches maximum for $V_{DS} = V_{DSsat}$. This maximum current is considered as the MOSFET saturation current I_{Dsat}. Putting Eq. (8.38) for V_{DSsat} in place of V_{DS} in the current equation (8.36), the saturation current I_{Dsat} is obtained as

$$I_{Dsat} = I_D(V_{DSsat}) = \frac{\beta}{2(1 + F_B)}(V_{GS} - V_T)^2 \tag{8.39}$$

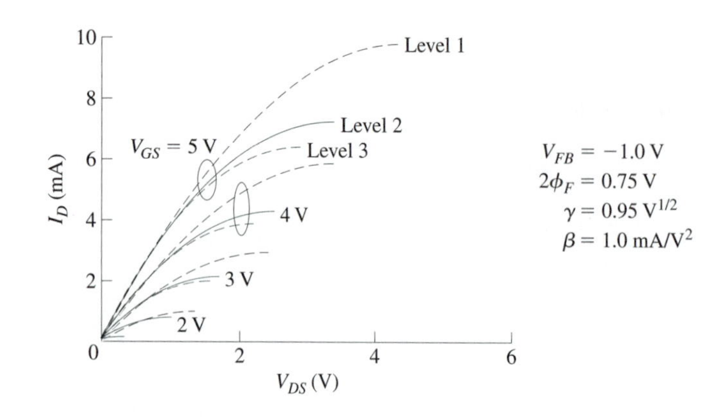

Figure 8.13 Comparison of SPICE LEVEL 1, LEVEL 2, and LEVEL 3 models.

The SPICE LEVEL 3 MOSFET model can be summarized as follows:

$$I_D = \begin{cases} \beta(V_{GS} - V_T)V_{DS} - (1 + F_B)\frac{V_{DS}^2}{2} & \text{if } 0 \le V_{DS} < V_{DSsat} \\ \frac{\beta}{2(1+F_B)}(V_{GS} - V_T)^2 & \text{if } V_{DS} \ge V_{DSsat} \end{cases} \tag{8.40}$$

where the drain saturation voltage is

$$V_{DSsat} = \frac{V_{GS} - V_T}{1 + F_B} \tag{8.41}$$

whereas the threshold voltage V_T, the gain factor β, and the factor F_B are given by Eqs. (8.9), (8.22), and (8.37), respectively.

Analogous equations apply to the case of P-channel MOSFETs. The form of these equations is presented in the tables summarizing SPICE models and parameters (Section 11.2.1).

Comparing LEVEL 3 and LEVEL 2 models, it can be concluded that the equations of the LEVEL 3 model are much simpler. Moreover, the LEVEL 3 equations are very similar to the simplest LEVEL 1 equations. The only difference between the LEVEL 3 and LEVEL 1 equations is in the factor $(1+F_B)$: replacing F_B by zero, the equations of LEVEL 3 model are reduced to the equations of LEVEL 1 model. In practice, F_B is not negligible compared to 1, so the accuracy of LEVEL 3 model is much higher. A comparison between LEVEL 1, LEVEL 2, and LEVEL 3 models is shown in Fig. 8.13. In this example, $F_B = 0.95/(2\sqrt{0.75}) = 0.55$, and as the results show, this leads to a significant difference between LEVEL 1 and LEVEL 3 models. Both the drain current and the saturation voltage are overestimated by the LEVEL 1 model. On the other hand, the difference between LEVEL 3 and LEVEL 2 models is much smaller. It should be noted that the same set of parameters is used to compare the characteristics in Fig. 8.13. Independent fitting of the LEVEL 2 and LEVEL 3 parameters would produce an even smaller difference between the two models.

In conclusion, the equations of the SPICE LEVEL 3 model resemble the simple LEVEL 1 equations that are typically used in circuit-design books. The only difference is due to the appearance of the factor $(1 + F_B)$. This factor, however, is needed to approach the accuracy of the LEVEL 2 model that is presented in many semiconductor-device books as the MOSFET model.

EXAMPLE 8.2 Drain-Current Calculations with the LEVEL 3 and LEVEL 1 Models

A depletion-type N-channel MOSFET has zero-bias ($V_{SB} = 0$) threshold voltage of $V_T = -2.5$ V. Calculate the drain current of this MOSFET if it is biased with $V_{GS} = 5$ V, $V_{DS} = 10$ V, and $V_{SB} = 0$ V, using the SPICE LEVEL 3 and LEVEL 1 models, and compare the results. The MOSFET channel-width-to-channel-length ratio is 25, $\gamma = 0.85\ \mathrm{V}^{1/2}$, $\phi_F = 0.35$ V, $C_{ox} = 7 \times 10^{-4}$ F/m^2, and $\mu_0 = 1000$ cm^2/V · s.

SOLUTION

The first step is to determine whether the MOSFET operates in the triode or the saturation region. To be able to calculate the saturation voltage V_{DSsat} using Eq. (8.41), the factor F_B needs to be determined:

$$F_B = \gamma/\left(2\sqrt{2\phi_F}\right) = 0.85/\left(2\sqrt{0.70}\right) = 0.51$$

Because the saturation voltage,

$$V_{DSsat} = (V_{GS} - V_T)/(1 + F_B) = [5 - (-2.5)]/(1 + 0.51) = 5.0\ \mathrm{V}$$

is smaller than the applied drain voltage $V_{DS} = 10$ V, the MOSFET is in saturation. By calculating the gain factor

$$\beta = \mu_0 C_{ox} W/L_{eff} = 0.1 \times 7 \times 10^{-4} \times 25 = 1.75 \times 10^{-3} A/V^2 = 1.75\ \mathrm{mA/V^2}$$

we obtain the saturation current, according to the LEVEL 3 model, as

$$I_D = \frac{\beta}{2(1 + F_B)}(V_{GS} - V_T)^2 = 32.6\ \mathrm{mA}$$

Because F_B is neglected in the LEVEL 1 model, the current according to the LEVEL 1 model is

$$I_D = \frac{\beta}{2}(V_{GS} - V_T)^2 = 49.2\ \mathrm{mA}$$

Obviously, the currents is significantly overestimated by the LEVEL 1 model.

EXAMPLE 8.3 MOSFET in the Linear Region

A MOSFET operating in its linear region can be used as a voltage-controlled resistor. Determine the sensitivity of the resistance on the gate voltage ($\partial R/\partial V_{GS}$) at $V_{GS} = 5$ V if the depletion-type MOSFET, considered in Example 8.2, is used as a voltage-controlled resistor.

SOLUTION

Based on the equation for drain current in the linear region,

$$I_D = \beta(V_{GS} - V_T)V_{DS}$$

the resistance can be expressed as

$$R = V_{DS}/I_D = \frac{1}{\beta(V_{GS} - V_T)}$$

Therefore, the sensitivity of this voltage-controlled resistor is

$$\frac{\partial R}{\partial V_{GS}} = \frac{\partial}{\partial V_{GS}}\left[\frac{1}{\beta(V_{GS} - V_T)}\right] = -\frac{1}{\beta}\frac{1}{(V_{GS} - V_T)^2} = -\frac{1}{1.75 \times 10^{-3}(5 + 2.5)^2} = 10.2\ \Omega/\mathrm{V}$$

8.3 SECOND-ORDER EFFECTS

This section describes the second-order effects included in the SPICE LEVEL 3 model. The term "second-order" should not be confused with "negligible," because some of these effects are very important in terms of simulation accuracy. The importance of some of the second-order effects would also depend on a particular application. Therefore, the decision to neglect a particular second-order effect can appropriately be made only if the effect is properly understood.

8.3.1 Mobility Reduction with Gate Voltage

This is a second-order effect that can rarely be neglected. The effect is related to the gate voltage and appears even at the smallest drain-to-source voltages. Figure 8.14a shows the transfer characteristic of a MOSFET in the linear region ($V_{DS} = 500$ mV). Due to the small V_{DS} value, the parabolic term ($V_{DS}^2/2$) in Eq. (8.40) can be neglected, which leads to the linear-region model given by Eq. (8.3). The linear I_D–V_{GS} dependence predicted by the model is plotted by the dashed line in Fig. 8.14a. However, experimental data frequently show a deviation from the predicted linear dependence, with the actual drain current falling increasingly bellow the predicted values as the gate voltage increases. The solid lines in Fig. 8.14 illustrate this effect.

This smaller-than-expected drain current is due to reduction of the channel-carrier mobility. The principal model assumes constant (gate-voltage-independent) mobility μ_0 of the carriers in the channel [refer to Eq. (8.16)]. In reality, the gate-voltage-induced vertical field influences the carrier-scattering mechanisms in the channel. The carrier-scattering

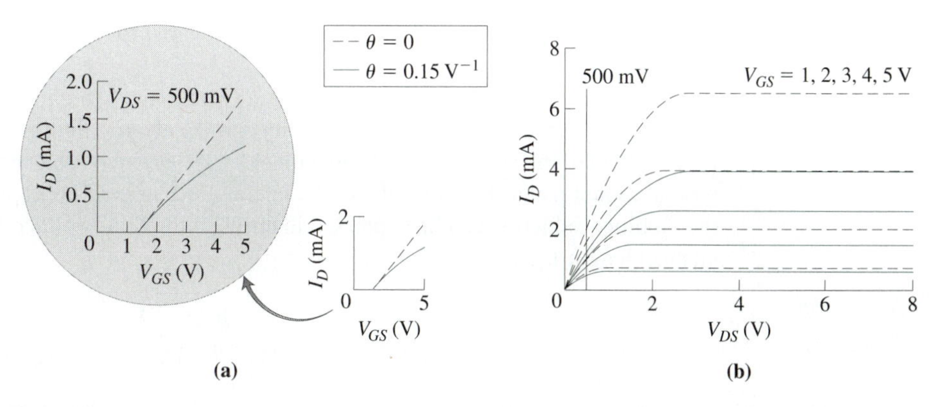

Figure 8.14 Influence of mobility reduction with gate voltage on (a) transfer and (b) output characteristics.

mechanisms in the very thin inversion layer are multiple and rather complex. However, a number of those scattering mechanisms depend on the inversion-layer thickness, and consequently on the applied gate voltage. Because the physically based equations of the mobility dependence on the gate voltage are complex, the following widely accepted semiempirical equation is used in SPICE:

$$\mu_s = \frac{\mu_0}{1 + \theta(V_{GS} - V_T)} \tag{8.42}$$

where the so-called surface mobility μ_s is now used instead of the low-field mobility μ_0 to calculate the transconductance parameter and consequently the gain factor [Eq. (8.22)]. In simple terms, KP is calculated as $KP = \mu_s C_{ox}$ instead of $KP = \mu_0 C_{ox}$. The parameter θ is a SPICE parameter that has to be experimentally determined. It is referred to as the *mobility modulation constant.*

Zero value of the θ parameter effectively eliminates this effect from the SPICE model, because $\mu_s = \mu_0$ in this case. Also, the added $\theta(V_{GS} - V_T)$ term does little in the near-threshold region (small $V_{GS} - V_T$ values). However, at moderate and high gate voltages, the effect becomes pronounced. Figure 8.14b illustrates the importance of this effect for accurate modeling of the output characteristics. Very frequently, it is impossible to achieve an acceptable agreement between the model and the experimental data without the help of the θ parameter.

8.3.2 Velocity Saturation (Mobility Reduction with Drain Voltage)

Channel-carrier mobility can also be reduced by a high *lateral* field in the channel. As explained in Section 3.3.2 [Eq. (3.23)], the carrier mobility relates the drift velocity to the electric field. At small electric fields $|E|$, the drift velocity v_d increases linearly with the electric field, which leads to the appearance of a lateral-field-independent carrier mobility $\mu_s = v_d/|E|$. Long-channel MOSFETs may operate in the low-field region, where the linear $v_d - |E|$ dependence is observed. At higher electric fields, the drift velocity deviates from linear dependence and even saturates, which is illustrated in Fig. 3.6. As explained in Section 8.1.4, this can be the mechanism responsible for the current saturation in short-channel MOSFETs.

The drain-current model is derived from Ohm's law, which assumes a linear dependence of the drift current (thus drift velocity) on the electric field [Eq. (8.14)]. To account for the effect of the velocity saturation with electric field increase, the mobility in the drain-current model has to be reduced. As the lateral electric field is proportional to V_{DS}/L_{eff}, the mobility reduction can be expressed in terms of the drain voltage V_{DS} and the effective channel length L_{eff}:

$$\mu_{eff} = \frac{\mu_s}{1 + \frac{\mu_s}{v_{max}} \frac{V_{DS}}{L_{eff}}} \tag{8.43}$$

Obviously, the surface mobility μ_s is divided by the $1+\mu_s V_{DS}/(v_{max} L_{eff})$ term to give the so-called *effective mobility* μ_{eff} that depends on V_{DS}/L_{eff}. It is the effective mobility μ_{eff}, and not μ_s, that is directly used to calculate the transconductance parameter *KP* and consequently the gain factor β, which appears in the drain-current equation. Therefore, the general form of Eq. (8.22) is

$$\beta = \frac{\mu_{eff} W C_{ox}}{L_{eff}} = KP \frac{W}{L_{eff}} \tag{8.44}$$

The strength of the considered effect is controlled by the v_{max} parameter, which has the physical meaning of *maximum drift velocity*. Although the typical value of v_{max} is 1–2 × 10^5 m/s in silicon, it can freely be adjusted in SPICE, like any other parameter. Setting $v_{max} = \infty$ would completely eliminate this effect from the SPICE MOSFET model because $\mu_{eff} = \mu_s$ in this case.

The velocity saturation also affects the saturation voltage V_{DSsat}. This effect is not covered by the effective mobility Eq. (8.43). To include this effect, the saturation voltage Eq. (8.41) is modified in SPICE. The modified equation is presented in Table 11.9 (Section 11.2.1).

8.3.3 Finite Output Resistance

The principal model assumes perfectly saturated current (horizontal I_D–V_{DS} characteristics in the saturation region), which means that it assumes that the dynamic output resistance of the MOSFET in saturation is infinitely large ($\Delta V_{DS}/\Delta I_D \rightarrow \infty$). Real MOSFETs exhibit finite output resistances. For some applications, it is very important to use the real value of the output resistance during circuit simulation. There are at least two effects that cause an increase of the drain current in the saturation region: (1) channel length modulation and (2) drain-induced barrier lowering (DIBL).

Channel-length modulation is basically the shortening of the actual channel when saturation happens due to the channel pinch-off at the drain end. The principal model is derived for the triode region, where the channel extends from the source to the drain, so that the average lateral field is $\approx V_{DS}/L_{eff}$. In the saturation, the voltage across the channel remains approximately constant (V_{DSsat}). However, the channel becomes shorter as V_{DS} "pushes" the pinch-off point away to a distance L_{pinch} from the drain. This leads to a higher field $V_{DSsat}/(L_{eff} - L_{pinch})$ in the shortened channel and therefore leads to a current increase with V_{DS}. This effect is modeled by changing (modulating) the channel length:

$$\beta = \frac{\mu_{eff} W C_{ox}}{L_{eff} - L_{pinch}} = KP \frac{W}{L_{eff} - L_{pinch}} \tag{8.45}$$

where the length of the pinch-off channel L_{pinch} depends on V_{DS} (the SPICE LEVEL 3 equation is given in Table 11.9). The equation for L_{pinch} is based on the depletion-layer width of an abrupt P–N junction [Eq. (6.45)], where $V_{DS} - V_{DSsat}$ is the voltage across the depletion layer and κ is a fitting parameter (SPICE input parameter). Because the channel-length shortening is applied abruptly when V_{DS} becomes larger than V_{DSsat}, the modeled current may not be "smooth" (a first-derivative discontinuity) around the saturation voltage.

Drain-induced barrier lowering (DIBL) is a more appropriate model for the finite output resistance in the case of short-channel MOSFETs. This effect is due to a strong lateral electric field. The principal model assumes that the gate voltage fully controls the surface potential in the channel region. In terms of the energy-band diagram, the assumption is that the channel current depends on how much is the barrier lowered by the gate voltage: in Fig. 8.4, the barrier has its full height, and there is no current flow; in Fig. 8.5b, the gate voltage lowers the barrier and the inversion-layer electrons flow from the source into the drain. In reality, the electric field from the drain can also cause some barrier lowering, which is called *drain-induced barrier lowering*. This is illustrated in Fig. 8.15. The DIBL adds to the barrier lowering due to the gate voltage; if pronounced, it can significantly enhance the electron injection from the source into the channel, leading to a noticeable increase in the channel current. Furthermore, this current increases with V_{DS}, which leads to the appearance of the finite dynamic resistance in the saturation region.

The "help" from the drain in injecting carriers from the source into the channel can be modeled by a drain-voltage-induced reduction of the threshold voltage. It was empirically established that the simplest linear relationship is quite satisfactory:

$$V_T = V_{FB} + 2\phi_F + F_s \gamma \sqrt{2\phi_F + V_{SB}} - \sigma_D V_{DS} \tag{8.46}$$

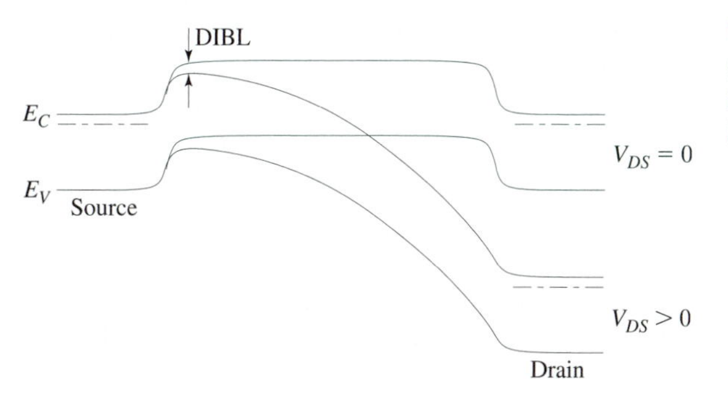

Figure 8.15 Energy-band diagrams along the channel of a MOSFET, illustrating drain-induced barrier lowering (DIBL).

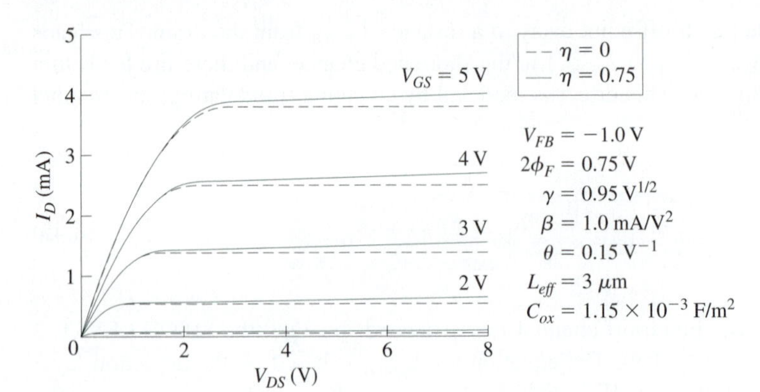

Figure 8.16 Output characteristics with (solid lines) and without (dashed lines) the influence of V_{DS} on V_T.

The coefficient expressing the strength of the V_T dependence on V_{DS}, σ_D, is not a SPICE parameter itself. It is calculated in SPICE by an equation that involves L_{eff}, C_{ox}, and a SPICE parameter η (frequently referred to as the coefficient of the static feedback). This equation is given in Table 11.10. Setting the parameter η to zero eliminates this effect from the model (the dashed lines in Fig. 8.16), whereas a larger η value expresses a stronger V_T dependence on V_{DS}, therefore a smaller output resistance (solid lines in Fig. 8.16).

8.3.4 Threshold-Voltage-Related Short-Channel Effects

The threshold voltage V_T of a MOSFET, as given by the principal model [Eq. (8.9)], is independent on the channel length or width. In reality, this is observed when the channel length and width are much larger than the channel region affected by the fringing field at the channel edges (edge effects). However, when the channel length or width is reduced to dimensions that are comparable to the edge-affected region, the threshold voltage experiences dependence on the channel length or width.

Figure 8.17 illustrates the edge effects by the electric field lines (the arrows in the figure) appearing in the MOSFET depletion layer (the clear area). The electric field lines originate at positive charge centers in the gate or the N-type drain/source regions, and

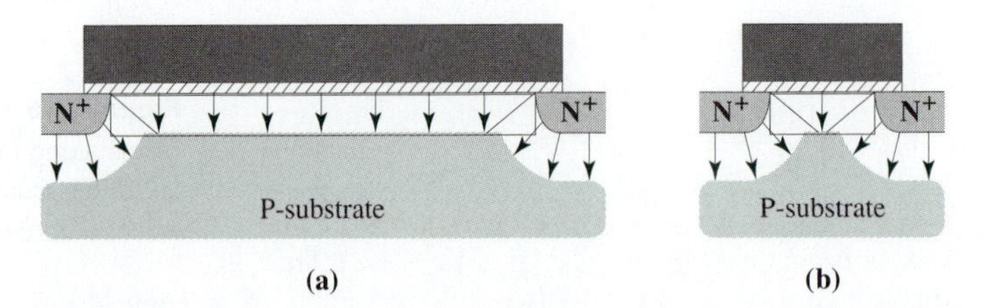

Figure 8.17 Illustration of the threshold-voltage-related short-channel effect. A part of the depletion-layer charge under the channel is created by the source and drain electric field (note the arrow origins). This helps the gate voltage to create the depletion layer under the channel, effectively reducing the threshold voltage. The effect is insignificant in (a) long-channel MOSFETs, but quite pronounced in (b) short-channel MOSFETs.

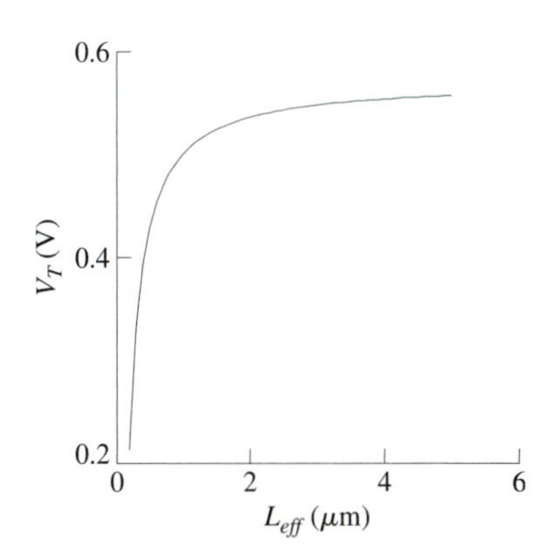

Figure 8.18 Threshold-voltage dependence on MOSFET channel length.

they terminate at negative charge centers (negative acceptor ions) in the depletion layer of the P-type body. It can be seen that the electric field lines at the ends of the channel originate from the source/drain regions and not from the gate. Consequently, some charge at the edges of the depletion layer is linked to source and drain charge and not to the gate charge. This means that less gate charge, and consequently less gate voltage, is needed to create the depletion layer under the channel.

The depletion-layer charge induced by the gate voltage is given by Q_d/C_{ox} and $\gamma\sqrt{2\phi_F + V_{SB}}$ terms in Eqs. (8.7) and (8.9), respectively, where C_{ox} is the gate-oxide capacitance per unit area, and Q_d is the depletion-layer charge per unit of channel area. Q_d is related to the total charge Q_d^* (in C) as $Q_d = Q_d^*/(L_{eff}W)$, where $L_{eff}W$ is the channel area. Noting that Q_d^* involves all the charge centers under the channel (the rectangular area in Fig. 8.17), it is easy to see that Q_d/C_{ox} (and consequently $\gamma\sqrt{2\phi_F + V_{SB}}$) overestimates the gate voltage needed to create the depletion layer under the channel. The gate voltage should be related to the charge inside the trapezoidal area, and not the whole (rectangular) area under the channel.

In long-channel MOSFETs, such as the one in Fig. 8.17a, the edge charge created by the source and drain is much smaller than the total charge Q_d^*. This can be seen by comparing the rectangular and the trapezoidal areas in Fig. 8.17a. Consequently, the $Q^*/(L_{eff}WC_{ox}) = Q_d/C_{ox} = \gamma\sqrt{2\phi_F + V_{SB}}$ term fairly correctly expresses the voltage needed to create the depletion layer under the channel. As the edge effects are negligible in the long-channel MOSFETs, the threshold voltage does not show a dependence on the gate length (Q_d/C_{ox} is independent of gate length). In Fig. 8.18, this is the case for MOSFETs with channels longer than approximately 2 μm.

In the case of short-channel MOSFETs (Fig. 8.17b), the charge enclosed in the trapezoidal area is significantly smaller than the charge inside the rectangular area. This is because the edge effect is now pronounced: the source- and drain-related fields are creating an observable portion of the depletion layer under the channel.

To model this effect, the depletion layer charge Q_d^* is multiplied by a *charge-sharing factor* F_s that can be obtained as the ratio between the trapezoidal and the rectangular areas; this is to convert the total charge in the depletion layer under the channel (Q_d^*) into the charge that is created by the gate (the charge enclosed in the trapezoidal area). As $F_s Q_d^*/(L_{eff}WC_{ox}) = F_s\, Q_d/C_{ox} = F_s\gamma\sqrt{2\phi_F + V_{SB}}$, the threshold-voltage equation

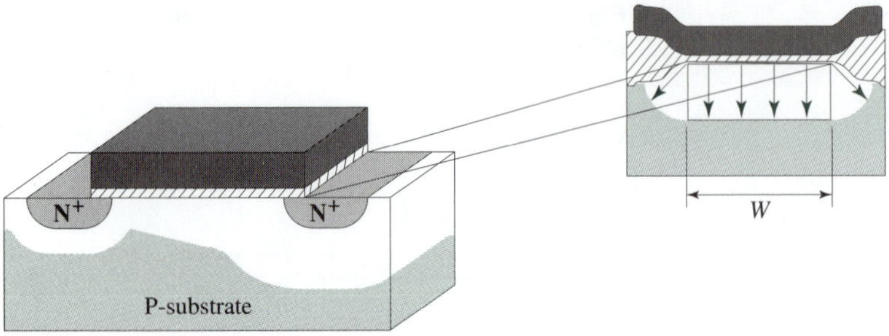

Figure 8.19 Illustration of the narrow-channel effect. Fringing electric field wastes the gate voltage, causing a threshold-voltage increase in narrow-channel MOSFETs.

[Eq. (8.9)] is modified to

$$V_T = V_{FB} + 2\phi_F + F_s \gamma \sqrt{2\phi_F + V_{SB}} \tag{8.47}$$

A number of different equations for the charge-sharing factor F_s have been developed. The equation that is used in SPICE is given in Table 11.10 (Section 11.2.1). SPICE uses the source and drain P–N junction depth x_j, the lateral diffusion $x_{j\text{-}lat}$, and the P–N junction built-in voltage V_{bi} as parameters to calculate the charge-sharing factor F_s.

8.3.5 Threshold-Voltage-Related Narrow-Channel Effects

Threshold voltage can also become dependent on the channel width W, if the channel width is reduced to levels that are comparable to the edge-effect regions. Figure 8.19 illustrates the edge effect at the channel ends that determine the channel width. As mentioned earlier, the principal threshold-voltage equation is derived under the assumption that the gate voltage creates the depletion-layer charge in the rectangular area under the channel. In reality, the gate voltage depletes a wider region, due to the fringing field effect (Fig. 8.19). This increases the threshold voltage.

To include this effect, the threshold-voltage equation is expanded again, its final form being as follows:

$$V_T = V_{FB} + 2\phi_F + \gamma F_s \sqrt{2\phi_F + V_{SB}} - \sigma_D V_{DS} + F_n(2\phi_F + V_{SB}) \tag{8.48}$$

The new term $F_n(2\phi_F + V_{SB})$ models the threshold-voltage increase in narrow-channel MOSFETs. The parameter F_n is calculated from the channel width W, the gate-oxide capacitance C_{ox}, and a SPICE parameter δ modulating the strength of this effect. The actual equation is given in Table 11.10. For $\delta = 0$, or wide channels (W large), the parameter F_n approaches zero, eliminating this effect from the threshold-voltage equation.

8.3.6 Subthreshold Current

The principal model assumes that the channel carrier density is zero for $V_{GS} \leq V_T$, appearing abruptly for gate voltages larger than the threshold: $Q_I = (V_{GS} - V_T)C_{ox}$. In reality, the transition from full depletion to strong inversion is gradual (the term *strong* inversion itself indicates that there could be a *moderate,* and even a *weak,* inversion). There

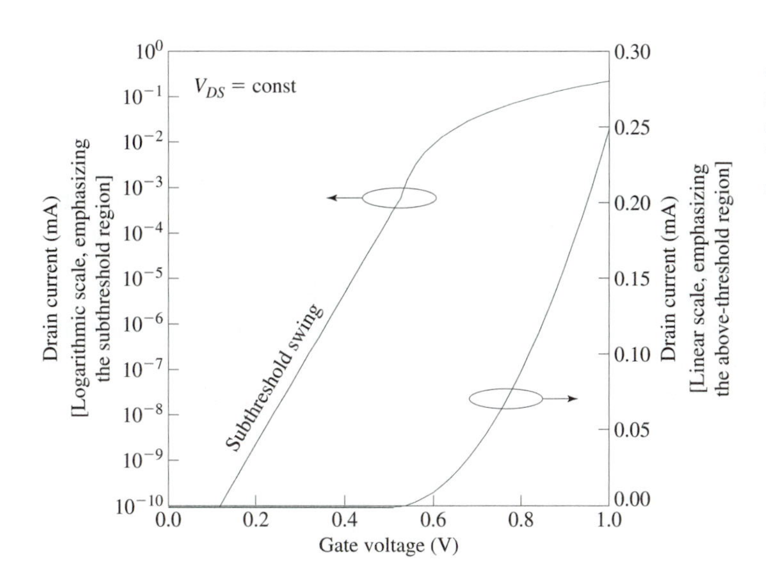

Figure 8.20 The use of a logarithmic drain axis emphasizes the subthreshold region of a transfer characteristic (the subthreshold current cannot be seen with a linear axis). The subthreshold swing is 60 mV/decade, corresponding to $n_s = 1$ in Eq. (8.49).

are mobile carriers in the channel, even for subthreshold gate voltages. Of course, their concentration is very small, and it rapidly decays as the gate voltage is reduced below V_T. Nonetheless, their effect is observable, because they account for the gradual decay of the drain current from the above-threshold levels toward zero.

The vertical field is very low in the subthreshold region, and it does not take a large drain voltage to fully remove the carriers from the drain end. This creates a concentration gradient. Because of this, and because of the low level of the carrier concentration, the diffusion current dominates over the drift current in the subthreshold region. Consequently, a diode-like, exponential current–voltage equation is used to model the subthreshold current:

$$I_{D-subth} = I_{D0} e^{V_{GS}/n_s kT} \tag{8.49}$$

Figure 8.20 shows drain-current plots in both the above-threshold and subthreshold regions. Because the subthreshold current cannot be observed on the plot corresponding to the linear current axis, a logarithmic current axis is used to present the subthreshold current. Given that the subthreshold current depends exponentially on the gate voltage [Eq. (8.49)], the semilogarithmic plot corresponding to the subthreshold current appears as a straight line. The slope of this straight line is referred to as the *subthreshold swing*. The subthreshold swing is the voltage increment needed to change the subthreshold current by an order of magnitude (one decade). Accordingly, the subthreshold swing is expressed in V/decade. The theoretical limit for the subthreshold swing at room temperature is 60 mV/decade (refer to Example 8.4). This value corresponds to $n_s = 1$ in Eq. (8.49). In reality, different leakage mechanisms add subthreshold current to the pure diffusion component, which increases the subthreshold swing (a slower current reduction with reducing gate voltage). In Eq. (8.49), this effect is modeled by setting n_s to a value that is larger than 1.

Together with the threshold voltage, the subthreshold swing fully characterizes the subthreshold region. It is a very important parameter for MOSFETs used as switches because it shows the voltage range that is needed to separate the *on* and *off* modes of the switch.

The form of Eq. (8.49) that is used in SPICE LEVEL 3 is shown in Table 11.7. The constant I_{D0} is selected so as to provide a continual transition from the subthreshold to the above-threshold drain current. I_D calculated by the above-threshold model is identical to the subthreshold current at $V_{GS} = V_T + n_s kT$. For smaller gate voltages, the subthreshold model is used, whereas the above-threshold model is used for larger gate voltages. The coefficient n_s is analogous to the emission coefficient n of diodes. The SPICE LEVEL 3 equation for n_s is given in Table 11.10. There is an input SPICE parameter, NFS, that can be used to control the value of n_s.

EXAMPLE 8.4 Subthreshold Swing

Show that 60 mV/decade is the theoretical limit for the subthreshold swing at room temperature. What value of the n_s parameter in Eq. (8.49) corresponds to the subthreshold swing of 100 mV/decade?

SOLUTION

The subthreshold swing is the voltage range ΔV_{GS} that corresponds to a tenfold increase in the current ($I_{D2}/I_{D1} = 10$). Applying Eq. (8.49) to two current levels (I_{D1} and I_{D2}) and dividing these current equations, we obtain

$$\frac{I_{D2}}{I_{D1}} = e^{(V_{GS2} - V_{GS1})/n_s kT} = e^{\Delta V_{GS}/n_s kT}$$

Taking logarithms with the base 10 of both sides leads to

$$\Delta V_{GS} = n_s kT \log\left(\frac{I_{D2}}{I_{D1}}\right) \Big/ \log(e)$$

For the case of $I_{D2}/I_{D1} = 10$ ($\log 10 = 1$),

$$\Delta V_{GS} = n_s kT / \log(e) \qquad \text{(in V/decade)}$$

The smallest ΔV_{GS} is for $n_s = 1$. This is the case when the subthreshold current is purely diffusion current, meaning that no additional current mechanisms exist. For $n_s = 1$, $\Delta V_{GS} = 0.026/\log(2.7183) = 60$ mV/decade. For $\Delta V_{GS} = 100$ mV/decade, $n_s = \Delta V_{GS} \log e / kT = 0.1 \log(2.7183)/0.026 = 1.67$.

†8.4 NANOSCALE MOSFETs

CMOS technology has undergone unmatched progress during the last five decades. In a way, this is illustrated by the names given to describe different levels of integration: small-scale integration (SSI), medium-scale integration (MSI), large-scale integration (LSI), very-large-scale integration (VLSI), and ultra-large-scale integration (ULSI). Although every one of these phases appears as an *era* in economic terms, it is a remarkable fact that the device and technology principles have not changed. The principles of modern MOSFETs are essentially the same, although critically important features have been added to enable aggressive dimension downscaling. This section describes the benefits of dimension reduction, the problems it opens, and the solutions that have been developed.

8.4.1 Downscaling Benefits and Rules

The switching speed (maximum operating frequency) was initially a disadvantage of the CMOS circuits. The v_{OUT} versus t diagram of Fig. 8.6b illustrates the times needed to achieve the high/low output level due to the charging/discharging of the load capacitance C_L. These times are determined by the value of the load capacitance and the value of the charging/discharging current supported by the MOSFETs. As illustrated by Example 11.6b (Section 11.2), the loading capacitance consists of the output and the input capacitances of the MOSFETs in the driving and the loading inverters, respectively. Both the loading capacitance and the charging/discharging current change favorably when the channel length L_g is reduced. A reduction of the channel length by a factor of S reduces the input capacitance of the CMOS cell S times, due to reduced gate area. It also increases the charging/discharging current S times as the MOSFET current increases proportionally with the channel-length reduction. This means a speed improvement by a factor of S^2. Additionally, this means a smaller cell area, thus the possibility of integrating more logic cells to create more powerful ICs. These are extremely motivating and economically extremely rewarding benefits.

However, a successful reduction of the channel length requires some other device parameters to be appropriately adjusted to avoid possible adverse effects. One of the things that may happen when the channel is reduced is that the drain field will start taking electrons directly from the source (punch-through effect). To prevent this from happening, the substrate doping is increased, which shortens the penetration of the drain field into the substrate. As Eq. (6.44) shows, to reduce a depletion-layer width S times, the doping concentration has to be increased S^2 times. However, the increased substrate doping increases the body factor γ and consequently the threshold voltage, as can be seen from Eqs. (7.22) and (8.9), respectively. A MOSFET that turns *on* at a voltage higher than 25% of the supply-voltage level is generally not acceptable. To keep the threshold voltage down, it is necessary to reduce the thickness of the gate oxide S times. Because this increases the input capacitance, the channel width is also reduced S times as a compensation. Table 8.1 summarizes these steps as a set of downscaling guidelines, more frequently referred to as downscaling rules, and their effects.

It has not been always possible to avoid all the important adverse effects of the scaling down by applying these general rules. The channel-length reduction, along with

TABLE 8.1 General Downscaling Rules and Their Effects

Rules			
Channel length	L	$\longrightarrow$	L/S
Channel width	W	$\longrightarrow$	W/S
Gate-oxide thickness	t_{ox}	$\longrightarrow$	t_{ox}/S
Substrate doping	$N_{A,D}$	$\longrightarrow$	$N_{A,D} \times S^2$
Effects			
Drain current	$I_D \propto W/(Lt_{ox})$	$\longrightarrow$	$I_D \times S$
Input capacitance	$C_{in} \propto WL/t_{ox}$	$\longrightarrow$	C_{in}/S
Maximum switching frequency	$f \propto I_D/C_{in}$	$\longrightarrow$	$f \times S^2$
Cell area	$A \propto WL$	$\longrightarrow$	A/S^2

the associated substrate doping increase, results in an increase of the lateral electric field in the channel. Equation (6.31) and Fig. 6.17c illustrate the electric-field dependence on the doping level in the case of an abrupt P–N junction. This increase in the electric field means that the breakdown voltage (Section 6.1.4) is reduced, reducing the maximum operating voltage. Reductions of the maximum operating voltage of CMOS integrated circuits have already occurred; initially, the CMOS integrated circuits could operate at voltages higher than 20 V, then the standard was 5 V, followed by 3.3 V, and nowadays we have limitations of 1.5 V and 1.0 V.

To reduce the problem of high electric field in the MOSFET channel, the abrupt-like N^+-drain and source–P-substrate junctions are modified. Lightly doped regions are introduced to linearize the junctions—that is, to reduce the maximum doping level at the junctions. An N-channel MOSFET with lightly doped drain and source extensions is shown in Fig. 8.21. As the doping concentration is reduced at the drain extension–substrate junction, the maximum lateral field appearing at the drain end of the channel is reduced as well. MOSFETs with this type of doping structure are referred to as *lightly doped drain* (LDD) MOSFETs.

The drain and source extensions of the LDD MOSFET are doped by implanting arsenic or antimony ions, which have lower diffusion constants than phosphorus. The lower diffusion constants make it possible to achieve shallow drain/source extensions, which

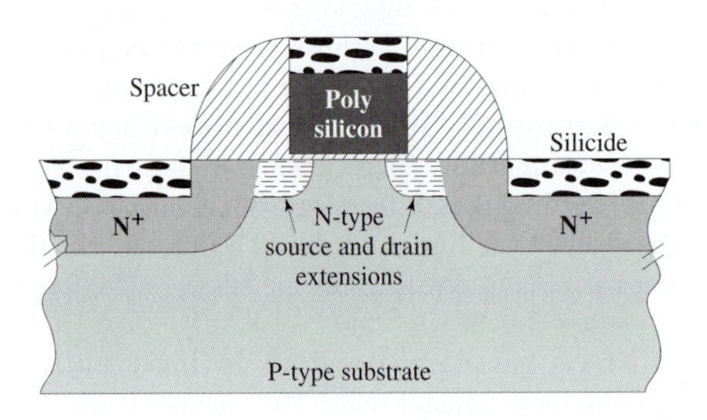

Figure 8.21 Deep-submicron MOSFET structure.

minimizes the penetration of the drain field under the channel (punch-through effect). The drain/source extensions are implanted with the polysilicon gate as a mask to achieve self-alignment between the gate and the drain/source areas. The spacer layer shown in Fig. 8.21 is created afterward to be used as a mask when the N^+ regions are created by phosphorus implantation. The spacer layer is created by deposition of either oxide or silicon nitride, which fills the polysilicon–substrate corner, and by subsequent etching to remove the deposited layer from the top of the polysilicon and outside the corner region of the silicon substrate.

Channel-length reduction leads to an undesirable increase in the resistance of the gate line. To minimize this effect, the MOSFET structure is further modified by creating a silicide layer at the top of the gate (Fig 8.21). The silicide can be created by depositing titanium and by subsequent annealing which leads to a reaction between the deposited titanium and the underlying silicon to create a layer of $TiSi_2$. The resistivity of the silicide is much smaller than the resistivity of the polysilicon, which reduces the gate resistance. The silicide is also created in the drain and the source regions to improve the contact between the metalization and the silicon—that is, to reduce the contact resistance.

8.4.2 Leakage Currents

MOSFETs with channel lengths less than 100 nm were experimentally demonstrated as early as 1987 by IBM researchers,[2] followed by a demonstration of sub-50-nm MOSFET by Toshiba researchers in 1993.[3] Although operational, and potentially applicable in a number of specific areas, these MOSFETs could not be used as building blocks for CMOS ULSI circuits.

Figure 8.22a shows simulated transfer characteristics of a 100-nm MOSFET with parameters and characteristics similar to the one published by the IBM researchers. Because the problems with such a short-channel MOSFET are mostly in the subthreshold region, the left-hand drain-current axis is presented in a logarithmic scale so that the subthreshold current can be seen over several orders of magnitude. According to the basic theory, the subthreshold current is due to diffusion and should not depend on V_{DS} voltage [Eq. (8.49)]. If that were the case, the subthreshold transfer characteristics for $V_{DS} = 0.1$ V and $V_{DS} = 1.0$ V in Fig. 8.22a would overlap one another. The observed shift of the transfer characteristic by V_{DS} indicates a short-channel effect due to too deep penetration of the drain-to-substrate depletion layer under the gate. The energy-band diagram and the equipotential contours in Fig. 8.23 illustrate that there is too strong influence from the drain voltage in the region that should be controlled by the gate.

---

[2]G. A. Sai-Halasz, M. R. Wordeman, D. P. Kern, E. Ganin, S. Rishton, D. S. Zicherman, H. Schmid, M. R. Polacri, H. Y. Ng, P. J. Restle, T. H. P. Chang, and R. H. Dennard, Design and experimental technology for 0.1-μm gate-length low-temperature FET's, *IEEE Electron Dev. Lett.*, vol. EDL-8, pp. 463–466 (1987).

[3]M. Ono, M. Saito, T. Yoshitomi, C. Fiegna, T. Ohguro, and H. Iwai, Sub-50 nm gate length n-MOSFET's with 10 nm phosphorus source and drain junctions, *IEDM Tech. Dig.*, pp. 119–122, (1993).

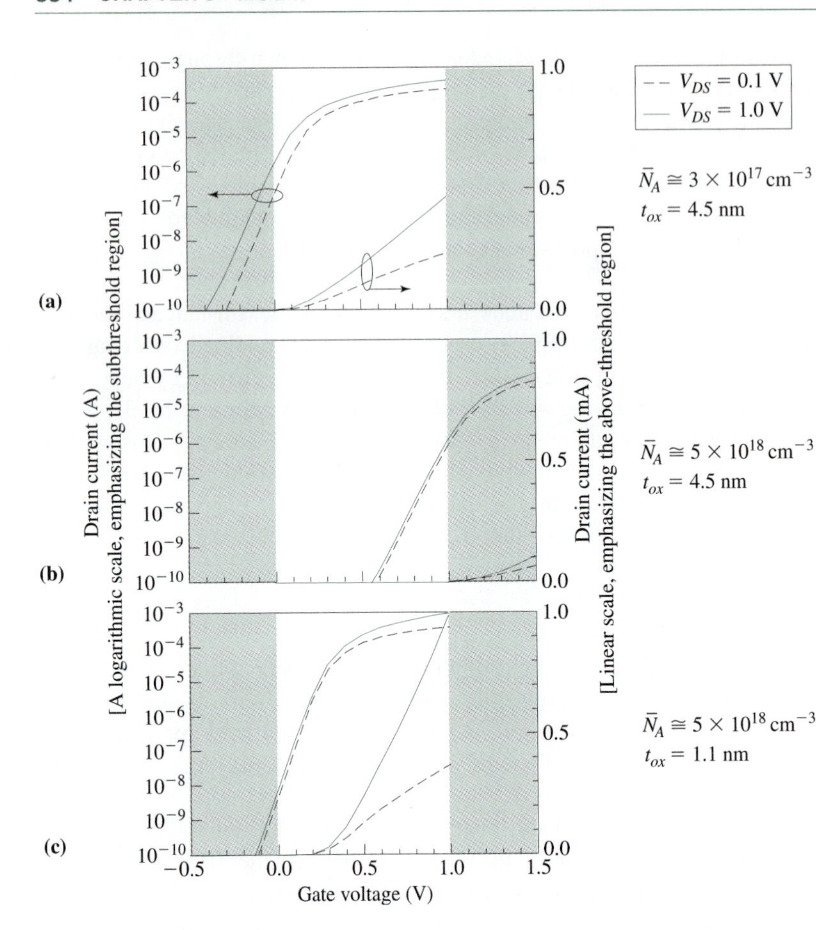

Figure 8.22 Transfer characteristics of 100-nm MOSFETs (the channel width is 1 μm). (a) Devices with similar characteristics were experimentally demonstrated as early as 1987, but they exhibit pronounced short-channel effects and very high *off* current. (b) As known from the downscaling rules, an increase in the substrate doping eliminates the short-channel effects, but also increases the threshold voltage, in this case above the 1.0-V supply voltage. (c) Downscaling rules require a 1.1-nm gate oxide that leaks a very high input current.

The most important problem with a too-high *off* current is that the CMOS cells would start consuming significant power, limiting the number of cells that can be integrated. The *off* current in Fig. 8.22a is $> 1\ \mu$A, which means that the leakage current of a 1-million-transistor IC would be at the level of > 1 A.

Figure 8.22b shows simulated transfer characteristics of the same MOSFET when the substrate doping is increased 16 times. The short-channel effects are all but eliminated; however, the threshold voltage is increased even beyond 1.0 V, which is set as the maximum supply voltage. Of course, the downscaling rules require a simultaneous reduction in the gate-oxide thickness by a factor of 4. Comparing the energy-band diagrams and the equipotential contours in Figs. 8.23 and 8.24, we can see how the increased substrate doping and reduced gate-oxide thickness eliminate the drain influence in the gate region. Figure 8.22c shows that this does result in improved transfer characteristics: lower *off* current and lower threshold voltage. The new problem here is that an oxide thickness of about 1 nm is needed. Tunneling of electrons through the 1-nm barrier that such a gate oxide could provide is quite significant. With a high leakage current through the gate oxide, the problem with current leakage and power dissipation reappears in just another form.

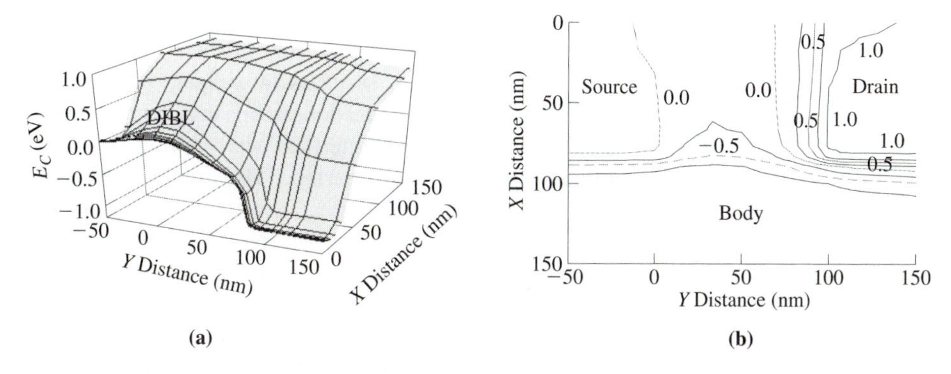

Figure 8.23 (a) Energy-band diagram and (b) equipotential contours for the MOSFET with the transfer characteristics shown in Fig. 8.22a ($V_{GS} = 0$ V, $V_{DS} = 1.0$ V).

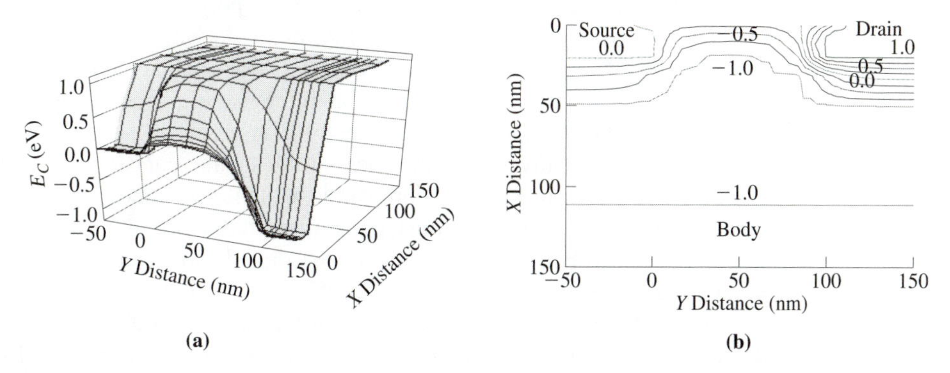

Figure 8.24 Energy-band diagram (a) and equipotential contours (b) for the MOSFET with the transfer characteristics shown in Fig. 8.22c ($V_{GS} = 0$ V, $V_{DS} = 1.0$ V).

8.4.3 Advanced MOSFETs

High-k Gate Dielectric

A possible solution to the problem of gate-oxide thickness is replacement of the oxide by another material with a higher permittivity. For the control over the channel region by the gate voltage, it does not matter whether the oxide thickness is reduced or the permittivity is increased. The strength of the field effect is related to the gate-dielectric capacitance per unit area:

$$C = \frac{\varepsilon_d}{t_d} \tag{8.50}$$

There are dielectric materials with permittivities hundreds and even thousands of times higher than the permittivity of silicon dioxide. These materials are commonly called high-k materials because k is sometimes used as a symbol for the dielectric constant: $\varepsilon_d = k\varepsilon_0$. Obviously, the high-$k$ materials enable the same strength of the field effect with much thicker dielectric layers. Although the increased thickness reduces the tunneling current exponentially, there are other leakage mechanisms that cause leakage-current problems with many potential gate dielectrics, notably ferroelectric materials. Some materials with high permittivity—in particular, some metal oxides such as HfO_2—exhibit acceptable leakage and breakdown field properties. A big disadvantage of all alternative materials has always been the much poorer interface between these films and silicon, compared to the native silicon–silicon dioxide interface. The excellent properties of SiO_2 (the quality of interface to Si, the dielectric strength, the large energy-band discontinuities with respect to both the conduction and the valence bands in silicon, hence large barrier heights), along with fair blocking of dopant diffusion, outweigh the comparative disadvantage of low dielectric permittivity. As a result, SiO_2 has remained the superior gate material so far. However, a solution will be needed for the tunneling limit, and that is where the metal oxides are expected to play a role.

Metal Gates

Another helpful possibility is to replace the polysilicon gates with a suitable metal so that two goals are achieved. Primarily, this is to help reduce the threshold voltage by a selection of a metal with a smaller work function compared to N^+ polysilicon. Because the effect of more negative work-function difference is a parallel negative shift of the transfer characteristics, this alleviates the need for aggressive scaling down of the gate dielectric thickness. Secondly, metal gate lines exhibit smaller parasitic gate resistances.

The gate metal has to be compatible with the semiconductor processing—in particular, the high ion-implant annealing temperatures (the *self-aligning* technology requires the gate to be formed before the implant for the drain/source extensions). Several high-melting-point refractory metals (such as Mo, W, Ti, Ta) and their nitrides (MoN, WN, TiN, TaN) have been investigated for use as MOSFET gates. In addition to the work-function value, a very important factor is how well a specific metal integrates in MOSFET technology. For example, some metals do not exhibit good adhesion to SiO_2 (or the gate dielectric used) and others present etching problems, making it difficult to create well-controlled small-dimension gates. Molybdenum (Mo) has emerged as the most attractive option for metal gates. An additional potential benefit with Mo is due to the possibility of controllably changing its work function. For example, the Mo work function is dependent on the energy and dose of the nitrogen implant.

Retrograde Substrate and Halo Doping Profiles

Quite a few modifications in the MOSFET structure have been developed to address the issue of the high leakage current due to penetration of the drain field through the depletion region under the channel. One approach relates to engineering of the substrate doping profile. For example, a high doping level is used below the surface region, where the gate field is not as strong, with a lower doping at the surface. This doping profile, illustrated in Fig. 8.25, is called *retrograde substrate doping*. The higher-doped region prevents the extension of the drain electric field through the lower part of the depletion region where the gate field is weak. The lower-doped region at the surface helps to prevent unacceptable increase in the threshold voltage and to prevent channel-mobility reduction that would be caused by higher surface doping.

Another useful alteration of the substrate doping profile is to increase the doping around the source and drain areas. This profile, also illustrated in Fig. 8.25, is referred

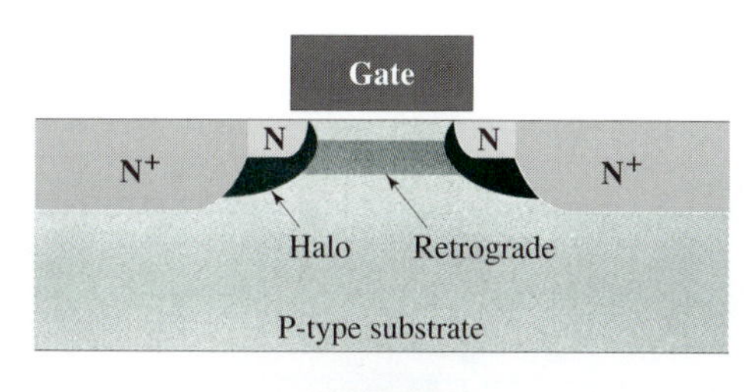

Figure 8.25 MOSFET with engineered doping profile in the substrate.

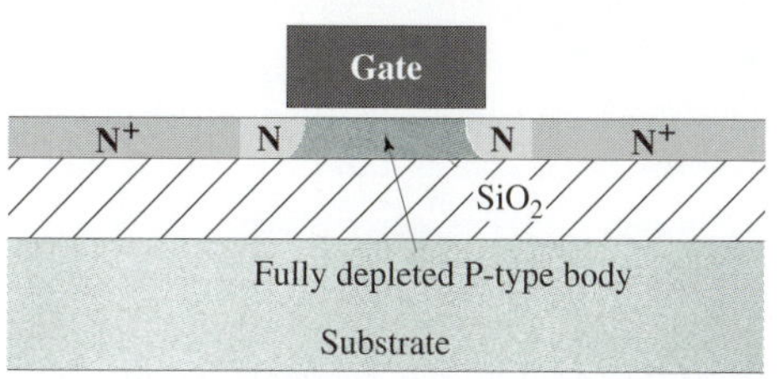

Figure 8.26 SOI MOSFET with ultrathin body.

to as *halo* doping. To create the halo profiles, ion implantation is performed at an angle and with energy high enough to ensure that the implanted atoms are outside the source/drain regions. The channel area is protected from this implantation by the already formed gate of the MOSFET. The increased doping around the source leads to an increased source-to-body barrier for carrier injection, which results in more effective suppression of the leakage current.

Silicon-on-Insulator (SOI) MOSFETs

Silicon-on-insulator (SOI) substrates have been under development for many years because of their potential for different applications. The SOI technology (refer to Section 16.2.4) provides a thin monocrystalline film of silicon on an insulating layer (typically, SiO_2). Because no current can flow through the buried SiO_2 layer (Fig. 8.26), much better control of the leakage current is possible in SOI MOSFETs. For this to work, however, the top silicon film has to be very thin so that it is fully depleted (MOSFETs are also made in partially depleted SOI films; but this provides no advantage for leakage control in short-channel MOSFETs). Notwithstanding fabrication difficulties in terms of achieving uniform layers, SOI MOSFETs with bodies as thin as 10 nm have been demonstrated. These MOSFETs are also called ultra-thin-body (UTB) MOSFETs.

Double-Gate MOSFETs: FinFET

Much better control of the MOSFET body in the gate area can be achieved by double-gate MOSFETs. In this approach, a thin MOSFET body is sandwiched between two gates. The second gate halves the body thickness; far more importantly, it helps to prevent penetration of the drain electric field under the channel. A big issue with the double-gate concept has been the inherent fabrication complexity. A double-gate structure that can be fabricated with a practical integrated-circuit technology is the so-called FinFET.[4] As Fig. 8.27 shows,

---

[4]L. Chang, Y.-K. Choi, D. Ha, P. Ranade, S. Xiong, J. Bokor, C. Hu, and T.-J. King, Extremely scaled silicon nano-CMOS devices, *Proc. IEEE,* Vol. 91, pp. 1860–1873 (2003).

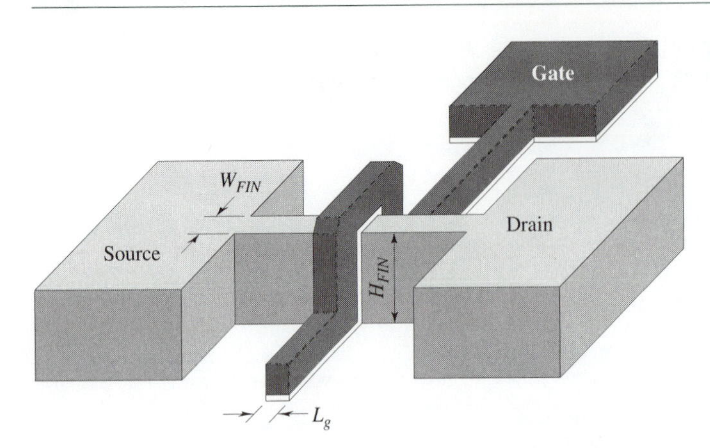

Figure 8.27 FinFET: the most promising structure for nanoscale MOSFETs.

the two MOSFET channels of this double-gate structure are along the vertical sides of the silicon area etched in the shape of a fin (the height of the fin is labeled by H_{FIN} and the width of the fin is labeled by W_{FIN}). To achieve source and drain regions that are self-aligned to the gate, the ion implantation for the source and drain is performed with the gate as a mask. Thus, the visible silicon areas are the source and drain regions, whereas the MOSFET body is under the gate.

The effective channel width of the FinFET is equal to $2H_{FIN}$. Given that H_{FIN} is a technological parameter, it may seem that it is not possible to freely adjust the effective channel width of a FinFET at the level of two-dimensional layout design. Moreover, it may seem that the channel width is restricted by the practical fabrication limit for the fin height. These difficulties are overcome by using MOSFETs with multiple parallel fins. For n fins, the effective channel width becomes $2nH_{FIN}$.

The most critical parameter is the fin width (W_{FIN}). The half-width, being equal to the effective body thickness, determines the strength of leakage-current suppression for zero-biased gates. For very wide fins—for example, for W_{FIN} larger than the depletion-layer width—there is no current-suppression advantage in comparison to ordinary MOSFETs. Thus, the FinFET structure is useful with very narrow fins, which by itself presents fabrication challenges. In single-gate MOSFET structures, the gate length (L_g) is made equal to the smallest dimension that can be achieved by a given technological process. FinFETs require $W_{FIN} < L_g$ to utilize the current suppression advantages of this structure. To set W_{FIN} as the minimum design dimension would mean to increase L_g, which would result in undesirable reduction of MOSFET current. To avoid this disadvantage, sophisticated "fin trimming" techniques have to be used to enable fabrication of fins with widths that are smaller than the minimum design dimension (L_g).

Assuming that fabrication of FinFETs with $W_{FIN} \approx 5$ nm is possible, and limiting the equivalent thickness of the gate dielectric to ≈ 1 nm, the desired leakage-current targets can be met at gate lengths down to ≈ 10 nm.[5] At these dimensions, quantum-mechanical effects will become important. In particular, carrier confinement in the thin body may cause

[5]L. Chang, S. Tang, T.-J. King, J. Bokor, and C. Hu, Gate-length scaling and threshold voltage control of double-gate MOSFETs, *IEDM Tech. Dig.*, pp. 719–722 (2000).

an intolerable shift in the threshold voltage. On the positive side, the carrier transport in the channel will approach the ballistic limit because the carrier scattering can be ignored.

*8.5 MOS-BASED MEMORY DEVICES

Information storage is a very important function in a variety of electronic systems and definitely of essential importance in computers. Although the memory function can be implemented with a number of nonsemiconductor devices, the progress made in the area of semiconductor technologies is having a very strong impact on the performance, cost, reliability, and physical size of semiconductor-based memory devices. Consequently, semiconductor memories are both expanding memory-based applications and replacing nonsemiconductor memory devices in a number of existing applications.

The essential characteristics of a good memory are (1) large number of fast reading/programming cycles, (2) long retention of the stored information, and (3) high memory capacity. No currently available memory is superior in terms of all the three essential characteristics. Accordingly, different types of semiconductor and nonsemiconductor memories are used in electronic systems. The semiconductor memories can be classified into three categories: static random-access memories (SRAMs), dynamic random-access memories (DRAMs), and flash memories. The main nonsemiconductor memories are optical disks (CD/DVD) and magnetic disks.

SRAMs are based on bistable electronic circuits, known as flip-flops. These memories provide a practically unlimited number of very fast reading/programming cycles. They retain the information for as long as an uninterrupted power supply is connected to the memory cells. The big disadvantage of these memories is a relatively low memory capacity. This results from the fact that the circuit of each memory cell consists of many transistors, leading to a relatively large area per cell. Although still used, SRAMs are not as widespread as DRAMs and flash memories. Given that DRAM and flash memory cells are specific MOS-based devices, this section describes these two types of memory cell.

8.5.1 1C1T DRAM Cell

Stored charge in a capacitor is utilized as information storage in DRAMs. To enable access to this charge for information reading and programming, one of the capacitor plates is connected to a MOSFET acting as a controlled switch. Thus, a capacitor connected to a MOSFET switch creates what is referred to as a 1C1T (one-capacitor, one-transistor) memory cell. Figure 8.28 shows that 1C1T cells can be connected in a memory array that enables an arbitrary (random) access to each individual cell. As Fig. 8.28 shows, the gates of the MOSFETs in one column of the array are connected to create a *word line*. Application of a positive voltage to a word line sets all the MOSFETs in that column in *on* mode. A single cell is selected from the word line by selecting a single *bit* line to either sense the charge stored in the capacitor of this cell (reading) or set the charge at desired level (programming). This memory array is referred to as *random-access* memory because an arbitrary cell can be selected at the cross between a word and a bit line.

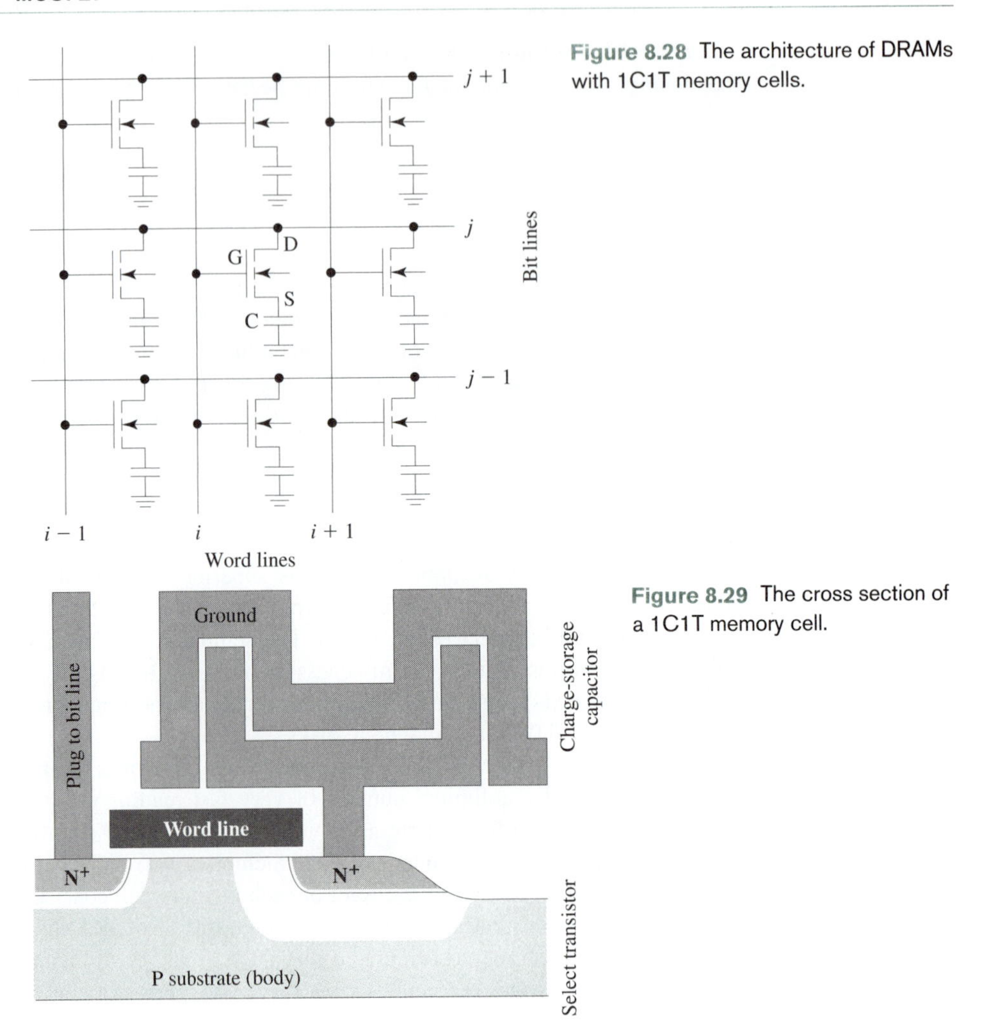

Figure 8.28 The architecture of DRAMs with 1C1T memory cells.

Figure 8.29 The cross section of a 1C1T memory cell.

The capacitor in a 1C1T cell can be implemented as a MOS capacitor. To increase the capacitance value without a proportional increase in the cell area, the MOS capacitors can be formed in trenches etched inside the silicon substrate. The alternative approach is to implement the capacitor as a metal–dielectric–metal capacitor. In the latter case, the capacitor is created on the top of the MOSFET, as illustrated in Fig. 8.29. Provided there is negligible leakage through the capacitor dielectric and negligible leakage through the MOSFET switch (set in *off* mode), the capacitor charge will be retained for an indefinite period of time. The leakage of practical MOSFETs, however, is not negligible. In fact it is so high that it discharges the capacitor within milliseconds. A millisecond may seem to be an impractically short time for a memory device, but it is at least hundreds of thousands times longer than the time needed to perform a single digital operation. This makes it practical to implement periodic refreshing of the memory cells. Because of the need to refresh the memory information, this type of memory is called *dynamic* RAM.

It may appear that the need for refreshing is a big disadvantage of DRAMs—in particular, in comparison to SRAMs, which do not need refreshing when operating with continuous power supply and are typically faster than DRAMs. The big advantage of DRAMs, however, is in their small cell size, thus large memory capacity. Although there are nonsemiconductor memories with capacity much larger than that of modern DRAMs, reading and/or programming of these memories is so slow that their use remains limited to *data storage*. DRAMs are the unique type of memory that is used for *data processing* in modern computers.

8.5.2 Flash Memory Cell

To create a nonvolatile memory, which will keep information stored even when the power supply is disconnected from the memory element, significantly modified devices have to be used. Semiconductor devices that enable electrical erasing and programming are based on trapping electrons in a deep potential well, so deep that the trapped electrons are very unlikely to gain enough thermal energy and escape from the potential well. Figure 8.30a shows the cross section of what is known as a *flash* memory cell. The deep potential well for electron trapping is created by inserting a floating polysilicon gate between the MOSFET gate (now referred to as the control gate) and the silicon substrate. The discontinuity of the bottom of the conduction band, appearing at the polysilicon–oxide interfaces, is used to create the trapping potential well. Because this discontinuity is more than 3 eV (Fig. 8.30c) and because the floating gate is completely surrounded by oxide, the electrons appearing in the floating gate are trapped.

Figure 8.30c shows the energy bands for the erased state (no trapped electrons in the floating gate). The electrons in the floating gate in this case are the normal doping-induced electrons in N-type polysilicon. The MOSFET is designed so that the application of V_{GS} reading voltage turns the MOSFET *on*, creating channel of electrons between the drain and the source. Figure 8.30b shows that the V_{GS} reading voltage is larger than the threshold voltage of the MOSFET in an erased state.

To program the MOSFET, a large V_{DS} voltage is applied to accelerate the channel electrons to kinetic energies larger than the energy barrier between the silicon and the oxide. Thus accelerated electrons are referred to as *hot electrons*. A number of these hot electrons will elastically scatter, changing direction toward the oxide, and with their high kinetic energy they will overpass the energy barrier created by the oxide to appear in the floating-gate area. With a sufficiently thick floating gate, most of these electrons will lose their energy inside the floating gate (through nonelastic scattering), becoming trapped in the potential well created by the floating gate. Although this process does not seem to be very efficient, with relatively long programming times (on the order of μs), enough electrons can be collected in the potential well of the floating gate to change the MOSFET state. Figure 8.30d shows the energy bands of the MOSFET with electrons trapped in the floating gate. The electric field of these electrons shifts the floating-gate energy bands upward; as a consequence, it changes the band bending (and the electric field) at the substrate surface. This increases the threshold voltage of the MOSFET. Figure 8.30b shows that the threshold voltage of the programmed MOSFET can be larger than the V_{GS} reading voltage.

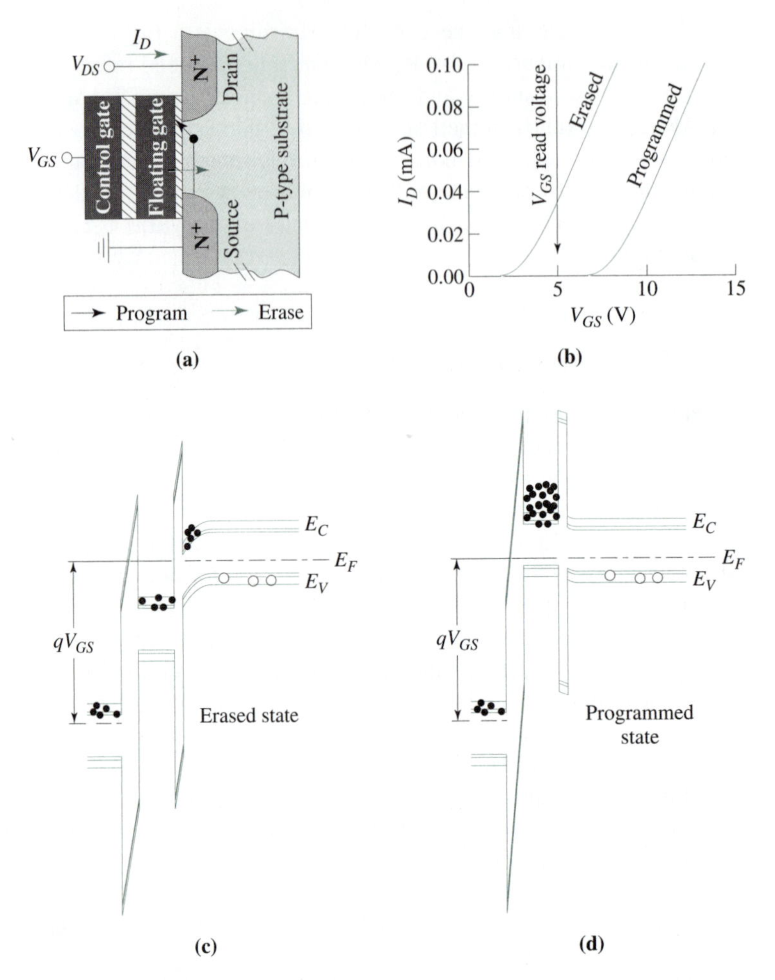

Figure 8.30 Flash memory MOSFET. (a) The cross section. (b) The transfer characteristics. (c) The energy bands in the erased state. (d) The energy bands in the programmed state.

To erase a programmed MOSFET, a large negative voltage is applied to the control gate, which forces the trapped electrons to tunnel back into the silicon substrate through the thin oxide separating the substrate and the floating gate (Fig. 8.30a).

The flash cell can be programmed and erased thousands of times before the erasing and programming mechanisms show any observable adverse impact on the MOSFET characteristics. Nonetheless, oxide stressing during programming and erasing creates oxide charge and interface traps. This accumulation of oxide charge and interface traps limits the number of programming and erasing cycles, ruling out any possible use of this nonvolatile memory for data processing (in addition, the programming and erasing are too slow for data-processing applications).

SUMMARY

1. The *inversion layer* of a MOS capacitor, with *source* and *drain* contacts at two ends, forms a switch in *on* mode. This is a voltage-controlled switch because the inversion layer (that is, the *channel* connecting the source and the drain) exists for as long as there is appropriate voltage at the *gate* electrode. In other words, the gate electric field can attract carriers of the same type as the source and drain regions (the switch in *on* mode), and can either deplete the channel region or attract opposite-type carriers, in which case the switch is in *off* mode (no conducting channel exists between the source and the drain terminals). This is known as the *field effect,* hence the name of the device: metal–oxide–semiconductor field-effect transistor (MOSFET).
2. There are N-channel MOSFETs (electrons in the inversion layer and N-type source/drain regions in a P-type substrate) and P-channel MOSFETs (holes in the channel and P-type source/drain regions in an N-type substrate). The N-channel MOSFETs are *on* when the gate voltage is larger, whereas the P-channel MOSFETs are *on* when the gate voltage is smaller than the *threshold voltage,*

$$V_T = V_{FB} \pm 2|\phi_F| \pm \gamma\sqrt{2|\phi_F| + |V_{SB}|}$$

   where the upper signs are for N-channel and the lower signs are for P-channel MOSFETs. A reverse-bias voltage across the source-to-substrate junction, $|V_{SB}|$, increases the threshold voltage (*body effect*).
3. N-channel and P-channel MOSFETs that are *off* for $V_{GS} = 0$ V are *normally off* MOSFETs, also called *enhancement-type* MOSFETs. MOSFETs with built-in channels (they can conduct current at $V_{GS} = 0$ V) are called *normally on,* or *depletion-type* MOSFETs. Enhancement-type N-channel and P-channel MOSFETs are used as voltage-controlled switches to create complementary MOS (CMOS) logic circuits.
4. The resistance of the channel (switch in *on* mode) is inversely proportional to $Q_I = C_{ox}(V_{GS} - V_T)$. Therefore, $R_{on} = 1/[\beta(V_{GS} - V_T)]$, which means that the current is

$$I_D = \beta(V_{GS} - V_T)V_{DS}$$

   At higher V_{DS} voltages, this linear dependence of I_D on V_{DS} is not maintained because the drain starts depleting the drain end of the channel, thereby reducing the current, which reaches its maximum at $V_{DS} = V_{DSsat}$. The region $0 \le |V_{DS}| \le |V_{DSsat}|$ is referred to as the *triode region*, and $|V_{DSsat}| < |V_{DS}| < |V_{BR}|$ is referred to as the *saturation region*. The saturation is reached due to either (1) full depletion of the drain end of the channel (channel "pinch-off") or (2) drift velocity saturation at high lateral electric fields in the channel. The saturation region enables the device to work as a current source whose current is controlled by the gate. Therefore, the MOSFET provides one possible implementation of the voltage-controlled current source used in analog circuits.
5. Modeling of the MOSFET current is based on Ohm's law, which accounts only for the linear region when applied literally (summary point 4). The rudimentary MOSFET model (SPICE LEVEL 1 model) modifies the basic Ohm relationship to include the reduction of channel-carrier density (Q_I) by the drain field. A more precise model is

TABLE 8.2 MOSFET Equations[a]

$$I_D = \begin{cases} \beta\left[(V_{GS}-V_T)V_{DS}-(1+F_B)\frac{V_{DS}^2}{2}\right], & \text{for } 0 \le |V_{DS}| < |V_{DSsat}| \\ \frac{\beta}{2(1+F_B)}(V_{GS}-V_T)^2, & \text{for } |V_{DS}| \ge |V_{DSsat}| \end{cases}$$

$$V_{DSsat} = \frac{V_{GS}-V_T}{1+F_B} \qquad \beta = \underbrace{\mu_{eff}\frac{\varepsilon_{ox}}{t_{ox}}}_{KP}\frac{W}{L_{eff}-L_{pinch}}$$

$$F_B = \frac{\gamma F_s}{2\sqrt{|2\phi_F| \pm V_{SB}}} + F_n \qquad \mu_{eff} = \frac{\mu_s}{1+\mu_s|V_{DS}|/(v_{max}L_{eff})}$$

$$\mu_s = \frac{\mu_0}{1+\theta|V_{GS}-V_T|}$$

$$V_T = V_{T0} \pm \gamma F_s\left(\sqrt{|2\phi_F| \pm V_{SB}} - \sqrt{|2\phi_F|}\right) - \sigma_D V_{DS} + F_n V_{SB}$$

Second-Order Effects	Principal Model
Mobility reduction due to vertical field (V_{GS} voltage)	$\theta = 0$
Mobility reduction due to lateral field (V_{DS} voltage)	$v_{max} = \infty$
Channel-length modulation in saturation region	$L_{pinch} = 0$
Short-channel charge sharing in the depletion layer	$F_s = 1$
Fringing-field widening of a narrow-channel depletion layer	$F_n = 0$
Drain-current increase in saturation due to DIBL	$\sigma_D = 0$

[a] Upper signs, N channel; lower signs, P channel.

obtained when the lateral nonuniformity of the depletion layer, which widens toward the reverse-biased drain-to-bulk junction, is included. This leads to a computationally inefficient model (SPICE LEVEL 2 model), involving $(V_{DS}+\cdots)^{3/2}$ terms. Simplifying this model by the linear and parabolic terms of Taylor series leads to what is referred to as the SPICE LEVEL 3 model. This model is structurally the same as the rudimentary model, with an added factor F_B to account for the effects of the depletion-layer variations. The LEVEL 3 model for the above-threshold drain current is summarized in Table 8.2.

6. A reduction of MOSFET dimensions (L, W, and t_{ox}), accompanied by an appropriate substrate-doping increase, results in faster CMOS cells that occupy a smaller area. Lower-doped drain and source extensions are used in small-dimension MOSFETs to reduce the electric field at the drain end of the channel. Deep-submicron MOSFETs also feature silicide gate layers and silicide source/drain contacts to decrease the gate and contact resistances. Better control over the channel region by the gate is needed to suppress excessive leakage current and to enable further scaling down of MOSFET dimensions. In response to this challenge, a number of potential techniques and advanced MOSFET structures are being considered, including high-k gate dielectrics, metal gates, engineered doping profiles in the substrate (retrograde and halo doping), ultra-thin-body SOI MOSFETs, and double-gate MOSFETs (FinFETs).
7. *DRAMs* and *flash* are two types of MOS-based semiconductor memory with a significant market impact. DRAMs are based on a 1C1T (one capacitor and one transistor) memory cell, where the capacitor is the charge-storage element and the transistor is the switch that enables fast and unlimited information reading and programming. Leakage

through modern MOSFETs, used for switching in DRAMs, is relatively high, so this type of memory has to be periodically refreshed. Flash is a nonvolatile semiconductor memory, however, with slow and limited programming cycles. In a flash memory cell, a polysilicon gate completely surrounded by oxide (floating gate) acts as a trap for electrons; thus, once it is charged, it remains in this state for a very long time. Charging (programming) and discharging (erasing) are achieved by injection of hot electrons and charge-tunneling through the gate oxide, respectively. These are relatively slow processes that also damage the gate oxide after a limited number of charge transfers.

PROBLEMS

8.1 Figure 8.31 shows five energy-band diagrams, drawn from the oxide–silicon interface into the silicon substrate, and the transfer characteristic of a MOSFET with four labeled points. Identify the four correct band diagrams and relate them to the four points on the transfer characteristic.

8.2 Figure 8.32 shows four energy-band diagrams, drawn from the source to the drain, along the silicon surface. Identify how the energy-band diagrams relate to each of the four points, labeled on the output characteristics of the MOSFET.

8.3 Which of the following statements, related to MOSFETs, are not correct?

(a) N-type substrate is used to make normally *on* P-channel MOSFETs.

(b) The net charge at the semiconductor surface is zero at $V_{GS} = V_T$.

(c) If a MOSFET is in the linear region, it is also in the triode region.

(d) Existence of a significant drain current at $V_{GS} = 0$ V indicates a faulty MOSFET.

(e) For a MOSFET in saturation, the channel carriers reach the saturation drift velocity at the pinch-off point.

(f) The threshold voltage of an enhancement-type P-channel MOSFET is negative.

(g) Positive gate voltage is needed to turn a normally *on* P-channel MOSFET off.

(h) A MOSFET cannot be in both the triode and the saturation region at the same time.

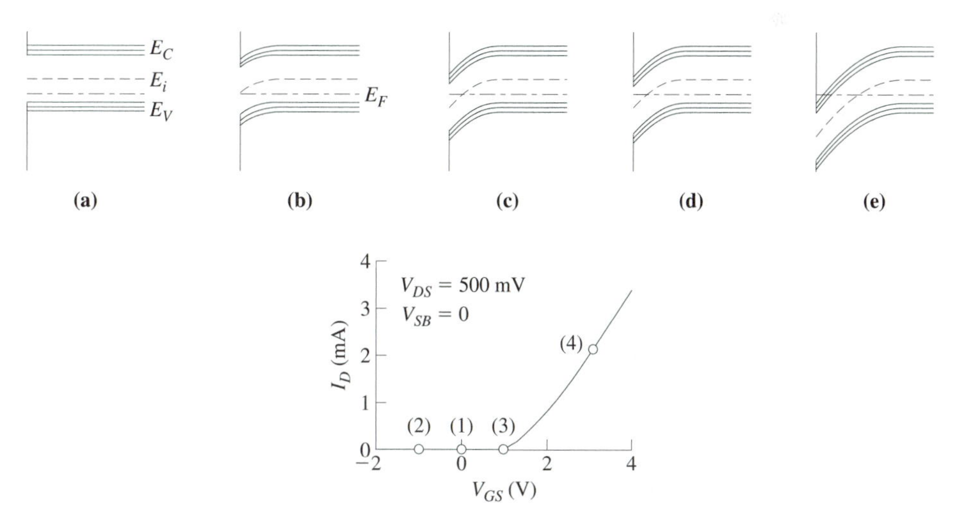

Figure 8.31 Energy-band diagrams and a MOSFET transfer characteristic.

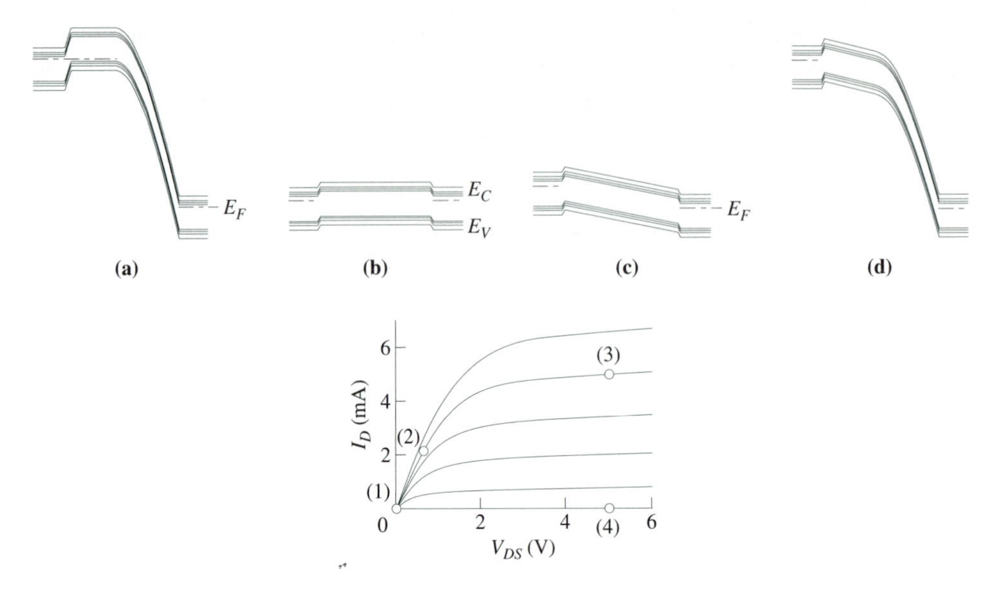

Figure 8.32 Energy-band diagrams and MOSFET output characteristics.

(i) The above-threshold current in the MOSFET channel is essentially due to diffusion.

(j) Both electrons and holes play significant roles in the flow of drain-to-source current.

8.4 For an N-channel MOSFET with uniform substrate doping of $N_A = 5 \times 10^{16}$ cm^{-3} and gate-oxide thickness of $t_{ox} = 5$ nm, determine the surface potential φ_s and the depletion-layer charge Q_d at:

(a) $V_{GS} = V_{FB} = -0.75$ V

(b) $V_{GS} = -0.5$ V and $V_{GS} = 0$ V (assume zero oxide charge and interface trap densities, so that $\varepsilon_{ox} E_{ox} = \varepsilon_s E_s = q N_A w_d$) **A**

(c) $V_{GS} = V_T = 0.2$ V

(d) $V_{GS} = 0.75$ V **A**

Plot $\varphi_s(V_{GS})$ and $Q_d(V_{GS})$. ($V_{SB} = 0$)

8.5 The transconductance of an N-channel MOSFET operating in the linear region ($V_{DS} = 50$ mV, $V_{SB} = 0$) is $g_m = dI_D/dV_{GS} = 2.5$ mA/V. If the threshold voltage is $V_T = 0.3$ V, what is the current at $V_{GS} = 1$ V?

8.6 Design an N-channel MOSFET, used as a voltage-controlled switch, so that the resistance in *on* mode is $R = 100$ Ω. The technological and circuit parameters are as follows: the threshold voltage is $V_T = 0.2$ V, the gate-oxide thickness is $t_{ox} = 3$ nm, the electron mobility in the channel is $\mu_n = 350$ cm^2/V · s, the gate voltage in *on* mode is $V_{GS} = 1.0$ V, and the minimum channel dimension is 0.2 μm.

8.7 The substrate doping and the body factor of an N-channel MOSFET are $N_A = 10^{16}$ cm^{-3} and $\gamma = 0.12$ V$^{1/2}$, respectively. If the threshold voltage, measured with $V_{SB} = 3$ V is $V_T = 0.5$ V, what is the zero-bias threshold voltage?

8.8 For the MOSFET of Problem 8.7, how many times is the channel resistance increased when V_{SB} is increased from 0 to 3.3 V? The gate and drain voltages are $V_{GS} = 3.3$ V and $V_{DS} = 50$ mV. **A**

8.9 Knowing the technological parameters $t_{ox} = 3.5$ nm, $N_D = 5 \times 10^{15}$ cm^{-3}, and $V_{FB} = 0.2$ V, determine the inversion-layer charge density at $V_{GS} = -0.75$ V, $V_{GS} = 0$ V, and $V_{GS} = 0.75$ V for

(a) $V_{BS} = 0$ V

(b) $V_{BS} = 0.75$ V **A**

8.10 The body factors of N-channel and P-channel MOSFETs are determined from body-effect measurements as 0.11 V$^{1/2}$ and 0.47 V$^{1/2}$, respectively. Determine the substrate doping levels in those MOSFETs. The gate-oxide capacitance is $C_{ox} = 1.726 \times 10^{-3}$ F/m^2.

8.11 The maximum operating voltage of an NMOS integrated circuit is 10 V and substrate doping level in the region between the MOSFETs (the field region) is $N_A = 5 \times 10^{17}$ cm^{-3}. Determine the minimum oxide thickness in the field region (the field-oxide thickness) needed to prevent current leakage between neighboring MOSFETs. Neglect the oxide charge, and consider aluminum gate (the worst-case scenario).

Hint: The field oxide can be considered as the gate oxide of a parasitic MOSFET that should be kept *off* (the maximum operating voltage should be below the threshold voltage) to prevent possible leakage.

8.12 The solid line in Fig. 8.33 (labeled by "N") is for an N$^+$-poly-N-channel MOSFET with the following parameters: $L = 2$ μm, $W = 2$ μm, $t_{ox} = 20$ nm, $N_A = 5 \times 10^{16}$ cm^{-3}, $x_j = 0.5$ μm, $x_{j-lat} = 0.4$ μm, $\mu_0 = 750$ cm^2/Vs, $\delta = 1$, $\theta = 0$, and $\eta = 0$. The other four characteristics are obtained by changing one of the listed parameters. State the altered parameter that relates to each of the transfer characteristics labeled by 1, 2, 3, and 4.

8.13 One set of output characteristics from Fig. 8.34 is for the nominal MOSFET parameters, as listed in the text of Problem 8.12, while the other three are for changed values of θ or η or for a specified v_{max} parameter. These parameters determine the strength of the following second-order effects: θ, mobility reduction with gate voltage; η, finite output resistance due to DIBL; v_{max}, the drift-velocity saturation. Relate each of the output characteristics to the appropriate set of parameters.

8.14 An N-channel MOSFET with $V_T = 0.25$ V is biased by $V_{GS} = 2.5$ V and $V_{DS} = 500$ mV. The gate-oxide capacitance is $C_{ox} = 2.5$ mF/m^2 and the effective channel length is $L = 1$ μm. Calculate:

(a) the average lateral field

(b) the average channel conductivity, assuming channel thickness $x_{ch} = 5$ nm

(c) current density

Assume $\mu_0 = 750$ cm^2/V · s for the channel-carrier mobility.

8.15 If the channel length of the MOSFET from Problem 8.14 is reduced to $L = 0.2$ μm, calculate the average lateral electric field for the same bias conditions and the same value of the threshold voltage. Assuming drift velocity of $v_d = 0.08$ μm/ps at this field (Fig. 3.6) and using the average carrier concentration in the channel ($x_{ch} = 5$ nm), calculate the current density. What is the channel-carrier mobility in this case? **A**

8.16 A P-channel MOSFET has $V_{T0} = 0.2$ V, $\gamma = 0.2$ V$^{1/2}$ (neglect F_B assuming $F_B \ll 1$), and $\beta = 5$ mA/V^2. If $V_{GS} = -1$ V, calculate V_{DSsat}, I_{Dsat}, and I_D at $V_{DS} = V_{DSsat}/5$ for

(a) $V_{SB} = 0$ V

(b) $V_{SB} = -4.1$ V, $|\phi_F| = 0.45$ V **A**

8.17 **(a)** Repeat the calculations of Example 8.2 using $\theta = 0.1$ V^{-1} to include the effect of mobility reduction with gate voltage. Use the SPICE LEVEL 3 model.

(b) Find the change in the drain current if the body of the MOSFET is biased at $V_{SB} = 5$ V. **A**

8.18 The technological parameters of an N$^+$-poly-gate P-channel MOSFET are $L/W = 10$, $t_{ox} = 4.5$ nm, and $N_D = 10^{16}$ cm^{-3}. Find the drain current at

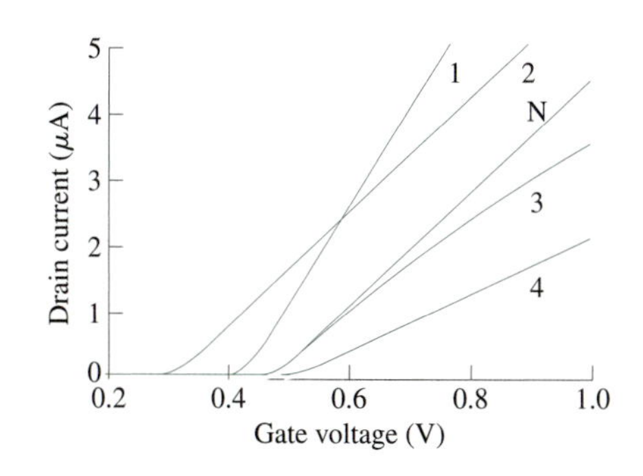

(a) $L = 1.5$ μm
(b) $W = 1$ μm
(c) $\theta = 0.5$ V^{-1}
(d) $N_A = 3 \times 10^{16}$ cm^{-3}

Figure 8.33 MOSFET transfer characteristics.

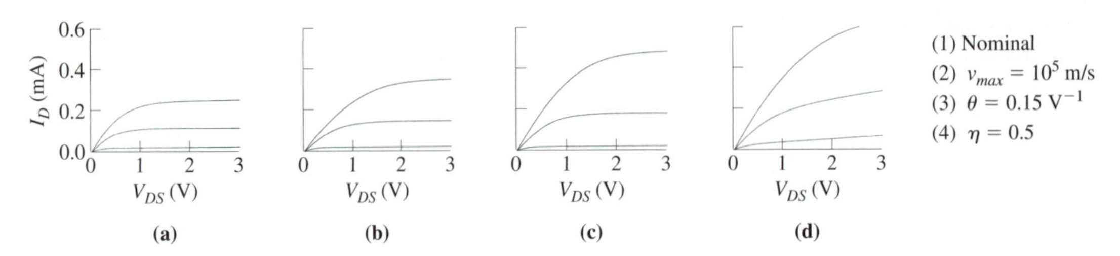

Figure 8.34 MOSFET output characteristics.

$V_{GS} = -2$ V, $V_{DS} = -50$ mV, and $V_{SB} = 0$ V. Assume $\mu_0 = 350$ cm^2/V · s and $N_{oc} = 0$.

8.19 Considered is an N-channel MOSFET with the gain factor $\beta = 600\ \mu$ A/V^2, the drain-bias factor $F_B = 0.7$, and the zero-bias threshold voltage $V_{T0} = 1.1$ V. If the coefficient of the influence of the drain bias on the threshold voltage is $\sigma_D = 0.01$, determine the dynamic output resistance ($r_o = dV_{DS}/dI_D$) of the MOSFET at $V_{GS} = V_{DS}/2 = 5$ V. Channel-length modulation can be neglected.

8.20 The output dynamic resistance of an N-channel MOSFET with $\sigma_D = 0.01$ and negligible channel-length modulation effect is $r_o = 1$ MΩ at $V_{GS} - V_T = 0.5$ V. What is the dynamic output resistance at $V_{GS} - V_T = 5$ V if

(a) $\theta = 0$ **A**

(b) $\theta = 0.05$ V^{-1}

8.21 If the saturation current of an N-channel MOSFET is 1 mA when measured at $V_{GS} = 2$ V, $V_{SB} = 0$ V, and $V_{DS} = 5$ V, what is the saturation current at $V_{GS} = 2$ V, $V_{SB} = 0$ V, and $V_{DS} = 10$ V? The following parameters are known: the gain factor $\beta = 3$ mA/V^2, the drain-bias factor $F_B = 0.5$, and the zero-bias threshold voltage $V_{T0} = 1.1$ V.

8.22 The length of the channel pinch-off region can be expressed as the depletion layer of an abrupt P–N junction, modulated by a fitting parameter κ:

$$L_{pinch} = \sqrt{\kappa \frac{2\varepsilon_s}{qN_A}(V_{DS} - V_{DSsat})}$$

What is the relative increase (expressed as percentage) of the drain current of a 1-μm MOSFET, when the voltage changes from $V_{DS} = V_{DSsat}$ to $V_{DS} = 5\text{ V} + V_{DSsat}$, if $N_A = 5 \times 10^{16}$ cm^{-3} and $\kappa = 0.2$? Assume constant threshold voltage ($\sigma_D = 0$).

8.23 The body of a nanoscale N-channel MOSFET is doped at $N_A = 5 \times 10^{18}$ cm^{-3}.

(a) Determine the maximum inversion-layer charge per unit area (Q_I) that can be reached so that the electric field in the oxide remains below $E_{ox} = 0.6$ V/nm. Neglect any interface-trap and gate-oxide charge.

(b) If the channel length and the channel width are $L = 50$ nm and $W = 500$ nm, respectively, determine how many electrons are creating this inversion layer.

(c) Calculate the threshold voltage of a MOSFET having gate-oxide thickness $t_{ox} = 1$ nm and flat-band voltage $V_{FB} = 1.0$ V.

(d) What gate voltage is needed to form the maximum inversion-layer charge?

8.24 The electron mobility in the channel of a nanoscale MOSFET is $\mu_n = 300$ cm^2/V · s. The channel length is $L = 50$ nm, and the applied voltage across the channel is $V_{DS} = 0.1$ V. Determine the scattering length (average distance between two scattering events) for the electrons in the channel. Assuming that an electron moves from the source toward the drain, calculate the kinetic energy that the electron gains between two scattering events. The effective mass of the electrons is $m^* = 0.19m_0$ and their thermal velocity is $v_{th} = 2 \times 10^7$cm/s.

REVIEW QUESTIONS

R-8.1 What type of substrate (N or P) is used to make normally *off* N-channel MOSFETs? Normally *on* N-channel MOSFETs?

R-8.2 What gate-to-source voltage, positive or negative, is needed to turn a normally *on* P-channel MOSFET *off*?

R-8.3 Can a single MOSFET be used as both a voltage-controlled switch (digital operation) and a voltage-controlled current source (analog operation)?

R-8.4 Typically, is the surface potential φ_s zero at $V_{GS} = 0$ V? What is the condition of $\varphi_s = 0$ called?

R-8.5 Can a normally *on* and a normally *off* N-channel MOSFET have the same flat-band voltage? Are the electrical conditions (energy bands) in the silicon of normally *off* and normally *on* MOSFETs equivalent at $\varphi_s = 0$?

R-8.6 Is there any charge at the semiconductor side of a MOS structure at $V_{GS} = V_T$? Is there any mobile carrier charge?

R-8.7 Is the threshold voltage of a normally *on* P-channel MOSFET positive or negative?

R-8.8 Why does source-to-bulk reverse-bias voltage ($V_{SB} > 0$) increase the threshold voltage? What is this effect called?

R-8.9 Can a MOSFET simultaneously be in both the linear and the triode region?

R-8.10 Do channel carriers face a negligible or infinitely large resistance between the channel pinch-off point and the drain of a MOSFET in saturation?

R-8.11 Is there analogy between the energy bands of a MOSFET in saturation and a waterfall?

R-8.12 Which SPICE model (LEVEL 1, 2, or 3) would you use to simulate a circuit with MOSFETs? Why?

R-8.13 The mobility reduction with gate voltage is a second-order effect. Can you, typically, neglect it?

R-8.14 What is the effect of neglecting the mobility reduction with the drain voltage?

R-8.15 What is the effect of neglecting the threshold-voltage dependence on V_{DS} voltage?

9 BJT

The bipolar junction transistor (BJT) was the first solid-state active electronic device. Before the BJT, electronic amplifiers were based on vacuum tubes. The BJT concepts were experimentally and theoretically established by Bardeen, Brattain, and Shockley at the Bell Telephone Laboratories during 1948. The era of semiconductor-based electronics, which has had an enormous influence on the way we live today, actually began with the invention of the BJT.

A number of alternative transistors have been developed since the first BJTs, notably MOSFETs and MESFETs. Nonetheless, the BJTs are still used because there are applications in which the BJTs still offer the best performance. In addition, there are applications in which they are combined with MOSFETs, even in integrated-circuit technology. It should also be noted that the BJT principles are frequently used in a number of specifically designed semiconductor devices.

9.1 BJT PRINCIPLES

Different from MOSFETs, the principal operation of a BJT (the normal active mode) relates to its function as a controlled source of constant current (an analog device). Nonetheless, BJTs can be operated as voltage-controlled switches (a digital device). Both digital circuits and semiconductor power-switching circuits were initially developed with BJTs.[1] This section introduces the BJT as a controlled current source, followed by a

[1] Resistor–transistor logic (RTL), transistor–transistor logic (TTL), and emitter-coupled logic (ECL) are all digital circuits based on BJTs.

description of its operation as a switch to present all the four modes of operation. At the end of the section, the BJT is compared to the MOSFET.

9.1.1 BJT as a Voltage-Controlled Current Source

The essential characteristic of an *ideal* current source is that its current does not depend on the voltage across the current source. In other words, it delivers a constant current at any voltage.

A reverse-biased P–N junction is a semiconductor implementation of a constant-current source, assuming that the current through a reverse biased P–N junction is due to the minority carriers. The energy-band diagram of Fig. 9.1 shows that the minority electrons easily roll down and the minority holes easily bubble up through the depletion layer, making the reverse-bias current. This current is limited by the number of minority electrons and holes appearing at the edges of the depletion layer per unit time, and not by the reverse-bias voltage V_{CB}, which sets the energy difference between the P-type and N-type regions. The flow of minority electrons through the depletion layer is analogous to a waterfall: the current of the falling water does not depend on the height of the fall but on the current of water coming to the edge of the fall.

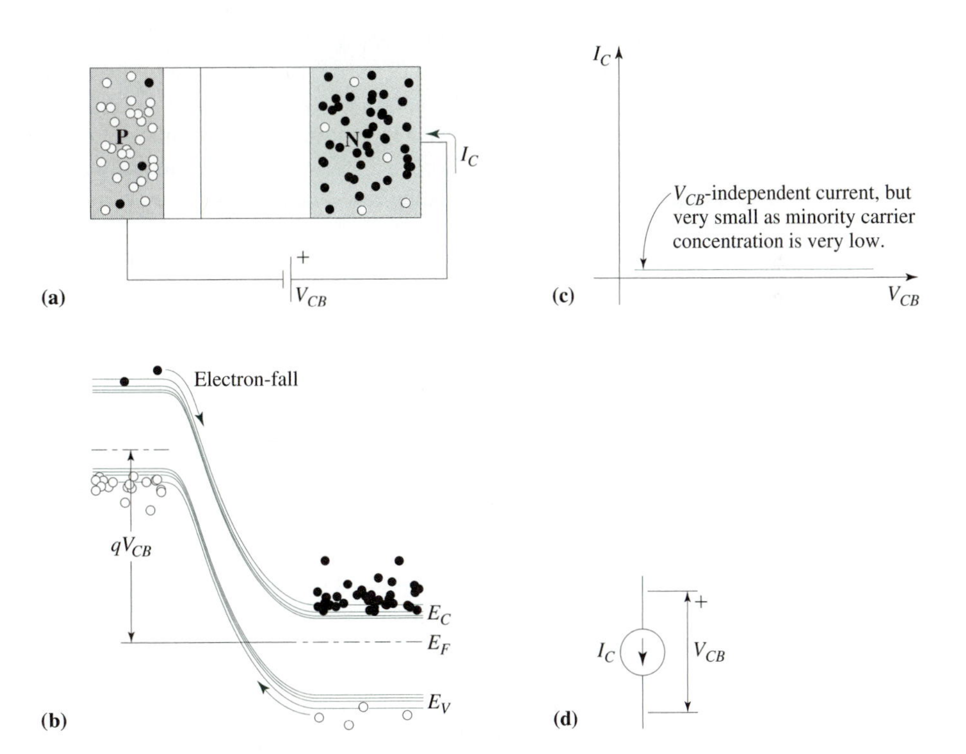

Figure 9.1 Reverse-biased P–N junction as a current source. (a) Cross section. (b) Energy-band diagram. (c) *I*–*V* characteristic. (d) Current-source symbol.

Being a device whose current does not change with the voltage across the device, the reverse-biased P–N junction exhibits the main characteristic of a constant-current source, at least theoretically. Figure 9.1c and 9.1d illustrates the I–V characteristic and the symbol of the reverse-biased P–N junction used as a current source. Of course, a very important question here is whether this current source is at all useful. It may seem that its current is too small for any realistic application.

It is true that the P–N junction reverse current is only a leakage current, usually negligible. However, what is important here is the principle of waterfall, or "electron-fall" as labeled in Fig. 9.1b. The current through the reverse-biased P–N junction can be increased to a significant level by providing more electrons, in the same way that the current of a waterfall increases after a heavy rainfall. In fact, it is necessary to have a way of controlling the number of electrons appearing at the edge of the "fall" so as to create a *controlled* current source.

More minority electrons in the P-type region can be created by increased temperature or exposure to light, which would break additional covalent bonds and generate additional electron–hole pairs. This would make a temperature-controlled or light-controlled current source. However, to have an electronic amplifying device, we need a current source that is electrically controlled, say a voltage-controlled current source.

Thinking of a *supply of electrons* that is *controlled by a voltage,* the forward-biased P–N junction appears as a possibility. As the forward bias causes a significant number of electrons to move from the N-type region into the P-type region, the forward-biased P–N junction could be used to supply electrons to the current source. Obviously, this can work only if the two P–N junctions, the forward-biased (the controlling junction) and the reverse-biased (the current source), share a common P-type region. This is the case in an NPN BJT structure, illustrated in Fig. 9.2a. The common P-type region is called a *base.* The N-type of the forward-biased P–N junction, which emits the electrons, is called an *emitter,* and the N type of the reverse-biased P–N junction is called a *collector* because it collects the electrons.

Figure 9.2 summarizes the operation of the NPN BJT as a voltage-controlled current source. The forward-bias voltage V_{BE} (the input voltage) controls the supply of electrons from the emitter to the depletion layer of the reverse-biased P–N junction ("electron-fall"). The output current depends on the input voltage V_{BE} (shown by the transfer characteristic), but it does not depend on the output voltage V_{CB} (horizontal lines of the output current–voltage characteristics).

The equivalent circuit in Fig. 9.2d is shown for a BJT in the *common-emitter* configuration. In this most frequently used configuration, the input voltage is between the base and the emitter, whereas the output voltage is between the collector and the emitter. Practically, there is no qualitative difference between V_{CE} and V_{CB}. Although V_{CE} involves the input V_{BE} voltage ($V_{CE} = V_{CB} + V_{BE}$), the changes in input voltage (the input-voltage signal) are practically very small and can almost always be neglected in comparison to other voltages in a BJT circuit. The total value of V_{BE} may be important, but it is almost always sufficient to approximate it by a constant value (the commonly used value in electronic-design books is $V_{BE} = 0.7$ V).

To be used as a signal amplifier, a BJT has to be connected to a power-supply circuit. BJTs cannot generate power to deliver to the amplified signal, but what they can do is

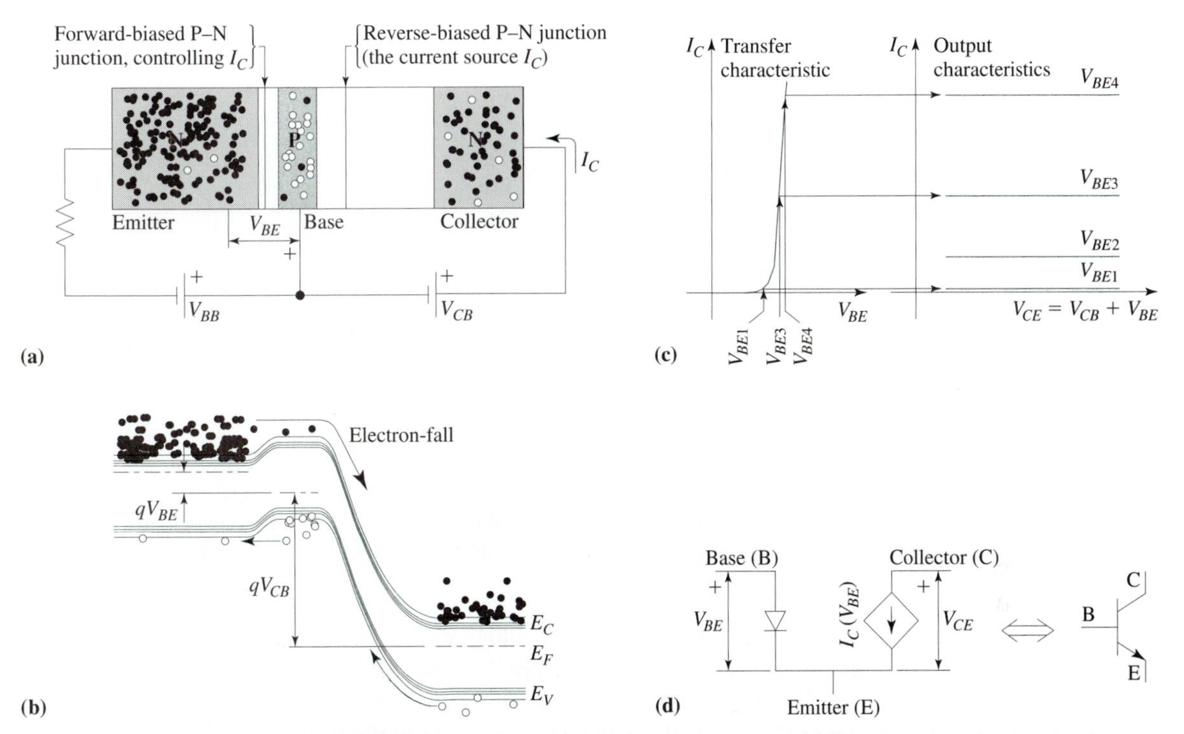

Figure 9.2 Summary of NPN BJT operation as a voltage-controlled current source. (a) Cross section showing the three regions, their names, the two junctions, and the biasing arrangement. (b) Energy-band model. (c) Main current–voltage characteristics. (d) An equivalent circuit (left) and the symbol (right).

convert supplied DC voltage/power into signal voltage/power. The DC voltage/power has to be supplied to the *active* device (the BJT in this case) through a *loading* element. Figure 9.3a shows the basic amplifying circuit, where the active device is the BJT, the loading device is the resistor R_C, and the power-supply voltage is labeled by V_+. Figure 9.3b provides a graphical analysis of this circuit to illustrate the principle of voltage amplification by a voltage-controlled current source. The output characteristics of the voltage-controlled current source (i_C–v_{CE} graph) are accompanied by the so-called *load line,* which is the i_C–v_{CE} dependence as determined by the loading resistor R_C and the DC power supply V_+. The load-line equation can be obtained from the fact that V_+ voltage is divided between the resistor ($R_C i_C$) and the BJT output (v_{CE}):

$$V_+ = R_C i_C + v_{CE} \tag{9.1}$$

The aim is to present this equation in the form of $i_C(v_{CE})$ so that it can be plotted on the same graph as the output characteristics of the BJT. From Eq. (9.1),

$$i_C = -\frac{1}{R_C} v_{CE} + \frac{V_+}{R_C} \tag{9.2}$$

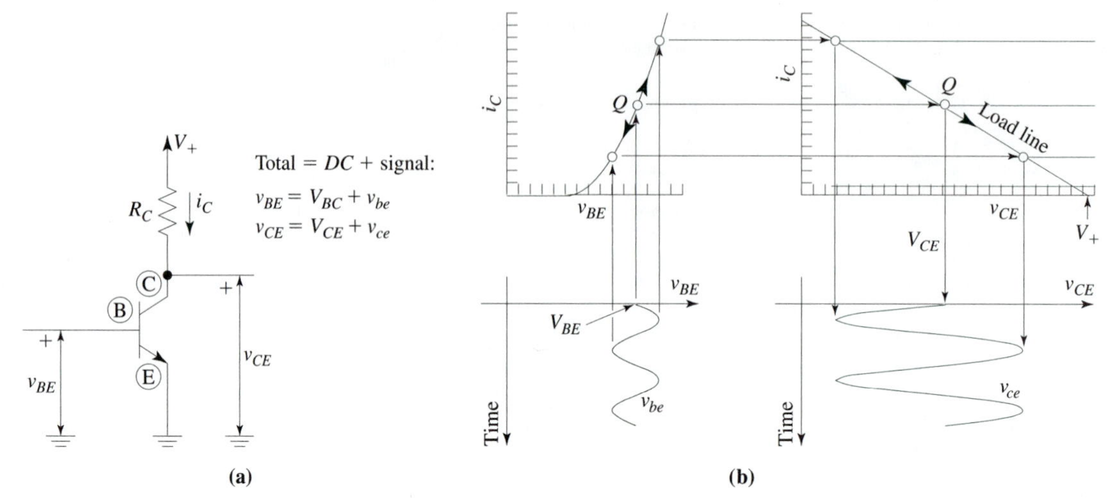

Figure 9.3 Principles of voltage amplification by a voltage-controlled current source. (a) A BJT is connected to a DC power supply through a loading resistor to create the principal amplifier circuit. (b) Graphic analysis of the amplifier circuit.

This is the i_C–v_{CE} characteristic of the loading R_C–V_+ circuit that together with the output i_C–v_{CE} characteristics of the BJT determines the output voltage (v_{CE}) and the output current (i_C). In other words, the actual v_{CE} and i_C values are found at the cross sections between the load line and the output characteristics of the BJT.

With this understanding, we can follow the transition of the small input signal into the large output signal that is illustrated in Fig. 9.3b. The sinusoidal input signal (v_{be}) oscillates the total input voltage around the quiescent point Q that is set in the middle of a linear-like segment of the transfer characteristic. The links between the transfer and the output characteristics show that a specific horizontal line of the output characteristics is "selected" by the input voltage. A horizontal line of the output characteristics means that a "selected" i_C current is possible for a range of v_{CE} voltages. The actual v_{CE} voltage is determined by the load line and is found at the cross section between the load line and the set i_C of the current source; thus, the current oscillations are converted into voltage oscillation. The amplitude of the sinusoidal output voltage is inversely proportional on the slope of the load line. Accordingly, it can be set at an arbitrarily high level, provided the oscillations remain within the operating regime of the output BJT characteristics. It should be noted that, for clarity, the slope of the transfer i_C–v_{BE} characteristics in Fig. 9.3 is reduced. In practice, the slope is so much higher that the amplitude of the input sinusoidal voltage is typically much smaller than the amplitude of the output signal. This means a very large voltage gain, v_{ce}/v_{be}.

9.1.2 BJT Currents and Gain Definitions

In general, four different currents flow through the two P–N junctions of a BJT. These currents will be labeled by I_{nE} and I_{pE} to denote the electron and hole currents through the emitter–base (E–B) junction and by I_{nC} and I_{pC} to denote the electron and hole currents

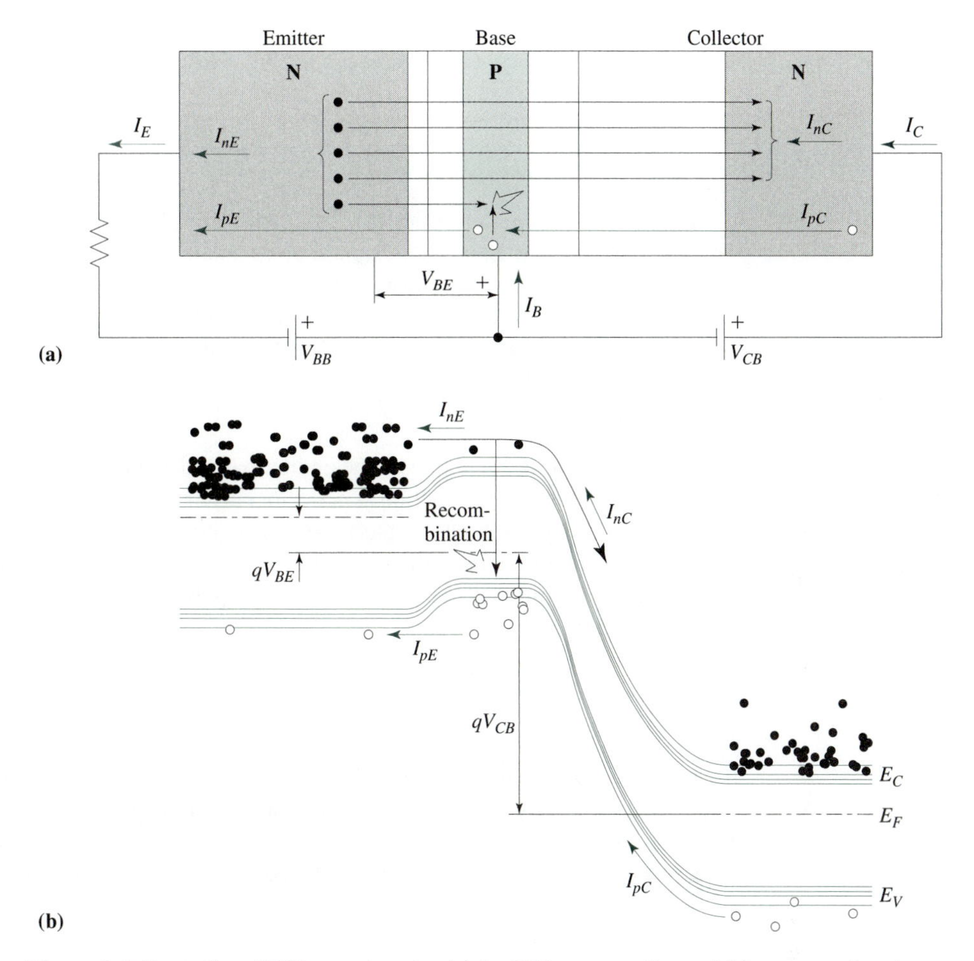

Figure 9.4 Illustration of BJT currents using (a) the BJT cross section and (b) an energy-band diagram.

through the collector–base (C–B) junction. Figure 9.4 illustrates the relationships between these currents.

Emitter Efficiency, γ_E

The N-type emitter of the forward-biased E–B junction emits electrons into the the P-type base, which is the I_{nE} current. The emission of these electrons is controlled by the input V_{BE} voltage, so it is these electrons that should be supplied to the reverse-biased B–C junction (the current source) to make the controlled (useful) transistor current. The other current through the E–B junction is due to the holes emitted from the base into the emitter (I_{pE} current). This current is not a useful transistor current because it is enclosed in the input circuit. The total emitter current is the sum of these two currents:

$$I_E = I_{nE} + I_{pE} \tag{9.3}$$

The ratio between the useful and the total emitter current is called *emitter efficiency* and is labeled by γ_E:

$$\gamma_E = \frac{I_{nE}}{I_E} \tag{9.4}$$

To maximize the emitter efficiency, I_{nE} (the useful current) should be as large as possible compared to I_{pE}, which should be as small as possible. Although both currents depend exponentially on the forward bias V_{BE}, they also depend on the majority-carrier concentrations—electrons in the emitter region and holes in the base region. To maximize I_{nE} and minimize I_{pE}, the doping level of the emitter should be as high as possible and the doping level of the base should be as low as possible.

Transport Factor, α_T

Most of the electrons emitted from the emitter pass through the base region to be collected by the reverse-biased C–B junction as collector current I_{nC}. However, some of the electrons are recombined by the holes in the P-type base, contributing to the base and not to the collector current. Obviously, the recombined electrons do not contribute to the transistor current. The ratio of electrons successfully transported through the base region is called *transport factor:*

$$\alpha_T = \frac{I_{nC}}{I_{nE}} \tag{9.5}$$

To maximize the transport factor, the recombination in the base has to be minimized, which is achieved by making the base region as thin as possible. This is well illustrated by a possible argument that two P–N junction diodes with connected anodes (P-type sides) electrically make the structure of the NPN BJT. The problem with such a BJT is that it is useless because of its zero transport factor: all the emitted electrons are recombined in the base, leaving any output (collector) current unrelated to the input current and voltage.

Transconductance, g_m

The collector current is not only due to the electrons arriving from the emitter. There is a small current due to the minority holes that move from the collector into the P-type base. This current, labeled I_{pC} in Fig. 9.4, is a part of the reverse-bias current of the C–B junction. The other part is the current of minority electrons that would exist even when no electrons are emitted from the emitter (zero- or reverse-biased E–B junction). The reverse-bias current of the C–B junction is usually labeled by I_{CB0}. It is a small leakage current, which can most frequently be neglected. It can be noticed only when the BJT is in *off* mode (both E–B and C–B junctions are zero- or reverse-biased) and is therefore used to characterize the leakage of a BJT in *off* mode. Neglecting I_{CB0} current, the terminal collector current becomes equal to the transistor current:

$$I_C \approx I_{nC} \tag{9.6}$$

The emitter current I_E depends exponentially on the input bias voltage V_{BE}, according to the diode relationship (6.4). The currents I_{nE}, I_{nC}, and eventually the output current I_C

maintain this exponential relationship:

$$I_C(V_{BE}) = I_C(0)e^{V_{BE}/V_t} \tag{9.7}$$

Equation (9.7) is the transfer characteristic shown in Fig. 9.2c and 9.3b. The slope of the transfer characteristic determines the gain that can be achieved by the BJT. For a voltage-controlled current source, the gain is defined as a *transconductance* (expressed in A/V):

$$g_m = \frac{dI_C}{dV_{BE}} = \frac{1}{V_t}\underbrace{I_C(0)e^{V_{BE}/V_t}}_{I_C} = \frac{I_C}{V_t} \tag{9.8}$$

Common-Base and Common-Emitter Current Gains, α and β

Alternatively, a BJT can be considered as a current-controlled current source. This is possible because the input voltage is related to the input current. By considering the BJT as a current-controlled current source, the BJT gain is defined as a unitless current gain. If we specify the collector as the output of a BJT used as an amplifier, the input can be either the emitter (in which case the base is common) or the base (in which case the emitter is common). These two configurations (common base and common emitter) lead to two possible current gain definitions:

common-base current gain:	common-emitter current gain:
$\alpha = I_C/I_E$	$\beta = I_C/I_B$
I_C—output	I_C—output
I_E—input	I_B—input

(9.9)

Being current ratios of terminal currents, α and β can be electrically measured, and because of this they are referred to as electrical parameters. The following equation shows that α is directly related to the emitter efficiency γ_E and the transport factor α_T, which are technological parameters:

$$\alpha = \frac{I_C}{I_E} \approx \frac{I_{nC}}{I_E} = \underbrace{\frac{I_{nC}}{I_{nE}}}_{\alpha_T}\underbrace{\frac{I_{nE}}{I_E}}_{\gamma_E} \tag{9.10}$$

$$\alpha = \alpha_T\gamma_E \tag{9.11}$$

The theoretical maximum for γ_E is 1 (no holes injected back into the emitter), and the theoretical maximum for α_T is 1 as well (no electrons recombined in the base). This means that the common-base current gain α cannot be larger than 1. Note that this does not mean that the common-base configuration is useless; it cannot provide a real current gain (the current gain is ≤ 1), but it can provide a power gain, for example.

The α and β factors are related. The following equations show the relationship:

$$\beta = \frac{I_C}{I_B} = \frac{I_C}{I_E - I_C} = \frac{1}{\frac{1}{\underbrace{I_C/I_E}_{\alpha}} - 1} = \frac{1}{\frac{1}{\alpha} - 1} \tag{9.12}$$

$$\beta = \alpha/(1-\alpha) \tag{9.13}$$

$$\alpha = \beta/(1+\beta) \tag{9.14}$$

Thus, if α is known, β can be calculated using Eq. (9.13), and vice versa; that is, if β is known, α can be calculated using Eq. (9.14). Typically, $\alpha > 0.99$ (but always <1), and $\beta > 100$.

Equation (9.10) shows that α and β are independent of any circuit biasing, provided the BJT is biased to operate as an amplifier—they depend only on α_T and γ_E. Once a BJT has been made, α and β are set and do not change with circuit conditions. As distinct from this, the transconductance gain depends on the biasing current I_C [Eq. (9.8)]. Consequently, α and β are used as the main BJT parameters.

EXAMPLE 9.1 BJT Currents

The common-emitter gain of a BJT operating as a voltage-controlled current source is $\beta = 450$. Calculate the base and the emitter currents if the collector current is 1 mA. What is the common-base current gain α?

SOLUTION

The base current is calculated from the definition of β [Eq. (9.9)]:

$$I_B = \frac{I_C}{\beta} = 2.22\ \mu\text{A}$$

The emitter current is obtained from Kirchhoff's first law applied to the BJT:

$$I_E = I_C + I_B = (\beta + 1)I_B = 1.002\ \text{mA}$$

The common-base current gain is

$$\alpha = \frac{\beta}{\beta + 1} = 0.9978$$

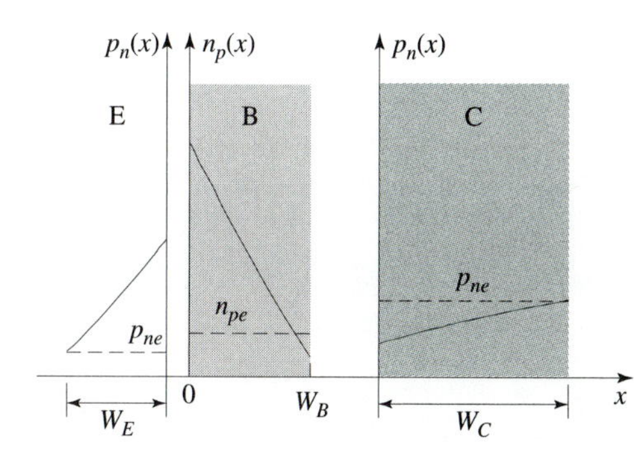

Figure 9.5 Profiles of minority carriers in a BJT (for clarity, the concentration scales in the base, the emitter, and the collector are different).

9.1.3 Dependence of α and β Current Gains on Technological Parameters

Equation (9.10) expresses the current gain(s) in terms of the emitter efficiency γ_E and the transport factor α_T, which are defined as ratios of electron and hole currents. These currents depend on technological and physical parameters that ultimately determine the current gains α and β.

Emitter Efficiency

The currents I_{nE} and I_{pE}, defining γ_E, are the two components of the current flowing through the forward-biased E–B junction. Consistent with the P–N junction theory (Chapter 6), I_{nE} is limited by the diffusion of electrons as the minority carriers in the base and I_{pE} is limited by the diffusion of holes as the minority carriers in the emitter. Figure 9.5 illustrates the concentration profiles of the minority carriers that determine the diffusion currents in a BJT. In general, these profiles are exponential [Eq. (5.18)], but assuming that the emitter, the base, and the collector widths are much smaller than the diffusion lengths, the exponential dependencies become close to the linear-like profiles shown in Fig. 9.5. Applying the diffusion-current equation [Eq. (4.5)] to the minority electrons in the base,

$$I_{nE} = qA_J D_B \frac{dn(x)}{dx} = qA_J D_B \frac{n_p(0) - n_p(W_B)}{W_B} \tag{9.15}$$

where A_J is the junction area and D_B is the diffusion constant of the minority carriers in the base.[2] The forward-bias voltage V_{BE} increases the minority-carrier concentration at the edge of the depletion layer exponentially above the equilibrium level. As a result, $n_p(0) \gg n_{pe}$. Moreover, $n_p(W_B) < n_{pe}$ because the reverse bias drops the minority-carrier concentrations below the equilibrium level at the C–B junction (Fig. 9.5). Therefore,

[2]Equation (9.15) is analogous to Eq. (6.5), which is the starting equation for derivation of the diode current. The main difference is in the replacement of the diffusion length L_n by the actual base width W_B because of the assumption that $W_B \ll L_n$. The minus sign is omitted in Eq. (9.15) because the chosen direction of the current I_{nE} is opposite to the direction of the x-axis in Fig. 9.5.

$n_p(W_B)$ can be neglected in comparison to $n_p(0)$:

$$I_{nE} = qA_J D_B \frac{n_p(0)}{W_B} \tag{9.16}$$

In analogy with Eq. (6.9):

$$n_p(0) = n_{pe} e^{V_{BE}/V_t} \tag{9.17}$$

With this, Eq. (9.16) becomes

$$I_{nE} = qA_J D_B \frac{n_{pe}}{W_B} e^{V_{BE}/V_t} \tag{9.18}$$

The equilibrium concentration of the minority carriers in the base is determined by $n_{pe} = n_i^2/N_B$, where N_B is the doping level in the base. Therefore,

$$I_{nE} = qA_J n_i^2 \frac{D_B}{N_B W_B} e^{V_{BE}/V_t} \tag{9.19}$$

The equation for I_{pE} can be written in analogy with Eq. (9.19):

$$I_{pE} = qA_J n_i^2 \frac{D_E}{N_E W_E} e^{V_{BE}/V_t} \tag{9.20}$$

where N_E is the doping level in the emitter and D_E is the diffusion constant of the minority holes in the emitter.

The ratio of these two currents,

$$\frac{I_{pE}}{I_{nE}} = \frac{D_E}{D_B} \frac{N_B W_B}{N_E W_E} \tag{9.21}$$

depends on the ratio of the diffusion constant, on the ratio of the base and emitter widths, and, most importantly, on the ratio between the base and emitter doping levels. This equation shows that $N_B \ll N_E$ will ensure that the unwanted current I_{pE} is much smaller than the useful current I_{nE}. The formal definition for the emitter efficiency can be expressed in terms of this ratio:

$$\gamma_E = \frac{I_{nE}}{I_E} = \frac{I_{nE}}{I_{nE} + I_{pE}} = \frac{1}{1 + (I_{pE}/I_{nE})} \tag{9.22}$$

Eliminating I_{pE}/I_{nE} from Eqs. (9.21) and (9.22), the emitter efficiency is expressed in terms of the relevant physical constants and technological parameters:

$$\gamma_E = 1 \bigg/ \left(1 + \frac{D_E}{D_B} \frac{N_B W_B}{N_E W_E}\right) \tag{9.23}$$

Transport Factor

Figure 9.4 illustrates that a small number of the injected electrons do not make it through the base because they get recombined by holes. The current associated with the flow of the recombined electrons, I_{rec}, is diverted to the base terminal:

$$I_{nE} = I_{nC} + I_{rec} \tag{9.24}$$

The difference between I_{nE} and I_{nC} relates to the deviation of the concentration profile $n_p(x)$ (Fig. 9.5) from a perfect straight line. The concentration gradient is somewhat smaller at $x = W_B$, showing that the diffusion current at the collector end of the base is somewhat smaller than the diffusion current at the emitter end of the base: $I_{nC} = \alpha_T I_{nE}$, where the transport factor $\alpha_T < 1$.

The transport factor can be expressed in terms of the ratio of the small recombination current I_{rec} and the diffusion current $I_{nC} \approx I_{nE}$:

$$\alpha_T = \frac{I_{nC}}{I_{nE}} = \frac{I_{nC}}{I_{nC} + I_{rec}} = \frac{1}{1 + (I_{rec}/I_{nC})} \approx \frac{1}{1 + (I_{rec}/I_{nE})} \tag{9.25}$$

The recombination current can be related to the number of excess electrons in the base and the lifetime of these electrons. The number of excess electrons is $A_J W_B [n_p(0) - n_p(W_B)]/2 \approx A_J W_B n_p(0)/2$. The charge associated with these electrons—the stored charge of minority carriers in the base—is $Q_S = q A_J W_B n_p(0)/2$. If the average lifetime of each electron is τ_n, the average time that it takes to recombine the stored charge Q_S is also τ_n. Under steady-state conditions, the charge Q_S remains constant over time because all the recombined electrons and holes are supplied by adequate currents of electrons from the emitter and holes from the base. This is the recombination current, I_{rec}. Therefore,

$$I_{rec} = \frac{Q_S}{\tau_n} = q A_J \frac{W_B n_p(0)}{2\tau_n} \tag{9.26}$$

The minority-carrier lifetime is related to the diffusion constant and the diffusion length [Eq. (5.15)]:

$$\tau_n = \frac{L_B^2}{D_B} \tag{9.27}$$

Using this relationship, Eq. (9.26) becomes

$$I_{rec} = q A_J \frac{W_B D_B n_p(0)}{2 L_B^2} \tag{9.28}$$

The ratio I_{rec}/I_{nE} can now be determined from Eqs. (9.28) and (9.16):

$$\frac{I_{rec}}{I_{nE}} = \frac{1}{2} \left(\frac{W_B}{L_B} \right)^2 \tag{9.29}$$

Accordingly, Eq. (9.25) for the transport factor becomes

$$\alpha_T = 1 \Bigg/ \left[1 + \frac{1}{2}\left(\frac{W_B}{L_B}\right)^2\right] \tag{9.30}$$

This equation shows that α_T becomes very close to unity in BJTs with a thin base ($W_B \ll L_B$).

Common-Base and Common-Emitter Current Gains

Given that $\alpha = \gamma_E \alpha_T$, Eqs. (9.23) and (9.30) for γ_E and α_T can be used to relate the common-base current gain directly to the relevant physical and technological parameters:

$$\alpha \approx \frac{1}{1 + \dfrac{D_E}{D_B}\dfrac{N_B W_B}{N_E W_E} + \dfrac{1}{2}\left(\dfrac{W_B}{L_B}\right)^2} \tag{9.31}$$

Because the terms with N_B/N_E and $(W_B/L_B)^2$ ratios in Eqs. (9.23) and (9.30) are small, their product is even smaller and it is neglected in Eq. (9.31).

The common-base current gain is obtained from Eqs. (9.13) and (9.31) as

$$\beta = \frac{1}{\dfrac{D_E}{D_B}\dfrac{N_B W_B}{N_E W_E} + \dfrac{1}{2}\left(\dfrac{W_B}{L_B}\right)^2} \tag{9.32}$$

EXAMPLE 9.2 Typical BJT Parameters

Typical technological parameters of a BJT are as follows: the emitter doping $N_E = 10^{20}$ cm^{-3}, the base doping $N_B = 2 \times 10^{18}$ cm^{-3}, the emitter width $W_E = 2\ \mu$m, and the base width $W_B = 1\ \mu$m.

(a) Assuming $\tau_n = \tau_p = 10\ \mu$s and estimating D_E and D_B from the mobility graphs (Fig. 3.8), determine L_E and L_B and compare them to the emitter and base widths. Are the conditions $W_B \ll L_B$ and $W_E \ll L_E$ satisfied?

(b) Determine the emitter efficiency and the transport factor.

(c) Determine the common-base and common-emitter current gains.

(d) Calculate the common-emitter gain by assuming (1) ideal emitter efficiency and (2) ideal transport factor. Compare these values to the result from part (c) and comment on the relative importance of the emitter efficiency and the transport factor values.

SOLUTION

(a) The mobilities of holes in the emitter and electrons in the base, for the given doping levels, are estimated from the graphs in Fig. 3.8 as follows: $\mu_E = 50$ cm^2/V · s and $\mu_B = 250$ cm^2/V · s. With this, the diffusion constants are

$$D_E = V_t \mu_E = 0.026 \times 50 = 1.30 \text{ cm}^2/\text{s}$$

$$D_B = V_t \mu_B = 0.026 \times 250 = 6.50 \text{ cm}^2/\text{s}$$

The diffusion lengths are [Eq. (5.15)]

$$L_E = \sqrt{D_E \tau_p} = \sqrt{1.30 \times 10^{-4} \times 10^{-5}} = 36.1\ \mu\text{m}$$

$$L_B = \sqrt{D_B \tau_n} = \sqrt{6.50 \times 10^{-4} \times 10^{-5}} = 80.6\ \mu\text{m}$$

Given that $W_E = 2$ μm and $W_B = 1$ μm, the conditions $W_B \ll L_B$ and $W_E \ll L_E$ are satisfied.

(b) Labeling the ratios of physical and technological parameters in Eq. (9.23) by R_γ,

$$R_\gamma = \frac{D_E}{D_B}\frac{N_B}{N_E}\frac{W_B}{W_E} = \frac{1.30}{6.50}\frac{2 \times 10^{18}}{10^{20}}\frac{1}{2} = 0.0020$$

the emitter efficiency is calculated as

$$\gamma_E = \frac{1}{1 + R_\gamma} = \frac{1}{1 + 0.0020} = 0.99800$$

Similarly, with the ratio

$$R_\alpha = \frac{1}{2}\left(\frac{W_B}{L_B}\right)^2 = \frac{1}{2}\left(\frac{1}{80.6}\right)^2 = 7.70 \times 10^{-5}$$

the transport factor [Eq. (9.30)] is

$$\alpha_T = \frac{1}{1 + R_\alpha} = \frac{1}{1 + 7.70 \times 10^{-5}} = 0.99992$$

(c) From Eq. (9.11)

$$\alpha = \alpha_T \gamma_E = 0.99800 \times 0.99992 = 0.99793$$

From Eq. (9.13),

$$\beta = \frac{\alpha}{(1 - \alpha)} = \frac{0.99793}{1 - 0.99793} = 481$$

(d) For ideal emitter efficiency ($\gamma_E = 1$), $\alpha = \alpha_T = 0.99992$, so

$$\beta = \frac{\alpha}{(1-\alpha)} = \frac{0.99992}{1-0.99992} = 12500$$

For ideal transport factor, $\alpha = \gamma_E = 0.99800$, so

$$\beta = \frac{\alpha}{(1-\alpha)} = \frac{0.99800}{1-0.99800} = 499$$

Comparing these values to $\beta = 481$ obtained in part (c), we see that the common-emitter current gain is largely determined by the value of the emitter efficiency.

9.1.4 The Four Modes of Operation: BJT as a Switch

The voltage-controlled current source is only one possible mode of BJT operation. This is the mode referred to as the *normal active mode*. Because each of the two P–N junctions (E–B and C–B) can be either forward- or reverse-biased, there are four bias possibilities for the BJT. This is illustrated in Fig 9.6.

Normal Active Mode

With a forward-biased E–B junction and a reverse-biased C–B junction, the BJT is set in the normal active mode. The reverse-biased C–B junction acts as a current source, and the forward-biased E–B junction supplies controlled current to the current source (the reverse-biased C–B junction). Accordingly, the BJT acts as a *controlled* current source. This is considered as the main mode of BJT operation and has been considered in detail in the previous sections.

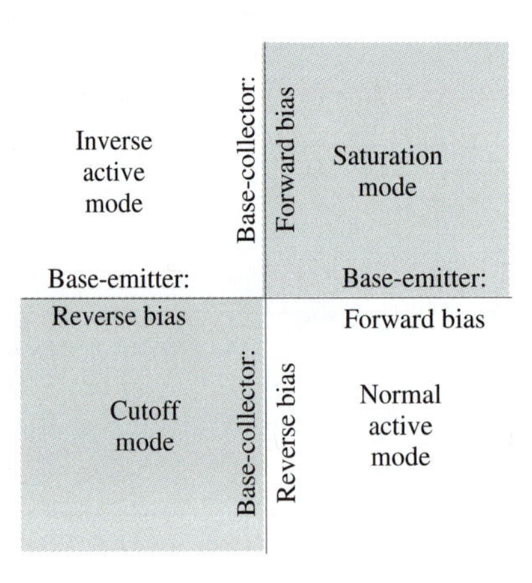

Figure 9.6 Two possible bias states of the two junctions lead to four possible modes of operations of the BJT.

Cutoff

If none of the two junctions is forward-biased, the BJT is in cutoff mode. All the terminal currents are zero (neglecting the reverse leakage) and the output is an open circuit. The BJT acts as a switch in *off* mode.

Taking the illustration in Fig. 9.6 literally, it may be concluded that the cutoff region requires $V_{BE} < 0$ and $V_{BC} < 0$ (in the case of the considered NPN structure). There is no doubt that a BJT would be in cutoff mode under these conditions. The question arises for biasing that is below the turn-on voltage ($V_F \approx 0.7$ V in silicon) but not below zero. Strictly speaking, reverse bias is for $V_{BE} < 0$ and forward bias exists when V_{BE} reaches V_F. The area between 0 and V_F is a gray area that requires special attention when one is determining the mode of a BJT operation. Nonetheless, it is clear that the E–B junction does not emit electrons into the base if $V_{BE} < V_F$. With this conclusion, it is clear that the BJT cannot be in the normal active mode. In addition, if $V_{BC} < V_F$, there is no emission of electrons or holes by the C–B junction either. With no current flowing through the device (neglecting reverse leakage), the BJT is classified as being in cutoff mode.

Saturation

If one junction is forward-biased and the other is not reverse-biased, the BJT is in saturation mode.

The C–B junction is reverse-biased in the normal active mode, thus $V_{CB} > 0$. The boundary between the normal active and the saturation modes is defined by the condition $V_{CB} = 0$. For $V_{CB} > 0$ (or, equivalently, $V_{BC} < 0$) the C–B junction is reverse-biased and for $V_{CB} < 0$ it is not.

The output characteristics of a BJT with common emitter are presented as the collector current versus the collector–emitter voltage. V_{CE} can be related to V_{CB} through the voltage loop (Kirchhoff's second law) applied to the BJT: $V_{CE} = V_{CB} + V_{BE}$. The condition $V_{CB} = 0$ is equivalent to the condition $V_{CE} = V_{BE}$. This is the straight line labeled by $V_{CB} = 0$ on the output characteristics shown in Fig. 9.7.

To explain the saturation mode, consider a transition from the normal active mode into saturation. The V_{CE}-independent collector current in the normal active mode is due to the "electron-fall" effect. The reverse C–B bias ($V_{CB} > 0$) can change the height of the fall but it does not cause a significant change in the current. For $V_{CB} < 0$ (the junction is not reverse-biased), the "electron-fall" structure no longer exists. The collector is no longer just collecting electrons diffusing from the base; rather, it starts emitting current in the opposite direction. As a result, the total collector current drops below the level in the normal active mode. This will happen when V_{CB} changes from positive to negative even if the input V_{BE} voltage and the input I_B current are kept constant. As a consequence, the ratio of the output and the input currents is no longer equal to the common-emitter current gain. Because of the reduction of I_C, the saturation region is also defined by the following condition:

$$I_C < \beta I_B \tag{9.33}$$

A decrease in V_{CB} beyond zero, which corresponds to a decrease of $V_{CE} = V_{CB} + V_{BE}$ on the output characteristics, pushes the BJT deeper into saturation. The decrease in V_{CB} toward the turn-on voltage of the C–B junction increases the carrier emission by the

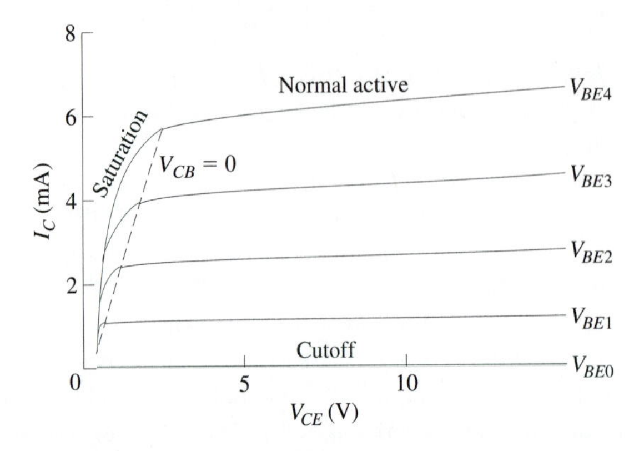

Figure 9.7 Output characteristics of a BJT. The normal active region corresponds to the characteristics of a voltage-controlled current source. The cutoff and the saturation regions correspond to the characteristics of a switch in *off* mode and the resistance of a switch in *on* mode, respectively.

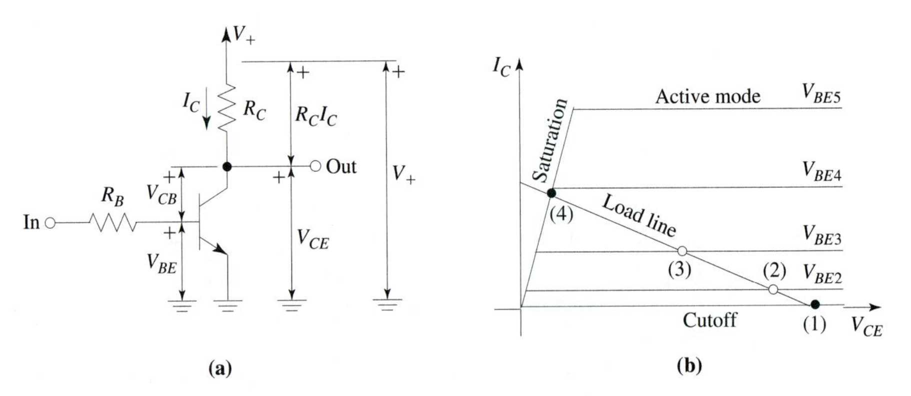

Figure 9.8 (a) The circuit of a BJT inverter with resistive load. (b) Graphic analysis showing that the BJT can be set in either cutoff, normal active, or saturation mode.

collector, which reduces the net current collected by the collector. In Fig. 9.7, this is seen as the I_C reduction toward zero with reduction of V_{CE} from V_{BE} (corresponding to $V_{CB} = 0$) to 0 (corresponding to $V_{CB} = -V_{BE}$). The "depth" of saturation can be characterized by the ratio $\beta I_B / I_C$, which shows how much is I_C reduced below the normal active level of βI_B.

The output characteristic in saturation is very close to the current–voltage characteristic of a small resistor. This is the parasitic resistance of a switch in *on* mode. Combined with the operation as a switch in *off* mode (the cutoff mode), the BJT can be used as a digital device. A representative digital circuit is the inverter. Figure 9.8a shows the basic circuit of the BJT inverter with resistive load. The loading circuit (R_C and V_+) is the same as for the voltage-amplifier circuit shown in Fig. 9.3. The difference between digital and analog operation can be explained by the graphical analysis provided in Fig. 9.8b.

For a small input voltage $V_{BE} < V_F$, the BJT is in the cutoff mode [point (1) on the output characteristics in Fig. 9.8b]. This is so because the E–B junction is not forward-biased and the C–B junction is reverse-biased: $V_{CB} = V_{CE} - V_{BE} > 0$. The collector current is $I_C = 0$ and the voltage across the resistor is $R_C I_C = 0$, which means that $V_+ = V_{CE}$. The small input voltage is *inverted* into a large output voltage ($V_+ = V_{CE}$).

When V_{BE} is increased, the BJT enters the normal active mode. These are points (2) and (3) on the output characteristics shown in Fig. 9.8b. If the input voltage is oscillated in the range that corresponds to the normal active mode, the circuit works as an amplifier—just the same as the circuit and the analysis shown in Fig. 9.3.

If V_{BE} is increased beyond a certain level—for example, to the level labeled as V_{BE4} in Fig. 9.8b—the BJT enters saturation mode. At this level the collector current is so high that

$$V_{CE} = V_+ - I_C R_C \tag{9.34}$$

is no longer larger than V_{BE}. With $V_{CE} < V_{BE}$, $V_{CB} = V_{CE} - V_{BE} < 0$, which is the condition for saturation when the E–B junction is forward-biased. The small output voltage V_{CE}, corresponding to point (4), will not change its value even if the input voltage is further increased to V_{BE5} and beyond. It is said that the output voltage and current are *saturated,* hence the name *saturation region.*[3] Point (4) shows that a large input voltage is inverted into a *small* output voltage—the circuit works as an inverter.[4]

Inverse Active Mode

If the C–B junction is forward-biased and E–B junction is reverse-biased, the BJT operates as a controlled current source but with swapped emitter and collector: the collector emits the carriers that are collected by the emitter. This mode is called inverse active mode.

If the NPN structure was symmetrical, the inverse mode of operation would be as good as the normal active mode. In real BJTs, the doping level of the collector is the lowest, which means that its efficiency (γ_E) is not good when used in the emitter role. Because of this, α and β values in the inverse active mode are small, and no good gain can be achieved in this mode of operation.

---

[3] The term "saturation" is not consistently used in the case of different types of transistors, namely the BJT and FETs (including the MOSFET). To avoid possible confusion, this fact should be noted and remembered. The MOSFET in saturation operates as a voltage-controlled current source (an analog device), whereas the BJT in saturation operates as a switch in *on* mode (a digital device). "Saturation" of the MOSFET output current means that it does not increase with the *output* (drain-to-source) voltage. In the case of the BJT, "saturation" means that the output voltage and output current do not change with the *input* (base-to-emitter) voltage.

[4] V_{BE} voltage cannot be increased to the digital high level $V_H = V_+$. The proper input voltage that changes from V_L to $V_H = V_+$ is connected through the input resistor R_B (Fig. 9.8a) to limit the base current to the levels that will not damage the BJT.

EXAMPLE 9.3 BJT Modes of Operation

Determine the mode of operation of an NPN BJT with $\beta \approx 450$, if it is known that:

(a) $V_{BE} = 0.7$ V, $V_{CE} = 5.2$ V
(b) $V_{BE} = 0.7$ V, $V_{CE} = 0.2$ V
(c) $V_{BE} = 0.8$ V, $V_{BC} = 0.3$ V
(d) $V_{BE} = 0.8$ V, $V_{BC} = -0.7$ V
(e) $V_{BE} = -0.8$ V, $V_{BC} = 0.7$ V
(f) $V_{BE} = 0.1$ V, $V_{BC} = -10$ V
(g) $I_C = 455$ mA, $I_B = 1$ mA
(h) $I_C = 455$ mA, $I_E = 502$ mA

SOLUTION

(a) $V_{BE} = 0.7$ V shows that the E–B junction is forward-biased. To conclude about the biasing of the C–B junction, V_{CB} voltage is needed. It is found as

$$V_{CB} = V_{CE} - V_{BE} = 4.5 \text{ V}$$

A positive collector-to-base voltage shows that this N–P junction is reverse-biased. With this combination, the BJT is in normal active mode.

(b) In this case, V_{CB} voltage is negative:

$$V_{CB} = V_{CE} - V_{BE} = -0.5 \text{ V}$$

which, in combination with the forward-biased E–B junction ($V_{BE} = 0.7$ V), sets the BJT in saturation.

(c) Again, a forward-biased E–B junction ($V_{BE} = 0.8$ V) and a negative collector-to-base voltage ($V_{CB} = -0.3$ V) bias the BJT in saturation mode.

(d) This time, the C–B junction is reverse-biased, which sets the BJT in normal active mode.

(e) This is the reverse situation: the E–B junction reverse-biased ($V_{BE} = -0.8$ V), while the C–E junction is forward-biased ($V_{CB} = -0.7$ V). The collector is emitting electrons, while the emitter is collecting them. Therefore, the BJT is in inverse active mode.

(f) Given that $V_{BE} = 0.1$ V is below the forward-bias level of the E–B junction and the C–B junction is reverse-biased, this BJT is in cutoff mode.

(g) In the normal active mode, the collector and the base currents are related through the gain factor β:

$$I_C = \beta I_B$$

This BJT satisfies this criterion.

(h) The base current in this case is

$$I_B = I_E - I_C = 47 \text{ mA}$$

and it is obvious that $I_C < \beta I_B$. This means the BJT is in saturation [refer to Eq. (9.33)].

9.1.5 Complementary BJT

Figure 9.9 shows the alternative possibility of making a BJT: the emitter and the collector are P-type, whereas the base is an N-type semiconductor. This type of transistor is called a PNP BJT.

To set a PNP BJT in the normal active mode, negative V_{BE} and positive V_{CB} voltages are needed, which is opposite to the case of an NPN BJT. The emitter region is at the highest potential and the collector is at the lowest potential, which causes holes from the emitter to be emitted and collected by the collector. With the holes making the transistor current, as opposed to electrons in the case of an NPN BJT, the emitter and the collector current directions are opposite to those in the NPN BJT. The same applies to the base current. Appearing as a mirror image of the NPN, the PNP BJT complements the NPN in some circuit applications.

9.1.6 BJT Versus MOSFET

Both the BJT and the earlier introduced MOSFET can perform equivalent principal functions: (1) voltage-controlled current source and (2) voltage-controlled switch. Some similarities exist even in the principle of operation. This is perhaps best illustrated by the fact that the energy-band diagram *along the channel* of a MOSFET in saturation (Fig. 8.9) is very similar to the energy-band diagram of a BJT (Fig. 9.2b). This certainly means that any electrical function implemented in MOSFET technology can in principle be achieved by BJTs, and vice versa. Extremely important differences exist, however, in the performances and efficiencies achieved by the two possible technologies. At the surface, these may seem like simple quantitative differences, but in practice they appear to be qualitative differences. Although it is theoretically possible to build a complex microprocessor in

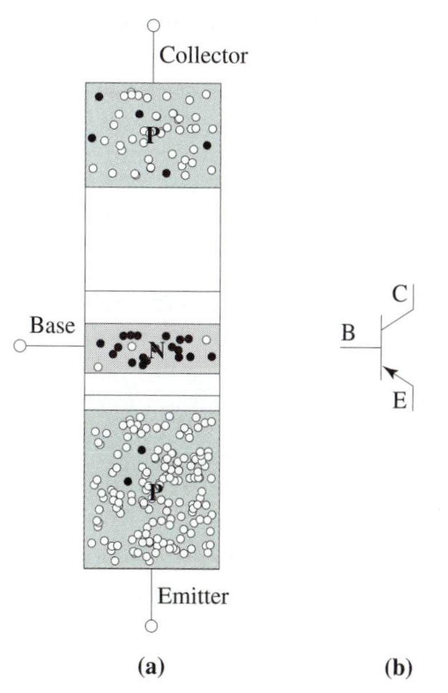

Figure 9.9 (a) The cross section and (b) the symbol of a PNP BJT.

BJT technology, the dramatic developments in the information technology would not have occurred with BJTs due to practical power-dissipation limits. This example illustrates the importance of understanding the differences between the two devices.

The following descriptions of BJT and MOSFET advantages highlight the differences.

BJT Advantages

1. The energy-band similarity does not apply to the same areas in both devices. The BJT energy-band diagram of Fig. 9.2b applies to any (x,z) point, assuming the y-axis in the direction of the electron flow (along the energy-band diagram). Because the whole emitter cross-sectional area A_J is effective, a sizable device current can be achieved. In the case of the MOSFET, the energy-band diagram that can be altered by the controlling electrode appears only along the channel (Fig. 8.9). The channel thickness (in x direction) is limited by the electric-field penetration into the semiconductor to a couple of nanometers. Consequently, the channel cross-sectional area $x_{ch}W$ is severely restricted. The BJT structure is advantageous in terms of achieving large device currents, which is important in power applications, both linear and switching.
2. The diode (P–N junction) used as a controlling device in the BJT offers an advantage over the capacitor (MOS structure) used in the MOSFET in terms of the sensitivity of the output current to input voltage. This is the concept of *transconductance,* mathematically expressed as $g_m = dI_{output}/dV_{input}$. A small change in the input V_{BE} voltage—for example, from 0.5 V to 0.8 V—is sufficient to drive the output current from practically zero to the maximum level. To achieve this with a MOSFET, an input voltage change of several volts may be necessary. Adding to these observations, the better current capabilities of the BJT, the picture of a superior transconductance becomes clear. The higher transconductance of the device relates not only to higher gains of amplifiers but also to shorter switching times and superior noise characteristics of both linear and digital circuits.

MOSFET Advantages

1. Here is the other side of the coin: the capacitor (MOS structure) used as the controlling device in MOSFETs, as opposed to the diode (P–N junction) used in BJTs, results in unmatched advantages for the MOSFET:
   - It enables a MOSFET operated as a switch to be maintained in *on* mode without any power consumption: no input current is needed to support the channel that creates the low-resistance path across the output. This adds to the fact that the other digital state, the switch in *off* mode, does not require any power consumption either. Using complementary MOSFETs, any logic function can be implemented, and no power consumption would be required to maintain any logic state. This was explained for the case of CMOS inverter in Section 8.1.2. The BJT dissipates significant power when in saturation (switch in *on* mode), because this state can be maintained only by significant input and output currents. The problem with the power consumption is not only the heat removal (big cooling elements, fans, etc. needed), but also an extremely low limit in the

number of logic cells that can be supported by the current that can be supplied to a geometrically small IC.

- With the capacitor at the input, MOSFETs do not require biasing resistors, which are necessary to limit the current through the input diode of BJTs (an example of an input-biasing resistor is R_B in Fig. 9.8a). Both digital and analog functions can be implemented by circuits consisting of complementary MOSFETs only; no resistors and capacitors are needed. Since large-value resistors and capacitors require enormous areas, compared to transistors, this makes the MOSFET technology much more efficient in terms of area usage, again enabling much more complex circuits to be integrated.

2. The MOSFET is a single-carrier transistor (also referred to as a *unipolar transistor*): only electrons matter in N-channel MOSFETs, and only holes matter in P-channel MOSFETs. As opposed to this, the holes do matter in NPN BJTs, even though the main transistor current is due to electron flow. The base current of an NPN BJT, which is the input current of the common-emitter transistor, is due to the holes. The fact that both types of carrier are active is reflected in the name of the device: *bipolar* junction transistor. The disadvantage of having both types of carrier in a single circuit (such as the base–emitter circuit) is that the recombination process, which links the two currents, is relatively slow. This is illustrated in the best way by the appearance of the stored charge (Section 6.4). Because the excess charge stored during the *on* period has to be removed by the recombination process before the diode (and therefore the BJT) turns *off*, the associated delay limits the maximum switching frequency to relatively low values.

Although no general rule can be established, it can be said that BJTs are more suitable for analog applications, especially when high output power is needed, whereas MOSFETs are much more suitable for digital circuits, especially in terms of achieving ICs able to perform extremely complex functions.

9.2 PRINCIPAL CURRENT–VOLTAGE CHARACTERISTICS: EBERS–MOLL MODEL IN SPICE

The characteristic of a forward-biased P–N junction in a BJT can be modeled by the Shockley equation for a forward-biased diode [Eq. (6.15)]. The characteristics of a reverse-biased P–N junction in a BJT correspond to the characteristics of a controlled current source. Consequently, combinations of diodes and controlled current sources can be used to create equivalent circuits that account for all four possible modes of operation. The BJT equations that are derived by the equivalent-circuit approach are known as the Ebers–Moll model.

The physically based version of the Ebers–Moll model, referred to as the *injection version,* is introduced first in this section. The equivalent circuit of this version uses two diodes and two controlled current sources to model the two junctions. The equivalent circuit of a BJT that is used in SPICE, as well as in circuit-design and analysis books, uses one effective controlled-current source. To link the physical effects in a BJT to the practically used equivalent circuit and related SPICE parameters, the injection version is transformed through what is called the *transport version* into the *SPICE version*.

9.2.1 Injection Version

In the normal active mode, the collector–base of the BJT plays the role of a current source controlled by a voltage, or equivalently by the corresponding current. The controlling emitter–base junction can be modeled by a diode. Therefore, the diode and the current source of the upper branch of the circuit in Fig. 9.10b ($V_{BE} > 0$ and $V_{BC} < 0$) make a proper equivalent circuit of the BJT in the normal active mode. The current of the controlling junction (I_F) is directly related to the current injected into the current-source junction. The current of the current source is labeled $\alpha_F I_F$ to express this fact, where $\alpha_F < 1$ due to the carrier losses related to nonideal emitter efficiency and nonideal transport factor. The current of the controlling junction I_F, of course, depends on the voltage applied to the base–emitter, V_{BE},

$$I_F = I_{ES}\left(e^{V_{BE}/V_t} - 1\right) \tag{9.35}$$

where I_{ES} is the saturation current of the base–emitter junction.

The BJT model should include all the possible bias arrangements, not only the normal active mode. In the inverse active mode, the roles of the P–N junctions are swapped, and the BJT can be modeled by a circuit that is a mirror image of the circuit modeling the BJT in the normal active mode. The current of the current source is analogously labeled $\alpha_R I_R$, where the controlling current I_R depends on the base–collector voltage,

$$I_R = I_{CS}\left(e^{V_{BC}/V_t} - 1\right) \tag{9.36}$$

where I_{CS} is the base–collector saturation current.

Adding the equivalent circuits for the inverse and normal active modes together, as shown in Fig. 9.10b, does not introduce any adverse effects. If the BJT is in the normal active mode, then we have $V_{BC} < 0$ and the current $I_R \approx -I_{CS} \ll \alpha_F I_F \approx$

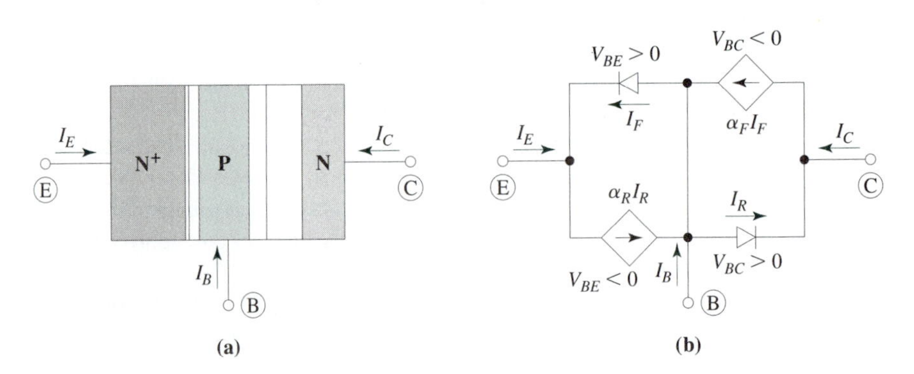

Figure 9.10 (a) SPICE definition of BJT current directions. (b) Injection version of static BJT equivalent circuit.

$I_{ES}\alpha_F \exp(V_{BE}/V_t)$. In fact, the corresponding equation for the collector terminal current, $I_C \approx I_{ES}\alpha_F \exp(V_{BE}/V_t) + I_{CS}$, properly includes the reverse-bias (leakage) current of the base–collector junction.

Moreover, the circuit of Fig. 9.10b automatically includes the two remaining biasing possibilities, the saturation and cutoff modes. In saturation, both V_{BE} and V_{BC} voltages are positive, and both I_F and I_R currents are significant. In the typical case of $V_{BC} < V_{BE}$, the terminal collector current retains the direction as in the case of the normal active mode; however, the current value is reduced because $I_C = \alpha I_F - I_R$. The voltage between the collector and emitter is very small, $V_{CE} = -V_{BC} + V_{BE}$. An increase in V_{BC} causes further reduction in V_{CE} voltage and I_C current, according to the I_C–V_{CE} characteristic in saturation (Fig. 8.9).

In cutoff, both P–N junctions are reverse-biased (V_{BE} and V_{BC} negative), allowing only the flow of the leakage currents I_{ES} and I_{CS}.

When the currents from the two branches of the equivalent circuit are added, the terminal collector and emitter currents can be expressed as

$$I_C = \alpha_F I_{ES}\left(e^{V_{BE}/V_t} - 1\right) - I_{CS}\left(e^{V_{BC}/V_t} - 1\right) \tag{9.37}$$

$$I_E = -I_{ES}\left(e^{V_{BE}/V_t} - 1\right) + \alpha_R I_{CS}\left(e^{V_{BC}/V_t} - 1\right) \tag{9.38}$$

whereas the base terminal current is the balance between the emitter and the collector current:

$$I_B = -I_C - I_E \tag{9.39}$$

This set of equations, which is the injection version of the Ebers–Moll model, relates all three terminal currents to the two terminal voltages (V_{BE} and V_{BC}) through the following four parameters: α_F, the common-base current gain of a BJT in the normal active mode; α_R, the common-base current gain of a BJT in the inverse active mode; I_{ES}, the emitter–base saturation current; and I_{CS}, the collector–base saturation current.

9.2.2 Transport Version

The equivalent circuit of the transport version is the same as the injection version of the Ebers–Moll model. The difference is in the way the internal currents are expressed: they are now based on the actual current source currents, labeled I_{EC} and I_{CC} in Fig. 9.11, rather than the currents injected by the P–N junctions as in Fig. 9.10b. Of course, the relationships between the currents of the controlling P–N junctions and the actual currents of the current sources have to be retained to correctly model the BJT. Consequently, the P–N junction currents in Fig. 9.11 cannot be considered as independent but have to be related to I_{EC} and I_{CC} through the corresponding common-base current gains α_F and α_R.

The transport version relates the I_{EC} and I_{CC} currents to the terminal voltages in the following way:

$$I_{CC} = I_S\left(e^{V_{BE}/V_t} - 1\right) \tag{9.40}$$

$$I_{EC} = I_S\left(e^{V_{BC}/V_t} - 1\right) \tag{9.41}$$

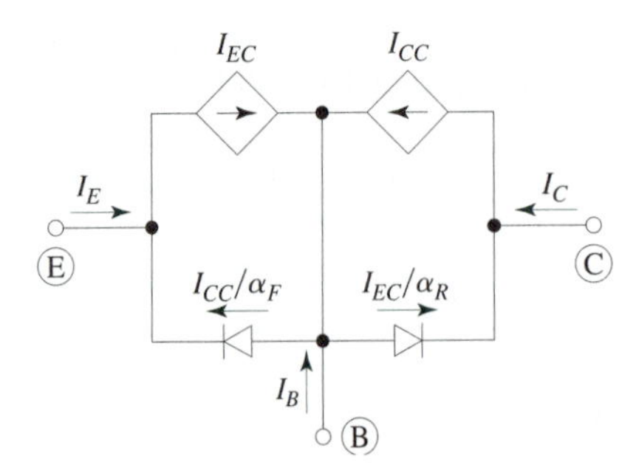

Figure 9.11 The transport version of static BJT equivalent circuit.

Therefore, the terminal currents are given as

$$I_C = I_S(e^{V_{BE}/V_t} - 1) - \frac{I_S}{\alpha_R}(e^{V_{BC}/V_t} - 1) \tag{9.42}$$

$$I_E = -\frac{I_S}{\alpha_F}(e^{V_{BE}/V_t} - 1) + I_S(e^{V_{BC}/V_t} - 1) \tag{9.43}$$

$$I_B = -I_C - I_E \tag{9.44}$$

Comparing the two models, it is obvious that the single saturation current I_S used in these equations is equivalent to neither the base–emitter saturation current I_{ES} nor the base–collector saturation current I_{CS}. The collector current equations [Eqs. (9.42) and (9.37)] become equivalent under the following conditions:

$$I_S = \alpha_F I_{ES} \tag{9.45}$$

$$I_S = \alpha_R I_{CS} \tag{9.46}$$

The same conditions also lead to the equivalence of the emitter current Eqs. (9.43) and (9.38). The base–emitter and base–collector saturation currents in real BJTs are different, due to different areas and different doping levels. Obviously, the single I_S current cannot realistically represent both saturation currents at the same time. For realistic simulations, the parameter I_S should be related to the base–emitter junction ($I_S = \alpha_F I_{ES}$) in the case of normal active mode and to the base–collector junction ($I_S = \alpha_R I_{CS}$) in the case of inverse active mode.

It appears that the choice of a single I_S parameter, instead of two parameters representing the two P–N junctions, complicates parameter measurement and reduces the generality of the model. However, it enables the more general and physically based equivalent circuit of Fig. 9.10b to be related to the equivalent circuit most frequently used in circuit-design and analysis books. This circuit is discussed in the following text.

9.2.3 SPICE Version

A single current source is used to model the BJT in the circuit-design and analysis books. The two current sources of Fig. 9.11 can be reduced to one while maintaining the same

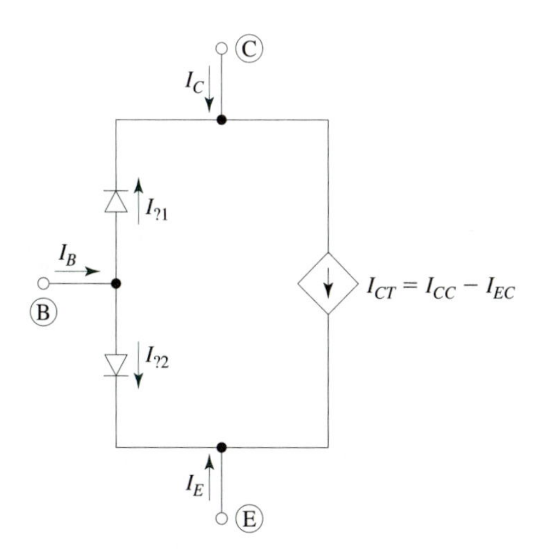

Figure 9.12 The SPICE version of static BJT equivalent circuit.

relationships between the terminal currents (I_C, I_E, and I_B) and the terminal voltages (V_{BE} and V_{BC}).

The circuit with a single current source is shown in Fig. 9.12. The currents $I_{?1}$ and $I_{?2}$ can be determined so that the terminal I_C and I_E currents are equivalent to the ones in Fig. 9.11:

$$\begin{aligned} I_C &= \underbrace{I_{CC} - \frac{I_{EC}}{\alpha_R}}_{\text{Fig. 9.11}} = \underbrace{I_{CC} - I_{EC} - I_{?1}}_{\text{Fig. 9.12}} \Rightarrow \\ I_{?1} &= I_{EC}\left(\frac{1}{\alpha_R} - 1\right) = I_{EC}\frac{1-\alpha_R}{\alpha_R}, \qquad I_{?1} = \frac{I_{EC}}{\beta_R} \end{aligned} \tag{9.47}$$

$$\begin{aligned} I_E &= \underbrace{-\frac{I_{CC}}{\alpha_F} + I_{EC}}_{\text{Fig. 9.11}} = \underbrace{-I_{CC} + I_{EC} - I_{?2}}_{\text{Fig. 9.12}} \Rightarrow \\ I_{?2} &= I_{CC}\left(\frac{1}{\alpha_F} - 1\right) = I_{CC}\frac{1-\alpha_F}{\alpha_F}, \qquad I_{?2} = \frac{I_{EC}}{\beta_F} \end{aligned} \tag{9.48}$$

According to Eq. (9.13), β_F and β_R are common-emitter current gains of the BJT in normal active and reverse active modes. With these values of $I_{?1}$ and $I_{?2}$, the terminal currents can be expressed as

$$\begin{aligned} I_C &= I_{CC} - I_{EC} - \frac{I_{EC}}{\beta_R} \\ I_E &= -I_{CC} + I_{EC} - \frac{I_{CC}}{\beta_F} \\ I_B &= -I_C - I_E \end{aligned} \tag{9.49}$$

When we replace I_{CC} and I_{EC} from Eqs. (9.40) and (9.41), the terminal currents are related to the terminal voltages:

$$\begin{aligned} I_C &= I_S(e^{V_{BE}/V_t}-1) - \left(1+\frac{1}{\beta_R}\right) I_S(e^{V_{BC}/V_t}-1) \\ I_E &= -\left(1+\frac{1}{\beta_F}\right) I_S(e^{V_{BE}/V_t}-1) + I_S(e^{V_{BC}/V_t}-1) \\ I_B &= \frac{1}{\beta_F} I_S(e^{V_{BE}/V_t}-1) + \frac{1}{\beta_R} I_S(e^{V_{BC}/V_t}-1) \end{aligned} \quad (9.50)$$

These are the final and general equations of the principal Ebers–Moll model. The three parameters, I_S, β_F, and β_R, are all SPICE parameters.

In the case of normal active mode, $V_{BE}/V_t \gg 1$ and $V_{BC}/V_t \ll -1$, which means $\exp(V_{BE}/V_t) \gg 1$ and $\exp(V_{BC}/V_t) - 1 \approx -1$. This simplifies the general equations to the following form:

$$\begin{aligned} I_C &= I_S e^{V_{BE}/V_t} \\ I_E &= -\left(1+\frac{1}{\beta_F}\right) I_S e^{V_{BE}/V_t} = -\frac{I_C}{\alpha_F} \\ I_B &= \frac{1}{\beta_F} I_S e^{V_{BE}/V_t} = \frac{I_C}{\beta_F} \end{aligned} \quad (9.51)$$

In the normal active mode, the output collector current I_C depends exponentially on the input voltage V_{BE} through the I_S parameter and the thermal voltage V_t. The emitter and the base currents, I_E and I_B, are related to the collector current through the current gains α_F and β_F, originally defined by Eq. (9.9). The minus sign in the I_E equation appears due to the fact that the I_E current direction in the SPICE models is defined into the BJT (Fig. 9.12), which is opposite to the actual direction of the conventional I_E current used in Section 9.1 (Fig. 9.4).

EXAMPLE 9.4 Ebers–Moll Model for a PNP BJT

In analogy with the Ebers–Moll model of an NPN BJT, draw the equivalent circuit and write down the general equations of the Ebers–Moll model for the case of a PNP BJT. Simplify these equations for the case of a PNP BJT in the normal active mode. In SPICE, the directions of the terminal currents of a PNP BJT are defined to be opposite to their NPN counterparts.

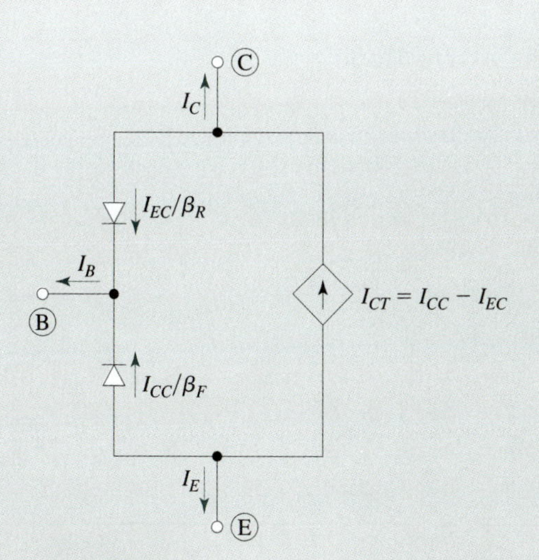

Figure 9.13 SPICE equivalent circuit of a PNP BJT.

SOLUTION

A PNP BJT is a mirror image of an NPN BJT in the sense that the diode (P–N junction) terminals are swapped, all the currents are in the opposite directions, and all the voltages are with the opposite polarities. When we swap the diode terminals and reverse the current directions in the circuit of Fig. 9.12, the SPICE equivalent circuit of a PNP BJT is obtained as in Fig. 9.13.

To avoid using negative voltages, the terminal voltages can be expressed as V_{EB} and V_{CB}, rather than $-V_{BE}$ and $-V_{BC}$. With these changes in Eq. (9.50), the Ebers–Moll model of a PNP BJT is obtained as

$$
\begin{aligned}
I_C &= I_S\left(e^{V_{EB}/V_t} - 1\right) - \left(1 + \frac{1}{\beta_R}\right) I_S\left(e^{V_{CB}/V_t} - 1\right) \\
I_E &= -\left(1 + \frac{1}{\beta_F}\right) I_S\left(e^{V_{EB}/V_t} - 1\right) + I_S\left(e^{V_{CB}/V_t} - 1\right) \\
I_B &= \frac{1}{\beta_F} I_S\left(e^{V_{EB}/V_t} - 1\right) + \frac{1}{\beta_R} I_S\left(e^{V_{CB}/V_t} - 1\right)
\end{aligned}
$$

The simplified equations for the case of normal active mode can be deduced in a similar way:

$$
\begin{aligned}
I_C &= I_S e^{V_{EB}/V_t} \\
I_E &= -\left(1 + \frac{1}{\beta_F}\right) I_S e^{V_{EB}/V_t} = -\frac{I_C}{\alpha_F} \\
I_B &= \frac{1}{\beta_F} I_S e^{V_{EB}/V_t} = \frac{I_C}{\beta_F}
\end{aligned}
$$

EXAMPLE 9.5 Ebers–Moll Model for Inverse Active Mode

Simplify Eq. (9.50) for the case of the inverse active mode.

SOLUTION

In this case, $\exp(V_{BC}/V_t) \gg 1$ and $\exp(V_{BE}/V_t) - 1 \approx -1$, which leads to

$$I_C = -\left(1 + \frac{1}{\beta_R}\right) I_S e^{V_{BC}/V_t} = -\frac{I_E}{\alpha_R}$$

$$I_E = I_S e^{V_{BC}/V_t}$$

$$I_B = \frac{1}{\beta_R} I_S e^{V_{BC}/V_t} = \frac{I_E}{\beta_R}$$

EXAMPLE 9.6 Fundamental BJT Parameters

The results of measurements performed on an NPN BJT are given in Table 9.1. Calculate the following SPICE parameters: I_S, β_F, and β_R.

TABLE 9.1 Measurement Data

V_{BE} (V)	V_{BC} (V)	I_B (μA)	I_C (mA)
0.80	−5.0	2.6	0.49
−5.0	0.72	353.3	0.90

SOLUTION

Voltages V_{BE} and V_{BC} indicate that the first raw of data is for the BJT in the normal active mode, whereas the second raw is related to the inverse active mode. According to Eqs. (9.51),

$$\beta_F = I_C/I_B = \frac{490}{2.6} = 188.5$$

and

$$\ln I_S = \ln I_C - V_{BE}/V_t = \ln 4.9 \times 10^{-5} - \frac{0.80}{0.02585} = -40.87 \Rightarrow I_S = 1.78 \times 10^{-17} \text{ A}$$

In the case of the inverse active mode, the results of Example 9.5 can be used to find β_R:

$$\beta_R = \frac{I_E}{I_B} = \frac{I_C - I_B}{I_B} = \frac{900.0 - 353.3}{353.3} = 1.55$$

Another value for the saturation current can be obtained for the case of the inverse active mode; however, this value is less relevant as the BJT normally operates in the normal active mode.

9.3 SECOND-ORDER EFFECTS

This section describes the most important second-order effects (again a distinction should be made between a *second-order effect* and a *negligible effect*). Different mathematical equations have been developed as models for the second-order effects. SPICE-based equations are selected for presentation in this section. Although lengthy arguments can be made about the advantages and disadvantages of particular models, there is no doubt that the SPICE-based equations are of unchallenged practical importance.

9.3.1 Early Effect: Finite Dynamic Output Resistance

The output collector current I_C in the normal active mode, as predicted by the principal Ebers–Moll model [Eq. (9.51)], does not depend on the output voltage. This is the case of ideal current source, illustrated by the dotted horizontal lines on the I_C–V_{CE} plot of Fig. 9.14a. The real BJTs, however, do not have perfectly horizontal I_C–V_{CE} characteristics: I_C always increases to some extent with an increase in V_{CE}. The reciprocal value of the slope of the output I_C–V_{CE} characteristic is defined as dynamic output resistance:

$$r_o = 1 \Big/ \left(\frac{dI_C}{dV_{CE}}\right) \tag{9.52}$$

The ideal current source has infinitely large r_o.

In the case of a BJT acting as a controlled current source, r_o is not the same for every input voltage/current. As the input voltage/current is increased, r_o is reduced, which is observed as a more pronounced slope on the corresponding I_C–V_{CE} line. This effect, known as the Early effect, is illustrated in Fig. 9.14a by the solid lines.

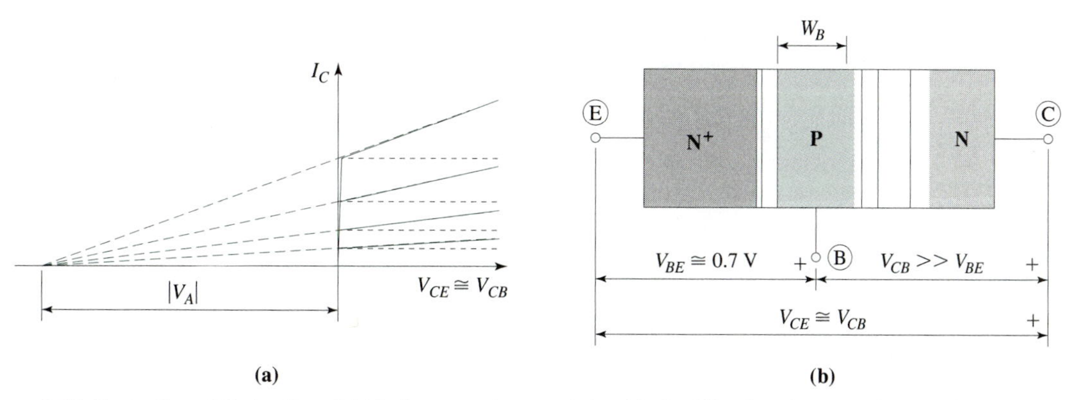

Figure 9.14 Illustration of Early effect. (a) Ideal output characteristics (dashed lines) and output characteristics with pronounced Early effect (solid lines). (b) BJT cross section illustrating base narrowing due to increased base–collector depletion-layer width by increased V_{CB} voltage.

Proper inclusion of the real dynamic output resistance is very important in the simulation and design of analog circuits. Consequently, the Early effect appears as the most important second-order effect.

Figure 9.14b illustrates that it is in fact the reverse voltage of the base–collector junction $V_{BC} = -V_{CB}$ that is directly related to the I_C current increase. Practically, however, the difference between V_{CE} and V_{CB}, which is $V_{BE} \approx 0.7$ V, is insignificant at the relatively large V_{CE} and V_{CB} voltages needed to observe the Early effect. The I_C current increase with an increase in V_{CE}, and therefore V_{CB} voltage, is due to effective shortening of the base width W_B, caused by the associated depletion-layer expansion. This is reflected in the alternative name for the Early effect, which is the *base modulation effect*. The narrower base leads to increased saturation current I_S, which causes the I_C increase [Eq. (9.50)].

A physical insight into this effect can be provided by referring to Fig. 9.5 and Eq. (9.19), which can be rewritten as

$$I_C \approx I_{nE} = I_S e^{V_{BE}/V_t} \tag{9.53}$$

where

$$I_S = qA_J n_i^2 \frac{D_B}{N_B W_B} \tag{9.54}$$

Therefore, the Early effect can be explained by the following sequence of effects, initiated by a V_{BC} (or V_{CE}) increase: (1) The depletion-layer width at the base–collector junction is increased, (2) W_B is reduced, and (3) the concentration gradient of minority carriers in the base is increased, which increases the diffusion current as shown by Eqs. (9.54) and (9.53) for I_S and I_C, respectively.

Early suggested a way of modeling the output resistance itself, and its variation with the level of output current, by a single parameter. Figure 9.14a shows that this is possible if it is assumed that the extrapolated I_C–V_{CE} characteristics (the dashed lines) intersect in a single point on the $V_{CE} \approx V_{CB}$ axis, which is known as Early voltage V_A. Obviously, a larger absolute value of the Early voltage means that the output resistance is higher (I_{CE}–V_{CE} lines are closer to the horizontal level), and vice versa. In the ideal case of $r_o \to \infty$, we have the following Early voltage: $V_A \to \infty$.

Using the rule of similar triangles, the following relationship can be written with the definition of $|V_A|$ as in Fig. 9.14a:

$$\frac{I_C(|V_{BC}| = 0)}{|V_A|} = \frac{I_C(|V_{BC}|)}{|V_A| + |V_{BC}|} \tag{9.55}$$

which, with regard to the related comments and Eq. (9.51), leads to the following equation for the saturation current:

$$I_S = I_{S0}\frac{|V_A| + |V_{BC}|}{|V_A|} = I_{S0}\left(1 + \frac{|V_{BC}|}{|V_A|}\right) \tag{9.56}$$

I_S becomes the V_{BC}-dependent saturation current, used in Eq. (9.50), the Ebers–Moll equation, and the zero-voltage saturation current I_{S0} becomes the SPICE parameter.

Analogous theory applies to the case of inverse active mode, when the base–emitter junction is reverse-biased, with the base–collector junction being forward-biased. The Early voltage in this case is denoted by V_B.

EXAMPLE 9.7 Early Effect

It has been found that the collector current of an NPN BJT increases from 1 mA to 1.1 mA if the collector-to-emitter voltage is increased from 5 V to 10 V. Calculate the Early voltage and the dynamic output resistance of this BJT.

SOLUTION

With the availability of two measurement points, the following set of two equations can be solved:

$$I_{C1} = I_C(0)\left(1 + \frac{|V_{CB1}|}{|V_A|}\right)$$

$$I_{C2} = I_C(0)\left(1 + \frac{|V_{CB2}|}{|V_A|}\right)$$

The solution can be expressed as

$$V_A = \left(V_{CB2} - V_{CB1}\frac{I_{C2}}{I_{C1}}\right) \Big/ \left(\frac{I_{C2}}{I_{C1}} - 1\right)$$

where $V_{CB1} = V_{CE1} - V_{BE}$ and $V_{CE2} = V_{CE2} - V_{BE}$. Therefore,

$$V_A = \frac{(10 - 0.7) - (5 - 0.7) \times 1.1}{0.1} = 45.7 \text{ V}$$

The reciprocal value of the dynamic output resistance is

$$\frac{1}{r_o} = \frac{dI_C}{dV_{CE}} \approx \frac{dI_C}{dV_{CB}}$$

The first derivative of the I_C–V_{CB} dependence leads to

$$\frac{1}{r_o} = \frac{d}{dV_{CB}}\left[\underbrace{I_{S0}e^{V_{BE}/V_t}}_{I_C(0)}\left(1 + \frac{|V_{CB}|}{|V_A|}\right)\right] = \frac{I_C(0)}{|V_A|}$$

which means that the output resistance is

$$r_o = \frac{|V_A|}{I_C(0)}$$

Find $I_C(0)$ from the first measurement point:

$$I_C(0) = I_{C1} \Big/ \left(1 + \frac{|V_{CB1}|}{|V_A|}\right) = 1 \Big/ \left(1 + \frac{4.3}{45.7}\right) = 0.914 \text{ mA}$$

The output resistance is now obtained as

$$r_o = \frac{45.7}{0.914} = 50 \text{ k}\Omega$$

9.3.2 Parasitic Resistances

Another second-order effect included in SPICE is due to the parasitic resistances. Very similarly to the case of the diode, series resistors r_E, r_B, and r_C are added to the emitter, base, and collector, respectively, to account for the contact resistances and the resistances of the respective regions in the silicon (Fig. 9.15). The resistances r_E, r_B, and r_C are direct SPICE parameters.

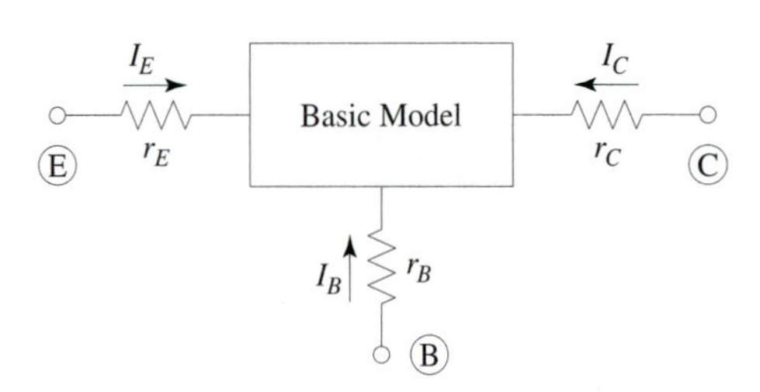

Figure 9.15 The equivalent circuit of the Ebers–Moll model is extended to include the parasitic resistances.

9.3.3 Dependence of Common-Emitter Current Gain on Transistor Current: Low-Current Effects

The common-emitter current gains β_F and β_R, which are SPICE parameters themselves, are constants in the Ebers–Moll model. However, Fig. 9.16 shows that the measured common-emitter current gain β_F of a BJT is different at different current levels. The common-emitter current gain increases with the collector current, slowly reaching the maximum value at medium currents and then rather rapidly decreasing at high currents. Noting that β_F is plotted versus the logarithm of I_C in Fig. 9.16, we can see that the Ebers–Moll assumption of constant β_F can satisfactorily be used in a range of I_C current about 2 orders of magnitude wide. However, if the BJT is operated in extreme conditions, very high or very low current levels, the changes of the common-emitter current gain cannot be neglected. There are equations in SPICE, based on what is known as the Gummel–Poon model, which account for this type of second-order effects.

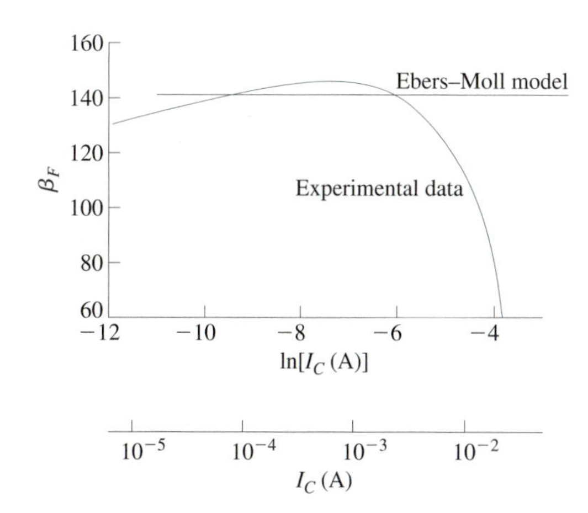

Figure 9.16 The common-emitter current gain at different levels of the collector current. The experimental data are measured on an NPN BJT from a 3086 IC.

According to the Ebers–Moll model, both I_C and I_B are proportional to $\exp(V_{BE}/V_t)$, which results in the expected constant $\beta_F = I_C/I_B$. The diffusion component of the base current does follow the $\exp(V_{BE}/V_t)$ dependence. However, at low biasing levels, the recombination of the carriers in the bulk and surface depletion layer, as well as other surface leakage mechanisms, lead to an increase of the base current. The increased base current is observed as the β_F reduction at low bias levels. To model this effect, Eq. (9.50) for the base current is modified in the following way:

$$I_B = \frac{I_{S0}}{\beta_{FM}}\left(e^{V_{BE}/V_t} - 1\right) + C_2 I_{S0}\left(e^{V_{BE}/(n_{EL}V_t)} - 1\right) + \frac{I_{S0}}{\beta_{RM}}\left(e^{V_{BC}/V_t} - 1\right) + C_4 I_{S0}\left(e^{V_{BC}/(n_{CL}V_t)} - 1\right) \tag{9.57}$$

Obviously, $C_2 I_S\left[\exp(V_{BE}/(n_{EL}/V_t) - 1)\right]$ and $C_4 I_S\left[\exp(V_{BC}/(n_{CL}/V_t) - 1)\right]$ terms are added to include the base-current increase for the cases of forward-biased base–emitter and forward-biased base–collector junctions, respectively. This introduces four SPICE parameters: C_2, the base–emitter leakage saturation current coefficient; n_{EL}, the base–emitter leakage emission coefficient; C_4, the base–collector leakage saturation current coefficient; and n_{CL}, the base–collector leakage emission coefficient. The current gains β_{FM} and β_{RM} are not additional SPICE parameters, they have the same values as β_F and β_R in the Ebers–Moll model. The subscript M is added to indicate that these are the constant mid-current values of the current gains and not the variable current gains.

For a BJT in the normal active mode and when the leakage current dominates, Eq. (9.57) is simplified to

$$I_B \approx C_2 I_{S0} e^{V_{BE}/n_{EL}V_t} \tag{9.58}$$

If plotted as $\ln I_B$–V_{BE}, straight line with the slope equal to $1/n_{EL}V_t$ is obtained. If the recombination current dominates, then $n_{EL} \approx 2$ (this is explained in Section 6.2.2). This slope is smaller than the slope for the pure diffusion current, which is $1/V_t$

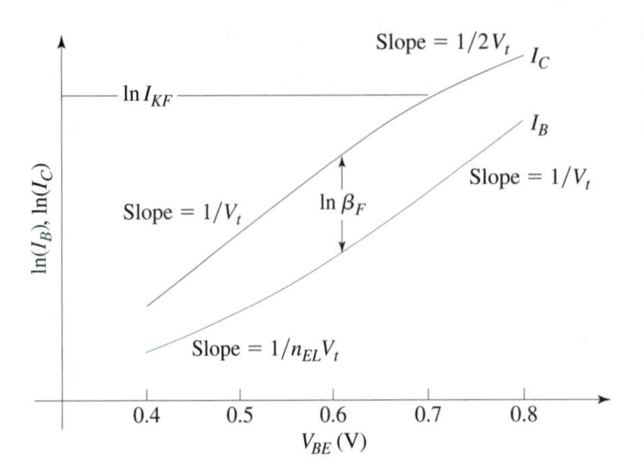

Figure 9.17 Semilogarithmic plots of the base and collector currents versus V_{BE} for an NPN BJT in the normal active mode.

[Eq. (9.51)]. Figure 9.17 shows that the leakage current dominates when the diffusion current, characterized by the section with the slope $1/V_t$, is very small.

9.3.4 Dependence of Common-Emitter Current Gain on Transistor Current: Gummel–Poon Model for High-Current Effects

To include the effects of high V_{CE} bias (the Early effect), the saturation current I_S was modified. Analogously, I_S can be modified to include the effects of high V_{BE} bias, which causes the collector current to fall below the $\exp(V_{BE}/V_t)$ level, and therefore causes the common-emitter current gain reduction at high-bias levels. Equation (9.54) shows that the saturation current I_S is inversely proportional to the two relevant technological parameters: the doping level in the base N_B and the base width W_B. The physical meaning of the $N_B W_B$ is the number of doping atoms in the base per unit of junction area; consequently, $Q_{B0} = qN_B W_B$ is the charge density due to the majority carriers (expressed in C/m^2). Q_{B0} is known as the *base Gummel number*.

Gummel and Poon suggested that the effects of both V_{BE} and V_{BC} bias can be included through a modification of the base Gummel number:

$$Q_{BT} = \underbrace{Q_{B0}}_{qN_B W_B} + \underbrace{C_{dE} V_{BE} + C_{dC} V_{BC} \frac{A_E}{A_C}}_{\text{depletion-layer charge}}$$

$$+ \underbrace{\frac{Q_{B0}}{Q_{BT}} \tau_F I_S \left(e^{V_{BE}/V_t} - 1\right) + \frac{Q_{B0}}{Q_{BT}} \tau_R I_S \left(e^{V_{BC}/V_t} - 1\right)}_{\text{stored charge}} \tag{9.59}$$

where A_E and A_C are the areas of the base–emitter and the base–collector junctions, respectively, and τ_F and τ_R are the normal mode and inverse mode transit times [refer to Eq. (6.58)]. The modification by the depletion-layer charge is related to the Early effect, and the modification by the stored charge is related to the drop in the current gain at high-injection levels.

Dividing I_S by Q_{BT}/Q_{B0} effectively replaces $N_B W_B$ by the modified Gummel number to include the contribution of the depletion-layer charge and the stored charge. Accordingly, the collector current I_C in the normal active mode is expressed as

$$I_C \approx \frac{I_{S0}}{q_b} e^{V_{BE}/V_t} \tag{9.60}$$

where

$$q_b = \frac{Q_{BT}}{Q_{B0}} \tag{9.61}$$

When we define the parameters

$$I_{KF} = \frac{Q_{B0}}{\tau_F}, \qquad I_{KR} = \frac{Q_{B0}}{\tau_R} \tag{9.62}$$

$$|V_A| = \frac{Q_{B0}}{C_{dC}} \frac{A_C}{A_E}, \qquad |V_B| = \frac{Q_{B0}}{C_{dE}} \tag{9.63}$$

the factor q_b can be expressed as

$$q_b = \frac{q_1}{2} + \frac{\sqrt{q_1^2 + 4q_2}}{2} \approx \frac{q_1}{2}\left(1 + \sqrt{1 + 4q_2}\right) \tag{9.64}$$

where

$$q_1 = 1 + \frac{V_{BE}}{|V_B|} + \frac{V_{BC}}{|V_A|} \approx \frac{1}{1 - \dfrac{V_{BE}}{|V_B|} - \dfrac{V_{BC}}{|V_A|}} \tag{9.65}$$

and

$$q_2 = \frac{I_{S0}}{I_{KF}}\left(e^{V_{BE}/V_t} - 1\right) + \frac{I_{S0}}{I_{KR}}\left(e^{V_{BC}/V_t} - 1\right) \tag{9.66}$$

The first equations of q_b and q_1 correspond directly to the original Gummel–Poon model, while the approximate equations correspond to those used in SPICE. Derivation of q_b is given in the following books: (1) R. S. Muller and T. I. Kamins, *Device Electronics for Integrated Circuits,* 2nd ed., Wiley, New York, 1986, pp. 359–362, and (2) G. Massobrio and P. Antognetti, *Semiconductor Device Modeling with SPICE,* 2nd ed., McGraw-Hill, New York, 1993, Chapter 2. Both books provide more detailed description of the Gummel–Poon model, while the second book also lists the equations used in SPICE. I_{KF} and I_{KR} are additional SPICE parameters, which are used to fit the high-level I_C current to the experimental data. The measurement of these parameters is described in Section 11.3.2. $|V_A|$ and $|V_B|$ are equivalent to the earlier described normal and reverse Early voltages.

Equations (9.61)–(9.66) are general: they cover all the possible BJT modes of operation. To generalize the collector current given by Eq. (9.60), the effects of V_{BC} voltage should be added in a way analogous to the case of Ebers–Moll model. The general I_C equation is given here along with I_B [Eq. (9.57)] and the corresponding I_E equations, to show the complete set of BJT equations at the Gummel–Poon level in SPICE:

$$\begin{aligned}
I_B &= \frac{I_{S0}}{\beta_{FM}}\left(e^{V_{BE}/V_t}-1\right) \qquad + C_2 I_{S0}\left(e^{V_{BE}/(n_{EL}V_t)}-1\right)\\
&\quad + \frac{I_{S0}}{\beta_R}\left(e^{V_{BC}/V_t}-1\right) \qquad + C_4 I_{S0}\left(e^{V_{BC}/(n_{CL}V_t)}-1\right)\\
I_C &= \frac{I_{S0}}{q_b}\left(e^{V_{BE}/V_t}-e^{V_{BC}/V_t}\right)\\
&\quad - \frac{I_{S0}}{\beta_{RM}}\left(e^{V_{BC}/V_t}-1\right) \qquad - C_4 I_{S0}\left(e^{V_{BC}/(n_{CL}V_t)}-1\right)\\
I_E &= -\frac{I_{S0}}{q_b}\left(e^{V_{BE}/V_t}-e^{V_{BC}/V_t}\right)\\
&\quad - \frac{I_{S0}}{\beta_{FM}}\left(e^{V_{BE}/V_t}-1\right) \qquad - C_2 I_{S0}\left(e^{V_{BE}/(n_{EL}V_t)}-1\right)
\end{aligned} \tag{9.67}$$

The Ebers–Moll equations are included in Eqs. (9.67) of the Gummel–Poon level in SPICE. Default values of the second-order effect parameters reduce Eqs. (9.67) to Eqs. (9.50). $I_{KF} = I_{KR} \to \infty$ turns q_2 into zero and $|V_A| = |V_B| \to \infty$ turns q_1 into unity, which means $q_b = 1$. The default value of C_2 and C_4 is zero, which, along with $q_b = 1$, clearly eliminates all the additions in Eqs. (9.67) compared to Eqs. (9.50).

It is useful to analyze the Gummel–Poon equation for the collector current in the normal active mode and without the Early effect [Eq. (9.60)]. At low and medium current levels, $I_{S0}\exp(V_{BE}/V_t) \ll I_{KF}$, which means that Eq. (9.64) reduces to $q_b \approx 1$ because $4q_2 \ll 1$. With $q_b \approx 1$, Eq. (9.60) becomes equivalent to the corresponding Ebers–Moll equation, which predicts the $1/V_t$ slope of the $\ln I_C$–V_{BE} line. However, at high current levels, $4q_2 \gg 1$ and $2\sqrt{q_2} \gg 1$. This means that Eqs. (9.64) and (9.60) can be simplified as

$$q_b \approx \sqrt{q_2} = \sqrt{\frac{I_{S0}}{I_{KF}} e^{V_{BE}/2V_t}} \tag{9.68}$$

and

$$I_C \approx \sqrt{I_{S0} I_{KF}}\, e^{V_{BE}/2V_t} \Rightarrow \ln I_C \approx \ln\sqrt{I_{S0} I_{KF}} + \underbrace{\frac{1}{2V_t}}_{\text{slope}} V_{BE} \tag{9.69}$$

The semilogarithmic plot of the collector current in Fig. 9.17 shows the two regions characterized by the two slopes: $1/V_t$ where $\beta_F \approx \beta_{FM}$, and $1/2V_t$, where $\beta_F < \beta_{FM}$.

9.4 HETEROJUNCTION BIPOLAR TRANSISTOR

In homojunctions, which are P–N junctions created in the same type of semiconductor, the bottom of the conduction band (E_C) and the top of the valence band (E_V) are parallel throughout the whole structure because they are always separated by the constant energy gap. As a result, the energy barrier at the P–N junction has the same value for both the electrons on the N type and the holes on the P-type side. For the case of zero bias (thermal equilibrium), this is illustrated in Fig. 6.3a where the barrier is labeled by qV_{bi}. With forward-bias V_D, the barrier height is reduced to $q(V_{bi} - V_D)$ for both the electrons and the holes, as shown in Fig. 6.6a. This means that the same fractions of both the electrons in the N-type and the holes in the P-type region possess sufficient energy to pass through the barrier. To focus on the effect of the barrier height, assume equal concentrations of majority electrons and holes (equal N-type and P-type doping levels) and neglect any differences in the physical parameters (such as the diffusion constants, the diffusion lengths, or the lengths of the neutral regions when shorter than the diffusion lengths) as second-order effects. With these assumptions, the currents of the electrons and the holes through the junction are equal. If this P–N junction was the base–emitter junction of a BJT, the emitter efficiency would be $\gamma_E = 0.5$, as can be seen from Eqs. (9.19), (9.20), and (9.22). To achieve emitter efficiency that is close to 1, BJTs with homojunctions are made with much lower doping in the base ($N_B \ll N_E$). Although this approach works, it has important limitations for some applications. An important limitation is due to the relatively high resistance of the low-doped base, which causes relatively large RC time constants and limits the high-frequency operation of the device.

The energy barriers for electrons and holes at a heterojunction are in general different, because of the existence of different band offsets. This enables us to create a heterojunction bipolar transistor that has both heavily doped base and excellent emitter efficiency.

A specific and frequently used heterojunction is the junction between N-type AlGaAs and P-type GaAs. Figure 9.18a shows the energy-band diagram of this junction under the flat-band condition. The figure shows that forward-bias voltage V_{FB} would have to be applied to split the Fermi levels by qV_{FB} that is necessary to create the flat-band condition. The energy-band diagram under the flat-band condition is drawn because it clearly illustrates the band offsets between the wider gap of AlGaAs and the narrower gap of GaAs. The energy-band diagram shown in Fig. 9.18a is in principle the same as the energy-band diagram at the interface between the wider energy gap of SiO_2 and the narrower energy gap of Si, shown previously in Fig. 7.4b. In practice, the flat-band condition cannot be reached at the N–P heterojunction because damagingly high forward current would be forced through the junction.

The energy gaps of AlGaAs and GaAs are 1.85 eV and 1.42 eV, respectively. These are material constants that cannot be changed by doping or biasing. The positions of the energy bands of AlGaAs and GaAs with respect to the vacuum level, and with respect to one another, are also material constants. The positions of the energy bands of these two materials are such that there are a conduction-band offset of $\Delta E_C = 0.28$ eV and a valence-band offset of $\Delta E_V = 0.14$ eV. Neither doping nor applied bias can change these band offsets.

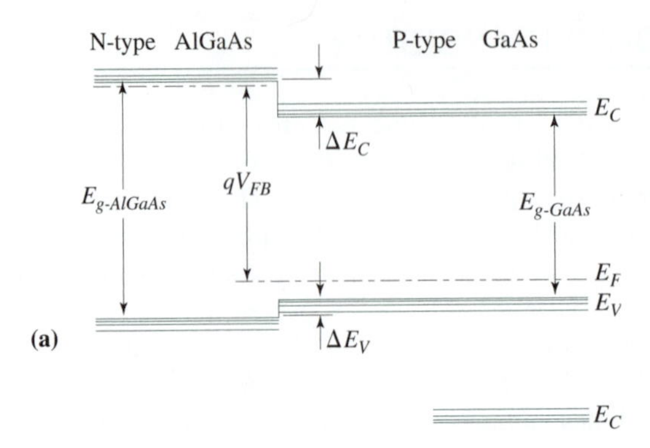

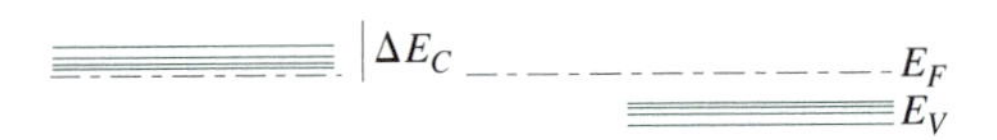

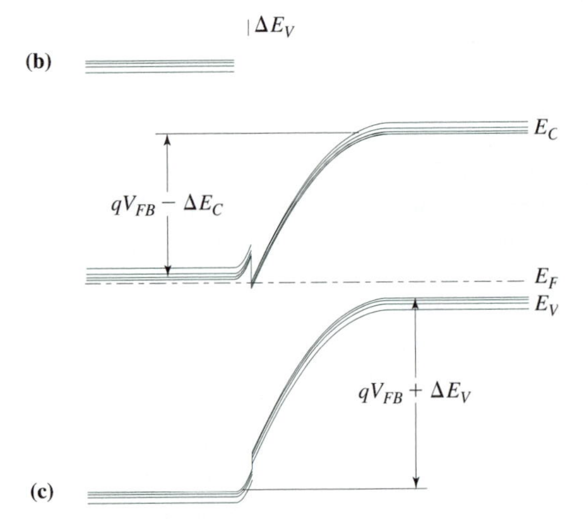

Figure 9.18 Energy-band diagrams of the AlGaAs–GaAs heterojunction. (a) The diagram under the flat-band condition, illustrating the band offsets ΔE_C and ΔE_V as material constants. (b) An incomplete diagram at thermal equilibrium, illustrating the process of diagram construction. (c) The complete diagram at thermal equilibrium, illustrating different barrier heights for the electrons in the N-type AlGaAs and the holes in the P-type GaAs.

The following steps can be used to construct the energy-band diagram of a heterojunction:

1. The constant Fermi-level (E_F) line, corresponding to the case of thermal equilibrium, is drawn first. Just as in the case of a homojunction, the energy bands of the two neutral regions are drawn away from the junction. In this case, the energy bands of N-type AlGaAs are drawn with the appropriate energy gap ($E_{g-AlGaAs}$) and are placed so that E_C is quite close to E_F to express the N-type doping. The energy bands of GaAs are drawn with the energy gap of E_{GaAs} and are placed so that E_V is close to E_F to express the P-type doping.

2. The energy-band discontinuities, ΔE_C and ΔE_V, are indicated at the junction (Fig. 9.18b). The exact energy positions of ΔE_C and ΔE_V depend on the doping levels in AlGaAs and GaAs, which leads to different degrees of field penetration and band bending. In this specific case, ΔE_V and ΔE_C are shifted slightly above E_V and E_C of AlGaAs to indicate the much smaller band bending in the heavier doped AlGaAs. The energy position of ΔE_C with respect to ΔE_V, however, corresponds to $E_{g-AlGaAs}$ and E_{g-GaAs} and is the same as in Fig. 9.18a.
3. To complete the diagram, the E_C levels from each material are connected to the ΔE_C ends and the E_V levels are connected to the ΔE_V ends by curved lines that indicate the built-in electric field in the depletion layer at the junction.

Figure 9.18c shows that the energy barrier for the electrons in the N-type AlGaAs is $qV_{FB} - \Delta E_C$, whereas the energy barrier for the holes in the P-type GaAs is $qV_{FB} + \Delta E_V$. With forward-bias V_D, both these barriers will be reduced by qV_D because E_F of the neutral AlGaAs region is moved up by qV_D, with respect to E_F in the neutral region of GaAs. Importantly, the difference between the energy barriers for holes and electrons remains the same: $\Delta E_C + \Delta E_V$. If this AlGaAs–GaAs heterojunction is the emitter–base junction of a BJT, the injection of electrons into the base (I_{nE} current) is much higher than the injection of holes into the emitter (I_{pE} current). To focus on this effect, we can again assume equal emitter and base doping and neglect any differences in the physical parameters as second order effects. With these assumptions, $I_{nE}/I_{pE} = \exp[(\Delta E_C + \Delta E_V)/kT]$. Even relatively small ΔE_C and ΔE_V offsets can lead to a very large I_{nE}/I_{pE} ratio and, according to Eq. (9.22), to emitter efficiency that is close to 1.

Given that there is no need to reduce the doping level of the base to achieve good emitter efficiency, it is possible to significantly reduce the resistance of the base in heterojunction bipolar transistors. This adds to a number of other favorable parameters of GaAs structures, such as higher electron mobility and reduced parasitic capacitances. As a result, heterojunction bipolar transistors are frequently used for high-frequency (microwave) and high-power applications.

SUMMARY

1. A BJT in the *normal active mode* acts as a voltage-controlled current source. In the case of an NPN BJT, the *forward-biased E–B junction* emits electrons into the base, which diffuse through the very narrow base region to be collected by the electric field of the *reverse-biased C–B junction*. Even the smallest positive collector voltage collects the electrons efficiently enough so that I_C does not increase with $V_{CB} \approx V_{CE}$ (a constant-current source). However, I_C strongly depends on the bias of the controlling E–B junction. A good analogy is a waterfall current, which does not depend on the height of the fall but on the amount of the incoming water.
2. In addition to the emitter electron current I_{nE}, there is also I_{pE}, the current due to the holes emitted from the base back into the emitter, which reduces the emitter efficiency: $\gamma_E = I_{nE}/(I_{nE} + I_{pE})$. Some of the electrons are recombined in the base, leading to a nonideal base transport factor: $\alpha_T = I_{nC}/I_{nE}$. The reverse-bias current of the C–B junction, I_{CB0}, can usually be neglected, so that emitter-to-collector and base-to-

collector current gains can be related to γ_E and α_T as follows:

$$\alpha = \frac{I_C}{I_E} = \alpha_T \gamma_E, \qquad \beta = \frac{\alpha}{1-\alpha}$$

3. To maximize emitter efficiency, the emitter of a good BJT will be much more heavily doped than the base:

$$\gamma_E = 1 \Bigg/ \left(1 + \frac{D_E}{D_B}\frac{N_B W_B}{N_E W_E}\right)$$

and to maximize the transport factor, it will have a very narrow base:

$$\alpha_T = 1 \Bigg/ \left[1 + \frac{1}{2}\left(\frac{W_B}{L_B}\right)^2\right]$$

4. When the collector and the emitter are swapped (*inverse active mode*), α and β are significantly smaller. If none of the junctions is forward-biased, no significant current flows through the BJT (*cutoff mode*). When one junction is forward-biased but the other is not reverse-biased, the BJT is in *saturation mode*. With a forward-biased E–B junction, the saturation mode in an NPN BJT is for $V_{CB} < 0$—the negatively biased collector does not collect the emitted electrons but rather begins to emit electrons on its own, reducing the I_C current. The cutoff and saturation modes enable a BJT to be used as a voltage-controlled switch in digital and power-switching circuits.

5. There is a complementary PNP BJT. It is a mirror image of the NPN: opposite doping types, opposite voltage polarities, and opposite current directions.

6. There are strong similarities between the energy-band diagrams and therefore between operation principles of BJTs and MOSFETs. However, there are also very important differences. The concentration of carriers in the MOSFET channel is controlled by an electric field (no input current is needed and no input power is wasted); a significant input current, and consequently input power, is needed to keep a BJT in the saturation mode (switch in *on* mode). On the other hand, a very shallow field penetration through the channel charge severely limits the cross-sectional area of the MOSFET channel; in the case of BJTs, the input control is over the entire P–N junction area—hence, superior current capability.

7. The general version (accounting for all the modes of operation) of the principal Ebers–Moll model for an NPN BJT is

$$\begin{aligned}
I_C &= I_S\left(e^{V_{BE}/V_t} - 1\right) - \left(1 + \frac{1}{\beta_R}\right) I_S\left(e^{V_{BC}/V_t} - 1\right) \\
I_E &= -\left(1 + \frac{1}{\beta_F}\right) I_S\left(e^{V_{BE}/V_t} - 1\right) + I_S\left(e^{V_{BC}/V_t} - 1\right) \\
I_B &= \frac{1}{\beta_F} I_S\left(e^{V_{BE}/V_t} - 1\right) + \frac{1}{\beta_R} I_S\left(e^{V_{BC}/V_t} - 1\right)
\end{aligned}$$

The assumed current directions are into the transistor—hence the reverse signs of the emitter current. The only difference in the PNP model is that the polarities of the voltages are opposite ($V_{BE} \rightarrow V_{EB}$, $V_{BC} \rightarrow V_{CB}$), and the current directions are assumed out of the transistor. In the normal active mode, $\exp(V_{BE}/V_t) \gg 1$ and $\exp(V_{BC}/V_t) - 1 \approx 1$, which simplifies the principal Ebers–Moll equations to

$$I_C = I_S e^{V_{BE}/V_t}$$

$$I_E = -\left(1 + \frac{1}{\beta_F}\right) I_S e^{V_{BE}/V_t} = -\frac{I_C}{\alpha_F}$$

$$I_B = \frac{1}{\beta_F} I_S e^{V_{BE}/V_t} = \frac{I_C}{\beta_F}$$

8. The principal model assumes a perfect current source (the collector current fully independent of the collector voltage). In reality, the expansion of the C–B depletion layer due to an increased reverse bias leads to narrowing of the effective base width (*base width modulation*), resulting in an increase of the collector current (a finite output dynamic resistance). This is known as the Early effect and is modeled through I_S,

$$I_S = I_{S0}\left(1 + \frac{|V_{BC}|}{|V_A|} + \frac{|V_{BE}|}{|V_B|}\right)$$

where I_{S0} is the SPICE parameter, and not I_S itself. The forward and reverse Early voltages $|V_A|$ and $|V_B|$ are also parameters.

9. Additional second-order effects relate to β dependence on the collector current: it has a maximum at medium currents, being smaller at small collector currents, but with a lot more dramatic reduction at high-level injection. The reduction at small currents is modeled by adding "leakage" components to the base-current equation:

$$\begin{aligned} I_B = {} & \frac{I_{S0}}{\beta_{FM}}\left(e^{V_{BE}/V_t} - 1\right) + C_2 I_{S0}\left(e^{V_{BE}/(n_{EL}V_t)} - 1\right) \\ & + \frac{I_{S0}}{\beta_{RM}}\left(e^{V_{BC}/V_t} - 1\right) + C_4 I_{S0}\left(e^{V_{BC}/(n_{CL}V_t)} - 1\right) \end{aligned}$$

where the maximum current gains β_{FM} and β_{RM} are parameters in the model, with the other parameters being C_2, C_4, n_{EL}, and n_{CL}. The reduction at high-level injection is modeled by equations that include a modified Gummel number in the base. The modifications include additions of the depletion-layer charge (the Early effect) and the stored charge (the reduction of β at high injection levels). The complete equations are referred to as the Gummel–Poon model. In the normal active mode and without the Early effect, the Gummel–Poon equation for the collector current is simplified to

$$I_C \approx \begin{cases} I_{S0} e^{V_{BE}/V_t} & \text{for } I_C < I_{KF} \\ \sqrt{I_{S0} I_{KF}}\, e^{V_{BE}/2V_t} & \text{for } I_C > I_{KF} \end{cases}$$

where the high-level knee current I_{KF} is a parameter. An analogous parameter I_{KR} relates to the inverse active mode.

10. The energy-barrier heights for injections of majority electrons and holes at a homojunction (a P–N junction with the P-type and N-type regions made of the same semiconductor) are equal. To achieve a high I_{nE}/I_{pE} ratio and good emitter efficiency in an NPN BJT with homojunctions, the base doping has to be much lower than the emitter doping. The energy-barrier heights for injections of majority electrons and holes at a heterojunction are different due to the energy-band discontinuities ΔE_C and ΔE_V. In some heterojunctions, such as AlGaAs–GaAs, the I_{nE}/I_{pE} ratio is proportional to $\exp[(\Delta E_C + \Delta E_V)/kT]$. This means that excellent emitter efficiency can be achieved even with a highly doped base.

PROBLEMS

9.1 Find the most suitable description for each of the concentration diagrams shown in Fig. 9.19. Electron concentrations are presented with solid lines, while the hole concentrations are presented with dashed lines.

9.2 Figure 9.20 shows four energy-band diagrams, drawn from the emitter to the collector. Explain how the energy-band diagrams relate to each of the four points, labeled on the output characteristics of the BJT.

9.3 Assign each of the energy-band diagrams from Fig. 9.21 to the proper description of the BJT type and mode of operation.

9.4 **(a)** List and briefly describe the voltage dependencies of the current of majority and minority carriers that flow through the emitter and the collector of an NPN BJT biased in the normal active mode.

(b) If the common-base current gain is 600, what is the common-emitter current gain?

(c) If the emitter efficiency is ≈ 1, what is the value of the transport factor through the base?

9.5 An NPN BJT with base–emitter and base–collector junction areas A_{JE} and A_{JC}, respectively, has the following parameters: the transport factor $\alpha_T = 0.9999$ and the emitter efficiency $\gamma_E = 0.9968$. The ratings of the transistor are as follows: the maximum collector current I_{max} is 10 mA, and the maximum collector–base voltage V_{max} is 25 V. What would the BJT parameters and ratings be if both junction areas (A_{JE} and A_{JC}) are doubled?

(a) $\beta = 604$, $I_{max} = 10$ A, $V_{max} = 25$ V
(b) $\beta = 302$, $I_{max} = 20$ mA, $V_{max} = 25$ V
(c) $\beta = 604$, $I_{max} = 20$ mA, $V_{max} = 50$ V
(d) $\beta = 151$, $I_{max} = 10$ mA, $V_{max} = 25$ V
(e) $\beta = 302$, $I_{max} = 20$ mA, $V_{max} = 12.5$ V
(f) $\beta = 604$, $I_{max} = 20$ mA, $V_{max} = 25$ V

9.6 Which of the following statements, related to BJTs, **are not** correct?

(a) The relationship $\alpha = \frac{\beta}{\beta+1}$ cannot be used when a BJT is in saturation.

(b) The emitter of an NPN BJT is grounded, and the collector is connected to V_+ through a loading resistor. V_{BE} voltage is such that the BJT is in the saturation mode. An increase in V_{BE} will increase I_{CE} and reduce V_{CE}.

(c) The transport factor α_T and the common-base current gain α are different and unrelated parameters.

(d) The emitter efficiency γ_E depends on the doping level in the base.

(e) The concentration of the minority carriers in the base is significantly higher compared to the equilibrium level when the BJT is in normal or inverse active mode.

(f) The minority-carrier lifetime in the base does not affect the transport factor α_T.

(g) The fact that the collector current I_C does not depend on the output voltage V_{CE} in the normal active mode is related to the fact that the saturation current of a diode does not depend on the voltage drop across the diode.

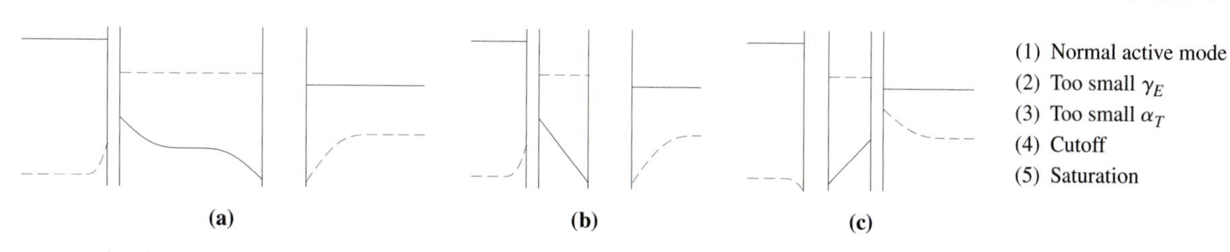

Figure 9.19 NPN concentration diagrams.

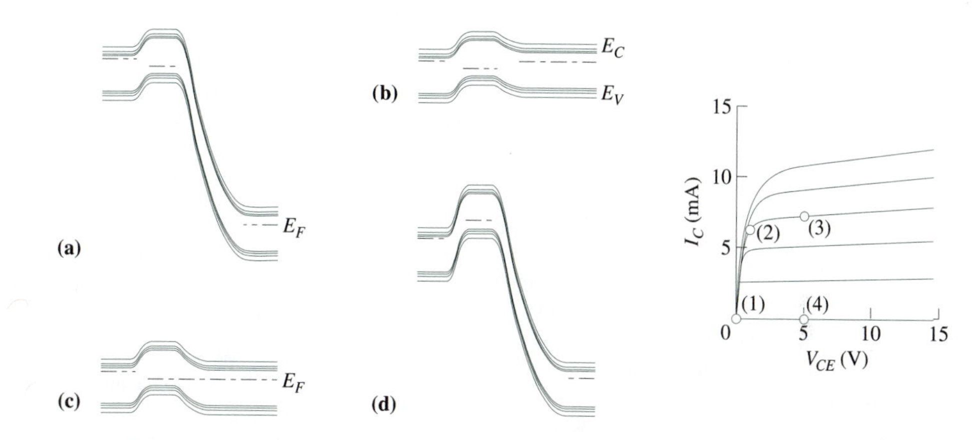

Figure 9.20 Energy-band diagrams and BJT output characteristics.

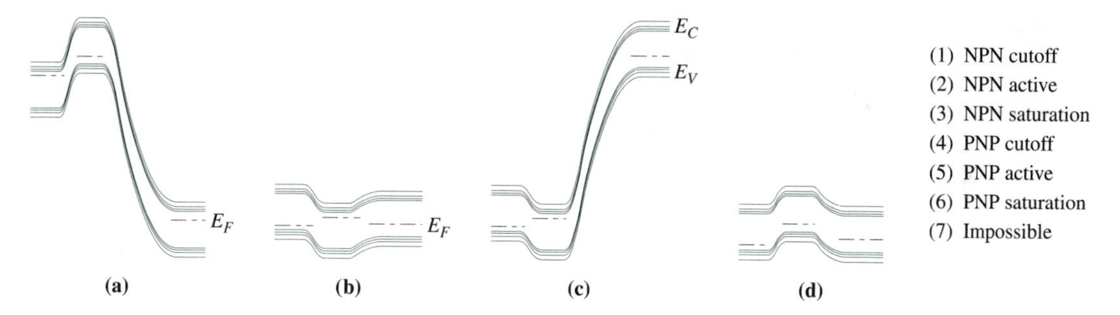

Figure 9.21 Energy-band diagrams of NPN and PNP BJTs.

9.7 **(a)** Calculate γ_E for an NPN and a PNP BJT using the following technological parameters: $N_E = 10^{19}$ cm^{-3}, $N_B = 10^{17}$ cm^{-3}, $W_E = W_B = 2$ μm, $D_n(10^{19}$ cm$^{-3}) = 2.5$ cm^2/s, D_p $(10^{19}$ cm$^{-3}) = 1.5$ cm^2/s, $\mu_n(10^{17}$ cm$^{-3}) = 800$ cm^2/V $\cdot$ s, $\mu_p(10^{17}$ cm$^{-3}) = 320$ cm^2/V $\cdot$ s.

(b) Assuming $\alpha_T = 1$, calculate the common-emitter current gain for both BJTs.

9.8 **(a)** Design the base–emitter area of an NPN BJT so that the current rating of the BJT operating in the normal active mode is 100 mA at $V_{BE} = 0.70$ V. The technological parameters are as follows: the emitter doping is $N_D = 10^{20}$ cm^{-3}, the base doping is $N_A = 5 \times 10^{17}$ cm^{-3}, the base width is $W = 1$ μm, and the electron mobility in the base is $\mu_n = 300$ cm^{2}V/s.

(b) What would the normal active current be at $V_{BE} = 0.7$ V if the BJT was made in GaAs with the same base width and doping levels? The electron mobility in the GaAs base is $\mu_n = 2500$ cm^2/V · s. **A**

9.9 **(a)** Design the base–emitter area of an NPN BJT so that its transconductance gain is $g_m = 2$ A/V at $V_{BE} = 0.70$ V. The technological parameters are as follows: the emitter is heavily doped, the base doping is 5×10^{17} cm^{-3}, the base width is $W = 0.5$ μm, and the electron mobility in the base is $\mu_n = 300$ cm^2/V · s.

(b) Room temperature (300 K) is assumed in part (a). What would be the transconductance gain at $T = 55$°C? Assume that the electron mobility does not change and that $n_i \approx 7.5 \times 10^{10}$ cm^{-3}. **A**

9.10 Table 9.2 lists the four possible combinations of two measured parameters (the common-emitter current gain and the collector current at $V_{BE} = 0.7$ V) for two NPN BJTs, one made in Si and the other in GaAs. Assuming that the doping levels and the geometric parameters of the two BJT are similar, determine the correct combination.

TABLE 9.2 Possible Data Combinations for Si and GaAs NPN BJTs (I_C Measured at $V_{BE} = 0.7$ V)

	Si		GaAs	
	β	I_C	β	I_C
1	500	0.1 μA	2000	300 mA
2	500	300 mA	2000	0.1 μA
3	2000	0.1 μA	500	300 mA
4	2000	300 mA	500	0.1 μA

9.11 Determine the modes of operation of a PNP BJT on the basis of the following sets of measurements:

(a) $V_{BE} = 0.7$ V, $V_{CE} = -5.2$ V
(b) $V_{BE} = 0.7$ V, $V_{BC} = -0.7$ V
(c) $V_{BE} = -0.7$ V, $V_{CE} = -0.2$ V
(d) $V_{BE} = 0.7$ V, $V_{CE} = -0.2$ V
(e) $V_{BE} = -5.2$ V, $V_{CE} = 0.7$ V **A**

9.12 The NPN BJT in the circuit of Fig. 9.8 has $\beta = 550$. The circuit parameters are $R_C = 1$ kΩ and $V_+ = 5$ V. Find the minimum base current i_B that ensures that the BJT is in saturation (switch in *on* mode). Assume $V_{BE} = 0.7$ V.

9.13 A PNP BJT with $\beta = 300$ is used in a circuit that is similar to the circuit of Fig. 9.8: the resistor is the same ($R_C = 1$ kΩ) but V_+ is replaced by $V_- = -5$ V. Find the maximum base current $|I_B|$ to ensure that the BJT is in the normal active mode. Assume $V_{BE} = -0.7$ V. **A**

9.14 NPN BJTs in integrated injection logic (I^2L) circuits operate in the inverse active mode. Calculate the common-emitter current gain (β_R) if the technological parameters are as follows: $N_C = 5 \times 10^{14}$ cm^{-2}, $N_B = 10^{16}$ cm^{-3}, $W_C = 10$ μm, $W_B = 2$ μm, $\mu_n = 1250$ cm^2/V · s, and $\mu_p = 480$ cm^2/V · s. Assume ideal transport factor of the base. **A**

9.15 The parameters of the Ebers–Moll model are $I_{S0} = 10^{-12}$ A, $\beta_F = 500$, and $\beta_R = 1$. Find the values of the emitter, base, and collector currents for

(a) $V_{BE} = 0.65$ V and $V_{CB} = 4.35$ V (the normal active mode)

(b) $V_{BE} = 0.65$ V and $V_{CE} = 0.1$ V (the saturation mode)

(c) $V_{EB} = 4.65$ V and $V_{BC} = 0.65$ V (the inverse active mode) **A**

9.16 For the BJT of Problem 9.15, calculate the transconductance g_m in the normal active mode.

9.17 The measured dynamic output resistance of an NPN BJT is $r_o = 35.7$ kΩ. Calculate the Early voltage if $I_{S0} = 1.016 \times 10^{-16}$ A and the measurement is taken at $V_{BE} = 0.8$ V. Estimate the resistance at $V_{BE} = 0.7$ V.

9.18 An increase in the reverse-bias base–collector voltage reduces the base width (base modulation effect). At what voltage would the base width be reduced to zero ("punch-through" breakdown), if the zero-bias base width is 0.937 μm and the doping levels are $N_{base} = 10^{16}$ cm^{-3} and $N_{collector} = 5 \times 10^{14}$ cm^{-3}? Assume that the junction is abrupt.

9.19 The breakdown voltage of the BJT considered in Problem 9.18 can be increased by increasing the zero-bias width of the base. What breakdown voltage can be achieved in this way before avalanche breakdown of the collector–base junction is reached? Assume that avalanching is triggered at $E_{max} = 30$ V/μm. **A**

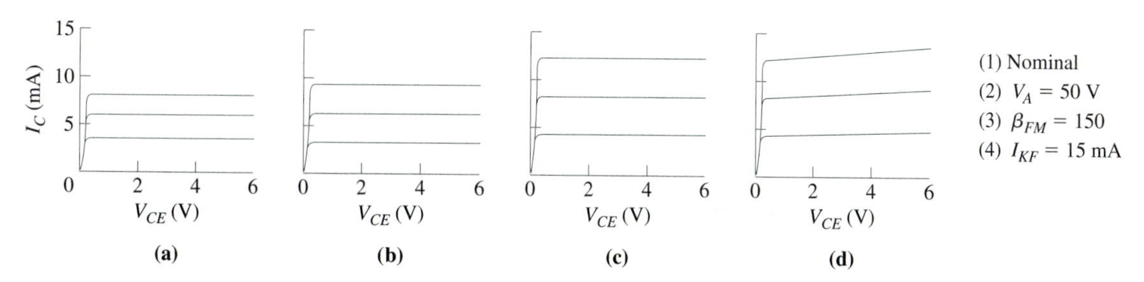

Figure 9.22 BJT output characteristics ($I_B = 0$, 20, 40, and 60 μA in every diagram).

TABLE 9.3 SPICE Model and Experimental Results

Model	Experiment
(a) $r_o = 10$ MΩ	$r_o = 80$ kΩ
(b) $I_C = 1$ mA, $I_B = 10$ μA	$I_C = 6.4$ mA, $I_B = 10$ μA
(c) $I_{C1} = 1.5$ mA, $I_{B1} = 10$ μA	$I_{C1} = 1.6$ mA, $I_{B1} = 10$ μA
$I_{C2} = 30.0$ mA, $I_{B2} = 200$ μA	$I_{C2} = 19.2$ mA, $I_{B2} = 200$ μA

9.20 One set of output characteristics from Fig. 9.22 is for $I_{S0} = 10^{-15}$ A and $\beta_{FM} = 200$ as nominal BJT parameters, whereas the other three are either for changed value of β_{FM} or for specified V_A or I_{KF} parameter. Relate each of the output characteristics to the appropriate set of parameters.

9.21 Which SPICE parameter should be changed in order to achieve better matching between the model and the experimental results in each of the cases shown in Table 9.3? The options are

(1) increase I_{S0}
(2) increase $C_2 I_{S0}$
(3) increase β_F
(4) decrease V_A
(5) decrease I_{KF}
(6) decrease n_{EL}

9.22 Are the following statements correct or incorrect?

(a) The current gain β_F is the only principal SPICE parameter for a BJT in the normal active mode.
(b) The saturation current I_{S0} is the only principal SPICE parameter for a BJT in the saturation mode.
(c) The parasitic resistances r_E and r_C are used to set the dynamic output resistance.
(d) I_{KF} parameter (the normal knee current) is important to model the effects at large V_{CE} voltages.

9.23 The emitter–base junction of a BJT is implemented as

(a) N–P homojunction in GaAs
(b) N–P AlGaAs–GaAs heterojunction

Determine the emitter efficiency in each case. The technological parameters satisfy the following conditions: $D_E/D_B = 0.1$, $N_B/N_E = 1$, and $W_B/W_E = 1$.

REVIEW QUESTIONS

R-9.1 Increased reverse-bias voltage increases the slope of energy bands—that is, the electric field—in the depletion layer of a P–N junction. Why does not this significantly increase the reverse-bias current?

R-9.2 The equilibrium concentration of electrons in the base of an NPN BJT is n_{pe}. Is the concentration of electrons significantly higher when this BJT is biased in the normal active mode? Is this concentration related to the collector current?

R-9.3 Does the doping level in the base affect the emitter efficiency γ_E? If so, why?

R-9.4 Does it affect the transport factor α_T? If so, why?

R-9.5 Is the transport factor α_T different from the common-base current gain α? Are they related? If so, how?

R-9.6 Is the common-base current gain α related to the common-emitter current gain β?

R-9.7 Does the relationship between α and β apply in saturation and cutoff?

R-9.8 A change in the base–emitter voltage V_{BE} drives an NPN BJT from the normal active mode into saturation. Is V_{BE} increased or decreased? If the BJT biasing circuit is as in Fig. 9.8, is V_{CE} increased or decreased? Is I_C increased or decreased?

R-9.9 The measured voltage between the base and collector of a BJT connected in a circuit is $V_{BC} = -5$ V. Provided the BJT is not broken down, can you say whether this is an NPN or a PNP BJT?

R-9.10 What parameters are essential for a proper SPICE simulation of a circuit with a BJT in the normal active mode? Inverse active mode? Saturation?

R-9.11 Which SPICE parameter is used to adjust the dynamic output resistance?

R-9.12 Is the I_{KF} parameter used to model the effects of a large V_{CE} voltage or the effects of a large I_C current?

10 Physics of Nanoscale Devices

The small size of modern semiconductor devices means that only a small number of electrons or holes appear in each device. For example, nonvolatile memories may leak only a few electrons over a period of 10 years. In fact, almost no modern semiconductor device has more than one minority carrier in the neutral regions. To be able to understand and deal with the effects associated with a small number of particles, we need to refine and upgrade the concepts of classical device physics, which are limited to continuous current flow, continuous carrier concentration, continuous generation/recombination rates, and continuous balance between generation and recombination in thermal equilibrium. Accordingly, Section 10.1 introduces the physics of single-carrier events.

The demand for an increase in the complexity of integrated electronic systems and the associated reduction in device dimensions have led to the development and use of devices with one or more dimensions in the nanoscale region. As this downscaling trend continues and device operation enters deeper into the quantum-mechanical world, the modern electronics engineer is exposed to confusing claims that vary from extreme gloom-and-doom scenarios to extremely optimistic promises that harnessing quantum effects will revolutionize not only electronics systems but technology in general. In spite of the inherent mystery, the relevant quantum-mechanical effects are well established and can be presented in a way that simply upgrades existing knowledge of semiconductor phenomena by incorporating unambiguous facts. Section 10.2 shows how to incorporate the effects of quantum confinement in the standard MOSFET models. Section 10.3, which is devoted to one-dimensional systems (nanowires and carbon nanotubes), includes the physics of transport without carrier scattering (ballistic transport) when the current is determined by the ultimate quantum-mechanical conductance limit.

10.1 SINGLE-CARRIER EVENTS

10.1.1 Beyond the Classical Principle of Continuity

Let us begin by recalling the specific question from Example 1.8b. In this case, take a typical neutral P-type silicon region with $p = 10^{18}$ cm^{-3} and calculated concentration of electrons of $n = n_i^2/p = 10^{20}/10^{18} = 100$ cm^{-3}. Let the size of the neutral P-type region be 100 nm in each dimension ($V = 10^{-15}$ cm^3), which means that the total number of electrons in this P-type region is $N = nV = 10^{-13}$. This result is clearly inconsistent with the well-established quantum fact that the number of electrons should be either zero or one but cannot be 10^{-13}. Yet, this result cannot be ignored because neither $N_{actual} = 0$ nor $N_{actual} = 1$ corresponds to the condition of thermal equilibrium, and the equilibrium condition cannot be ignored because the carrier balance is not affected by external forces. To address the question about the implications of this result, we have to establish a workable meaning of the calculated numbers.

The meaning of $N = 10^{-13}$ cannot be *the number of electrons in this device*; however, it can be the *average* number of these electrons. One way to define the average number of electrons is to consider 10^{15} of these devices, which will provide the total volume of 1 cm^3. Then we can have $M_1 = 100$ devices with $N_{actual} = 1$ in each of their P-type regions. Most of the remaining devices will be with $N_{actual} = 0$, although it is possible to have M_2 devices with $N_{actual} = 2$. According to this scheme, the average number of electrons is

$$\overline{N} = \frac{0 \times M_0 + 1 \times M_1 + 2 \times M_2 + \cdots}{M_0 + M_1 + M_2 + \cdots} = \sum_{N_{actual}=0}^{N_{max}} N_{actual}\, p(N_{actual}) \qquad (10.1)$$

where $p(N_{actual})$ is the probability that the P-type region in a device will contain N_{actual} electrons. Probability $p(N_{actual})$ defined in this way relates to the statistics of a large number of devices; that is, it does not establish the physics of a single device. Furthermore, this approach does not address the issue that none of these devices is in thermal equilibrium with $N_{actual} = 0, 1, 2, \ldots$.

Focusing on the physics of a single device, we may think of $N = 10^{-13}$ as the average number of electrons over a period of time. Referring to Fig. 10.1, consider a period of time t_0 when the actual number of electrons in the P-type region is $N_{actual} = 0$, which implies actual concentration of $n_{actual} = 0$. Thermal generation and recombination are not balanced during this period because the probability of a recombination event is zero (there are no electrons to be recombined) and the probability of a generation event is higher. This period will last until a generation event occurs, setting the beginning of a period of time t_1 when the actual number of electrons is $N_{actual} = 1$ and the implied actual concentration is $n_{actual} = N_{actual}/V = 10^{15}$ cm^{-3}. During t_1, the recombination probability becomes much higher than the generation probability, so we can assume that this period will be terminated by a recombination event. This sets the state of $N_{actual} = 0$ again, which will last until another generation event occurs. In principle, the state with $N_{actual} = 1$ may also be changed by another generation event, leading to the state with $N_{actual} = 2$, which will last for the period t_2 until one of the electrons is recombined. With a significant number of

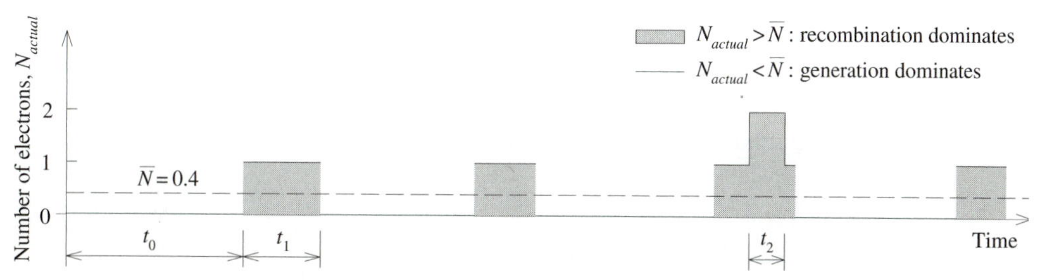

Figure 10.1 Time balance of nonequilibrium generation and recombination events: a refined equilibrium concept that can meaningfully account for the result that the average number of electrons can be $\overline{N} < 1$.

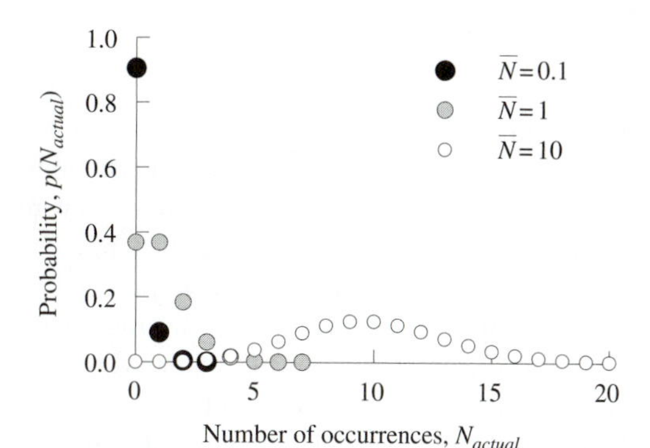

Figure 10.2 The distributions of the *actual* number of occurrences (N_{actual}) for three different *average* numbers of occurrences: $\overline{N}=0.1$, $\overline{N}=1$, and $\overline{N}=10$.

generation and recombination events, we can establish the average values $\tau_0 = \overline{t_0}$, $\tau_1 = \overline{t_1}$, $\tau_2 = \overline{t_2}$, ... to define the average number of electrons in the following way:

$$\overline{N} = \frac{0 \times \tau_0 + 1 \times \tau_1 + 2 \times \tau_2 + \cdots}{\tau_0 + \tau_1 + \tau_2 + \cdots} = \sum_{N_{actual}=0}^{N_{max}} N_{actual} \, p(N_{actual}) \tag{10.2}$$

If the external conditions that impact the generation and recombination events do not change, the probability that N_{actual} electrons will appear at an instant of time, $p(N_{actual})$, is given by the Poisson distribution:

$$p(N_{actual}) = \frac{\overline{N}^{N_{actual}} e^{-\overline{N}}}{N_{actual}!} \qquad (N_{actual} = 0, 1, 2, \ldots) \tag{10.3}$$

This distribution is plotted in Fig. 10.2 for three different values of $\overline{N}$. The Poisson distribution can also be applied in space to determine the probability that N_{actual} particles will appear in a single device if the average number of particles per device is $\overline{N}$ and the particles are uniformly distributed (i.e., the space conditions that impact the appearance of a particle are identical across the considered space). In general, this distribution provides the probability for an integer number of occurrences, N_{actual} ($N_{actual} = 0, 1, 2, 3, \ldots$), for a given average number of occurrences, $\overline{N}$, that is in general a noninteger number. The mean of this distribution (or the expected number of occurrences) is $E(N_{actual}) = \overline{N}$, whereas the standard deviation is $\sigma_{N_{actual}} = \sqrt{\overline{N}}$. For large values of $\overline{N}$, this distribution can be approximated by the Gaussian function with the same values for the mean and the standard deviation:

$$p(N_{actual}) = \frac{1}{\sqrt{2\pi\overline{N}}} e^{-(N_{actual}-\overline{N})^2/2\overline{N}} \tag{10.4}$$

The approximation of the Poisson distribution by Gaussian function for a large number of particles is useful because it becomes difficult to calculate $N_{actual}!$ for large values of N_{actual}.

Let us focus again on the case of a single P-type silicon region with the doping level of $N_A = 10^{18}$ cm^{-3} and the volume of $V = 10^{-15}$ cm^3. In this example, the number of holes in the P-type region is $P = pV = 10^{18} \times 10^{-15} = 1000$. This number is large enough to make the continuous approximation of a "sea" of holes and to allow us to conclude that τ_1 is equal to the minority-carrier lifetime. This time will not change if the size of the P-type region is increased, provided τ_1 is not dominated by surface recombination.

As distinct from the minority-carrier lifetime, the average generation time τ_0 does depend on the size of the considered P-type region. If the volume of the P-type region doubles in this example, the average number of electrons will double to $nV = \overline{N} = 2 \times 10^{-13}$, which will halve the time needed to generate an electron because $\overline{N} \approx \tau_1/(\tau_0 + \tau_1) \approx \tau_1/\tau_0$.[1] This is consistent with the understanding that the time constant in the generation process is the average time required for a single R–G center to generate an electron–hole pair (τ_t), as discussed in Section 5.3.3. If the number of R–G centers in the volume V is M_t, the time τ_0 is obtained as the total probability per unit time that a generation event will occur:[2]

$$\frac{1}{\tau_0} = \sum_{i=1}^{M_t} \frac{1}{\tau_t} = \frac{M_t}{\tau_t} \tag{10.5}$$

If the volume is doubled, M_t is doubled and τ_0 is halved.

Equation (10.2) provides the meaning for $\overline{N} < 1$. It can also be applied for the case of $\overline{N} > 1$. For example, the state of $\overline{N} = 99.3$ corresponds to the Gaussian function of N_{actual} with the mean of $\overline{N} = 99.3$ and the standard deviation of $\sigma_{N_{actual}} = \sqrt{99.3} = 10.0$.[3] Although the absolute variation in the actual number of occurrences is now increased to

[1]The times $\tau_2, \tau_3, \ldots$, are neglected. The probability for the occurrence of $N_{actual} = 2$ electrons, according to Eq. (10.3), is $p(2) = 5 \times 10^{-27}$. This is a much smaller value than $p(1) = 10^{-13}$, so $\tau_2 \ll \tau_1$.

[2]This assumes $M_t > 1$, which corresponds to a concentration of R–G centers exceeding $> 10^{15}$cm^{-3}. The case of $M_t < 1$ is considered in Section 10.1.4.

[3]The Poisson distribution given by Eq. (10.3) can be approximated by Gaussian function with the same mean and standard deviation for large values of $\overline{N}$.

about $\sigma_{N_{actual}} = 10.0$, the relative variation is reduced to $\sigma_{N_{actual}}/\overline{N} = 0.10$. Furthermore, the times that correspond to different N_{actual} values, $\tau_{N_{actual}}$, are reduced. Take the example of $N_{actual} = 100$ electrons in a P-type silicon region. The state of $N_{actual} = 100$ will change to the state of $N_{actual} = 99$ if any of the 100 available electrons is recombined, and the time needed to recombine any of the available 100 electrons is 100 times shorter than the minority-carrier lifetime. The state of $N_{actual} = 100$ can also change to the state of $N_{actual} = 101$ if an additional electron is generated. Once the times $\tau_{N_{actual}}$ have been reduced well below the shortest period of time that we are interested in, Δt, it becomes more convenient to work with the classical concepts of generation rate (number of generation events/Δt) and recombination rate (number of recombination events/Δt) and with the classical concept of thermal equilibrium, which is defined by equal generation and recombination rates.

It should be noted that the use of the thermal equilibrium equation $np = n_i^2$ to calculate $N = nV < 1$ can be considered only as an extrapolation-based estimate of $\overline{N}$. This is similar to the practice of extrapolating equilibrium-based generation or recombination rates to calculate nonequilibrium generation or recombination currents. Section 10.1.5 will introduce the idea of direct modeling of single generation and recombination events without the assumption of thermal equilibrium.

EXAMPLE 10.1 Generation Time in a Neutral Semiconductor

The minority-carrier lifetime in the P-type silicon region considered in this section is 1 μs. Using the estimate for the average number of electrons in this P-type region ($\overline{N} = 10^{-13}$), obtained from the thermal equilibrium equation $np = n_i^2$, calculate the average time that it takes to generate an electron.

SOLUTION

The minority-carrier lifetime is the average time needed to recombine an excess minority carrier in the "sea" of majority carriers. Accordingly, τ_1 is equal to the minority-carrier lifetime. Based on Eq. (10.2), the average time to generate an electron can be expressed in terms of τ_1:

$$\overline{N} \approx \frac{\tau_1}{\tau_0 + \tau_1}$$

$$\tau_0 \approx \frac{1 - \overline{N}}{\overline{N}} \tau_1 \approx \frac{\tau_1}{\overline{N}} = \frac{10^{-6}}{10^{-13}} = 10^7 \text{ s} = 115.7 \text{ days}$$

This time is much longer than the time a single defect in a depletion layer takes to generate an electron–hole pair (Example 5.7) because most levels in the energy gap of a P-type semiconductor are occupied by holes, which significantly reduces the availability of electrons at the R–G levels, hence the probability for an electron to be emitted into the conduction band.

10.1.2 Current–Time Form of the Uncertainty Principle

The demand for increased complexity of modern electronic systems requires faster access to smaller devices conducting smaller currents. However, the simultaneous reduction in both access time and current value is ultimately limited by the following condition:

$$I\Delta t \gg q \tag{10.6}$$

This condition is a form of uncertainty principle expressing the fact that it is not possible to indefinitely reduce both the observation time and the current value, while maintaining the certainty in the current value, because electrons behave as quantum particles and not as a continuous fluid. The limiting condition $I\Delta t \gg q$ emerges from the fundamental frequency–time form of the uncertainty principle, establishing that the frequency of a repeating event, f, cannot be determined within a time interval Δt that is shorter than the period between two events, τ; therefore, $\Delta t > \tau$, which is the same condition as $f\Delta t > 1$. In the case of electric current, the period between two events can be defined as the time between the collection of two electrons by a device contact, $\tau = q/I$, which means that the relevant frequency is $f = I/q$ and the condition $f\Delta t > 1$ can be expressed as $I\Delta t > q$. The general condition $I\Delta t > q$ is specified as $I\Delta t \gg q$ because the period between two events in this case is not a set constant, and to reduce the uncertainty in the current value, a large number of electrons must be collected. This will be explained and quantified shortly, using the examples of diffusion current in both reverse- and forward-biased P–N junctions.

The P-type region, considered in Section 10.1.1 and Example 10.1, will dominate the reverse-bias current of a P–N junction if the N-type doping is much higher than the P-type doping. For a start, ignore the generation current in the depletion layer to focus on the reverse-bias current due to minority carriers, which is dominated by the minority electrons in the P-type region (the effect related to the dominance of the generation current is considered in the next section). Using the basic equation for P–N junction saturation current [Section 6.2.1, Eq. (6.16)], the reverse-bias current is

$$I_S = qA_J n_i^2 \frac{D_n}{L_n N_A} \tag{10.7}$$

where $n_i = 10^{10}$ cm^{-3}, $D_n = 9$ cm^2/s (from the solution of Example 6.3), $N_A = 10^{18}$ cm^{-3}, $A_J = (100 \text{ nm})^2 = 10^{-10}$ cm^2, and $L_n = \sqrt{D_n \tau_n}$. Taking the value of $\tau_1 = 10^{-6}$ s for the minority-carrier lifetime τ_n, the diffusion length is $L_n = 30$ μm. With this, the calculated saturation current is $I_S = 4.800 \times 10^{-24}$ A. Such a current value is not possible if our observation time is 1 s because it corresponds to the flow of $I_S/q = 3.0 \times 10^{-5}$ electron/s. This shows that the concept of continuously flowing current can lose its meaning. The number 4.800×10^{-24} A maintains its meaning if the shortest period of time that we are interested in is $\Delta t = 1$ year $= 3.15576 \times 10^7$ s. If a current measurement takes $\Delta t = 1$ year, the average number of collected electrons during each measurement is $\overline{N} = I_S\Delta t/q = 946.7$. This means that some measurements will show $947 \times q/\Delta t = 4.801 \times 10^{-24}$ A, and other measurements will show $946 \times q/\Delta t = 4.796 \times 10^{-24}$ A, which is a small variation. However, if the shortest period of time that we are interested in is $\Delta t = 1$ h, which corresponds to $\overline{N} = I_S\Delta t/q = 0.108$ electron,

89.8% of the measurements will show no current, 9.7% will show $1.6 \times 10^{-19}/3600 = 4.44 \times 10^{-23}$ A, and 0.5% will show $2 \times 1.6 \times 10^{-19}/3600 = 8.89 \times 10^{-23}$ A.[4] To avoid these complications when we use the concept of current, we can simply specify that the average time between the flow of minority electrons through the P–N junction is $\tau = q/I_S = 3.33 \times 10^4$ s $= 9.26$ h.

It may be observed that the diffusion length, $L_n = 30\ \mu$m, is much longer than the assumed width of the P-type region, which is $W_{anode} = 100$ nm. In a case like this, the reverse-bias current would be calculated from the saturation-current equation for a P–N junction with a short anode [Section 6.2.1, Eq. (6.17)],

$$I_S = qA_J n_i^2 \frac{D_n}{W_{anode} N_A} \tag{10.8}$$

The result is $I_S = 1.44 \times 10^{-21}$ A, which corresponds to $\tau = q/I_S = 111.1$ s, still a fairly long average time between the flows of two electrons.

Consider now that this P–N junction is forward-biased with $V_D = 0.8$ V. Given that the built-in voltage is $V_{bi} = V_t \ln(N_D N_A)/n_i^2 = 1.0$ V, this forward-bias voltage corresponds to a very small barrier height of $q(V_{bi} - V_D) = 0.2$ eV. Because the width of the neutral P-type region ($W_{anode} = 100$ nm) is smaller than the diffusion length of electrons ($L_n = 30\ \mu$m), the result obtained by Eq. (10.8) should be used as the value for I_S when calculating the forward-bias current:

$$I_D = I_S e^{V_D/V_t} = 1.44 \times 10^{-21} \times e^{0.8/0.02585} = 39.7 \text{ nA} \tag{10.9}$$

The current $I_D \approx I = 40$ nA means that the average time between two events of electron collection by the anode contact is $\tau = q/I = 4$ ps. If this current is measured over $\Delta t = 10$ ps, the average number of collected electrons is $\overline{N} = \Delta t/\tau = 2.5$. In this case, the actual current values appear in multiples of $q/\Delta t = 16$ nA with probabilities in accordance with the Poisson distribution [Eq. (10.3)]; this scenario is illustrated in Fig. 10.3. The probability distributions for $\Delta t = 100$ ps, 1 ns, and 10 ns, also shown in Fig. 10.3, are obtained from the Gaussian function [Eq. (10.4)] and the following conversion of particle numbers into currents: $I_{actual} = qN_{actual}/\Delta t$.

Based on the relationship between the current and the number of particles ($I_{actual} = qN_{actual}/\Delta t$), and knowing that the standard deviation of N_{actual} is $\sigma_{N_{actual}} = \sqrt{\overline{N}}$, we can obtain the standard deviation of the current:

$$\sigma_I = \frac{q}{\Delta t}\sigma_{N_{actual}} = \frac{q}{\Delta t}\sqrt{\overline{N}} = \frac{q}{\Delta t}\sqrt{\frac{I\Delta t}{q}} = \sqrt{\frac{qI}{\Delta t}} \tag{10.10}$$

[4]This is according to the Poisson distribution for $\overline{N} = 0.108$ and the probabilities for $N_{actual} = 0, 1,$ and 2 electrons.

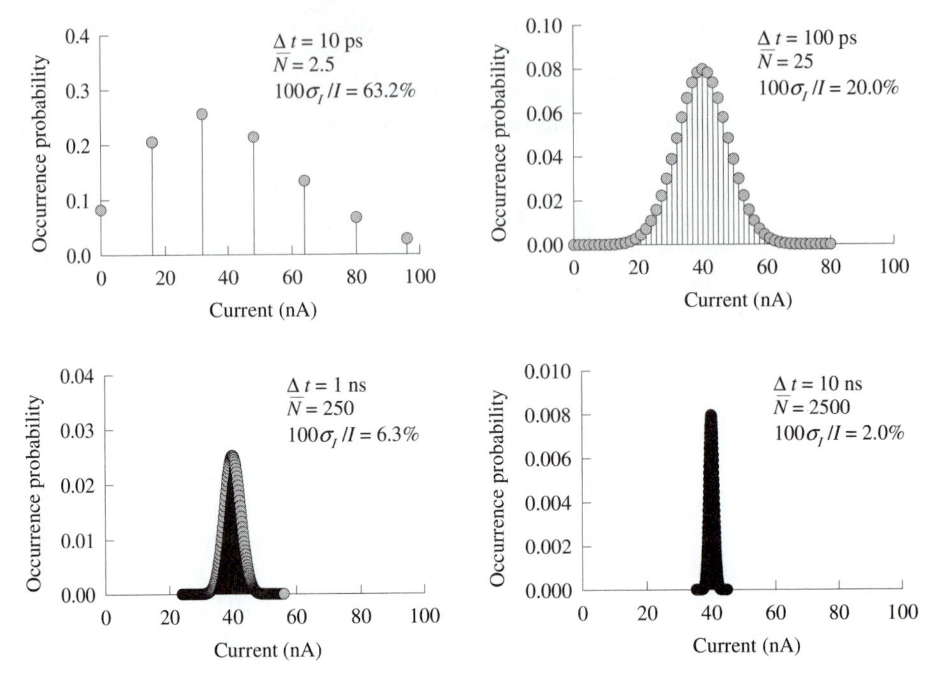

Figure 10.3 Distributions of 40-nA current for observation times from $\Delta t = 10$ ps to $\Delta t = 10$ ns. The current exhibits discrete and random values for short observation times; with the increase in observation time, the distribution of current values becomes almost continuous and narrows down toward the expected value of 40 nA.

In this equation, I is the average current ($I = q\overline{N}/\Delta t$). The coefficient of current variation, which is the relative error in the terminology of current measurements, is

$$c_v \equiv \frac{\sigma_I}{I} = \sqrt{\frac{q}{I\Delta t}} \tag{10.11}$$

This quantifies the uncertainty relationship $I\Delta t \gg q$ in the following way:

$$I\Delta t = \frac{q}{c_v^2} \tag{10.12}$$

The variability of measured/observed current that is quantified by the coefficient of variation c_v and illustrated in Fig. 10.3 is known as *shot noise*. For the case of $I = 40$ nA, the variation of $100 \times \sigma_I/I = 2.0\%$ ($c_v = 0.02$) requires $\Delta t = 10$ ns, which means that the operating frequency of this device has to be smaller than $1/\Delta t = 100$ MHz. Clearly, the designers of modern electronic systems need to check the validity of the assumption of continuous current before implementing a solution obtained by any standard equation for device current.

10.1.3 Carrier-Supply Limit to Diffusion Current

The equations used to calculate both the forward- and the reverse-bias currents in the preceding section are derived from the diffusion-current equation. The parameter in this equation, the diffusion constant $D_{n,p}$, can be expressed in terms of thermal velocity and the average scattering length of the current carriers. This is because the random thermal motion of the current carriers is the driving force of diffusion current. The derivation of the Einstein relationship in Section 4.2.1 shows that

$$D_{n,p} = \frac{v_{th} l_{sc}}{3} \tag{10.13}$$

Replacing the diffusion constant in the equation for the forward-bias current,

$$I_D = qA_J n_i^2 \frac{D_n}{W_{anode} N_A} e^{V_D/V_t} \tag{10.14}$$

leads to the following equation for the forward-bias current as determined by the diffusion of minority electrons in the P-type region of a P–N junction:

$$I_D = qA_J \frac{n_i^2 v_{th}}{6N_A} \frac{2l_{sc}}{W_{anode}} e^{V_D/V_t} \tag{10.15}$$

The minority electrons, supplied to the diffusion process in the neutral P-type region, arrive from the N-type region of the P–N junction. *Because the diffusion current cannot exceed the rate of electron supply from the N-type region, there is a limit to the applicability of the diffusion-current equation.* The rate of electron supply is basically the thermionic emission current from the N-type region. To estimate this current, we focus on the electrons as majority carriers in the N-type region that hit the depletion layer at the P–N junction. The number of these hits per unit area and unit time can be determined by the reasoning applied in the derivation of Einstein relationship (Section 4.2.1). In analogy with Eq. (4.6), the number of these hits per unit area and unit time is $N_D v_{th}/6$, where N_D is equal to the concentration of electrons in the N-type region. The number of hits per unit time is $A_J N_D v_{th}/6$. The barrier height at the P–N junction is $qV_B = q(V_{bi} - V_D)$ and the probability that an electron will possess energy higher than the barrier height is

$$T_e = \frac{\int_{qV_B}^{\infty} e^{-E/kT}}{\int_0^{\infty} e^{-E/kT}} = e^{-qV_B/kT} = e^{-(V_{bi}-V_D)/V_t} \tag{10.16}$$

Therefore, the thermionic emission current from the N-type region is

$$I_{th} = qA_J \frac{N_D v_{th}}{6} T_e = qA_J \frac{N_D v_{th}}{6} e^{-(V_{bi}-V_D)/V_t} \tag{10.17}$$

Replacing the built-in voltage by $V_{bi} = V_t \ln(N_D N_A / n_i^2)$, we obtain

$$I_{th} = qA_J \frac{N_D v_{th}}{6e^{V_{bi}/V_t}} e^{V_D/V_t} = qA_J \frac{n_i^2 v_{th}}{6N_A} e^{V_D/V_t} \tag{10.18}$$

Equations (10.15) and (10.18) show that the diffusion current can be expressed in terms of the thermionic emission current as follows:

$$I_D = I_{th} \frac{2l_{sc}}{W_{anode}} \tag{10.19}$$

Equation (10.19) shows that the diffusion current I_D limits the current through the P–N junction to a fraction of the thermal-supply current I_{th} when $W_{anode} > 2l_{sc}$. In fact, when $W_{anode} > L_n$, the diffusion length L_n should be used in Eq. (10.19). However, for $W_{anode} < 2l_{sc}$, the diffusion-current equation predicts current values that are larger than the current of electron supply to the diffusion process. Clearly, this is not possible. This problem with the diffusion-current equation occurs under the assumption of continuous electron fluid. This assumption does not provide any limit on the increase in the concentration gradient, and consequently does not limit the diffusion current when the width of the neutral P-type region (W_{anode}) is reduced toward zero. However, when particle scattering and the concept of scattering length are included in the analysis of diffusion current, Eq. (10.19) shows that reducing the dimensions of the device below the scattering length cannot increase the diffusion current because the scattering events cannot be reduced below zero. When scattering is eliminated, the forward-bias current through a P–N junction is limited by the thermionic current given by Eq. (10.17).

The average scattering length can be estimated from the carrier mobility and the effective mass in the following way:

$$l_{sc} = v_{th} \tau_{sc} \tag{10.20}$$

$$\mu_n = \frac{q\tau_{sc}}{m^*} \tag{10.21}$$

$$\frac{m^* v_{th}^2}{2} = \frac{3}{2} kT \tag{10.22}$$

$$v_{th} = \sqrt{\frac{3kT}{m^*}} \tag{10.23}$$

$$l_{sc} = \frac{\mu_n}{q} \sqrt{3kTm^*} \tag{10.24}$$

Assuming that the electron mobility in the considered P-type region is $\mu_n = 300\ \text{cm}^2/\text{V}\cdot\text{s}$ and that the effective mass is $m^* = 0.26m_0$, the average scattering length is $l_{sc} = 10.2$ nm. Given that $W_{anode} > 2l_{sc}$, we conclude that the diffusion-current equation can be applied in the considered example. We can also find that the thermionic current limit is $I_{th} = W_{anode} I_D / 2l_{sc} = 196$ nA, which is not too far from the diffusion current of 40 nA.

Let us analyze now the reverse-bias currents due to minority electrons, calculated by Eqs. (10.7) and (10.8) in the preceding section. The result in Example 10.1 shows that the average appearance of a single minority electron in the neutral P-type region is once in $\tau_0 = 10^7$ s = 115.7 days. Assume that each minority electron that is generated in the neutral P-type region crosses the P–N junction to contribute to the minority-electron current. This assumption shows that the current of minority electrons is limited to

$$I_S = \frac{q}{\tau_0} = \frac{1.6 \times 10^{-19}}{10^7} = 1.6 \times 10^{-26}\ \text{A} \tag{10.25}$$

Comparison of this value to the numbers obtained by the diffusion-current equations ($I_S = 4.80 \times 10^{-24}$ A and $I_S = 1.44 \times 10^{-21}$ A) shows that the assumption of continuous electron fluid significantly overestimates the minority-carrier current in P–N junctions; it results in current values that are higher than the carrier-supply limit. However, this issue is not practically important because the reverse-bias current is dominated by the generation current. Although this was shown earlier, in Section 6.2.2, it is useful to estimate the generation current for the considered example to complete the analysis of the reverse-bias current. With $N_D \gg N_A$, and assuming reverse bias of $V_R = 1$ V, the depletion-layer width is

$$w_d = \sqrt{\frac{2\varepsilon_s(V_R + V_{bi})}{qN_A}} = \sqrt{\frac{2 \times 11.8 \times 8.85 \times 10^{-12}(1+1)}{1.6 \times 10^{-19} \times 10^{24}}} = 51.1\ \text{nm} \tag{10.26}$$

The volume of the depletion layer is $V_d = A_J w_d = 5.1 \times 10^{-16}\ \text{cm}^3$. Assuming that the concentration of the fastest R–G centers is $N_t = 10^{16}\ \text{cm}^{-3}$, the average number of these centers in the depletion layer is $M_t = N_t V_d = 5.1$ (the case of $M_t < 1$ will be considered in Section 10.1.4). If the time a single R–G center takes to generate an electron–hole pair is labeled by τ_t, the average time between generation events in this depletion layer [according to Eq. (10.5)] is $\tau_{0d} = \tau_t/M_t$. With the assumption that the depletion-layer field sweeps away every generated electron–hole pair, the generation current is

$$I_G = \frac{q}{\tau_{0d}} = M_t \frac{q}{\tau_t} \tag{10.27}$$

Equations for the direct calculation of τ_t are introduced in Section 10.1.5. At this stage, we can use the relationship between τ_t and the minority-carrier lifetime that is derived in Section 5.3.3: $\tau_t \approx 2\tau_1(N_t/n_i)$, assuming that the minority-carrier lifetimes for the electrons (τ_n) and holes (τ_p) are the same ($\tau_1 = \tau_n = \tau_p$).[5] With this equation, $\tau_t = 2 \times 10^{-6} \times 10^{16}/10^{10} = 2$ s. This means that the average time between electron–hole generation events in the depletion layer is $\tau_{0d} = 2\ \text{s}/5.1 = 0.39$ s and the average generation current is $I_G = 4.1 \times 10^{-19}$ A. This is a much larger value than any of the estimated I_S values for the minority-carrier current.

[5]With this relationship, Eq. (10.27) can be transformed into the standard generation-current equation, shown in Section 6.2.2: $I_G = qM_t n_i/2\tau_1 N_t = (qn_i/2\tau_1)w_d A_J$.

10.1.4 Spatial Uncertainty

It is quite possible that the average number of fixed particles, such as doping atoms and R–G centers, will become smaller than unity when the device dimensions are reduced. For the doping level of 10^{15} cm^{-3}, this effect will occur when the dimensions of that region are reduced below $1/10^{15}$ cm^{-3} $= 10^{-15}$ cm^{-3} $=$ 100 nm $\times$ 100 nm $\times$ 100 nm. In fact, significant randomness appears even when the average number of particles is larger than unity. The data plotted in Fig. 10.2 show that even for an average number of particles as large as $\overline{N} = 10$, only 12.5% of devices contain $N_{actual} = 10$ particles, and as many as 9% of devices contain $N_{actual} = 7$ particles (which is a variation of 30%). Given that the standard deviation of the actual number of particles is $\sigma_{N_{actual}} = \sqrt{\overline{N}}$, the normalized variability in the number of particles drops with the square root of the average number of particles: $\sigma_{N_{actual}}/\overline{N} = \sqrt{\overline{N}}/\overline{N} = 1/\sqrt{\overline{N}}$.

It should be clarified that there is neither spatial localization nor spatial variation of free electrons in crystals, even at distances that are comparable to the distance between the doping atoms. Take the example of $N_D = 10^{15}$ cm^{-3} when the average distance between the doping ions is $(1/N_D)^{1/3} = 100$ nm. When the electron of this donor appears in the conduction band, it appears as a uniformly spread wave across the whole crystal, as described in Section 2.2.2 (Fig. 2.8), and no fluctuation in the electron concentration/density can be observed even at 50 nm from the donor ion.

Spatial uncertainty in small-size devices causes device-to-device fluctuations in the number of fixed particles even in the case of uniformly distributed particles. This effect is illustrated in Fig. 10.4. Take again the doping of $N_D = 10^{15}$ cm^{-3}, which is uniform across many devices in a large silicon wafer. Assume that the volume of individual N-type regions, belonging to individual devices and separated by P-type areas and/or dielectric, is $V = 215$ nm $\times$ 215 nm $\times$ 215 nm $= 10^{-14}$ cm^3. With this, the average number of donor ions per individual N-type region is $\overline{N} = N_D V = 10$. Based on the results shown in Fig. 10.2, only 12.5% of these N-type regions will have the expected doping concentration of $10/10^{-14} = 10^{15}$ cm^{-3}; as many as 9% of the N-type regions will have a doping concentration 30% lower: $7/10^{-14} = 7 \times 10^{14}$ cm^{-3}.

Turning to the R–G centers as fixed particles, let us consider the impact of a reduction of the P–N junction area on the reverse-bias current, which is the generation current given

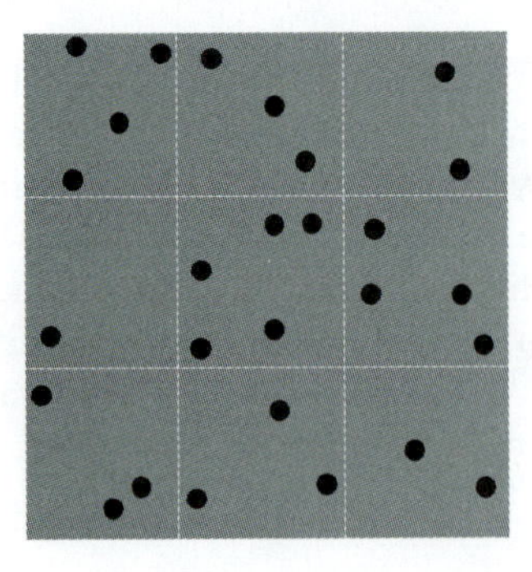

Figure 10.4 Illustration of the uncertainty in the actual number of particles per device (square) for the average value of $\overline{N} = 3.0$ particles/device.

by Eq. (10.27). The generation-current density is

$$j_G = \frac{I_G}{A} = \frac{M_t}{A}\frac{q}{\tau_t} = N_t w_d \frac{q}{\tau_t} \tag{10.28}$$

where w_d is the depletion-layer width, Aw_d is the depletion-layer volume, and $N_t = M_t/(Aw_d)$ is the concentration of the fastest R–G centers. By its definition and purpose, the current density should not depend on the device area. This is correct when the device area is large enough that $M_t = N_t A w_d \gg 1$. If we take the same specific values for the doping levels and the depletion layer as in the section on reverse-bias current ($w_d = 51.1$ nm) and the same type of R–G center with $\tau_t = 2$ s, but a much cleaner crystal with $N_t = 10^{14}$ cm^{-3}, the generation-current density will be $j_G = 4.1 \times 10^{-7}$ A/m^2. This current density will be observed for P–N junctions with areas of 1 mm $\times$ 1 mm, 100 μm $\times$ 100 μm, 10 μm $\times$ 10μm, ..., because the average number of R–G centers in the depletion layer remains much greater than 1: 5.1×10^6, 5.1×10^4, 511, ..., respectively. There is also insignificant device-to-device variation.

If the area of the diode is reduced to 100 nm $\times$ 100 nm, however, the average number of R–G centers in the depletion layer drops to $M_t = 0.05$. According to Eq. (10.3), 95.1% of P–N junctions will be free of this type of R–G center ($M_t = 0$), 4.8% of devices will have $M_t = 1$, and 0.1% of devices will have $M_t = 2$ R–G centers in the depletion layer. Equation (10.28) can no longer be applied to calculate the current density. To determine what current densities would be measured in the case of 4.8% of devices with $M_t = 1$ and 0.1% of devices with $M_t = 2$, we can first use Eq. (10.27) to calculate the corresponding currents and then divide the result by the area. The results are $j_G = 8 \times 10^{-6}$ A/m^2 and 1.6×10^{-5} A/m^2 for $M_t = 1$ and $M_t = 2$, respectively. These current densities are 20 and 40 times larger, respectively, than the large-area current density of $j_G = 4.1 \times 10^{-7}$ A/m^2. Although the percentage of devices with increased current densities seems small, it is these devices that may determine the design of an electronic system with a large number of P–N junctions.

Regarding the 95.1% of P–N junctions with $M_t = 0$, note that the reverse-bias current in these devices will not be zero because there are likely to be R–G centers of other types with an average generation time of $\tau_t > 2$ s. The current density of these devices will be reduced in comparison to the large-area value; indeed, in a large fraction of these devices it may be significantly reduced because of the exponential dependence of τ_t on the energy level of the responsible R–G center.

In conclusion, the current density in small-area devices is not constant but appears as a distribution when a large number of devices are considered. The tail of this distribution exhibits a significant current-density increase compared to the constant large-area value.

10.1.5 Direct Nonequilibrium Modeling of Single-Carrier Events

Irreversible processes can be modeled directly and independently of equilibrium conditions if we assume that the irreversible processes consist of irreversible events. The following specific examples can clarify the meaning of this statement: (1) collections of electrons by the collector contact of a BJT are irreversible events that create the process of collector current; (2) electron–hole generations in a depletion layer are irreversible events that make the process of generation current; and (3) electron-scattering events are irreversible events that can be modeled to determine average scattering time, mobility, and ultimately resistivity, which is the main model parameter for the process of drift current.[6] Irreversible events define discrete intervals of time, τ_i ($i = 1, \ldots, M$), corresponding to each of the M events that form a process. A process that results in an irreversible outcome becomes an event, which will be referred to as a composite event to distinguish it from the constituent events. If a process consists of consecutive events, then the time needed for the composite event is simply

$$\tau = \sum_{i=1}^{M} \tau_i \tag{10.29}$$

If a process consists of simultaneous and independent events, then the probabilities per unit time for each of these events are added to obtain the probability per unit time for the composite event:

$$\frac{1}{\tau} = \sum_{i=1}^{M} \frac{1}{\tau_i} \tag{10.30}$$

This approach is applicable to both macroscopic and microscopic events and processes. Furthermore, if the time intervals for the constitutent events are not known, this approach can be applied hierarchically to simpler levels until events with known times have been identified.

Carrier Capture Time, Minority-Carrier Lifetime, and Recombination Rate

An event of indirect recombination, or recombination through an R–G center, can be considered as a composite event consisting of two consecutive carrier capture events: either electron capture by an R–G center followed by a hole-capture event by the same center or a hole capture followed by an electron capture. This is according to the illustration in Fig. 5.1b. The time needed to capture an electron that has been emitted in the conduction band can be obtained by considering the probability per unit time for the electron capture by an R–G center. The probability that a moving electron will hit an R–G center per unit length, with a chance of being captured, is proportional to the concentration of R–G centers (N_t) and the capture cross section of the R–G centers (σ_n). Given that the length traveled per unit time is the thermal velocity (v_{th}), the probability that the electron will hit an R–G center per unit time is $v_{th}\sigma_n N_t$. Multiplying this probability by $(1 - f_t)$, the probability that the R–G center is empty, gives the probability that a single electron is captured per unit time. The reciprocal value of this probability is the average

[6]It may be more convenient to take measured values of resistivity, but even so, it is scattering events that determine the value of drift current.

electron-capture time:

$$\tau_{c,n} = \frac{1}{v_{th}\sigma_n N_t(1 - f_t)} \tag{10.31}$$

The average capture time can be used to establish the time needed to recombine an excess electron in a P-type semiconductor, which is the minority-carrier lifetime for excess electrons. In Eq. (10.31), $1 - f_t \approx 1$ for $(E - E_F)/kT \gg 1$, which is most of the energy gap in a P-type semiconductor. Therefore, for the dominant R–G centers, $\tau_{c,n} = 1/(v_{th}\sigma_n N_t)$.

An event of electron capture by an R–G center, which takes the average time of $\tau_{c,n}$, can be followed by one of two events: (1) emission of this electron back into the conduction band or (2) capture of a hole by this R–G center to complete the composite recombination event. The average times for emission of electrons into the conduction band (the first possible event) are very short for R–G centers with energy levels close to the conduction band. This means that these R–G centers will not be contributing to recombination, which sets the upper energy boundary for the R–G centers that contribute to recombination. The R–G centers with energy levels between this upper boundary and the lower boundary around E_F are the effective recombination centers, as shown in Fig. 5.4. When an electron is captured by one of these centers, the average time that a hole will be captured by this center can be determined in a similar way as for the electron capture, except that in this case the concentration of R–G centers (N_t) is replaced by the concentration of holes (p) because the focus is on a single R–G center, and the capture by any of the available holes is a relevant capture event that will complete the recombination process. Therefore, $\tau_{c,p} = 1/(v_{th}\sigma_p p)$. If we assume sufficiently clean crystal so that the position of the Fermi level is set by p and not by N_t, then $p \gg N_t$ and $\tau_{c,n} \gg \tau_{c,p}$. Accordingly, the minority-carrier lifetime τ_n is

$$\tau_n \approx \tau_{c,n} = \frac{1}{v_{th}\sigma_n N_t} \tag{10.32}$$

This is the same result we found earlier in Section 5.3.3. The analysis in this section provides deeper insight into the recombination process, the constituent events, and accordingly clarifies the inherent assumptions for the validity of this result.

Recombination of any existing electrons is an event that contributes to the recombination process. If there are N electrons, the average time between two recombination events, according to Eq. (10.30), is τ_n/N. In other words, the number of electrons recombined per unit time is N/τ_n. Furthermore, the number of electrons recombined per unit time and unit volume, which is the effective recombination rate, is

$$U = \frac{N}{V}\frac{1}{\tau_n} = \frac{n}{\tau_n} = v_{th}\sigma_n N_t n \tag{10.33}$$

Again, this result is consistent with the result shown in Section 5.3.3, except that it is obtained by the bottom-up approach.

Carrier Emission Time, Generation Time, and Generation Rate

Carrier emission is usually not modeled independently of recombination. As shown in Section 5.3.1, the modeling of carrier emission and generation is based on the carrier emission rate that is determined from the conditions of thermal equilibrium with the recombination rate. The considerations in this chapter show that the meaning of the concepts associated with this approach can be lost in small-dimension devices where the single-carrier events become apparent. It is obvious from the considerations in Section 10.1.1 that the concept of balanced emission and capture rates loses its meaning if it takes 1 μs to capture the only electron in a P-type region and then, on average, 115.7 days to emit another electron (Example 10.1). In addition, this approach relates the emission and generation events to the parameters of independent capture events, specifically to the capture cross section and the thermal velocity of a free carrier. Although this logical problem is believed to be resolved by the mathematical symmetry of the detailed balance principle, specific examples show that the meaning of capture parameters is lost when they are used to explain emission events. For example, the capture cross section for electrons is larger for positively charged traps than for neutral traps because of Coulomb attraction. A positively charged trap can capture an electron more easily, but this does not mean that the electron can be released more easily from that trap, as implied by Eq. (5.28) in Section 5.3.1. Likewise, it is harder to emit an electron from a deeper trap, but this does not mean that it is harder for an electron to fall into a deeper trap. In general, it is easier to fall into a wider trap and it is harder to get out of a deeper trap; however, it is not easier to get out of a wider trap and it is not harder to fall into a deeper trap.

The problem with the direct modeling of the emission time has resulted from the difficulty of granulating an emission event to the point at which the times of the constituent events are known. It has recently been suggested that the transfer of energy from a phonon to the electron to be emitted can be taken as an elementary event that limits further time granulation. Specifically, it has been observed that the time needed for the event of energy transfer from a phonon to the R–G center that holds the electron to be emitted cannot be shorter than the period of the phonon considered as a wave: h/E. Based on this and considerations of the related phonon statistics, the following equation has been established for the average electron emission time:[7]

$$\frac{1}{\tau_{e,n}} = \frac{\eta(E_C - E_t + kT)}{h} e^{-(E_C - E_t)/kT} \tag{10.34}$$

In Eq. (10.34), $E_C - E_t$ is the energy difference between the bottom of the conduction band and the energy level of the captured electron, h is the Planck constant, and η is a parameter whose value is shown to be very close to the reciprocal value of the average

[7] S. Dimitrijev, Irreversible event-based model for thermal emission of electrons from isolated traps, *J. Appl. Phys.*, vol. 105, p. 103706-1 (2009).

number of phonons needed to emit a single electron from a midgap R–G center in silicon ($\eta = 0.07$).[8]

An electron–hole generation is a composite event that consists of two consecutive emission events: an electron emission and a hole emission in either order. According to Eq. (10.29), the times taken by each consecutive event ($\tau_{e,n}$ for the emission of an electron and $\tau_{e,p}$ for the emission of a hole) are added to obtain the time for the composite event (τ_t):

$$\tau_t = \tau_{e,n} + \tau_{e,p} \tag{10.35}$$

For M_t R–G centers that act simultaneously, Eq. (10.30) is applied to obtain the average time for the generation of a single electron–hole pair: τ_t/M_t. Looking at this result another way, M_t/τ_t is the number of electron–hole pairs generated per unit time. The number of electron–hole pairs generated per unit time and unit volume, which is the generation rate, is then

$$|U| = \frac{M_t}{V}\frac{1}{\tau_t} = \frac{N_t}{\tau_t} = \frac{N_t}{\tau_{e,n} + \tau_{e,p}} \tag{10.36}$$

With the equation for $\tau_{e,p}$ that is analogous to Eq. (10.34) for $\tau_{e,n}$, the generation rate is expressed as

$$|U| = \frac{N_t}{A_e e^{(E_C - E_t)/kT} + A_h e^{(E_t - E_V)/kT}} \tag{10.37}$$

where $A_e = h/\eta(E_C - E_t)$, $A_h = h/\eta_h(E_t - E_V)$, E_V is top of the valence band, and η_h is analogous parameter to the parameter η. The only difference between Eq. (10.37) and the widely used Shockley-Read-Hall equations, which are derived from the equilibrium model (Section 5.3.3), is in the parameters A_e and A_h. This difference can be reconciled by adjusting the values of the capture cross sections in the Shockley-Read-Hall equations. In other words, the widely established experimental evidence for the Shockley-Read-Hall equation is just as valid for Eq. (10.37).

The Thermal Equilibrium Condition and the Degeneracy Factor

Application of the equilibrium-independent equations for the emission and capture events to equilibrium conditions can provide a deeper insight into the equilibrium parameters. Take the case of thermal equilibrium where the number of electron emission events per unit time (the emission rate, $r_{e,n}$) is balanced by the number of captured electrons per unit time (the capture rate, $r_{c,p}$). This is the condition that is used to derive the emission time in the absence of the nonequilibrium Eq. (10.34), as shown in Section 5.3.1. In the spirit

---

[8] In silicon, the maximum energy of phonons is $E_{p-max} = 0.066$ eV (this energy corresponds to the minimum phonon wavelength set by the crystal-lattice constant). The value of $1/\eta = 14.3$ and the energy of $E_C - E_t = 0.593$ eV correspond to the average phonon energy of $E_p = (E_C - E_t)/(1/\eta) = 0.041$ eV.

of Eq. (10.30), the capture rate is $r_{c,n} = n_0/\tau_{c,n}$, where n_0 is the equilibrium electron concentration. Therefore,

$$r_{c,n} = \frac{n_0}{\tau_{c,n}} = v_{th}\sigma_n N_t(1 - f_t)n_0 \tag{10.38}$$

This equation is the same as Eq. (5.22) in Section 5.3.1. Similarly, the emission rate is

$$r_{e,n} = \frac{N_t f_t}{\tau_{e,n}} \tag{10.39}$$

Based on Eqs. (10.38), (10.39), and (10.34), the equilibrium condition $r_{c,n} = r_{e,n}$ becomes

$$v_{th}\sigma_n N_t(1 - f_t)n_0 = N_t f_t \frac{\eta(E_C - E_t + kT)}{h} e^{-(E_C - E_t)/kT} \tag{10.40}$$

With $n_0 = N_C \exp-(E_C - E_F)/kT$, Eq. (10.40) is transformed into

$$v_{th}\sigma_n N_C(1 - f_t) = f_t \frac{\eta(E_C - E_t + kT)}{h} e^{(E_t - E_F)/kT} \tag{10.41}$$

The dependent variable in this equation is the occupancy probability of the R–G center, f_t. From Eq. (10.41), the following equation is obtained for f_t:

$$f_t = \frac{1}{1 + \eta(E_C - E_t + kT)/(hv_{th}\sigma_n N_C)e^{(E_t - E_F)/kT}} \tag{10.42}$$

A comparison to the Fermi–Dirac distribution for donors, given in Example 2.15, shows that this is the Fermi–Dirac distribution with the degeneracy factor

$$g = \frac{hv_{th}\sigma_n N_C}{\eta(E_C - E_t + kT)} \tag{10.43}$$

According to Eq. (10.42), f_t depends on both capture and emission parameters and, for example, is higher for larger capture cross sections. This result is logical because a larger capture cross section and an increased capture rate should shift the equilibrium balance toward a higher population of R–G centers. Therefore, the nonequilibrium equation for the emission probability clarifies the need for the degeneracy factor in the Fermi–Dirac distribution applied to trapped electrons (including electrons on donor and R–G centers).

EXAMPLE 10.2 Emission, Capture, and Generation Times in a Neutral Semiconductor

Consider neutral P-type silicon with $p = 10^{18}$ cm^{-3}, $V = 100$ nm $\times$ 100 nm $\times$ 100 nm, and a single donor-type R–G center with $E_C - E_t = 0.6$ eV, $\eta = 0.07$, $\eta_h = \eta = 0.07$, $\sigma_p = 10^{-15}$ cm^2, and $\sigma_n = 10^{-13}$ cm^2. Assume that $v_{th} = 10^7$ cm/s.

(a) Using the nonequilibrium emission equation, calculate the average electron and hole emission times from the R–G center.

(b) Calculate the average time this R–G center requires to capture a hole after a hole emission event.

(c) Use the values for hole capture and hole emission times to determine the probability that an electron will occupy the R–G center.

(d) Based on the electron occupancy probability and the electron emission time, determine the average time required for an electron–hole pair to be generated by the R–G center (τ_0). Compare the result with the result obtained in Example 10.1.

(e) Assuming that the considered R–G center is the only defect in the P-type region (including the boundaries) that can capture an electron from the conduction band, calculate the average electron capture time. Based on the result, determine the average number of electrons in the P-type region.

(f) Use the equations for the hole capture and hole emission times to determine the degeneracy factor in the Fermi–Dirac function for hole occupancy of the R–G center.

SOLUTION

(a) Based on Eq. (10.34), the average time needed to emit an electron from the R–G center is

$$\tau_{e,n} = \frac{h}{\eta(E_c - E_t + kT)} e^{(E_c - E_t)/kT}$$
$$= \frac{6.62 \times 10^{-34}}{1.6 \times 10^{-19} \times (0.6 + 0.026)} \times e^{0.52/0.02585} = 1.1 \text{ ms}$$

The distance of the R–G center from the valence band is $E_t - E_V = E_g - (E_C - E_t) = 1.12 - 0.6 = 0.52$ eV. The hole emission time is

$$\tau_{e,p} = \frac{h}{\eta_h(E_t - E_V + kT)} e^{(E_t - E_V)/kT}$$
$$= \frac{6.62 \times 10^{-34}}{1.6 \times 10^{-19} \times (0.52 + 0.026)} \times e^{0.52/0.02585} = 59.0\ \mu\text{s}$$

(b)

$$\tau_{c,p} = \frac{1}{v_{th}\sigma_p p} = 100 \text{ ps}$$

(c) An electron appears at the R–G center when a hole is emitted from the center. This electron appears at the R–G center for the period of time needed to capture a hole, $\tau_{c,p}$. Therefore, the probability that the R–G center is occupied by an electron is

$$f_t = \frac{\tau_{c,p}}{\tau_{c,p} + \tau_{e,p}} = \frac{10^{-10}}{10^{-10} + 5.9 \times 10^{-5}} = 1.69 \times 10^{-6}$$

(d) This question is not about generation in a depletion layer but about generation in a neutral P-type semiconductor. Because of that, τ_0 is not equal to $\tau_t = \tau_{e,n} + \tau_{e,p}$. There are two possible events that can occur when a hole is emitted from this R–G center: (1) electron emission to complete the electron–hole generation process and (2) hole capture. There are many holes in this P-type region, so the capture time for this R–G center is very short: $\tau_{c,p} = 100$ ps. Accordingly, hole capture and hole emission are the dominant events. These events determine the probability that the R–G center is occupied by an electron, as determined in part (c). The probability per unit time that an electron–hole pair is generated by the R–G center is equal to the probability that the R–G center is occupied by an electron (f_t) multiplied by the probability per unit time that the electron will be emitted into the conduction band $(1/\tau_{e,n})$:

$$\frac{1}{\tau_0} = \frac{f_t}{\tau_{e,n}}$$

Therefore,

$$\tau_0 = \frac{\tau_{e,n}}{f_t} = \frac{1.1 \times 10^{-3}}{1.69 \times 10^{-6}} = 649 \text{ s}$$

Although this time is much shorter than the result obtained in Example 10.1, it is still much longer than the time that this type of R–G center would need to generate an electron–hole pair in a depletion layer: $\tau_t = \tau_{e,n} + \tau_{e,p} = 1.1$ ms.

(e) The capture cross section for electrons of this donor-type R–G center is $\sigma_n = 10^{-13}$ cm^2. This corresponds to the radius of 1.8 nm, which is much smaller than the dimensions of the P-type region. The probability per unit length for a free electron to hit the R–G center is σ_n/V. This is equivalent to $\sigma_n N_t$, given that N_t for the case of one R–G center inside volume V is $N_t = 1/V = 10^{15}$ cm^{-3}. Because the electron travels with the thermal velocity v_{th}, the probability of hitting the R–G center per unit time is $v_{th}\sigma_n N_t$. This is also the probability per unit time that the electron is captured, because the R–G center is not occupied by an electron. Therefore,

$$\tau_{c,n} = \frac{1}{v_{th}\sigma_n N_t} = \frac{1}{10^5 \times 10^{-17} \times 10^{-21}} = 1.0 \text{ ns}$$

The average number of electrons is

$$\overline{N} = \frac{\tau_{c,n}}{\tau_{c,n} + \tau_{e,n}} = \frac{10^{-9}}{10^{-9} + 649} = 1.5 \times 10^{-12}$$

Using the equilibrium equation $n = n_i^2/p = 100\ \text{cm}^{-3}$, the average number of electrons is $\overline{N} = nV = 10^{-13}$.

(f) The R–G center is occupied by a hole after a hole capture event. This hole appears at the R–G center for time needed to emit it into the valence band, $\tau_{e,p}$. Therefore, the probability for hole occupancy is

$$f_{ht} = \frac{\tau_{e,p}}{\tau_{e,p} + \tau_{c,p}} = \frac{1}{1 + \tau_{c,p}/\tau_{e,p}}$$

With the equations for $\tau_{c,p}$ and $\tau_{e,p}$, given in parts (a) and (b), we obtain

$$\frac{\tau_{c,p}}{\tau_{e,p}} = \frac{\eta_h(E_t - E_V + kT)}{h v_{th} \sigma_p p \exp(E_t - E_V)/kT}$$

Taking into account that $p = N_V \exp[-(E_F - E_V)/kT]$, we write

$$\frac{\tau_{c,p}}{\tau_{e,p}} = \frac{\eta_h(E_t - E_V + kT)}{h v_{th} \sigma_p N_V} e^{(E_F - E_t)/kT}$$

$$f_{ht} = \frac{1}{1 + \eta_h(E_t - E_V + kT)/(h v_{th} \sigma_p N_V) e^{(E_F - E_t)/kT}}$$

Given that the Fermi–Dirac distribution for occupancy of isolated centers is

$$f_{ht} = \frac{1}{1 + (1/g) e^{(E_F - E_t)/kT}}$$

the degeneracy factor is

$$g = \frac{h v_{th} \sigma_p N_V}{\eta_h(E_t - E_V + kT)} = \frac{6.62 \times 10^{-34} \times 10^5 \times 10^{-19} \times 3.1 \times 10^{25}}{0.07 \times 1.6 \times 10^{-19} \times (0.52 + 0.026)} = 0.034$$

10.2 TWO-DIMENSIONAL TRANSPORT IN MOSFETs AND HEMTs

The classical electron-gas model, which considers electrons as mobile particles, can be applied as long as the electron wavelengths are much smaller than the dimensions of the space in which the electrons can freely move. When the size of the potential-energy well that contains the electrons becomes comparable to the electron wavelength, the electrons do not move as free particles but appear as standing waves. This effect is called *quantum confinement*. With one-dimensional quantum confinement, the electrons are free to move in the other two directions, and they are referred to as a two-dimensional electron gas.

This situation appears in the channel of MOSFETs and other field-effect devices such as the high-electron mobility transistor (HEMT). With two-dimensional confinement, the electron gas becomes one-dimensional. This is the case with nanowire transistors and carbon nanotubes, which are considered in Section 10.3. The situation with the holes—the second current carrier in semiconductors—is fully analogous to the case of electrons, so we deal with the concepts of two-dimensional and one-dimensional hole gases. The effect of one-dimensional quantum confinement is not so dramatic that it completely replaces the classical understanding of carrier transport, and in fact the classical device equations can be used with some adaptation of the model parameters. The effects of quantum confinement are introduced in this section to the level that is necessary for a modern engineer to understand the meaning of the terminology used, to learn how to incorporate these effects into the existing MOSFET equations, and to learn how to model HEMTs with these equations.

10.2.1 Quantum Confinement

The current carriers in field-effect transistors appear in approximately triangular potential wells, such as the one illustrated in Fig. 10.5. This is the case with the ordinary MOSFETs and also with the HEMT. In the case of silicon MOSFETs, the wide energy gap of SiO_2 grown on Si creates the conduction-band discontinuity (offset) that forms the triangular potential-energy well. This effect is achieved by what is known as *energy-gap engineering* in the case of compound semiconductors such as GaAs and GaN. Figure 10.6 illustrates the cross section and the energy-band diagram of the AlGaAs–GaAs heterojunction.[9] The energy gap of AlGaAs is wider, and it creates a conduction-band offset at the interface with GaAs. This leads to the formation of a triangular potential well.

Because the width of the triangular potential well approaches zero at the tip, we have a device dimension (the width of the potential well) that must become smaller than the electron wavelength. As a consequence, the wave properties of electrons become important. In the direction along the width of the potential well, which is the x-direction in Fig. 10.5, electrons cannot move freely and appear as standing waves. The first standing wave that can be formed corresponds to half of the electron wavelength and appears at the energy

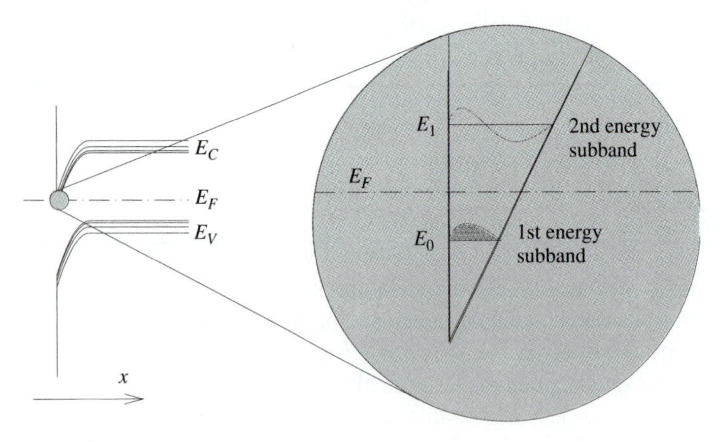

Figure 10.5 Illustration of energy level quantization that creates two-dimensional electron gas (2DEG).

[9] Section 9.4 describes the drawing of the energy-band diagram of the AlGaAs–GaAs heterojunction.

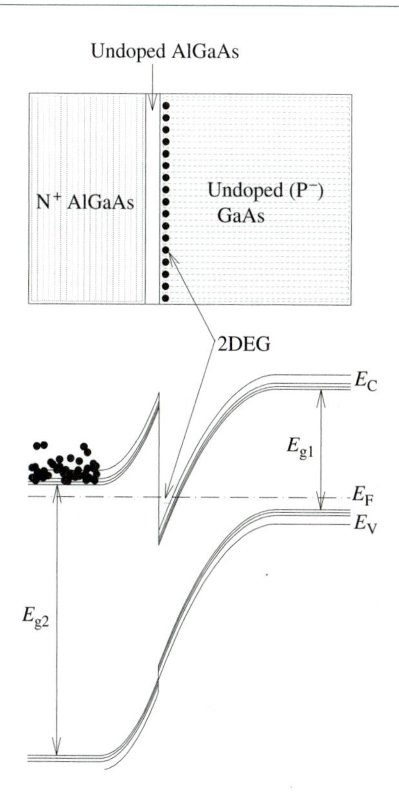

Figure 10.6 Cross section and energy-band diagram of an AlGaAs–GaAs system.

level that is above the bottom of the conduction band by a value labeled by E_0 (Fig. 10.5). The second standing wave corresponds to the full electron wavelength and appears at energy level E_1. As distinct from the quantized energy levels in the x-direction, the electron energy in the y- and z-directions can still take almost continuous values that depend on the p_y and p_z components of the electron momentum. Accordingly, the total electron energy can be expressed as

$$E = E_n + \frac{p_y^2}{2m_{n,y}} + \frac{p_z^2}{2m_{n,z}} \qquad (n = 0, 1, 2, \ldots) \tag{10.44}$$

where $m_{n,y}$ and $m_{n,z}$ are the electron effective masses in y and z directions, respectively. Therefore, what is seen as a single energy level in the x-direction appears as a two-dimensional energy continuum in the y and z dimensions. Each energy level E_n ($n = 0, 1, 2, \ldots$) represents the bottom of a two-dimensional energy continuum that is called an energy subband. Each of these subbands is two-dimensional, which means that the electrons can move and be scattered like particles in two dimensions and that the *electron-gas* concept can be applied only to these two dimensions. Consequently, the electrons in a two-dimensional subband are referred to as the two-dimensional electron gas (2DEG).

The values of the electric field forming the triangular potential well, which corresponds to the slope of the E_C line in Fig. 10.5, and the electron wavelength determine the quantized energy levels E_n ($n = 0, 1, 2, \ldots$). The energy levels E_n can be obtained by solving the Schrödinger equation, analogously to the case of electrons confined in a rectangular potential well presented in Example 2.2. Because of fairly involved mathematics in the case of triangular potential well, it is sufficient to present the final result:[10]

$$E_n \approx \left(\frac{h^2 q^2}{8\pi^2 m_x}\right)^{1/3} \left[\frac{3\pi}{2}\left(n+\frac{3}{4}\right)\right]^{2/3} E_{eff}^{2/3} \qquad (n = 0, 1, 2, \ldots) \qquad (10.45)$$

In Eq. (10.45), m_x is the electron effective mass in the x direction and E_{eff} is the effective electric field defining the triangular potential well (the slope of E_C lines in Figs. 10.5 and 10.6), which can be approximated by

$$E_{eff} = \frac{Q_I/2 + Q_d}{\varepsilon_s} \qquad (10.46)$$

where Q_I is the mobile charge per unit area in the two-dimensional electron gas (equivalent to the inversion-layer charge in MOSFETs), Q_d is the fixed depletion-layer charge per unit area, and ε_s is the semiconductor permittivity.

The effective mass of electrons in GaAs is isotropic and very small: $m_x = m_y = m_z = 0.067m_0$ (Fig. 2.13 and Section 2.3.1). With this effective mass and $Q_I/2 + Q_d = 2 \times 10^{12}$ cm^{-2}, which corresponds to $E_{eff} = 1.4 \times 10^5$ V/cm, the values of E_n for $n = 0$ and 1 are $E_0 = 112$ meV and $E_1 = 197$ meV, respectively. This means that these two two-dimensional subbands are separated by $E_1 - E_0 = 85$ meV, which is more than three times higher than the thermal energy at room temperature. For sufficiently high doping level [high Q_d in Eq. (10.46)] the Fermi level can appear between E_0 and E_1 for low and moderate values of Q_I (as shown in Fig. 10.5) so that most of the electrons are confined in the first subband. For higher values of Q_I, electrons will also populate the second two-dimensional subband (see Example 10.4). We refer to this situation as a quasi-two-dimensional electron gas.

The effective mass of electrons in Si is anisotropic, corresponding to six ellipsoidal E–k minima, as shown in Fig. 2.14 (Section 2.3.1). Projections of these six ellipsoids onto a two-dimensional plane that corresponds to two-dimensional subbands lead to two circular and four elliptical E–k minima. In the third dimension, the electrons appear as standing waves with the following effective masses in Eq. (10.45): $m_x = m_l = 0.98m_0$ for the two circular subbands and $m_x = m_t = 0.19m_0$ for the four elliptical subbands. With $m_x = 0.98m_0$ and $Q_I/2 + Q_d = 2 \times 10^{12}$ cm^{-2}, the bottoms of the first and the second circular subbands are obtained from Eq. (10.45) as $E_0 = 86$ meV and $E_1 = 49$ meV, respectively. In this case, the separation between these two subbands (37 meV) is just above the thermal energy, which means that a considerable fraction of the electrons occupies the second circular subband. As shown in Example 10.4, the first elliptical subband is also occupied by electrons because its bottom is very close to the bottom of the second circular subband. Again, this is the case of quasi-two-dimensional electron gas.

To obtain the actual electron-charge density in a two-dimensional subband, we multiply the density of the electron states $[D(E_{kin})]$ by the probability that an energy

[10] T. Ando, A. B. Fowler, and F. Stern, Electronic properties of two-dimensional systems, *Rev. Mod. Phys.*, vol. 54, pp. 437–672 (1982).

level E_{kin} is occupied by an electron $[f(E_{kin})]$ and then integrate to include all the energy levels in the subband:

$$Q_{In} = -q\int_0^{\infty} D(E_{kin})f(E_{kin})dE_{kin} \tag{10.47}$$

The two-dimensional density of electron states can be determined by a procedure that is analogous to the three-dimensional case described in Example 2.5. The result is (Problem 2.15):

$$D(E_{kin}) = M\frac{4\pi m^*}{h^2} \tag{10.48}$$

where M is the degeneracy factor that accounts for the number of equivalent E–k minima ($M = 1$ for the electrons in GaAs, $M = 2$ for the electrons associated with the circular subbands in Si, and $M = 4$ for the electrons associated with the elliptical subbands in Si) and m^* is the density-of-states effective mass ($m^* = 0.067m_0$ for the electrons in GaAs, $m^* = 0.19m_0$ for the electrons associated with the circular subbands in Si, and $m^* = \sqrt{m_l m_t} = 0.417m_0$ for the electrons associated with the elliptical subbands in Si). With $f(E_{kin}) = 1/\{1 + \exp[(E_{kin} - E_F)/kT]\}$ and $D(E_{kin})$ given by Eq. (10.48), the integration in Eq. (10.47) results with the following equation:

$$Q_{I,n} = -M\frac{4\pi q m^* kT}{h^2}\ln\left[1 + e^{(E_F - E_n)/kT}\right] \tag{10.49}$$

The energy E_n in Eq. (10.49) is the bottom of the two-dimensional subband with respect to the bottom of the conduction band at the surface, E_{C-s}—the tip of the triangular potential well in Fig. 10.5. To establish the Fermi level with respect to E_{C-s}, we observe that E_{C-s} is below the bottom of the conduction band in the body of the crystal (E_C) by the band bending due to the surface potential ($q\psi_s$):

$$E_F - E_{C-s} = E_F - (E_C - q\psi_s) = q\psi_s - (E_C - E_F) \tag{10.50}$$

Next we recall that E_F in the body of a P-type semiconductor is further away from the intrinsic Fermi-level position ($E_C - E_i$) by the value of the Fermi potential ($q\phi_F$):

$$E_C - E_F = E_C - E_i + q\phi_F \tag{10.51}$$

From Eqs. (10.50) and (10.51) we obtain

$$E_F - E_{C-s} = q\psi_s - (E_C - E_i) - q\phi_F \tag{10.52}$$

Given that E_n in Eq. (10.49) is with respect to E_{C-s}, $E_F - E_n$ in Eq. (10.49) is equal to $q\psi_s - (E_C - E_i) - q\phi_F - E_n$:

$$Q_{I,n} = -M\frac{4\pi q m^* kT}{h^2}\ln\left\{1 + e^{[q\psi_s - (E_C - E_i) - q\phi_F - E_n]/kT}\right\} \tag{10.53}$$

EXAMPLE 10.3 Separations Between Two-Dimensional Subbands in GaAs and Si

If the effective electric field that forms the triangular potential well is $E_{eff} = 5 \times 10^5$ V/cm in both an Si MOSFET and a GaAs HEMT, calculate E_0 and E_1 in (a) GaAs, (b) Si for circular subbands and (c) Si for elliptical subbands.

SOLUTION

(a) For the case of GaAs, where $m_x = 0.067 \times 9.1 \times 10^{-31} = 6.097 \times 10^{-32}$ kg, Eq. (10.45) gives the following results:

$$E_0 = \left[\frac{(6.62 \times 10^{-34})^2 \times (1.6 \times 10^{-19})^2}{8\pi^2 \times 6.097 \times 10^{-32}}\right]^{1/3} \left(\frac{9\pi}{8}\right)^{2/3} \times (5 \times 10^7)^{2/3}$$

$$= 4.424 \times 10^{-20} \text{ J} = 276.5 \text{ meV}$$

$$E_1 = \left[\frac{(6.62 \times 10^{-34})^2 \times (1.6 \times 10^{-19})^2}{8\pi^2 \times 6.097 \times 10^{-32}}\right]^{1/3} \left(\frac{9\pi}{8}\right)^{2/3} \times (5 \times 10^7)^{2/3}$$

$$= 7.782 \times 10^{-20} \text{ J} = 486.4 \text{ meV}$$

(b) For the case of circular subbands in Si, $m_x = 0.98m_0$, which gives $E_{0-c} = 113.2$ meV and $E_{1-c} = 199.1$ meV.

(c) For the case of elliptical subbands in Si, $m_x = 0.19m_0$, which gives $E_{0-e} = 195.4$ meV and $E_{1-e} = 343.8$ meV.

EXAMPLE 10.4 Position of Fermi Level and Population of Subbands

Assuming that $Q_I \gg Q_d$ in Example 10.3 and that all the electrons occupy the lowest subband, determine the corresponding positions of the Fermi level for both GaAs and Si. The dielectric constant of GaAs is 12.9. Is the assumption that only the lowest subband is occupied correct?

SOLUTION

From Eq. (10.46) we can obtain the electron density in 2DEG:

$$N_I = \frac{Q_I}{q} = \frac{2\varepsilon_s E_{eff}}{q}$$

The results are $N_I = 2 \times 12.9 \times 8.85 \times 10^{-12} \times 5 \times 10^7/1.6 \times 10^{-19} = 7.13 \times 10^{16}$ m^{-2} for GaAs and $N_I = 2 \times 11.8 \times 8.85 \times 10^{-12} \times 5 \times 10^7/1.6 \times 10^{-19} = 6.53 \times 10^{16}$ m^{-2} for Si.

Assuming that $\exp[(E_F - E_0)/kT] \gg 1$ in Eq. (10.49),

$$N_I = M\frac{4\pi m^*}{h^2}(E_F - E_0)$$

$$E_F - E_0 = \frac{N_I h^2}{4\pi M m^*}$$

In the case of GaAs ($M = 1$ and $m^* = 0.067m_0$),

$$E_F - E_0 = \frac{7.13 \times 10^{16} \times (6.62 \times 10^{-34})^2}{4\pi \times 0.067 \times 9.1 \times 10^{-31}} = 4.08 \times 10^{-20}\ \text{J} = 255.1\ \text{meV}$$

In the case of Si ($M = 2$ and $m^* = 0.19m_0$),

$$E_F - E_0 = \frac{6.53 \times 10^{16} \times (6.62 \times 10^{-34})^2}{4\pi \times 2 \times 0.19 \times 9.1 \times 10^{-31}} = 6.59 \times 10^{-21}\ \text{J} = 41.2\ \text{meV}$$

We can see that for these conditions E_F would be above the bottom of the second subband in GaAs ($E_1 - E_0 = 486.4 - 276.5 = 209.9$ eV), which means that the population of the second subband cannot be ignored. In the case of Si, E_F is close to both E_{1-c} ($E_{1-c} - E_{0-c} = 199.1 - 113.2 = 85.9$ meV) and E_{0-e} ($E_{0-e} - E_{0-c} = 195.4 - 113.2 = 82.2$ meV), meaning that the populations of the second circular and the first elliptical subbands cannot be ignored.

10.2.2 HEMT Structure and Characteristics

As shown in Fig. 10.6, the 2DEG in HEMTs results from conduction-band discontinuity at the AlGaAs–GaAs interface. The quality of the AlGaAs–GaAs interface is extraordinary. It is created by a continuous process of molecular-beam epitaxy, which enables us to change from GaAs to AlGaAs within a single atomic layer by changing the gas composition inside the epitaxial chamber. In addition, no doping atoms appear in the area of 2DEG, and a thin AlGaAs layer can be inserted to separate the 2DEG from the heavily doped N^+ AlGaAs. Consequently, both interface roughness scattering and Coulomb scattering from the doping ions are virtually eliminated. Thus any form of scattering is eliminated in one of the three spatial dimensions. The reduction in scattering means high electron mobility in the 2DEG, and it also means reduced noise. The properties of high electron mobility and low noise associated with the 2DEG are very useful for high-frequency and low-noise applications.

The 2DEG created at the AlGaAs–GaAs heterojunction makes the HEMT channel. As Fig. 10.7 illustrates, the channel is contacted at the ends to create source and drain terminals. Also a gate in the form of Schottky contact is created between the source and drain to provide a means of controlling the device current. Negative gate voltages reduce

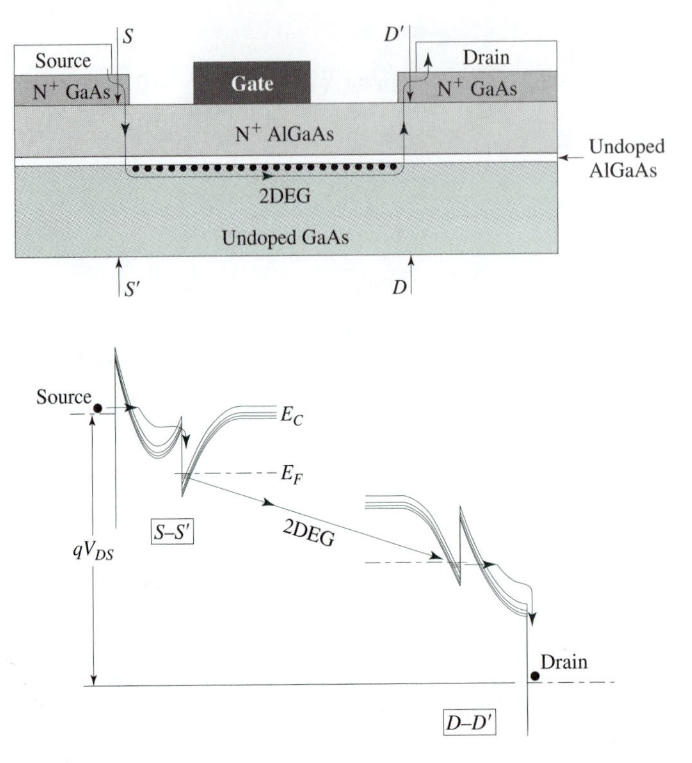

Figure 10.7 HEMT structure and energy bands at source and drain contacts.

the electron concentration in the channel and can completely repel the electrons, turning the device off. Therefore, the HEMT appears as a depletion-type FET. The transfer and output characteristics are analogous to those previously described for the depletion-type MOSFET.

The electron path from the source to the drain is indicated by the line and the arrows in Fig. 10.7. Electrons pass through a couple of heterojunctions between the source terminal and the 2DEG, as well as between the 2DEG and the drain terminal. The energy-band diagrams illustrate the barrier shapes associated with these heterojunctions at typical drain-to-source bias V_{DS}. It can be seen that electrons have to tunnel from the N^+ GaAs source into the N^+ AlGaAs layer and then tunnel again from the drain end of the 2DEG back into the N^+ AlGaAs layer. This introduces significant source and drain resistances, which adversely affect the high-speed and noise performance of the HEMT. Consequently, numerous modifications to source and drain contacts have been introduced to minimize the adverse effects of the source and drain resistances. The gate resistance is also very important, especially in terms of noise performance. The gate is typically made in a mushroom shape to cut its resistance without increasing the input capacitance, which improves the noise figure of the HEMT.

10.2.3 Application of Classical MOSFET Equations to Two-Dimensional Transport in MOSFETs and HEMTs

The threshold voltage (V_T) is a defining concept in MOSFETs used as switches and is therefore the central parameter in the classical MOSFET equations. The threshold voltage determines whether mobile carriers do or do not exist in the MOSFET channel,

$$Q_I = \begin{cases} 0 & \text{for } V_{GS} \leq V_T \\ -(V_{GS} - V_T)C_{ox} & \text{for } V_{GS} \geq V_T \end{cases} \tag{10.54}$$

where Q_I is the density of mobile charge in the channel, C_{ox} is the gate-oxide capacitance, and V_{GS} is the gate voltage.

Assuming that most of the mobile charge is located in the lowest two-dimensional subband, the density of mobile charge based on the model of two-dimensional electron gas is given by Eq. (10.53) for $n = 0$:

$$Q_I \approx Q_{I,0} = -M\frac{4\pi q m^* kT}{h^2} \ln\left\{1 + e^{[q\psi_s - (E_C - E_i) - q\phi_F - E_0]/kT}\right\} \tag{10.55}$$

For $q\psi_s \ll (E_C - E_i) + q\phi_F + E_0$, $\exp\{[q\psi_s - (E_C - E_i) - q\phi_F - E_0]/kT\} \approx 0$ and $Q_{I,0} \approx 0$. For $q\psi_s \gg (E_C - E_i) + q\phi_F + E_0$, $\exp\{[q\psi_s - (E_C - E_i) - q\phi_F - E_0]/kT\} \gg 1$ and $Q_{I,0} \approx (4M\pi q m^*/h^2)[q\psi_s - (E_C - E_i) - q\phi_F - E_0]$. The boundary between these two regions can be defined as $\psi_s = (E_C - E_i)/q + \phi_F + E_0/q$ and Q_I expressed analogously to Eq. (10.54):

$$Q_I = \begin{cases} 0 & \text{for } \psi_s \leq (E_C - E_i)/q + \phi_F + E_0/q \\ (4M\pi q^2 m^*/h^2)\left[\psi_s - (E_C - E_i)/q - \phi_F - E_0/q\right] & \text{for } \psi_s \geq (E_C - E_i)/q + \phi_F + E_0/q \end{cases} \tag{10.56}$$

Clearly, in the case of a two-dimensional gas, the threshold condition of the surface potential, ψ_{s-th}, can be defined as

$$\psi_{s-th} = (E_C - E_i)/q + \phi_F + E_0/q \tag{10.57}$$

As distinct from this, the threshold condition of the surface potential in the case of classical three-dimensional model is

$$\psi_{s-th} \approx 2\phi_F \tag{10.58}$$

Based on Eqs. (7.18) and (7.19) and labeling the surface potential at the threshold condition by ψ_{s-th}, the following form of the threshold-voltage equation accounts for both values of ψ_{s-th}:

$$V_T = V_{FB} + \psi_{s-th} + \gamma\sqrt{\psi_{s-th}} \tag{10.59}$$

In nondegenerate semiconductors, $\phi_F < (E_C - E_i)/q$, which means that the 2DEG value of ψ_{s-th} is higher than the classical value of $\psi_{s-th} = 2\phi_F$. Nonetheless, $2\phi_F$ is a parameter in the classical MOSFET model and can be adjusted so that the classical threshold-voltage equation corresponds to the 2DEG threshold condition.

Equation (10.59) can also be used to determine the threshold voltage in HEMTs, with a note that C_{ox} in $\gamma = \sqrt{2\varepsilon_s q N_A}/C_{ox}$ is now the capacitance per unit area of the AlGaAs layer: ε_s/t_{AlGaAs}.

Similarly, the channel-carrier mobility (μ_0) or the transconductance parameter ($KP = \mu_0 C_{ox}$) in the gain factor β [Eq. (8.22)] can be adjusted so that the classical MOSFET models provide a good fit even when there is a pronounced carrier confinement so that the electron gas is purely two dimensional. This is because the classical equations are derived from the application of Ohm's law to the inversion layer, which is already assumed to have negligible thickness; hence the density of the inversion-layer charge Q_I is expressed per unit area. Equations (8.13) to (8.16) show how the differential form of Ohm's law leads to the following general relationship between the current through the inversion layer I_D and the voltage that drives this current V_{DS}:

$$I_D = \frac{\mu_0 W}{L_{eff}} \overline{Q_I} V_{DS} \tag{10.60}$$

In this equation, $\overline{Q_I}$ is the average charge density along the channel. The need to use the average value is because V_{DS} impacts the charge density so that Q_I varies along the channel. For the case of small V_{DS}, this impact is insignificant: Q_I is given by Eq. (10.54), and I_D is linearly dependent on V_{DS}. For larger V_{DS} values, however, Eq. (10.54) for Q_I has to be modified to include the impact of V_{DS}. Because the electric potential due to V_{DS} varies between source and drain, changing from 0 V at the source end of the channel to V_{DS} itself at the drain end, a position-dependent electric-potential term $V(y)$ has to be included in the equation for Q_I:

$$Q_I(y) = \begin{cases} 0 & \text{for } V_{GS} \le V_T - V(y) \\ -[V_{GS} - V_T - V(y)]C_{ox} & \text{for } V_{GS} \ge V_T - V(y) \end{cases} \tag{10.61}$$

As a result, Q_I becomes position-dependent and has to be averaged so that Eq. (10.60) can be used to determine the total drain current:

$$\overline{Q_I} = \frac{1}{L} \int_0^L Q_I(y)dy \tag{10.62}$$

Many different approaches and assumptions are utilized to perform the averaging shown in Eq. (10.62), which results in many different MOSFET models. Nonetheless, they are all able to predict the departure from the linear I_D–V_{DS} relationship and eventual saturation of I_D. These models can also be used when the quantum confinement is pronounced, resulting in 2DEG, because the effects of the quantum confinement can be included by adjustments of the model parameters, in particular the threshold voltage V_T and the channel-carrier mobility μ_0.

EXAMPLE 10.5 Comparison of 2DEG and 3DEG Threshold Voltages in Si MOSFET

The body doping of a silicon N-channel MOSFET is $N_A = 5 \times 10^{17}$ cm^{-3}.

(a) Determine the 3DEG and 2DEG threshold values of the surface potential (ψ_{s-th}).
(b) If $t_{ox} = 2$ nm and $V_{FB} = 0.2$ V, determine the 2DEG and 3DEG threshold voltages.

SOLUTION

(a) Equation (10.58) specifies the threshold condition for the surface potential in the case of the classical three-dimensional model:

$$\psi_{s-th} \approx 2\phi_F$$

where the Fermi potential [see Eq. (2.88)] is

$$\phi_F = V_t \ln \frac{N_A}{n_i} = 0.02585 \ln \frac{5 \times 10^{17}}{10^{10}} = 0.46 \text{ V}$$

Therefore, the threshold value of the surface potential with the 3DEG model is $\psi_{s-th} =$ 0.92 V.

Equation (10.57) specifies the threshold condition for the surface potential in the case of two-dimensional gas,

$$\psi_{s-th} = E_g/2q + \phi_F + E_0/q$$

where $E_g/2 = E_C - E_i$ in the case of Si, E_0 is given by Eq. (10.45) for $n = 0$,

$$E_0 = \left(\frac{h^2 q^2}{8\pi^2 m_x}\right)^{1/3} \left(\frac{9\pi}{8}\right)^{2/3} E_{eff}^{2/3}$$

and E_{eff} is given by Eq. (10.46) with $Q_I = 0$ at the threshold condition:

$$E_{eff} = \frac{Q_d}{\varepsilon_s} = \frac{1}{\varepsilon_s} q N_A w_d = \sqrt{\frac{2qN_A \psi_{s-th}}{\varepsilon_s}}$$

It can be seen that the calculation of ψ_{s-th} requires E_0 and E_{eff}, which depend on ψ_{s-th} itself. An iterative process has to be applied to solve these equations. We can begin the iterative process with $\psi_{s-th} = 1$ V as the input parameter in the equation for E_{eff} and then calculate E_0 and ψ_{s-th} from the other two equations, which provide us with a new value for ψ_{s-th} to begin a new iteration. This process converges to a stable value of ψ_{s-th} very quickly:

Input		Output	
$\psi_{s\text{-}th}$ (V)	E_{eff} (V/m)	E_0 (eV)	$\psi_{s\text{-}th}$ (V)
1.0000	3.91×10^7	0.1336	1.1518
1.1518	4.20×10^7	0.1400	1.1538
1.1583	4.21×10^7	0.1403	1.1586

Therefore, the threshold value of the surface potential with the 2DEG model is $\psi_{s-th} = 1.16$ V.

(b) The threshold voltage is given by Eq. (10.59) where the body factor γ is

$$\gamma = \frac{\sqrt{2\varepsilon_s q N_A}}{C_{ox}} = \frac{t_{ox}\sqrt{2\varepsilon_s q N_A}}{\varepsilon_{ox}} = 0.24\ \mathrm{V}^{1/2}$$

For the case of 3DEG,

$$V_T = V_{FB} + \psi_{s-th} + \gamma\sqrt{\psi_{s-th}} = 0.2 + 0.92 + 0.24\sqrt{0.92} = 1.35\ \mathrm{V}$$

For the case of 2DEG,

$$V_T = V_{FB} + \psi_{s-th} + \gamma\sqrt{\psi_{s-th}} = 0.2 + 1.16 + 0.24\sqrt{1.16} = 1.62\ \mathrm{V}$$

EXAMPLE 10.6 The Question of Ballistic Transport

The standard MOSFET equations cannot be applied for ballistic transport, which occurs when the carriers are not scattered as they pass through the MOSFET channel. To check the applicability of the standard MOSFET equations, estimate the number of scattering events for each electron in the channel of a nanoscale MOSFET having a channel length of $L = 50$ nm and an applied voltage between drain and source of $V_{DS} = 0.1$ V. The mobility of the electrons in the channel is $\mu_n = 300\ \mathrm{cm^2/V\cdot s}$, and their effective mass for conductivity is $m^* = 0.19m_0$.

SOLUTION

The time between two scattering events (τ_{sc}) can be determined from the mobility value:

$$\mu_n = \frac{q\tau_{sc}}{m^*}$$

$$\tau_{sc} = \frac{m^*}{q}\mu_n = 3.24 \times 10^{-14}\ \mathrm{s}$$

The scattering length is related to the scattering time and the thermal velocity,

$$l_{sc} = v_{th}\tau_{sc}$$

The thermal velocity can be determined from the following condition:

$$\frac{m^* v_{th}^2}{2} = kT$$

Note that the average thermal energy for the two-dimensional case is used (kT) instead of the usual three-dimensional term ($3kT/2$). From the preceding equations, the following equation can be obtained for the relationship between scattering length and mobility:

$$l_{sc} = \frac{\mu_n}{q}\sqrt{2kTm^*} = 7.1 \text{ nm}$$

The scattering length is smaller than the channel length. For an electron that moves along the channel only, the average number of scattering events would be L/l_{sc}. However, this estimate gives the smallest number of scattering events because the electrons scatter in random directions, meaning that they travel a longer path than the channel length. Following the insights from Section 10.1, the number of scattering events (N_{sc}) can be estimated as the ratio between the average time that an electron travels from the source to the drain τ_T and the scattering time: $N_{sc} = \tau_T/\tau_{sc}$. The time τ_T can be determined from the drift velocity, which is obtained from the mobility and the electric field:

$$v_{dr} = \mu_n E = \mu_n \frac{V_{DS}}{L} = 6.0 \times 10^4 \text{ m/s}$$

$$\tau_T = \frac{L}{v_{dr}} = 8.33 \times 10^{-13} \text{ s}$$

$$N_{sc} = \frac{\tau_T}{\tau_{sc}} = \frac{8.33 \times 10^{-13}}{3.24 \times 10^{-14}} = 25.7 \text{ scattering events}$$

This is a sufficient number of scattering events to allow us to conclude that the MOSFET is not in ballistic mode and that the standard MOSFET equations, based on Ohm's law, are applicable. From the analysis presented, it is obvious that the number of scattering events reduces with an increase in mobility:

$$N_{sc} = \frac{\tau_T}{\tau_{dr}} = \frac{qL}{\mu_n^2 m^* E}$$

This means that ballistic transport is most likely to occur in materials with high carrier mobility.

10.3 ONE-DIMENSIONAL TRANSPORT IN NANOWIRES AND CARBON NANOTUBES

Recent progress in terms of techniques for growth of semiconductor nanowires and carbon nanotubes has generated considerable interest in their potential to enable the ultimate size reduction of electronic devices. The atomic structure of semiconductor nanowires is the same as in the common three-dimensional crystals. The nanowires exhibit distinct properties, however, because their diameter is smaller than the electron wavelength so that the electrons are under the condition of two-dimensional quantum confinement. This results in one-dimensional electron gas, the only dimension for electron transport being along the wire. The atomic structure of carbon nanotubes is described in Section 1.1.3; it is different from that of the nanowires because carbon nanotubes are hollow cylinders whose walls have the two-dimensional graphene structure. Another difference having practical importance is due to significantly reduced carrier scattering in the two-dimensional graphene-type crystal, which results in very high carrier mobilities. Yet another difference is due to the fact that carbon nanotubes can appear as both metallic and semiconductive, depending on their helicity. Similar to nanowires, however, carbon nanotubes are so small in diameter that the electrons appear as a one-dimensional electron gas due to the two-dimensional quantum confinement.

Since the electron transport in semiconductor nanowires and carbon nanotubes is one-dimensional, the standard MOSFET equations must be adapted for application to nanowire and carbon-nanotube FETs. This issue is considered in Section 10.3.1, under the assumption that the ohmic nature of electron transport is maintained. Because semiconductor nanowires and carbon nanotubes are of practical interest due to their potential for the ultimate size reduction, it is very important to consider the case of *ballistic transport*, which emerges when the length of the wires or the tubes becomes smaller than the carrier-scattering length, with the result that ohmic resistance no longer exists. Because of the absence of ohmic resistance, ballistic transport has generated considerable hype in terms of potential applications. Section 10.3.2 describes the ultimate limit in terms of resistance reduction (quantum conductance limit), since a modern engineer cannot properly consider potential applications of nanowires and nanotubes without this effect of quantum resistance.

10.3.1 Ohmic Transport in Nanowire and Carbon-Nanotube FETs

To enable continuous reduction in channel length while avoiding short-channel effects due to penetration of the electric field from the drain to the source regions, MOSFET structures that enable better gate control of the channel region are needed. The SOI MOSFET with ultrathin body (Fig. 8.26) and the FinFET (Fig. 8.27) demonstrate that the evolution of these structures is moving toward the thinnest possible body, fully surrounded by the gate electrode, as shown by the conceptual diagram in Fig. 10.8.[11]

In the nanowire implementation of a coaxially gated FET, the body is a semiconductor nanowire exhibiting two-dimensional carrier confinement. This means that the electron gas is one-dimensional (1DEG) and Q_I in these FETs is the channel charge per unit length, expressed in C/m. In the carbon-nanotube implementation of the coaxially gated FETs from Fig. 10.8, the body is a carbon nanotube. This enables the FET to have the thinnest possible body. The atomic thickness of the tube enables us to reduce the length between the source and drain contact to several nanometers while avoiding the short-channel effects.

Q_I can be determined by the quantum-mechanical approach described in Section 10.2.1 for the case of 2DEG. In the case of 1DEG, the density of states in Eq. (10.47)

[11] J. Appenzeller, Carbon nanotubes for high-performance electronics—Progress and prospect, *Proc. IEEE*, vol. 96, pp. 201–211 (2008); *IEEE Trans. Electron Devices*, Special Issue on Nanowire Transistors: Modeling, Device Design, and Technology, vol. 55 (2008).

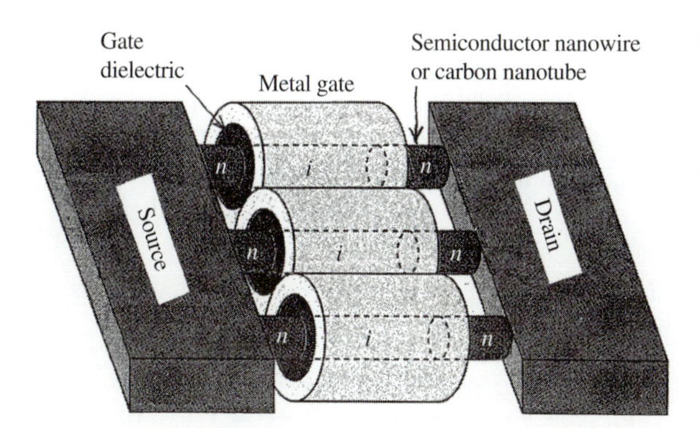

Figure 10.8 Conceptual diagram of an array of coaxially gated FETs where the body is either semiconductor nanowire (nanowire FET) or carbon nanotube (carbon-nanotube FET).

is one-dimensional, expressed in eV/m, which leads to Q_I in C/m. Similar to the 2DEG, the one-dimensional Q_I can also be related to applied voltage and the capacitance through Eq. (10.54), provided the threshold-voltage value is adjusted to account for the carrier-confinement effects. In the case of 1DEG, however, C_{ox} in Eq. (10.54) is the capacitance per unit length (expressed in F/m). This is consistent with the fact that the resultant Q_I should be in C/m.

Because the free-carrier density (Q_I) in 1DEG becomes charge per unit length, there is no channel width (W) in Eq. (10.60). In the case of transistor arrays (N nanowires or carbon nanotubes running in parallel between the source and the drain), W in Eq. (10.60) is replaced by the number of nanowires or carbon nanotubes N:

$$I_D = N\frac{\mu_0}{L_{eff}}\overline{Q_I}V_{DS} \tag{10.63}$$

Apart from the use of C_{ox} as capacitance per unit length in Eq. (10.54) and the replacement of W by N in Eq. (10.63), the classical MOSFET equations can still be used to model the current–voltage characteristics of nanowire and carbon-nanotube FETs, but only if the current remains limited by carrier scattering and the concept of mobility can be applied. Because the reason for considering nanowires and carbon nanotubes is to reduce the channel length, it may happen that the free carrier path between two scattering events becomes longer than the channel length. When this occurs, the current carriers moving from the source to the drain do not experience ohmic resistance and the concept of mobility cannot be applied. In this case, we deal with ballistic transport of carriers between the source and the drain, which is considered in the next section.

EXAMPLE 10.7 Resistance of Semiconductor Nanowire

Determine the resistance of a coaxially gated semiconductor nanowire with length $L = 20$ nm, diameter $d = 3$ nm, gate-dielectric thickness $t_{ox} = 2$ nm, and dielectric permittivity $\varepsilon = 10\varepsilon_0$ if the effective gate voltage is $V_{GS} - V_T = 0.1$ V and the carrier mobility is $\mu_0 = 100$ cm^2/V · s. The capacitance per unit length of a coaxial cable is given by $C = 2\pi\varepsilon/\ln(D/d)$, where d is the outside diameter of the inner conductor and D is the inside diameter of the shield.

SOLUTION

Based on the equation for capacitance per unit length of a coaxial cable, the gate-dielectric capacitance of the nanowire is

$$C_{ox} = \frac{2\pi\varepsilon}{\ln(d + t_{ox})/d} = \frac{2\pi \times 10 \times 8.85 \times 10^{-12}}{\ln(3+2)/3} = 1.1 \times 10^{-9} \text{ F/m}$$

The charge per unit length is obtained from Eq. (10.54):

$$Q_I = (V_{GS} - V_T)C_{ox} = 0.1 \times 1.1 \times 10^{-9} = 1.1 \times 10^{-10} \text{ C/m}$$

The resistance can be determined from Eq. (10.63):

$$R = \frac{V_{DS}}{I_D} = \frac{L_{eff}}{\mu_0 Q_I} = \frac{20 \times 10^{-9}}{100 \times 10^{-4} \times 1.1 \times 10^{-10}} = 18.4 \text{ k}\Omega$$

10.3.2 One-Dimensional Ballistic Transport and the Quantum Conductance Limit

The term *ballistic transport* describes the transport of current carriers that do not experience scattering. Such transport occurs when the distance between two contacts, which is the channel length between the source and drain regions in FETs, is shorter than the free path between two scattering events. This situation can practically occur in nanowire and carbon-nanotube FETs because these devices enable a very aggressive reduction in channel length. In addition to the channel length reduction, carbon nanotubes enable significant reduction in carrier scattering, which means a significant increase in the mean free path between two scattering events. Therefore, the most likely devices to exhibit ballistic transport are those based on carbon nanotubes.

The absence of carrier scattering, which corresponds to zero ohmic resistance, does not mean that huge currents can be achieved by applying almost negligible voltages. The factors that limit the current in the absence of scattering are (1) limited density of electron states, which means limited electron concentration, and (2) limited electron velocity. In the case of a one-dimensional electron gas, the concentration of electrons is the number of electrons per unit length, $n_{\rightarrow}$ (the arrow in the subscript indicates that we are focusing on the electrons moving in the positive x-direction only). If the electrons move with velocity v_x along the x-direction, we can determine the current of these electrons as the number of electrons that reach a selected point per unit time. The electrons that can reach the selected point during time Δt are within the distance $L = v_x \Delta t$. The number of electrons within the distance L is equal to $n_{\rightarrow} L$ and the number of electrons reaching the selected point per unit time is $n_{\rightarrow} L/\Delta t = n_{\rightarrow} v_x$. This is the particle current; the electric current of the electrons moving in the x-direction is

$$I_{\rightarrow} = -qn_{\rightarrow} v_x \tag{10.64}$$

Consider first the ultimate quantum-mechanical limit to the number of electrons per unit length ($n_{\rightarrow}$). This limit relates to the finite one-dimensional density of possible electron states. The concept of density of states, and their inherent limitation, was introduced in Section 2.3.3. Nonetheless, full derivation of the one-dimensional density of states is provided in the following text so that the real reason for the ultimate conductance limit can be clearly seen. Furthermore, this derivation begins with the fundamental reason for this conductance limit, which is that electron size is finite. If we label the smallest possible electron size in one dimension by Δx, the maximum number of electrons per unit length is $C = 2/\Delta x$ because only two electrons with different spins can share a single state defined by Δx. Although Δx itself is not a physical constant, the product between Δx and the smallest possible momentum granulation, Δp_x, is equal to the Planck constant:

$$\Delta x \Delta p_x = h \tag{10.65}$$

This is, of course, the fundamental quantum-mechanical principle expressed in the form of the Heisenberg uncertainty relationship between particle position and particle momentum. Including this relationship, the maximum possible number of electrons per unit length becomes

$$C = \frac{2}{h} \Delta p_x \tag{10.66}$$

Particle momentum is directly related to the particle velocity ($p_x = m^* v_x$), which is directly related to the kinetic energy ($E_{kin} = m^* v_x^2/2$). This leads to the following relationship between the momentum and the kinetic energy:

$$p_x = \pm\sqrt{2m^* E_{kin}} \tag{10.67}$$

The first derivative of this equation establishes the following relationship between a small change in the momentum (dp_x) and a small change in the kinetic energy (dE_{kin}):

$$dp_x = \pm\sqrt{\frac{m^*}{2}} E_{kin}^{-1/2} dE_{kin} \tag{10.68}$$

Given that the momentum change (dp_x) cannot be smaller than Δp_x, Eqs. (10.66) and (10.68) can be used to express the maximum possible number of electrons per unit length in terms of the kinetic energy:

$$C_{\rightarrow}(E_{kin}) = \underbrace{\frac{\sqrt{2m^*}}{h} E_{kin}^{-1/2}}_{D_{\rightarrow}(E_{kin})} dE_{kin} \tag{10.69}$$

Note that only electrons with positive p_x (velocities along x direction) are included in Eq. (10.69), which is indicated by the arrows in the symbol subscripts. This means that $D_{\rightarrow}$ in Eq. (10.69) is half the value of the one-dimensional density of electron states per unit energy (to obtain the total density of states, as required in Problem 2.16, $D_{\rightarrow}$ should be doubled to include $D_{\leftarrow}$, which corresponds to the negative momenta). Both $D_{\rightarrow}(E_{kin})$ and $C_{\rightarrow}(E_{kin})$ express the maximum possible number of electrons per unit length, the difference being that $C_{\rightarrow}(E_{kin})$ is the total number within a small energy range dE_{kin}, whereas $D_{\rightarrow}(E_{kin})$ is the number per unit energy. The *actual* number of electrons is smaller than the *maximum possible* number when some of the electron states are not occupied by electrons. This means that the actual number of electrons per unit length (moving in the x-direction) is $C_{\rightarrow}(E_{kin})f(E_{kin})$, where $f(E_{kin})$ is the probability that an electron state is actually occupied by an electron. It should also be noted that this is the actual number of electrons with kinetic energies in the small energy range between E_{kin} and $E_{kin} + dE_{kin}$; in other words, this is a fraction of the actual number of electrons per unit length that will be labeled by $dn_{\rightarrow}$:

$$dn_{\rightarrow}(E_{kin}) = \frac{\sqrt{2m^*}}{h} E_{kin}^{-1/2} f(E_{kin}) dE_{kin} \tag{10.70}$$

Having established that the number of electrons per unit length depends on the kinetic energy, we can modify Eq. (10.64) to account for this effect:

$$dI_{\rightarrow}(E_{kin}) = -qv_x dn_{\rightarrow}(E_{kin}) = -q\sqrt{\frac{2E_{kin}}{m^*}} dn_{\rightarrow}(E_{kin}) \tag{10.71}$$

Eliminating $dn_{\rightarrow}$ from Eqs. (10.70) and (10.71), we obtain

$$dI_{\rightarrow}(E_{kin}) = -\frac{2q}{h} f(E_{kin}) dE_{kin} \tag{10.72}$$

This equation gives the fraction of the current due to the electrons with kinetic energies in the range between E_{kin} and $E_{kin} + dE_{kin}$. To obtain the total current, we integrate the current fractions along the whole range of kinetic energies:

$$I_{\rightarrow} = -\frac{2q}{h} \int_0^{\infty} f(E_{kin}) dE_{kin} \tag{10.73}$$

This current is illustrated by the shaded area in Fig. 10.9 corresponding to the source end of a nanowire or nanotube. To simplify this integration, we will assume that $f(E_{kin}) = 1$

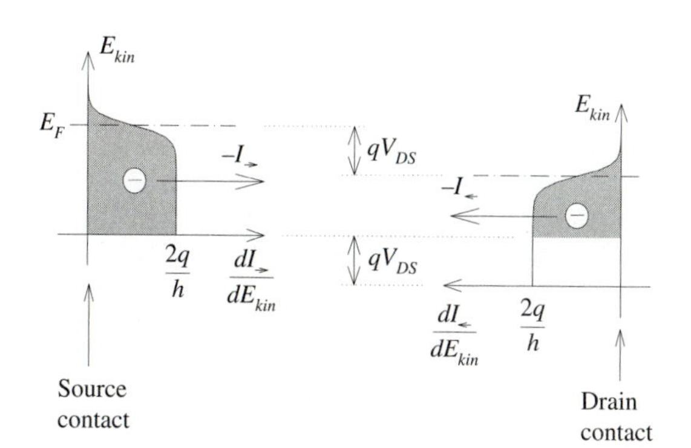

Figure 10.9 Illustration of energy distributions of ballistic electron currents in a nanowire/nanotube, as defined by Eq. (10.72). $I_{\rightarrow}$ originates at the source end, whereas $I_{\leftarrow}$ originates at the drain end of the nanowire/nanotube.

for $E_{kin} \leq E_F$ and $f(E_{kin}) = 0$ for $E_{kin} > E_F$,[12] where E_F is the Fermi level. Taking the bottom of the one-dimensional subband ($E_{kin} = 0$) as the reference energy level for E_F, we obtain

$$I_{\rightarrow} = -\frac{2q}{h}\int_0^{E_F} dE_{kin} = -\frac{2q}{h}E_F \tag{10.74}$$

This is the ultimate limit for the current of electrons moving from the source toward the drain (assuming transport through a single one-dimensional channel). This is also the limit for the effective current that could be observed if no electrons were moving in the opposite direction.

The current of electrons that are moving in the negative x-direction, from the drain contact toward the source contact, can be obtained by fully analogous derivation. In this case, however, the energy distribution of $I_{\leftarrow}$ is integrated over a different energy range. As illustrated in Fig. 10.9 (the diagram for the drain contact), the integration should begin at $E_{kin} = qV_{DS}$, where V_{DS} is the applied voltage between the drain and source contacts. It can be seen from this diagram that the electrons with energies below qV_{DS} at the drain end cannot ballistically move to the source contact because there are no electron states below $E_{kin} = 0$ at the source end of the nanotube/nanowire. Analogously to the source end, the upper limit for the integration is assumed to be the Fermi level. With this, we obtain the following result:

$$I_{\leftarrow} = -\frac{2q}{h}\int_{qV_{DS}}^{E_F} dE_{kin} = -\frac{2q}{h}(E_F - qV_{DS}) \tag{10.75}$$

The effective current in the positive x direction is

$$I = I_{\rightarrow} - I_{\leftarrow} = -\frac{2q^2}{h}V_{DS} \tag{10.76}$$

[12]This is the abrupt Fermi–Dirac distribution that is strictly correct for $T = 0$ K.

The minus sign in Eq. (10.76) simply indicates that the electric current is in the negative x-direction when the electrons are moving in the positive x-direction (from the source to the drain). It should be stressed that Eq. (10.76) can be used when $0 \leq qV_{DS} \leq E_F$. For larger applied voltages, the current from the drain toward the source drops to zero because $E_F - qV_{DS}$ drops below zero and, as a consequence, there are no electron states below $E_{kin} = 0$ at the source end to accept the electrons from the drain end (refer to Fig. 10.9). This means that the current saturates at the maximum level given by Eq. (10.74).

The dependence between the current and the applied voltage in Eq. (10.76) is linear, which enables us to define a corresponding quantum resistance $R_Q = V_{DS}/(-I)$:

$$R_Q = \frac{h}{2q^2} = 12.93\ \text{k}\Omega \tag{10.77}$$

The derivation that leads to the result expressed by Eq. (10.77) assumes that the electrons populate a single one-dimensional subband. If the subband is degenerated because of multiple E–k minima or because more subbands are populated, degeneracy factor M is introduced to account for the multiple transport channels:

$$R_Q = \frac{h}{2Mq^2} \tag{10.78}$$

This is the absolute possible minimum resistance of a one-dimensional wire or tube, which can also be expressed as the absolute conductance limit ($G_Q = 1/R_Q$):

$$G_Q = M\frac{2q^2}{h} \tag{10.79}$$

This result comes from a modeling approach first introduced by R. Landauer[13] and is known as the Landauer conductance formula.

The quantum resistance R_Q does not depend on the length of the carbon nanotube or nanowire. Because of that, it is frequently referred to as *contact resistance*. Ohmic resistance, contact or otherwise, is associated with carrier scattering, which causes carriers to give their kinetic energy away as heat. Each carrier reaching the drain contact dissipates its kinetic energy of $E_{kin} = qV_{DS}$ as heat inside the drain region. If N carriers reach the drain contact per unit time, the total energy dissipated per unit time is $qV_{DS}N/t = V_{DS}I$, where we take into account that $qN/t = I$. This means that the power is dissipated inside the drain region, or at the drain contact itself if every carrier is scattered and forced to dissipate its kinetic energy at the contact. It may be argued that the whole V_{DS} voltage drops across the contact so that $V_{DS}I = R_QI^2$ correctly reflects that the power is dissipated at the contact. However, thinking of R_Q as an ohmic contact resistance that limits the current is misleading. If R_Q were ohmic contact resistance, we should get symmetrical power dissipation at the source contact, and it should be possible to alter this resistance by technological means. In fact, the quantum resistance R_Q is a physical constant determined

[13] R. Landauer, Spatial variation of currents and fields due to localized scatterers in metallic conduction. *IBM J. Res. Dev.* vol. 1, p. 233 (1957).

by the Planck constant, since the Planck constant sets the limit to the number of electrons that can move through a one-dimensional wire. This should be quite obvious from the foregoing derivation of R_Q, which did involve the quantum limit on electron size, hence the number of electrons, but no carrier scattering.

EXAMPLE 10.8 Quantum Limit for Nanowire Resistance

Compare the quantum resistance limit with the nanowire resistance calculated in Example 10.7. What is implied if the nanowire resistance in Example 10.7 is calculated for $V_{GS} - V_T = 0.5$ V?

SOLUTION

The result in Example 10.7 shows a nanowire resistance of 18.4 kΩ, which is not much higher than the quantum limit of 12.93 kΩ. If Example 10.7 is solved for $V_{GS} - V_T = 0.5$ V instead of 0.1 V, the result would be a resistance five times smaller: $18.4/5 = 3.68$ kΩ. This is much smaller than the quantum limit for a single one-dimensional channel. The nanowire resistance could drop below 12.93 kΩ only if there were multiple one-dimensional subbands and/or E–k minima to contribute to the current flow [$M > 1$ in Eq. (10.78)]. It is quite possible, however, that the number of electrons per unit length is overestimated by the capacitance–voltage equation, $n = (V_{GS} - V_T)C_{ox}/q$.

EXAMPLE 10.9 Carbon Nanotubes as Interconnect Material: Myth or Future Reality?

A bundle of densely packed, single-wall carbon nanotubes is considered as interconnect material. Assuming ballistic transport through two one-dimensional subbands, determine the required density of carbon nanotubes to match the resistance of 1-μm-long copper-interconnecting track with the same cross-sectional area (the resistivity of copper is 17.2 nΩ·m).

SOLUTION

The resistance of the copper interconnect is

$$R_{Cu} = \rho \frac{L}{A}$$

The resistance of a bundle of N carbon nanotubes is

$$R_{CNT} = \frac{R_Q}{N}$$

where R_Q is the quantum resistance,

$$R_Q = \frac{h}{2Mq^2} = \frac{h}{4q^2} = 6.46\ \text{k}\Omega$$

From the condition $R_{Cu} = R_{CNT}$, we obtain

$$\frac{N}{A} = \frac{R_Q}{\rho L} = \frac{6.46 \times 10^3}{17.2 \times 10^{-9} \times 10^{-6}} = 0.376\ \text{nm}^{-2}$$

This analysis shows that carbon nanotubes with the currently demonstrated density of 0.1 CNT/nm^2 are not competitive as interconnect material. Obviously, the calculation will begin to favor CNTs for much longer interconnects; when the length is significantly increased, however, ballistic transport may not be maintained.

SUMMARY

1. The concept of continuous carrier concentration leads to noninteger electron and/or hole numbers in small devices. Not only this is impossible but also instantaneous thermal equilibrium conditions that require noninteger carrier numbers are not possible. Noninteger carrier numbers can be understood as average values over a sufficiently long period of time, and the thermal equilibrium concept can be refined as a long-term balance between nonequilibrium carrier generation and recombination events.
2. The *actual* number of occurrences N_{actual} for a given *average* number of occurrences $\overline{N}$ follows the Poisson probability distribution:

$$p(N_{actual}) = \frac{\overline{N}^{N_{actual}} e^{-\overline{N}}}{N_{actual}!} \qquad (N_{actual} = 0, 1, 2, \ldots)$$

   The mean and the standard deviation of this distribution are $\overline{N}$ and $\sqrt{\overline{N}}$, respectively. For a large number of occurrences, the Poisson distribution can be approximated by the Gaussian function with the same mean and standard deviation. These distributions can be used to model the actual number of occurrences in both time and space under uniform time/space conditions.
3. The concept of continuous current loses its meaning at low current levels when the average time between two events of single-carrier collections at the device contact (q/I) becomes longer than the shortest observation time (Δt). In that case, it is more meaningful to express the average or expected time between two carrier events than

the average current flow. Even if $\Delta t > q/I$, but not $\gg q/I$, there is significant uncertainty in the current value. The needed observation time (Δt) for a given current (I) and specified coefficient of current variation ($c_v = \sigma_I/I$) can be determined from the following relationship:

$$I\Delta t = \frac{q}{c_v^2}$$

4. The diffusion-current models may overestimate the actual current in nanoscale devices when the apparent gradient of the carrier concentration, $(n_1 - n_2)/W$, becomes large because of the small device dimensions, W. The diffusion current is due to a difference in thermal flow of carriers in two opposite directions:

$$I_{\rightarrow} - I_{\leftarrow} = A_J q v_{th}(n_1 - n_2)$$

Irrespective of how small the device dimensions become (how small W is), the diffusion current cannot exceed the carrier-supply level $I_{\rightarrow}$.

5. The concept of continuous concentrations of fixed particles—in particular, doping atoms and R–G centers—also loses its meaning in small devices that correspond to a noninteger number of particles. As distinct from the time effects in the case of current flow, the small-size effects in the case of fixed particles are observed as spatial uncertainties.
6. Electron and hole capture and emission events can be modeled as individual events, separate from any equilibrium assumptions. This provides proper insight into the processes of carrier generation and recombination in small devices, where the detailed-balance principle fails because it necessitates dealing with instantaneous balance of noninteger number of particles.
7. Electrons in crystals are held in potential wells. When one dimension of the potential well becomes comparable to the electron wavelength, the separation between the allowed energy levels becomes larger than the thermal energy of the electrons, and the electrons can no longer jump freely from one energy level to another. It is said that the electrons are confined as standing waves in this dimension, whereas the model of free electrons as particles that form an electron gas can still be applied in the other two dimensions. Accordingly, the one-dimensional quantum confinement results in a two-dimensional electron gas (2DEG). The concepts and equations of free electrons in the conduction band now apply to two-dimensional subbands whose bottoms are positioned at the energy levels E_n set by the allowed energy levels in the confined direction:

$$E = E_n + \underbrace{\frac{p_y^2}{2m_{n,y}} + \frac{p_z^2}{2m_{n,z}}}_{\text{2D subbands}} \qquad (n = 0, 1, 2, \ldots)$$

Analogous concepts apply to the case of holes.

8. Two practically important devices that exhibit pronounced one-dimensional quantum confinement are the MOSFET and the HEMT. The 2DEG in HEMTs is formed due to the conduction-band discontinuity at the AlGaAs–GaAs heterojunction and appears even when no voltage is applied. As a result, HEMTs are typically normally *on* FETs, used in applications that can utilize high mobility of the electrons in the 2DEG.

9. The classical MOSFET equations are valid even with pronounced one-dimensional confinement. This is because they are derived with the assumption of a two-dimensional electron sheet whose charge density per unit area is given by

$$Q_I = \begin{cases} 0 & \text{for } V_{GS} \leq V_T \\ -(V_{GS} - V_T)C_{ox} & \text{for } V_{GS} \geq V_T \end{cases}$$

The threshold voltage (V_T) is a parameter in this equation that can be adjusted to account for the quantum-mechanical effects. As long as the carrier-scattering length is smaller than the device-channel length (L_{eff}), Ohm's law can be applied and the device current can be modeled by the following equation:

$$I_D = \frac{\mu_0 W}{L_{eff}} \overline{Q_I} V_{DS}$$

The carrier mobility μ_0 is the second important device parameter that can be adjusted to account for numerous physical effects (including the wave properties of carriers). The average value of charge density ($\overline{Q_I}$) is used so that the effects of the drain voltage on the charge density can be incorporated. Different methods are used for this averaging, which is the reason for the existence of different MOSFET models.

10. Semiconductor nanowires and carbon nanotubes confine electrons and holes in two dimensions, resulting in one-dimensional electron/hole gases. Coaxially gated FETs based on either semiconductor nanowires or carbon nanotubes provide the best possible gate control of the channel carriers; this reduces the short-channel effects due to the drain voltage and enables the most aggressive reduction of the channel length. As long as the channel length does not drop below the carrier-scattering length and Ohm's law can be applied, the device current can be modeled by the following equation:

$$I_D = \frac{\mu_0 N}{L_{eff}} \overline{Q_I} V_{DS}$$

There is no channel width W in the case of 1DEG, which is replaced by the number of nanowires or nanotubes (N) connected in parallel. In this equation Q_I, the channel charge per unit length (in C/m), can also be calculated as $(V_{GS} - V_T)C_{ox}$, with a note that C_{ox} is the coaxial capacitance per unit length.

11. When the channel length becomes shorter than the length between two scattering events and the carriers no longer experience scattering in the channel, the transport is said to be ballistic. In spite of the absence of ohmic resistance due to carrier scattering, the channel conductance is not infinite. The conductance in this case is limited by the quantum limit on the number of carriers per unit length (this limit relates to the finite size of electrons) and the finite carrier velocity. This ultimate conductance limit is determined by the Planck constant and the number of populated subbands M (conducting channels):

$$G_Q = M\frac{2q^2}{h}$$

Using G_Q, which is also called the Landauer conductance, the current can be calculated as $I = G_Q V$, but only for small applied voltages V. For large voltages, the current saturates at the value set by the limited number of carriers per unit length and the finite carrier velocity. For conduction through a single channel (a single populated subband), the current limit is $I = (2q/h)E_F$.

PROBLEMS

10.1 Consider a silicon wafer with diameter D = 300 mm and thickness t_w = 1 mm that is N-type doped with $N_D = 10^{20}$ cm^{-3}. Calculate the number of holes in the entire wafer.

10.2 A test structure consists of 10^9 P–N junctions connected in parallel. The measured reverse-bias current of this test structure is I_R = 10.0 pA. Assuming that the devices are identical, estimate the time that elapses between the flow of two electrons in a single device.

10.3 Measured DC current through a device is I = 1 pA. If the electrons flowing through this device are collected during pulsed (repeated) time intervals of Δt = 1 μs, determine the average number of collected electrons. What is the probability of collecting $N = 10$ electrons?

10.4 The current through $N_{dev} = 10^6$ devices in parallel, measured over the period of $\Delta t = 1$ s, is $I = 1$ fA. Determine the percentages of devices that contribute with $N_{el} = 0$, $N_{el} = 1$, and $N_{el} = 2$ electrons to this current measurement, respectively. **A**

10.5 Two types of R–G center are uniformly distributed across reverse-biased P–N junctions. The average numbers of R–G centers appearing in the depletion layer of a P–N junction are $\overline{M_{t1}} = 2.1$ for the first type and $\overline{M_{t2}} = 0.1$ for the second type of R–G center. The average times between two generation events are $\tau_{t1} = 25$ ms and $\tau_{t2} = 1$ ms for the first and second type of R–G center, respectively.

(a) Determine probabilities $p(M_{t1}, M_{t2})$, where M_{t1} and M_{t2} are the actual numbers of R–G centers, as indicated in the following table.

	M_{t1}			
M_{t2}	0	1	2	3
0	$p(0,0)$	$p(1,0)$	$p(2,0)$	$p(3,0)$
1	$p(0,1)$	$p(1,1)$ **A**	$p(2,1)$	$p(3,1)$

(b) Calculate the generation currents that correspond to the probabilities calculated in part (a).

10.6 There are two R–G centers in the depletion layer of a reverse-biased P–N junction with the following electron and hole emission times: $\tau_{e,n-1} = \tau_{e,p-1} = 1$ ms for the first R–G center; $\tau_{e,n-2} = 20$ μs and $\tau_{e,p-2} = 48$ ms for the second center. Determine the average time for each of these R–G centers to generate electron–hole pairs and the total average time between two generation events in this depletion layer.

10.7 Consider a single R–G center in a neutral N-type silicon. If the average time for electron capture by this R–G center is $\tau_{c,n}$ = 1 ps and the average time for the electron emission from this R–G center is $\tau_{e,n}$ = 1 ms, determine the probability that the R–G center is empty of an electron (occupied by a hole). If the average time for hole emission is equal to the average time for electron emission, what is the average time this R–G center needs to generate an electron–hole pair?

10.8 The electron density in a 2DEG formed at the heterojunction between AlGaAs and undoped GaAs is $N_I = 10^{12}$ cm^2. Assuming that only one two-dimensional subband is occupied by electrons, determine the position of the Fermi level with respect to the bottom of the conduction band of GaAs at the heterojunction. Based on the determined Fermi-level position, calculate the electron densities in the first and the second subbands to verify the assumption that the population of higher subbands can be neglected.

10.9 Determine the electric potential at the heterojunction between AlGaAs and undoped GaAs with respect to the potential of the neutral undoped GaAs if the bottom of the lowest subband is at $E_0 = 70$ meV with respect to the bottom of the conduction band of GaAs and the Fermi level is $E_F \approx E_0$.

10.10 The following are the parameters of an AlGaAs/GaAs HEMT: the thickness of the AlGaAs film $t_{AlGaAs} = 100$ nm, the dielectric constant of both AlGaAs and GaAs $\varepsilon_s/\varepsilon_0 = 13$, the electron mobility in the 2DEG $\mu_0 = 10{,}000$ cm^2/V·s, the channel length $L_{eff} = 100$ nm, and the channel width $W = 100$ μm.

(a) If the density of electrons in the 2DEG is $N_I = 10^{12}$ cm^{-2} at zero gate voltage, determine the threshold voltage of this HEMT.

(b) Determine the HEMT current for $V_{GS} = 5$ V and $V_{DS} = 50$ mV. **A**

10.11 Determine the channel resistance of a HEMT if the electron density in the 2DEG is $N_I = 10^{12}$ cm^{-2}, the electron mobility is $\mu_0 = 10{,}000$ cm^2/V·s, the channel length is $L_{eff} = 100$ nm, and the channel width is $W = 100$ μm.

10.12 The density of electrons and the electron mobility in a graphene sheet are $N_I = 5 \times 10^{12}$ cm^{-2} and $\mu_0 = 20{,}000$ cm^2/Vs, respectively.

(a) Determine the sheet resistance of this graphene film.

(b) How many graphene sheets are needed to match the sheet resistance of a 100-nm thick copper sheet (the resistivity of copper is 17.2 nΩ·m)? What distance between the graphene sheets is required so that their effective thickness is also 100 nm? **A**

10.13 The following parameters are identical for a MOSFET with a 2DEG channel and a coaxially gated FET with a 1DEG channel: threshold voltage $V_T = 0.2$ V, gate-oxide thickness $t_{ox} = 3$ nm, gate dielectric permittivity $\varepsilon_{ox} = 3.9\varepsilon_0$, channel length $L_{eff} = 100$ nm, and electron mobility $\mu_0 = 100$ cm^2/V·s. The channel width of the 2DEG MOSFET is $W = 1$ μm and the nanowire diameter of the 1DEG nanowire FET is $d = 3$ nm. Identical gate voltage is applied to both devices: $V_{GS} = 1$ V.

(a) Determine the channel resistance of the 2DEG MOSFET. **A**

(b) Determine the channel resistance of the 1DEG nanowire FET.

(c) How many nanowire FETs are needed to match the resistance of the 2DEG MOSFET? Could the width of the array of coaxially gated nanowire FETs (refer to Fig. 10.8) match the channel width of the 2DEG MOSFET?

(d) The advantage of coaxially gated nanowire FETs is that they enable far more aggressive scaling of L_{eff} without the detrimental short-channel effects. Assuming that ohmic conduction is maintained when the channel length of the 1DEG nanowire FET is reduced to $L_{eff} = 10$ nm, determine the number of required nanowire FETs to match the channel resistance of the 2DEG MOSFET. **A**

10.14 Use the relationship between the scattering length and scattering time, $l_{sc} = v_{th}\tau_{sc}$, to determine the scattering length at room temperature in a graphene sheet with electron mobility of $\mu_0 = 100{,}000$ cm^2/V·s and effective electron mass $m^* = 0.06m_0$.

10.15 The resistance of a semiconductor nanowire is $R = 12.93$ kΩ for wire lengths up to 100 nm and increases by $dR/dL = 30$ Ω/nm for wire lengths exceeding 100 nm, measured at room temperature. If the effective electron mass is $m^* = 0.2m_0$, what is the number of electrons per unit length in this nanowire?

10.16 The current through a semiconductor nanowire increases linearly with the applied voltage V for $0 < V < 0.25$ V and saturates at $I = 19.3$ μA for $V > 0.25$ V.

(a) Determine the position of the Fermi level with respect to the bottom of the lowest one-dimensional subband.

(b) Assuming that the electron states below the Fermi level are occupied and the electron states above the Fermi level are empty, determine the number of electrons per unit length that move toward the positively biased contact.

(c) In the current-saturation region ($V > 0.25$ V), the average electron velocity is set at a saturation-velocity level, $\overline{v_x} = v_{sat}$. What is the value of v_{sat}? **A**

REVIEW QUESTIONS

R-10.1 If the minority-hole concentration is $p = 0.01$ cm^{-3}, can we assume that the number of holes is equal to zero because the volume of a semiconductor device is $\ll 1$ cm^3? If so, does that mean that the actual hole concentration can be rounded down to $p = 0$ cm^{-3}?

R-10.2 If no current is detected through a small device for 10 s, even with a measurement method that can detect a single electron, does that mean that no current will be detected for the next 1000 s under the same conditions?

R-10.3 If device is designed so that the average current during a switching cycle corresponds to 10 electrons, should we expect the actual current to follow the Poisson distribution? That is, will the current correspond to 10 electrons exactly in only 12.5% of the cycles?

R-10.4 Considering electrons as particles in a semiconductor device, can we assume that their size is zero irrespective of the device dimensions?

R-10.5 Consider a triangular potential well for electrons, noting that the width of the well reduces to zero at the bottom of the well (the triangle tip). Can any electron fit at the bottom of this well?

R-10.6 An electron appearing in a triangular potential well forms a standing wave at the energy level E_0 needed for the width of the potential well to become equal to the half-wavelength of the electron. Is E_0 related to the kinetic energy of this electron?

R-10.7 Do channel electrons in Si MOSFETs and AlGaAs–GaAs HEMTs appear in triangular potential wells? If so, does that mean that they appear as standing waves? If so, does that mean that they cannot move as free particles?

R-10.8 The triangular potential well is formed by a built-in electric field in HEMTs and contains electrons even at zero gate voltage (normally *on* FET). Can threshold voltage (V_T) be associated with this device? If so, can the threshold voltage be defined as the gate voltage (V_G) needed to repel the electrons from the triangular potential well? If so, can the electron density be calculated as $N_I = (V_G - V_T)C/q$? If so, what dielectric thickness should be used to calculate C?

R-10.9 Can the classical MOSFET equations be used to model HEMT? If yes, how do these equations account for the structural, technological, and physical differences?

R-10.10 Can the threshold-voltage equation be used in the case of pronounced one-dimensional quantum confinement?

R-10.11 Can we define channel length (L) and channel width (W) in the case of coaxially gated nanowire FETs?

R-10.12 What is the unit for charge density in a carbon-nanotube FET?

R-10.13 In the case of ballistic transport, the ohmic resistance due to carrier scattering becomes zero. Does that mean infinite conductance? If not, what does limit the conductance?

R-10.14 Is power dissipated in a carbon nanotube due to its quantum resistance R_Q? If not, where is the power VI dissipated?

11 Device Electronics: Equivalent Circuits and SPICE Parameters

Semiconductor diodes and transistors are complex nonlinear devices. An elegant way of presenting their overall electrical characteristics is by use of equivalent circuits. Basically, a nonlinear voltage-controlled current source represents the main function of a considered device, rendering the additional components in the equivalent circuit as *parasitic elements*. Importantly, it is the capacitors in the equivalent circuit that model the dynamic characteristics of the considered device. Although considered as *parasitic elements,* their role in high-frequency analyses and design is quite fundamental.

Large-signal equivalent circuits provide general models, and these are the equivalent circuits used in SPICE. They include nonlinear I–V and C–V dependencies to provide adequate models across the whole range of current and voltage values. *Small-signal* equivalent circuits and their parameters are extensively used for analyzing and designing electronic circuits that provide a *linear* response to a *small* input signal. Consequently, the small-signal equivalent circuits are linear circuits themselves, consisting of linear elements (resistors, capacitors, and inductors). They can be derived by adequate simplification and transformation from the more general large-signal equivalent circuits.

This chapter is devoted to parameters and equations used in SPICE for both the main functions and the complete equivalent circuits of diodes, MOSFETs, and BJTs. There is a special emphasis on parameter measurement or, more specifically, on the graphic method for determination of the initial parameter values that can be used for nonlinear fitting. In addition to general large-signal equivalent circuits, simplified small-signal equivalent circuits for diodes, MOSFETs, and BJTs and simple digital model for MOSFETs are also described. This enables us to link the device physics and the SPICE models and parameters to the simple and frequently used equivalent circuits in circuit analysis and design books.

11.1 DIODES

11.1.1 Static Model and Parameters in SPICE

The SPICE model for the static I–V diode characteristic is summarized in Table 11.1. The diagram in Table 11.1 shows that the diode is modeled as a nonlinear current source, controlled by the voltage across the current source itself (V_{D0}), connected in series to a resistor r_S. The resistor r_S represents the parasitic resistances, whereas the current source describes the I–V characteristic of the P–N junction. The $I_D(V_{D0})$ equation appears in three parts. The first part is for $V_{D0} > -BV$, and it expresses the normal I–V diode characteristic, whereas the second and the third parts are for $V_D \leq -BV$, expressing the breakdown characteristic of the diode.

TABLE 11.1 Summary of the SPICE Diode Model: Static *I–V* Characteristic

Symbol	Usual SPICE Keyword	Parameter Name	Typical Value/Range	Unit
Static Parameters				
I_S	IS	Saturation current		A
n	N	Emission coefficient	1–2	
r_S	RS	Parasitic resistance		Ω
BV	BV	Breakdown voltage (positive number)		V
	IBV	Breakdown current (positive number) Note: $\mathtt{IBV} = \mathtt{IS}\frac{\mathtt{BV}}{V_t}$		A
Temperature-Related Parameters				
E_g	EG	Energy gap	1.12 for Si	eV
p_t	XTI	Saturation-current temperature exponent	3	

Static Diode Model

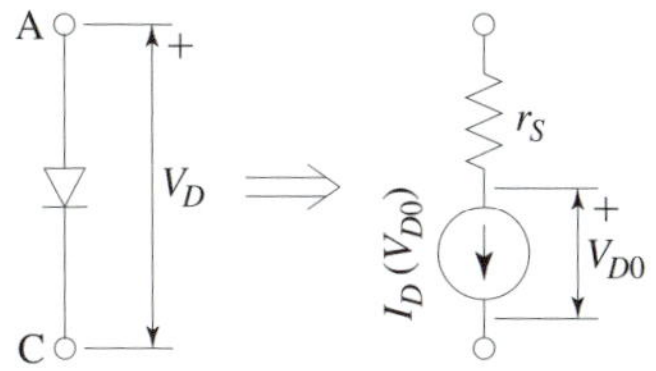

$$I_D(V_{D0}) = \begin{cases} I_S(T)\left(e^{V_{D0}/\mathtt{N}V_t} - 1\right) + V_{D0}G_{MIN} & \text{if } V_{D0} > \text{-BV} \\ -\mathtt{IBV} & \text{if } V_{D0} = \text{-BV} \\ -I_S(T)\left[e^{-(\mathtt{BV} + V_{D0})/V_t} - 1 + \frac{\mathtt{BV}}{V_t}\right] & \text{if } V_{D0} < \text{-BV} \end{cases}$$

$$I_S(T) = \mathtt{IS}\left(\frac{T}{T_{nom}}\right)^{\mathtt{XTI/N}} \exp\left[-\frac{q\mathtt{EG}}{kT}\left(1 - \frac{T}{T_{nom}}\right)\right]$$

The first part of the SPICE $I_D(V_{D0})$ equation is a variation of the previously derived Eq. (6.22). An important change is that the voltage across the P–N junction (V_{D0}) is used instead of the terminal diode voltage V_D (obviously, this is to include the effects of the parasitic resistance r_S). Therefore, the parameters in this equation are I_S, n, and r_S. The additional term $V_{D0}G_{MIN}$ is not important in terms of the diode characteristic description: it is added to enhance computational efficiency. G_{MIN} is set to a small value (typically 10^{-15} A/V) and normally does not show any observable influence on the diode characteristic. G_{MIN} is a program parameter (not a device parameter), and it can be altered by the user.

The second and the third parts of the SPICE $I_D(V_{D0})$ equation given in Table 11.1 model the breakdown characteristic of the diode. Figure 6.13 shows the diode characteristic in breakdown (V_D around $-BV$). Due to the junction breakdown, the current sharply rises when the reverse-bias voltage $-V_D$ is increased beyond the breakdown voltage $-BV$. This sharp rise in the current is modeled by the exponential dependence shown in the third part of the SPICE $I_D(V_{D0})$ equation of Table 11.1. The second part of this equation defines IBV as the current at $V_{D0} = -BV$. BV and IBV appear as device parameters. SPICE would accept independently set IBV parameter; however, this may lead to numerical problems if there is a large discontinuity in the current around the $-BV$ point. The third part of the SPICE $I_D(V_{D0})$ equation is used for any voltage that is smaller than $-BV$. For the points very close to $-BV$ (thus $V_{D0} \approx -BV$), the exponential term $\exp{-(BV + V_{D0})/V_t}$ is approximately 1, which means that the current is approximately $-I_S\frac{BV}{V_t}$. To avoid discontinuity in the current (and therefore possible numerical problems), the parameter IBV should be set at the value $I_S\frac{BV}{V_t}$.

To include the temperature effects, the saturation current I_S is multiplied in SPICE by a semiempirical factor, as shown by the $I_S(T)$ equation in Table 11.1. There is an additional parameter p_t that is called the saturation-current temperature exponent and whose value is typically $p_t = 3$ for silicon diodes. The operating temperature T is set in SPICE in the same way as the device parameters, with a limitation that a unique operating temperature has to be used for the whole circuit. The nominal temperature is set to $T_{nom} = 27°\text{C}$, although this value can also be changed by the user.

11.1.2 Large-Signal Equivalent Circuit in SPICE

The effects of the depletion-layer and stored-charge capacitances (described in Sections 6.3 and 6.4) can be incorporated by adding capacitors in parallel with the current source. Therefore, the large-signal equivalent circuit of a P–N junction consists of (1) the nonlinear voltage-controlled current source, modeling the DC I–V characteristic, (2) the series resistor, modeling the parasitic resistances, and (3) the parallel capacitor, modeling the capacitance-related effects. This equivalent circuit is shown in Table 11.2.

The reverse-bias voltage $V_R > 0$ in Eq. (6.57) can be converted into V_{D0}. In the reverse-bias region, $V_{D0} \approx V_D = -V_R$ (the difference $V_D - V_{D0} = r_S I_D$ is insignificant due to the very small I_D current). Thus, the only difference between Eq. (6.57) and the SPICE equation shown in Table 11.2 is in the use of $-V_{D0}$ instead of V_R. In addition, due to the insignificant I_D current for $V_{D0} < 0.5V_{bi}$, the concepts applied to derive the equations for the depletion-layer width and capacitance are still valid. Therefore, the validity of $C_d(V_R)$ equation is expanded by the transformation $V_R = -V_{D0}$, where V_{D0}

TABLE 11.2 Summary of the SPICE Diode Model: Dynamic Characteristics

Dynamic Parameters

Symbol	Usual SPICE Keyword	Parameter Name	Typical Value/Range	Unit
$C_d(0)$	CJO	Zero-bias junction capacitance		F
V_{bi}	VJ	Built-in (junction) voltage	0.65–1.25	V
m	M	Grading coefficient	$\frac{1}{3}$–$\frac{1}{2}$	
τ_T	TT	Transit time		s

Large-Signal Diode Model

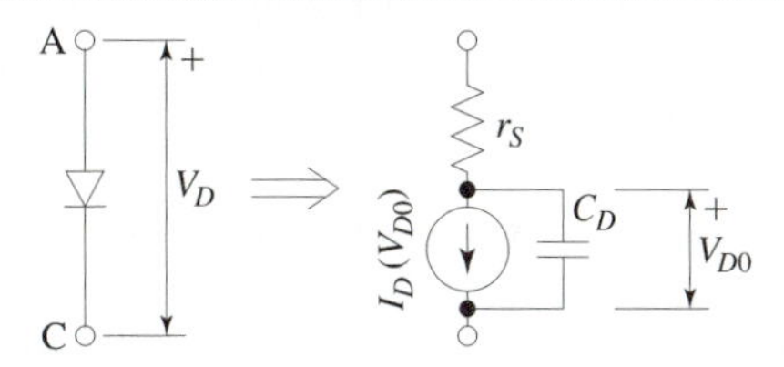

$I_D(V_{D0})$ is given in Table 11.1

$$C_D = C_d + C_s$$

$$C_d = C_d(0,T)\left[1 - (V_{D0}/V_{bi}(T))\right]^{-\mathtt{M}} \qquad (\text{for} V_{D0} < 0.5V_{bi})$$

$$C_d(0,T) = \mathtt{CJO}\left\{1 + \mathtt{M}\left[400\times10^{-6}(T - T_{nom}) - \frac{V_{bi}(T)-\mathtt{VJ}}{\mathtt{VJ}}\right]\right\}$$

$$V_{bi}(T) \approx \frac{T}{T_{nom}}\mathtt{VJ} - 2\frac{kT}{q}\ln\left(\frac{T}{T_{nom}}\right)^{1.5}$$

$$C_s = \mathtt{TT}\ \ (dI_D/dV_{D0})$$

is allowed to take values up to $0.5V_{bi}$. SPICE has a different equation to calculate C_d at voltages $V_{D0} > 0.5V_{bi}$. However, the depletion-layer capacitance in that region is not very important as the total diode capacitance is dominated by the stored-charge capacitance.

The depletion-layer capacitance depends on temperature, however, this temperature dependence is less significant than the dependence of the I–V characteristic. The temperature dependence of C_d is mainly due to the temperature dependence of the built-in voltage V_{bi} [refer to Eq. (6.2)], although $C_d(0)$ also depends slightly on the temperature. Table 11.2 shows the SPICE equations used to calculate the values of V_{bi} and $C_d(0)$ at temperature T, which is different from the nominal temperature T_{nom}.

The stored-charge capacitance is related to the stored charge, as defined by Eq. (6.59). This is the equation used in SPICE, as shown in Table 11.2. Both the stored-charge capacitance C_s and the depletion-layer capacitance C_d appear across the P–N junction. Therefore, the total diode capacitance C_D is expressed as a parallel connection of C_s and C_d:

$$C_D = C_d + C_s \tag{11.1}$$

11.1.3 Parameter Measurement

Very sophisticated fitting algorithms and software exist that can be used to fit the nonlinear SPICE equations to experimental data. These algorithms, however, require a set of initial values for the fitted parameters to be specified. In some cases, the nonlinear fitting depends very critically on the initial parameter values. Simple graphic methods can be used to determine the values of SPICE parameters. The parameter values obtained in this way can be used as initial values for nonlinear fitting. Importantly, the graphic methods provide a visual demonstration of how good fit can be achieved between selected SPICE equations and specific experimental data.

Measurement of I_S, n, and r_S

The three device parameters involved in the current–voltage equation for $V_{D0} > -BV$ are I_S, n, and r_S. Proper values of these SPICE parameters need to be set to ensure correct simulation. The default values of these parameters (typically, $I_S = 10^{-14}$ A, $n = 1$, $r_S = 0$) cannot guarantee an acceptable agreement between the model and the real characteristic of any possible type of diode. Figures 6.11 and 6.12 demonstrate the importance of properly setting the values of n and r_S, respectively.

Figure 11.1 illustrates a graphic method for determination of I_S, n, and r_S parameters from experimental I_D–V_D data. Given that the current depends exponentially on the voltage, a $\ln I_D$–V_{D0} graph is used to linearize the problem. The open symbols show the raw experimental data—obtained when the measured V_D voltage served as the P–N junction voltage V_{D0}. Because the voltage across r_S (which is $r_S I_D$) is neglected in this case, the voltage V_{D0} is effectively overestimated by $r_S I_D$ (refer to Fig. 6.12). This effect is not pronounced at small currents as $r_S I_D \ll V_{D0}$ (the linear part of the graph); however, it becomes observable at high currents. A good initial guess for r_S can be obtained by judging the maximum deviation $r_S I_D$ of the raw experimental data from the straight line

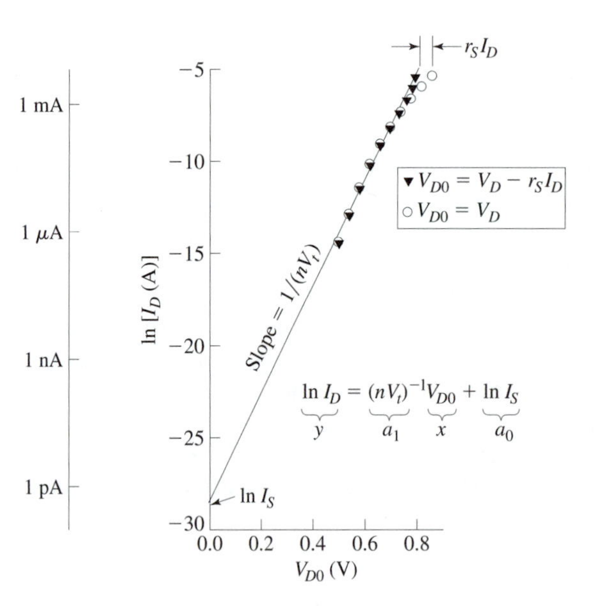

Figure 11.1 Measurement of diode static SPICE parameters, I_S, n, and r_S. The experimental data (symbols) and the fitting (line) are also shown in Fig 6.12.

extrapolated from the low-current linear portion of the I_D–V_D dependence. The maximum deviation has to be, obviously, divided by the maximum current I_D to obtain r_S. Using the estimated value of r_S, the experimental diode voltage points V_D are transformed into P–N junction voltage points as $V_{D0} = V_D - r_S I_D$. If a straight line is obtained, the value of r_S is taken as the final value. Alternatively, r_S is altered and the process repeated until a straight line is obtained.

The closed symbols in Fig. 11.1 show the straight line obtained after the extraction of voltage effect of the parasitic resistance, $r_S I_D$, from the raw experimental data. That is, the closed symbols represent the experimental characteristic of the current source in the SPICE diode model (Table 11.1). Because the experimental data are collected in the forward-bias region, where $\exp(V_{D0}/nV_t) \gg 1$, the SPICE $I_D(V_{D0})$ equation is reduced to

$$I_D = I_S e^{V_{D0}/nV_t} \tag{11.2}$$

Therefore, the logarithm of the current $\ln I_D$ linearly depends on V_{D0}:

$$\ln I_D = \frac{1}{nV_t} V_{D0} + \ln I_S \tag{11.3}$$

Figure 11.1 illustrates that the parameters I_S and n are obtained from the coefficients a_0 and a_1 of the linear $\ln I_D$–V_{D0} dependence as

$$\begin{aligned} I_S &= e^{a_0} \\ n &= 1/a_1 V_t \end{aligned} \tag{11.4}$$

EXAMPLE 11.1 Measurement of Static SPICE Parameters

A set of measured I_D–V_D values for a P–N junction diode are given in Table 11.3. Obtain SPICE parameters I_S, n, and r_S for this diode.

SOLUTION

Let us assume that the parasitic resistance r_S is on the order of 10 Ω. In that case, the voltage across the parasitic resistance is $\leq 1 \times 10^{-3} \times 10 = 0.01$ V for currents ≤ 1 mA. This means that the parasitic resistance effect can be neglected (0.01 V is much smaller than ≈ 0.7 V appearing across the P–N junction) for currents ≤ 1 mA. The measured diode current I_D can then be directly related to the measured voltage V_D as $I_D = I_S \exp(V_D/nV_t)$. This exponential equation can be

TABLE 11.3 Current–Voltage Measurements

V_D (V)	0.67	0.70	0.73	0.76	0.80	0.84	0.91	1.00	1.26	1.65
I_D (mA)	0.1	0.2	0.5	1.0	2.0	5.0	10.0	20.0	50.0	100.0

TABLE 11.4 Linearization of Current–Voltage Data

I_D (mA)	$y = \ln V_D$ ln(mA)	$x = V_D$ (V)
0.1	−2.303	0.67
0.2	−1.609	0.70
0.5	−0.693	0.73
1.0	0.000	0.76

linearized in the following way:

$$\ln I_D = \ln I_S + \frac{1}{nV_t} V_D$$

that is,

$$y = a_0 + a_1 x$$

where $y = \ln I_D$, $x = V_D$, $a_0 = \ln I_S$, and $a_1 = 1/nV_t$. The results of this linearization, applied to the first four experimental points ($I_D \leq 1$ mA) from Table 11.3, are given in Table 11.4.

The graphic method, explained in the previous section, can be used to find the coefficients a_0 and a_1 of this linear relationship. Alternatively, these coefficients can be calculated using the numerical linear regression method. For the case of a one-variable linear equation (and two parameters, a_0 and a_1), the following system of two linear equations has to be solved:

$$\begin{aligned} na_0 \quad &+ \left(\textstyle\sum_{i=1}^{n} x_i\right) a_1 = \textstyle\sum_{i=1}^{n} y_i \\ \left(\textstyle\sum_{i=1}^{n} x_i\right) a_0 &+ \left(\textstyle\sum_{i=1}^{n} x_i^2\right) a_1 = \textstyle\sum_{i=1}^{n} x_i y_i \end{aligned} \tag{11.5}$$

where n is the number of experimental points used for the linear fitting. Applying the system of equations (11.5) to the data of Table 11.4, one obtains

$$4a_0 + 2.86a_1 = -4.605$$
$$2.86a_0 + 2.0494a_1 = -3.1752$$

The solution of the foregoing system of equations is $a_0 = -19.80$, and $a_1 = 26.08$. The parameters I_S and n can now be calculated as $I_S = \exp(a_0) = 2.52 \times 10^{-9}$ mA $= 2.52 \times 10^{-12}$ A, $n = 1/a_1 V_t = 1.48$.

If the parasitic resistance r_S was zero, the voltage V_D at the highest current $I_D = 100$ mA would be $V_D = 1.48 \times 0.02585 \times \ln(100/2.52 \times 10^{-9}) = 0.93$ V. It can be seen from Table 11.3 that the measured voltage is 1.65 V. The difference $1.65 - 0.93 = 0.72$ V is due to the voltage across r_S: $r_S I_D = 0.72$ V. Using this difference, the parasitic resistance is estimated as $r_S = 0.72\ \text{V}/I_D = 0.72/100\ \text{mA} = 7.2\ \Omega$. If $r_S = 7.2\ \Omega$ is a proper value, the voltage across the P–N

TABLE 11.5 Transformed Current–Voltage Data

I_D (mA)	$V_{D0} = V_D - r_s I_D$ (V)	$V_{D0} = nV_t \ln(I_D/I_S)$ (V)
	$I_S = 2.5\times10^{-12}$A, $n = 1.48$, $r_S = 7.2\ \Omega$	
0.1	0.67	0.67
0.2	0.70	0.70
0.5	0.73	0.73
1.0	0.76	0.76
2.0	0.79	0.78
5.0	0.80	0.82
10.0	0.84	0.84
20.0	0.86	0.87
50.0	0.90	0.91
100.0	0.93	0.93

junction V_{D0}, calculated as $V_D - r_S I_D$, should closely match the values calculated from the diode equation $nV_t \ln(I_D/I_S)$. The results of these calculations are presented in Table 11.5. It can be seen that the theoretical values (the third column) closely match the transformed experimental values (the second column). Therefore, we conclude that $I_S = 2.5 \times 10^{-12}$ A, $n = 1.48$, and $r_S = 7.2\ \Omega$ represent a good set of SPICE parameters for the considered diode. If the matching was not good, the value of r_S would be altered to try to improve the matching.

Measurement of $C_d(0)$, V_{bi}, and m

It is not possible to completely linearize the model for the reverse-biased P–N junction capacitance, given by Eq. (6.57). Therefore, the graphic or the linear regression methods cannot be directly applied. The situation is further complicated by the fact that the measured data contain an additional, parasitic capacitance component. The P–N junction capacitance can be measured in different ways, perhaps most suitably by means of a bridge. The measurement frequency can be set low enough that the parasitic series resistance becomes negligible compared to the impedance of the capacitor. However, the parasitic capacitance, caused mainly by pin capacitance, stray capacitance, and pad capacitance, cannot be avoided. Assuming that the parasitic capacitance C_p does not depend on the voltage applied, the measured capacitance can be expressed as

$$C_{meas} = C_d(0)\left(1 + \frac{V_R}{V_{bi}}\right)^{-m} + C_p \qquad (11.6)$$

Although the parameter C_p in Eq. (11.6) is not needed as a SPICE parameter, it has to be extracted from the experimental data.

Curve fitting can be the most effective way of parameter measurement in this case, provided the initial parameter values are properly determined. There are four parameters in Eq. (11.6): $C_d(0)$, V_{bi}, m, and C_p. The built-in voltage V_{bi} depends on the doping levels in the P- and N-type regions, as given by Eq. (6.2). It is useful to estimate the likely extreme values of this parameter. The lowest value is obtained when the lowest doping levels are assumed; let this be $N_A N_D = 10^{15} \times 10^{16}$ cm^{-3} $\times$ cm^{-3}. With this, $V_{bi}(min) = 0.02585 \ln 10^{31}/(1.02 \times 10^{10})^2 = 0.65$ V. The highest value is obtained when the highest doping levels are assumed, say $N_A N_D = 10^{20} \times 10^{21}$ cm^{-3} $\times$ cm^{-3}. With this, $V_{bi}(max) = 1.25$ V. If the value of $V_{bi} = 0.9$ V is assumed, the maximum error cannot be much bigger than 0.3 V, which is $0.3/0.9 \times 100 = 33\%$.

Assuming a constant value for the parameter V_{bi} enables linearization of the equation for the reverse-biased P–N junction capacitance in the following way:

$$\log(C_{meas} - C_p) = \log C_d(0) - m \log\left(1 + \frac{V_R}{V_{bi}}\right) \tag{11.7}$$

where $C_{meas} - C_p$ represents the reverse-biased P–N junction capacitance (the depletion-layer capacitance C_d). If the parasitic capacitance C_p were zero and V_{bi} were correctly assumed, plotting the logarithm of the measured capacitances versus $\log(1 + V_R/V_{bi})$ would give a straight line. The parameters of the linear relationship $y = a_0 + a_1 x$ would be related to the parameters m and V_{bi}: $\log C_d(0) = a_0$ and $-m = a_1$. This would enable the measurement of m and $C_d(0)$ by the graphical or linear regression method. Although the value of C_p cannot be neglected, $C_p = 0$ can still be used as the initial value to enable the measurement of initial values of m and $C_d(0)$.

To illustrate this technique, consider the example of experimental data given in Table 11.6. Assuming $V_{bi} = 0.9$ V and $C_p = 0$, $\log(1 + V_R/V_{bi})$ and $\log(C_{meas} - C_p)$ can be calculated as shown in Table 11.6 as well. The $\log(C_{meas} - C_p)$ versus $\log(1 + V_R/V_{bi})$ graph (for $V_{bi} = 0.9$ V and $C_p = 0$) is shown in Fig. 11.2a by the squared symbols. The dashed line represents the best linear fit for these data. The parameters of the dashed line are found to be $a_1 = -0.25$, and $a_0 = 0.639$. This means that $m = -a_1 = 0.25$,

TABLE 11.6 Example of Experimental Data Used to Obtain Measured Values of the Capacitance-Model Parameters

V_R (V)	C_{meas} (pF)	$\log(1 + \frac{V_R}{0.9})$	$\log C_{meas}$	$\log(C_{meas} - 1.26$ pF)
0.0	4.45	0.0000	0.6484	0.5038
1.0	3.59	0.3245	0.5551	0.3686
2.0	3.21	0.5082	0.5065	0.2981
3.0	2.98	0.6368	0.4742	0.2379
4.0	2.82	0.7360	0.4502	0.1963
5.0	2.70	0.8166	0.4314	0.1625
6.0	2.61	0.8846	0.4166	0.1340
7.0	2.54	0.9434	0.4048	0.1094
8.0	2.47	0.9951	0.3927	0.0876
9.0	2.42	1.0414	0.3838	0.0682
10.0	2.37	1.0832	0.3747	0.0507

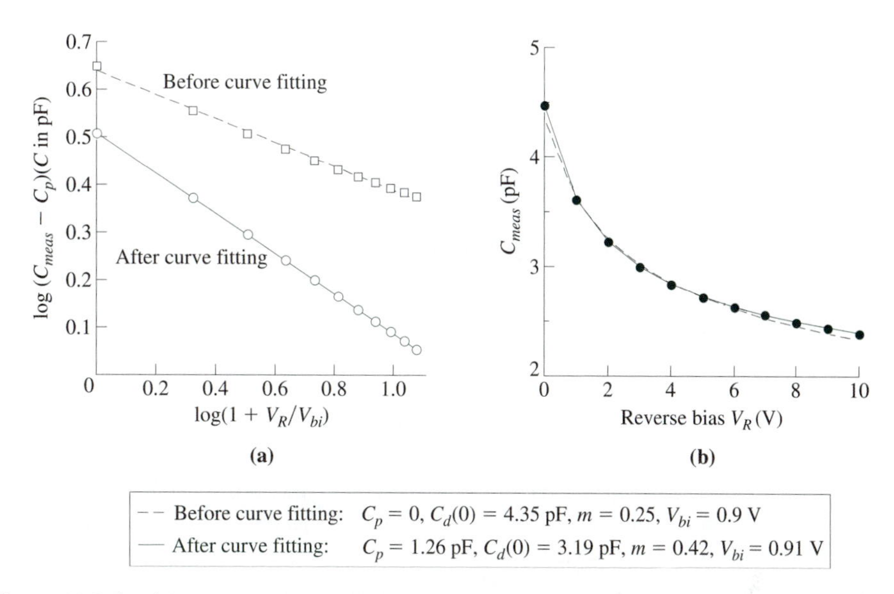

Figure 11.2 (a, b) Measurement of the parameters of the capacitance model: (1) $V_{bi} = 0.9$ V and $C_p = 0$ is assumed, and the linear regression is applied to estimate m and $C_d(0)$—the square symbols and the dashed line in (a); (2) curve fitting is performed using the estimated values as the initial parameters, which improves the fit (the solid lines) but more importantly provides parameters that are physically justified.

and $C_d(0) = 10^{0.639} = 4.35$ pF. This completes the set of possible values of the four parameters. Using these values, the corresponding theoretical curve (dashed line) for C_{meas} versus V_R is compared to the experimental data (symbols) in Fig. 11.2b.

The fit between the dashed line and the symbols in Fig. 11.2b appears as quite reasonable. However, the value of the parameter m is questionable. It has been shown that m is the smallest for the extreme case of the linear junction, in which case it is $m = 0.33$. The value of $m = 0.25$ is not physically justified. The good fit is achieved due to the compensating effect of the assumption $C_p = 0$. We should not accept this situation. The value of C_p can be as high as 2 pF, which means that the measured capacitance can be more than twice as high as the real junction capacitance.

This procedure can be repeated assuming, for example, $C_p = 2$ pF. If a better fit is achieved, it would indicate that the real C_p is closer to 2 pF than 0 pF. Making yet another guess and performing an additional iteration may improve the fit further. This is in essence the iterative process used for the curve fitting. As mentioned, there are a number of software packages that automatically perform the iterative process used in the curve fitting. The initially extracted values of the four parameters can be attempted as the initial parameters for the curve-fitting procedure. Using $C_p = 0$, $C_d(0) = 4.35$ pF, $m = 0.25$, and $V_{bi} = 0.9$ V as the initial values, the following final values are obtained by the curve fitting available in *SigmaPlot* scientific graphing software: $C_p = 1.26$ pF, $C_d(0) = 3.19$ pF, $m = 0.42$, $V_{bi} = 0.91$ V. The solid lines in Fig. 11.2 represent the model with these parameters. It can be seen that the fitting is improved (the fact that the iterative process used for the curve fitting converged indicated that the best fit was achieved). More importantly, the parameter $m = 0.42$ is now within the extreme limits of 0.33 and 0.5, and the value $C_p = 1.26$ pF appears as practically reasonable. When $\log(C_{meas} - C_p)$ is plotted versus $\log(1 + V_R/V_{bi})$ (the solid line in Fig. 11.2a), a straight line is obtained, which confirms that the measurement procedure is successfully completed.

EXAMPLE 11.2 Linear Regression Analysis

Use the data given in columns 3 and 5 of Table 11.6 to apply linear regression analysis to obtain the parameters m and $C_d(0)$ in Eq. (11.6).

SOLUTION

To be able to apply linear regression analysis, the linear form of Eq. (11.6) has to be used, as given by Eq. (11.7). If the numbers from the third and fifth columns of Table 11.6 are used as the experimental x_i and y_i data, respectively, the system of two linear equations (11.5) can be applied directly. Upon transforming the parameters as $\log C_d(0) = a_0$ and $-m = a_1$, the system of two equations can be written as

$$11a_0 + 7.9698a_1 = 2.2108$$

$$7.9698a_0 + 6.8982a_1 = 1.1314$$

In solving this system, it is found that $a_0 = 0.504$ and $a_1 = -0.42$. Therefore, $C_d(0) = 10^{a_0} = 10^{0.504} = 3.19$ pF, and $m = -a_1 = 0.42$.

11.1.4 Small-Signal Equivalent Circuit

The term *small signal* is applied to indicate that a device is used in a linear region; the signals have to be small enough to permit approximately linear relationships to exist between all the *signal* voltages and currents in the considered device/circuit. Accordingly, a small-signal equivalent circuit of a device has to consist of linear elements. A frequency-dependent linear ratio between a signal voltage and a signal current is called *small-signal impedance;* a frequency-dependent ratio between signal current and signal voltage is called *small-signal admittance*. In the specific case of no pronounced frequency dependence, these ratios are called *small-signal resistance* and *small-signal conductance,* respectively. If possible, it is convenient to express a small-signal impedance (or admittance) as a circuit consisting of small-signal resistances, capacitances and, if necessary, inductances.

Small-Signal Resistance

Figure 11.3 illustrates how linear behavior can be obtained from an essentially nonlinear characteristic. In this example, the nonlinear characteristic is the exponential i_D–v_D dependence of the diode. Referring to Fig. 11.3, let us assume that a signal voltage v_d is applied across the diode and that the signal voltage is a sinusoid with a small amplitude. If the voltage amplitude is small enough, the corresponding signal current oscillates within a nearly linear segment of the i_D–v_D characteristic. As a result, the signal current is

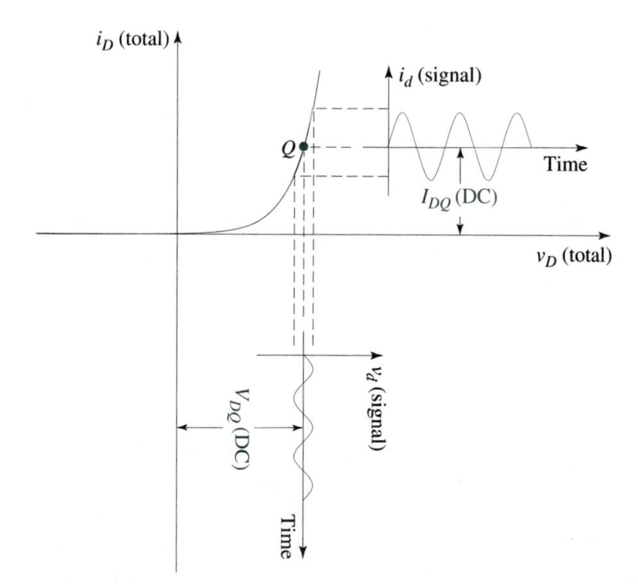

Figure 11.3 Small enough signals are confined in a linear-like segment of a nonlinear characteristic; DC offset is used to set the quiescent point (Q).

also a sinusoid. Assuming that the voltage and the current sinusoids are in phase, the ratio between the signal current (i_d) and the signal voltage (v_d) defines the small-signal conductance:[1]

$$g_d = \frac{i_d}{v_d} \tag{11.8}$$

Given that this ratio depends on the slope of the linear i_D–v_D segment, the small-signal conductance is equal to the slope of the i_D–v_D characteristic at the quiescent/offset voltage V_{DQ}:

$$g_d = \left.\frac{di_D}{dv_D}\right|_{@V_{DQ}} \tag{11.9}$$

The current–voltage characteristic of the diode is given by Eq. (6.22). In the forward-bias region, this equation can be expressed in the form

$$i_D = I_S \exp(v_D/nV_t) \tag{11.10}$$

to indicate that it applies to the total variable current i_D and voltage v_D. Therefore,

$$g_d = \left.\frac{d\left(I_S e^{v_D/nV_t}\right)}{dv_D}\right|_{@V_{DQ}} = \frac{1}{nV_t}\left[I_S e^{v_D/nV_t}\right]_{@V_{DQ}} = \frac{1}{nV_t}\underbrace{I_S e^{V_{DQ}/nV_t}}_{I_{DQ}} = \frac{I_{DQ}}{nV_t} \tag{11.11}$$

[1]Small-signal quantities are usually labeled with lowercase letters.

Given that resistance is equal to the reciprocal value of conductance, the small-signal resistance is

$$r_d = \frac{1}{g_d} = n\frac{V_t}{I_{DQ}} \tag{11.12}$$

Therefore, the small-signal equivalent circuit of a forward-biased diode can be as simple as a resistor with the resistance r_d. Of course, this equivalent circuit can be used when the assumption is satisfied that the signal current and the signal voltage are in phase. At higher frequencies, the capacitances (and perhaps even inductances) associated with the diode structure cause a phase shift between the signal current and voltage, so the small-signal capacitances/inductances have to be added to the small-signal equivalent circuit.

Equation (11.12) shows that the small-signal resistance depends on the DC current, I_{DQ}. The current I_{DQ} is the quiescent/offset current that corresponds to the quiescent/offset voltage V_{DQ}, and they together define the quiescent point (Q) on the current–voltage characteristic. In absence of any signal, the applied voltage across the diode is equal to the DC level V_{DQ} and the corresponding current is equal to the DC level I_{DQ}. When a signal voltage is superimposed on the DC voltage, the total voltage $v_D = V_{DQ} + v_d$ and the total current $i_D = I_{DQ} + i_d$ starts oscillating around the quiescent point (V_{DQ}, I_{DQ}).

The fact that r_d depends on the quiescent point Q [Eq. (11.12)] can easily be visualized by keeping in mind that r_d is equal to the reciprocal value of the slope of the i_D–v_D characteristic at Q. From Fig. 11.3 we can see that the slope of the characteristic is smaller for smaller currents and voltages, so a shift of Q toward the knee of the characteristic would increase the small-signal resistance. Therefore, r_d can be electrically controlled by the value of the applied DC offset (V_{DQ}).

Small-Signal Capacitances

The two capacitances associated with a P–N junction are described in Section 6.3 (the depletion-layer capacitance) and Section 6.4 (the stored-charge capacitance).

The depletion-layer capacitance dominates for reverse-biased P–N junctions, so the large-signal equivalent circuit becomes simply a voltage-dependent capacitor. The C–V dependence was illustrated in Figs. 6.16 and 11.2 and is modeled by Eq. (6.57). For a small enough signal, the capacitance variation with signal voltage can be neglected, so that the reverse-biased P–N junction acts as a constant-value capacitor with capacitance equal to

$$c_d = C_d(0)\left(1 - \frac{V_{DQ}}{V_{bi}}\right)^{-m} \tag{11.13}$$

Analogously to the case of small-signal resistance, the value of this capacitor can be set by the DC offset voltage $V_R = -V_{DQ}$ (the Q point). This enables us to use reverse-biased P–N junctions as *varactors* (Example 6.6).

The stored-charge capacitance becomes pronounced in forward-biased junctions when the flowing current causes significant levels of stored charge. Given that this capacitance is inseparable from a significant current flow through a forward-biased diode, it appears in parallel with the small-signal resistance in any small-signal equivalent circuit for a forward-biased diode. The stored-charge capacitance is also voltage-dependent, so for small-signal analyses, the value of the capacitance at the quiescent point is used. Based on Eqs. (6.59), (6.65), and (11.9),

$$c_s = \tau_T \left. \frac{di_D}{dv_D} \right|_{@V_{DQ}} = \tau_T g_d = \begin{cases} \frac{\tau_n}{2} g_d & \text{for N}^+\text{–P junction} \\ \frac{\tau_p}{2} g_d & \text{for P}^+\text{–N junction} \end{cases} \quad (11.14)$$

where τ_n and τ_p are the minority-carrier lifetimes.

11.2 MOSFET

The current–voltage characteristics of a MOSFET can be represented by a voltage-controlled current source, $I_D(V_{GS}, V_{DS}, V_{SB})$, where the I_D dependence on V_{GS}, V_{DS}, and V_{SB} is according to the DC model considered in Section 8.2. This voltage-controlled current source is added to the parasitic elements in the structure of a MOSFET to obtain an equivalent circuit that models the electrical response to both large and small signals at both low and high frequencies.

11.2.1 Static Model and Parameters: LEVEL 3 in SPICE

Regarding the voltage-controlled current source, there is a need both to summarize the MOSFET equations used in SPICE and to present the hierarchy of MOSFET parameters. This section provides a series of tables designed to enable an intuitive reference to MOSFET parameters. Multiple tables are used because it is not very useful to present the MOSFET SPICE parameters in a single list. Not only would the list be rather long, but the relationships and (in)compatibilities between the parameters could not be clearly expressed. The following hierarchy is used to classify the MOSFET parameters presented in this section:

- Static LEVEL 3 Model (Table 11.7)
  - Principal effects (Table 11.8)
  - Channel-related second-order effects (Table 11.9)
  - Depletion-layer-related second-order effects (Table 11.10, parts I and II)
- Dynamic model (the complete large-signal equivalent circuit, including the parasitic elements) (Table 11.11, parts I and II)

The first variables that need to be specified when using a MOSFET are the gate length and width (referred to as geometrical variables in Table 11.7). Different MOSFETs can have different gate lengths and widths even in the integrated circuits, where the MOSFETs are made by the same technological process and all the other MOSFET parameters are identical. In SPICE, the gate length and width are considered as device attributes, and they are typically stated for every individual MOSFET. However, the gate length and width can also be specified as MOSFET parameters, together with all the other device parameters.

Generally, there is a difference between the *gate length* L_g and the *channel length* L_{eff}. The MOSFET diagram given in Table 11.7 illustrates this difference. Although a self-

TABLE 11.7 Summary of SPICE LEVEL 3 Static MOSFET Model

Geometric Variables

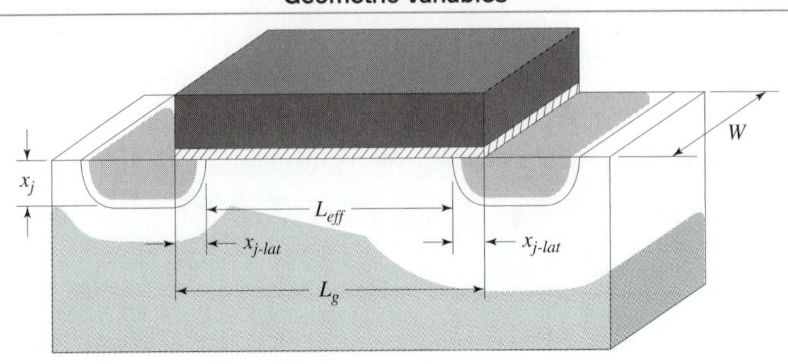

$L_{eff} = L_g - 2x_{j-lat}$ (x_{j-lat} is a parameter; refer to Table 11.10)

Symbol	SPICE Keyword	Variable Name	Default Value	Unit
L_g	L	Gate length	100×10^{-6}	m
W	W	Channel width	100×10^{-6}	m

Note: L and W can also be specified as parameters.

Static LEVEL 3 Model

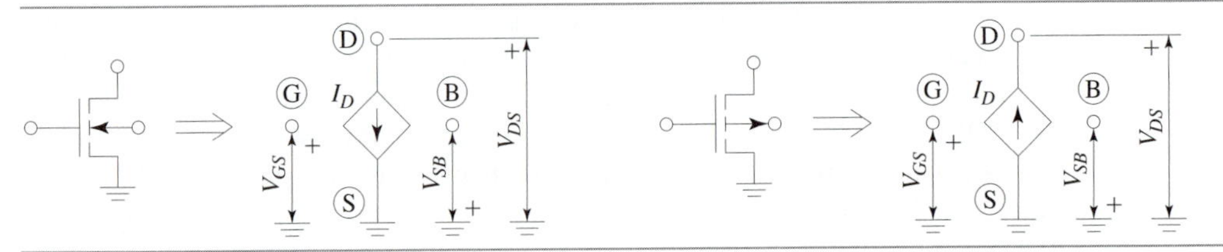

NMOS ($V_{Ts} = V_T + n_s kT/q$)		**PMOS ($V_{Ts} = V_T - n_s kT/q$)**	
sub-V_T:	$V_{GS} \le V_{Ts}$	sub-V_T:	$V_{GS} \ge V_{Ts}$
triode:	$V_{GS} > V_{Ts}$, and $0 < V_{DS} < V_{DSsat}$	triode:	$V_{GS} < V_{Ts}$, and $0 > V_{DS} > V_{DSsat}$
satur.:	$V_{GS} > V_{Ts}$, and $V_{DS} \ge V_{DSsat} > 0$	satur.:	$V_{GS} < V_{Ts}$, and $V_{DS} \le V_{DSsat} < 0$

$$I_D = \begin{cases} f(V_{GS}) = \begin{cases} \beta\left[(V_{GS} - V_T)V_{DS} - (1 + F_B)\frac{V_{DS}^2}{2}\right] & \text{triode region} \\ \frac{\beta}{2(1+F_B)}(V_{GS} - V_T)^2 & \text{satur. region} \end{cases} \\ f(V_{GS} = V_{Ts}) \times e^{-qV_{subth}/n_s kT}, \text{ sub-}V_T \text{ region} \end{cases}$$

NMOS:

$$V_{subth} = V_{Ts} - V_{GS} \ge 0$$

$$F_B = \frac{\gamma F_s}{2\sqrt{|2\phi_F| + V_{SB}}} + F_n^a$$

PMOS:

$$V_{subth} = V_{GS} - V_{Ts} \ge 0$$

$$F_B = \frac{\gamma F_s}{2\sqrt{|2\phi_F| - V_{SB}}} + F_n^a$$

	Principal Effects	Second-Order Effects: Channel-Related	Second-Order Effects: Depletion-Layer-Related	Second-Order Effects: All
β, V_{DSsat}	Table 11.8	Table 11.9	Table 11.8	Table 11.9
V_T, $\lvert 2\phi_F \rvert$, γ, F_s, F_n, n_s	Table 11.8	Table 11.8	Table 11.10	Table 11.10

[a] By error, Berkeley SPICE, PSPICE, and HSPICE use factor 4 instead of 2 in front of the square root. (*Source:* D. Foty, *MOSFET Modeling with SPICE: Principles and Practice,* Prentice-Hall, Upper Saddle River, NJ, 1997, p. 173).

TABLE 11.8 Summary of SPICE LEVEL 3 Static Parameters: Principal Effects

Principal Static Parameters

Symbol	SPICE Keyword	Parameter Name	Typical Value NMOS	Typical Value PMOS	Unit
KP (or	KP	Transconductance parameter[a]	1.2×10^{-4}		A/V^2
μ_0 and	Uo	Low-field mobility[b]	700		cm^2/V · s
t_{ox})	Tox	Gate-oxide thickness[b]	20×10^{-9}		m
V_{T0}	Vto	Zero-bias threshold voltage	1	−1	V
$\lvert 2\phi_F \rvert$	Phi	Surface potential in strong inversion	0.70		V
γ	Gamma	Body-effect parameter	>0.3		V$^{1/2}$

β, V_T, V_{DSsat}, F_s, F_n, and n_s Equations

NMOS	PMOS

$$\beta = \begin{cases} \mathrm{KP}\dfrac{\mathrm{W}}{L_{eff}} & \text{if KP is specified} \\ \mu_0 \dfrac{\varepsilon_{ox}}{t_{ox}} \dfrac{\mathrm{W}}{L_{eff}} & \text{if KP is not specified} \end{cases}$$

$$L_{pinch} = 0 \ (\mathrm{KAPPA} = 0)$$

$$V_T = \mathrm{Vto} + \gamma(\sqrt{\mathrm{Phi} + V_{SB}} - \sqrt{\mathrm{Phi}}), \quad V_T = \mathrm{Vto} - \gamma(\sqrt{\mathrm{Phi} + V_{BS}} - \sqrt{\mathrm{Phi}})$$

$$V_{DSsat} = \frac{V_{GS} - V_T}{1 + F_B}$$

$$F_s = 1 \ (\mathrm{Xj} = 0)$$

$$F_n = 0 \ (\mathrm{DELTA} = 0)$$

$$n_s = 1 + \frac{\gamma F_s (\mathrm{Phi} + |V_{SB}|)^{-1/2}}{2C_{ox}} \ (\mathrm{NFS} = 0)$$

Constant: $\varepsilon_{ox} = 3.9 \times 8.85 \times 10^{-12}$ F/m

[a,b]Incompatible parameters.

aligned technique is typically used to define the channel as the source and drain regions are created, the lateral-diffusion effect leads to the difference between the masking gate and the effective channel. The SPICE input variable is the gate length L_g, whereas the current–voltage equations use the effective channel length L_{eff}. The relationship between L_g and L_{eff} is shown in Table 11.7. If the lateral-diffusion parameter x_{j-lat} is not specified (set to zero), the gate and channel lengths become equal.

To model the static characteristics, the MOSFET is considered as a voltage-controlled current source. The current is I_D, whereas the controlling voltages are the voltage across the current source V_{DS} and the two separate voltages V_{GS} and V_{SB}. Zero current is assumed between the G and S as well as between the B and S terminals. Table 11.7 shows a compact form of the LEVEL 3 $I_D(V_{GS}, V_{DS}, V_{SB})$ equation that appears in three parts corresponding to the three different modes of operation: subthreshold, triode (including the linear mode), and saturation. It also shows the equations used to calculate the factor F_B. The equations for β, V_T, V_{DSsat}, $|2\phi_F|$, γ, F_s, F_n, and n_s that are obviously needed to

TABLE 11.9 Summary of SPICE LEVEL 3 Static Parameters: Channel-Related Second-Order Effects

Channel-Related Static Parameters

Symbol	SPICE Keyword	Parameter Name	Typical Value	Unit
KP (or	KP	Transconductance parameter[a]	1.2×10^{-4}	A/V^2
μ_0 and	Uo	Low-field mobility[b]	700	$cm^2/V \cdot s$
t_{ox})	Tox	Gate-oxide thickness[b]	20×10^{-9}	m
θ	THETA	Mobility modulation constant	0.1	—
v_{max}	Vmax	Maximum drift velocity	10^5	m/s
κ	KAPPA	Channel-length modulation coefficient (needs Nsub)	0.2	—
N_A, N_D	Nsub	Substrate doping concentration	10^{15}	cm^{-3}

β and V_{DSsat} Equations

NMOS **PMOS**

$$\Rightarrow \quad \beta = \mu_{eff} \frac{\varepsilon_{ox}}{\mathtt{Tox}} \frac{\mathtt{W}}{L_{eff} - L_{pinch}}$$

$$\mu_{eff} = \frac{\mu_s}{1 + \mu_s |V_{DS}| / (\mathtt{Vmax} L_{eff})}$$

$$\mu_s = \frac{\mu_0}{1 + \mathtt{THETA} |V_{GS} - V_T|}$$

$$\mu_0 = \mathtt{KP} \frac{\mathtt{Tox}}{\varepsilon_{ox}} \quad \text{if KP is specified;} \quad \text{else } \mu_0 = \mathtt{Uo}$$

$$L_{pinch} = \begin{cases} L_a = \sqrt{\mathtt{KAPPA} \frac{2\varepsilon_s}{q\mathtt{Nsub}} |V_{DS} - V_{DSsat}|} & \text{if Vmax is not specified}^c \\ \left[\left(\frac{\varepsilon_s}{q\mathtt{Nsub}} \frac{V_{DSsat}}{L_{eff}} \right)^2 + L_a^2 \right]^{1/2} - \frac{\varepsilon_s}{q\mathtt{Nsub}} \frac{V_{DSsat}}{L_{eff}} & \text{if Vmax is specified} \end{cases}$$

$$\Rightarrow \quad V_{DSsat} = \begin{cases} \frac{V_{GS} - V_T}{1 + F_B} & \text{if Vmax is not specified}^c \\ V_{DSsat-corr} & \text{if Vmax is specified} \end{cases}$$

$$V_{DSsat\text{-}corr} = V_a + V_b - \sqrt{V_a^2 + V_b^2},^d \qquad V_{DSsat\text{-}corr} = V_a - V_b + \sqrt{V_a^2 + V_b^2}^{\,d}$$

$$V_a = \frac{V_{GS} - V_T}{1 + F_B}, \qquad V_b = \frac{\mathtt{Vmax} L_{eff}}{\mu_s}^{\,d}$$

Constant: $\varepsilon_{ox} = 3.9 \times 8.85 \times 10^{-12}$ F/m

[a,b]Incompatible parameters.
[c]D. Foty, *MOSFET Modeling with SPICE: Principles and Practice,* Prentice-Hall, Upper Saddle River, NJ, 1997, p. 599.
[d]G. Massobrio and P. Antognetti, *Semiconductor Device Modeling with SPICE,* 2nd ed., McGraw-Hill, New York, 1993, p. 208.

calculate $I_D(V_{GS}, V_{DS}, V_{SB})$ are given in Tables 11.8, 11.9, and 11.10. Different equations (given in different tables) are used at different levels of complexity.

Table 11.8 represents the simplest choice, covering the principal effects only. All the parameters shown in Table 11.8 are considered as essential. Although any version of SPICE is expected to have sensible default values of these parameters, no simulation should be trusted unless the values of these parameters are checked and properly specified. Note that the MOSFET gain factor β can be influenced in two ways: one is to specify the transconductance parameter KP, and the other is to specify the low-field mobility μ_0 and the gate-oxide thickness t_{ox}. Because these two options are mutually exclusive, SPICE will ignore μ_0 and/or t_{ox} when calculating β if KP is specified.

The parameters and equations are common for the enhancement-type (normally *off*) and depletion-type (normally *on*) MOSFETs. The typical values of the zero-bias threshold voltage, shown in Table 11.8, are for the enhancement-type MOSFETs. If a negative V_{T0} is specified for an N-channel MOSFET, it automatically becomes a depletion-type MOSFET (refer to Fig. 8.2). Analogously, a positive V_{T0} indicates a depletion-type P-channel MOSFET.

As described in Section 8.3, some second-order effects significantly influence MOSFET characteristics. In particular, the mobility reduction with the gate voltage (Fig. 8.14) is so important that it can rarely be neglected. Table 11.9 summarizes the three channel-related second-order effects; they all influence the gain factor β. Additionally, V_{DSsat} is modified for the case of pronounced velocity-saturation effect.

The depletion-layer-related static parameters (Table 11.10) involve the finite output resistance (η parameter), the gate-oxide charge influence on V_T (N_{oc} parameter), the short-channel (x_j, x_{j-lat}, and V_{bi} parameters), and the narrow-channel (δ parameter)

TABLE 11.10 Summary of SPICE LEVEL 3 Static Parameters: Depletion-Layer-Related Second-Order Effects

PART I
Depletion-Layer-Related Static Parameters[a]

Symbol	SPICE Keyword	Parameter Name	Typical Value	Unit
t_{ox}	Tox	Gate-oxide thickness	20×10^{-9}	m
η	ETA	Static feedback	0.7	—
		Note: This parameter can be used with V_{T0}, $2\|\phi_F\|$, and γ; t_{ox} should also be specified.		
N_A, N_D	Nsub	Substrate doping concentration	10^{15}	cm^{-3}
		Note: This parameter has to be specified to include the parameters below.		
N_{oc}	Nss	Oxide-charge density	10^{10}	cm^{-2}
	TPG	Gate material type		—
		Same as drain/source: TPG = 1		
		Opposite of D/S: TPG = −1		
		Metal: TPG = 0		
x_j	Xj	P–N junction depth	0.5×10^{-6}	m
x_{j-lat}	Ld	Lateral diffusion	$0.8 \times x_j$	m
V_{bi}	PB	P–N junction built-in voltage	0.8	V
δ	DELTA	Width effect on threshold voltage	1.0	—
	NFS	Subthreshold current-fitting parameter	10^{11}	cm^{-2}

[a]Incompatible parameters: V_{T0}, $|2\phi_F|$, and γ.

TABLE 11.10 (Continued)

PART II
V_T, $2|\phi_F|$, γ, F_s, F_n, and n_s Equations

	NMOS	PMOS
	$C_{ox} = \varepsilon_{ox}/\mathtt{Tox}$	
⇒	$V_T = V_{T0} + \gamma F_s\left(\sqrt{\lvert 2\phi_F\rvert + V_{SB}} - \sqrt{\lvert 2\phi_F\rvert}\right) - \sigma_D V_{DS} + F_n(V_{SB} + 2\phi_F)$	$V_T = V_{T0} - \gamma F_s\left(\sqrt{\lvert 2\phi_F\rvert + V_{BS}} - \sqrt{\lvert 2\phi_F\rvert}\right) - \sigma_D V_{DS} - F_n(V_{BS} + \lvert 2\phi_F\rvert)$
	$V_{T0} = \phi_{ms} - \frac{q\mathtt{Nss}}{C_{ox}} + \lvert 2\phi_F\rvert + \gamma F_s\sqrt{\lvert 2\phi_F\rvert}$	$V_{T0} = \phi_{ms} - \frac{q\mathtt{Nss}}{C_{ox}} - \lvert 2\phi_F\rvert - \gamma F_s\sqrt{\lvert 2\phi_F\rvert}$
	$\phi_{ms} = \begin{cases} -\frac{E_g}{2q} - \lvert\phi_F\rvert & \text{if TPG} = 1 \\ \frac{E_g}{2q} - \lvert\phi_F\rvert & \text{if TPG} = -1 \\ \phi_* - \lvert\phi_F\rvert & \text{if TPG} = 0 \end{cases}$	$\phi_{ms} = \begin{cases} \frac{E_g}{2q} + \lvert\phi_F\rvert & \text{if TPG} = 1 \\ \frac{-E_g}{2q} + \lvert\phi_F\rvert & \text{if TPG} = -1 \\ \phi_* + \lvert\phi_F\rvert & \text{if TPG} = 0 \end{cases}$
	$\sigma_D = 8.15 \times 10^{-22}\mathtt{ETA}/(C_{ox}L_{eff}^3)^a$	
⇒	$\gamma = \frac{1}{C_{ox}}\sqrt{2\varepsilon_s q\mathtt{Nsub}}$	
⇒	$\lvert 2\phi_F\rvert = 2\frac{kT}{q}\ln\frac{\mathtt{Nsub}}{n_i}$	
⇒	$F_s = 1 - \frac{\mathtt{Xj}}{L_{eff}}\left(\frac{\mathtt{Ld}+w_c}{\mathtt{Xj}}\sqrt{1 - \frac{w_p}{\mathtt{Xj}+w_p}} - \frac{\mathtt{Ld}}{\mathtt{Xj}}\right)^a$	
	$w_p = \sqrt{\frac{2\varepsilon_s}{q\mathtt{Nsub}}(V_{bi} + V_{SB})}$	$w_p = \sqrt{\frac{2\varepsilon_s}{q\mathtt{Nsub}}(V_{bi} + V_{BS})}$
	$w_c = 0.0631353\mathtt{Xj} + 0.8013929 w_p - 0.0111077 w_p^2/\mathtt{Xj}\ ^a$	
⇒	$F_n = \mathtt{DELTA}\,\varepsilon_s\pi/(4C_{ox}\mathtt{W})^{(1)}$	
⇒	$n_s = 1 + \frac{q\mathtt{NFS}}{C_{ox}} + \frac{\gamma F_s(\lvert 2\phi_F\rvert + \lvert V_{SB}\rvert)^{-1/2} - F_n}{2C_{ox}}{}^b$	

Constants:

$\varepsilon_{ox} = 3.45 \times 10^{-11}$ F/m, $k = 8.62 \times 10^{-5}$ eV/K, $n_i = 1.4 \times 10^{10}$ cm^{-3}

$q = 1.6 \times 10^{-19}$ C, $\phi_* = \phi_m - 4.61$ V, $\varepsilon_s = 1.044 \times 10^{-10}$ F/m

[a] G. Massobrio and P. Antognetti, *Semiconductor Device Modeling with SPICE,* 2nd ed., McGraw-Hill, New York, 1993, pp. 205–206.

[b] D. Foty, *MOSFET Modeling with SPICE: Principles and Practice,* Prentice-Hall, Upper Saddle River, NJ, 1997, p. 597.

effects. Because the gate-oxide charge and the short-channel and narrow-channel effects modify the zero-bias threshold voltage V_{T0}, the parameters associated with these effects are incompatible with V_{T0} as well as with $|2\phi_F|$ and γ. To properly include these effects, $N_{A,D}$ must be specified, while V_{T0}, $|2\phi_F|$ and γ must not be specified (they are calculated by SPICE). When $N_{A,D}$ is specified, a proper "type of gate" (**TPG** parameter) should be used to ensure that the zero-bias threshold voltage V_{T0} is properly calculated by SPICE. The static feedback parameter η (modeling the effect of finite output resistance) can be used with either group of parameters, provided t_{ox} is specified.

11.2.2 Parameter Measurement

This section describes graphic methods for measurement of the most important MOSFET parameters. As mentioned before, the graphic method is a valuable tool for establishing the initial values of the parameters needed in any nonlinear fitting algorithm.

Measurement of V_{T0} and KP

The parameters V_{T0} and *KP* can be obtained from the linear part of transfer characteristic I_D–V_{GS}. The MOSFET is in the linear mode for small V_{DS} voltages, when the quadratic term $(1 + F_B)V_{DS}^2/2$ is negligible compared to $(V_{GS} - V_T)V_{DS}$, and for small V_{GS}–V_T voltages, when $\mu_{eff} \approx \mu_0$ because the mobility modulation factor $\theta(V_{GS} - V_T) \ll 1$. To simplify the following considerations, define a low-field gain factor β_0

$$\beta_0 = \mu_0 C_{ox} \frac{W}{L_{eff}} \tag{11.15}$$

that is a constant, as opposed to the generally voltage-dependent gain factor β used in Tables 11.7 and 11.9:

$$\beta = \mu_{eff} C_{ox} \frac{W}{L_{eff}} \tag{11.16}$$

Therefore, the MOSFET model in the linear region can be written as

$$I_D \approx \beta_0(V_{GS} - V_T)V_{DS} \tag{11.17}$$

Note that the range of V_{GS} voltages in which I_D–V_{GS} characteristic is approximately linear can be expressed as $V_{DS}(1 + F_B) \ll V_{GS} - V_T \ll 1/\theta$.

Figure 11.4 provides an example of a MOSFET transfer characteristic measured at $V_{DS} = 50$ mV. The linear part appears approximately between $V_{GS} = 0.8$ V and $V_{GS} = 2.0$ V. This part of the transfer characteristic is modeled by Eq. (11.17) and can be used to obtain $V_{T0} = V_T(V_{SB} = 0)$ and *KP*. The zero-bias threshold voltage V_{T0} is obtained at the intersection between the V_{GS} axis and the straight line extrapolating the linear part of the transfer characteristic. It can also be obtained analytically, applying Eq. (11.17) to two different measurement points (I_{D1}, V_{GS1}) and (I_{D2}, V_{GS2}) to obtain a system of two linear equations. Eliminating β_0 from these two equations, the zero-bias threshold voltage

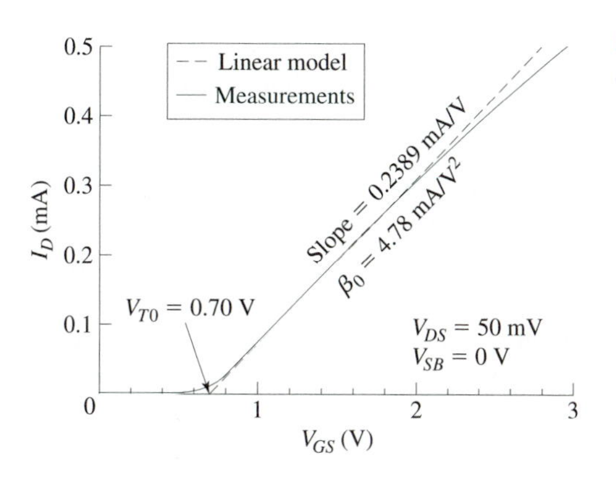

Figure 11.4 MOSFET transfer characteristics in the linear region.

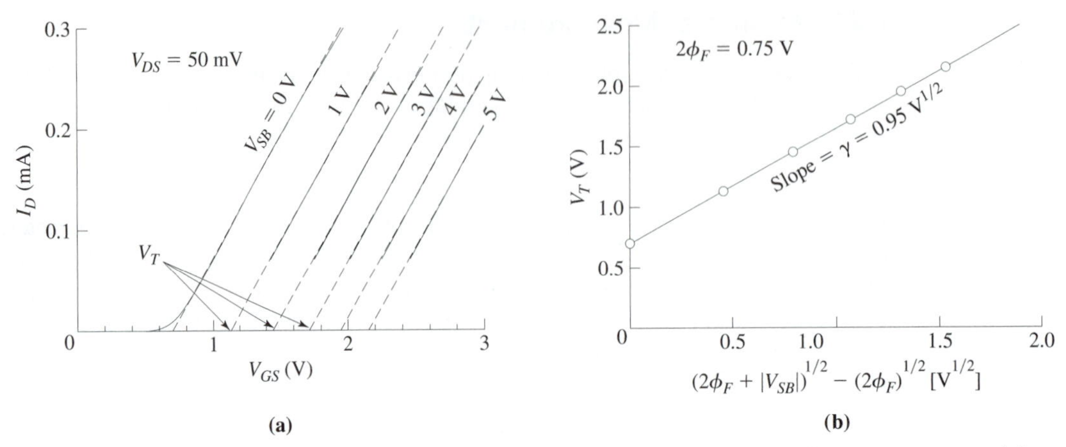

Figure 11.5 (a) Graph used to obtain $V_T(V_{SB})$ data. (b) Graphical extraction of the body-effect parameter γ and the surface-inversion potential $2\phi_F$.

is obtained as

$$V_{T0} = \frac{V_{GS1} - (I_{DS1}/I_{DS2})V_{GS2}}{1 - I_{DS1}/I_{DS2}} \tag{11.18}$$

The slope of the linear part of the transfer characteristic is, according to Eq. (11.17), $\beta_0 V_{DS}$. Therefore, factor β_0 is obtained when the slope is divided by the voltage V_{DS}. The transconductance parameter KP can then be calculated as $KP = \beta_0 L_{eff}/W$.

Measurement of γ and $2\phi_F$

Measurement of the parameters γ and $2\phi_F$ is based on the dependence of the threshold voltage V_T on the source-to-substrate voltage V_{SB} (the body effect). To collect experimental V_T-versus-V_{SB} data, the threshold voltage is measured by the previous procedure with a difference that the MOSFET is biased by different V_{SB} voltages. Figure 11.5a gives an example of transfer characteristics measured with different V_{SB} voltages that are used to obtain the corresponding V_T voltages.

The equation modeling $V_T(V_{SB})$ dependence is shown in Table 11.8. It can be seen that for a properly chosen $2\phi_F$, V_T versus $(\sqrt{2\phi_F + V_{SB}} - \sqrt{2\phi_F})$ exhibits linear dependence having the slope γ. Therefore, making an initial guess for $2\phi_F$, the V_T-versus-$(\sqrt{2\phi_F + V_{SB}} - \sqrt{2\phi_F})$ plot can be used to verify the validity of the assumed $2\phi_F$ value. If the plotted line is not straight, a second guess for $2\phi_F$ is made and the plot is redone. Note that a concave curve indicates that $2\phi_F$ should be increased, whereas a convex curve indicates that $2\phi_F$ should be decreased. This process is continued until an appropriate straight line is obtained, as illustrated in Fig. 11.5b. The slope of this line is γ.

When the second-order effects from Table 11.10 are to be employed, the substrate doping $N_{A,D}$ and the gate-oxide-thickness t_{ox} have to be specified instead of V_{T0}, $2\phi_F$, and γ. If the gate-oxide thickness is not known, it can be obtained from gate-oxide-capacitance (C_{ox}) measurements. The gate-oxide-capacitance can be measured using a large-area MOSFET biased in accumulation. With the gate-oxide-capacitance value and the already obtained body factor γ, the doping level is calculated using the equation given in Table 11.10.

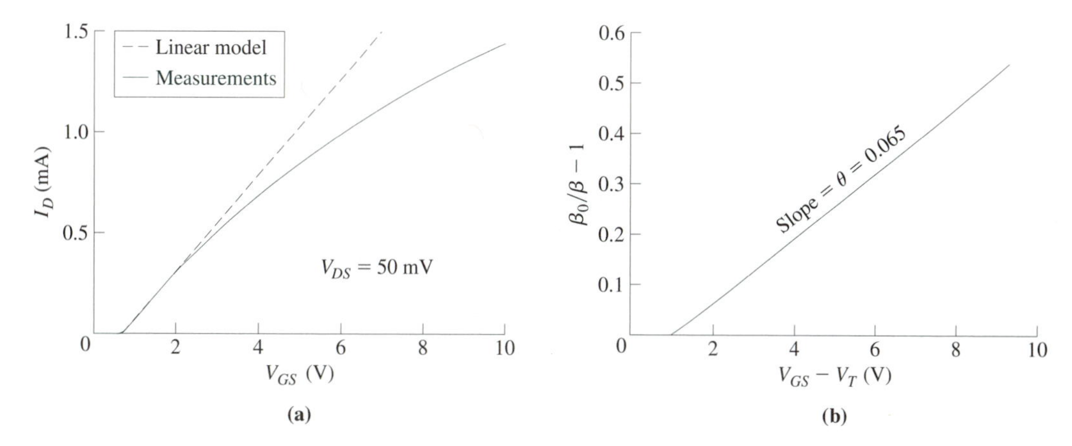

Figure 11.6 MOSFET transfer characteristics at (a) higher V_{GS} voltages and (b) extraction of θ parameter.

Measurement of θ

Figure 11.6a illustrates that the linear model [Eq. (11.17)] overestimates the actual current at higher V_{GS} voltages. The deviation of the measured I_D current from the linear dependence is due to the mobility-reduction effect and is taken into account by the mobility-modulation coefficient θ in the SPICE LEVEL 3 model. Therefore, to properly describe the drain current I_D at larger V_{GS} voltages, but still small V_{DS} voltages $V_{DS}(1 + F_B) \ll V_{GS} - V_T$, an equation more general than Eq. (11.17) is used:

$$I_D \approx \beta(V_{GS} - V_T)V_{DS} \tag{11.19}$$

The gain factor β in Eq. (11.19) is related to β_0 as

$$\beta = \frac{\beta_0}{1 + \theta(V_{GS} - V_T)} \tag{11.20}$$

By using Eq. (11.19), a set of β values can be calculated from the experimental I_D points measured at different V_{GS} voltages. Of course V_{DS} is known, and the threshold voltage V_{T0} needs to be obtained first. Having the set of β values for different V_{GS} values and having measured β_0 and V_{T0} as previously described, the θ parameter can be obtained from Eq. (11.20). To enable application of the graphical method, Eq. (11.20) is transformed to

$$\frac{\beta_0}{\beta} - 1 = \theta(V_{GS} - V_{T0}) \tag{11.21}$$

It is obvious that plotting $\beta_0/\beta - 1$ versus V_{GS}–V_T should produce a straight line with the slope equal to the parameter θ, as illustrated in Fig. 11.6b.

Measurement of Effective Length and Parasitic Resistances

Lateral diffusion leads to a difference between the *gate length* L_g and the *effective channel length* L_{eff}. This is illustrated in Table 11.7. Although L_g is the SPICE input parameter, the gate length is not necessarily equal to the *nominal gate length,* specified at the design

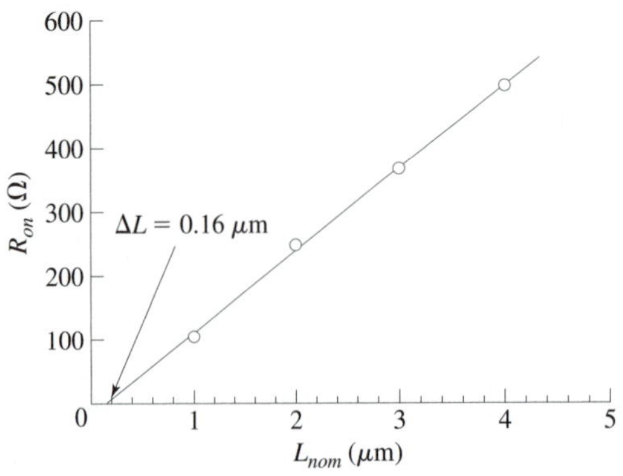

Figure 11.7 Measurement of the difference between the nominal and effective channel lengths.

level. This is due to imperfections of the manufacturing process (over- or underexposed photoresist, over- or underetched polysilicon, etc.).

The difference between the nominal and the effective channel lengths,

$$\Delta L = L_{nom} - L_{eff} \tag{11.22}$$

can electrically be measured if special test structures consisting of MOSFETs with equal widths and scaled lengths are available. Because the MOSFET current in the linear region is given by Eq. (11.17), the *on* resistance of the MOSFET can be defined and expressed as follows:

$$R_{on} \equiv \frac{V_{DS}}{I_D} = \frac{1}{\beta_0(V_{GS} - V_T)} \tag{11.23}$$

Replacing β_0 by $KP(W/L_{eff}) = KPW/(L_{nom} - \Delta L)$ shows that the *on* resistance is linearly dependent on L_{nom}:

$$R_{on} = \frac{L_{nom} - \Delta L}{KPW(V_{GS} - V_T)} \tag{11.24}$$

Figure 11.7 provides an example of measured *on* resistances for MOSFETs with four different channel lengths and all the other parameters identical. Zero *on* resistance corresponds to $L_{eff} = 0$ or, equivalently, to $L_{nom} = \Delta L$ [refer to Eq. (11.24)]. The intersection between the extrapolated linear dependence and the L_{nom}-axis (this is $R_{on} = 0$) directly shows ΔL. Once ΔL has been determined, the effective channel length is obtained as $L_{eff} = L_{nom} - \Delta L$ and the gate length can be specified as $L_g = L_{eff} + 2x_{j-lat}$, where x_{j-lat} is also specified as an input parameter.

The method illustrated in Fig. 11.7 is based on the assumption that the *on* resistance is equal to the channel resistance, which implicitly assumes that the parasitic series resistance is zero. When the parasitic series resistance is not negligible, a measured V_{DS}/I_D is not equal to the channel resistance but to the sum of the channel resistance and the parasitic series resistance: $V_{DS}/I_D = R_{ch} + R_{par}$. Obviously, using V_{DS}/I_D data as R_{ch} when R_{par} is not negligible leads to errors; in this case, the following, extended version of Eq. (11.24)

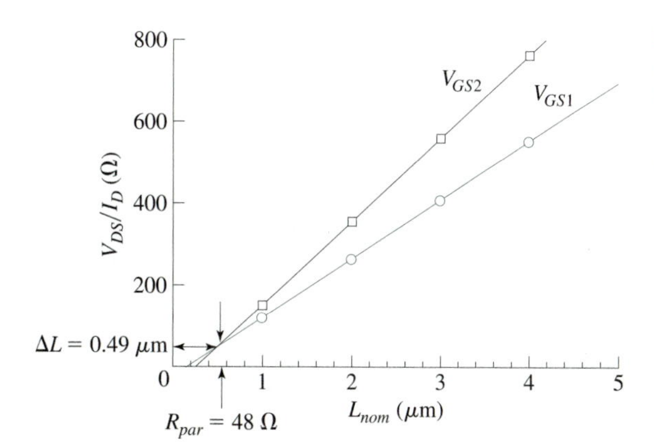

Figure 11.8 Simultaneous measurement of the difference between the nominal and effective channel lengths and the parasitic resistance.

should be used:

$$V_{DS}/I_D = \underbrace{\frac{L_{nom} - \Delta L}{KPW(V_{GS} - V_T)}}_{R_{ch}} + R_{par} \tag{11.25}$$

The channel resistance [the first term of Eq. (11.25)] depends on the gate voltage, whereas the parasitic resistance does not. This helps to distinguish between the contributions of the channel resistance and the parasitic resistance to V_{DS}/I_D values. It is necessary to measure the set of V_{DS}/I_D-versus-L_{nom} data for one or more additional gate-to-source voltages V_{GS}. Figure 11.8 shows an additional set of data points (labeled as V_{GS2}) added to the plot previously shown in Fig. 11.7. According to Eq. (11.25), different R_{ch} resistances at different V_{GS} voltages correspond to different V_{DS}/I_D values. However, there is one point where the influence of V_{GS} on R_{ch} does not exist and the V_{DS}/I_D value is the same for any V_{GS} voltage. This point is $L_{nom} = \Delta L$ because it turns R_{ch} into zero for any V_{GS} voltage. Therefore, the straight lines of the linear V_{DS}/I_D versus L_{nom} dependencies, measured at different V_{GS} voltages, intersect at a single point that is defined by $L_{nom} = \Delta L$ and $V_{DS}/I_D = R_{par}$.

When ΔL has been determined, the gate length L_g can accurately be specified. In addition, the determined value of the parasitic series resistance R_{par} shows the combined effect of the source and drain parasitic resistances R_S and R_D. Assuming symmetrical MOSFET, these two parameters can be specified as $R_S = R_D = R_{par}/2$.

EXAMPLE 11.3 Effective Channel Length

Two adjacent MOSFETs have the following design dimensions: $L_{nom1} = 1$ μm, $L_{nom2} = 2$ μm, and $W_1 = W_2$. The drain currents, measured at $V_{GS} - V_T = 2.5$ V and $V_{DS} = 50$ mV, are $I_{D1} = 495$ μA and $I_{D2} = 180$ μA. Neglecting the parasitic series resistance, determine the effective channel lengths.

SOLUTION

The measurement conditions ($V_{GS} - V_T = 2.5$ V and $V_{DS} = 50$ mV) indicate that the currents I_{D1} and I_{D2} are measured in the linear region. Corresponding *on* resistances are

$R_{on1} = V_{DS}/I_{D1} = 101.0\Omega$ and $R_{on2} = 277.8\ \Omega$. Because the *on* resistance is given by Eq. (11.24),

$$R_{on} = \underbrace{\frac{1}{KP\,W(V_{GS} - V_T)}}_{a}(L_{nom} - \Delta L)$$

the following system of two equations and two unknowns (a and ΔL) can be written:

$$R_{on1} = a(L_{nom1} - \Delta L)$$

$$R_{on2} = a(L_{nom2} - \Delta L)$$

To find ΔL, these two equations are divided:

$$R_{on1}/R_{on2} = (L_{nom1} - \Delta L)/(L_{nom2} - \Delta L)$$

and ΔL expressed as

$$\Delta L = \left(L_{nom2}\frac{R_{on1}}{R_{on2}} - L_{nom1}\right) \Big/ \left(\frac{R_{on1}}{R_{on2}} - 1\right) = 0.43\ \mu\text{m}$$

Therefore, the effective channel lengths are $L_{eff1} = L_{nom1} - \Delta L = 1 - 0.4 = 0.6\ \mu$m, and $L_{eff2} = 2 - 0.4 = 1.6\ \mu$m.

EXAMPLE 11.4 Static Feedback on the Threshold-Voltage (η)

To extract the static feedback on the threshold voltage η (SPICE parameter), the dependence of the threshold voltage on the drain-to-source voltage has been measured and the following data obtained: $V_{DS} = 2$ V, $V_T = 0.68$ V; $V_{DS} = 3$ V, $V_T = 0.66$ V; $V_{DS} = 4$ V, $V_T = 0.63$ V; $V_{DS} = 5$ V, $V_T = 0.61$ V. Determine the static feedback on the threshold voltage η for this MOSFET. The effective channel length is 2 μm, and the oxide capacitance is 1.726×10^{-3} F/m^2.

SOLUTION

According to Table 11.10 (Part II), the threshold-voltage dependence on the drain-to-source voltage is given by

$$V_T = V_{T0} - \sigma_D V_{DS}$$

where V_{T0} is the zero-bias threshold voltage, and σ_D is a coefficient that can be determined as the slope of the linear $V_T - V_{DS}$ dependence. Figure 11.9 shows that σ_D is found to be 0.024.

The relationship between σ_D and η is also given in Table 11.10 (Part II). Using that equation, η is calculated as

$$\eta = \sigma_D C_{ox} L_{eff}^3 / 8.15 \times 10^{-22} = 0.024 \times 1.726 \times 10^{-3} (2 \times 10^{-6})^3 / 8.15 \times 10^{-22} = 0.41$$

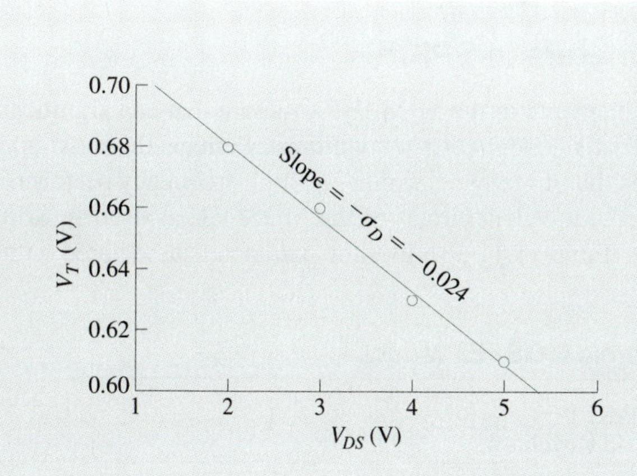

Figure 11.9 Threshold-voltage dependence on drain-to-source voltage.

EXAMPLE 11.5 Effective Channel Width

The MOSFET channel width can also differ from the nominal value, which can be significant in narrow-channel MOSFETs. To obtain this difference, the channel conductance $G_{on} = I_D/V_{DS}$ is measured in the linear region for MOSFETs having different channel widths and all the other parameters identical. Determine $\Delta W = W_{nom} - W$, using the following results: $W_{nom} = 4\ \mu m$, $G_{on} = 0.40\ \Omega^{-1}$; $W_{nom} = 6\ \mu m$, $G_{on} = 0.58\ \Omega^{-1}$; $W_{nom} = 0.8\ \mu m$, $G_{on} = 0.76\ \Omega^{-1}$.

SOLUTION

The channel conductance G_{on} in the linear region is

$$G_{on} = KP\frac{W}{L_{eff}}(V_{GS} - V_T) = \frac{KP}{L_{eff}}(V_{GS} - V_T)(W_{nom} - \Delta W)$$

The G_{on}–W_{nom} plot of Fig. 11.10 shows the linear dependence predicted by this equation. As the zero channel conductance appears at $W_{nom} = \Delta W$, ΔW is found at the intersection between the linear G_{on}–W_{nom} dependence and the W_{nom}-axis. From Fig. 11.10, $\Delta W = -0.44\ \mu m$.

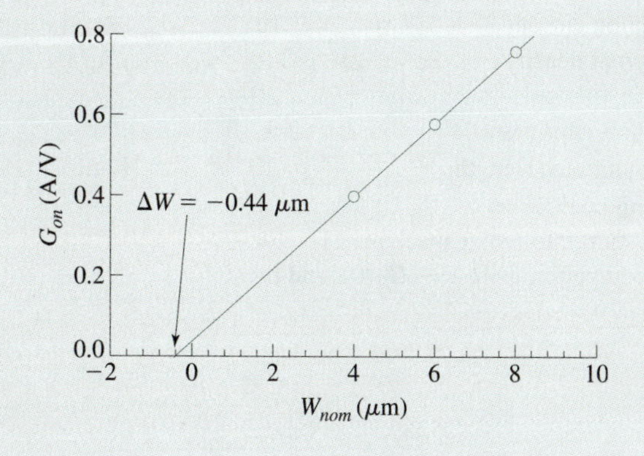

Figure 11.10 Dependence of the channel conductance on channel width.

11.2.3 Large-Signal Equivalent Circuit and Dynamic Parameters in SPICE

There are a number of parasitic elements in the MOSFET structure that can significantly influence the MOSFET characteristics under certain conditions. Perhaps the most important are the parasitic capacitances that directly determine the high-frequency performance of the MOSFET. The large-signal equivalent circuit of the MOSFET, as used in SPICE, is shown in Table 11.11 (Part II). Table 11.11 also lists the parameters associated with all

TABLE 11.11 Summary of SPICE Dynamic MOSFET Model

PART I
Geometric Variables

Symbol	SPICE Keyword	Variable Name	Default Value	Unit
A_D; P_D	AD; PD	Drain diffusion area; . . . perimeter	0; 0	m^2; m
A_S; P_S	AS; PS	Source diffusion area; . . . perimeter	0; 0	m^2; m

Parasitic-Element-Related Parameters

Symbol	SPICE Keyword	Related Parasitic Element	Parameter Name	Typical Value	Unit
R_D	Rd	R_D	Drain resistance	10	Ω
R_S	Rs	R_S	Source resistance	10	Ω
R_G	Rg	R_G	Gate resistance	10	Ω
R_B	Rb	R_B	Bulk resistance	10	Ω
	Rds	Not shown	Drain–source leakage resistance	∞	Ω
t_{ox}	Tox	C_{S1}; C_{D1}	Gate-oxide thickness	20×10^{-9}	m
$\|2\phi_F\|$ (or $N_{A,D}$)	Phi (or Nsub)	C_{S1}; C_{D1}	Surface potential (substrate doping)	0.7 (10^{15})	V (cm^{-3})
C_{GDO}	Cgdo	C_{D2}	Gate–drain overlap capacitance per channel width	4×10^{-11}	F/m
C_{GSO}	Cgso	C_{S2}	Gate–source overlap capacitance per channel width	4×10^{-11}	F/m
C_{GBO}	Cgbo	Not shown	Gate–bulk overlap capacitance per channel length	2×10^{-10}	F/m
I_S (or J_S)	IS (or JS)	D_B	Saturation current (current density)	10^{-14} (10^{-8})	A (A/m^2)
V_{bi}	PB/PBSW	D_B/D_P	Built-in voltage	0.8	V
$C_d(0)$	Cj/Cjsw	D_B/D_P	Zero-bias capacitance per unit area (length)	2×10^{-4} 10^{-9}	F/m^2 (F/m)
m	Mj/Mjsw	D_B/D_P	Grading coefficient	$\frac{1}{3}$–$\frac{1}{2}$	—
C_{BD}; C_{BS}	Cbd; Cbs	D_B/D_P	Drain/source-to-bulk capacitance [incompatible with V_{bi}, $C_d(0)$, and m]		F

TABLE 11.11 (Continued)

PART II
Large-Signal Equivalent Circuit

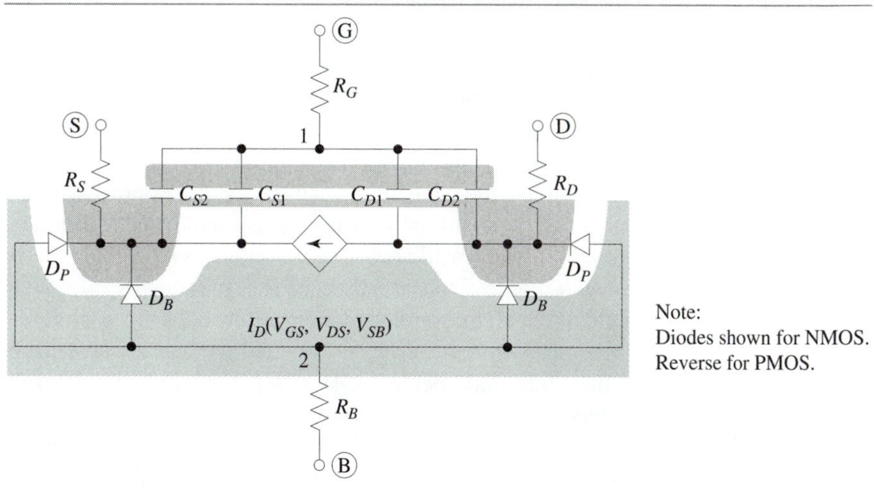

$I_D(V_{GS}, V_{DS}, V_{SB})$ is given in Table 11.7
D_B/D_P is according to diode model of Table 11.2
$C_{S2} = C_{GSO}W$
$C_{D2} = C_{GDO}W$
$C_{GB} = C_{GBO}L_{eff}$; C_{GB} appears between points 1 and 2 (not explicitly shown)
C_{S1} and C_{S2} calculated by SPICE from the terminal voltages, and t_{ox} and $|2\phi_F|$ (or $N_{A,D}$) parameters $\left(|2\phi_F| = \frac{kT}{q} \ln \frac{N_{A,D}}{n_i}\right)$
$I_S = J_S A_D$ (drain–bulk)
$I_S = J_S A_S$ (source–bulk)

the elements of the equivalent circuit but the current source, which is the only nonparasitic element.

The pairs of gate-to-source and gate-to-drain capacitors in Table 11.1 have different origins, and consequently different models and parameters are associated with these capacitors. The gate-to-source capacitance C_{S2} and the gate-to-drain capacitance C_{D2} are due to overlap between the gate and source/drain regions. SPICE parameters that include these capacitances are C_{GSO} and C_{GDO} (overlap capacitances per unit width). They have to be specified in F/m, and SPICE then multiplies the specified values by the channel width W, to convert them into capacitances expressed in F. Assuming overlap of l_{olp}, these parameters can be estimated as $C_{GSO,GDO} = l_{olp}\varepsilon_{ox}/t_{ox}$. Gate-to-body overlap capacitance is not shown explicitly in the figure; however, it exists, and it is connected between points 1 and 2 in the equivalent circuit. The MOSFET cross section along the channel width is shown in Fig. 8.19. The gate-to-body overlap capacitance is due to the gate extension outside the effective channel width (W). Assuming overlap of l_{olp}, the parameter of this capacitance can also be estimated as $C_{GBO} = l_{olp}\varepsilon_{ox}/t_{f-ox}$. This parameter, however, has the meaning of capacitance per unit length, and SPICE multiplies it by the effective channel length L_{eff} to convert the capacitance into farads.

The gate-oxide capacitance inside the active channel area is included by the capacitors C_{S1} and C_{D1}. These capacitances vary with the applied voltages, and SPICE calculates them accordingly. In the linear region, when the channel expands from the source to the drain, C_{S1} and C_{D2} each makes half of the total gate-oxide capacitance $C_{ox}WL_{eff}$. In saturation, the channel is pinched off at the drain side, and C_{D1} capacitance is smaller. Although no specific parameters are needed to calculate these capacitances, the gate-oxide thickness t_{ox} and the doping level ($|2\phi_F|$ or $N_{A,D}$) have to be specified.

In addition to the parasitic capacitors, there are parasitic resistors as well. Although the origins can be different (contact resistances and/or neutral-body resistances), they can be expressed by four parasitic resistors associated with each of the four terminals. The values of these parasitic resistors are direct SPICE parameters.

Finally, the source-to-body and drain-to-body P–N junctions create parasitic diodes. Although these diodes are normally off, they can create leakage currents; more importantly, they introduce depletion-layer capacitances. As described in Section 6.3, these capacitances depend on the reverse-bias voltage. The SPICE MOSFET model includes the full diode model (equivalent circuit) for these two P–N junctions. Additionally, each of these junctions is represented by two independent diodes, the body and perimeter diode. The zero-bias capacitance (as diode parameter) can be specified per unit area (for the case of the body diode D_B) or per unit length (for the perimeter diode D_P). Of course, these parameters necessitate properly specified geometrical variables (drain and source diffusion area/perimeter). As an alternative to the complete P–N junction capacitance models, the body-to-drain and body-to-source capacitances can directly be specified (C_{BD} and C_{BS} parameters). In this case, however, the capacitance dependence on the reverse-bias voltage is not included.

11.2.4 Simple Digital Model

Both digital and analog circuits with MOSFETs can be analyzed by the SPICE simulator. As described in Section 11.2.1, the MOSFET is represented in SPICE by its large-signal equivalent circuit with the associated equations and parameters. Assuming properly set parameters, SPICE simulations enable the fairly precise electrical characterization of a circuit. The simulation results can be utilized in many different ways during the design process. The question that is important for this section relates to the need for simplified equivalent circuits. If the large-signal equivalent circuit used in SPICE can be used to solve and analyze digital circuits, why do we need a simple digital model?

Precise SPICE simulations, obtained by the trial-and-error method, are not very helpful for a number of the very important decisions that a circuit/system designer has to make in the early stages of the design process. For this purpose, the designer has to rely on insights and hand calculations based on the simplest possible models. In the case of digital circuits, the simplest model for a MOSFET is the voltage-controlled switch. An ideal switch has zero resistance in *on* mode, infinite resistance in *off* mode, and zero parasitic capacitances. The concept of ideal switch is an abstraction that can help with the functional design of a digital circuit but is totally inadequate when it comes to the reality of *implementation* constraints. The parasitic resistance and the parasitic capacitances should be added to the ideal switch to enable considerations related to power/energy

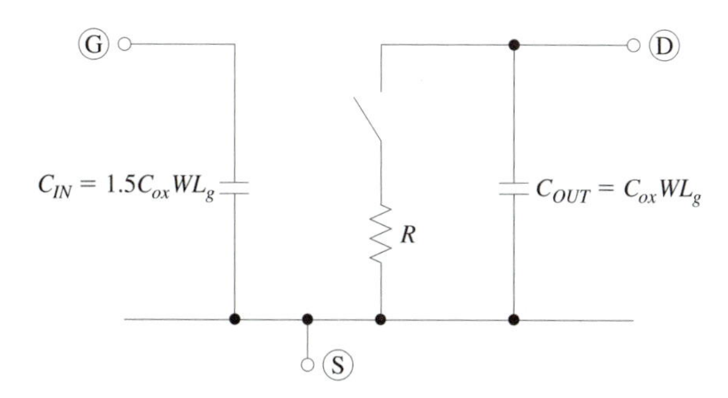

Figure 11.11 Simple digital MOSFET model.

dissipation and time response (energy and time are the two fundamental concepts that impose implementation constraints).

Figure 11.11 shows the equivalent circuit of a simple digital model that includes the parasitic resistance of the switch in *on* mode, the input capacitance, and the output capacitance.[2] The resistance of the switch in *off* mode is assumed to be infinite, which is appropriate for most applications where the MOSFET is used in series with much smaller resistances. Naturally, the input and output capacitances are ignored when DC or low-frequency analyses are performed.

The question now is how to relate the components of this simple equivalent circuit to the physically based components of the large-signal equivalent circuit. Starting with the resistor (R), consider a "sudden" change in the gate voltage of an N-channel MOSFET from 0 to V_+. In a digital circuit, such as the CMOS inverter shown in Fig. 8.6a, this should cause the drain-to-source voltage of the MOSFET to change from V_+ to 0. Looking at the output characteristics (Fig. 11.12), the instantaneous current–voltage point has to move from ($V_{DS} = V_+$, $I_D = 0$) to ($V_{DS} = 0$, $I_D = 0$). The actual path of the current–voltage point is also illustrated in Fig. 11.12. Because of the parasitic capacitance between the drain and the source, V_{DS} voltage cannot be changed instantly while the "sudden" change in input voltage from $V_{GS} = 0$ to $V_{GS} = V_+$ changes the current from 0 to I_{Dsat}. Accordingly, this transition is shown by the vertical arrow. The saturation-drain current, corresponding to $V_{GS} = V_+$, starts discharging the drain-to-source capacitance. This reduces the drain-to-source voltage; when the voltage reaches the V_{DSsat} value at time t_1, the discharging path enters the triode region.

The actual discharging path is nonlinear, which is not consistent with the linear equivalent circuit (Fig. 11.11) whose resistance R we need to establish. Nonetheless, the arguments for the usefulness of even a coarse but simple digital model remain valid. As discussed earlier, simple equivalent circuits would not be used to achieve the precise calculations that would be needed for circuit optimization. SPICE simulations would be used for this type of analysis. The simple equivalent circuits are needed at the conceptual level, where many important decisions are typically made. Given that *simplicity* and not *precision* is of ultimate importance at the conceptual level, we use the best *linear path* as the model for the voltage–current changes associated with the discharging of the drain-to-source capacitance. This linear path is also illustrated in Fig. 11.12. Based on this linear

---

[2]R. J. Baker, H. W. Li, and D. E. Boyce, *CMOS: Circuit Design, Layout, and Simulation,* IEEE Press, New York, 1998.

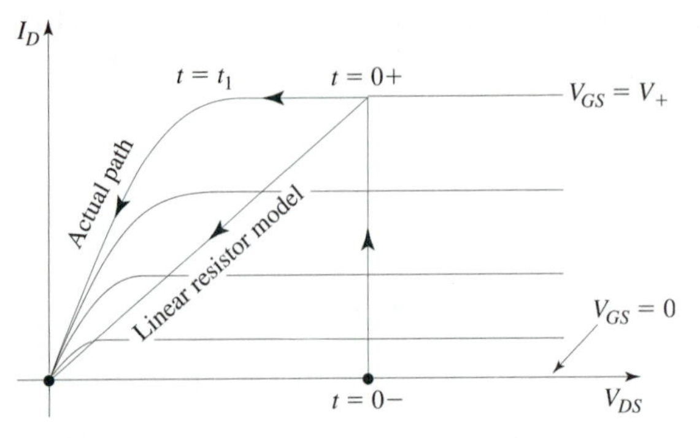

Figure 11.12 Transitions of NMOS operating point from $V_{DS} = V_+$ and $I_D = 0$ to $V_{DS} = 0$, illustrating the actual path and the linear path that provide the best model for the resistance R in the equivalent circuit of Fig. 11.11.

path and the MOSFET equation for saturation current, the resistance R in the equivalent circuit of Fig. 11.11 can be expressed as

$$R = \frac{V_+}{I_{Dsat}(V_{GS} = V_+)} = R_{Seff}\frac{L}{W} \tag{11.26}$$

where the kind of effective sheet resistance R_{Seff} includes the voltage V_+ and all the relevant technological parameters. In the case of the LEVEL 3 SPICE model, $R_{Seff} = 2(1 + F_B)V_+/KP(V_+ - V_T)^2$.

The remaining two components of the simple digital MOSFET model are the input (C_{IN}) and output (C_{OUT}) capacitances. Basically, these capacitances have to replace and represent the capacitances of the large-signal equivalent circuit shown in Table 11.11 (Part II). Labeling the gate length and width by L_g and W, respectively, and the gate-oxide capacitance per unit area by C_{ox}, we can clearly see that one component of C_{IN} is $C_{S1} + C_{S2} = C_{ox}WL_g/2$. The other component of C_{IN} is due to $C_{D1} + C_{D2}$. Although $C_{D1} + C_{D2} = C_{ox}WL_g/2$, its contribution to the effective input capacitance is different from this value because the drain is not short-circuited to the source and V_{GD} voltage is not equal to the voltage across C_{IN}. To turn the MOSFET on, V_{GS} has to change from 0 to V_+ while V_{DS} has to change from V_+ to 0. The change of V_{GD} is $\Delta V_{GD} = \Delta V_{GS} - \Delta V_{DS} = (V_+ - 0) - (0 - V_+) = 2V_+$. Assuming linear transitions, the current through the gate-to-drain capacitance $C_{D1} + C_{D2}$ is

$$I_{G\to D} = \frac{C_{ox}WL_g}{2}\frac{\Delta V_{GD}}{\Delta t} = \frac{C_{ox}WL_g}{2}2V_+ = C_{ox}WL \underbrace{\frac{\Delta V_{GS}}{\Delta t}}_{=-\Delta V_{DS}/\Delta t} \tag{11.27}$$

Equation (11.27) means that the gate-to-drain capacitance $C_{D1} + C_{D2}$ can be split into components from the gate to ground and from the drain to ground, each of them of value $(C_{D1} + C_{D2}) = C_{ox}WL_g$.[3] Therefore, the total input capacitance (due to both the gate-to-

[3]The effective input and output capacitances are larger than the real value of the capacitance connected between the input and the output. This effect is due to increased voltage changes across the bridging capacitance. In linear circuits, this effect is known as the *Miller* effect. In this case, the voltage change of $2V_+$ across $C_{D1} + C_{D2} = C_{ox}WL_g/2$ is equal to twice as small voltage change (V_+) across twice as large a capacitance $[2(C_{D1} + C_{D2}) = C_{ox}WL_g]$.

source and drain-to-gate components) is $C_{IN} = C_{ox}WL_g/2 + C_{ox}WL_g = 3C_{ox}WL_g/2$, whereas the output capacitance is $C_{OUT} = C_{ox}WL_g$. These are the values shown in Fig. 11.11.

EXAMPLE 11.6 Time Response of CMOS Inverter

The saturation current of the NMOS transistor in a CMOS technology is $I_{Dsat} = 8$ mA. The following parameters are also known: $L_{g-NMOS} = L_{g-PMOS} = 0.25\ \mu$m, $W_{NMOS} = 0.5\ \mu$m, $W_{PMOS} = 1.5\ \mu$m, $t_{ox} = 5$ nm, and $V_+ = 3.3$ V. A sharp 0-to-V_+ voltage transition is achieved at the inverter input. Using the simple digital MOSFET model, calculate the fall time if the inverter output is

(a) open (intrinsic fall time)

(b) connected to the input of an identical inverter (cascade connection)

Note: The fall time is defined as the time that is needed for the output voltage to drop from $0.9V_+$ to $0.1V_+$, as illustrated in Fig. 11.13.

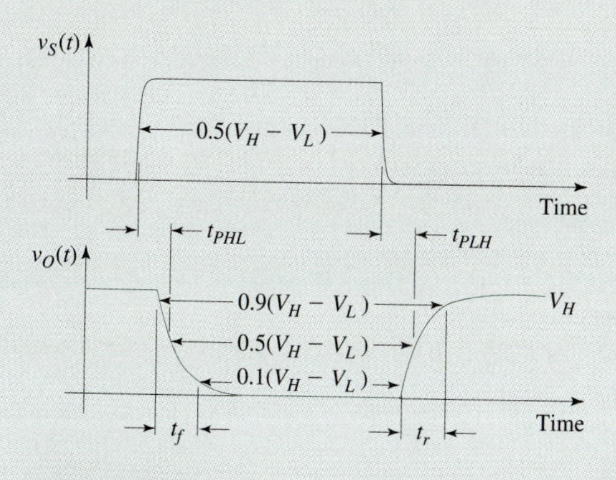

Figure 11.13 Practical definitions of rise (t_r), fall (t_f), and propagation (t_{PHL} and t_{PLH}) times associated with an inverter (v_S is the input voltage, v_O is the output voltage, V_H is the logic high level, and V_L is the logic low level).

SOLUTION

(a) Although the inverter output is open, the MOSFET capacitances have to be charged/discharged as the inverter changes states. The output capacitance of the NMOS transistor is connected between the inverter output and ground, which means that its voltage has to change from V_+ to zero as the inverter changes the output level from *high* to *low*. The output capacitance of the PMOS transistor is connected between the output and V_+, which means that the voltage across this capacitance has to change from 0 V to $-V_+$. The charging/discharging current of both these capacitances flows through the NMOS transistor (the PMOS transistor is *off*). This means that the capacitances are effectively connected in parallel, with the total capacitance being

$$C_{OUT-INV} = 6.9 \times 10^{-3} \times 0.25 \times 10^{-6} \times (0.5 + 1.5) \times 10^{-6} = 3.45 \text{ fF}$$

As the voltage between the drain and source of the NMOS transistor changes from V_+ to 0, the current changes from I_{Dsat} to 0. This means the average resistance is

$$R = \frac{V_+}{I_{Dsat}} = \frac{3.3}{8 \times 10^{-3}} = 412.5\ \Omega$$

The capacitance $C_{OUT-INV}$ and the resistance R create the simple R–C discharging circuit. In this circuit, the voltage drops exponentially with time constant $\tau = RC_{OUT-INV}$:

$$v_O = V_+ e^{-t/\tau}$$

Applying the definition of fall time, we have

$$t_1 = -\tau \ln 0.9$$

$$t_2 = -\tau \ln 0.1$$

$$t_f = t_2 - t_1 = \tau(-\ln 0.1 + \ln 0.9) = \tau \ln \frac{0.9}{0.1} = 2.2\tau$$

$$t_f = 2.2 \times 412.5 \times 3.45 \times 10^{-15} = 3.13 \text{ ps}$$

(b) In this case the input capacitances of the NMOS and PMOS transistors creating the inverter connected to the output of the first inverter have to be added to the intrinsic loading capacitance $C_{OUT-INV}$:

$$C_L = C_{OUT-INV} + \frac{3}{2} C_{ox} L_{g-NMOS} W_{NMOS} + \frac{3}{2} C_{ox} L_{g-PMOS} W_{PMOS}$$

$$C_L = 3.45 \times 10^{-15} + 1.5 \times 6.9 \times 10^{-3} \times 0.25 \times 10^{-6} \times (0.5 + 1.5) \times 10^{-6} = 8.625 \text{ fF}$$

$$t_f = 2.2 \times RC_L = 2.2 \times 412.5 \times 8.625 \times 10^{-15} = 7.8 \text{ ps}$$

EXAMPLE 11.7 Design of CMOS Transistors for Large Capacitive Loads

The parameters of the simple digital models for N-channel and P-channel MOSFETs in a minimum-size CMOS inverter are $R_{NMOS} = R_{PMOS} = 400\ \Omega$, $C_{IN-PMOS} = 2C_{IN-NMOS} = 10$ fF, and $C_{OUT-PMOS} = 2C_{OUT-NMOS} = 7$ fF.

(a) Estimate the total delay time $t_{delay} = t_{PHL} + t_{PLH}$ if this inverter drives external load of equivalent capacitance $C_L = 10$ pF. (*Note:* The propagation times t_{PHL} and t_{PLH} are defined in Fig. 11.13.)

(b) Estimate t_{delay} if C_L is driven by the cascade connection of three inverters with scaled-up channel widths, so that W_{NMOS} and W_{PMOS} are increased by $S = 10$ times in each subsequent inverter. Compare the result to the case of single-inverter driver of part (a).

SOLUTION

(a) To estimate t_{PHL} we note that C_L is discharging from $V_H = V_+$ to $V_L = 0$ through R_{NMOS}. The output voltage is decaying exponentially, $V_+ \exp(-t/\tau_{PHL})$, and t_{PHL} is defined as the time when the output voltage drops to $0.5V_+$:

$$0.5V_+ = V_+ e^{-t_{PHL}/\tau_{PHL}}$$

$$t_{PHL} = \tau_{PHL} \ln 2$$

where

$$\tau_{PHL} = R_{NMOS}(C_{OUT-NMOS} + C_{OUT-PMOS} + C_L)$$

Similarly, we can find that

$$t_{PLH} = (\ln 2) R_{PMOS}(C_{OUT-NMOS} + C_{OUT-PMOS} + C_L)$$

Therefore,

$$t_{delay} = (\ln 2)(R_{NMOS} + R_{PMOS})(C_{OUT-NMOS} + C_{OUT-PMOS} + C_L)$$

$$t_{delay} = 0.693 \times 800 \times (3.5 \times 10^{-15} + 7 \times 10^{-15} + 10 \times 10^{-12}) = 5.55 \text{ ns}$$

(b) As channel widths in the cascaded inverters are increased by factor S toward the load, the capacitances are also increased by factor S, whereas the resistances are reduced by factor S compared to the previous inverter. To obtain the total delay time, we have to add the delay times after each of the inverters:

$$\begin{aligned} t_{delay} &= (\ln 2)(R_{NMOS} + R_{PMOS}) \\ &\quad \times C_{OUT-NMOS} + C_{OUT-PMOS} + S(C_{IN-NMOS} + C_{IN-PMOS}) && \text{1st inverter} \\ &\quad + (\ln 2)\frac{R_{NMOS} + R_{PMOS}}{S} \\ &\quad \times S(C_{OUT-NMOS} + C_{OUT-PMOS}) + S^2(C_{IN-NMOS} + C_{IN-PMOS}) && \text{2nd inverter} \\ &\quad + (\ln 2)\frac{R_{NMOS} + R_{PMOS}}{S^2} \\ &\quad \times S^2(C_{OUT-NMOS} + C_{OUT-PMOS}) + C_L && \text{3rd inverter} \end{aligned}$$

$$\begin{aligned} t_{delay} &= \ln 2 \times 800 \times 10.5 \times 10^{-15} + \ln 2 \times 800 \times 10 \times 15 \times 10^{-15} \\ &\quad + \ln 2 \times 800 \times 10.5 \times 10^{-15} + \ln 2 \times 800 \times 10 \times 15 \times 10^{-15} \\ &\quad + \ln 2 \times 800 \times 10.5 \times 10^{-15} + \ln 2 \times 800 \times \tfrac{10\times 10^{-12}}{10^2} \\ &= 239.3 \text{ ps} \end{aligned}$$

It can be seen that the triple-inverter driver provides much shorter delay (0.24 ns $\ll$ 5.55 ns), even though the pulse has to propagate through three inverter stages.

11.2.5 Small-Signal Equivalent Circuit

In analogy with the arguments presented in Section 11.2.4 for the need of a simple equivalent circuit for digital applications, there is a need for a simplified equivalent circuit applicable to analog applications. As described in Section 11.1.4, the term *small signal* is used to designate that a device is used in a small linear-like region of its characteristics.

The general small-signal equivalent circuit of a MOSFET is shown in Fig. 11.14. The components of this circuit can be related to the components of the large-signal equivalent circuit, shown in Table 11.11 (Part II). The resistances r_G, r_S, and r_D are equivalent to their large-signal counterparts, R_G, R_S, and R_D, respectively. These are the parasitic contact and body resistances of the gate, source, and drain regions. These resistances are small and can usually be neglected.

The small-signal capacitance C_{gs} is equal to the large-signal capacitance $C_{S1} + C_{S2}$ at the specific quiescent point (DC bias). In general, $C_{S1} + C_{S2}$ capacitance depends on V_{GS} voltage, but in a linear circuit the DC component of the gate-to-source voltage is fixed at a selected quiescent point V_{GSQ}. Because the variations of the signal voltage v_{gs} around the quiescent point V_{GSQ} are small, it is assumed that the small-signal capacitance C_{gs} has a constant value that is equal to the value of $C_{S1} + C_{S2}$ at $V_{GS} = V_{GSQ}$. Analogously, the small-signal capacitance C_{gd} is equal to the value of the large-signal capacitance $C_{D1} + C_{D2}$ at $V_{GD} = V_{GDQ}$.

Typically, no signals are applied between the body and source terminals of MOSFETs used in linear circuits. This means that the body and the source terminals are effectively

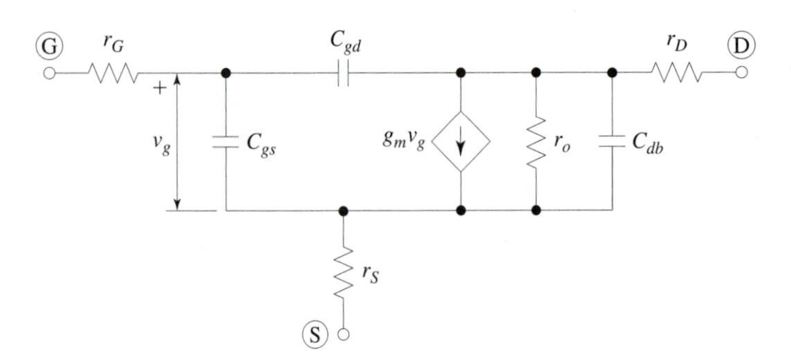

Figure 11.14 The general small-signal equivalent circuit of a MOSFET.

short-circuited as far as the signals are concerned.[4] Accordingly, the small-signal capacitance C_{db} should be related to the parasitic capacitance between the drain and body terminals (any capacitance between the body and the source terminals is short-circuited). As can be seen from the large-signal equivalent circuit shown in Table 11.11 (Part II), the capacitance between the drain and the body terminals is due to the reverse-biased diodes D_B and D_P. Therefore, the small-signal capacitance is equal to the value of the depletion-layer capacitance of the reverse-biased drain-to-body P–N junction at the quiescent voltage V_{DSQ}.

As in the case of the simple digital model (Section 11.2.4), the input–output bridging capacitance can be split into effective input and output components. Assuming the common-source configuration, the input is between the gate and source terminals and the output is between the drain and source terminals. In this case, C_{gd} is the bridging input–output capacitance. Analogously to the digital case, the effective input capacitance will appear larger than C_{gd} to account for the fact that the voltage changes between the gate and drain (the output signal) are larger than the voltage changes between the gate and source (the input signal). This is the Miller effect. Defining the ratio between the output and input signals as the gain, $A = v_{ds}/v_{gs}$, the effective input component of C_{gd} is equal to $(1 + A)C_{gd}$. The effective output component of the bridging capacitance is approximately equal to C_{gd} itself, under the assumption that $v_{ds} \gg v_{gs}$ for a high gain ($A \gg 1$). The effective input and output components of C_{gd} have to be added to C_{gs} and C_{db}, respectively, to obtain the total effective input and output capacitances of a MOSFET used as an amplifier with common source:

$$\begin{aligned} C_i &= C_{gs} + (1 + A)C_{gd} \\ C_o &= C_{db} + C_{gd} \end{aligned} \tag{11.28}$$

The capacitances are important for high-frequency analyses. If the frequencies of interest are small enough, the impedances associated with the parasitic capacitances become so large that the capacitances can be neglected (replaced by open circuits in the equivalent circuit). This leaves us with the two essential components of the small-signal equivalent circuit: the voltage-controlled current source ($g_m v_g$) and the output resistance (r_o). As described in Section 11.1.4, a small-signal resistance is different from the static resistance.[5] In this case the small-signal resistance v_{ds}/i_d is different from the static resistance V_{DSQ}/I_{DQ}. The static resistance relates to the slope of a line connecting a selected quiescent point (I_{DQ}, V_{DSQ}) to the (0, 0) point (the origin of the output characteristics). An example is the line labeled as "linear resistor model" in Fig. 11.12. The small-signal resistance relates to the slope of the actual I_D–V_D line at the selected quiescent point. The mathematical definition is

$$r_o = \left(\left. \frac{dI_D}{dV_{DS}} \right|_{@V_{DSQ}} \right)^{-1} \tag{11.29}$$

In the ideal case, the saturation current of a MOSFET does not change with V_{DS}, so $r_o \to \infty$. This means that the MOSFET acts as a perfect voltage-controlled current source.

[4]Even when a DC bias is applied between the body and the source, the voltage source providing this DC bias acts as a short circuit in terms of any signal analysis.

[5]The *small-signal* resistance is also called *dynamic* resistance.

In practical MOSFETs, however, there is always a slight increase of I_D with V_{DS}, so the small-signal output resistance is not infinite. Section 8.3 describes the most important second-order effects that lead to the finite output resistance.

The single most important component of the small-signal equivalent circuit is the voltage-controlled current source. The controlling factor is the transconductance g_m. The transconductance relates the signal drain current to the signal gate-to-source voltage. In other words, it shows how much change in the drain current is caused by a certain change in the gate-to-source voltage. Therefore, the transconductance relates to the transfer I_D–V_{GS} characteristic in the following way:

$$g_m = \left.\frac{dI_D}{dV_{GS}}\right|_{@V_{GSQ}} \tag{11.30}$$

EXAMPLE 11.8 Small-Signal Transconductance

Determine the transconductance of the MOSFET from Example 8.2 at the following bias points:

(a) $V_{GSQ} = 5$ V, $V_{DSQ} = 10$ V, and $V_{SBQ} = 0$ V
(b) $V_{GSQ} = 0$ V, $V_{DSQ} = 10$ V, and $V_{SBQ} = 0$ V

Compare and comment on the results.

SOLUTION

Using the LEVEL 3 equation in the saturation region and Eq. (11.30), we obtain

(a)

$$g_m = \frac{\partial}{\partial V_{GS}}\left[\frac{\beta}{2(1+F_B)}(V_{GS} - V_T)^2\right]_{@V_{GSQ}}$$

$$= \frac{\beta}{1+F_B}(V_{GSQ} - V_T)$$

$$g_m = \frac{1.75}{1+0.51}(5+2.5) = 8.7 \text{ mA/V}$$

(b)

$$g_m = \frac{1.75}{1+0.51}(0+2.5) = 2.9 \text{ mA/V}$$

The transconductance is smaller for the smaller V_{GS} because the slope of the transfer characteristic is smaller.

11.3 BJT

11.3.1 Static Model and Parameters: Ebers–Moll and Gummel–Poon Levels in SPICE

The Ebers–Moll and Gummel–Poon levels in SPICE are two levels of model complexity that differ in the way the common-emitter current gains β_F and β_R are treated. At the Ebers–Moll level, they are considered as constants whose values can be specified by the user. At the Gummel–Poon level, the specified values are considered as the maximum values only (β_{FM}, and β_{RM}), while modified I_C and I_B equations account for the variation of the current gain I_C/I_B at different bias conditions. The more complex equations at the Gummel–Poon level include the simpler Ebers–Moll equations. Consequently, and unlike the MOSFET case, the BJT levels do not need to be explicitly specified by the user.

The Ebers–Moll parameters and equations are summarized in Table 11.12. They include the principal effects and the most important second-order effect, which is the Early effect. The Early voltages can be ignored if precise modeling of the output dynamic resistance is not important. For completeness, the equations are given for the case of both NPN and PNP BJT. If the approximation of constant common-emitter current gain is satisfactory and Table 11.12 is used, then the Gummel–Poon parameters and equation in Table 11.13 should be ignored. However, Table 11.15 is still relevant, because it summarizes the parasitic elements that may need to be added to the Ebers–Moll parameters for more precise simulation.

In the extreme biasing conditions, when the current gain reduction at very low or high current levels cannot be neglected, the parameters and equations of the more complex Gummel–Poon level, summarized in Table 11.13, should be used instead of the parameters and equations from Table 11.12.

11.3.2 Parameter Measurement

This section uses the example of a 3086 NPN BJT biased in the normal active mode to show practical techniques of BJT parameter measurement. Analogous techniques are employed for measurement of the parameters related to the inverse active mode. The techniques described are useful for obtaining estimated parameter values, which can then be used as the initial values for nonlinear parameter fitting.

Measurement of the Saturation Current and the Current Gain

To find the saturation current I_{S0} and the current gain β_F, a set of measured I_C and I_B values over a range of V_{BE} voltages is needed. The voltage V_{CE} is kept constant, and it is set at neither too low a value (to avoid the saturation region) nor too high a value (to avoid the influence of the Early effect).

In the normal active mode, I_C and I_B dependencies on V_{BE} are given by Eqs. (9.51). Because these are exponential relationships, they can be linearized in the following way:

$$\ln I_C = \ln I_{S0} + \frac{1}{V_t} V_{BE} \tag{11.31}$$

$$\ln I_B = \ln I_{S0} - \ln \beta_F + \frac{1}{V_t} V_{BE} \tag{11.32}$$

TABLE 11.12 Summary of SPICE BJT Model: Static Ebers–Moll Level

Ebers–Moll Parameters

Symbol	Usual SPICE Keyword	Parameter Name	Typical Value	Unit
I_{S0}	IS	Saturation current	10^{-16}	A
β_F	BF	Normal common-emitter current gain	150	—
β_R	BR	Inverse common-emitter current gain	5	—
V_A	VA	Normal Early voltage	>50	V
V_B	VB	Inverse Early voltage		V

Ebers–Moll Model

NPN BJT

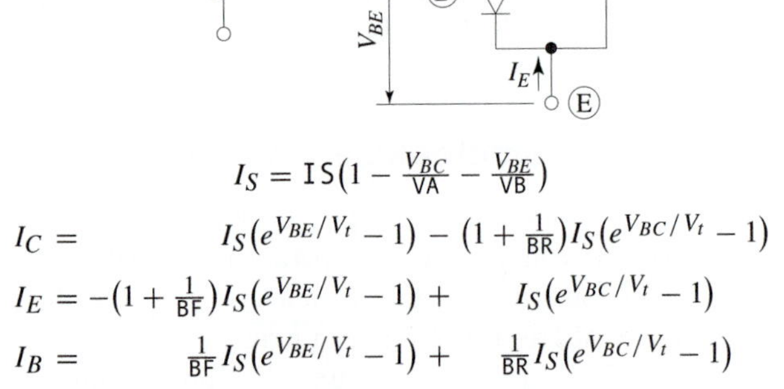

$$I_S = \mathtt{IS}\left(1 - \frac{V_{BC}}{\mathsf{VA}} - \frac{V_{BE}}{\mathsf{VB}}\right)$$

$$\begin{aligned}
I_C &= I_S\left(e^{V_{BE}/V_t} - 1\right) - \left(1 + \frac{1}{\mathsf{BR}}\right) I_S\left(e^{V_{BC}/V_t} - 1\right)\\
I_E &= -\left(1 + \frac{1}{\mathsf{BF}}\right) I_S\left(e^{V_{BE}/V_t} - 1\right) + I_S\left(e^{V_{BC}/V_t} - 1\right)\\
I_B &= \frac{1}{\mathsf{BF}} I_S\left(e^{V_{BE}/V_t} - 1\right) + \frac{1}{\mathsf{BR}} I_S\left(e^{V_{BC}/V_t} - 1\right)
\end{aligned}$$

PNP BJT

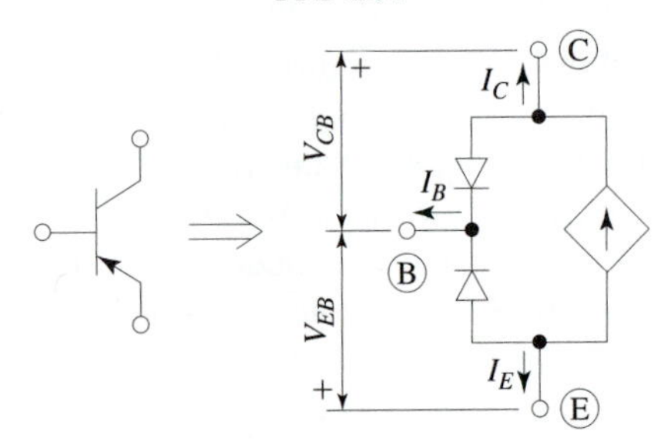

$$I_S = \mathtt{IS}\left(1 - \frac{V_{CB}}{\mathsf{VA}} - \frac{V_{EB}}{\mathsf{VB}}\right)$$

$$\begin{aligned}
I_C &= I_S\left(e^{V_{EB}/V_t} - 1\right) - \left(1 + \frac{1}{\mathsf{BR}}\right) I_S\left(e^{V_{CB}/V_t} - 1\right)\\
I_E &= -\left(1 + \frac{1}{\mathsf{BF}}\right) I_S\left(e^{V_{EB}/V_t} - 1\right) + I_S\left(e^{V_{CB}/V_t} - 1\right)\\
I_B &= \frac{1}{\mathsf{BF}} I_S\left(e^{V_{EB}/V_t} - 1\right) + \frac{1}{\mathsf{BF}} I_S\left(e^{V_{CB}/V_t} - 1\right)
\end{aligned}$$

TABLE 11.13 Summary of SPICE BJT Model: Static Gummel–Poon Level

Gummel–Poon Parameters

Symbol	Usual SPICE Keyword	Parameter Name[a]	Typical Value	Unit
I_{SO}	IS	Saturation current	10^{-16}	A
β_{FM}	BF	Maximum normal current gain	150	—
β_{RM}	BR	Maximum inverse current gain	5	—
V_A	VA	Normal Early voltage	>50	V
V_B	VB	Inverse Early voltage		V
I_{KF}	IKF	Normal knee current	$>10^{-2}$	A
I_{KR}	IKR	Inverse knee current		A
$C_2 I_{SO}$	ISE	B–E leakage saturation current	$<I_{SO}$	A
n_{EL}	NE	B–E leakage emission coefficient	2	—
$C_4 I_{SO}$	ISC	B–C leakage saturation current	$<I_{SO}$	A
n_{CL}	NC	B–C leakage emission coefficient	2	—

Gummel–Poon Model

NPN BJT (equivalent circuit as in Table 11.12)

$$\lambda_{BE} = e^{V_{BE}/V_t} - 1, \qquad \lambda_{BC} = e^{V_{BC}/V_t} - 1$$

$$\lambda_{BEL} = e^{V_{BE}/(\mathrm{NE}V_t)} - 1, \qquad \lambda_{BCL} = e^{V_{BC}/(\mathrm{NC}V_t)} - 1$$

$$q_1 = \left(1 - \frac{V_{BC}}{\mathrm{VA}} - \frac{V_{BE}}{\mathrm{VB}}\right)^{-1}, \qquad q_2 = \frac{\mathrm{IS}}{\mathrm{IKF}}\lambda_{BE} + \frac{\mathrm{IS}}{\mathrm{IKR}}\lambda_{BC}, \qquad q_b = 0.5q_1\left(1 + \sqrt{1 + 4q_2}\right)$$

$$I_B = \frac{\mathrm{IS}}{\mathrm{BF}}\lambda_{BE} + \mathrm{ISE}\lambda_{BEL} + \frac{\mathrm{IS}}{\mathrm{BR}}\lambda_{BC} + \mathrm{ISC}\lambda_{BCL}$$

$$I_C = \frac{\mathrm{IS}}{q_b}(\lambda_{BE} - \lambda_{BC}) - \frac{\mathrm{IS}}{\mathrm{BR}}\lambda_{BC} - \mathrm{ISC}\lambda_{BCL}$$

$$I_E = -\frac{\mathrm{IS}}{q_b}(\lambda_{BE} - \lambda_{BC}) - \frac{\mathrm{IS}}{\mathrm{BF}}\lambda_{BE} - \mathrm{ISE}\lambda_{BEL}$$

PNP BJT (equivalent circuit as in Table 11.12)

$$\lambda_{EB} = e^{V_{EB}/V_t} - 1, \qquad \lambda_{CB} = e^{V_{CB}/V_t} - 1$$

$$\lambda_{EBL} = e^{V_{EB}/(\mathrm{NE}V_t)} - 1, \qquad \lambda_{CBL} = e^{V_{CB}/(\mathrm{NC}V_t)} - 1$$

$$q_1 = \left(1 - V_{CB}\mathrm{VA} - \frac{V_{EB}}{\mathrm{VB}}\right)^{-1}, \qquad q_2 = \frac{\mathrm{IS}}{\mathrm{IKF}}\lambda_{EB} + \frac{\mathrm{IS}}{\mathrm{IKR}}\lambda_{CB}, \qquad q_b = 0.5q_1\left(1 + \sqrt{1 + 4q_2}\right)$$

$$I_B = \frac{\mathrm{IS}}{\mathrm{BF}}\lambda_{EB} + \mathrm{ISE}\lambda_{EBL} + \frac{\mathrm{IS}}{\mathrm{BR}}\lambda_{CB} + \mathrm{ISC}\lambda_{CBL}$$

$$I_C = \frac{\mathrm{IS}}{q_b}(\lambda_{EB} - \lambda_{CB}) - \frac{\mathrm{IS}}{\mathrm{BR}}\lambda_{CB} - \mathrm{ISC}\lambda_{CBL}$$

$$I_E = -\frac{\mathrm{IS}}{q_b}(\lambda_{EB} - \lambda_{CB}) - \frac{\mathrm{IS}}{\mathrm{BF}}\lambda_{EB} - \mathrm{ISE}\lambda_{EBL}$$

[a] B–E, base–emitter; B–C, base–collector.

Figure 11.15 shows that the linear ln I_C–V_{BE} and ln I_B–V_{BE} dependencies, with the slope of $1/V_t$, are observed over a wide range of output currents. As Eq. (11.31) shows, the logarithm of the saturation current is obtained as ln I_C at $V_{BE} = 0$.

It is obvious from Eqs. (11.31) and (11.32) that $\ln I_C - \ln I_B = \ln \beta_F$. Therefore, the logarithm of the current gain is obtained as the difference between the ln I_C–V_{BE} and ln I_B–V_{BE} lines.

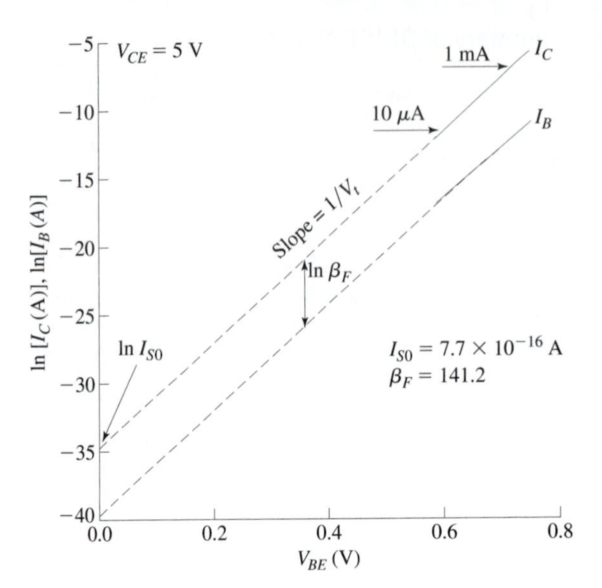

Figure 11.15 Measurement of I_{SO} and β_F.

Measurement of the Early Voltage

It may seem from Fig. 9.14 that V_A measurement is as simple as the extrapolation of several I_C–V_{CE} lines. However, a number of important points are not immediately obvious. To begin with, Eq. (9.55) cannot directly be used to calculate V_A because the $I_C(V_{BC} = 0)$ point is not in the active region. Using another reference point and assuming that $V_{BC} \approx V_{CE}$, for convenience, we modify Eq. (9.55) as follows:

$$\frac{I_{C-ref}}{V_{CE-ref} + V_A} = \frac{I_C}{V_{CE} + V_A} \tag{11.33}$$

This equation can further be transformed into the following form:

$$V_{CE} = \underbrace{\frac{V_{CE-ref} + V_A}{I_{C-ref}}}_{\text{slope}} I_C - V_A \tag{11.34}$$

Only two points, (V_{CE-ref}, I_{C-ref}) and (V_{CE}, I_C), are needed to construct the line defined by Eq. (11.34). It is important, however, that the two points span the entire operating range of V_{CE} voltages.

The range of input voltages V_{BE}, or alternatively input currents I_B, should also cover the entire operating range. The importance of this point is the best illustrated by Table 11.14, which shows significant differences between Early voltages obtained by

TABLE 11.14 Early Voltages Obtained by Extrapolation of Different I_C–V_{CE} Lines

I_B (μA)	V_A (V)
20	200.3
40	146.4
60	115.7
80	98.9
100	86.0

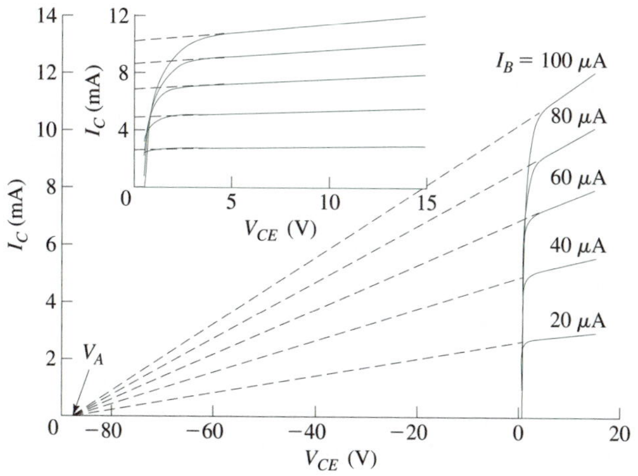

Figure 11.16 Early voltage V_A of a 3086 NPN BJT.

extrapolation of I_C–V_{CE} lines corresponding to different I_B currents. Obviously, if the input current I_B is restricted to a too small value, the Early voltage will be overestimated.

Of course, the results from Table 11.14 raise the question of whether it is possible to establish a unique V_A, which will properly represent the complete set of I_C–V_{CE} characteristics. Given that the Early effect is the most pronounced at the highest I_B current, the corresponding Early voltage can be used as the first estimate. Using $V_A = 86.0$ V as the initial value, Eq. (11.34) was fitted to the experimental data shown in Fig. 11.16. The best fit was achieved with $V_A = 87.7$ V, which is a very close value to the initial one. Figure 11.16 illustrates that this unique value can properly represent the complete set of output characteristics.

Measurement of the High-Level Knee Current and the Leakage Parameters

The first problem that appears at high current levels is the difference between the applied V_{BE} voltage and the voltage that actually appears across the P–N junction. This difference is due to the voltage drop across the parasitic resistances, as described in Section 6.2.2.

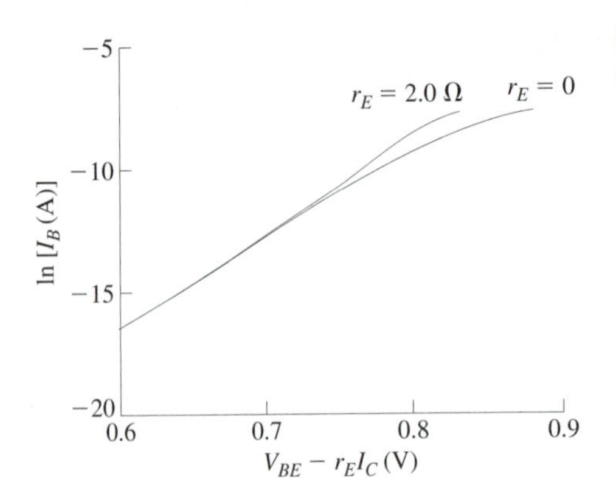

Figure 11.17 The effect of r_E on $\ln I_B$–V_{BE} dependence.

In the case of the BJT, the voltage across r_B is negligible because the base current is very small; however, the voltage across r_E can become pronounced. Although this voltage is $r_E I_E$, it can be approximated by $r_E I_C$, since $I_C \approx I_E$. Following the procedure described in Section 6.2.2, the value of r_E is determined to calculate the effective base–emitter voltage $V_{BE} - r_E I_C$ that linearizes the high-current part of the $\ln I_B$–V_{BE} dependence (Fig. 11.17).

Plotting $\ln I_C$ versus the effective base–emitter voltage $V_{BE} - r_E I_C$, as in Fig. 11.18, shows that the actual $\ln I_C$ data still depart from the linear dependence at high current levels. The reduced slope of the $\ln I_C$–V_{BE} dependence causes the reduction in the current gain observed at high I_C currents in Fig. 9.16.

The related parameter in the Gummel–Poon equations [Eqs. (9.60) and (9.64)–(9.66)] is I_{KF}. To estimate the value of this parameter, $\ln I_C$–V_{BE} measurements are extended into the high-current region. Again, the measurements are performed with properly selected V_{CE} value so that the Early effect is avoided, simplifying Eq. (9.65) to $q_1 \approx 1$. Equation (9.69) for the collector current at high current levels shows that a plot of $\ln I_C$ versus V_{BE} should be a straight line with the slope $1/2V_t$. Figure 11.18 shows that I_{KF} can be determined in one of the following two ways: (1) as the logarithm of the current at which the high-current line with the slope of $1/2V_t$ intersects the low-current line with the slope of $1/V_t$ and (2) from $\ln \sqrt{I_{S0} I_{KF}}$ that is determined as value of the high-current line at $V_{BE} = 0$, using the earlier determined value of I_{S0}.

At very low current levels, a significant reduction of the current gain β_F can occur due to the increase of base leakage current, which is modeled by Eq. (9.57) and the associated parameters C_2 and n_{EL}. The leakage component results in a deviation of the $\ln I_B$–V_{BE} dependence from the line with the $1/V_t$ slope. This situation is not observed in Fig. 11.17, and therefore the parameter C_2 is assumed to be zero. When the effect does appear, the changed slope of the line and its value at $V_{BE} = 0$ are used to estimate n_{EL} and $C_2 I_{S0}$, respectively. This is equivalent to the procedure for the measurement of n and I_S, as described in Section 11.1.3.

The validity of the measured parameters is checked in Fig. 11.19, where the model predictions (the dashed lines) are compared to the experimental data (the solid lines). As mentioned earlier, the estimated parameter values can be used as initial values for nonlinear curve fitting. Figure 11.19 also shows that the curve fitting changed slightly

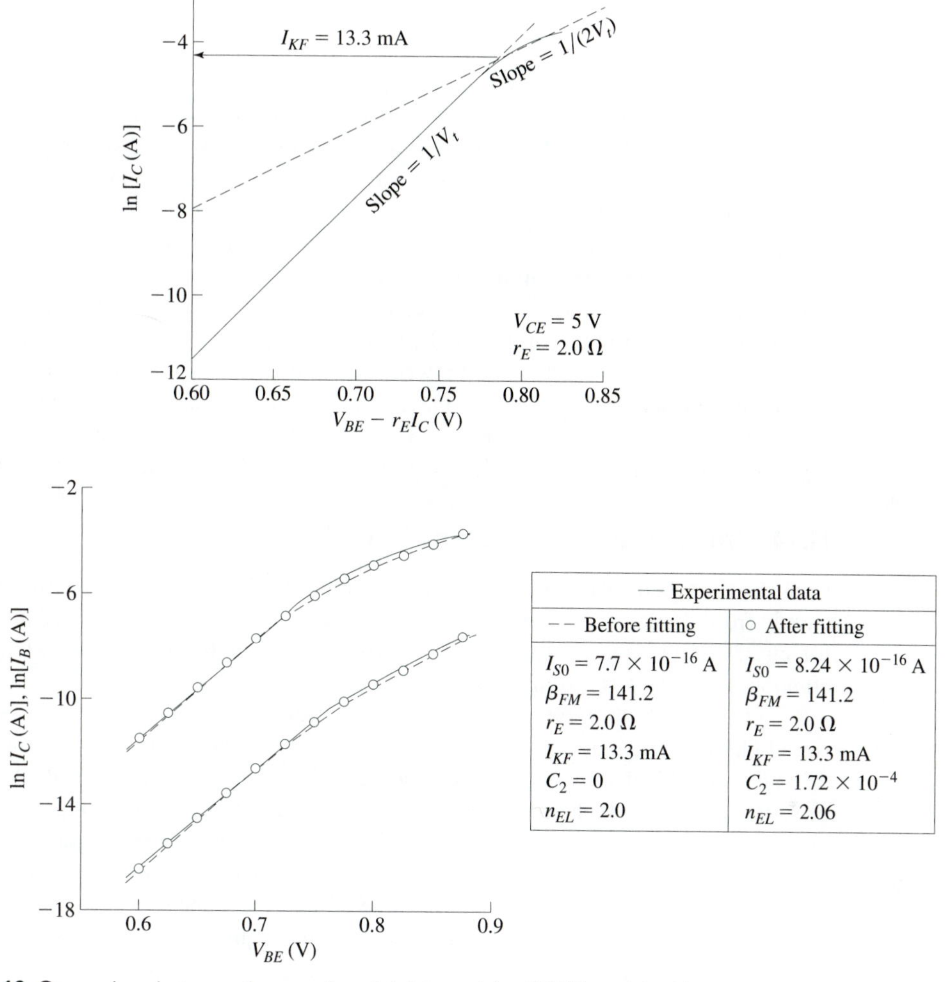

Figure 11.18 Measurement of I_{KF}.

Figure 11.19 Comparison between the experimental data and the SPICE model with parameter values as estimated by the described techniques (before fitting), as well as with parameter values after nonlinear curve fitting.

some parameter values to ensure the best fit between the model (open symbols) and the experimental data (the solid lines).

11.3.3 Large-Signal Equivalent Circuit and Dynamic Parameters in SPICE

Two P–N junctions are inherently present in any BJT structure, and they bring along the associated capacitances. As described in Chapter 6, there are two capacitances associated with every P–N junction: the depletion-layer capacitance C_d and the stored-charge capacitance C_s, the latter of which is important only in the case of forward-biased mode when significant current flows through the P–N junction. The depletion-layer and the stored-charge capacitances of the base–emitter and base–collector junctions appear in parallel with the respective diodes, as shown by the equivalent circuit in Table 11.15. Subscripts E and C are used for the base–emitter and the base–collector junctions, respectively. All these capacitances are voltage-dependent, as described in Chapters 6.

There is an additional P–N junction in the standard bipolar IC structure of the BJT, which is the isolating N-epi–P-substrate, or collector–substrate junction (shown later: Fig. 16.15). This junction is always reverse-biased, which means that the stored-charge capacitance never becomes important. However, the depletion-layer capacitance is always there, and it can definitely influence the high-frequency characteristics of the BJT. The depletion-layer capacitance of the collector–substrate junction is included in the large-signal equivalent circuit used in SPICE, which is C_{dS} in the equivalent circuit shown in Table 11.15. It appears connected between the collector and the lowest potential in the circuit (V_-), which is effectively zero level for the signal voltages and currents.

The equivalent circuit in Table 11.15 also includes the parasitic resistances in the base (r_B), the emitter (r_E), and the collector (r_C), which are all direct SPICE parameters.

11.3.4 Small-Signal Equivalent Circuit

Consider the specific case of the normal active mode to simplify the large-signal equivalent circuit from Table 11.15 into a small-signal equivalent circuit that is frequently used for circuit design and analysis. Table 11.16 summarizes the relationships between the components of the large-signal and the small-signal equivalent circuits.

To begin with, the base–collector junction is reverse-biased, so the diode D_C and the stored-charge capacitance C_{sC} can be removed because no diode current flows through this junction. What remains between the base and the collector is the depletion-layer capacitance C_{dC}, which in general depends on the voltage across the junction V_{CB} (analogous to the equation given in Table 11.2). However, as we are considering the small-signal situation, we are interested in the voltage V_{CB} that corresponds to the DC bias point (quiescent point) Q. Therefore, a single value of the capacitance between the base and the collector can be obtained and used in the small-signal equivalent circuit. This capacitance, labeled as C_μ in the circuit of Fig. 11.20, is known as Miller capacitance. Its importance lies in the fact that it makes a feedback between the output (collector) and the input (base) when the BJT is used in the common-emitter configuration.

The base–emitter junction is forward-biased, and both capacitances are important. The values of the depletion-layer capacitance C_{dE} and the stored-charge capacitance C_{sE} at the DC bias voltage V_{BE} (quiescent point Q) are summed to obtain the capacitance C_π.

The I–V characteristic of the forward-biased diode D_E is not linear; however, in the small range of the small input voltage change, it can be approximated by a linear segment, which is basically the slope of I–V characteristic at the operating point. The reciprocal value of this slope is equivalent to the resistance that the small input signal is facing. Therefore, for the small signals, the diode D_E is replaced by its small-signal resistance at the quiescent point, r_π.

TABLE 11.15 Summary of SPICE BJT Model: Parasitic Elements

Parasitic-Element-Related Parameters

Symbol	Usual SPICE Keyword	Related Parasitic Element	Parameter Name[a]	Typical Value/Range	Unit
r_B	RB	r_B	Base resistance	10	Ω
r_E	RE	r_E	Emitter resistance	2	Ω
r_C	RC	r_C	Collector resistance	15	Ω
$C_{dE}(0)$	CJE	C_{dE}	Zero-bias B–E capacitance		F
V_{biE}	VJE	C_{dE}	B–E built-in voltage	0.8	V
m_E	MJE	C_{dE}	B–E grading coefficient	$\frac{1}{3}$–$\frac{1}{2}$	—
τ_F	TF	C_{sE}	Normal transit time	10^{-9}	s
$C_{dC}(0)$	CJC	C_{dC}	Zero-bias B–C capacitance		F
V_{biC}	VJC	C_{dC}	B–C built-in voltage	0.75	V
m_C	MJC	C_{dC}	B–C grading coefficient	$\frac{1}{3}$–$\frac{1}{2}$	—
τ_R	TR	C_{sC}	Inverse transit time	10^{-9}	s
$C_{dS}(0)$	CJS	C_{dS}	Zero-bias C–S capacitance		F
V_{biS}	VJS	C_{dS}	C–S built-in voltage	0.7	V
m_S	MJS	C_{dS}	C–S grading coefficient	$\frac{1}{3}$–$\frac{1}{2}$	—

Large-Signal Equivalent Circuit

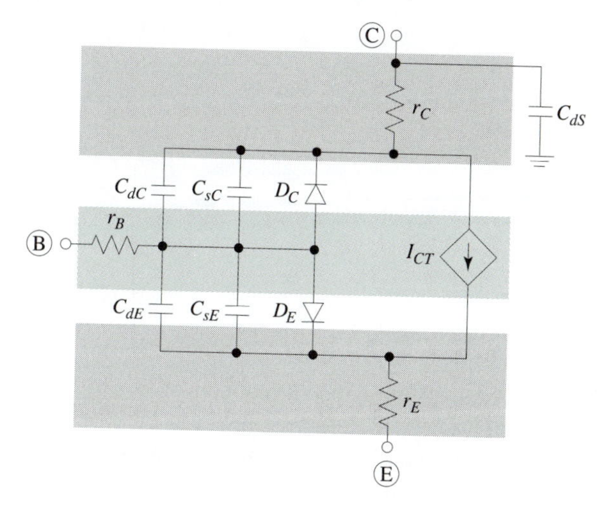

Note: The diodes and the current-source direction are shown for NPN BJT. Reverse current direction and diode polarities apply in the case of PNP BJT.

C_{dE}, C_{dC}, C_{dS} — According to C_d equation of Table 11.2

C_{sE}, C_{sC} — According to C_s equation of Table 11.2

[a] B–E, base–emitter; B–C, base–collector; C–S, collector–substrate.

TABLE 11.16 Relationship Between the Components of Large-Signal and Small-Signal Equivalent Circuits

Large-Signal Equivalent Circuit (General case, Table 11.15)	Small-Signal Equivalent Circuit (Normal active mode, Fig. 11.20)
D_C	Neglected
C_{sC}	Neglected
C_{dC} at V_{CB}	C_μ
$(C_{dE} + C_{sE})$ at V_{BE}	C_π
D_E at V_{BE}	r_π
I_{CT} at V_{BE}	$g_m v_\pi$ and r_o
C_{dS} at V_{CS}	C_o (common emitter)
r_B	r_B
r_E	r_E
r_C	r_C

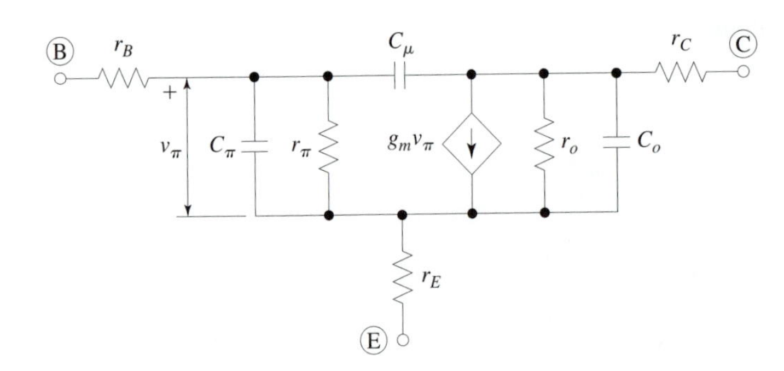

Figure 11.20 Small-signal equivalent circuit of a BJT.

Similarly, the current of the current source I_{CT} is a nonlinear function of V_{BE}, and through the Early effect, a nonlinear function of V_{CE}. For small signals, however, the I_{CT}–V_{BE} dependence can be approximated by a linear segment $i_c = g_m v_\pi$, where i_c is the small-signal output current, v_π is the small-signal input voltage, and g_m is the slope of the I_{CT}–V_{BE} characteristic at the quiescent point. The concept of g_m, which is called *transconductance*, is explained in Section 11.2.5. The dependence of I_{CT} on V_{CE} shows through the dynamic, or small-signal, output resistance r_o. The relationship between r_o and $I_{CT} \approx I_C$ is given by Eq. (9.52). Again, the derivative dI_C/dV_{CE} is calculated at the quiescent point to obtain the value for the small-signal equivalent circuit.

The capacitance C_o is due to any parasitic capacitances between the collector and the emitter, such as the depletion-layer capacitance C_{dS}. Finally, the parasitic resistances r_B, r_E, and r_C are the same in both the large-signal and small-signal equivalent circuits.

EXAMPLE 11.9 g_m and r_π

For the BJT of Example 9.6, calculate the transconductance g_m and the small-signal input resistance r_π at $V_{BE} = 0.80$ V and $V_{BC} = -5$ V.

SOLUTION

This BJT is in normal active mode:

$$g_m = \frac{dI_C}{dV_{BE}} = \frac{d}{dV_{BE}}\left(I_S e^{V_{BE}/V_t}\right) = \frac{1}{V_t}\underbrace{I_S e^{V_{BE}/V_t}}_{I_C}$$

$$g_m = \frac{I_C}{V_t}$$

$$g_m = \frac{0.49 \times 10^{-3}}{0.02586} = 18.95 \text{ mA/V}$$

$$r_\pi = \frac{dV_{BE}}{dI_B} = \beta_F \frac{dV_{BE}}{dI_C} = \beta_F \frac{1}{\underbrace{\frac{dI_C}{dV_{BE}}}_{1/g_m}} = \frac{\beta_F}{g_m}$$

$$r_\pi = \frac{188.5}{18.95} = 9.95 \text{ k}\Omega$$

SUMMARY

1. Equivalent circuits utilize elementary components (resistors, capacitors, controlled sources, etc.) to model the characteristics of more complex devices such as the diode. In the large-signal equivalent circuit of the diode, the nonlinear current–voltage characteristic is represented by a voltage-controlled current source. Two parallel capacitors represent the depletion-layer and stored-charge capacitances, whereas a series resistor represents the parasitic (contact and body) resistances.
2. The small-signal equivalent circuit is a linear circuit, applicable to a narrow voltage/current range that is centered at a DC bias (or quiescent) point. The values of the small-signal elements—the resistance and the parallel capacitances—depend on the applied DC bias. For example, the small-signal resistance is $r_d = nV_t/I_{DQ}$, where I_{DQ} is the DC-bias (quiescent) current.
3. The first estimation of SPICE parameters can be obtained by applying mathematical transforms that "linearize" the model equations. This enables the use of the graphic method, where the parameters or their mathematical transformations are related to the coefficients a_0 and a_1 of the general linear dependence: $y = a_0 + a_1 x$. For the case of forward-bias current, $I_D = I_S \exp(V_D/nV_t)$ is linearized as $\ln I_D = \ln I_S + V_D/nV_t$, which means $a_0 = \ln I_S$ and $a_1 = 1/nV_t$ for $x = V_D$ and $y = \ln I_D$. For the case of reverse-biased capacitance, $C_d = C_d(0)(1 - V_D/V_{bi})^{-m}$ is linearized as $\log C_d = \log C_d(0) - m \log(1 - V_D/V_{bi})$, which means $a_0 = \log C_d(0)$ and $a_1 = -m$ for $x = \log(1 - V_D/V_{bi})$ and $y = \log C_d$.

4. The threshold voltage V_T and the transconductance parameter KP of a MOSFET are obtained from the transfer characteristic in the linear region: $I_D = KP(W/L_{eff})(V_{GS} - V_T)$. The transconductance parameter is determined from the slope $KP(W/L_{eff})$, and the threshold voltage is determined from the intercept with the V_{GS}-axis. A particularly useful technique employs measurements of V_T for different V_{SB} voltages (body effect) to obtain the body factor γ (it involves the substrate doping concentration and the gate-oxide thickness).
5. The MOSFET structure has a number of inherent parasitic elements: source-to-body and drain-to-body diodes, resistors in series with all the four terminals, gate–source overlap capacitor, gate–drain overlap capacitor, and the central MOS capacitor that is modeled by two parts: gate–source ($C_{ox}/2$ in strong inversion) and gate–drain ($C_{ox}/2$ in series with C_d, with its value becoming smaller as the drain voltage expands the depletion layer). The parasitic capacitances are especially important because they determine the high-frequency behavior.
6. The base–emitter and base–collector P–N junctions introduce the associated parasitic capacitances in a BJT, both the depletion-layer and stored-charge capacitances. Also, there are parasitic resistances in series with all the BJT terminals. It is these parasitic elements that determine the high-frequency linear and switching performance of BJTs.

PROBLEMS

11.1 Determine the static SPICE parameters (the saturation current I_S, the emission coefficient n, and the contact resistance r_S) from the I_D–V_D data shown in Table 11.17.

TABLE 11.17 Current–Voltage Data

V_D (V)	0.65	0.70	0.76	0.81	0.89
I_D (μA)	79	264	879	2925	9685

11.2 Find the appropriate set of parameters for each of the lines in Fig. 11.21.

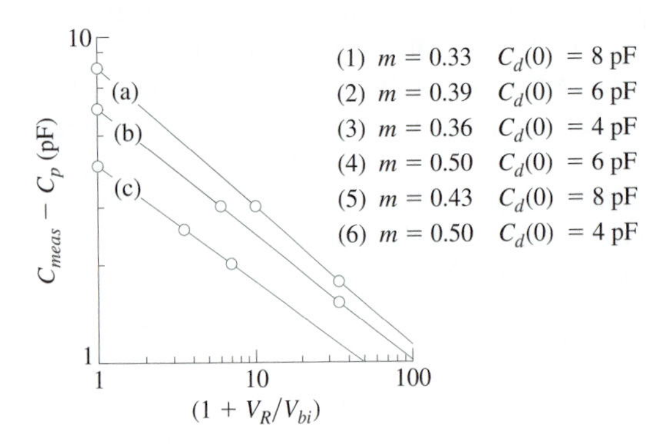

Figure 11.21 Three sets of capacitance–voltage measurements.

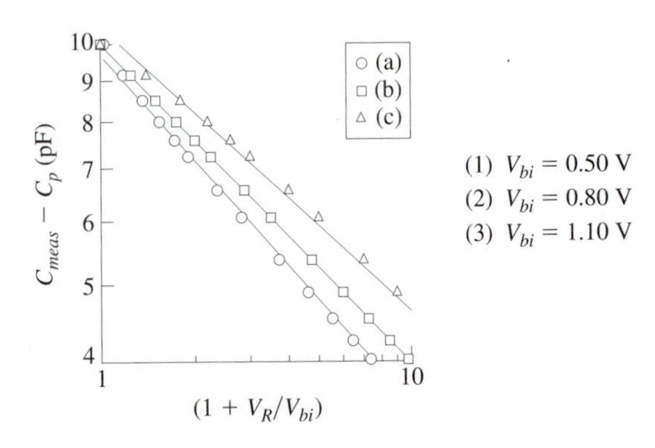

Figure 11.22 A set of measured C_d–V_R points (symbols) is plotted with three different assumptions for V_{bi}. The lines show the best linear fits.

11.3 A set of C_d–V_R data is plotted in Fig. 11.22 with three different assumptions for V_{bi}.

(a) Identify the set of data that corresponds to each of the V_{bi} values.

(b) Determine the grading coefficient m.

11.4 Using the data from Table 11.18 and assuming parallel parasitic capacitance $C_p = 1$ pF and built-in voltage $V_{bi} = 0.9$ V, obtain the best estimate of the grading coefficient m and the zero-bias capacitance $C_d(0)$.

TABLE 11.18 Capacitance–Voltage Measurements

P–N Junction Capacitance (pF)	Voltage (V)
3.33	−1
2.72	−3
2.44	−5
2.28	−7
2.16	−9

11.5 The SPICE equation for the temperature dependence of the saturation current of a P–N junction is

$$I_S(T) = I_S\left(\frac{T}{T_{nom}}\right)^{p_t/n} \exp\left[-\frac{qE_g}{kT}\left(1-\frac{T}{T_{nom}}\right)\right]$$

Using the following theoretical dependencies for the intrinsic carrier concentration and the mobility on temperature:

$$N_C = A_C T^{3/2}, \qquad N_V = A_V T^{3/2},$$

$$n_i = \sqrt{N_C N_V} e^{-E_g/2kT} \qquad \mu_{n,p} = C_{n,p} T^{-3/2}$$

and assuming $n = 1$, find the theoretical value of the parameter p_t (saturation-current temperature exponent). **A**

11.6 The saturation current of a Schottky diode can be expressed as [Eq. (7.4)]

$$I_S = A_J A^* T^2 \exp{-q\phi_B/(kT)}$$

where A_J is the junction area and A^* is effective Richardson constant. Assuming $n = 1$, find the theoretical value of the parameter p_t (saturation-current temperature exponent), so that the SPICE equation given in Problem 11.5 can be used for this Schottky diode. **A**

11.7 Obtain the zero-bias threshold voltage V_{T0} and the transconductance parameter KP of an N-channel MOSFET using the data given in Table 11.19. The channel-width-to-channel-length ratio of the transistor is 100.

TABLE 11.19 I_D–V_{GS} Data

Drain Current (mA)	Gate Voltage (V)
$V_S = 0$, $V_B = 0$, $V_{DS} = 50$ mV	
0.18	1
0.50	1.5

11.8 To determine the value of the body factor γ of an N-channel MOSFET, the threshold voltage dependence on $\sqrt{2\phi_F + |V_{SB}|} - \sqrt{2\phi_F}$ is analyzed. This dependence becomes linear for $2\phi_F = 0.82$ V. Two points of this dependence are given in Table 11.20. Determine the body factor of this MOSFET.

TABLE 11.20 Body-Effect Data

V_T (V)	$(\sqrt{2\phi_F + \lvert V_{SB}\rvert} - \sqrt{2\phi_F})$ ($V^{1/2}$)
1.0	0.0
2.0	1.0

11.9 For the N-channel MOSFET considered in Problem 11.7, obtain the best estimate of the drain current at $V_{GS} = 5$ V, to complete Table 11.21. **A**

TABLE 11.21 I_D–V_{GS} Data

Drain Current (mA)	Gate Voltage (V)
$V_S = 0$, $V_B = 0$, $V_{DS} = 50$ mV	
0.18	1
0.50	1.5
?	5.0
2.49	8.0

11.10 A set of measurements of β versus V_{GS}–V_T is given in Table 11.22. Determine the mobility-modulation coefficient θ, used to express the mobility reduction with the gate voltage in the SPICE LEVEL 3 MOSFET model.

TABLE 11.22 β–(V_{GS}–V_T) Data

V_{GS}–V_T (V)	0.5	1.0	3.0	5.0	7.0	9.0
β (mA/V^2)	455	455	385	333	294	263

11.11 The gate of a MOSFET overlaps the source and drain regions by 100 nm each, and it overlaps the field oxide by 500 nm. The gate-oxide thickness is $t_{ox} = 10$ nm, whereas the field-oxide thickness is $T_{ox} = 100$ nm. Determine the following SPICE parameters: C_{GDO} (gate–drain overlap capacitance per channel width), C_{GSO} (gate–source overlap capacitance per channel width), and C_{GBO} (gate–bulk overlap capacitance per channel length). (**A** for C_{GDO})

11.12 The source–bulk and drain–bulk junction depth is $x_j = 100$ nm and the lateral diffusion is $x_{j-lat} = 0.8x_j$. What are the gate–drain and the gate–source overlap capacitances per channel width? The gate-oxide thickness is 8 nm, and the gate itself is used as a mask for source/drain implantation (self-aligned structure). What is the total gate capacitance if $L_{gate} = 0.3\ \mu$m and $W = 3\ \mu$m? (Ignore any gate–bulk overlap capacitance.)

11.13 To extend the result obtained in Example 11.7b for the case of N cascaded inverters that drive load capacitance C_L, we assume that $C_L = S^N(C_{IN-NMOS} + C_{IN-PMOS})$. Express the delay time in terms of the simple-model parameters of the minimum-size inverter and generalized number of scaled-up and cascaded inverters, N. Plot t_{delay} versus N and discuss the result.

11.14 The maximum operating frequency, also called cutoff frequency, of a FET is defined by

$$f_{max} = \frac{g_m}{2\pi(C_{gs} + C_{gd})}$$

where g_m is the transconductance and C_{gs} and C_{gd} are small-signal gate–source and gate–drain capacitances. Find f_{max} at the onset of saturation and $V_{GS} - V_{T0} = 1$ V for an N-channel MOSFET with $L_{gate} = 250$ nm, $L_{eff} = 200$ nm, $W = 20\ \mu$m, $t_{ox} = 5$ nm, $\mu_{eff} = 350$ cm^2/V·s, and $F_B \ll 1$.

11.15 Using the data from Table 11.23, estimate the following parameters: I_{S0}, β_F, and I_{KF}. (Use $V_t = 26$ mV).

TABLE 11.23 Input and Transfer Characteristics Data

V_{BE} (V)	I_B (A)	I_C (A)	V_{BE} (V)	I_B (A)	I_C (A)
0.60	1.2×10^{-7}	1.2×10^{-5}	0.78	1.3×10^{-4}	7.3×10^{-3}
0.62	2.6×10^{-7}	2.6×10^{-5}	0.80	2.8×10^{-4}	1.2×10^{-2}
0.64	5.5×10^{-7}	5.6×10^{-5}	0.82	6.0×10^{-4}	2.0×10^{-2}
0.66	1.2×10^{-6}	1.2×10^{-4}	0.84	1.3×10^{-3}	3.1×10^{-2}
0.68	2.7×10^{-6}	2.6×10^{-4}	0.86	2.8×10^{-3}	4.8×10^{-2}
0.70	5.8×10^{-6}	5.5×10^{-4}	0.88	6.1×10^{-3}	7.3×10^{-2}
0.72	1.2×10^{-5}	1.1×10^{-3}	0.90	1.3×10^{-2}	1.1×10^{-1}
0.74	2.7×10^{-5}	2.2×10^{-3}	0.92	2.9×10^{-2}	1.6×10^{-1}
0.76	5.9×10^{-5}	4.1×10^{-3}	0.94	6.2×10^{-2}	2.4×10^{-1}

TABLE 11.24 Output Characteristics Data

V_{CE}	I_C (mA)				
(V)	@$I_B = 20\ \mu$A	@$I_B = 40\ \mu$A	@$I_B = 60\ \mu$A	@$I_B = 80\ \mu$A	@$I_B = 100\ \mu$A
3.0	2.73	5.05	7.07	8.82	10.2
6.0	2.79	5.21	7.34	9.24	11.0
9.0	2.85	5.34	7.54	9.53	11.4
12.0	2.90	5.46	7.74	9.81	11.7
15.0	2.95	5.58	7.93	10.1	12.1

11.16 Using the data from Table 11.24, estimate the Early voltage V_A.

11.17 I_B–V_{BE} measurements taken at low bias level are given in Table 11.25. Determine the related parameters $C_2 I_{S0}$ and n_{EL} ($V_t = 26$ mV).

TABLE 11.25 Transfer Characteristic Data

V_{BE} (V)	I_B (nA)	V_{BE} (V)	I_B (nA)
0.60	5.0	0.63	9.50
0.61	6.3	0.64	12.0
0.62	7.6	0.65	15.0

REVIEW QUESTIONS

R-11.1 What are the two essential SPICE parameters that determine the current–voltage characteristic of a diode?

R-11.2 How is the SPICE model related to the theoretical equations for depletion-layer capacitances of abrupt and linear junctions?

R-11.3 Ideally, what is the small-signal resistance of a reverse-biased P–N junction? What if the reverse-biased P–N junction is operated in the breakdown region?

R-11.4 What are the origins of the two parasitic capacitances in the diode structure? Are there any other parasitic elements? How are they connected in the large-signal equivalent circuit?

R-11.5 What are the parasitic capacitances in the large-signal equivalent circuit of a MOSFET? Relate these capacitances to the MOSFET structure.

R-11.6 Draw the equivalent circuit of the simple digital MOSFET model. How do the circuit elements relate to the MOSFET structure and the physical parameters?

R-11.7 Draw the small-signal equivalent circuit of a MOSFET for high frequencies (capacitors included) and low frequencies (capacitors replaced by open circuits).

R-11.8 How is the high-level knee current (I_{KF}) determined from measured I_C–V_{BE} data?

R-11.9 What are the capacitances in the large-signal equivalent circuit of a BJT?

R-11.10 Draw the small-signal equivalent of a BJT and compare it to the equivalent circuit of a MOSFET. Are there similarities between these two circuits? What are the differences?

12 Photonic Devices

Electrons can interact with light through the following fundamental mechanisms: (1) *spontaneous light emission,* (2) *light absorption,* and (3) *stimulated light emission*. Each of these three effects is exploited for very useful devices: light-emitting diodes (LED), photodetectors/solar cells, and lasers, respectively. A common name for these devices is *optoelectronic* devices or *photonic* devices. These devices cover a range of very important applications: displays, sensors, optical communications, control, and so on.

Photons are quanta of light energy: $h\nu$, where h is Planck's constant, and ν is the light frequency. To emit a photon, an electron has to give away energy equal to $h\nu$, whereas after a photon absorption an electron gains energy equal to $h\nu$. In semiconductors, electrons lose and gain energy through the processes of *recombination* with the holes and electron–hole *generation,* respectively. The generation and recombination mechanisms are described in Chapter 5. This chapter provides specifics on the applications and structure of diodes used as LEDs, photodetectors/solar cells, and lasers.

12.1 LIGHT-EMITTING DIODES (LED)

The energy-band diagram of a forward-biased diode, shown previously in Fig. 6.6a, is reproduced again in Fig. 12.1b to specifically illustrate the light emission by recombination of the minority carriers. As illustrated in this figure, the electrons with high enough energy to overpass the energy barrier at the depletion layer and appear on the P-type side as minority carriers will sooner or later recombine with majority holes. Analogously, the holes appearing in the N-type region are recombined with the majority electrons. In the process of electron–hole recombination, the electrons change their energy status from the high energy levels in the conduction band to the low energy levels in the valence band. The energy difference between a free electron and a recombined electron must be released, and

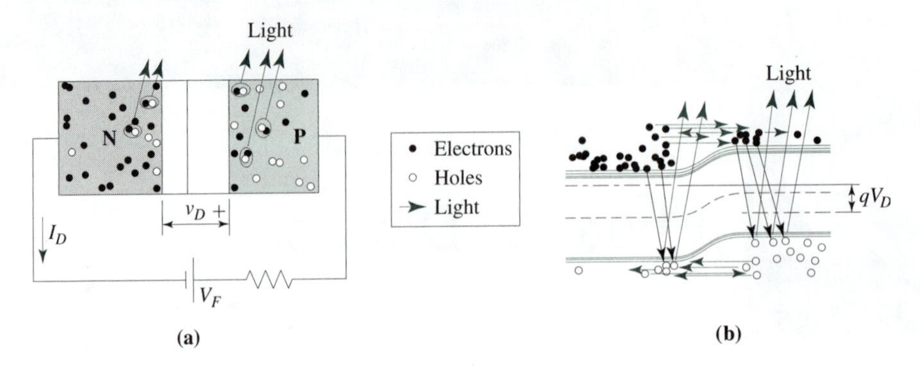

Figure 12.1 (a) Cross section and (b) energy-band diagram of a forward-biased LED, illustrating emission of photons due to the electron–hole recombination.

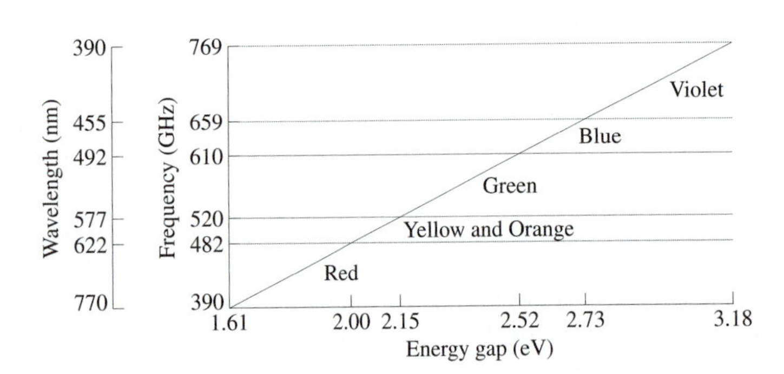

Figure 12.2 Different energy gaps are needed to produce LEDs emitting light of different colors.

it can be released in the form of a photon. As mentioned earlier, a photon is a quantum of light energy, given as

$$E_{photon} = h\nu = \frac{hc}{\lambda} \tag{12.1}$$

where h is Planck's constant ($h = 6.626 \times 10^{-34}$ J · s), ν is the light frequency, c is the speed of light ($c = 3 \times 10^8$ m/s), and λ is the light wavelength.

Figure 12.1 illustrates the photons produced by electron–hole recombination. It is obvious that the photon energy $h\nu$ is approximately equal to the energy gap E_g. There are a variety of compound semiconductor materials that provide different energy-gap values suitable for *different colors* of the visible light. Figure 12.2 shows the energy gap values needed to produce different colors of light.

Ternary and quaternary compound semiconductors, such as the GaAsP and InAlGaP systems, exhibit energy-gap values that correspond to visible light. In addition, the energy gap changes with the composition, enabling adjustments for specific light colors. For example, $GaAs_{0.6}P_{0.4}$ is typically used for red LEDs ($\approx$1.9 eV). The InAlGaP system is useful for yellow and green LEDs. Blue LEDs have been the hardest to develop. GaN, with its energy gap of about 3.4 eV, emits blue light very efficiently, but it has not been possible to develop monocrystalline GaN wafers. The recently developed blue LEDs use GaN deposited on either SiC or sapphire wafers.

Another important characteristic of LEDs is the *light intensity,* which is the number of photons emitted per unit time. It is directly related to the optical power, $P_{opt}A_J$, where P_{opt} is the optical power density in W/m^2 and A_J is the junction area. By observing that the optical power is equal to the energy of photons emitted per unit time and by knowing that the energy of each photon is $h\nu$, we find that the number of photons emitted per unit time—the light intensity—is equal to $P_{opt}A_J/h\nu$. If every recombined minority carrier was emitting a photon, $P_{opt}A_J/h\nu$ would be equal to the number of injected (and consequently recombined) minority carriers per unit time, I_D/q. However, only a fraction η_Q of the recombination events will result in light emission, so that

$$\frac{P_{opt}}{h\nu}A_J = \frac{\eta_Q}{q}I_D \tag{12.2}$$

The parameter η_Q expresses the efficiency of an LED and is referred to as *radiative recombination efficiency*.

The radiative recombination efficiency is not equal to 1 because some of the recombination events inevitably release the energy in the form of phonons (heat). Section 5.1 describes that the ratio of energy released as photons (light) and phonons (heat) depends to a large extent on whether the semiconductor is direct or indirect. In direct semiconductors, the top of the valence band and the bottom of the conduction band appear for the same wave vector $\boldsymbol{k}$, so the radiative band-to-band recombination is very likely. In the case of indirect semiconductors, the top of the valence band and the bottom of the conduction band appear for different wave vectors $\boldsymbol{k}$. Requiring changes in both the energy and the wave vector, recombination in indirect semiconductors typically occurs through R–G centers (Fig. 5.1) and involves phonons so that the energy is typically released to the phonons.

Given a certain radiative recombination efficiency, the light intensity is directly proportional to the concentration of electron–hole pairs recombined per unit time, which is the recombination rate. The recombination rate is directly proportional to the concentration of available electrons and the concentration of available holes. In the P-type region, there are plenty of available holes, so the recombination rate is basically limited by the excess concentration of minority electrons. In the N-type region, there are plenty of electrons, so the recombination rate is limited by the excess concentration of holes (again, the minority carriers). The excess concentrations of the minority carriers make the stored charge discussed in Section 6.4 and illustrated in Fig. 6.19. In a steady-state situation (constant diode current I_D), the stored-charge density does not change in time. This is because the recombined minority carriers are replaced by the new minority carriers that are injected over the junction barrier as the forward-bias current I_D. Therefore, the light intensity is directly proportional to the diode current I_D.

Figure 12.3a shows a simple LED driving circuit, whereas Fig. 12.3b indicates the operating point on the I_D–V_D characteristic. If $v_{IN} \gg V_{DO}$, the light intensity becomes directly proportional to the input voltage v_{IN}:

$$\text{Light intensity} \propto I_{DO} = \frac{v_{IN} - V_{DO}}{R} \approx \frac{v_{IN}}{R} \tag{12.3}$$

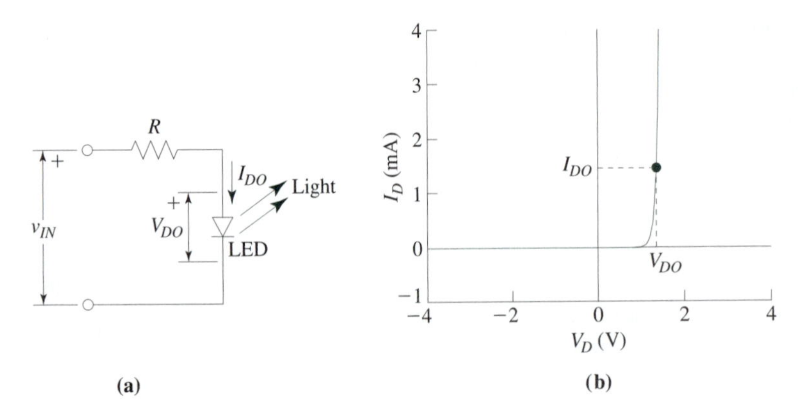

Figure 12.3 (a) LED driving circuit. (b) Operating point.

This enables us to use LEDs as binary indicators, that is, to visualize the two (*yes* and *no*) logic states. In this case the diodes are operated in two points: zero current (light off) and the optimum (recommended) operating current I_{DO} (light on). The resistor R of the driving circuit is used to adjust the *on* level of the input voltage to the optimum operating current I_{DO}.

Given that visible LEDs are made of semiconductor materials with larger energy gaps than silicon, they exhibit proportionally larger turn-on voltages. This is because the energy barriers at LED junctions are significantly higher than is the case with silicon P–N junctions. This means that the 0.7 V, which is the typical turn-on voltage in silicon diodes, cannot reduce the barrier height sufficiently for the electrons and holes to be able to overpass it, and in turn produce the forward-bias current.

12.2 PHOTODETECTORS AND SOLAR CELLS

In addition to emitting light, diodes can absorb light to generate electrons and holes. This process of light conversion into electric current is useful not only for electronic light detection, but also for conversion of solar power into electric power, using specifically designed P–N junctions called solar cells.

12.2.1 Biasing for Photodetector and Solar-Cell Applications

Photodetector diodes, or photodiodes, are biased in the reverse-bias region, as illustrated in Fig. 12.4. In the dark, the current–voltage characteristics of photodiodes are the same as the characteristics of rectifying diodes. This means that only the leakage current flows in the reverse-bias region. When exposed to light, the reverse current of the photodiode increases proportionally to the light intensity. This current is referred to as *photocurrent*. Figure 12.4b illustrates that the photocurrent, similar to the normal reverse-bias current, does not depend on the reverse-bias voltage.

The circuit of Fig. 12.4a converts the light intensity into voltage V_o. In the dark, the current through the circuit is approximately zero, and therefore the voltage across R is zero as well. The load line in Fig. 12.4b intersects the diode characteristics at 0 mA and −7 V,

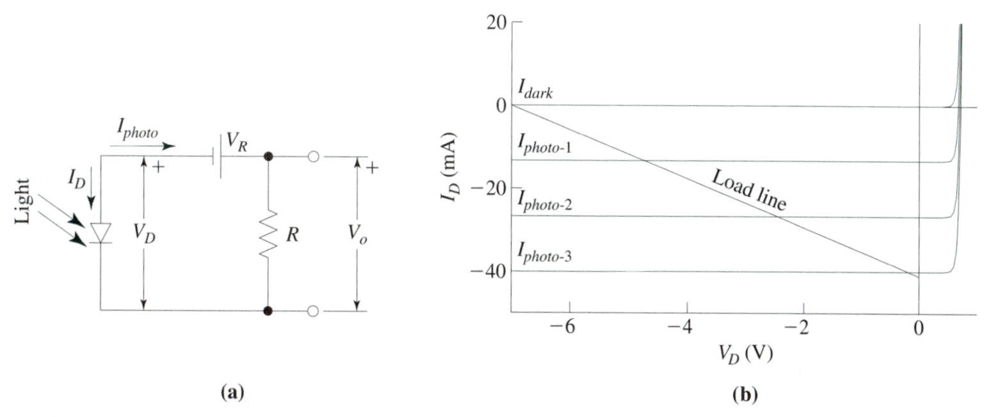

Figure 12.4 (a) A simple circuit biasing a photodetector diode. (b) The current–voltage characteristics of the photodiode and the load line provide graphical analysis of the circuit.

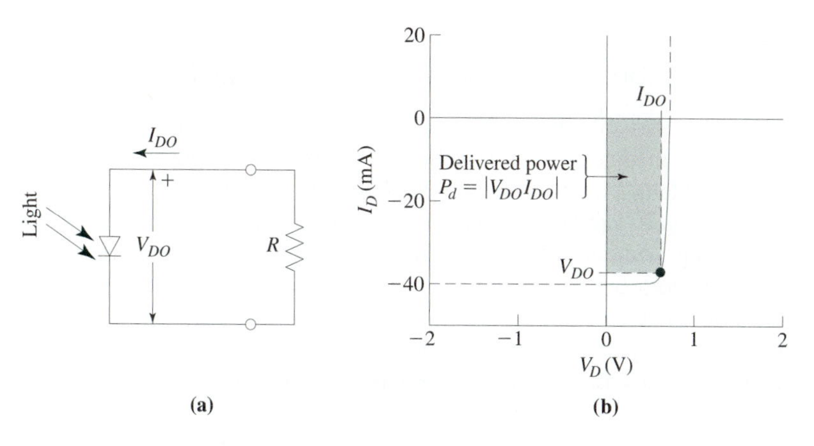

Figure 12.5 A solar-cell diode is directly connected to a loading element. (a) The electric circuit. (b) The operating point (I_{DO}, V_{DO}) is in the quadrant of negative currents and positive voltages (forward bias).

which is the assumed reverse-bias voltage V_R; therefore, $V_o = V_R - V_D = 7 - 7 = 0$. With an increase in the light intensity and consequently the photocurrent, the voltage across the resistor—that is, the output voltage V_o—increases as well. This is accompanied by a corresponding reduction in the reverse bias of the photodiode. The load line in Fig. 12.4b illustrates that the maximum output voltage is approximately limited to the reverse-bias voltage V_R. If the light intensity is increased beyond this point, the diode is pushed toward the forward-bias region, where the appearance of the normal forward-bias current, which flows in the opposite direction, limits the output voltage increase.

As opposed to photodetectors, the diodes used as solar cells operate in the forward-bias region. Figure 12.5a shows that the diode as a solar cell is directly connected to a loading element (resistor R).

Two extreme biasing conditions of the solar-cell diode are short circuit ($R = 0$) and open circuit ($R = \infty$). At short circuit ($V_{DO} = 0$), the only current flowing through the diode is the photocurrent. Although this condition is useful to measure the value of the photocurrent, it produces no power as $V_{DO}I_{DO} = 0$.

With some load resistance R, the voltage across the diode V_{DO} becomes positive, whereas the current I_{DO} still remains negative. The negative power value $P_d = V_{DO}I_{DO}$ indicates that the diode acts as a power generator. The power generated by the diode is indicated by the shaded area in Fig. 12.5b.

The current I_{DO} in the circuit of Fig. 12.5a remains negative because the photocurrent is larger than the normal forward-bias current. The total current I_{DO} can never become positive because the source of the forward bias is the photocurrent itself. In the extreme case, the normal forward-bias current can become equal to the photocurrent making the total current I_{DO} equal to zero, which is the open-circuit condition. Although the open-circuit condition provides the maximum voltage drop V_{DO}, it produces no power because, again, $V_{DO}I_{DO} = 0$.

It is obvious that the value of the load resistance R directly influences the power that is delivered to the loading element. The maximum delivered power corresponds to a single value of the load resistance R. However, any change in the photocurrent (that is, the light intensity) will change the value of the load resistance that is needed to maximize the power.

12.2.2 Carrier Generation in Photodetectors and Solar Cells

The photodetector and solar-cell diodes operate by the mechanism of light-induced electron–hole generation. In this process, which is the opposite of light emission, the photon energy $h\nu$ is used to destroy a covalent bond, liberating an electron and creating a hole. If the generated electron and hole diffuse to or are generated in the depletion layer of the diode, the existing electric field sweeps them away before they get a chance to recombine, creating the photocurrent.

The photodetector and the solar-cell application circuits shown in Fig. 12.4a and Fig. 12.5a, respectively, are presented in Fig. 12.6a and 12.6b with the diode symbol replaced by cross sections illustrating electron–hole generation by light. It is shown that the electrons and holes generated in the depletion layer move toward their respective majority-carrier regions, like the minority carriers, due to the direction of the electric field (E) in the depletion layer. Therefore, the current due to the light-generated electrons and holes adds up to the thermal reverse-bias current I_S. The light also generates electrons and holes in the neutral N-type and P-type regions that can diffuse to the depletion layer, contributing to the photocurrent.

Clearly, to maximize photodiode sensitivity, the depletion layer should be as wide as possible. The depletion-layer width depends on the doping level: $w_{depl} \propto 1/\sqrt{N_{A,D}}$ in the case of an abrupt P–N junction (Section 6.3.2). Therefore, the lowest doping level that can technologically be achieved is the most favorable in terms of maximizing depletion-layer volume. The most common photodetectors are made with such a layer between the P-type and N-type regions, which are needed as the diode terminals (the anode and the cathode). To distinguish this very lightly doped, almost *intrinsic* region from the N-type and P-type regions, it is labeled I; consequently, the diode is specifically referred to as a *PIN photodiode*. The thickness of the I region is such that it is completely depleted.

The energy-band diagrams of the photodetector and the solar-cell diodes, also given in Fig. 12.6, provide deeper insight into the exploitation of the carrier-generation mechanism for light detection and solar-power conversion. Remembering that electrons roll down and the holes bubble up along the energy bands, we can clarify the first important point, which is the direction of the photocurrent flow. This is a simple question as far as the

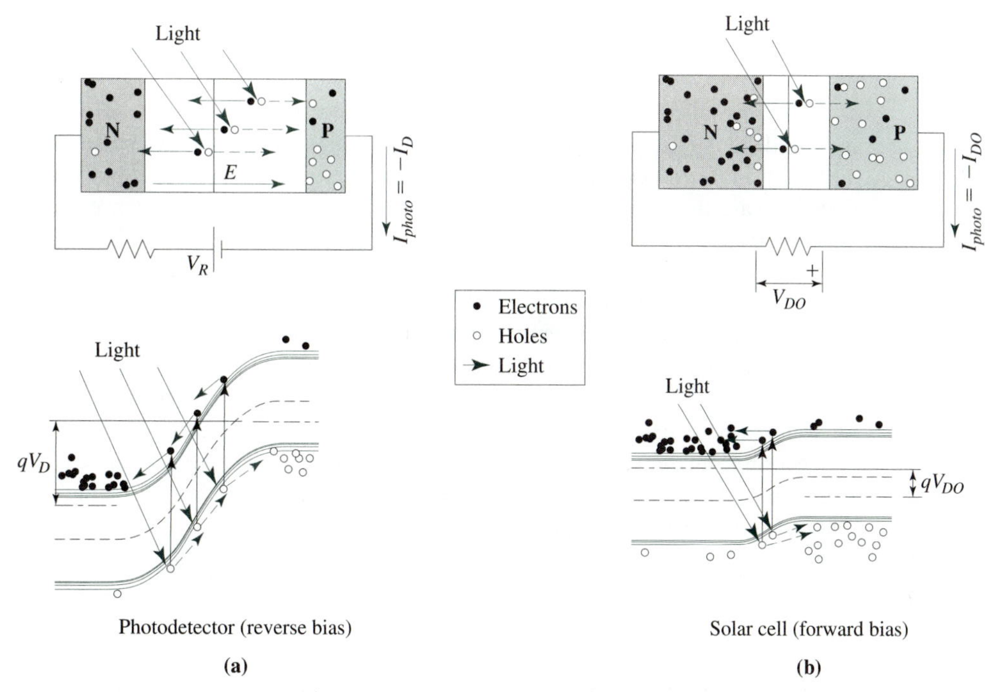

Figure 12.6 Energy-band diagrams of (a) a photodetector and (b) a solar cell, accompanied by their respective application circuits, where the diode symbols are replaced by diode cross sections.

photodetector circuit is concerned, and the energy bands are not necessary to answer it: the reverse-bias voltage V_R (that is, the associate electric field E) drives the electrons generated in the depletion layer toward the neutral N-type region and the holes toward the neutral P-type region. The situation is not as obvious in the case of the solar-cell circuit. The question is: What drives the light-generated electrons and holes if there is only a resistor connected to the diode?

The energy-band diagram of Fig. 12.6b shows that there is a slope in the energy bands; hence the electrons generated in the depletion layer roll down toward the N-type region and holes bubble up toward the P-type region. This energy-band slope and the associated field are due to the ionized doping atoms in the depletion layer (the built-in electric field), as explained in Section 6.1.1. The splitting of the Fermi levels in the photodetector and the solar-cell diodes (Figs. 12.6a and 12.6b, respectively) is in the opposite directions, indicating the reverse bias of the photodetector and the forward bias of the solar cell. However, the energy-band slopes in the depletion layers are in the same direction, which means that the photocurrents are in the same direction as well. This is consistent with the basic application circuits. The photodiode is reverse-biased by the DC voltage source V_R, and the photocurrent flows in the same direction as the reverse-bias current. There is no DC bias in the solar-cell circuit, but the photocurrent causes voltage $I_{photo}R$ across both the resistor and the diode itself. It can be seen from the circuit in Fig. 12.5a that the negative photocurrent, flowing in the direction from the anode through the resistor into the cathode, causes a forward-biased voltage (positive voltage between the anode and the cathode).

As a result, the power dissipated by the diode is negative, which means that the diode acts as a generator of electrical power (in fact it converts the solar power into electrical power).

The second point that can easily be explained by the energy-band diagrams is the independence of the photocurrent on the reverse-bias voltage applied (refer to the current voltage characteristics of Fig. 12.4a). Although an increase in the reverse-bias voltage does increase the steepness of the energy bands in the depletion layer, the photocurrent is not increased because it is not limited by the steepness of the energy bands. The built-in bending of the bands (the built-in electric field) alone is good enough for every generated electron or hole to easily roll down or bubble up through the depletion layer. What determines the photocurrent is the rate of electron–hole pairs generated by the light. This number is increased by the light intensity but not by an increase in the reverse-bias voltage.

The third point that becomes obvious from the energy-band diagrams is that light with photon energies smaller than the energy gap E_g of the semiconductor cannot possibly move an electron from the valence band into the conduction band to generate free electron–hole pairs. From the condition $E_g = h\nu_{min} = hc/\lambda_{max}$, the maximum wavelength of the light that can generate electrons and holes is found as

$$\lambda_{max} = \frac{hc}{E_g} \tag{12.4}$$

The energy gap of silicon corresponds to $\lambda_{max} = 1.1\ \mu\text{m}$. Because almost the complete spectrum of the solar radiation is below this maximum wavelength, silicon appears to be an excellent material for solar cells. The fact that silicon is an indirect semiconductor does not prevent light absorption, so solar cells are most frequently made of silicon. Photodetectors, similarly to LEDs, are made of different materials to maximize device sensitivity to the light of a designated color while minimizing sensitivity to the other colors.

An additional difference is in the design of the diode structure. To provide a quick response, the capacitance of the photodetector is minimized by minimizing the diode area and increasing the depletion-layer width by inserting the "I" region. Solar cells are essentially large-area P–N junctions, designed to maximize the current generated by the incoming light.

12.2.3 Photocurrent Equation

In this section, a photocurrent equation is derived, assuming uniform carrier generation in the area of interest (around the P–N junction). Let us first deal with the photocurrent due to carrier generation in the depletion layer. Given that the electric field in the depletion layer immediately separates the generated electrons and holes, we basically need to convert the external generation rate G_{ext} into photocurrent. The generation rate G_{ext} is the *number* of electron–hole pairs generated in a unit of the depletion-layer volume per second. The generation rate multiplied by the unit charge, qG_{ext}, expresses the *charge* generated in a unit of the depletion-layer volume per second. As the electrons generated in the entire volume of the depletion layer contribute to the photocurrent, qG_{ext} is multiplied by the depletion-layer volume $A_J w_d$, which gives the electric charge generated per unit time: $qG_{ext} w_d A_J$. The charge generated per unit time is the photocurrent (C/s = A):

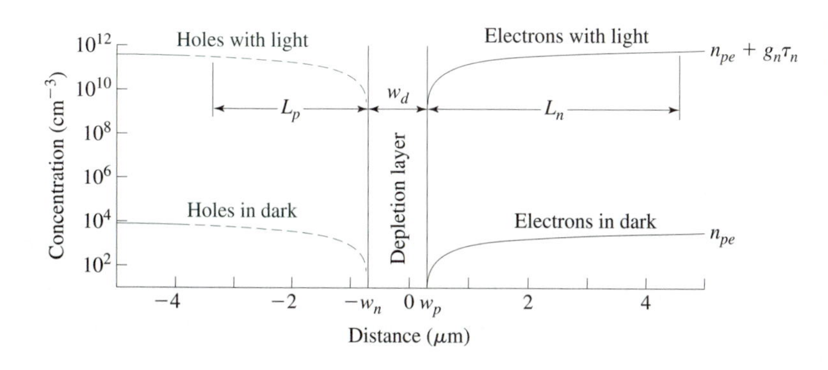

Figure 12.7 Minority-carrier concentration diagrams for a reverse-biased photodiode.

$$I_{photo-d} = qA_J G_{ext} w_d \tag{12.5}$$

The carriers generated in the neutral regions, far from the P–N junction, recombine because there is neither electric field nor concentration gradient to produce drift or diffusion currents. With $\partial j_n/\partial x = 0$ and $\partial n_p/\partial t = 0$, we find from the continuity equation [Eq. (5.1)] that $G_{ext} = U$. The effective thermal generation–recombination rate U is related to the excess concentration of minority carriers and the minority-carrier lifetime. In analogy with Eq. (5.8), $U = \delta n/\tau_n$ in the P-type region. This means that the steady-state level of excess electron concentration $\delta n = n_p - n_{pe}$, due to a uniform generation rate G_{ext}, is equal to $G_{ext}\tau_n$. However, at the edge of the depletion layer, the concentration is $n_p(w_p) = n_{pe}\exp(V_D/V_t) \ll n_{pe}$, because $V_D/V_t \ll -1$. This means that there is a concentration gradient around the P–N junction, as illustrated in Fig. 12.7. In the dark, the concentration gradient is small, leading to a small reverse-bias diffusion current I_s, which is the already well-known saturation current of the diode. The carrier generation lifts the concentration in the neutral region to $G_{ext}\tau_n + n_{pe}$, increasing the concentration gradient and therefore the reverse-bias diffusion current. To determine the component of the photocurrent due to the diffusion of excess electrons, $I_{photo-n}$, we can again assume a linear concentration gradient in the diffusion equation [Eq. (4.5)]. In analogy with Eq. (6.5), $\partial n_p/\partial x \approx (G_{ext}\tau_n + n_{pe})/L_n$. Clearly, this leads to the following result:

$$I_n = \underbrace{qA_J D_n \frac{n_{pe}}{L_n}}_{I_{S-n}} + \underbrace{qA_J D_n \frac{G_{ext}\tau_n}{L_n}}_{I_{photo-n}} \tag{12.6}$$

Given that $L_n^2 = D_n\tau_n$, the diffusion component of the photocurrent can be expressed as

$$I_{photo-n} = qA_J G_{ext} L_n \tag{12.7}$$

An analogous equation would be obtained for the diffusion current of holes in the N-type region: $I_{photo-h} = qA_J G_{ext} L_p$.

Upon adding the three components of the photocurrent together, we have the total photocurrent:

$$I_{photo} = qA_J G_{ext}(w_d + L_n + L_p) \tag{12.8}$$

It should be noted that the drift photocurrent $I_{photo-d}$ responds almost instantly to changes in light intensity. On the other hand, the response of the diffusion photocurrents $I_{photo-n}$ and $I_{photo-p}$ is limited by the rate of establishing the concentration profiles. For a fast-response photodetector, it is desirable to have $w_d \gg L_n + L_p$, so that the drift photocurrent dominates. To achieve this, the PIN structure is used. The very low doping of the "intrinsic" region enables very wide depletion layers, which not only helps satisfy the condition $w_d \gg L_n + L_p$ but also increases the magnitude of the photocurrent. The width of the "intrinsic" region is designed so that it is fully depleted at very small reverse-bias voltages. Because the total depletion-layer width is dominated by the "intrinsic" region, this means that the widening of the depletion layer with the reverse-bias voltage is negligible, and so the photocurrent is approximately voltage-independent (a light-controlled constant-current source).

EXAMPLE 12.1 Derivation of the Link Between Optical Power Density and the External Generation Rate

The absorption of light in a semiconductor material is characterized by the optical absorption coefficient, α. The absorption coefficient is a wavelength-dependent material constant that expresses the fraction of absorbed photons per unit length.

(a) If the incident optical power density is $P_{opt}(0)$, derive the equation for the change of the optical power density $P_{opt}(x)$ as the light is absorbed inside the material ($x > 0$).

(b) Derive the link between the absorbed optical power density and the external-generation rate.

SOLUTION

(a) Considering a distance interval dx, the product αdx shows the fraction of absorbed light across dx. Therefore, $P_{opt}(x)\alpha\, dx$ is the reduction of the optical power density $[dP_{opt}(x)]$ across dx:

$$dP_{opt}(x) = -P_{opt}(x)\alpha\, dx$$

$$\frac{dP_{opt}(x)}{P_{opt}(x)} = -\alpha\, dx$$

Integrating both sides of this equation and assuming the integration constant in the form $C = +\ln A$, we obtain

$$\ln P_{opt}(x) = -\alpha x + \ln A$$

$$\ln \frac{P_{opt}(x)}{A} = -\alpha x$$

$$P_{opt}(x) = Ae^{-\alpha x}$$

where the integration constant A is determined from the boundary condition ($x = 0$) to be equal to the incident power density $P_{opt}(0)$:

$$P_{opt}(x) = P_{opt}(0)e^{-\alpha x}$$

(b) The generation rate is the number of generated electron–hole pairs per unit volume and unit time. Dividing the absorbed optical power density by the energy of each photon, $-dP(x)/h\nu$, we obtain the number of photons absorbed per unit area and unit time. This is equal to the number of electron–hole pairs generated per unit area and unit time. To obtain the generation rate, we divide $-dP(x)/h\nu$ by dx:

$$G_{ext}(x) = -\frac{1}{h\nu}\frac{dP(x)}{dx}$$

It was shown in part (a) that $dP_{opt}(x)/dx = -\alpha P_{opt}(x)$. Therefore,

$$G_{ext}(x) = \frac{\alpha P_{opt}(x)}{h\nu}$$

EXAMPLE 12.2 Light Absorption and Drift Photocurrent

Due to the absorption of light in the semiconductor material, the assumption of a uniform carrier generation is not always acceptable. Assuming that $w_N = 10\ \mu\text{m}$ of N-type region is at the surface of the PIN diode, calculate the photocurrent generated in the depletion layer. The surface generation rate is $G_{ext}(0) = 10^{19}\ \text{cm}^{-3}\,\text{s}^{-1}$, and the optical absorption coefficient is $\alpha = 0.01\ \mu\text{m}^{-1}$. The width of the fully depleted "intrinsic layer" is $w_d = 100\ \mu\text{m}$ and the junction area is $A_J = 1\ \text{mm}^2$.

SOLUTION

Based on the results from Example 12.1, the nonuniform generation rate can be expressed as

$$G_{ext}(x) = G_{ext}(0)e^{-\alpha x}$$

In this case, $G_{ext}(x)$ cannot simply be multiplied by qA_Jw_d to obtain the photocurrent. However, we can find the average generation rate inside the depletion layer:

$$\overline{G_{ext}} = \frac{1}{w_d}\int_{w_N}^{w_N+w_d} G_{ext}(x)\,dx = \frac{1}{w_d}G_{ext}(0)\int_{w_N}^{w_N+w_d} e^{-\alpha x}\,dx$$

Note that the depletion-layer edges are at w_N and $w_N + w_d$, which are the integration limits. Solving the integral leads to

$$\overline{G_{ext}} = \frac{G_{ext}(0)}{w_d\alpha}\left[e^{-\alpha w_N} - e^{-\alpha(w_N+w_d)}\right]$$

The photocurrent is then

$$I_{photo} = qA_Jw_d\overline{G_{ext}} = qA_J\frac{G_{ext}(0)}{\alpha}\left[e^{-\alpha w_N} - e^{-\alpha(w_N+w_d)}\right]$$

$$I_{photo} = 91.5\ \mu\text{A}$$

EXAMPLE 12.3 Concentration Diagrams for a Reverse-Biased Photodiode and Diffusion Photocurrents

Solve the continuity equation to obtain the equations for the concentration gradients plotted in Fig. 12.7, and then use the result to derive the diffusion photocurrents.

SOLUTION

For the steady-state case, $\partial n_p/\partial t = 0$. With this and $U = n_p(x) - n_{pe}/\tau_n$, the continuity equation [Eq. (5.1)] becomes

$$0 = \frac{1}{q}\frac{dj_n}{dx} + G_{ext} - \frac{n_p(x) - n_{pe}}{\tau_n}$$

where $j_n = qD_n dn_p/dx$ [Eq. (4.5)]. Given that $L_n^2 = D_n\tau_n$ [Eq. (5.15)], we have

$$\frac{d^2\left[n_p(x) - n_{pe}\right]}{dx^2} - \frac{n_p(x) - n_{pe}}{L_n^2} + \frac{G_{ext}}{D_n} = 0$$

The only difference between this equation and Eq. (5.14) is the added constant G_{ext}/D_n. Because of that, the general solution of this equation can be expressed as the solution of Eq. (5.14), plus an additional constant:

$$n_p(x) - n_{pe} = A_1e^{x/L_n} + A_2e^{-x/L_n} + A_3$$

As in Section 5.2.3, A_1 has to be zero to have a finite concentration for $x \to \infty$. A_3 is not an independent constant: it has to be set so to ensure that $n_p(x) - n_{pe} = A_2 \exp(-x/L_n) + A_3$ is a solution of the differential equation. Replacing the solution expressed in this form into the differential equation, we obtain

$$\underbrace{\frac{A_2}{L_n^2} e^{-x/L_n}}_{d^2[n_p(x)-n_{pe}]/dx^2} \quad \underbrace{- \frac{A_2}{L_n^2} e^{-x/L_n} - \frac{A_3}{L_n^2}}_{[n_p(x)-n_{pe}]/L_n^2} + \frac{G_{ext}}{D_n} = 0$$

$$-\frac{A_3}{L_n^2} + \frac{G_{ext}}{D_n} = 0$$

$$A_3 = G_{ext} \frac{L_n^2}{D_n} = G_{ext} \tau_n$$

Therefore,

$$n_p(x) - n_{pe} = A_2 e^{-x/L_n} + G_{ext} \tau_n$$

where A_2 has to be determined so that the boundary condition $n_p(w_p)$ is satisfied:

$$n_p(x) - n_{pe} = \left[n_p(w_p) - n_{pe} - G_{ext}\tau_n\right] e^{-(x-w_n)/L_n} + G_{ext}\tau_n \qquad (x \geq w_p)$$

In the case of a reverse-biased P–N junction, $n_p(w_p) = \exp(V_D/V_t) \ll n_{pe}$:

$$n_p(x) = (n_{pe} + G_{ext}\tau_n)\left[1 - e^{-(x-w_n)/L_n}\right] \qquad (x \geq w_p)$$

This is the equation for the concentration profile of minority electrons, plotted in Fig. 12.7. Clearly, if there is no external generation ($G_{ext} = 0$), the equation is simplified to

$$n_p(x) = n_{pe}\left[1 - e^{-(x-w_n)/L_n}\right] \qquad (x \geq w_p)$$

Using this result in the diffusion-current equation

$$I_n = qA_J D_n \left.\frac{dn_p(x)}{dx}\right|_{x = w_n}$$

the following equation is obtained:

$$I_n = qA_J D_n \frac{n_{pe} + G_{ext}\tau_n}{L_n} = \underbrace{qA_J \frac{D_n n_{pe}}{L_n}}_{I_{S-n}} + \underbrace{qA_J G_{ext} L_n}_{I_{photo-n}}$$

which is the same result for the photocurrent as Eq. (12.7).

The corresponding equations for the minority holes are

$$p_n(x) = (p_{ne} + g_p\tau_p)\left[1 - e^{(x+w_p)/L_p}\right] \qquad (x \leq -w_n)$$

$$I_p = -qA_J D_p \frac{p_{ne} + g_p\tau_p}{L_p} = \underbrace{qA_J \frac{D_p p_{ne}}{L_p}}_{I_{S-p}} + \underbrace{qA_J g_p L_p}_{I_{photo-p}}$$

12.3 LASERS

The word *laser* is an acronym for "light amplification by stimulated emission of radiation." The distinguishing characteristic of lasers is emission of strong narrow beams of monochromatic light.[1] Lasers are widely used, with many applications being familiar to almost everybody. An important application of semiconductor lasers is the generation of the monochromatic light that carries information through optical-fiber communication systems.

12.3.1 Stimulated Emission, Inversion Population, and Other Fundamental Concepts

The following text introduces the semiconductor lasers in a way that does not require a preliminary study of gas lasers. The semiconductor lasers are most similar to the LEDs introduced in Section 2.2. Similar to LEDs, the current of a forward-biased P–N junction causes recombination of the excess minority carriers, leading to light emission. The difference is that laser light is monochromatic, having resulted from the process of *stimulated emission,* as distinct from the spontaneous emission in the case of LEDs.

To better understand the process of stimulated emission, it is helpful to know some of the fundamental properties of photons, which distinguish them as particles from the electrons. It was mentioned in Section 2.2 that no more than one electron can occupy a single electron state (Pauli exclusion principle). Therefore, if there is an electron in a particular state, the probability that another electron will get into that state is 0: electrons "shy away" from each other. As opposed to this behavior, photons "flock" into a single state. When there are n identical photons, the probability that one more photon will enter the same state is *enhanced* by the factor $(n + 1)$. *The probability that an atom will emit a photon with particular energy* $h\nu$ *is increased by the factor* $(n + 1)$ *if there are*

[1] Monochromatic light is single-wavelength light, or, practically, a very narrow band of wavelengths.

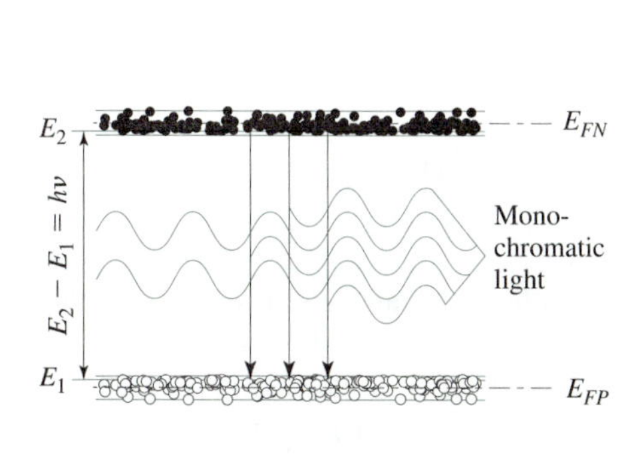

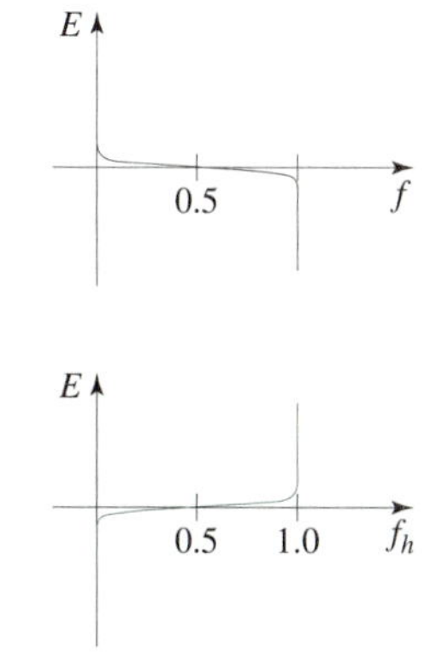

Figure 12.8 Stimulated emission: incoming photons trigger electron–hole recombination to generate more photons with the same energy $h\nu$. Population inversion: electron and hole distributions (f and f_h) relate to separate quasi-Fermi levels (E_{FN} and E_{FP}), which appear inside the bands to indicate that the concentrations of both electrons and holes are very high.

already n photons with this energy.[2] According to this property, electrons are classified as Fermi particles (being described by the Fermi–Dirac distribution), whereas the photons are classified as Bose particles and obey a different, Bose–Einstein, distribution.

Let us put mirrors at the two ends of a P–N junction that emits light due to a forward-bias current. The emitted light will reflect from the mirrors, so that the intensity of the light in the direction normal to the mirrors becomes dominant. More importantly, the presence of this light with frequency ν will increase the probability of minority excess electrons falling from the conduction band down to the valence band emitting light with the same frequency ν. As the presence of more $h\nu$ photons causes further increase in the emission of light with frequency ν, a chain reaction is triggered, leading to what is known as *stimulated emission*. Clearly, the stimulated emission can amplify a small-intensity incoming light beam to produce a large-intensity light beam, as illustrated in Fig. 12.8. This is called *optical amplification*.

To make use of the generated light, one of the mirrors is made slightly transparent so that a highly directional, monochromatic beam of light can exit the device. One more problem needs to be solved, and the laser is operational. The concentration of excess electrons in the conduction band needs to be maintained at a high level; otherwise, as the photons get out, the excess electrons will be spent and the light that was generated will die away. This is achieved by maintaining the forward-bias current above the needed level (*threshold current*), so that the concentration of the minority electrons injected into the P-type region is sufficiently higher than the equilibrium level. This condition is referred to as *population inversion*.

In nonequilibrium case, the concentrations of electrons and holes are expressed by two separate quasi-Fermi levels (E_{FN} and E_{FP}). The example of Fig. 12.8 illustrates the case of very high concentrations of both electrons and holes, so high that both E_{FN} and E_{FP} appear inside the energy bands. It can be shown that this is necessary for the stimulated emission to exceed the absorption. There are three necessary factors for a stimulated recombination of an electron–hole pair: (1) existence of photons with energy $h\nu$, (2) existence of an electron

[2]R. Feynman, R. Leighton, and M. Sands, *The Feynman Lectures on Physics—Quantum Mechanics,* Addison-Wesley, Reading, MA, 1965, pp. 4-7.

at energy level E_2 in the conduction band, and (3) existence of a hole at level E_1 in the valence band, where $E_2 - E_1 = h\nu$. Therefore, the rate of stimulated recombination can be expressed as

$$R_{(st)} \propto f(E_2, E_{FN}) f_h(E_1, E_{FP}) I(h\nu) = f_2(1 - f_1) I(h\nu) \tag{12.9}$$

where $I(h\nu)$ is the density of $h\nu$ photons, and $f(E_2, E_{FN}) = f_2$ and $f_h(E_1, E_{FP}) = 1 - f_1$ are the probabilities of having an electron and a hole at E_2 and E_1, respectively. The presence of photons, however, can cause electron–hole generation due to absorption by electrons at E_1, which subsequently move to E_2. The generation rate G_{ext} is proportional to the probability of having an electron at E_1, $f_1 = 1 - f_h(E_1, E_{FP})$, and a hole at E_2, $1 - f_2 = 1 - f(E_2, E_{FN})$:

$$g \propto f_1(1 - f_2) I(h\nu) \tag{12.10}$$

The condition for stimulated emission to exceed absorption is

$$R_{(st)} > G_{ext} \tag{12.11}$$

which according to Eqs. (12.9) and (12.10), and assuming equal proportionality coefficients, can be expressed as

$$f_2(1 - f_1) > f_1(1 - f_2) \tag{12.12}$$

This condition can be transformed as follows:

$$f_2 - f_2 f_1 > f_1 - f_1 f_2 \quad \rightarrow \quad f_2 > f_1 \tag{12.13}$$

$$\frac{1}{1 + \exp\left(\dfrac{E_2 - E_{FN}}{kT}\right)} > \frac{1}{1 + \exp\left(\dfrac{E_1 - E_{FP}}{kT}\right)} \tag{12.14}$$

$$\frac{E_1 - E_{FP}}{kT} > \frac{E_2 - E_{FN}}{kT} \tag{12.15}$$

and finally,

$$E_{FN} - E_{FP} > h\nu = E_2 - E_1 \tag{12.16}$$

Therefore, the stimulated emission can exceed the absorption only if strong nonequilibrium concentrations of electrons and holes are maintained, such that the difference between the respective quasi-Fermi levels is larger than the energy of the emitted photons.

12.3.2 A Typical Heterojunction Laser

Similar to LEDs, semiconductor lasers are made of direct semiconductors, typically III–V and II–VI compound semiconductors. In addition, it is practically impossible to reach the condition given by Eq. (12.16) by an ordinary P–N junction. Also, there is a need to confine the emitted light beam inside the active laser region. All these requirements can be met by using a structure with layers of different III–V and/or II–VI semiconductors—a structure with heterojunctions.

Different materials have different energy gaps, creating energy-band discontinuities (offsets) at the heterojunction, such as the band offsets at the AlGaAs–GaAs N–P junction discussed in Section 9.4. These offsets can help to achieve the condition given by Eq. (12.16). Also, different materials have different refractive indices, which can help achieve total beam reflection at the parallel interfaces, so that the light is confined inside the active region. Molecular-beam epitaxy is used to deposit one semiconductor layer over another, creating heterojunctions. To achieve the necessary high-quality interfaces, the wafers remain inside a high-vacuum chamber, while the reacting gases are changed as a transition from one layer to another is to be made.

The choice of the semiconductor that will emit the light (the active layer) depends on what light frequency (color) is needed. Figure 12.9 shows a laser with GaAs as the active layer,[3] as an illustration of a typical semiconductor laser. To create appropriate energy-band discontinuities, layers with wider energy gaps are needed at each side of the active layer. In this example, N-type and P-type AlGaAs are used for this purpose.[4] Let us concentrate on the N-AlGaAs–P-GaAs heterojunction. If the doping of N-AlGaAs is high enough, the electron quasi-Fermi level (E_{FN}) is close enough to the bottom of the conduction band that the conduction-band offset brings the bottom of the conduction band in the GaAs region below E_{FN}. Analogously, the valence-band offset at the P-AlGaAs–P-GaAs heterojunction places E_{FP} below the top of the valence band. This provides the condition for population inversion when appropriate forward bias (V_F) is applied. The emitted light is reflected backward and forward by the mirror surfaces capturing enough photons to trigger the stimulated emission. The photons that get through the partially transparent mirror (the useful laser beam) have to be replaced by new electrons and holes provided by the laser forward current (I_F).

The conduction-band offsets at both heterojunctions (N-AlGaAs–P-GaAs and P-GaAs–P-AlGaAs) are important. They create a potential well that traps the minority carriers (electrons) so that they cannot diffuse away from the active region, as would happen in the case of an ordinary P–N junction. This *carrier confinement* is important for maintaining the population inversion and maximizing the stimulated recombination and light emission. On the other hand, *light confinement* is achieved by the fact that the refractive index of AlGaAs is smaller than the refractive index of GaAs.

The outside GaAs layers, the P-type capping layer and the N-type substrate layer, also create heterojunctions with the adjacent P-type and N-type AlGaAs layers. These also lead to energy-band offsets; since, however, the associated depletion layers are narrow because of the high doping, they do not present a practical problem for the current flow.

---

[3]The energy gap of the GaAs (1.42 eV) corresponds to 900-nm light, which is in the infrared region.

[4]The energy gap of $Al_xGa_{1-x}As$ varies between 1.42 eV for pure GaAs ($x = 0$) and 2.9 eV for pure AlAs ($x = 1$).

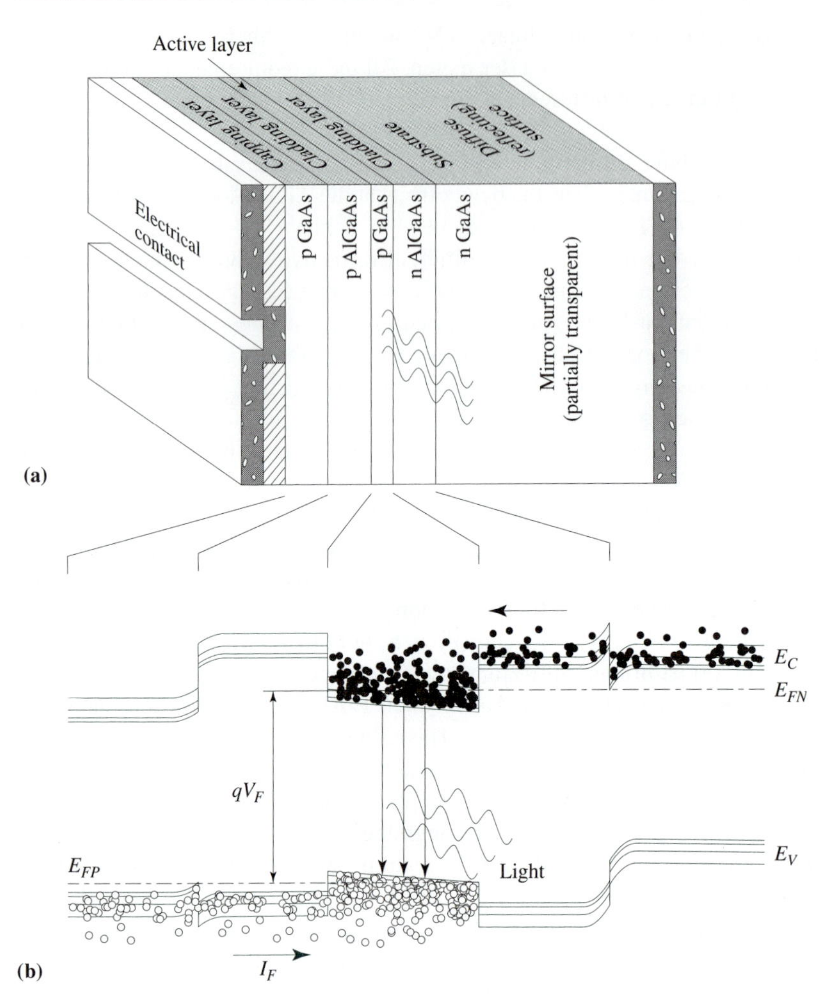

Figure 12.9 A typical heterojunction laser. (a) Cross section. (b) Energy-band diagram.

SUMMARY

1. In direct semiconductors, the energy released due to *recombination* of an electron–hole pair may be in the form of a photon. The opposite process, a photon absorption by an electron, results in *generation* of an electron–hole pair. During these processes the total energy is conserved: the electron must lose energy $h\nu$ when a photon is emitted and the electron's energy is increased by $h\nu$ when a photon is absorbed.
2. Electron–hole recombination is utilized in *light-emitting diodes* (LED). LEDs are operated in the forward-bias mode, so that the forward-bias current injects minority carriers into the neutral regions, which are then recombined by the majority carriers to emit light.
3. Electron–hole generation due to absorbed light is utilized in *photodetectors* and *solar cells*. The reverse-bias current of a photodetector diode or a solar cell is increased from the normal saturation current I_S to $I_S + I_{photo}$, where for the case of uniform generation

rate G_{ext} we have

$$I_{photo} = \underbrace{qA_J G_{ext} w_d}_{\text{drift}} + \underbrace{qA_J G_{ext} L_n + qA_J G_{ext} L_p}_{\text{diffusion}}$$

4. Photodetector diodes are operated in the reverse-bias mode, as light-controlled current sources. A solar cell is directly connected to a load resistance, which results in a positive voltage across the cell while the current remains negative—the *negative power* means that the solar cell is delivering power to the load.
5. As opposed to electrons, which obey the Pauli exclusion principle, the probability that a photon with energy $h\nu$ will be emitted is increased by the factor $(n + 1)$ if there are already n photons with this energy. This "behavior" leads to *stimulated light emission.* If *population inversion* is reached, that is, so many electrons and holes are injected into a semiconductor material that both electron and hole quasi-Fermi levels enter the conduction and valence bands, respectively:

$$E_{FN} - E_{FP} > h\nu = E_2 - E_1$$

the stimulated emission exceeds the recombination rate. The emitted light is reflected by parallel mirrors at the ends of the semiconductor so that the intensity of photons, and therefore the probability of stimulated emission, is enhanced. These principles are utilized by *lasers* to generate highly directional beams of monochromatic light. The light that exits the laser is compensated by maintaining the forward-bias current of the laser above the needed *threshold current.*

6. To practically achieve the inversion-population condition, a narrower-energy-gap semiconductor is sandwiched by N^+ and P^+ semiconductors with wider energy gaps. The energy-band discontinuities at the *heterojunctions,* "force" the quasi-Fermi levels inside the conduction/valence band of the active semiconductor. The energy-band discontinuities also help confine the carriers inside the active region so that they will recombine rather than diffuse away. Also, the active layer in the middle has a higher refraction index, which helps confine the generated light inside the active region.

PROBLEMS

12.1 Three semiconductor samples have distinct appearances: sample A is like a colorless glass, sample B is like a yellow glass, and sample C is like a metallic mirror. Knowing that one of these samples is Si ($E_g = 1.12$ eV), the other is 4H SiC ($E_g = 3.2$ eV), and the third is 3C SiC ($E_g = 2.4$ eV), determine

(a) the minimum wavelengths of light that can be transmitted through each of these semiconductors without absorption

(b) the sample labels that correspond to each of these semiconductors

12.2 The input bias voltage in the circuit of Fig. 12.3 is $v_{IN} = 5$ V and the current flowing through the LED is $I_{D0} = 10$ mA.

(a) If half of the recombined carriers emit light, how many photons are emitted per unit time?

(b) If the energy of each photon is $h\nu = 2.0$ eV, what is the optical power of the emitted light?

(c) What is the efficiency of the conversion of electrical power to light power?

12.3 The top P-type layer of a P–N^+ GaAs LED can be considered much wider than the diffusion length of the minority carriers, which is $L_n = 5\ \mu$m. The other technological parameters and constants are as follows: $N_A = 5 \times 10^{16}$ cm^{-3}, $D_n = 30$ cm^2/V · s, $n_i = 2.1 \times 10^6$ cm^{-3}, the diode area $A_J = 4$ mm^2, and the radiative recombination efficiency $\eta_Q = 0.7$. If the forward bias of the diode is $V_D = 1$ V, calculate

(a) the number of photons emitted per unit time, $P_{opt} A_J / h\nu$

(b) the optical power of the emitted light, assuming that the photon energy is equal to the energy gap ($E_g = 1.42$ eV) **A**

12.4 Exposure to light increases the reverse-bias current of a photodiode 10^6 times. What would be the reading of a voltmeter if the reverse bias is removed and the voltmeter is attached to the terminals of the diode as it remains exposed to light?

12.5 A PIN photodiode has junction area $A_J = 1$ mm^2 and fully depleted "intrinsic" region $W_I = 100\ \mu$m. Calculate the photocurrent, assuming that light of a certain intensity generates on average 10^{19} electron–hole pairs per second in 1 cm^3 of the photodiode material.

12.6 Assuming that the 100-μm intrinsic layer of a PIN diode is fully depleted, and neglecting any light absorption in the very thin P-type region at the top, calculate the photocurrent for surface generation rate of $G_{ext}(0) = 5.1 \times 10^{19}$ cm^{-3} s^{-1} and absorption coefficient $\alpha = 0.05\ \mu$m^{-1}. The junction area of the diode is $A_J = 1$ mm^2.

12.7 The power density of a 0.5-μm light is $P_{opt} = 900$ W/m^2.

(a) Calculate the number of photons per unit area and unit time that would hit the surface of a semiconductor exposed to this light. **A**

(b) Due to absorption by the semiconductor, the photon density decays exponentially down to zero, as does the generation rate: $G_{ext}(x) = G_{ext}(0)\exp(-\alpha x)$. Neglecting reflection losses and assuming that every absorbed photon of this light generates an electron–hole pair, calculate the surface generation rate. The absorption coefficient is $\alpha = 1\ \mu$m^{-1}.

(c) What would be the photocurrent if the PIN diode described in Problem 12.6 were exposed to this light? **A**

12.8 Design the area of a PIN photodiode so that its photocurrent is 10 mA when illuminated by 0.1 W/cm^2 of a 0.5-μm light. The technological parameters are as follows: the thickness of the top P-type layer is negligible, the thickness of the fully depleted I region is 10 μm, and the absorption coefficient is $\alpha = 0.06\ \mu$m^{-1}.

12.9 The short-circuit current of a silicon P$^+$–N solar cell at room temperature is $I_{sc} \approx I_{photo} = 100$ mA. The technological parameters of the cell are $A_J = 4$ cm^2, $N_D = 5 \times 10^{16}$ cm^{-3}, $\mu_p = 380$ cm^2/$V \cdot s$, $L_p = 10\ \mu$m, and $n \approx 1$.

(a) Derive the equation for the open-circuit voltage, V_{oc}, and calculate V_{oc}. **A**

(b) Calculate the maximum power that can be obtained from this cell. What load resistance R_L is needed to extract this power?

(c) How many cells, operating at the maximum power, have to be connected in series to obtain a voltage of 12 V? How many 12-V cells have to be connected in parallel so that the maximum power is 5.7 W? **A**

12.10 The technological parameters of a solar cell are as follows: $N_A = 3 \times 10^{16}$ cm^{-3}, $N_D = 8 \times 10^{15}$ cm^{-3}, $D_n = 20$ cm^2/s, $D_p = 10$ cm^2/s, $\tau_n = 0.3\ \mu$s, $\tau_p = 0.1\ \mu$s, and $A_J = 1$ cm^2. The generation rate of electron–hole pairs due to the absorbed light can be considered constant around the P–N junction: $G_{ext} = 5 \times 10^{19}$ cm^{-3}.

(a) Calculate the short-circuit current.

(b) Is the drift or diffusion photocurrent dominant, and what fraction of the total current is due to the dominant mechanism?

(c) What are the maximum concentrations of the minority electrons and holes?

12.11 Due to manufacturing problems and increased defect levels, the actual lifetimes of the solar-cell considered in Problem 12.10 are $\tau_n = 3$ ns and $\tau_p = 1$ ns. Repeat the calculations of Problem 12.10. **A**

12.12 The short-circuit current of a solar cell is $I_S = 10^{-11}$ A and $I_{photo} = 100$ mA in dark and when exposed to light, respectively. The load resistance that extracts the maximum power is $R_L = 5.2\ \Omega$. If the actual load resistance is 20% higher, what is the relative reduction in extracted power?

REVIEW QUESTIONS

R-12.1 How can the light intensity in LEDs be varied?

R-12.2 How are different colors achieved with LEDs?

R-12.3 Does the photocurrent in P–N junctions depend on the reverse-bias voltage? Why?

R-12.4 How is the P–N junction biased (forward or reverse) in photodetector circuits, and how is it biased in solar-cell circuits? Why?

R-12.5 Both LEDs and semiconductor lasers are basically P–N junctions that emit light. What is the difference?

R-12.6 Why do photons stimulate electron–hole recombination? Do they stimulate emission of photons with the same or with different energy $h\nu$?

R-12.7 What is optical amplification? In principle, can it be used to amplify a weakened optical signal carrying a communication signal through an optical fiber?

R-12.8 If the P–N junction is in equilibrium or close to equilibrium ($E_{FN} \approx E_{FP}$), will stimulated emission or absorption dominate? Would the stimulated emission and absorption rates be the same, given that both are proportional to the light intensity (photon density)? Do they depend on anything else?

R-12.9 What is needed to achieve a higher rate of stimulated emission compared to the rate of absorption? What is this condition called?

R-12.10 What is achieved by the heterojunctions in a typical semiconductor laser? Are they related to population inversion? Minority-carrier confinement? Light confinement?

13 JFET and MESFET

Two transistor structures that are very similar to the depletion (normally *on*) MOSFET are the junction field-effect transistor (JFET) and the metal–semiconductor-field-effect transistor (MESFET). The difference in the case of JFET is that reverse-biased P–N junction is used for DC isolation of the JFET gate, as distinct from the isolation by gate oxide in the case of the depletion MOSFET. JFETs are usually made as discrete devices in Si and recently in SiC, mostly for power applications. Regarding the MESFET, the gate is created as a Schottky diode (contact) rather than as a P–N junction diode. The MESFET is the most frequent FET implementation in GaAs.

This chapter provides descriptions and the energy-band diagrams of the JFET and the MESFET. This complements the detailed description of the enhancement (normally *off*) MOSFET, given in Chapter 8. SPICE does include JFET and MESFET models, and this chapter also provides a systematized reference to the SPICE models and parameters.

13.1 JFET

13.1.1 JFET Structure

The JFET is usually used as a discrete device, although it can be implemented in integrated-circuit technology as well. Figure 13.1 shows the principal JFET structure. This is the case of *N-channel JFET,* where the device current is due to flow of electrons from the source N^+ region, through the N-type layer called the channel, into the N^+ drain. The N-channel is sandwiched between top and bottom P^+ layers, whose separation determines the channel thickness. The top and bottom P^+ layers create P–N junctions with the N-type channel. The depletion layers associated with these junctions predominantly expand into the N-layer because its doping level is much lower than the doping of the P^+ regions. The dotted lines

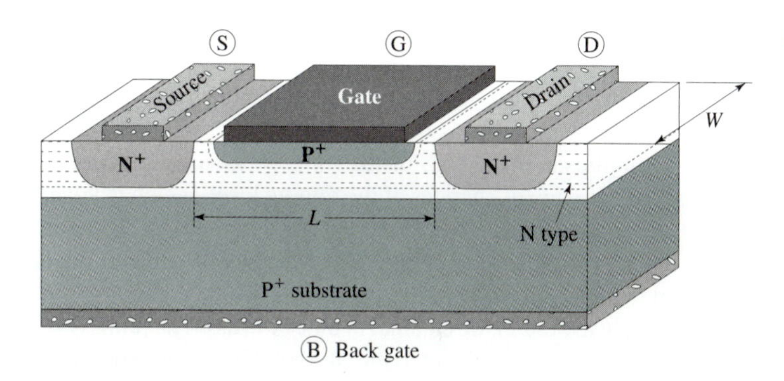

Figure 13.1 JFET structure.

in Fig. 13.1 show the edges of the depletion layers associated with the upper and lower P–N junctions. It is the distance between the two depletion layers, not the separation between the P–N junctions themselves, that is equal to the electrically effective channel thickness.

The upper and the lower P–N junctions can be reverse-biased, which increases the depletion-layer widths, thereby reducing the channel thickness and thus reducing the current that can flow through the channel. In the extreme case, the depletion layers extend over the whole thickness of the N-type layer, thereby reducing the electrically effective channel thickness to zero and thus reducing the JFET current to zero. This shows that the current can be controlled by the value of negative voltage applied to the upper and/or lower P^+ layer. Either of the two P^+ layers can be used as a gate. To distinguish between them, they are labeled *gate* and *back gate* in Fig. 13.1.

The JFET structure is similar to the depletion-type MOSFET. N^+ source and drain are the same, the built-in N-type channel that connects the drain and the source is the same, and the bottom P-type substrate that creates the isolation junction is the same. The only difference is the top P–N junction, which appears in place of the metal–oxide–semiconductor (MOS) structure. These different gate structures, however, control the current in a similar way: in both cases, the current flows through built-in channels at zero gate voltage (normally *on* FETs), and negative voltage is needed to reduce the current to zero by removing the electrons from the channel.

Differences between the JFET and the depletion-type MOSFET are important in some applications. In the case of the MOSFET, positive voltage can be applied to the gate without adverse effects; in fact it will only increase the channel current by attracting new electrons into the channel. In the case of the JFET, positive voltage at the gate is not desirable because it sets the gate–channel P–N junction in forward mode, effectively short-circuiting the gate and the channel. This appears as a disadvantage of the JFET structure. However, the lack of gate oxide in the JFET has a positive side as well: because no part of the gate-to-source voltage is wasted across the gate oxide, the current is more efficiently controlled; that is, the transconductance $g_m = dI_D/dV_{GS}$ is higher. Also, the gate can be shorter than the channel, and the depletion layer larger than the oxide thickness, leading to a smaller input capacitance.

This comparison relates only to the relationship between the device structures and the electrical performance. In both cases, development cost and device availability are strongly influenced by a complex set of manufacturing issues.

13.1.2 JFET Characteristics

Figure 13.2a and 13.2b shows the cross section and the energy-band diagram along the channel, respectively, for the case of zero gate-to-source voltage and a small drain-to-source voltage. The small discontinuities in the energy bands at the N^+-source–N-channel and N-channel–N^+-drain transitions are due to changes of the doping level, which is lower in the channel than in the source and drain regions. There are quite enough electrons in the channel to make a significant current when the energy bands are tilted by the applied V_{DS} voltage. This current depends linearly on V_{DS}, because it is basically limited by the slope of the energy bands (or the electric field in the channel in other words). The point Q in Fig. 13.2c is in the linear region of the I_D–V_{DS} characteristic (the solid line) corresponding to $V_{GS} = 0$.

The dotted lines in Fig. 13.2a show the edges of the depletion layers. The widths of the depletion layers at every point along the channel depend on the actual reverse bias at that point. Although the heavy doping of P^+ layers maintains approximately the same potential inside those regions (zero in the example of Fig. 13.2a), the voltage applied between the drain and the source distributes along the lower-doped N-type channel. Taking the potential of the source as reference, the energy-band diagram of Fig. 13.2b shows that the potential energy difference increases along the channel and reaches the value of qV_{DS} at the drain. The depletion-layer widths follow this trend: the narrowest depletion layers correspond to the zero reverse bias and appear at the source end of the channel; the depletion-layer widths increase along the channel and reach the maximum at the drain end of the channel. The

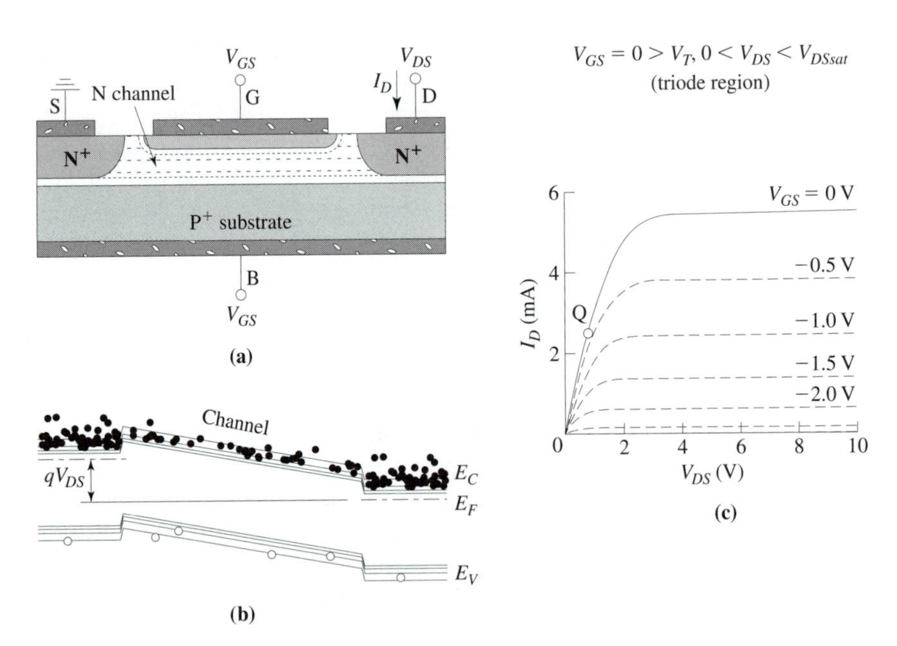

Figure 13.2 JFET in triode region. (a) Cross section, (b) energy bands, and (c) point Q showing the bias conditions on I_D–V_{DS} characteristics.

electrically effective channel thickness follows the opposite trend: it is the thickest at the source end, and the thinnest at the drain end of the channel.

The channel-thickness variation is not pronounced at small V_{DS} voltages. However, at higher V_{DS} voltages, the thickness reduction is reflected in a smaller current, which is seen as a departure of the actual I_D–V_{DS} characteristic from the linear trend followed at smaller V_{DS} voltages.

At sufficiently high V_{DS} voltage it can happen that the electrically effective channel thickness becomes zero (the channel is pinched off). The channel pinch-off first occurs at the drain end of the channel, expanding toward the source as V_{DS} is increased. *The V_{DS} voltage that causes channel pinch-off at the drain end is called saturation voltage and is labeled V_{DSsat}.*

Figure 13.3 illustrates the situation of channel pinch-off caused by a high V_{DS} voltage. Analogously to the MOSFET case, the voltage across the channel (between the source and the pinch-off point) remains constant and equal to V_{DSsat}, because any additional V_{DS} increase drops across the laterally expanding depletion layer. Because the current is limited by the conditions in the channel, and not in the depletion layer (the waterfall analogy with the energy-band diagram of Fig. 13.3b), this leads to saturation in the current increase with V_{DS} voltage.

It is now interesting to consider the case of the application to the gate of sufficient negative voltage (V_{GS}) to turn the JFET off. The negative voltage applied to the P^+ region adds up to the positively biased channel, increasing the total reverse-bias voltage. This can lead to expansion of the depletion layers over the entire N-layer thickness, even at the source end of the channel. Although Fig. 13.4a clearly illustrates this situation, it is not obvious from this cross-sectional diagram that the current through the channel is reduced to zero. The cross-sectional diagrams do not show any principal difference between the depletion layer in Fig. 13.4a and the depletion layer in Fig. 13.3a, but there is a significant difference. The depletion layer at the source end of the channel in Fig. 13.4a is controlled by the gate(s), and the electric-field lines terminate at the gates. Electrons taken from the source or any remaining part of the channel would follow the field lines only to hit the potential barriers of the reverse-biased P–N junctions. Consequently, no electron current can flow through this depletion layer. As opposed to this, the depletion layer in Fig. 13.3a is controlled by the drain, and the electric-field lines terminate at the drain. Electrons taken by the field from the end of the N-channel (the pinch-off point) are quickly transported to the drain.

Energy-band diagrams can clearly show this difference. The negative voltage at the gates increases the potential energy in the N-type layer (remember, negative electric potential corresponds to positive potential energy). Therefore, to modify the energy-band diagram of Fig. 13.3b (the case of $V_{GS} = 0$) so to represent the case of $V_{GS} < 0$, the energy bands between the source and the drain should be lifted by the value that corresponds to the reduced electric potential by the negative gate voltage. Figure 13.4b shows the energy-band diagram in this case. It can clearly be seen that the electrons from the source cannot flow into the drain because of the high potential barrier created by the gate voltage.

It is useful to compare the energy-band diagrams presented in this section for the JFET with the corresponding MOSFET energy-band diagrams. Figure 13.2 (a JFET in the triode region) shows an energy-band diagram similar to the energy-band diagram of a MOSFET in the channel region (the cross section of the two-dimensional energy-band diagram in Fig. 8.5b along the channel). The energy-band diagrams for the saturation and the cutoff modes of a JFET (Figs. 13.3 and 13.4) are also similar to the energy-band diagrams along the channel of a MOSFET in the saturation and the cutoff modes (Figs. 8.9 and 8.5a).

13.1.3 SPICE Model and Parameters

SPICE parameters and mathematical equations modeling the dependence of the output JFET current I_D on the terminal voltages V_{GS} and V_{DS} are given in Table 13.1. The equations modeling the JFET are very similar to the SPICE LEVEL 1 model of the MOSFET. In fact, the triode-region equations become equivalent if the gain factor β of the JFET is taken to be half of the MOSFET value. It should be noted that the threshold voltage (also called pinch-off voltage in the case of JFET) is negative for N-channel devices

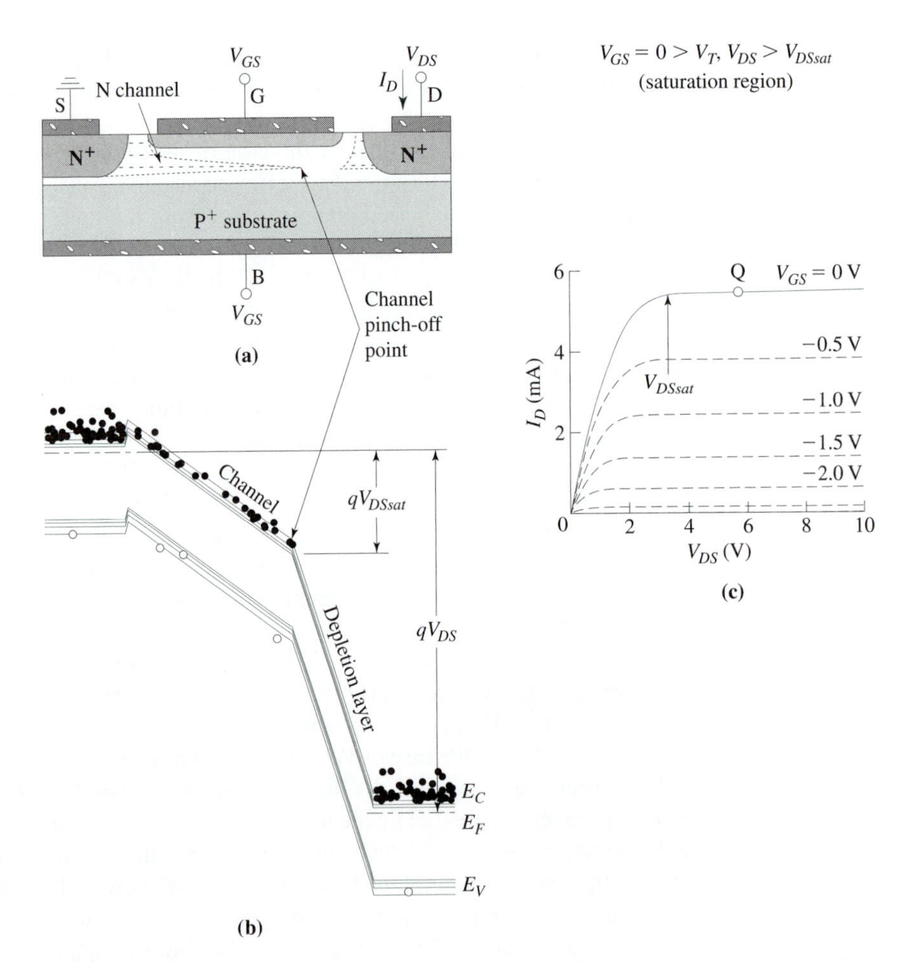

Figure 13.3 JFET in the saturation region. (a) Cross section, (b) energy bands, and (c) point Q showing the bias conditions on I_D–V_{DS} characteristics.

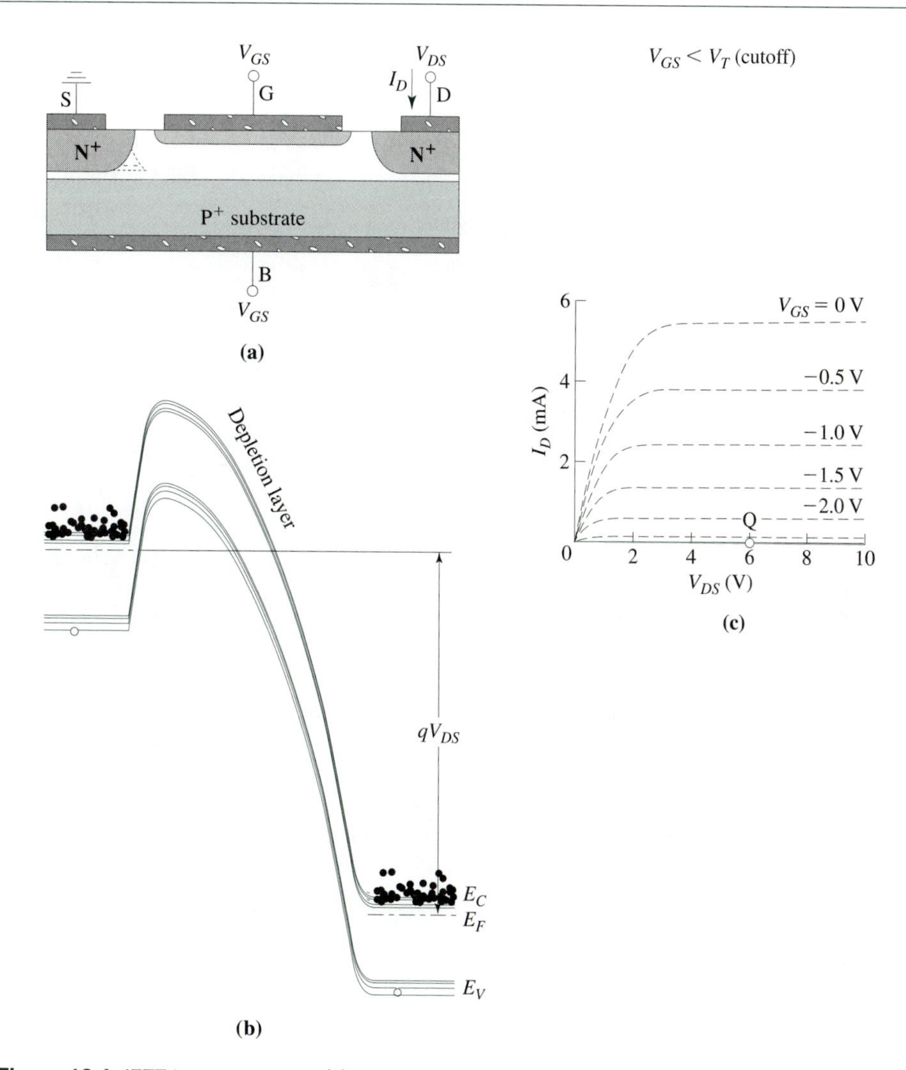

Figure 13.4 JFET in cutoff region. (a) Cross section, (b) energy bands, and (c) point Q showing the bias conditions on I_D–V_{DS} characteristics.

because the JFET and the depletion-type MOSFET are normally *on* FETs. Therefore, $V_{GS} - V_T$ is positive for any $V_{GS} > V_T$. The case of the P-channel devices is opposite, the threshold voltage is positive, $V_{GS} - V_T$ is negative when the FETs are *on* ($V_{GS} < V_T$), and V_{DS} voltage is negative as well. This gives a positive number for the current I_D in both cases; however, the current is assumed in the opposite direction, as shown in Table 13.1. As for measurement of the two key parameters, the threshold voltage V_T and the gain factor β, a procedure analogous to the one described in Section 11.2.2 (Fig. 11.4) can be used.

The multiplier $\lambda(V_{DS}) = 1 \pm \lambda V_{DS}$ is to include the effect of a small current increase with V_{DS} in the saturation region, or the finite dynamic output resistance in other words. The SPICE model for this effect is analogous to the Early model used in BJTs. Noting that

TABLE 13.1 Summary of SPICE JFET Model: The Principal Equivalent-Circuit Elements

Static Parameters

Symbol	SPICE Keyword	Parameter Name	Typical Value N Channel	Typical Value P Channel	Unit
V_T	Vto	Threshold voltage (pinch-off voltage)	-3	3	V
β	Beta	Gain factor (transconductance coefficient)	5×10^{-3}		A/V^2
λ	Lambda	Reciprocal Early voltage (channel-length modulation)	0.002		V^{-1}

Voltage-Controlled Current-Source Model

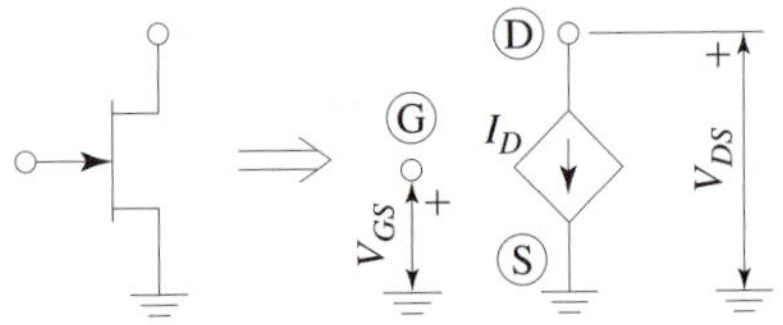

N-channel JFET

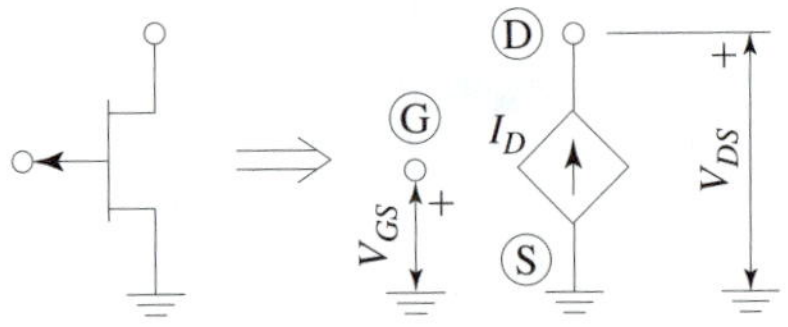

Off: $V_{GS} \le \text{Vto}$

Triode: $V_{GS} > \text{Vto}$, and $0 < V_{DS} < V_{DSsat}$

Satur.: $V_{GS} > \text{Vto}$, and $V_{DS} \ge V_{DSsat} > 0$

P-channel JFET

Off: $V_{GS} \ge \text{Vto}$

Triode: $V_{GS} < \text{Vto}$, and $0 > V_{DS} > V_{DSsat}$

Satur.: $V_{GS} < \text{Vto}$, and $V_{DS} \le V_{DSsat} < 0$

$$V_{DSsat} = V_{GS} - \text{Vto}$$

$$I_D = \begin{cases} 0, & \text{off region} \\ \text{Beta}\ \left[2(V_{GS} - \text{Vto})V_{DS} - V_{DS}^2\right], & \text{triode region} \\ \text{Beta}\ (V_{GS} - \text{Vto})^2\, \lambda(V_{DS}), & \text{satur. region} \end{cases}$$

$\lambda(V_{DS}) = 1 + \text{Lambda}\ V_{DS}$ (N-channel) $\qquad \lambda(V_{DS}) = 1 - \text{Lambda}\ V_{DS}$ (P-channel)

the output-voltage-independent current is multiplied by $(1 + |\lambda V_{DS}|)$ and $(1 + |\frac{1}{V_A} V_{CE}|)$ in the cases of JFET and BJT, respectively, we can say that λ has the meaning of the reciprocal Early voltage $1/V_A$. Therefore, the parameter measurement procedure described in Section 11.3.2 can be applied to obtain the parameter λ. Physically, the slight current increase in the saturation region is due to channel-length shortening as the pinch-off point moves toward the source. Consequently, the parameter λ is frequently called *channel-length modulation coefficient*.

Although the JFET can have two separate gates, the SPICE model assumes a single V_{GS} voltage. This is sufficient to cover the two most frequent application arrangements: (1) the two gates are connected together, electrically forming a single gate, and (2) one of the gates is connected to a constant voltage or grounded. These configurations are included

TABLE 13.2 Summary of SPICE JFET Model: Parasitic Elements

Parasitic-Element Related Parameters

Symbol	SPICE Keyword	Related Parasitic Element	Parameter Name	Typical Value	Unit
R_D	Rd	R_D	Drain resistance	10	Ω
R_S	Rs	R_S	Source resistance	10	Ω
I_S	IS	D_S, D_D	Saturation current	1×10^{-14}	A
n	N	D_S, D_D	Emission coefficient	1	—
$C_{GD}(0)$	Cgd	D_D	Gate–drain zero-bias capacitance	4×10^{-12}	F
$C_{GS}(0)$	Cgs	D_S	Gate–source zero-bias capacitance	4×10^{-12}	F
V_{bi}	PB	D_S, D_D	Built-in potential	0.8	V
m	M	D_S, D_D	Grading coefficient	$\frac{1}{3}-\frac{1}{2}$	—

Large-Signal Equivalent Circuit

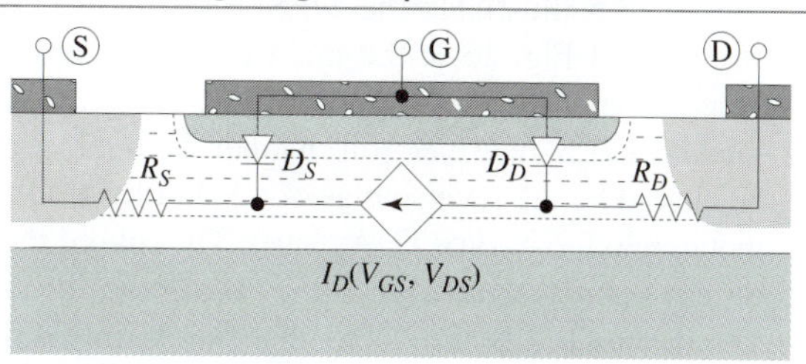

Note: Diodes and the current direction shown for N-channel JFET. Reverse for P-channel JFET.

$I_D(V_{GS}, V_{DS})$ is given in Table 13.1

D_S, D_D } according to diode model of Table 11.2

through the JFET parameters: (1) when the gates are connected together, the gain factor is increased (doubled in the case of symmetrical gates) and the absolute value of the threshold voltage (or pinch-off voltage) is reduced, and (2) when a constant reverse-bias voltage is applied to one of the gates, the absolute value of the threshold voltage is also reduced, as it helps the other gate to pinch the channel off.

Table 13.2 summarizes the parameters that are available to include some of the parasitic elements associated with the JFET structure, and gives the complete large-signal equivalent circuit used in SPICE. As always, the parasitic capacitances are of special importance because they determine the high-frequency behavior of the device. In the equivalent circuit of Table 13.2, the capacitances are included through the diode model given in Table 11.2. Therefore, the parameter measurement techniques are the same as described in Sections 11.1.2 and 11.1.3. It should be mentioned, however, that the stored-charge capacitance is neglected (no τ_T parameter appears in the list), which is justified by the fact that the gate-to-channel diodes D_S and D_D are normally reverse-biased.

The equivalent circuit of Table 13.2 can properly include all the parasitic capacitances for the case of mutually connected gates. In the case of one of the gates being grounded or being biased with a constant voltage, the top and bottom P–N junctions are not connected in parallel and cannot precisely be modeled by a single set of D_S and D_D diodes. Nonetheless, the parameters $C_{GS}(0)$ and $C_{GD}(0)$ can be adjusted to provide close fitting in that case. If more precise fitting of the capacitance–voltage dependencies is required, it will be necessary to add a set of diodes to the JFET, to include the effects of the second gate-to-channel junction.

13.2 MESFET

13.2.1 MESFET Structure

Figure 13.5 illustrates the MESFET structure. Although it is very similar to the JFET structure of Fig. 13.1, the main difference is that neither the upper nor the lower P–N junction appears in the MESFET structure. Semiinsulating GaAs substrate defines the thickness of N-type electron channel. The electrically effective thickness of the channel can be altered by the reverse bias of a Schottky diode, created by deposition of an appropriate metal onto the N-type GaAs layer. The role of the Schottky diode is equivalent to the role of the upper P–N junction in the case of the JFET.

It should also be noted that the MESFET is surrounded by semiinsulating GaAs. Although this does not affect the principal characteristics of the device, it is an important factor in terms of circuit speed.

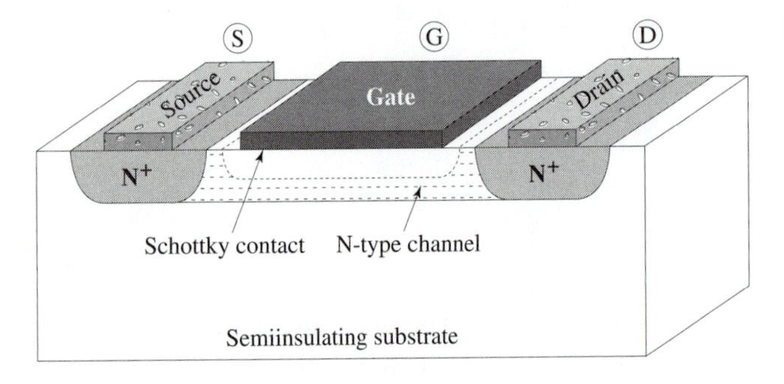

Figure 13.5 MESFET structure.

13.2.2 MESFET Characteristics

Device operation, energy-band diagrams, and consequently current–voltage characteristics are similar to those previously introduced with the MOSFET and JFET. Section 8.1.4 describes two mechanisms of current saturation: channel pinch-off and drift velocity saturation. The mechanism responsible for the current saturation in JFETs is typically the channel pinch-off. As opposed to this, the current in a typical GaAs MESFET saturates

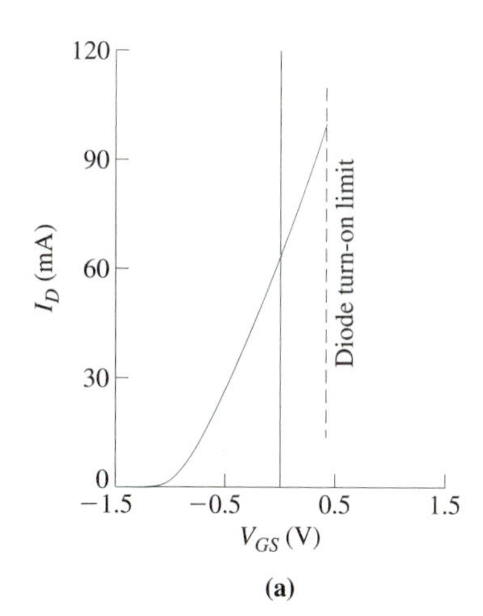

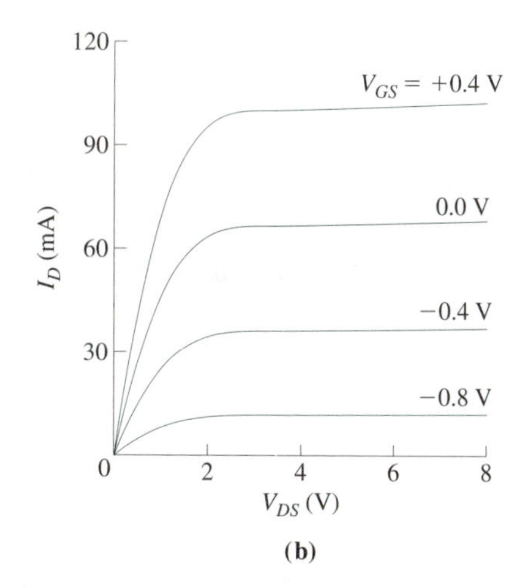

Figure 13.6 Transfer and output characteristics of a MESFET.

due to the velocity saturation. This difference influences the device modeling; however, this problem will be considered in the next section.

In this section, we show the transfer (Fig. 13.6a) and the output (Fig. 13.6b) characteristics, with the purpose of more carefully discussing the previously mentioned limitation to V_{GS} voltage.

The characteristics shown in Fig. 13.6 are for a normally *on* (depletion-type) MESFET. A significant current flows at $V_{GS} = 0$, and a negative $V_{GS} = V_T$ is needed to turn the device off. It was stated in Section 13.1.1 that positive V_{GS} is undesirable because it may turn the input diode on. Strictly speaking, some positive V_{GS} voltage can be applied before the input diode is turned on. In the example of Fig. 13.6a, the diode turn-on limit is shown to be about $V_{GS} = 0.4$ V.

The fact that some positive V_{GS} voltage can be applied makes it possible to design a normally *off* (enhancement-type) MESFET. To achieve this, the N-type GaAs layer is made thinner than the depletion-layer width of the Schottky diode at $V_{GS} = 0$ V. Therefore, the N channel is pinched off by the depletion layer, and no current flows at $V_{GS} = 0$ V. The current can start flowing if positive V_{GS} voltage is applied to narrow down the depletion-layer width. The V_{GS} voltage at which the current starts flowing, which is the threshold voltage V_T, is positive in this case. A practical problem with this type of normally *off* FET is that the threshold voltage, no matter how close to zero, is too close to the diode turn-on limit for V_{GS}. This not only makes the manufacturing requirements very strict, it also limits the input voltage range and consequently input noise margin that circuits made of these devices can handle. Nonetheless, so many circuits perform much better with normally *off* FETs that the enhancement-type MESFETs are used, especially in high-frequency digital circuits.

13.2.3 SPICE Model and Parameters

The current saturation in a typical MESFET is due to the drift-velocity saturation, which occurs before the channel can be pinched off by the drain voltage. The drift-velocity–electric-field curve of GaAs, is very complex. This leads to a complicated distribution of the lateral electric field in the channel, which is not possible to model by simple and

TABLE 13.3 Summary of SPICE MESFET Model

Static Parameters

Symbol	SPICE Keyword	Parameter Name	Typical Value	Unit
	LEVEL	Model type (1 = Curtice, 2 = Raytheon)		
V_T	Vto	Threshold voltage (pinch-off voltage)	−2	V
β	Beta	Gain factor (transconductance coefficient)	0.1	A/V^2
λ	Lambda	Reciprocal Early voltage (channel-length modulation)	0.005	V^{-1}
α_{sat}	Alpha	Saturation coefficient	1	V^{-1}
b	B	β reduction coefficient (LEVEL 2 only)	1	V^{-1}

Voltage-Controlled Current-Source Model[a]

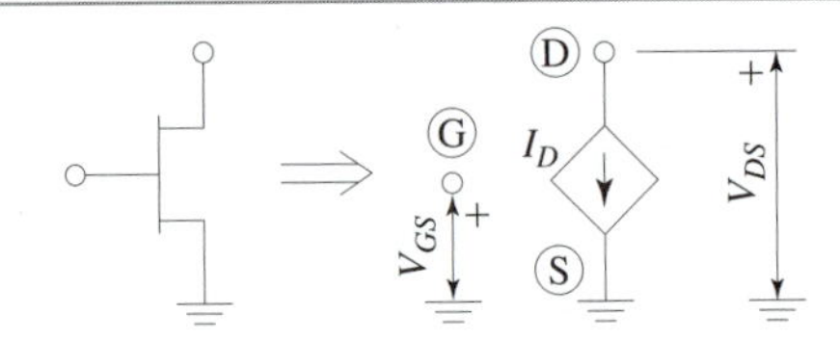

LEVEL = 1 *(Curtice Model)*

$$I_D = \begin{cases} 0 & \text{for } V_{GS} \le \texttt{Vto} \\ \texttt{Beta}\ (V_{GS} - \texttt{Vto})^2(1 + \texttt{Lambda}\ V_{DS})\tanh(\texttt{Alpha}\ V_{DS}) & \text{for } V_{GS} > \texttt{Vto} \end{cases}$$

LEVEL = 2 *(Raytheon Model)*

$$I_D = \begin{cases} 0 & \text{for } V_{GS} \le \texttt{Vto} \\ \frac{\texttt{Beta}}{1 + \texttt{B}\ (V_{GS} - \texttt{Vto})}(V_{GS} - \texttt{Vto})^2(1 + \texttt{Lambda}\ V_{DS})t(\alpha_{sat} V_{DS}) & \text{for } V_{GS} > \texttt{Vto} \end{cases}$$

$$t(\alpha_{sat} V_{DS}) = \begin{cases} 1 - \left(1 - \frac{\texttt{Alpha}\ V_{DS}}{3}\right)^3 & \text{for } \texttt{Alpha}\ V_{DS} \le 3 \\ 1 & \text{for } \texttt{Alpha}\ V_{DS} > 3 \end{cases}$$

[a] The parasitic elements are as in Table 13.2.

yet physically based equations. Consequently, SPICE uses empirical equations, which are presented in Table 13.3.

Two levels of MESFET model are available in SPICE: LEVEL 1 is an earlier developed Curtice model, and LEVEL 2 is a newer Raytheon model. The levels are selected by specifying the LEVEL as an input parameter.

Three essential parameters of the Curtice model are the threshold voltage V_T, the gain factor β, and the saturation coefficient α_{sat}. Although the meaning of V_T and β is the same as for a MOSFET, the saturation effect is modeled differently [the $\tanh(\alpha_{sat} V_{DS})$ term], which makes it easier to measure V_T and β in the saturation rather than linear region. As $\tanh(\alpha_{sat} V_{DS}) \approx 1$ in the saturation region, and assuming $1 + \lambda V_{DS} \approx 1$, the drain current becomes

$$I_D \approx \beta(V_{GS} - V_T)^2 \tag{13.1}$$

Plotting the square root of I_D versus V_{GS} produces a line that intersects the V_{GS}-axis at V_T and whose slope is equal to $\sqrt{\beta}$:

$$\sqrt{I_D} = \sqrt{\beta}(V_{GS} - V_T) \tag{13.2}$$

The measurement of the α_{sat} parameter is not as easy. The best way is to assume an initial value that is close to $3/V_{DSsat}$ (V_{DSsat} being the saturation voltage), and then use nonlinear curve fitting to more precisely determine the values of all the parameters, including α_{sat}.

The term $(1 + \lambda V_{DS})$ accounts for the finite output dynamic resistance, in the same way as for a JFET. This parameter has analogous meaning to the Early voltage in BJTs and is therefore measured in analogous way.

The Curtice model predicts parabolic increase of the drain current with gate-to-source voltage [Eq. (13.1)], which frequently fails to properly fit the experimental data. The newer Raytheon model (LEVEL 2 in SPICE) introduces an additional parameter to correct this problem. The new parameter b, called here β reduction coefficient, is analogous to the mobility-modulation constant θ of MOSFETs. The measurement of this parameter is analogous to the measurement of θ described in Section 11.2.2, with a difference that the measurements are taken in the saturation region. Noting that $t(\alpha_{sat} V_{DS}) = 1$ in the saturation region and assuming $1 + \lambda V_{DS} \approx 1$, the Raytheon equation of Table 13.3 can be written as

$$\underbrace{\frac{I_D}{(V_{GS} - V_T)^2}}_{S} = \frac{\beta}{1 + b(V_{GS} - V_T)} \tag{13.3}$$

This equation can further be modified to the form analogous to Eq. (11.21):

$$\frac{\beta}{S} - 1 = b(V_{GS} - V_T) \tag{13.4}$$

Obviously, the parameter b is the slope of $(\frac{\beta}{S} - 1)$-versus-$(V_{GS} - V_T)$ plot.

Another difference introduced in the Raytheon model is the replacement of $\tanh(\alpha_{sat} V_{DS})$ by the computationally more effective terms $1 - (1 - \frac{\alpha_{sat} V_{DS}}{3})^3$ in the triode and 1 in the saturation region.

The large-signal equivalent circuit of the JFET, shown in Table 13.2, can be applied to the MESFET case. Although Schottky diodes rather than P–N junction diodes should appear in the case of MESFET, there is no difference from the modeling point of view. The difference is taken into account by specifying appropriate parameter values. The same list of parameters applies, with a difference that the usual SPICE keyword for the built-in voltage is `VBI` rather than `PB`. For the sake of completeness, it should be mentioned that the SPICE equivalent circuit of the MESFET includes a resistor in series with the gate (R_G) and a capacitor in parallel with the current source (C_{DS}). The usual SPICE keywords for these parameters are `RG` and `CDS`, respectively. Because MESFETs are frequently used in high-frequency applications, these parameters are helpful for more precise simulation.

SUMMARY

1. The input capacitance of Si JFETs is smaller than that of a comparative MOSFET, making the JFET a superior device for high-frequency analog applications.
2. GaAs MESFET can be used at higher frequencies than Si MOSFETs, JFETs, and BJTs due to
   - **(a)** higher low-field mobility of GaAs electrons, which relates to a higher transconductance
   - **(b)** wider energy gap, which enables semiinsulating substrates, eliminating the capacitor-based isolation structures used in Si

PROBLEMS

13.1 Design the doping level in the channel region of the JFET shown in Fig. 13.1, so that the threshold (pinch-off) voltage is $V_T = -5$ V when the thickness of the channel region is 2 μm. The back gate is grounded and the JFET is to be implemented in GaAs ($\varepsilon_s = 13.2 \times 8.85 \times 10^{-12}$ F/m). Assume $V_{bi} = 1.35$ V.

13.2 The technological parameters of a GaAs JFET are as follows: the thickness of the channel region between the two junctions is $T_N = 2$ μm, the channel length is $L = 5$ μm, the channel width is $W = 50$ μm, the doping of the channel region is $N_D = 10^{16}$ cm^{-3}, the doping of the P^+ regions is $N_A = 10^{20}$ cm^{-3}, $\varepsilon_s = 13.2 \times 8.85 \times 10^{-12}$ F/m, and the electron mobility in the channel is $\mu_n = 7000$ cm^2/V·s.

(a) Determine the resistance of the channel when both the front and the back gates are grounded.

(b) Based on the channel resistance calculated in part (a), determine the parameter β in the SPICE equation for the triode region,

$$I_D = \beta\left[2(V_{GS} - V_T)V_{DS} - V_{DS}^2\right]$$

if the front and the back gates are short-circuited.

13.3 For the JFET from Problem 13.2, calculate the saturation voltage V_{DSsat} and the saturation current I_{Dsat} at $V_{DS} = V_{DSsat}$ for $V_{GS} = 0$. **A**

13.4 The threshold (pinch-off) voltage and the gain factor of a JFET are $V_T = -5$ V and $\beta = 10$ mA/V^2, respectively. The measured dynamic output resistance in saturation and for $V_{GS} = 0$ is $r_o = \Delta V_{DS}/\Delta I_D = 500$ Ω. Determine the value of SPICE parameter `Lambda` to match the dynamic output resistance.

13.5 The channel region of a JFET can be considered as a resistor for small V_{DS} voltages. The following are the technological parameters of an N-channel JFET that has symmetrical front and back P^+–N junctions and connected front and back gates: the thickness of the N-type channel $T_N = 1.6$ μm, the channel width $W = 1$ mm, the channel length $L = 10$ μm, the N-type doping in the channel $N_D = 10^{16}$ cm^{-3}, the P$^+$-type doping in the gate regions $N_A = 10^{20}$ cm^{-3}, and the electron mobility in the channel $\mu_n = 1250$ cm^2/V·s.

(a) Determine the channel resistance at $V_{GS} = 0$ V.

(b) Determine the value of SPICE parameter `Beta` to match this resistance.

13.6 The wide energy gap of 4H SiC can be utilized to design normally *off* N-channel JFETs ($V_T > 0$).

(a) Determine the thickness of the N-channel region between two symmetrical P$^+$ gate regions that will result in a threshold voltage of the JFET of $V_T = +1.0$ V (the front and the back gates are connected together).

(b) Determine the resistance of the channel for the maximum gate voltage $V_{GS} = 2.5$ V and $V_{DS} \approx 0$.

(c) Determine `Beta` in the SPICE model to match this resistance at $V_{GS} = 2.5$ V for the same V_T.

(d) Determine the maximum saturation current.

The following parameters are known: the channel width $W = 1$ mm, the channel length $L = 10\ \mu$m, the doping levels of the N-channel and the P$^+$ gate regions are $N_D = 10^{16}$ cm^{-3} and $N_A = 10^{10}$ cm^{-3}, respectively, the intrinsic carrier concentration $n_i = 2 \times 10^{-8}$ cm^{-3}, the electron mobility $\mu_n = 500$ cm^2/V·s, and the semiconductor permittivity $\varepsilon_s = 10\varepsilon_0$.

13.7 The dimensions of the front gate of an N-channel silicon JFET are $W = 1$ mm and $L = 5\ \mu$m; the back gate is grounded. The doping levels in the gate and in the channel are $N_A = 10^{20}$ cm^{-3} and $N_D = 10^{15}$ cm^{-3}, respectively. Determine the input/gate capacitance at $V_{GS} = 0$ V and $V_{DS} = 0$ V. Determine the values of SPICE `Cgs` and `Cgd` parameters.

13.8 A GaAs MESFET has a nickel gate electrode with the work function $q\phi_m = 5.15$ eV and $T_N = 200$ nm thick N-type channel with the doping level $N_D = 10^{17}$ cm^{-3}, deposited on a semiinsulating substrate. Calculate (a) the thickness of the active N channel at zero bias and (b) the threshold (pinch-off) voltage of this MESFET. The semiconductor affinity of GaAs is $q\chi_s = 4.07$ eV and the permittivity is $\varepsilon_s = 13.2\varepsilon_0$.

13.9 Table 13.4 gives data for the transfer characteristic of a GaAs MESFET in saturation. Determine the value of the β reduction coefficient b (SPICE keyword `B`) in the Raytheon model (SPICE LEVEL=2 MESFET model). The data correspond to V_{DS} values small enough to allow us to neglect any I_D increase due to V_{DS}.

TABLE 13.4 Current–Voltage Data for a MESFET

V_{GS} (V)	≤ -2.0	-1.5	-1.0	-0.5	0.0
I_D (mA)	0.0	2.3	8.3	17.3	28.6

13.10 **(a)** Design the doping level in the channel of the GaAs MESFET shown in Fig. 13.5 so that the *on* resistance is minimized and the MESFET is the normally *off* type with $V_T = 0.2$ V. The minimum channel thickness is limited to 100 nm by technological issues. Assume that $V_{bi} = 0.80$ V. The permittivity of GaAs is $\varepsilon_s = 13.2 \times 8.85 \times 10^{-12}$ F/m.

(b) Calculate the channel resistance for the maximum gate voltage $V_{GS} = 0.4$ V, channel-width-to-channel-length ratio $W/L = 100$, and the electron mobility of $\mu_n = 5000$ cm^2/V·s.

13.11 Repeat the design from Problem 13.10 without the assumption for V_{bi}, but knowing that nickel is to be used for the gate metal ($q\phi_m = 5.15$ eV) and that the electron affinity of GaAs is $q\chi_s = 4.07$ eV. What is the built-in voltage in this case, and what is the channel resistance at $V_{GS} = 0.4$ V? **A**

REVIEW QUESTIONS

R-13.1 Is the input capacitance of a depletion-type FET larger or smaller than the input capacitance of a comparable MOSFET? Is this advantageous for high-frequency analog applications?

R-13.2 Which material provides better high-frequency performance of depletion-type FETs, Si or GaAs? Why?

R-13.3 What is the main difference between the MOSFET and the JFET?

R-13.4 Is the JFET a normally *on* or normally *off* device?

R-13.5 What happens if positive gate voltage is applied to the gate of an N-channel JFET? What if this voltage is smaller than 0.7 V?

R-13.6 What is the mechanism of current saturation in silicon JFETs?

R-13.7 The threshold voltage, also called pinch-off voltage, is the most important SPICE parameter of the JFET. Is the threshold voltage of an N-channel JFET positive or negative?

R-13.8 What are the most important parasitic elements inherently present in the JFET structure, and what are the related SPICE parameters?

R-13.9 Can the MESFET structure be implemented in silicon? If so, what would be the difference from the JFET structure?

R-13.10 Can a GaAs MESFET be designed as a normally *off* (enhancement-type) device? If not, why not; if so, are there any application constraints?

R-13.11 What is the mechanism of current saturation in GaAs MESFETs?

14 Power Devices

Power-electronic circuits are used to convert electrical energy from the form supplied by a source to the form required by the load. A typical example of power conversion is the rectification and filtering of AC line voltage to provide a constant DC voltage. Other types of conversion include DC–AC, DC–DC, AC–AC, and combinations of the previous types. Diodes can be used to rectify an AC voltage (AC–DC conversion). Plain rectification produces "wavy" voltage, even after filtering with large-value capacitors. *Voltage regulation* is needed to obtain the needed "smooth" voltage. Such regulation is almost invariably based on *switching* techniques.[1] The other types of power conversion also involve switching techniques. Devices such as BJT, MOSFET, JFET, IGBT (insulated-gate bipolar transistor), and thyristor are used as controlled switches in these circuits.

Two fundamental characteristics of a semiconductor switch (either a diode or a controlled switch) are (1) the voltage that can be sustained by a switch in *off* mode (*blocking voltage*), which is determined by the breakdown voltage of the device, and (2) the parasitic resistance of the switch in *on* mode (*on resistance*), which relates to the current capability of the device.

In addition to switches and capacitors, the power circuits typically involve inductors and transformers. Large inductance values are necessary at lower switching frequencies, making the inductors and transformers inconveniently large and heavy. High switching frequencies are the only solution to this problem. Consequently, the switching characteristics of power devices are almost as important as the *on* resistance and the blocking voltage.

Sections 14.1 and 14.2 describe power-related specifics of devices that have already been introduced, the diode and the MOSFET, respectively. Sections 14.3 and 14.4

[1] It is possible to regulate the voltage by allowing the excess voltage to drop across the controlled-variable resistance of a BJT or MOSFET. This, so-called *linear regulation,* is not efficient, as it inevitably involves power dissipation by the regulating resistance.

introduce the IGBT and the thyristor, respectively, as alternative controlled switches to the MOSFET and BJT.

14.1 POWER DIODES

Both P–N junctions and Schottky contacts are used for power diodes. The principles of P–N junctions and Schottky contacts are introduced in Chapter 6 and Section 7.1, respectively.

14.1.1 Drift Region in Power Devices

Power diodes are used in circuits where relatively high voltages have to be rectified. The breakdown voltage of a silicon-based P–N junction can be higher than 1000 V, provided that at least one side (P or N) is very lightly doped. Equations (6.31) and (6.44)–(6.45) and Fig. 6.17c show that the maximum electric field, which appears right at the P–N junction, is proportional to $\sqrt{N_{A,D}}$, where $N_{A,D}$ is the concentration of the lower-doped region. For the case of $N_D \ll N_A$, the following equation can be obtained by eliminating w_n from Eqs. (6.31) and (6.45):

$$E_{cr} = \sqrt{2qN_DV_B/\varepsilon_s} \tag{14.1}$$

In Eq. (14.1), the maximum field E_{max} is set at the critical breakdown field of the semiconductor, E_{cr}, and the reverse bias V_R is set at the blocking voltage $V_B \approx V_R + V_{bi}$. The critical (breakdown) electric field in silicon is in the order of several tens of V/μm. It can be seen from Eq. (14.1) that lower doping levels (N_D) correspond to higher blocking voltages (V_B).

The forward voltage for the nominal *on* current is also very important in power-electronic circuits. As an example, assume that a diode-based rectifier is to provide $V_{OUT} = 5$ V and that the forward voltage across the diode is $V_F = 1$ V. V_F is comparable to V_{OUT}, and it will cause a significant power loss.

The low-doped region needed for the high blocking voltage inevitably increases the series resistance and therefore increases the forward voltage. This would be especially dramatic if the power diode were created by diffusion of P-type layer into a very-low-doped N-type substrate. However, there is no need to limit the doping level of the whole substrate, since its significant thickness would introduce a large series resistance. In power diodes as discrete devices, the relatively thick substrate (needed for mechanical strength only) can be heavily doped. The high breakdown voltage is achieved by depositing a very-low-doped epitaxial layer. A heavily doped layer at the top completes the structure, as shown in Fig. 14.1. The low-doped region, sandwiched between the N^+ and P^+ layers, is commonly referred to as *drift layer*. It is sometimes labeled “I” (“insulator”); accordingly, this type of diode is referred to as a *PIN diode*. In reality, the “I” layer is either P-type or N-type with a very low doping concentration. Another difference from the diode in integrated circuits (Fig. 16.1) is that the curved sections of the P–N junction are avoided. This is because the field is stronger at the sharpest sections of the curve, reducing the breakdown voltage that could be achieved by a planar P–N junction. A variety of etching and surface passivation techniques are used to avoid this problem (not illustrated in Fig. 14.1).

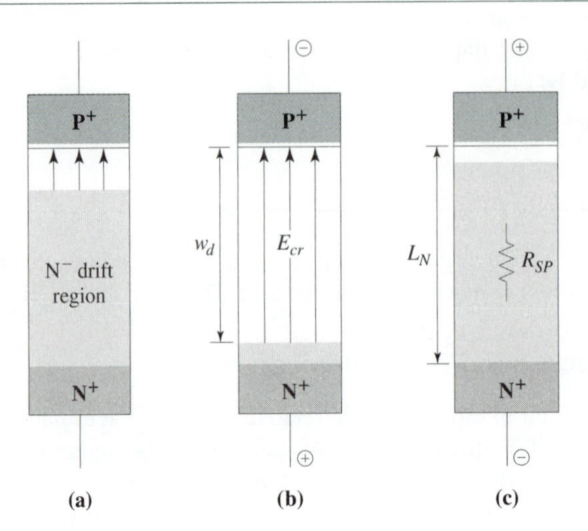

Figure 14.1 (a) The drift region in power devices is needed to support (b) the electric field at the blocking voltage; but it introduces (c) a series resistance in *on* mode.

The parameters of the drift region (its length and doping level) are determined so to achieve a desired blocking voltage and to minimize the *on* resistance. The blocking voltage and the *on* resistance, however, are related to one another: in general, a higher blocking voltage means a higher *on* resistance. For a set blocking voltage, the length of the drift region can be set to be approximately equal to the depletion-layer width[2] at $V_R = V_B$ (Fig. 14.1b and 14.1c):

$$L_N \geq w_d = \sqrt{\frac{2\varepsilon_s V_B}{qN_D}} \tag{14.2}$$

The resistance of the drift region is then

$$R = \underbrace{\rho L_N}_{R_{SP}} \frac{1}{A} \tag{14.3}$$

where A is the cross-sectional area and ρ is the resistivity. The area A is a geometric-design parameter, with the technological parameter being the resistance per unit area. This resistance is called specific resistance: $R_{SP} = \rho L_N$. The specific resistance depends on the doping level, given that $\rho = q\mu_n n \approx q\mu_n N_D$:

$$R_{SP} = \frac{L_N}{q\mu_n N_D} \tag{14.4}$$

[2]The depletion-layer width is given by Eq. (6.45).

With Eq. (14.2) for L_N, and eliminating N_D from Eq. (14.1) for the critical field, the specific resistance becomes

$$R_{SP} = \frac{4V_B^2}{\varepsilon_s \mu_n E_{cr}^3} \tag{14.5}$$

This relationship between the specific resistance and the blocking voltage involves only material parameters. Accordingly, the ratio V_B^2/R_{SP} is a figure of merit of a semiconductor material, known as the Baliga figure of merit:

$$\text{FOM} = \frac{V_B^2}{R_{SP}} = \varepsilon_s \mu_n E_{cr}^3/4 \tag{14.6}$$

Clearly, the figure of merit depends very strongly on the critical (breakdown) field. It is this figure of merit that has motivated the development of SiC as a material for power devices. The critical field is about five times higher in SiC than in Si, which for approximately the same μ_n and ε_s means a figure of merit more than a hundred times higher. This means a hundred times smaller *on* resistance for the same blocking voltage, or it also means that higher blocking voltages with reduced *on* resistances can be achieved.

14.1.2 Switching Characteristics

The switching characteristics of the diode become very important in switching power circuits. The diode can neither be turned on nor off instantly. The times needed for both *forward recovery* (t_{fr}) and *reverse recovery* (t_{rr}) are very important, especially because higher switching frequencies are needed to reduce the size of power-electronic circuits. Figure 14.2 shows the voltage and current waveforms for a power PIN diode.

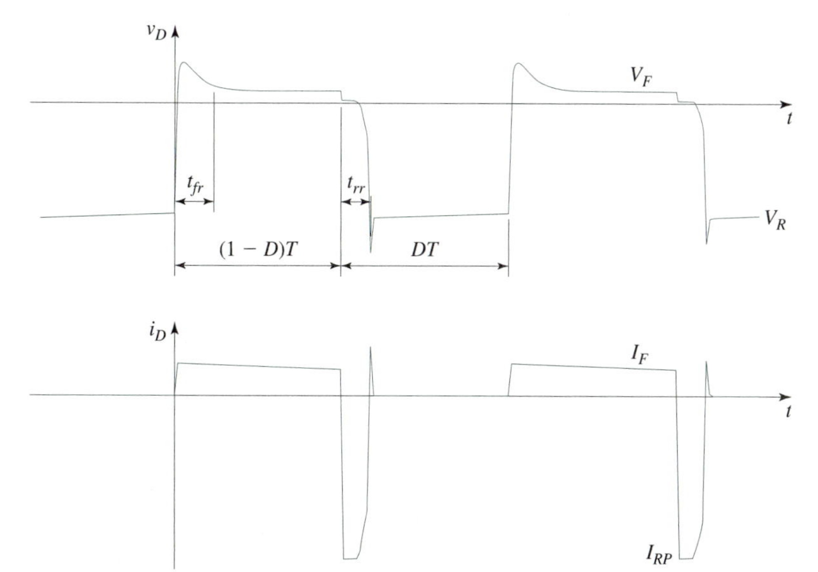

Figure 14.2 Switching response of a power diode: t_{fr}, forward-recovery time; t_{rr}, reverse-recovery time.

The reverse-recovery time was discussed in Section 6.4.3; the same reverse-recovery mechanisms influence the switching response of both the signal and power diodes. There is, however, a difference between the switching responses of signal diodes (Fig. 6.20) and power diodes (Fig. 14.2), which is due to the forward recovery time. This is especially the case when the power diode is connected to an inductor in a switching circuit so that the constant inductor current is forced through the diode ($I_F = i_L$). In a steady state, the forward current of minority carriers is due to diffusion. Earlier, Fig. 6.10 showed that an appropriate gradient of minority-carrier concentration is needed so that the diffusion current can transport the minority carriers injected over the barrier. The concentration gradient of minority carriers is not established instantly. As a consequence, diffusion is not the dominant mechanism of minority-carrier transport at the beginning. A significant electric field is established in the lightly doped neutral region to sustain the forced current by drifting the minority carriers, as well as injecting more majority carriers over the P–N junction barrier. This electric field adds to the electric field in the depletion layer to result in the voltage overshoot observed during the turn-on period of the diode. The time needed to establish the steady-state profile of the minority carriers (the stored charge), and therefore to reach the steady-state forward condition, is the *forward recovery time* (t_{fr} in Fig. 14.2a). The voltage overshoot is undesirable because it results in increased power dissipation ($>V_F I_F$) during the forward recovery period. The SPICE model of the diode does not include this effect. A parallel L–R circuit, added in series to the diode, can be used as the simplest equivalent circuit of this effect.

EXAMPLE 14.1 Stored-Charge Removal–t_{rs} Time

In the buck DC–DC converter of Fig. 14.3, $V_{IN} = 6$ V, $\bar{i}_L = 3.5$ A, and the transit time of the diode (SPICE parameter) is $\tau_T = 5\ \mu$s. Calculate the peak reverse current of the diode I_{RP} and determine the time it takes for the removal of the stored charge t_{rs}, for two values of the *on* resistance of the switch:

(a) $R_{on} = 0.5\ \Omega$
(b) $R_{on} = 1.0\ \Omega$

SOLUTION

(a) When the switch is off, forward-bias current $I_F = i_L$ flows through the diode. The stored charge caused by this current [Eq. (6.58)] is

$$Q_s = \tau_T I_F = 5 \times 10^{-6} \times 3.5 = 1.75 \times 10^{-5}\ \mathrm{C}$$

When the switch is *on*, the minority holes stored at the cathode side of the junction move toward the positive V_{IN} terminal and the electrons stored at the anode side move toward the negative V_{IN} terminal. This is the reverse peak current I_{RP} that removes the stored charge. The value of this current is set by the value of the input voltage (neglecting the small voltage across the diode itself) and the switch resistance:

$$I_{RP} = V_{IN}/R_{on} = 6/0.5 = 12\ \mathrm{A}$$

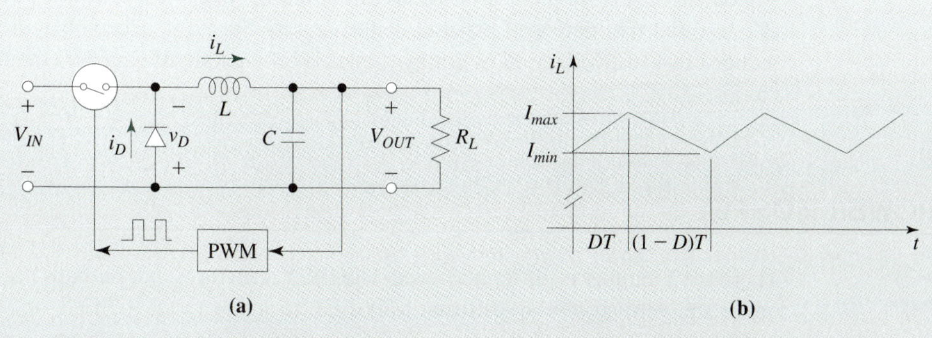

Figure 14.3 (a) Step-down (buck) DC–DC converter. (b) Current waveform.

The diode is not turned off as long as there is stored charge and the discharge current flows through the circuit. The time that is needed for stored-charge removal by this current is

$$t_{rs} = Q_s/I_{RP} = 1.75 \times 10^{-5}/12 = 1.46\ \mu s$$

(b) Answer: $I_{RP} = 6$ A, $t_{rs} = 2.92\ \mu$s.

14.1.3 Schottky Diode

Schottky diodes, introduced in Section 7.1, are single-carrier devices. Although there are minority carriers in the semiconductor, they play an insignificant role as far as the current flow through the metal–semiconductor contact is concerned. Consequently, the effects associated with the stored charge of minority carriers do not exist in Schottky diodes. This means that there is no forward voltage overshoot and $t_{rs} = 0$ (no reverse-recovery time is spent on a stored-charge removal). Clearly, these devices have superior switching characteristics, and consequently they are very useful in modern high-frequency power circuits.

The avalanche breakdown of a Schottky diode depends on the parameters of the semiconductor layer that creates the metal–semiconductor contact. Therefore, it is essentially the same as in the case of PIN diode. Similar to the PIN diode, the very lightly doped drift region is needed to take the high reverse-bias voltage. Also, the heavy doping of the substrate is used in discrete Schottky diodes to reduce the series resistance.

Typically, Pt, W, Cr, or Mo is used as the metal electrode (the anode), whereas the N^-–N^+ Si, GaAs, or SiC structure is used on the cathode side. The metal selection influences the forward voltage of the diode. Equation (7.1) shows that the barrier height at the junction is equal to the difference between the work functions of the metal and the semiconductor, and this barrier determines the forward current [Eqs. (7.4) and (7.6)]. In general, much higher currents at lower forward voltages are possible with Schottky diodes. Equation (7.6) shows that this is achieved at the expense of increased I_S, which is the reverse-bias current.

Nonetheless, it is frequently far more important to reduce the forward power dissipation ($V_F I_F$), and the increased leakage of the reverse-biased is acceptable. Accordingly, the reduced forward voltage of Schottky diodes is another significant advantage over the PIN diodes.

14.2 POWER MOSFET

There are a number of different power MOSFET structures, but perhaps the most accepted one is the **v**ertical **d**ouble-diffused MOSFET structure (VDMOSFET or DMOS), shown in Fig. 14.4. There is only one way of achieving a significant current by a field-effect transistor, and this is to significantly increase the effective channel width of the MOSFET while minimizing or at least maintaining the channel length to a small value. Tens of centimeters of effective channel width and micrometers of channel length are needed to achieve amperes of current. To facilitate a MOSFET with a channel tens of centimeters wide and only several micrometers long, a multiple cell structure is used. The top view of the VDMOSFET (Fig. 14.4a) shows the packing of hexagonal MOSFET cells to minimize

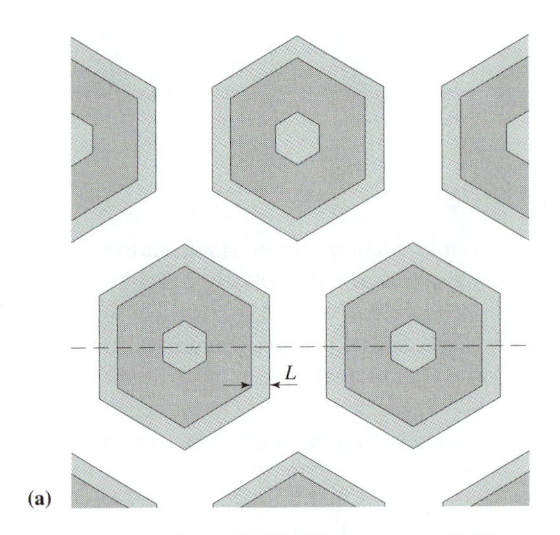

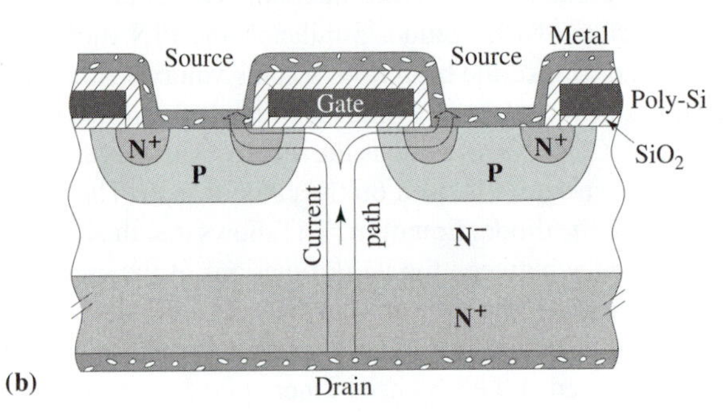

Figure 14.4 VDMOSFET: (a) top view and (b) cross section.

the occupied area. As Fig. 14.4 indicates, all the MOSFET cells are effectively connected in parallel: (1) the silicon substrate is used as the common drain of the MOSFET cells, (2) all the N^+-source and P-well regions are connected by the top metalization, and (3) although windows are etched in the polysilicon layer to enable the creation of the P-type and N^+-type regions, the polysilicon gates remain electrically connected to each other. This means that the effective channel width of the MOSFET is equal to the perimeter of a single cell multiplied by the number of cells used in the structure.

The P-type region is created by using the polysilicon layer as the mask, and the same edge of the polysilicon layer is used to define one side of the N^+ region. The difference between the edges of the P-type and N^+-type regions at the surface is due to different P-type and N^+-type lateral diffusions. This difference determines the channel length L, as illustrated in Fig. 14.4. On one side of the P-type channel region is the N^+ source, and on the other is the N^- drain. The surface region of the drain is very-low-doped (N^-); this is the *drift region*. Analogously to the drift region in the case of diodes, this region becomes depleted at high V_{DS} voltages to take the voltage across itself, enabling the needed forward blocking capability. It is created by epitaxial deposition of low-doped silicon onto a heavily doped N^+ substrate. The heavy doping of the substrate is needed to reduce the parasitic resistance inside the drain-neutral region.

Clearly, a sufficiently wide and low-doped drift region is needed to achieve the minimum acceptable breakdown voltage. Any increase in the width or reduction in the doping concentration, however, results in an increase in the *on* resistance. It may not be possible to find an acceptable trade-off between the needed blocking capability and *on* resistance by adjusting the doping level and width of the drift region. The *on* resistance can be reduced, while maintaining the desired breakdown voltage, by an increase in the channel width. This is a quite simple and quite efficient approach, which can even be applied after the device manufacture, because discrete MOSFETs can be paralleled to achieve the needed *on* resistance. A related effect is *positive temperature coefficient*: if a cell (or a discrete MOSFET operating in parallel combination with others) conducts more current, its *on* resistance increases because of the mobility reduction, which drops the current down to the stable shared value. The importance of this effect is seen when MOSFETs are compared to BJTs. BJTs have a negative temperature coefficient, which means that a current density increase leads to *on* resistance reduction, this causing further current density increase, and so on. This effect, known as *thermal runaway,* can destroy the device.

Is there a practically important limit to the reduction of *on resistance* by paralleling MOSFETs or MOSFET cells? In the case of BJTs, even if thermal runaway is ignored, the high input current needed to maintain the *on*-switch state would be a significant problem. The input of a MOSFET is a capacitor—no DC current is needed to maintain either *on* or *off*-switch states. But what about the current that is needed to charge or discharge the input capacitance to open or close the switch? Because of the large number of cells used, the input capacitance of a power MOSFET is significant. Take as an example $C_{in} = 10$ nF, to perform an estimation of the order of magnitude of the charging/discharging current. If this capacitance is to be charged from 0 to $Q = CV = 10$ nF $\times$ 10 V in $\Delta t = 0.1$ μs by a constant current, the charging current has to be $I = \Delta Q/\Delta t = 1$ A. This is not a small current, and it increases proportionally with the switching frequency and the input capacitance.[3] Clearly, the performance of the circuits driving the power MOSFET used as a switch becomes extremely important. However, no driving circuit will be able to provide unlimited charging/discharging current at unlimited frequency. The input capacitance of the MOSFET poses an ultimate limit to the increase of switching frequency.

As in Fig. 14.4, the source and the body of a power MOSFET are internally connected. Consequently, the capacitance of the drain–body N–P junction appears as a fairly large output capacitance. This can also be a limiting factor, although it can conveniently be utilized as a part of resonant circuits. The same junction acts as an integral power diode, connected across the output. This limits the voltage across the switch to positive values, which can also be a convenient feature. However, if a switch with reverse blocking capability is required, it would appear as a limitation.

The power capability of a power MOSFET is, clearly, another essential characteristic. To have the maximum rated current of a power MOSFET at 5 A while the blocking voltage is rated at 200 V does not mean that this MOSFET can dissipate $200 \times 5 = 1$ kW of power. The maximum current is stated for a fully switched-on MOSFET. If the *on* resistance is $R_{on} = 0.1\ \Omega$, the voltage across the MOSFET is $v_{DS} = R_{on}I_D = 0.1 \times 5 = 0.5$ V, and the dissipated power is $0.5 \times 5 = 0.25$ W. When the MOSFET is in the off mode, no power is dissipated because $I_D = 0$. Ideally, the MOSFET would be in one of these two states, and the dissipated power would never exceed $R_{on}I_D^2$. However, during transitions between those two states, a significantly larger power is dissipated because neither v_{DS} nor i_D is small. To avoid overheating and failure of the MOSFET, i_D and v_{DS} should stay inside the so-called *safe operating area (SOA)*, which is defined by I_{max}, V_{DS-max}, and i_D–v_{DS} points corresponding to the maximum power P_{max}.[4] The maximum power depends on how efficiently the heat is removed. MOSFETs can withstand short pulses of higher power compared to the steady-state level. Because the power MOSFETs are mainly used in switching circuits, the maximum power is typically expressed for a given pulse duration.

14.3 IGBT

The comparison of MOSFETs and BJTs, given in Section 9.1.6, shows that the main advantage of BJTs is the fact that the whole cross section is utilized for current flow. This makes the BJT structure superior in terms of achievable output currents. However, to maintain a power BJT switch in the *on* state, an input base current as high as one-fifth of the output current may be needed. This is a serious drawback because a 100-A device may need 20 A of base current, which significantly complicates the input-drive circuits and reduces the power efficiency. MOSFETs as field-effect devices do not need any current to maintain the *on* state. Still, the input currents needed to charge/discharge the

[3]"Proportional to input capacitance" effectively means "inversely proportional to the *on* resistance."

[4]The i_D–v_{DS} points corresponding to the maximum power P_{max} define a hyperbola on the i_D–v_{DS} graph.

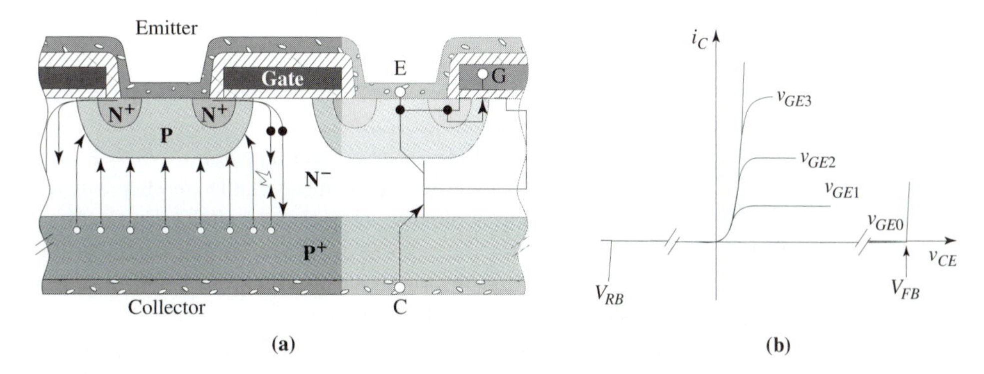

Figure 14.5 IGBT. (a) Cross section of a cell with MOSFET and BJT symbols to illustrate the IGBT components. (b) Current–voltage characteristics.

input capacitance may become quite significant if too many MOSFET cells are paralleled to achieve a high drain current (this is described in the previous section).

It can be said that the input characteristics of MOSFETs and output characteristics of BJTs are needed to create a device capable of switching high currents. It is possible to combine a MOSFET and a BJT to create such a device. When the drain of an N-channel MOSFET is connected to the collector while the source is connected to the base of an NPN BJT, the power MOSFET is utilized as a driver device supplying the base current to the BJT. Developing this principle further, an integrated device with field-effect input control and bipolar output action was created. Reflecting its principal features, this device is called an *insulated-gate bipolar transistor (IGBT)*. A shorter name, *insulated-gate transistor (IGT)*, is used as well, and it has also been called *conductivity-modulated field-effect transistor (COMFET)*. Commercial IGBTs with blocking voltage exceeding 500 V and able to switch hundreds of amperes of currents have been developed. IGBTs are replacing power BJTs in many applications.

The cross sections of IGBT and VDMOSFET, given in Fig. 14.5a and 14.4b, respectively, show that these devices are structurally very similar. The only difference is that the N^+ substrate is replaced by a P^+ layer in the case of IGBT. This creates the PN^-P^+ structure of the BJT. It is very easy to explain the operation of the device in the *off* state (zero gate voltage). In this case, the N^- layer is floating, and the IGBT appears as a BJT with unconnected base; therefore it behaves as two P–N junction diodes connected back to back. The applied voltage drops across the reverse-biased P–N junction, and no significant current flows through the device. In the case of positive v_{CE}, the breakdown voltage of the upper P–N^- junction (V_{FB}) determines the forward blocking capability. In the case of negative v_{CE}, the breakdown voltage of the lower N^-–P^+ junction (V_{RB}) determines the reverse blocking capability. The existence of the reverse blocking capability is a significant difference from the power MOSFET. In applications that use the integrated diode of the MOSFET, the reverse blocking of the IGBT is a disadvantage. However, in the applications that require reverse blocking (in addition to forward blocking), this feature of the IGBT appears as a very significant advantage.

When sufficient gate voltage is applied, a strong-inversion layer connects the N^- layer to the N^+ regions, and then it connects the N^- layer to the electrode labeled *emitter*. A significant part of the applied collector-to-emitter voltage drops across the lower N^-–P^+ junction, setting it in forward-bias mode. As a result, holes are injected into the N^- base region, most of them finishing at the *emitter terminal* through the P-type region (collector of the PNP BJT).[5] These holes make most of the *on*-state current. For as long as the inversion layer is strong enough to neglect any voltage drop across it, the device appears as a diode with a significant current capability, because almost the whole cross section is utilized. Figure 14.5a illustrates that the MOSFET supplies electrons to maintain the forward bias of the lower N^-–P^+ junction.

When the v_{CE} voltage is increased, its electric field opposes the gate field, reducing the concentration of electrons in the inversion layer. This can result in channel pinch-off, enabling a significant voltage drop across the depleted channel area. The forward bias of the junction injecting the holes, and therefore the device current, does not increase with any further increase of v_{CE}. The device is in its *saturation region,* which is illustrated in Fig. 14.5b for the sake of completeness, although this operation region is not useful for a device used as a switch.

In conclusion, it can be reiterated that the IGBT integrates the superior input and output current performances of MOSFET and BJT, respectively. However, the "selection" of superior characteristics is limited to the current capability. Being a BJT-like device from the output, it inevitably suffers from the effects associated with storage of minority carriers. As discussed in the sections on the power diode and the power MOSFET, this results in inferior switching performance comparing to the one-carrier devices such as the MOSFET. As far as SPICE simulation is concerned, there is no IGBT device model with specific model parameters. The device is modeled by its equivalent MOSFET–BJT circuit, as shown in Fig. 14.5a. Consequently, the MOSFET and BJT model parameters introduced in Chapters 8 and 9 are used.

14.4 THYRISTOR

Thyristors are four-layer PNPN devices that are capable of blocking thousands of volts in the *off* state and conducting thousands of amperes of current in the *on* state. They work on the principle of internal regenerative mechanism that leads to so-called *latch-up* effect. Clearly, basic understanding of the latch-up effect is necessary to understand the specifics of thyristors used as power switches.

Although thyristors utilize the latch-up effect, this effect is a potential problem in any other device that involves four layers (PNPN). An example of such a device is the IGBT described in the previous section. Also, PNPN structures are inherently present in the very

[5]The adopted labeling of the terminals as *emitter* and *collector* comes from the previously described connection of a MOSFET and NPN BJT. It can be confusing in the case of the IGBT because the *collector terminal* is connected to the *emitter* of the inside PNP BJT, while the *emitter terminal* is connected to the *collector* of the PNP BJT.

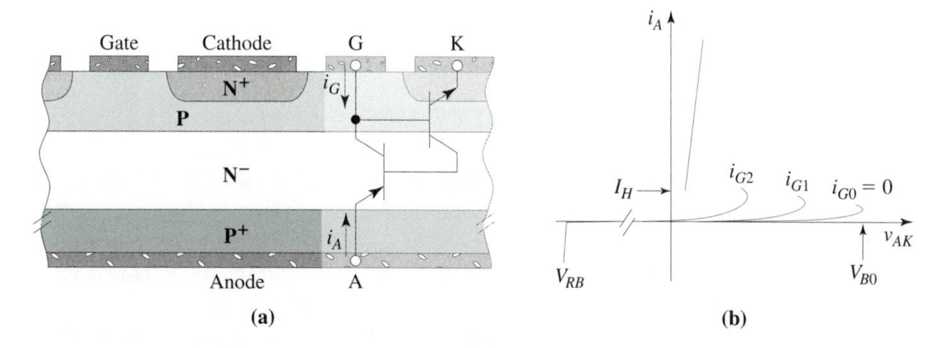

Figure 14.6 SCR. (a) Cross section of a cell with two-transistor model. (b) Current–voltage characteristics.

common CMOS structures. Unwanted latch-up of the parasitic thyristor structure may lead to permanent damage of these devices.

The most common thyristor type is the *silicon-controlled rectifier (SCR)*. Its cross section is shown in Fig. 14.6, which also shows the two-transistor model that is used to explain the regenerative mechanism involved in the thyristor operation. It can be seen that the PNP BJT collector is the same region as the NPN BJT base, whereas the NPN collector is the same as the PNP base. A small gate current can trigger closed-loop amplification (the BJTs amplify each others collector current) until both transistors enter saturation, providing a low-resistance path between the anode and the cathode.

The following mathematical analysis provides the basis for a more detailed discussion. The anode current i_A, being in fact the emitter current of the PNP BJT, is given by

$$i_A = (1 + \beta_{pnp}) I_{B-pnp} \tag{14.7}$$

I_{B-pnp} is the same current as the collector current of the NPN BJT, and it can therefore be expressed as

$$I_{C-npn} \equiv I_{B-pnp} = \beta_{npn}\overbrace{(i_G + \underbrace{\beta_{pnp} I_{B-pnp} + I_{CB0-pnp}}_{I_{C-pnp}})}^{I_{B-npn}} + I_{CB0-npn} \tag{14.8}$$

In Eq. (14.8), $I_{CB0-pnp}$ and $I_{CB0-npn}$ are the reverse-bias collector–base leakage currents of the PNP and NPN BJTs, respectively. Because of the closed loop, I_{B-pnp} appears on both the left-hand and right-hand sides of Eq. (14.8). Extracting I_{B-pnp},

$$I_{B-pnp} = \frac{\beta_{npn}(i_G + I_{CB0-pnp}) + I_{CB0-npn}}{1 - \beta_{pnp}\beta_{npn}} \tag{14.9}$$

and putting it into Eq. (14.7), the following equation for the anode current is obtained:

$$i_A = \frac{1+\beta_{pnp}}{1-\beta_{pnp}\beta_{npn}}\left[\beta_{npn}(i_G + I_{CB0\text{–}pnp}) + I_{CB0\text{–}npn}\right] \tag{14.10}$$

Assuming $i_G = 0$ and negligible leakage currents ($I_{CB0\text{–}pnp} \approx 0$ and $I_{CB0\text{–}npn} \approx 0$), the anode current is negligible, and the thyristor behaves as a switch in *off* mode even for $v_{AK} > 0$. This is because the base–collector N^-–P junction is reverse-biased and the base currents are ≈ 0, so both BJTs are in the cutoff mode. However, if v_{AK} is increased, both $I_{CB0\text{–}npn}$ and $I_{CB0\text{–}pnp}$ are increased, leading to the appearance of some i_A current. At very small transistor currents, the current gains (βs) are small as well, but their values increase as the transistor current is increased.[6] At some voltage $v_{AK} = V_{B0}$, the transistor currents increase the current gains to the point where $\beta_{pnp}\beta_{npn} \rightarrow 1$. According to Eq. (14.10), $i_A \rightarrow \infty$ at this condition. The regenerative mechanism has occurred, the BJTs enter the saturation mode, and Eq. (14.10) no longer applies. All the P–N junctions are forward-biased, which means that the voltage between the anode and the cathode is very small, and the current flowing between the anode and the cathode is limited by the external circuit. The thyristor is latched up, and it behaves as a switch in *on* mode with small parasitic resistance. The latch-up happens quickly, and no trace connecting V_{B0} to the resistor-like characteristic of the *on* thyristor is shown in Fig. 14.6b.

The latching just described occurred at voltage V_{B0} due to the internal leakage currents, in the absence of any gate current. The voltage V_{B0} is the forward blocking voltage of the thyristor. If some gate current is provided, the critical condition $\beta_{pnp}\beta_{npn}$ is reached at a smaller v_{AK} voltage, enabling latch-up with smaller anode voltages, as shown in Fig. 14.6b.

Once latched up, the SCR cannot be switched off by setting the gate current to zero, as the regenerative process is self-sustainable. The thyristor can be turned off by reducing the current to a level where the current–gain product is $\beta_{pnp}\beta_{npn} < 1$. The minimum current at which the thyristor is still on is called the *holding current* (I_H in Fig. 14.6b). Another type of thyristor, called a *MOS-controlled thyristor,* integrates a MOSFET whose gate can be used to control the switching of the thyristor. When latched up, the thyristor current creates stored charge at the forward-biased P–N junctions, which means that the switching characteristics exhibit the P–N junction-type delay, due to the need for stored-charge removal.

For negative v_{AK}, both base–emitter junctions (P^+N^- and PN^+) are reverse-biased, which means that no current flows because both BJTs are in cutoff. Therefore, this type of thyristor exhibits reverse blocking capability. The depletion layer is the widest in the lowest-concentration N^- layer, which therefore supports most of the reverse v_{AK} voltage. The breakdown voltage of this junction determines the reverse blocking voltage V_{RB}.

This type of reverse blocking capability is unwanted in some applications: more precisely, it is desirable that the thyristor conduct current in both directions once latched up. In this case, one must use TRIAC: a thyristor that, in a sense, is a pair of SCRs connected in antiparallel.

---

[6]This effect is shown in Fig. 9.16 and described in Section 9.3.

SUMMARY

1. Power-electronic circuits use switches to convert electrical energy from one form to another. P–N junction (including PIN) and Schottky diodes are used as two-terminal switches (rectifiers). BJTs, MOSFETs, IGBTs, and thyristors are most frequently used as controlled switches.
2. PIN diodes are, in principle, P–N junction diodes with an inserted low-doped ("insulating"—I) region, called a *drift region,* which takes the reverse-bias voltage across its depletion layer. PIN diodes exhibit good reverse blocking (*off* state) and current (*on* state) capabilities. Relatively high barrier height and the voltage across the drift region contribute to relatively high forward voltage (>1 V). In a 5-V power supply, this alone can lead up to 25% power-efficiency loss.
3. Minority carriers, involved in the current conduction of a P–N junction diode, cause two undesirable switching effects: (1) forward voltage overshoot due to limited rate of minority-carrier accumulation to the forward-bias level of stored charge and (2) delay and current flow needed to remove the stored charge so that the diode is switched off again.
4. Schottky diodes exhibit superior switching characteristics because the current conduction involves majority carriers only, and no stored-charge effects appear. The choice of metal work function can adjust the barrier height so that the forward voltage is reduced, compared to PIN diodes. This inevitably leads to increase in reverse-bias current, which is a limiting factor for the forward-voltage reduction. Analogously to PIN diodes, power Schottky diodes involve a drift (low-doped) region to achieve the desired reverse blocking capability.
5. Power BJTs, as controlled switches, share the positive (reverse blocking and good current capabilities) and negative (stored-charge-related switching delays) characteristics of PIN diodes. In addition, a large input base current is needed to maintain the BJT in on state.
6. Tens of thousands of MOSFET cells can be paralleled to compensate for the inherently inferior current capability and to enable small *on* resistance. Being a majority-carrier field-effect device, the power MOSFET exhibits superior switching performance, and no steady-state input current is needed to maintain either *on* or *off* states. Still, high transient currents are needed to charge/discharge the input capacitance to switch the device on/off, imposing a practical limit to input capacitance increase due to paralleling MOSFET cells. Analogously to diodes and BJTs, a drift region is used to provide good forward-blocking capability ("forward" refers to positive drain-to-source voltage). In the *on* state, the drift region contributes to, and even dominates, the *on* resistance. The body and the source are short-circuited in a power MOSFET, which short-circuits the internal body-to-source power diode—no reverse blocking capability.
7. IGBTs combine input MOSFET and output BJT to create a device with superior input control and output current capabilities. Its BJT-like output renders it inferior to the MOSFET in terms of high-frequency switching performance.
8. Thyristors are four-layer PNPN structures that can be latched into the conduction (*on*) state by a regenerative mechanism: the PNPN layers form PNP and NPN BJTs, with collectors connected to each other's base, amplifying and pushing the current in a closed loop until both BJTs enter the saturation mode. This mechanism works when $\beta_{pnp}\beta_{npn} \rightarrow 1$, which does not happen if the collector currents are below a certain level.

In this state, the thyristor is in the blocking (off) mode, either reverse or forward. A gate electrode is used to provide the triggering level of current—the gate is used to turn the thyristor on, but it cannot be used to turn the basic thyristor (SCR) off. The SCR turns off when the current is dropped below the *holding current* so that $\beta_{pnp}\beta_{npn} < 1$. There is an accumulation of minority carriers, and consequently the switching performance is limited by stored-charge effects. However, thyristors have superior blocking and current capabilities: they can conduct thousands of amperes of current in the *on* state and block thousands of volts in the *off* state.

PROBLEMS

14.1 In the boost DC–DC converter of Fig. 14.7, $\bar{i}_L = 5$ A, $V_{OUT} = 5$ V, the transit time of the diode is $\tau_T = 5\ \mu$s, and the on resistance of the switch is $R_{on} = 0.5\ \Omega$. Calculate the reverse diode current I_{RP} that discharges the stored charge, and find the time t_{rs} needed for the stored charge to be removed.

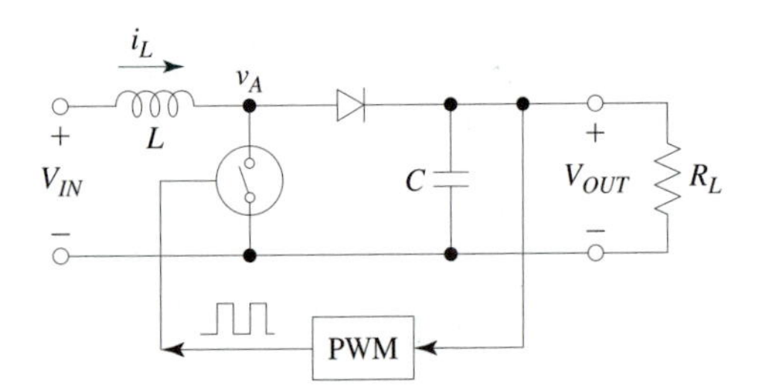

Figure 14.7 Step-up (boost) DC–DC converter.

14.2 Design the N-type drift region in a silicon diode so that the avalanche breakdown voltage is $V_{BR} = 12$ V. The breakdown electric field of silicon is $E_{cr} = 60$ V/μm. Assume that $V_{bi} = 0.9$ V. The design should specify the doping level (N_D) and the minimum length of the N region (L_N), which ensures that the depletion layer does not become wider than the drift region.

14.3 The same doping level of the N-type drift region, as obtained for the Si diode designed in Problem 14.2, is used for a diode implemented in SiC ($\varepsilon_s = 9.8 \times 8.85 \times 10^{-12}$ F/m, $E_{cr} = 320$ V/μm, $V_{bi} = 2.8$ V).

(a) Determine the breakdown voltage if the length of the drift region L_N is made larger than the maximum depletion-layer width. **A**

(b) Determine the breakdown voltage if the minimum L_N, as calculated in Problem 14.2, is used (in this case, define the breakdown voltage as the reverse-bias voltage that fully depletes the drift region—punch-through breakdown).

(c) What L_N is needed to achieve the breakdown voltage calculated in part (a) of this problem? **A**

14.4 Design the N-type drift region of a power diode so that the avalanche breakdown voltage is $V_{BR} = 1000$ V, if the diode is to be implemented in

(a) Si ($E_{cr} = 60$ V/μm and $V_{bi} = 0.9$ V)

(b) 4H SiC ($E_{cr} = 320$ V/μm, $V_{bi} = 2.8$ V, and $\varepsilon_s = 9.8 \times 8.85 \times 10^{-12}$ F/m)

The design should minimize the resistance of the drift region. Determine the specific resistance of the drift region in each case if electron mobilities are 1400 cm^2/V · s and 800 cm^2/V · s for Si and 4H SiC, respectively.

14.5 The doping concentrations of a silicon PIN diode are $N_A = 10^{20}$ cm^{-3}, $N_{D-l} = 10^{15}$ cm^{-3}, and $N_{D-h} = 10^{20}$ cm^{-3}, respectively, and the length of the drift region is $L_N = 5\ \mu$m. At what voltage does the depletion layer width become equal to L_N (punch-through breakdown)? Is this voltage larger or smaller than the avalanche breakdown voltage ($E_{cr} = 60$ V/μm)?

14.6 The following are the dimensions of a hexagonal VDMOSFET (such as the one in Fig. 14.4): the length of the polysilicon gate $L_g = 20\ \mu$m, the distance between two P-layers (under the gate) $L_{P-P} = 10\ \mu$m, the distance between two N^+ layers (under the gate) $L_{N^+-N^+} = 18\ \mu$m, the gate-oxide thickness $t_{ox} = 80$ nm, and the thickness of the insulating oxide between the polysilicon gate and the source metalization $t_I = 1\ \mu$m. If the MOSFET consists of 15,000 hexagonal cells, calculate

(a) the gate-to-drain (C_{GD}) capacitance
(b) the gate-to-source (C_{GS}) capacitance A

14.7 A power MOSFET is switched on and off by a 1-MHz pulse signal, with 0 V and 10 V as the low- and high-voltage levels, respectively. Assuming a constant input capacitance of 7 nF and a total series resistance in the charging/discharging circuit of 5 Ω, calculate the average dissipated power by the input-control circuit.

REVIEW QUESTIONS

R-14.1 Is it better to use higher or lower switching frequency in terms of minimizing the size of inductors (and transformers if present) in a switching power circuit?

R-14.2 Can the PIN diode impose a limit to the switching frequency in a power circuit?

R-14.3 Do power Schottky diodes exhibit the stored-charge-related effects (forward voltage overshoot and constant-current reverse recovery)?

R-14.4 Are the switching characteristics of a PIN diode (e.g., forward voltage overshoot and reverse recovery) related to the power efficiency?

R-14.5 Is the reverse blocking voltage related to the power efficiency? Indirectly?

R-14.6 Low doping concentration and sufficient width of the drift region are needed to achieve the desired blocking capability. Does the drift region affect the forward voltage?

R-14.7 If a PIN and a Schottky diode have identical drift regions, will the forward voltages be the same?

R-14.8 Is the forward voltage of a diode (either PIN or Schottky) related to the power efficiency?

R-14.9 Given better switching characteristics and smaller forward voltage, is the superiority of Schottky diodes over PIN diodes total? In other words, is there any important disadvantage of power Schottky diodes? If so, what is it?

R-14.10 Is the I_S parameter in the I_D–V_D equation related to the forward voltage? If so, how?

R-14.11 If two controlled switches are connected in parallel and one of them takes a larger share of the current, it will heat more, reaching higher operating temperatures. In the case of BJTs the temperature increases the current, whereas in the case of MOSFETs it decreases the current. How does this difference affect the stability of the parallel connection?

R-14.12 The channel thickness of MOSFETs is inherently limited to several nanometers (the penetration of the electric field creating the channel). A channel that is tens of centimeters in width is necessary to compensate for this limitation, and therefore necessary to achieve an acceptable channel resistance. How is it possible to place a tens-of-centimeters-wide power MOSFET in a package not larger than 1 cm in diameter?

R-14.13 Paralleling MOSFETs or MOSFET cells increases the input capacitance. Can this represent a practical problem, given that no DC current flows through the input capacitance?

R-14.14 Charging and discharging of the input capacitance has obvious implications for the maximum switching frequency. If there is no leakage through the capacitor, do charging and discharging affect the power efficiency?

R-14.15 Is the forward blocking voltage of a MOSFET related to the *on* resistance? If yes, what is the relationship?

R-14.16 The drain current of a power MOSFET increases as the gate voltage is increased above the threshold voltage, but then quickly "saturates." Does that mean that the resistance of the drift region dominates the *on* resistance? If not, why does this happen?

R-14.17 IGBT integrates a MOSFET and a BJT to achieve the advantages of MOSFET input controllability and BJT output current capability. However, MOSFETs can easily be paralleled, which increases the current capability. Do IGBTs exhibit a real advantage in terms of current capability? If so, there must be a limit/problem with paralleling the MOSFETs. What would it be?

R-14.18 For an IGBT in the *on* state, minority carriers are injected into the drift region. Does that mean that the *on* resistance is smaller than it would be if an equivalent drift region were used in a power MOSFET?

R-14.19 IGBT retains all the stored-charge-related effects found in BJTs. Does this make them inferior to MOSFETs in terms of high-frequency switching performance?

R-14.20 Thyristors are controlled switches that consist of four layers (PNPN). With either voltage polarity across the switch, there is always a reverse-biased P–N junction. How can a conductive path ever be established through the PNPN structure to set the switch in on state?

R-14.21 What happens if the collector current of transistor "A" is amplified by transistor "B" and the amplified current is fed back into the base of transistor A? Is there a limit to this closed-loop amplification? Can a finite "loop gain" be reached if the maximum $\beta_{pnp}\beta_{npn}$ product is <1? What if $\beta_{pnp}\beta_{npn} > 1$?

R-14.22 Consider the PNPN structure of an IGBT. If $\beta_{pnp-max}\beta_{npn-max}<1$, does this mean that this structure can never be latched up?

R-14.23 Assume that a thyristor conducts 1000 A at 1 V of forward voltage, which means that the *on* resistance is 1 mΩ. Can this thyristor be used as a small-resistance switch for low-current applications (say <1 A) to achieve negligible forward voltage drop (<1 mV)? A negative answer implies a minimum-current limit. Why should there be a "minimum-current limit"?

15 Negative-Resistance Diodes

Negative-resistance diodes are active microwave devices that can be used as amplifiers and oscillators at frequencies up to 100 GHz (despite the continuous improvement of FETs, their performance is still inferior or even inadequate at these frequencies). As the name suggests, a common characteristic of negative-resistance diodes is the negative-resistance (or negative-conductance) phenomenon. More precisely, we are dealing here with *negative dynamic (or differential) resistance (NDR)* or, alternatively, negative dynamic (differential) conductance. This means that a voltage increase causes a current decrease ($r = dv/di < 0$), which is different from the fact that the ratio between the instantaneous voltage and current is still positive ($R = v/i > 0$). The negative dynamic resistance causes current and voltage to be 180° out of phase with each other. As a consequence, the signal power is negative, which means that a negative-resistance diode does not dissipate but, rather, generates signal power. Again, this should not be confused with the total power, which is positive, expressing the fact that the device efficiency is $<100\%$.

Signal amplification by two-terminal negative-resistance diodes is conceptually different from the amplification by the voltage-controlled current sources (three-terminal BJTs and FETs). Section 15.1 describes the principles of amplification and oscillation by negative-resistance devices, followed by Sections 15.2 and 15.3, which introduce Gunn and IMPATT diodes as representatives of *transferred-electron devices* (TED) and *avalanche transit time diodes,* respectively. Finally, Section 15.4 briefly describes the tunnel diode.

15.1 AMPLIFICATION AND OSCILLATION BY NEGATIVE DYNAMIC RESISTANCE

A part of the current–voltage characteristic of a negative-resistance diode exhibits a *negative slope*. This region of the current–voltage characteristic can be used to produce signal voltage and current 180° out of phase with each other, thus to generate signal power. To achieve this, the loading element (G_L) is connected in parallel with the negative-

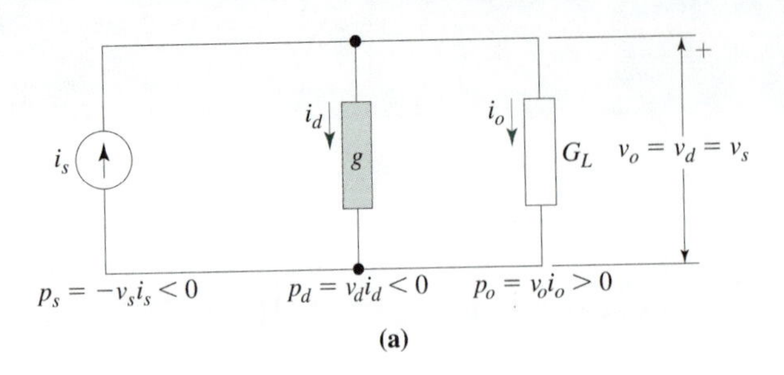

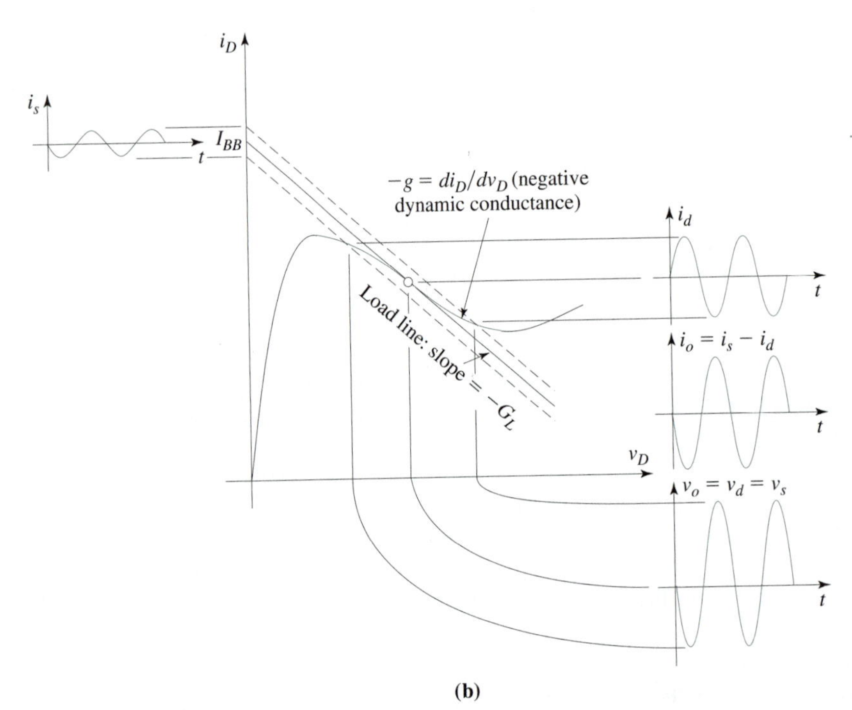

Figure 15.1 Signal amplification by a negative-conductance diode. (a) Fundamental small-signal equivalent circuit. (b) Graphic load-line analysis.

conductance diode (g), as illustrated in Fig. 15.1a. This does not mean that the diode can deliver signal power to the load out of nothing—it still needs a DC power source; but because of the parallel connection, the power source is effectively in the form of a current source I_{BB} that is connected in parallel to the signal source i_s.[1]

Figure 15.1b shows a typical i_D–v_D characteristic of a negative conductance diode, with the negative-conductance part of the curve highlighted. To use this characteristic for a graphic analysis of the circuit, the load line representing the load (G_L) and the DC bias (I_{BB}, not shown in Fig. 15.1a) is needed. When the source current i_s equals zero (the DC

[1]The actual implementation of the DC current source (I_{BB}) is a separate issue; but typically, an ordinary voltage source separated by a lossy low-pass filter is used.

bias point), we have

$$i_D = I_{BB} - G_L v_D \tag{15.1}$$

to satisfy the Kirchhoff current law at the parallel connection of g and G_L. Equation (15.1) is the needed load line, which can be drawn by observing the following two facts: (1) $i_D = I_{BB}$ for $v_D = 0$, which means that the line intersects the i_D axis at I_{BB}, and (2) the slope of the line is $-G_L$. The solution is found at the intersection between the load line and the diode characteristic. The absolute values of i_D and v_D are both positive, which means that the diode dissipates (uses) instantaneous power supplied by the DC source.

When a signal current i_s is added to the DC bias current I_{BB}, the load line is shifted up or down (depending on the sign of the signal current), but the slope is not changed. This is illustrated by the dashed lines in Fig. 15.1b. It can be seen that a small decrease/increase in the incoming current, caused by the superimposed signal current (i_s) on the bias current (I_{BB}), causes a rather large increase/decrease in the diode current. Therefore, a current gain is achieved. More importantly, the increase in i_D is accompanied by a decrease in v_D, which means that the signal voltage is 180° out of phase with the signal current flowing through the diode. This means negative signal power $p_d = v_d i_d$, or signal power generation in other words. On the other hand, the signal current flowing through the load (i_o) is in phase with the voltage, which means that the load is using power $p_o = v_o i_o$.

The current gain ($A = i_o/i_s$) and the power gain ($A = p_o/p_s$) are identical to each other, because $v_o = v_s$. Figure 15.1b shows that the magnitude of i_d current depends on the slopes of both the negative-conductance part of the diode curve (g) and the load line (G_L). This further means that the magnitude of the output current ($i_o = i_s - i_d$) and hence the current/power gain depends on both g and G_L. To find this dependence, let us start from Kirchhoff's current law,

$$\frac{i_o}{A} = i_d + i_o \tag{15.2}$$

where the source current i_s is expressed in terms of the output current, using the gain definition $A = i_o/i_s$. Furthermore, express the currents in terms of the conductances and the unique voltage $v_o = v_d = v_s$ by applying Ohm's law:

$$\frac{G_L}{A} v_o = -|g| v_o + G_L v_o \tag{15.3}$$

When one is applying Ohm's law to the negative-conductance element, care is taken to express that the current i_d is 180° out of phase with the voltage $v_d = v_o$. This leads to the following equation for the gain:

$$A = \frac{G_L}{G_L - |g|} \tag{15.4}$$

We can conclude from this equation that

$$G_L > |g| \tag{15.5}$$

is a condition for stable amplification. With this condition, $p_s = -v_o i_s = -v_o^2 G_L/A < 0$ (power generated), $p_d = v_o i_d = -|g|v_o^2 < 0$ (power generated), and $p_o = v_o i_o = G_L v_o^2 > 0$ (power used). If $G_L < |g|$, the gain A would be negative, and v_d and i_d inverted. With this, the input signal source consumes power ($p_s > 0$) while the negative-conductance diode still generates power ($p_d < 0$). Here the signal power generated by the negative-conductance diode is unrelated to the signal source. Is this situation possible?

To clarify the question of power generation unrelated to any excitation, let us remove the signal source ($i_s = 0$), but not the supply current I_{BB}. In addition to that, let G_L be adjusted so that

$$G_L = |g| \tag{15.6}$$

In this case, $A \to \infty$, which does mean that no input signal is needed to have a finite output signal current ($i_o = Ai_s = \infty \times 0 =$ finite value). An electronic system that produces an output signal without any input excitation, thereby generating the signal, is called *oscillator*. Referring to Fig. 15.1b, the condition $G_L = |g|$ means that the load line overlaps the diode curve in the negative-conductance section. There is not a unique i_D–v_D solution (intersection between the load line and the diode curve) in this case, meaning that the circuit is unstable. The operating point will oscillate, as governed either by internal physical processes in the diode (unresonant mode) or by a resonant circuit if such a circuit is connected (resonant mode).

EXAMPLE 15.1 Negative-Resistance Oscillator

The current–voltage characteristic of a negative-resistance diode is shown in Fig. 15.2, where $i_A = 200\ \mu$A, $i_B = 60{,}200\ \mu$A, $v_A = 16$ V, and $v_B = 10$ V.

(a) Find the value of the load conductance G_L so that this diode is used as an oscillator.

(b) Assuming maximum signal amplitude, calculate the *power conversion efficiency* of this oscillator ($\eta = \bar{p}_o/\bar{p}_S$, where $\bar{p}_o$ is the average output signal power, and $\bar{p}_S$ is the average instantaneous power supplied by the DC source).

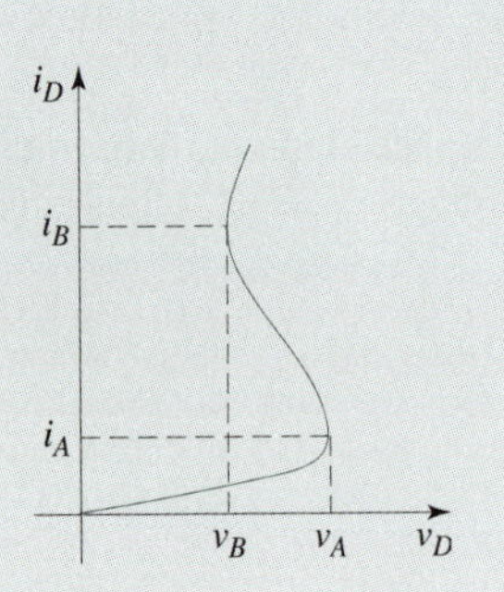

Figure 15.2 Current–voltage characteristic of a negative-resistance diode.

SOLUTION

(a) The oscillation condition is given by Eq. (15.6) as $G_L = |g|$. Assuming linear i_D–v_D dependence between i_A and i_B, $|g|$ can be estimated as

$$|g| \approx (i_B - i_A)/(v_A - v_B) = \frac{60{,}200 - 200}{16 - 10} = 10 \text{ mA/V} = 10 \text{ mS}$$

Therefore, the load conductance should be

$$G_L = 10 \text{ mS}$$

(b) To achieve maximum signal amplitude, the signal current and voltage should oscillate around the central (I_D, V_D) point, where I_D and V_D are $(60.2 + 0.2)/2 = 30.2$ mA and $(16 + 10)/2 = 13$ V, respectively. The peak amplitudes of the signal are then $I_m = i_B - I_D = (I_B - I_A)/2 = 30$ mA and $V_m = (V_A - V_B)/2 = 3$ V. Therefore, the voltage and current signals can be expressed as $v_o = v_d = V_m \sin(\omega t)$ and $i_o = i_s - i_d = -i_d = I_m \sin(\omega t)$. With this, the average output signal power is found as

$$\bar{p}_o = \frac{1}{T}\int_0^T v_o i_o \, dt = \frac{V_m I_m}{\pi}\int_0^{\pi} \sin^2(\omega t)\, d(\omega t) = V_m I_m/2$$

The instantaneous power delivered by the DC source is $p_S = I_{BB} v_D$, where I_{BB} is the current of the DC current source supplying the power and v_D is the voltage across the current source, which is equal to the voltage across the diode, as well as the output. Because $v_D = V_D - V_m \sin(\omega t)$, the average power delivered by the source can be expressed as

$$\bar{p}_S = \frac{I_{BB}}{2\pi}\int_0^{2\pi} v_D \, d(\omega t) = I_{BB} V_D - \frac{I_{BB} V_m}{2\pi}\underbrace{\int_0^{2\pi} \sin(\omega t)\, d(\omega t)}_{=0} = I_{BB} V_D$$

where I_{BB} is the current at which the load line intersects the i_D-axis. Because the load line passes through the central (I_D, V_D) point and has a slope of $-G_L$, the following relationship can be established:

$$\frac{I_{BB} - I_D}{V_D} = G_L$$

which further means that

$$I_{BB} = G_L V_D + I_D = 160.2 \text{ mA}$$

The power conversion efficiency is, therefore,

$$\eta = \frac{\bar{p}_o}{\bar{p}_S} = \frac{V_m I_m}{2 V_D I_{BB}} = \frac{3 \times 30}{2 \times 13 \times 160.2} = 0.0216 = 2.16\%$$

15.2 GUNN DIODE

"Gunn diode" represents a group of devices called *transferred-electron devices (TED)*. TEDs are used as oscillators and amplifiers, covering the frequency range from 1 to 100 GHz, with output power capabilities greater than 1 W. Principally, these devices are made of plain n-type semiconductor pieces (no P–N junctions), with ohmic contacts at two opposite sides. The name *Gunn diode* is typically used for GaAs-based diode, while the other options are InP and CdTe.

The appearance of negative dynamic resistance in such a simple structure is due to the specific E–k dependence, which has two close minima, as shown in Fig. 2.13a for GaAs. As illustrated in Fig. 15.3, the electrons in the lower E–k valley have much smaller effective mass, and consequently they possess much higher mobility μ_l (note that the subscript l refers to either *light* effective mass or *lower* E–k valley but not to *low* mobility). Because the higher E–k valley is much wider, electrons appear to be much heavier there, and consequently their mobility μ_h is much lower. At small voltages, thus small electric fields inside the semiconductor, all the electrons are in the lower E–k valley. The mobility of all the electrons is high (μ_l), and the conductance $G = qAn\mu_l/L$ is large. This is illustrated by the large-slope dashed line in Fig. 15.3. At very large voltages, thus very large electric fields, most of the electrons gain sufficient energy to appear in the higher E–k valley. The mobility of the electrons is now low (μ_h), and the conductance $G = qAn\mu_h/L$ is small, as shown by the small-slope dashed line in Fig. 15.3. In the medium-voltage range, a voltage increase causes a transfer of a number of electrons from the lower to the higher E–k valley, which reduces the average drift velocity of the electrons (Fig. 15.4) and therefore reduces the current. This current reduction due to a voltage increase is the effect of negative dynamic conductance $g = dI/dV < 0$.

The semiconductor is not stable in the negative-conductance region because it cannot establish a unique electric field when biased so that V/L exceeds the critical electric field (the critical electric field corresponds to maximum drift velocity in Fig. 15.4). To explain this, let us consider a packet of electrons injected by a negative cathode into GaAs biased beyond the critical V/L. As the electrons move toward the anode, the electric field back toward the cathode is reduced, while the electric field toward the anode is slightly increased. This is illustrated in Fig. 15.5a. In normal circumstances, the packet of electrons

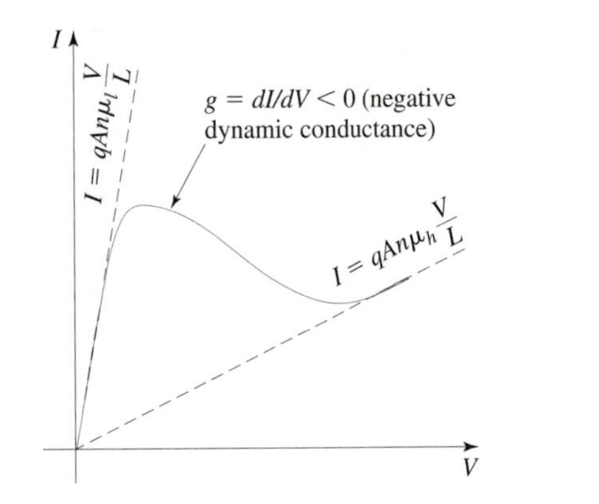

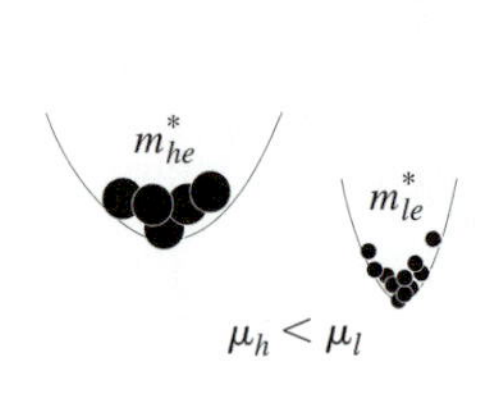

Figure 15.3 Illustration of the negative-dynamic conductance/resistance due to the Gunn effect.

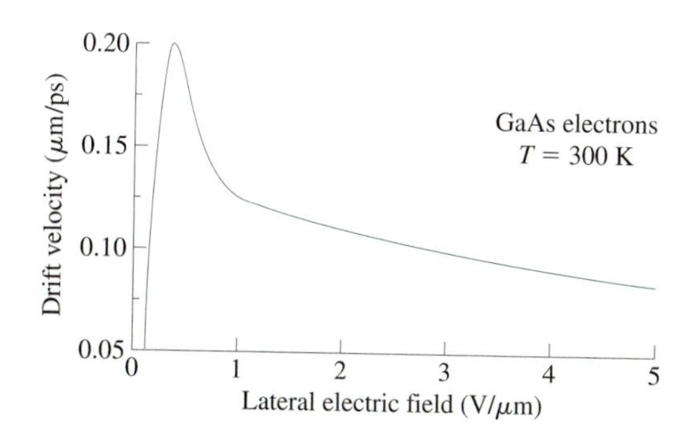

Figure 15.4 Drift velocity versus electric field in GaAs.

will disperse itself, as the electrons leaving the packet toward the anode move faster in the region of the stronger field. These electrons are not adequately replaced by electrons that come from the cathode at a slower rate. As a result, the semiconductor will stabilize itself. In the case of negative v_d–E slope (negative mobility), the situation is exactly opposite: the number of electrons that leave the packet moving toward the anode is lower than the number of electrons coming from the cathode. As a result, the packet of electrons grows.

To facilitate visualization of this effect by the energy bands, the region of strong electric field, such that the v_d–E slope is negative (negative mobility), is shaded in the conduction bands shown in Fig. 15.5. Normally, we think of the electrons as rolling down faster when the slope of the conduction band is larger. The shaded area is not like this, and the simple model breaks down here. Still, we can gain some graphic insight if we think of the shaded area as a very "dense" medium, so that the particles move faster in the clear than in the shaded area.[2] With this, we can visualize the effect of continuous electron accumulation as the accumulation layer moves down toward the anode. The increasing electron concentration in the accumulation layer increases the electric field toward the anode, which reduces the drift velocity of the electrons in this region, reducing the anode current. Therefore, *a voltage increase results with a current decrease,* as illustrated in Fig. 15.5b.

As the accumulation process continues and the accumulation layer moves toward the anode, further increasing the electric field, the drift-velocity reduction approaches saturation. At some point the increased electron concentration starts dominating the anode current, which leads to a current increase (Fig. 15.5c). The increased current causes a voltage reduction on the negative-slope load line, which only helps the current increase by causing a smaller field in the negative-mobility region than would be present at a higher voltage.

As the electron concentration in the accumulation layer drops down due to the electrons terminating at the anode, the electric field back toward the cathode is being increased, eventually above the critical level (Fig. 15.5d). By the time all the electrons from the accumulation layer have been collected by the anode, a new electron "packet"

---

[2]To be quite precise, we need to assume that this imaginary medium gets "denser" as the slope of the bands increases; this is to include the fact that v_d reduces with E.

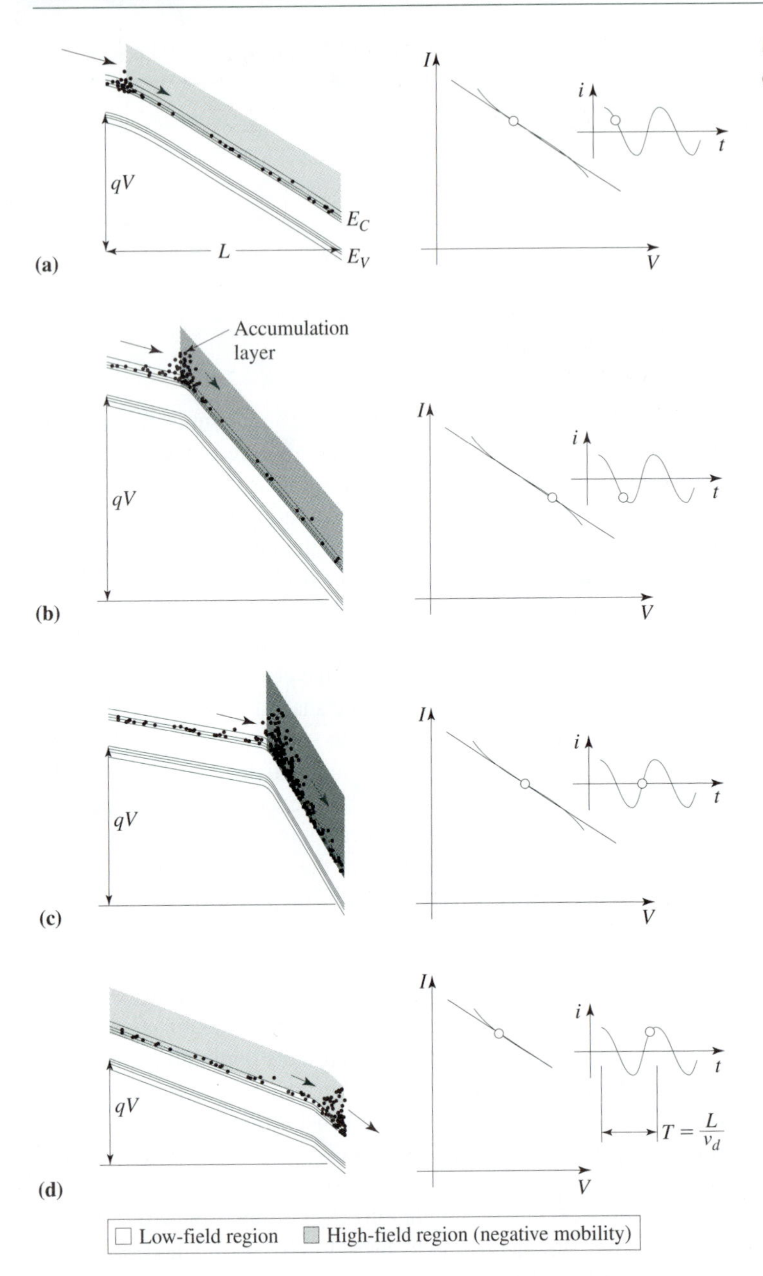

Figure 15.5 Illustration of a Gunn-effect oscillator.

is being formed at the cathode end of the semiconductor,[3] and the cycle is back to the situation represented by Fig. 15.5a.

---

[3]The creation of this electron "packet," or the initiation of the accumulation layer, is explained by the fact that the rate of electrons injected into the semiconductor is higher than their drift velocity in the high-field region.

This explanation shows that the oscillation period is approximately equal to the time needed for the accumulation layer to drift from the cathode to the anode. Assuming average drift velocity $\bar{v}_d$, the oscillation frequency is given by

$$f = \frac{\bar{v}_d}{L} \tag{15.7}$$

where L is the length of the sample.

This oscillation mechanism is referred to as either an *accumulation-layer mode,* a *transit-time mode,* or a *Gunn-oscillation mode*. It is a nonresonant mode because the oscillation frequency is set by the physical parameters of the diode itself (length and electron velocity). A number of more complex modes of operation are possible. For example, at higher doping concentration, or with longer samples, dipoles of accumulation and depletion layers are formed: that is, the diode is operated at a so-called *transit-time dipole-layer mode*. Additional oscillation modes include *quenched dipole-layer mode* and *limited-space-charge accumulation (LSA) mode*. At very low doping levels, or with very short samples, the device can operate in *stable amplification mode,* also called *uniform-field mode*. In this case, there are not enough electrons to create an accumulation layer. Consequently, the electric field is uniform so that I–V characteristic with a negative-resistance part (Fig. 15.3) results from a simple scaling of the velocity-field characteristic (Fig. 15.4). Because the spontaneous oscillation is avoided in this mode, the device can be used as a stable amplifier of signals with frequencies near the transit-time frequency. The other types of transferred-electron devices, namely, InP and CdTe diodes, are similar, but with different physical parameters.

15.3 IMPATT DIODE

The IMPATT diode represents *avalanche transit-time devices,* which can operate at frequencies higher than 100 GHz, providing the highest continuous power of all semiconductor microwave devices. IMPATT stands for **imp**act ionization **a**valanche **t**ransit **t**ime. These devices can be implemented in Si as well as GaAs.

The IMPATT diode is a resonant device because it requires a resonant circuit for its operation. The parallel connection of an inductor and a capacitor is the resonant circuit in Fig. 15.6a. Structurally, the IMPATT diode is typically a PIN diode. In Fig. 15.6, the "intrinsic" region is fully depleted and is sandwiched by heavily doped N^+ and moderately doped P-type regions that serve as contacts. Practically, the "intrinsic" region would be a very lightly doped P-type region. The DC reverse-bias voltage V_{BB} sets the diode very close to avalanche breakdown. The resonant circuit is designed so that the positive oscillating voltage v, superimposed on the bias voltage V_B, takes the diode into avalanche mode. The electric field in the structure is strongest at the N^+–I junction, and only in a narrow region around the N^+–I junction does the field exceed the critical breakdown value. This leads to generation of electron and hole avalanches, as illustrated in Fig. 15.6b. The electrons are quickly neutralized by the positive charge from the nearby anode, whereas the holes start drifting through the depletion layer toward the cathode. Assuming drift velocity v_d and sample length L, it will take time of $\tau = L/v_d$ for the holes to reach the cathode, causing

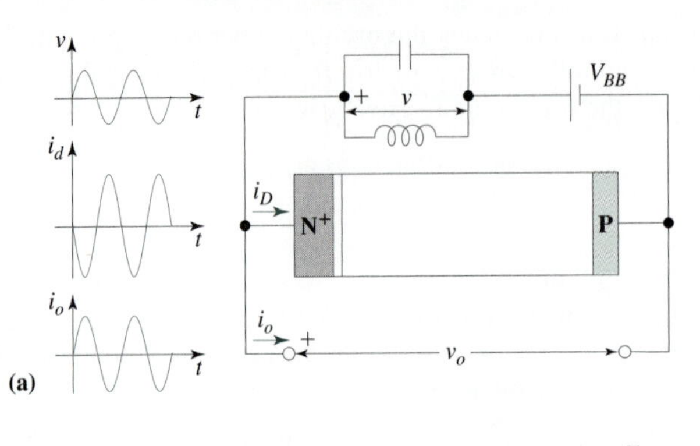

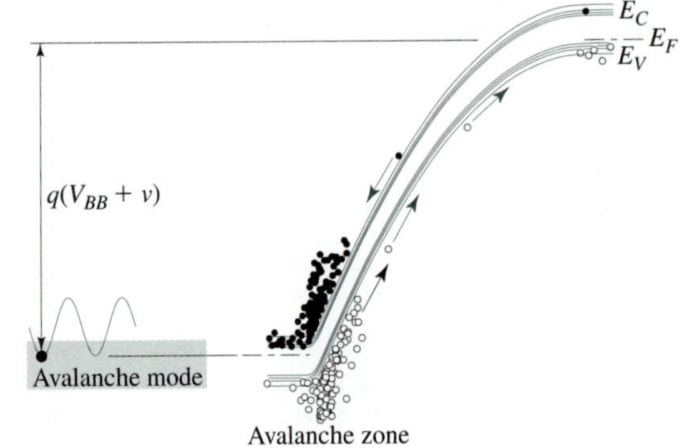

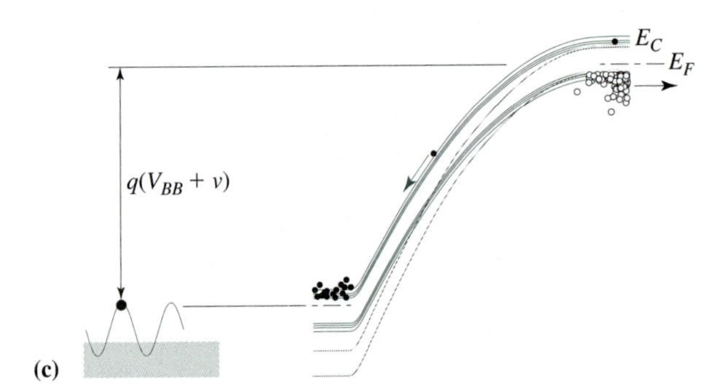

Figure 15.6 The principle of IMPATT diode operation.

maximum cathode current. If the resonant circuit is designed so that its frequency is

$$\frac{T}{2} = \tau \quad \Rightarrow \quad f = \frac{v_d}{2L} \tag{15.8}$$

the voltage is at its minimum when the current reaches its maximum. The voltage and current are 180° out of phase with each other, which means that negative dynamic resistance has been established.

The I–V characteristic of the avalanche diodes is S-shaped, as in Fig. 15.2, which is different from the N shaped I–V characteristic of the Gunn diode, shown in Fig. 15.3. This is because very small levels of current relate to any voltage between 0 and the breakdown.[4] In a situation such as the one illustrated in Fig. 15.5, the negative dynamic resistance causes a voltage decrease as the current is increased. However, if the diode is set in a continuous-breakdown mode, the structure behaves as an ordinary small-value resistor because of the abundance of current carriers.

15.4 TUNNEL DIODE

Tunnel diodes are small-power small-voltage devices that can be used as microwave amplifiers and oscillators, although they are being displaced by other semiconductor devices. The tunnel diodes, also known as Esaki diodes, have a historic importance related to the Nobel–prize winning discovery of the tunnel effect by L. Esaki in 1958. The tunnel diode is a P–N junction diode with heavy doping (on the order of 10^{20} cm^{-3}) in both P-type and N-type regions. The heavy doping is related to two important facts for the operation of the tunnel diode: (1) The Fermi level is inside the conduction band in the N-type material, reflecting the extreme concentration of electrons, and inside the valence band in the P-type region, to reflect the extreme concentration of holes. (2) The depletion layer separating the N-type and P-type regions is very narrow (<10 nm). As a result, (1) the electrons at the bottom of the conduction band in the N-type region are energetically aligned with the holes in the valence band of the P-type region (Fig. 15.7) and (2) the space separation between them is comparable to the characteristic length of the electron wave function. This enables the electrons to tunnel through the depletion-layer barrier.

Referring to Fig. 15.7, we see that an increasing tunneling current flows as an increasing forward-bias voltage is applied. However, the forward bias splits the quasi-Fermi levels reducing the energy overlap between the N-type conduction and P-type valence bands. This leads to a tunneling current decrease as the voltage is increased. This is the negative-resistance region that is utilized by microwave oscillators and amplifiers. As the energy overlap disappears, the tunneling current drops to zero. At this point, however, the normal diode current already dominates the total current. This current is due to the electrons (and holes) being able to go over the depletion-layer barrier, and it increases as the increasing voltage reduces the barrier height. This explains the N-shaped I–V curve of the tunneling diode, shown in Fig. 15.7.

SUMMARY

1. The maximum operating frequency of any transistor is not limited by the time it takes for the electrons to travel through the transistor (transit time), but by the input parasitic capacitance. Negative-resistance diodes operate at or near the transit-time frequencies. Negative dynamic resistance can be employed to convert DC supply power into signal

---

[4]This is opposite to the Gunn diode, where a small voltage is related to a very small resistance, due to the high mobility of the conducting electrons.

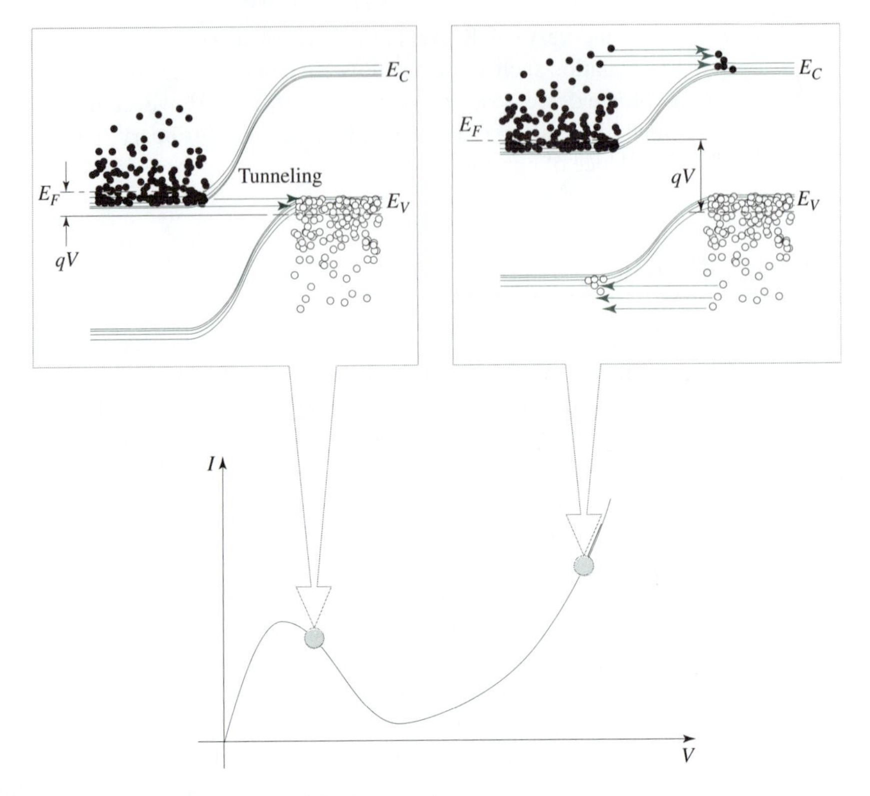

Figure 15.7 The principle of tunnel diode operation.

power ($p_d = gv_d^2 < 0$ as $g < 0$). This enables amplifiers and oscillators to operate at higher frequencies than any other solid-state system.

2. Negative dynamic resistance in a Gunn diode appears when an increasing voltage (electric field) converts an increasing number of high-mobility electrons into low-mobility electrons. This happens due to a transfer of electrons from an energetically lower E–k valley into a higher E–k valley, associated with a heavier effective mass of the electrons.
3. Negative dynamic resistance of an IMPATT diode is achieved by synchronizing the transit time of carriers created by avalanche breakdown to the half-period of a resonant voltage. The positive peak of the resonant voltage adds to a DC bias voltage to set the diode in avalanche mode. The created carriers take time τ (transit time) to come to the opposite electrode (maximum signal current); the half-period of the voltage is set to τ, which means that it is most negative when the current is at the positive peak.
4. The negative resistance of a tunnel diode is due to diminishing overlap (in energy terms) of N-type conduction and the nearby P-type valence bands, caused by Fermi-level splitting by an increasing forward-bias voltage. The diminishing "energy alignment" of N-type electrons to the P-type holes means fewer electrons can tunnel through the very narrow depletion layer at the P–N junction.

PROBLEMS

15.1 Power gain that can be achieved by a negative-conductance diode is given by Eq. (15.4). Derive this equation using the power balance principle.

15.2 Referring to Example 15.1:

(a) Sketch v_d, i_d, and i_s, positioning them appropriately with respect to I_D and V_D.

(b) Shade the rectangle defined by the axes and V_D and I_{BB} lines. What does the area $V_D I_{BB}$ represent?

(c) Observe the triangle defined by i_B and V_D lines and the diode characteristic. Compare it to the triangle defined by i_A, V_D, and the diode characteristic. How does the area of each of these triangles relate to the signal power?

15.3 The negative-resistance region of a Gunn diode is defined by the following two voltage–current points: $(I_1, V_1) = (8 \text{ V}, 12 \text{ mA})$ and $(I_2, V_2) = (12 \text{ V}, 2 \text{ mA})$. Calculate G_L to maximize the power gain of a sinusoidal signal with peak current $I_{mi} = 0.5$ mA.

15.4 The voltage across a Gunn diode oscillates between $V_{min} = 5.5$ V and $V_{max} = 10$ V. Knowing that the effective length of the diode is $L = 3.5\ \mu$m, and the drift velocity is $\bar{v}_d = 0.10\ \mu$m/ps, estimate the oscillation frequency. **A**

15.5 Find the resonant frequency for a silicon N^+P^-P IMPATT diode with effective length $L_{P^-} = 10\ \mu$m. The saturation velocity is $v_{sat} = 0.07\ \mu$m/ps.

REVIEW QUESTIONS

R-15.1 Can an analog signal be amplified by a two-terminal device? If so, how?

R-15.2 Is the signal power $p_d = g v_d^2$ positive or negative if the dynamic conductance of the device g is negative? What does "negative signal power" mean, anyway?

R-15.3 What happens if a negative-resistance diode is biased so that the load line overlaps the negative-resistance region (in other words, there is no unique intersection between the load line and the diode characteristic)?

R-15.4 Are Gunn diodes based on P–N junctions? Schottky contacts?

R-15.5 Is the Gunn effect possible in Si?

R-15.6 Is it possible to have a Gunn-diode-based oscillator without an external resonator circuit?

R-15.7 Is the electron drift velocity related to oscillation frequency that can be achieved by a Gunn diode? If so, why and how?

R-15.8 Is the electron drift velocity related to the oscillation frequency in the case of IMPATT diode? Is there any difference from the case of Gunn diode?

R-15.9 Does the IMPATT diode need an external resonator?

R-15.10 Do the N-shaped I–V characteristics of Gunn and tunnel diodes mean that their physical principles are similar, while quite different from those of avalanche diodes, which exhibit S-shaped I–V characteristics?

R-15.11 Is the tunnel diode based on a P–N junction? If so, is the P–N junction forward- or reverse-biased in the negative-resistance mode?

R-15.12 Is the operating voltage of a tunnel diode higher or lower than IMPATT diode? Is this related to the output power capabilities?

16 Integrated-Circuit Technologies

In comparison with circuits implemented with discrete devices, integrated-circuit (IC) technology offers unmatched advantages, including sophisticated functions that are enabled by mere circuit complexity, increased speed due to reduced parasitic capacitances and resistances, reduced cost, reduced size, and improved reliability. In terms of the performance of individual semiconductor devices, however, the various IC technologies impose severe limitations. A device structure may appear superior in isolation, but its practical significance will be diminished if integration is difficult. The fundamental device structures, presented in Part II, satisfy the integration criterion. Even so, integration may become impossible, or at the very least its advantages will be lost, if the technological parameters are optimized for discrete semiconductor devices.

This chapter introduces the principles of IC technology. The emphasis is on presenting new concepts that are associated with integration: (1) the need to make all the devices with a single optimized sequence of technology steps, (2) the need to provide electrical isolation between individual semiconductor devices, and (3) the possibility of physically merging regions of individual devices (layer-merging principle).

16.1 A DIODE IN IC TECHNOLOGY

This section provides an introduction and a brief review of essential processing steps, using the example of the P–N junction diode.

16.1.1 Basic Structure

The basic structure of a diode in integrated circuits is shown in Fig. 16.1. The diode terminals, the anode and the cathode, are labeled by A and C, respectively. A P–N junction used as a diode in integrated circuits has to be isolated from the other components built in

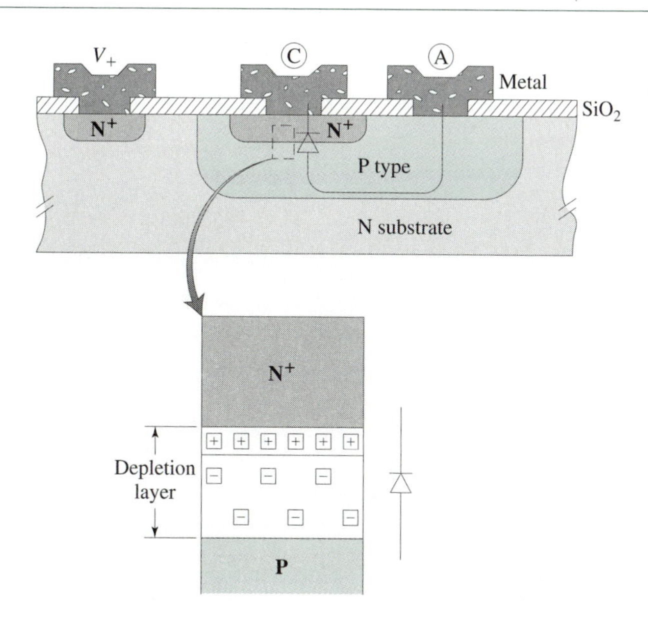

Figure 16.1 Structure of a P–N junction diode in integrated circuits.

the same substrate. Figure 16.1 illustrates that an additional reverse-biased P–N junction can be used to electrically isolate the diode from the other components of the integrated circuit. The principle of utilizing a reverse-biased P–N junction for electrical isolation relies on the fact that the reverse-bias current is negligible. The N-type substrate in Fig. 16.1 is contacted to the highest potential in the circuit (V_+) to ensure that the isolating P–N junction, formed at the junction between the N-type substrate and the P-type region (the anode of the diode), is reverse-biased for any anode voltage. In the worst-case scenario, the anode voltage is also equal to the highest potential in the circuit, V_+. In this case the voltage across the isolating P–N junction is equal to zero. For any anode potential lower than V_+, the isolating P–N junction is reverse-biased.

16.1.2 Lithography

The diode structure shown in Fig. 16.1 illustrates that semiconductor devices utilize N-type and P-type doped regions with certain depths and widths. If the doping is performed by diffusion, the depth of the doped regions is controlled by the time and the temperature during the diffusion. As far as the width is concerned, it is specified by the width of the window in the protective "wall," as illustrated in Fig. 1.20. Typically, the "wall" is a layer of SiO_2 (a thin glass layer), although Si_3N_4 as well as some other materials are sometimes used. If the semiconductor in question is silicon, the SiO_2 layer at the surface can be created by thermal oxidation of the silicon. Alternatively, it can be deposited on the surface. Either way, the window has to be created by *selective etching* of the protective layer ("wall").

The set of processing steps used to prepare the substrate for selective etching is referred to as *lithography*. Lithography plays a crucial role in semiconductor fabrication because it is utilized to create the device and IC patterns at almost any level. Etching of the window in the SiO_2 layer, to enable selective diffusion, is an example of the need for selective etching of layers. Another example is etching of a metal layer to define metal lines that interconnect the devices created in the semiconductor.

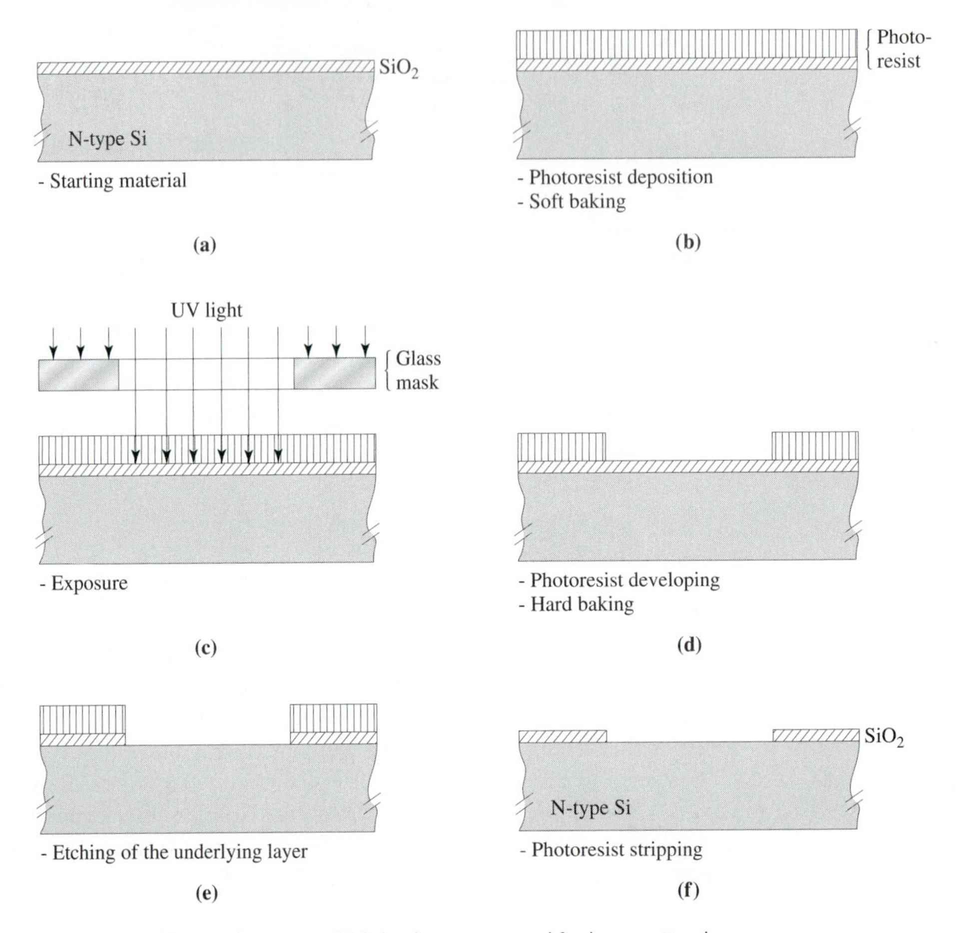

Figure 16.2 Lithography–a set of fabrication steps used for layer patterning.

Figure 16.2 illustrates the set of processes referred to as lithography. The starting material in this example is an N-type silicon wafer with a thin film of silicon dioxide grown at the surface. To etch a window in the silicon dioxide film, the areas of the oxide that are not to be etched need to be protected. To this end, a special film, referred to as a *photoresist,* is deposited onto the oxide surface.

The photoresist has to fulfill a double role: (1) it should be light-sensitive to enable the transfer of the desired pattern by way of exposure/nonexposure to the light, and (2) it should be resistant to chemicals used to etch the underlying layer. Generally, there are two types of photoresist: negative and positive. The exposed areas of the positive photoresist are washed away by a developing chemical, whereas the unexposed areas remain at the surface as shown in Fig. 16.2d. In the case of the negative photoresist, the unexposed areas are removed during the developing. The photoresist is originally in liquid form, which is spun onto the substrate surface to create a thin film. The deposited photoresist is soft-baked (80–100°C) before the exposure, to evaporate the solvents, and hard-baked (120–150°C) after the developing, to improve the adhesion to the underlying layer.

The photoresist is exposed through a glass mask by ultraviolet (UV) light (Fig. 16.2c). The pattern that is to be transferred onto the substrate is provided on the glass mask in the form of clear and opaque fields. The mask itself is made by computer-controlled exposure of a thin film deposited on the glass mask and sensitive to either UV light or an electron beam; the clear and opaque fields are obtained after film development. Figure 16.2c illustrates so-called *contact lithography,* which requires a $1\times$ image on the mask. In *projection lithography*, which is now typically used, the mask is positioned above the substrate. When light passes through the mask, it is focused onto the substrate, which results in image reduction. This has several advantages: (1) the image on the mask can be much larger than the image on the substrate, (2) the substrate can be exposed part by part (chip by chip), which enables better focusing, and (3) the mask is not contaminated due to contact with the photoresist.

In principle, it is possible to expose an electron-beam resist deposited directly on the substrate by a computer-controlled electron beam, in which case there would be no need for the glass mask. Although this may seem advantageous, there is a practical problem. There can be more than 10^{10} picture elements in today's ICs, which are exposed simultaneously using the glass mask. It would be absurdly slow to use an electron beam to directly write these 10^{10} picture elements, and of course repeat the process for every IC. Because there is a limit in terms of the finest pattern that can be obtained by the UV light (the limit imposed by the wavelength of the UV light), the research of nanometer structures is carried out using direct electron-beam writing.

Once the photoresist has been developed and hard-baked, the underlying layer can be etched (Fig. 16.2e). This etching can be wet (etching by an appropriate chemical solution) or dry (plasma etching). After the etching is completed, the photoresist is removed to obtain the desired structure as shown in Fig. 16.2f.

16.1.3 Process Sequence

The IC diode shown in Fig. 16.1 can be made by combining lithography and diffusion. The process sequence is illustrated in Fig. 16.3. In this example, the initial material is the N-type silicon with a window in the overlying oxide layer, prepared as illustrated previously in Fig. 16.2. The P–N junction shown in Fig. 16.3a is the isolating junction: it will have to be reverse-biased to electrically isolate the diode from the common N-type substrate. This can be achieved if the N-type substrate is connected to the most positive voltage in the circuit (V_+). A good contact to N-type silicon can be achieved only with high concentrations of donor atoms ($N_D > 10^{19}$ cm^{-3}). Because of this, it is necessary to provide a highly doped N-type region (labeled as N$^+$) for the contact with the metal terminal where the positive voltage is to be connected. To this end, the oxide mask (the oxide layer with the window that enabled the P-type layer to be formed by boron diffusion) has to be stripped, and a new oxide layer grown to create a diffusion mask with a window over the area where the N$^+$ region is to be formed. This obviously involves an oxide deposition/growth and a lithography to open the window. After this, diffusion of phosphorus is performed to create the N$^+$ region needed for the contact, as shown in Fig. 16.3b.

Analogously to the creation of the P-type layer by diffusing a higher acceptor concentration into the N-type substrate, the cathode of the diode can be created by diffusing donors with concentration N$^+$ that is even higher than the acceptor concentration in the P-type region. The doping concentration achieved by the N$^+$ diffusion for the contact to

N-type substrate is higher than the maximum acceptor concentration in the P-type region. Therefore, the same diffusion process can be used to create both the N^+ contact and the N^+ cathode of the diode. Figure 16.3b shows that this is achieved by opening windows in the oxide layer over the N-type substrate (for the contact) and over the P-type region (for the cathode). The resulting diffusion profiles along the N^+PN structure (cathode–anode–substrate) are illustrated in Fig. 16.4.

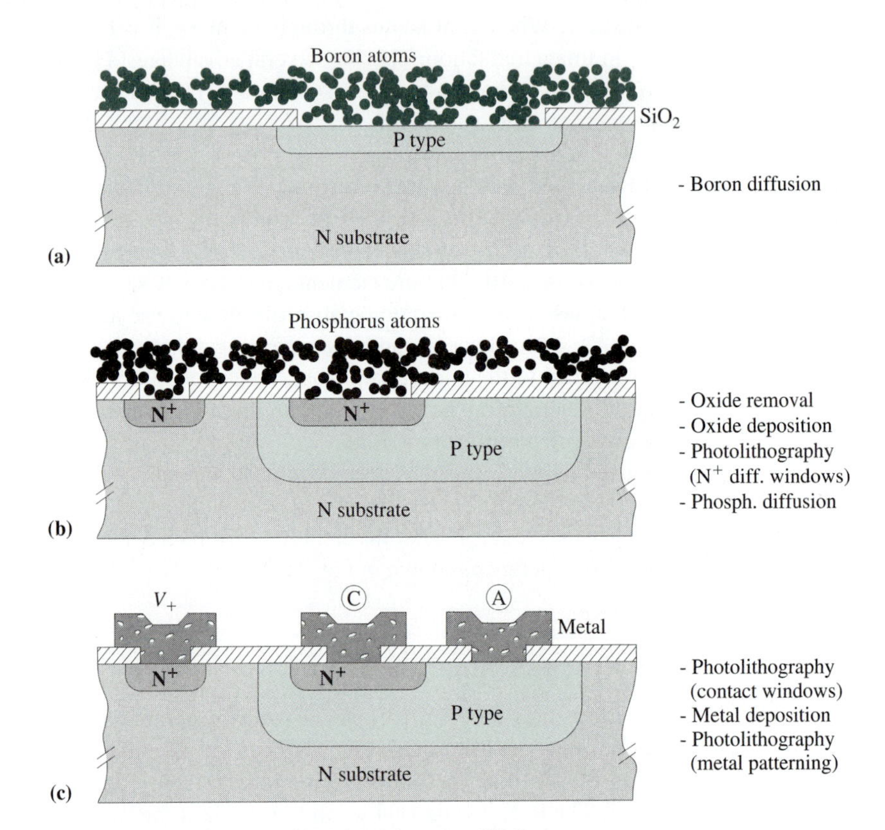

Figure 16.3 Process steps used in fabrication of an IC diode.

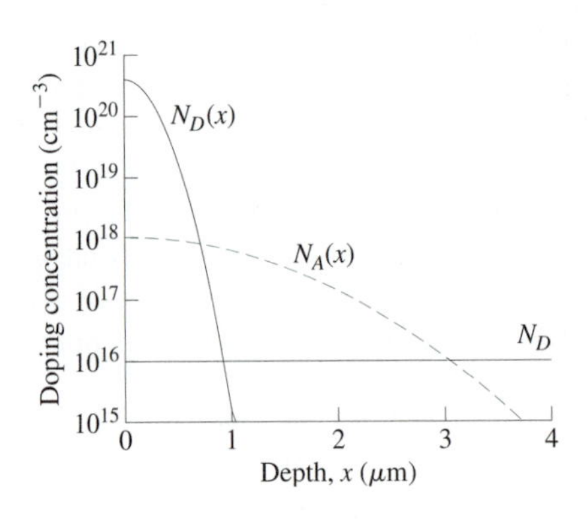

Figure 16.4 Diffusion profiles associated with the N^+PN structure (cathode–anode–substrate), shown in Fig. 16.3b.

Finally, the diode terminals and the metal track for the positive voltage (V_+) have to be made. The areas where the metal should contact the silicon are defined as contact windows (another lithography process) in a freshly grown/deposited oxide layer. The oxide is a good insulator, so it will electrically isolate the metal layer from the silicon in all other areas. After the metal is deposited, it appears all over the chip and has to be removed from the areas where it is not wanted to create the desired metal tracks. Again, a lithography process is needed to enable selective etching of the metal layer, as shown in Fig. 16.3c. At this stage, the diode is electrically functional.

16.1.4 Diffusion Profiles

Each of the diffusion profiles shown in Fig. 16.4 is achieved by controlling the time and temperature of a two-step diffusion process. The principal equations that can be used as models for these diffusion profiles are obtained as solutions of the so-called Fick equation. This second-order partial differential equation is obtained by eliminating the diffusion current from the diffusion-current equation [Eq. (4.3)] and the continuity equation [Eq. (4.24)]:

$$\frac{\partial N(x,t)}{\partial t} = D\frac{\partial^2 N(x,t)}{\partial x^2} \tag{16.1}$$

If we solve Fick's equation with appropriate boundary and initial conditions, the function $N(x,t)$ modeling the time-dependent distribution (profile) of the doping atoms in the semiconductor can be obtained. The temperature dependence is through the diffusion coefficient D—the Arrhenius equation [Eq. (4.21)] shows that D increases exponentially with temperature.

Integrated-circuit diffusions are typically performed in two steps, the predeposition diffusion and the drive-in diffusion. The first is a constant-source diffusion, which means that a constant doping concentration N_0 is maintained at the surface of the semiconductor substrate. The second is a redistribution of the doping atoms in the semiconductor by heating the substrate while not providing any additional doping atoms to the surface of the substrate.

If the predeposition is carried out in an atmosphere sufficiently rich in doping atoms, then the surface doping concentration N_0 is at the solid-solubility limit (the maximum concentration of doping atoms that the host semiconductor crystal can accommodate). The solid-solubility limit depends on the temperature; however, in silicon it is approximately equal to 4×10^{20} cm^{-3} for boron, 8×10^{20} cm^{-3} for phosphorus, 1.5×10^{21} cm^{-3} for arsenic, and 4×10^{19} cm^{-3} for antimony. When one is solving Fick's diffusion equation for constant-source diffusion, the following initial and boundary conditions apply:

$$N(x,0) = 0, \qquad N(0,t) = N_0, \qquad N(\infty,t) = 0 \tag{16.2}$$

The solution of Eq. (16.1) that satisfies the boundary and initial conditions [Eq. (16.2)] is

$$N(x,t) = N_0 \operatorname{erfc}\frac{x}{2\sqrt{Dt}} \tag{16.3}$$

where erfc is the complementary error function defined by

$$\operatorname{erfc}(z) = \frac{2}{\sqrt{\pi}} \int_z^{\infty} e^{-\alpha^2} d\alpha \tag{16.4}$$

This function is plotted in Fig. 16.5, whereas the normalized doping profiles for three different predeposition times/temperatures (more precisely Dt products) are plotted in Fig. 16.6a. Note that the doping profiles depend on the time–diffusion coefficient product Dt, where the diffusion coefficient depends exponentially on the temperature, as given by Eq. (4.21). A typical predeposition temperature is 950°C and a typical time is 30 min. The relatively low temperatures and times result in shallow doping profiles, which serve as the initial condition for the following drive-in diffusion.

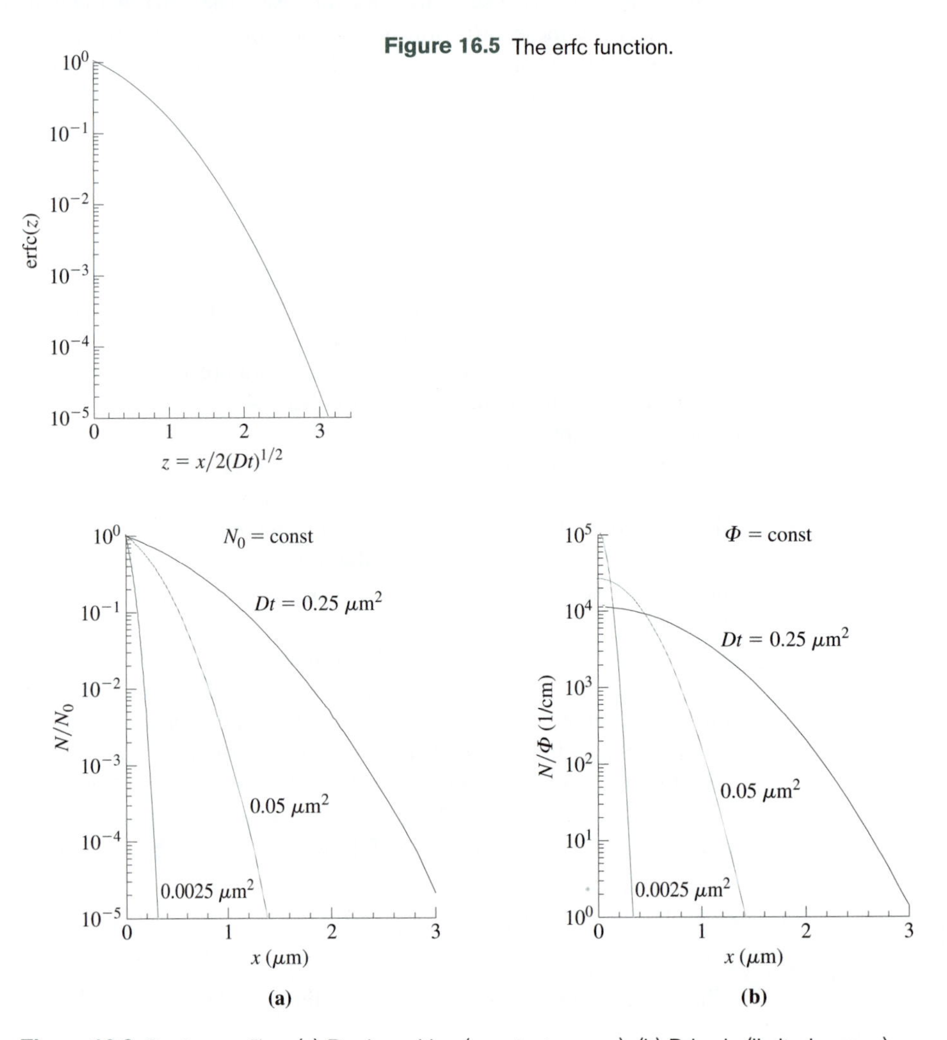

Figure 16.5 The erfc function.

Figure 16.6 Doping profiles. (a) Predeposition (constant source). (b) Drive-in (limited source).

The drive-in is done in an atmosphere containing oxygen or nitrogen, but not dopants. Because of that, the following so-called limited-source boundary conditions will apply:

$$\int_0^\infty N(x,t)\,dx = \Phi, \qquad N(\infty,t) = 0 \tag{16.5}$$

where Φ is called the dose of doping atoms that are incorporated into the semiconductor during the predeposition. *The dose expresses how many doping atoms per unit area are in the semiconductor, regardless of their depth in the semiconductor, so its unit is* $1/\mathrm{m}^2$. The concentration of the doping atoms, $N(x,t)$, expressing their in-depth distribution in the semiconductor is integrated along the x-axis to express the overall number of doping atoms per unit area—that is, the dose Φ [Eq. (16.5)]. The dose does not change during the drive-in diffusion [$\Phi \neq \Phi(t)$], as there are no new doping atoms that are incorporated into the semiconductor, but the already existing ones are being redistributed diffusing deeper into the semiconductor. The solution of Eq. (16.1) that satisfies boundary conditions (16.5) is given by

$$N(x,t) = \frac{\Phi}{\sqrt{\pi D t}} e^{-x^2/4Dt} \tag{16.6}$$

The doping profiles obtained after drive-in diffusion are illustrated in Fig. 16.6b for three different time–diffusion-coefficient products.

If the predeposition profile given by Eq. (16.3) is integrated to obtain Φ,

$$\Phi = N_0 \int_0^\infty \operatorname{erfc}\frac{x}{2\sqrt{Dt}}\,dx \tag{16.7}$$

and the obtained result is used in Eq. (16.6), the following equation is obtained as a model for the two-step (predeposition and drive-in) diffusion process:

$$N(x) = \underbrace{\frac{2N_0}{\pi}\sqrt{\frac{D_1 t_1}{D_2 t_2}}}_{N_s} e^{-x^2/4D_2t_2} = N_s e^{-x^2/4D_2t_2} \tag{16.8}$$

In Eq. (16.8), D_1 and t_1 refer to the predeposition, D_2 and t_2 refer to the drive-in diffusion, and $N_s = N(0)$ obviously expresses the doping concentration at the surface after the drive-in.

EXAMPLE 16.1 Constant-Source Diffusion

Constant-source boron diffusion is carried out at 1050°C onto an N-type silicon wafer with uniform background doping N_B, where $N_B = 10^{16}$ cm^{-3}. The surface concentration is maintained at 4×10^{20} cm^{-3}, which is the solid-solubility limit. A junction depth of 1 μm is required. What should be the diffusion time?

SOLUTION

Calculate first the diffusion coefficient at $T = 1050°C$. Repeating the procedure described in Example 4.2, we find that $D = 5.1 \times 10^{-14}$ cm^2/s.

At the P–N junction, the boron concentration $N(x)$ is equal to the substrate concentration N_B. Referring to Eq. (16.3) we obtain:

$$N_B = N(x_j) = N_0 \text{erfc}(z), \qquad z = x_j/\left(2\sqrt{Dt}\right)$$

It is easy to find that $\text{erfc}(z) = N_B/N_0 = 10^{16}/4 \times 10^{20} = 2.5 \times 10^{-5}$. From Fig. 16.5, we can find that $\text{erfc}(z) = 2.5 \times 10^{-5}$ for $z = 3.0$. Using $z = x_j/(2\sqrt{Dt})$, the time required is obtained as

$$t = x_j^2/(4z^2 D) = 5447 \text{ s} = 1.51 \text{ h}$$

EXAMPLE 16.2 Drive-in Diffusion

How will the junction depth from Example 16.1 change if a drive-in diffusion is performed after the constant-source diffusion? Assume that the time and temperature of the drive-in diffusion are the same as the time and the temperature of the constant-source diffusion. What is the surface doping concentration after the drive-in diffusion?

SOLUTION

The doping profile after a drive-in, which follows a constant-source diffusion, is given by Eq. (16.8). At the P–N junction, ($x = x_j$), the boron concentration $N(x)$ is equal to the substrate concentration N_B (refer to Fig. 1.21):

$$N_B = N(x_j) = N_s e^{-x_j^2/(4D_2 t_2)}$$

From this equation we calculate x_j as

$$x_j = \sqrt{4Dt \ln 2N_0/(\pi N_B)} = 1.1 \times 10^{-4} \text{ cm} = 1.1\ \mu\text{m}$$

The surface concentration after the constant-source diffusion and the drive-in is also given by Eq. (16.8):

$$N_s = 2N_0/\pi = 2.55 \times 10^{20} \text{ cm}^{-3}$$

During the drive-in, the junction depth is increased while the surface concentration is reduced.

EXAMPLE 16.3 Designing Two-Step Diffusion Process

The doping concentration of an N-type substrate is $N_B = 10^{16}$ cm^{-3}. Determine the times and temperatures of the predeposition and drive-in diffusions of boron that will achieve the following parameters in the resulting P-type region: the surface concentration $N_s = 2 \times 10^{18}$ cm^{-3} and the P–N junction depth $x_j = 2$ μm. Assume that the surface concentration during the predeposition is equal to the solid-solubility limit, $N_0 = 4 \times 10^{20}$ cm^{-3}. The activation energy and the frequency factor for the boron diffusion are $E_A = 3.46$ eV and $D_0 = 0.76$ cm^2/s, respectively.

SOLUTION

The doping profile after a two-step boron diffusion is given by Eq. (16.8). At the P–N junction,

$$N_B = N(x_j) = N_s e^{-x_j^2/(4D_2t_2)}$$

The product of a drive-in time t_2 and diffusion coefficient D_2 that is necessary to achieve the desired x_j and N_s can be calculated using the stated conditions:

$$D_2t_2 = \frac{x_j^2}{4}\frac{1}{\ln(N_s/N_B)} = 0.189\ \mu\text{m}^2 = 1.89 \times 10^{-9}\ \text{cm}^2$$

Theoretically, any combination of t_2 and D_2 that gives this calculated product is a correct solution. However, the time t_2 and the diffusion temperature T_2, which determines the value of D_2 through Eq. (4.21), should have practically meaningful values. The times should be neither too long nor too short, and the temperature should not be higher than 1200°C. If we assume the drive-in time of 1 hour—that is, $t_2 = 3600$ s—the diffusion coefficient should be $D_2 = 1.89 \times 10^{-9}/3600 = 5.25 \times 10^{-13}$ cm^2/s. Using Eq. (4.21), the drive-in temperature is calculated as

$$T_2 = \frac{E_A}{k}\frac{1}{\ln(D_0/D_2)} = 1433.5\ \text{K} = 1160°\text{C}$$

This is an acceptable temperature; therefore, $t_2 = 3600$ s and $T_2 = 1160$°C can be used as a practically acceptable set of values.

The predeposition time t_1 and temperature T_1 (i.e., diffusion coefficient D_1) are determined from Eq. (16.8):

$$D_1t_1 = \left(\frac{\pi N_s}{2N_0}\right)^2 D_2t_2 = 1.17 \times 10^{-13}\ \text{cm}^2$$

If the predeposition temperature is set to $T_1 = 900$°C, then $D_1 = 1.05 \times 10^{-15}$ cm^2/s, and $t_1 = 1.17 \times 10^{-13}/1.05 \times 10^{-15} = 111$ s. Therefore, $T_1 = 900$°C and $t_1 = 111$ s is a possible set of values for the predeposition temperature and time.

16.2 MOSFET TECHNOLOGIES

There are essentially two categories of MOSFET-based IC technology: (1) NMOS technology is based on the two types of N-channel MOSFET (the enhancement type and the depletion type), and (2) CMOS technology is based on the complementary N-channel and P-channel enhancement-type MOSFETs. Both NMOS and CMOS technologies utilize electronic isolation by field oxide created by local oxidation of silicon (LOCOS), in addition to the already introduced isolation by a reverse-biased P–N junction that is also used in CMOS technology. Following an introduction to LOCOS, this section presents the basic NMOS, the basic CMOS, and the silicon-on-insulator (SOI) CMOS technologies.

16.2.1 Local Oxidation of Silicon (LOCOS)

Silicon nitride (Si_3N_4) can be used to protect parts of the silicon surface against oxidation because silicon nitride represents a very efficient barrier against the diffusion of oxygen and water molecules. Figure 16.7 illustrates the sequence of process steps used to locally oxidize a silicon surface—that is, to create thick oxide surrounding the active gate-oxide area. The metal/polysilicon creating the gates of the MOS capacitors are usually formed as tracks, so it is the windows in the thick oxide that define the area of the MOS capacitors. The thick oxide surrounding the gate-oxide areas is called *field oxide*. A metal/polysilicon track running over the field oxide would in principle create a parasitic MOS capacitor; however, the thickness of the field oxide makes this capacitance much smaller in comparison with the gate-oxide capacitance.

The process sequence known as LOCOS begins with deposition or growth of very thin buffer oxide. Then, a silicon nitride layer is deposited and patterned to open windows in the areas that are to be oxidized (Fig. 16.7a). After that, thermal oxidation is performed to create the field oxide of desired thickness, as illustrated in Fig. 16.7b, which also shows that approximately half the created oxide grows at the expense of the silicon, whereas the other half builds up at the top of the original surface. This is because silicon atoms are used to create the oxide (silicon is consumed), and oxygen atoms are incorporated into the oxide film (oxide rises above the original silicon surface). The precise ratio between the depth of the consumed silicon and the total oxide thickness is 0.46. This is a useful property because a thick isolation field oxide can be created without the need to have large oxide-to-silicon steps that are hard to reliably cover by thin metal films. In addition, some lateral oxidation occurs, which smooths the step, making it easier to cover by the metal film used for contacts and interconnections.

Once the field oxide has grown, the silicon nitride and the buffer oxide are removed. In preparation for the thermal oxidation that will create the thin gate oxide, the surface is thoroughly cleaned; Fig. 16.7c illustrates the structure just before the gate oxidation. Immediately after the gate oxidation, polysilicon is deposited, and then it is patterned to define the gate. To provide a contact to the semiconductor terminal of the MOS capacitor, boron is implanted through the thin oxide to create the P^+ region illustrated in Fig. 16.7d. The final steps are contact window opening, metal deposition, and patterning by appropriate photolithography processing to obtain the structure shown in Fig. 16.7e.

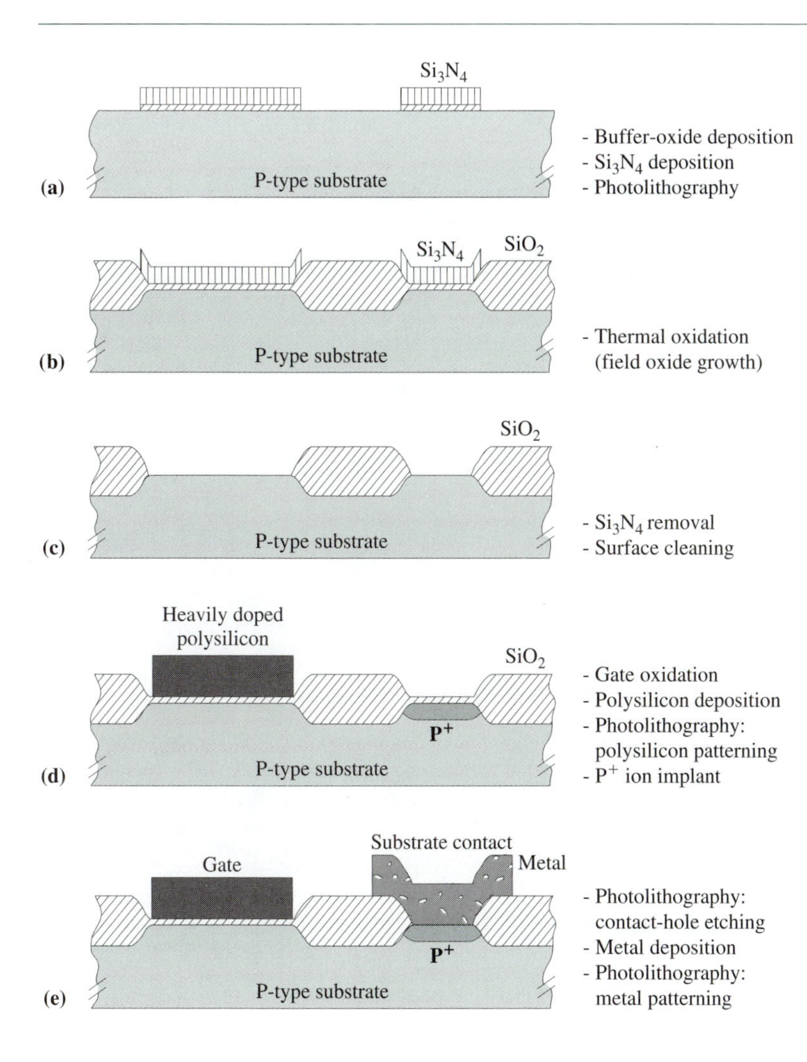

Figure 16.7 LOCOS process used to create the field oxide (the thick oxide surrounding the thin gate oxide) that defines the active area of the MOS capacitor.

The structure shown in Fig. 16.7e is just a MOS capacitor. To obtain a MOSFET, the source and drain regions are added as illustrated in Fig. 16.8 for a typical N-channel MOSFET in IC technology. A cross section along the channel width was shown previously in Fig. 8.19. It can be seen that the thick field oxide terminates the channel, providing the necessary electronic isolation on the common IC substrate.

16.2.2 NMOS Technology

Enhancement and depletion N-channel MOSFETs alone are sufficient to create both digital and analog circuits. The technology that provides integration of the two types of N-channel MOSFET is called NMOS technology. Figure 16.9 shows the circuit of an NMOS inverter, which is the elementary and the representative circuit.

The input signal is applied to the gate of the enhancement MOSFET because it is this MOSFET that acts as the voltage-controlled switch. A high voltage level (V_H) applied to

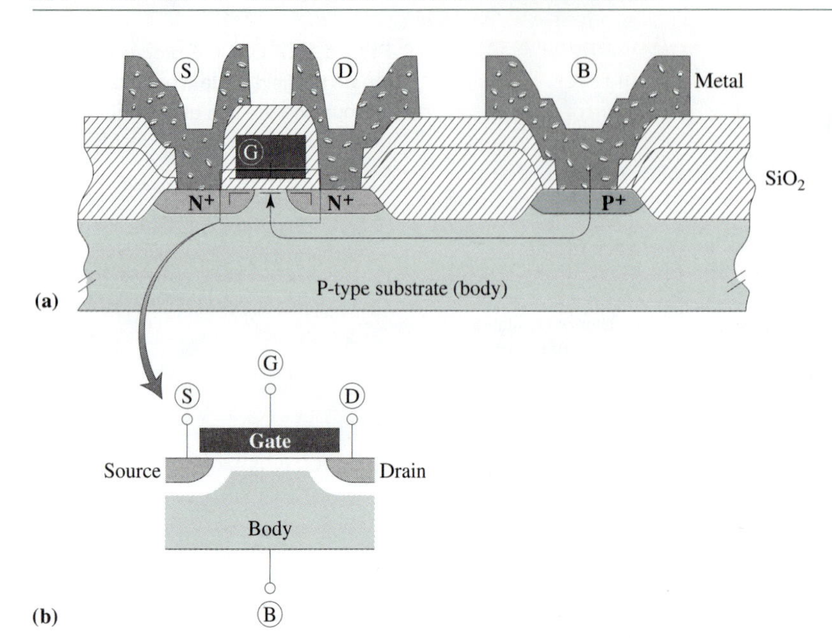

Figure 16.8 (a) Typical integrated-circuit structure of a MOSFET. (b) Schematic cross section used earlier (Chapter 8).

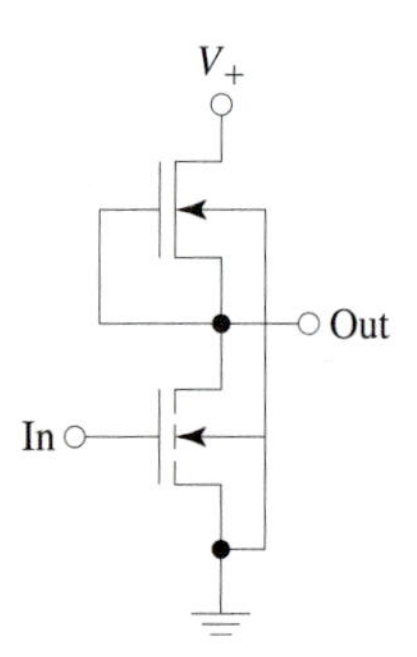

Figure 16.9 The circuit of an NMOS inverter.

the input sets the enhancement MOSFET in *on* mode, connecting the output to ground; V_H at the input corresponds to V_L at the output. The drain of the enhancement-type MOSFET has to be biased by a positive voltage. The simplest way of achieving this is through a resistor. Of course, the resistance of the loading resistor has to be much larger than the resistance of the MOSFET in *on* mode, so that the output voltage is close to zero. In integrated-circuit technology, however, it is far more efficient to replace the loading resistor by the channel resistance of a MOSFET (active load). The depletion MOSFET, connected as in Fig. 16.9, quite efficiently plays the role of the loading element. The short-circuited gate and source set its V_{GS} voltage to zero; but because this is a depletion-type MOSFET and the channel is formed at $V_{GS} = 0$, it is the resistance of this channel that is utilized as the loading element.

For the case of a low voltage level (V_L) at the input, the enhancement MOSFET acts as a switch in *off* mode, disconnecting the output from the ground. The output is now connected through the channel resistance of the depletion MOSFET to V_+; V_L at the input corresponds to $V_H = V_+$ at the output. Therefore, the circuit inverts the input voltage level.

Neglecting leakage currents through the enhancement MOSFET in *off* mode, the NMOS inverter does not conduct any current in the state of low input and high output levels. This is because the output of the inverter is connected to a capacitive load (for example, the input gates of subsequent NMOS circuits) that does not conduct any DC current. Importantly, the NMOS inverter has to conduct current to maintain the other possible state, which is the input at V_H and the output at V_L. In this case, the voltage across the channel resistance of the depletion MOSFET is $V_+ - V_L$. This means that for a channel resistance of R_{ch}, the current flowing through the inverter is $(V_+ - V_L)/R_{ch}$. The importance of this fact is due to the power dissipation. The power dissipation may not be of a real concern for integrated circuits with relatively small number of transistors, but it limits the development of sophisticated digital functions that require integrated circuits of very high complexity.

The cross section and composite layout of the NMOS inverter are shown in Fig. 16.10. They show that the active region of the NMOS inverter is surrounded by the thick field

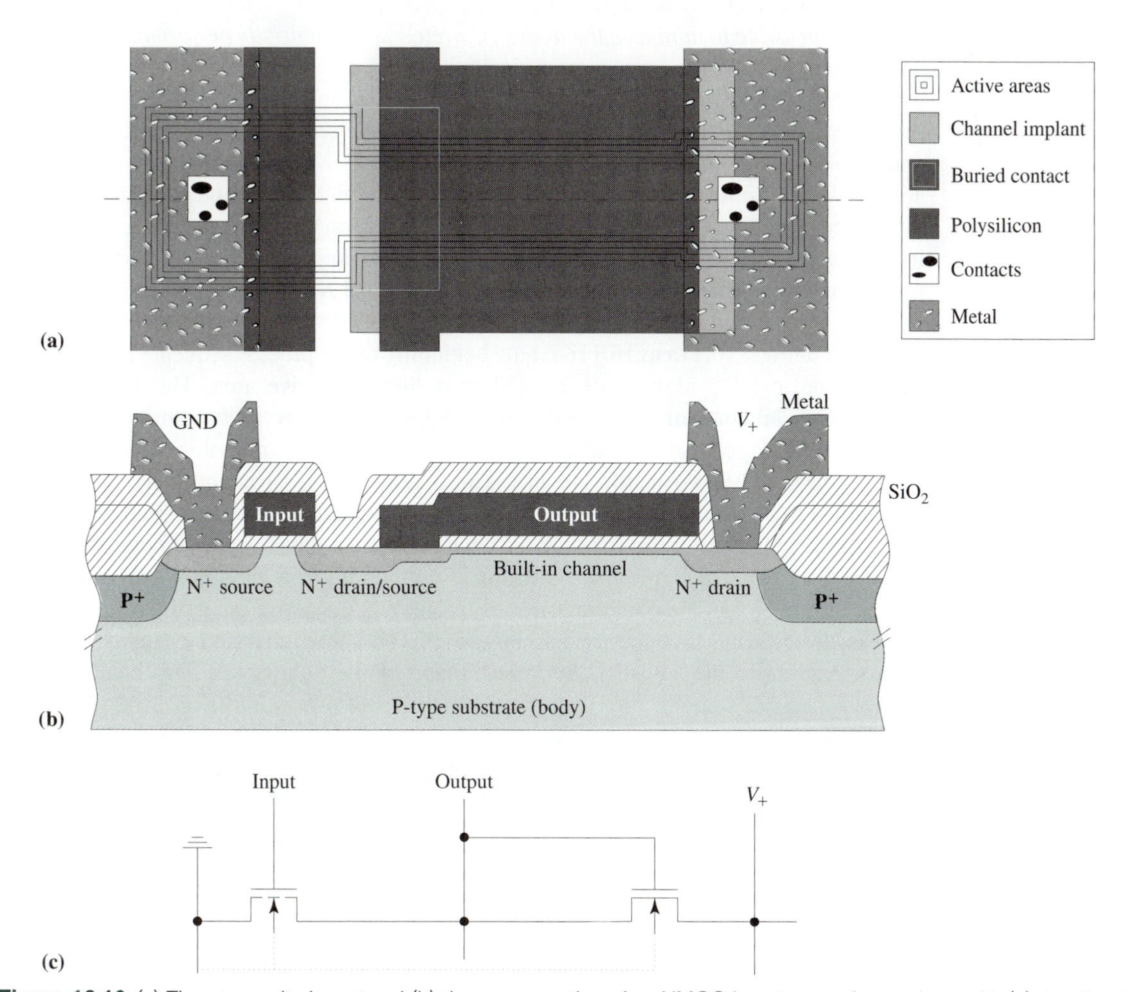

Figure 16.10 (a) The composite layout and (b) the cross section of an NMOS inverter are shown along with (c) the circuit diagram.

oxide and a P^+ diffusion region, which electrically isolate the inverter from the rest of the circuit. Imagine an N^+ region of a neighboring device adjacent to the N^+ drain of the depletion MOSFET, along with a V_+ metal line running over the space that separates them. If the field oxide and the P^+ region were not there, the two N^+ regions and the metal over the thin oxide between them would comprise a turned-on parasitic MOSFET, causing current leakage between the two devices. The thick field oxide and the increased substrate concentration underneath (P^+ region) increase the threshold voltage of this parasitic MOSFET over the value of the supply voltage V_+, ensuring that it remains off.

Figure 16.10 also shows that the N^+ drain of the enhancement MOSFET and the N^+ source of the depletion MOSFET are *merged* into a single N^+ region. This is done to minimize the area of the inverter. The circuit would not operate better if separate N^+ regions were made for the drain of enhancement and the source of depletion MOSFETs, respectively, and then connected by a metal line from the top. Quite the opposite, this approach would introduce additional parasitic resistance and capacitance, adversely affecting circuit performance. *This principle of IC layer merging is employed whenever possible to minimize the active IC area and maximize its performance.*

In the spirit of the layer-merging principle, the gate of the depletion MOSFET is extended to directly contact the N^+ source/drain region. The polysilicon gate area is further extended to serve as the output line of the inverter. In Fig. 16.10, both the input and output lines are implemented by polysilicon stripes, whereas the V_+ and ground rails appear as metal lines. This illustrates that the polysilicon layer can be used as the second interconnection level. When necessary, the N^+ diffusion areas can be used as the third interconnection level. The contact between the metal and the polysilicon levels can be made in a way similar to the contacts between metal and N^+ regions, shown in Fig. 16.10.

The technology sequence used to fabricate the NMOS inverter of Fig. 16.10 is presented in Figs. 16.11a to 16.11i. At the beginning of the process sequence, the isolation (field oxide and P^+ region) are created to define the active area. The active area is protected by silicon nitride deposited over a thin, thermally grown buffer oxide. The silicon nitride layer and the buffer oxide are patterned by a photolithography process, using a mask as shown in Fig. 16.11a. The opaque and transparent areas of the mask correspond to photolithography with positive photoresist. Silicon wafers prepared in such a way are exposed to boron implantation, which creates the P^+-type doping outside the active region. After this, thermal oxidation is applied to grow the thick field oxide. The silicon nitride blocks the oxidizing species, protecting the active area; this is the LOCOS process described in Section 16.2.1. Figure 16.11b shows that some lateral oxidation occurs, leading to the so-called *bird beak* shape of the field oxide. This has beneficial effects because it smooths the oxide step, which would otherwise be too sharp and high, thereby leading to gaps in the metal layer deposited subsequently. After the field oxide growth, the silicon nitride and the buffer oxide are removed (Fig. 16.11c).

After the active areas have been defined, the second photolithography process is applied to provide selective implantation of phosphorus, creating the built-in channel of the depletion MOSFET (Fig. 16.11d). In this case, the photoresist itself is used to protect the channel area of the enhancement-type MOSFET. In the following steps, the photoresist is removed and the surface thoroughly cleaned to prepare the wafers for gate-oxide growth. Once the gate oxide has been grown, the third photolithography is used to etch a hole in the gate oxide (Fig. 16.11e), where the subsequently deposited polysilicon layer should contact the silicon (this is the contact between the gate of depletion MOSFET, its source, and the drain of the enhancement MOSFET). The deposited polysilicon is patterned by the fourth photolithography process and associated polysilicon and gate-oxide etching (Fig. 16.11f). The photolithography mask used for the polysilicon patterning defines the gate lengths of the MOSFETs (the width of the polysilicon areas) and the width of N^+ diffusion regions (the separation between the polysilicon areas).

The alignment of the N^+ source/drain regions to the MOSFET gate is very important. Although no gap between the gate and the N^+ source/drain regions is desirable, large gate-to-N^+ source/drain overlaps would create large parasitic gate-to-source/drain capacitances, adversely affecting the high-speed performance of the device. If the N^+ source/drain regions were to be created by a separate mask, the alignment of this mask would create significant problems and/or limitations. The use of polysilicon as a gate material, instead of the initially used aluminum, enables a self-aligning technique to be employed. Figure 16.11g shows that the N^+ source/drain regions are obtained by diffusion from

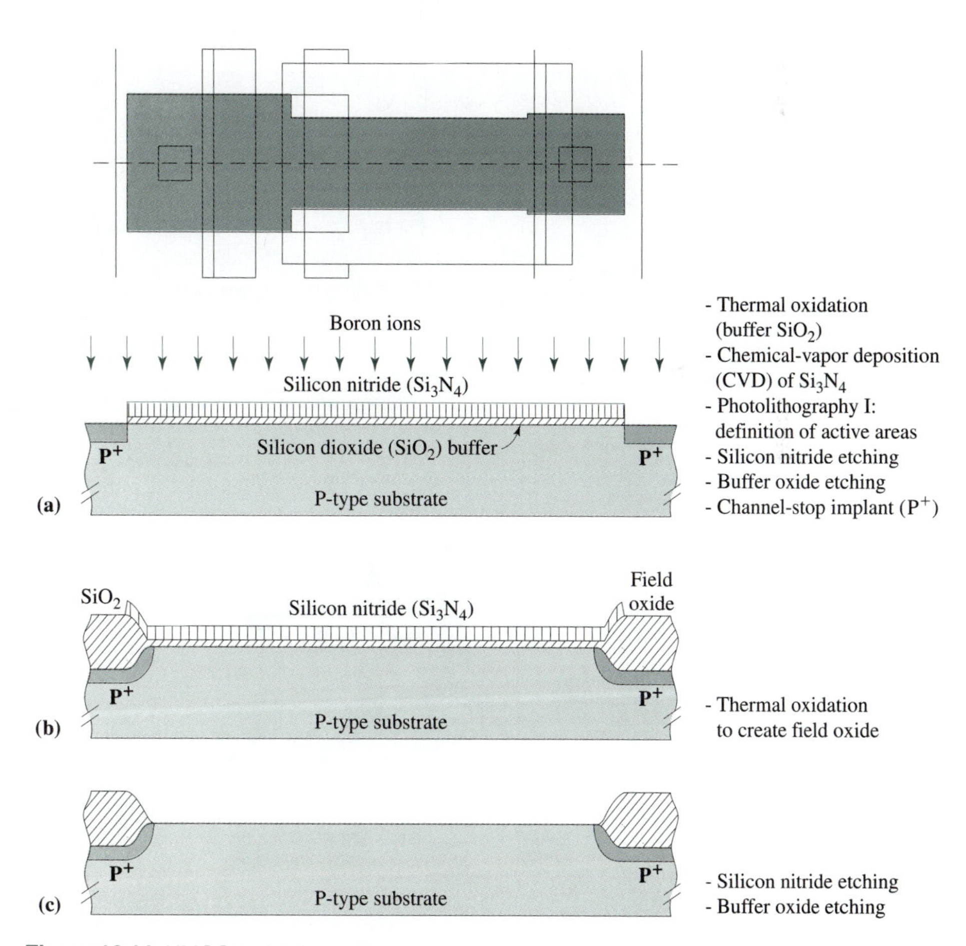

Figure 16.11 NMOS technology process.

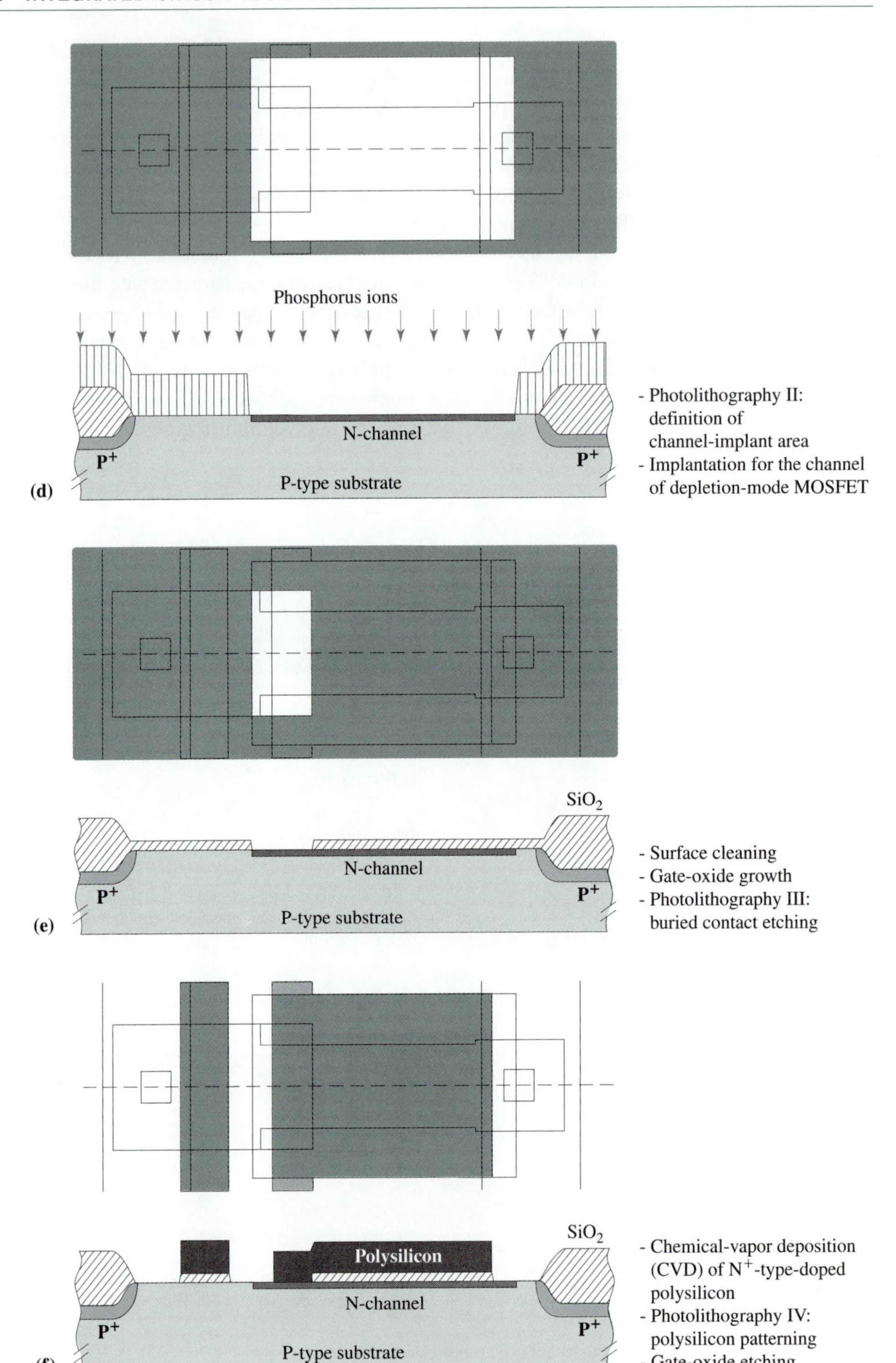

Figure 16.11 (Continued)

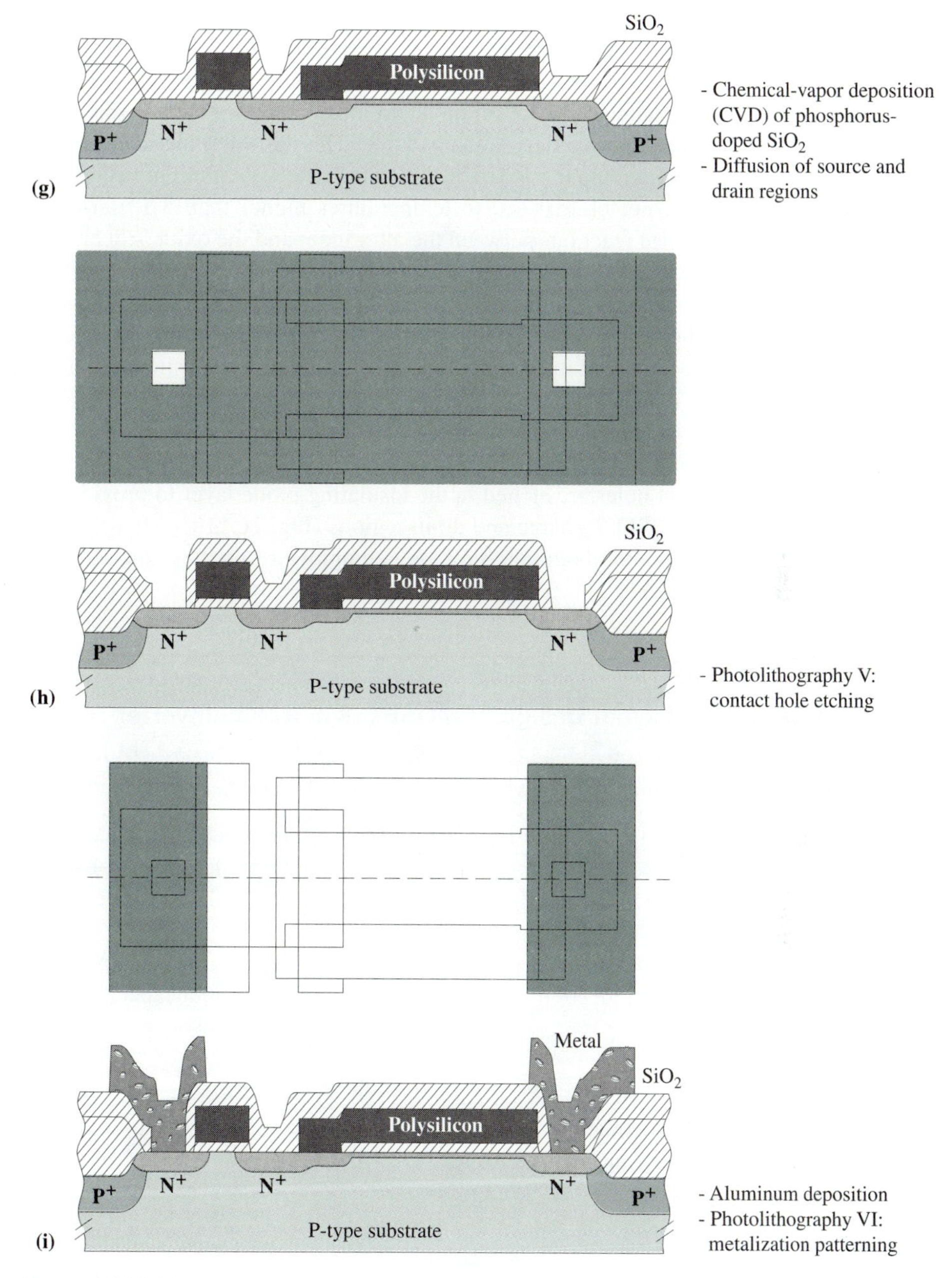

Figure 16.11 (Continued)

phosphorus-doped oxide, which is deposited onto the wafer. The polysilicon gates protect the area underneath against phosphorus diffusion because the phosphorus diffuses into the polysilicon, and the substrate areas not covered by the polysilicon (these are the drain/source regions). Some lateral diffusion does occur, which means that some gate-to-source/drain overlap is unavoidable. This self-aligning technique cannot be implemented with aluminum instead of the polysilicon because aluminum deposited over oxide or silicon should not be exposed to temperatures higher than 570°C (this would lead to adverse chemical reactions between the aluminum and the oxide/silicon) and the diffusion process requires much higher temperatures. Aluminum has smaller resistivity than polysilicon, and initially was considered a better choice for the gate material. However, the beneficial effects of the described self-aliging technique are so important that they led to almost complete replacement of the originally used aluminum by polysilicon as the gate material.

The phosphorus-doped oxide, deposited to provide the N^+ source/drain diffusion, is also used as the insulating layer between the polysilicon and the subsequently deposited aluminum layer for device interconnection. However, before the aluminum is deposited, contact holes are etched in the insulating oxide layer to provide the necessary contacts to the MOSFET source and drain regions (Fig. 16.11h). After the aluminum deposition, the sixth photolithography process is employed to pattern the aluminum layer, as shown in Fig. 16.11i.

EXAMPLE 16.4 Layout Design of MOSFETs in NMOS Inverter

When the input of the active-load NMOS inverter (Fig. 16.9) is biased at $V_H = 5.0$ V, the driving MOSFET is in the linear region, whereas the loading MOSFET is in the saturation region.

(a) Using the simplest MOSFET equations ($F_B \approx 0$ in the SPICE LEVEL 3 model), derive an equation for the ratio $r = \frac{W_d/L_d}{W_l/L_l}$. If the threshold voltages of the driving and the loading MOSFETs are $V_{Td} = 1.0$ V and $V_{Tl} = -1.0$ V, respectively, determine r so that the low-output-voltage level is $V_L = 0.2$ V.

(b) If the minimum channel dimension is limited to 0.25 μm, specify the channel lengths and channel widths of the two MOSFETs so that the input capacitance of the inverter is minimized.

SOLUTION

(a) The simplest equations for MOSFETs in the linear and the saturation regions are given by Eqs. (8.3) and (8.26), respectively. Therefore, the currents of the driving and the loading MOSFETs can be expressed as

$$I_{D-d} = \beta_d(V_{GS} - V_{Td})V_{DS}$$

$$I_{D-l} = \frac{\beta_l}{2}(\underbrace{V_{GS-l}}_{0} - V_{Tl})^2$$

where

$$\beta_d = KP\frac{W_d}{L_d} \quad \text{and} \quad \beta_l = KP\frac{W_l}{L_l}$$

For $V_{GS} = V_H$, the driving MOSFET is *on,* taking the constant current of the saturated loading MOSFET to ground. From the condition that the two currents are equal, $I_{D-d} = I_{D-l}$, we can find the requested equation for the r ratio:

$$KP\frac{W_d}{L_d}(V_H - V_{Td})V_L = \frac{KP}{2}\frac{W_l}{L_l}V_{Tl}^2$$

$$r = \frac{W_d/L_d}{W_l/L_l} = \frac{V_{Tl}^2}{2(V_H - V_{Td})V_L}$$

For the numerical values provided in the text of the example, we obtain

$$r = \frac{(-1)^2}{2 \times (5-1) \times 0.2} = 0.625$$

(b) To minimize the input capacitance of the inverter, the area of the driving MOSFET ($W_d L_d$) should be as small as possible. Given that $W_d/L_d < W_l/L_l$, we select $W_d = L_d = L_l = 0.25\ \mu$m and obtain $W_l = 0.25/r = 0.25/0.625 = 0.4\ \mu$m.

16.2.3 Basic CMOS Technology

The issue of high static power dissipation, identified in Section 16.2.2 for the NMOS technology, can be removed if the loading depletion MOSFET is replaced by a complementary enhancement-type P-channel MOSFET. This is the complementary MOS (CMOS) technology. The circuit of a CMOS inverter was shown in Fig. 8.6a and discussed in Section 8.1.2.

The composite layout and the cross section of the CMOS inverter implemented in basic N-well technology are shown in Fig. 16.12. The technology is referred to as N-well technology because the N-type body needed for the PMOS transistor is implemented as the N-well region diffused into the P-type substrate. The NMOS transistors are created in the P-type substrate itself. Given that the body of the NMOS transistors (the P-type substrate) is grounded and the body of the PMOS transistors (the N-well) is connected to the most-positive potential V_+, the N-well–P-substrate junction is reverse-biased. This is the electrical isolation by a reverse-biased P–N junction, utilized in CMOS technology. In addition, to ensure that no surface leakage occurs, the devices are separated by thick field oxide (the LOCOS isolation). The N^+-type layer, used to create the source and the drain of the NMOS, is also used to provide contact to the N-well region, which is directly connected to the source of the PMOS. Although no direct connection between the P-type substrate and the source of the NMOS is shown in Fig. 16.12b, the P-type substrate is connected to ground, typically by a rail enclosing the whole IC area.

There is also P-well CMOS technology, in which PMOS transistors are placed in an N-type silicon substrate, and P wells are created to place NMOS transistors. The P-well technology was developed before the N-well technology because it was easier to achieve

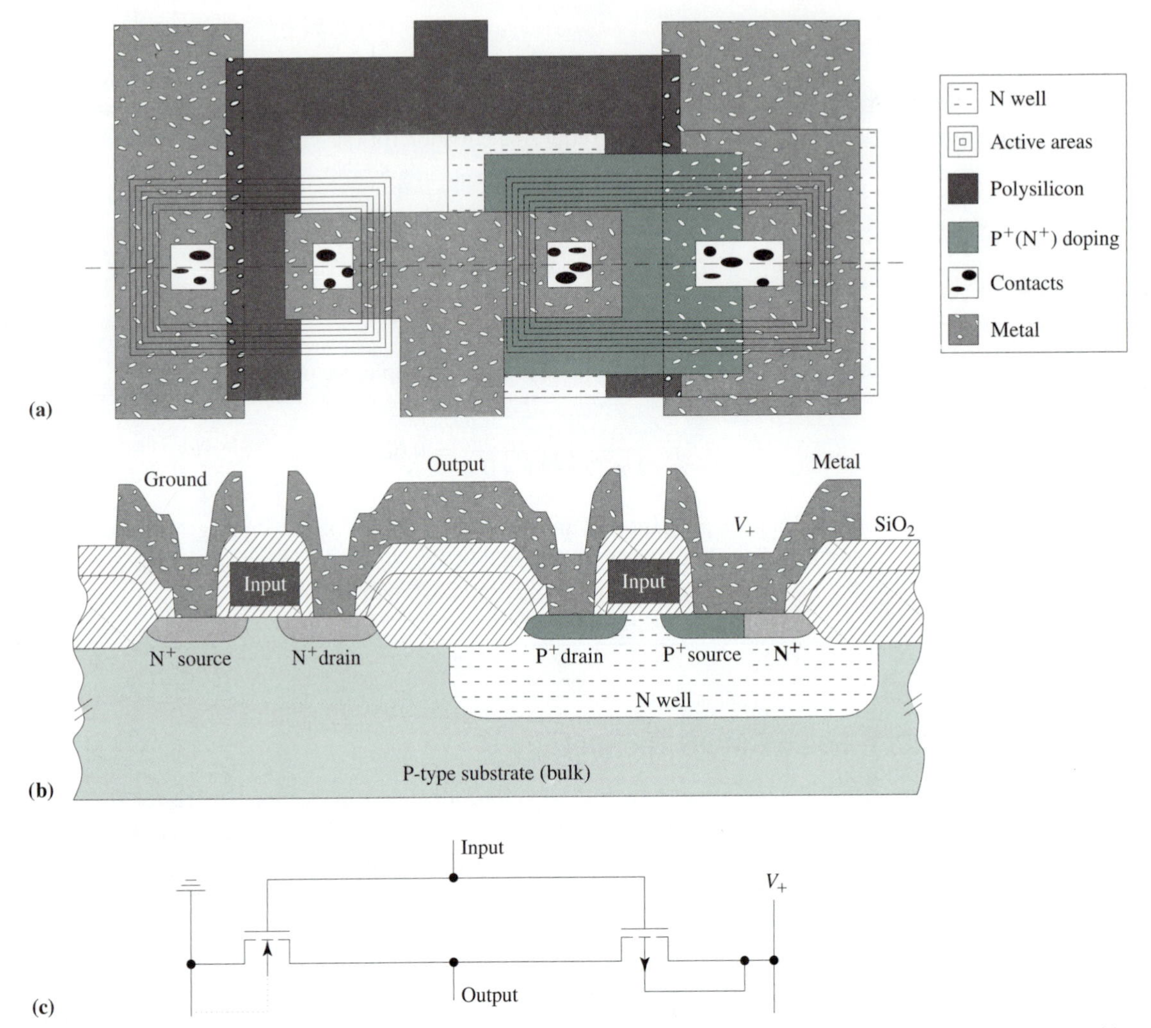

Figure 16.12 (a) The composite layout and (b) the cross section of a basic N-well CMOS inverter are shown along with (c) the circuit diagram.

the desired threshold voltages of NMOS and PMOS transistors. When the use of ion implantation made it possible to adjust the threshold voltages, N-well technology became more popular because the use of the P-type substrate made it compatible with NMOS and bipolar technology (described in Section 16.3).

The basic technology sequence that can be used to fabricate a CMOS inverter is presented in Fig. 16.13a to 16.13i. At the beginning of the process, N wells have been created by ion implantation of phosphorus through an appropriately patterned SiO_2 masking layer and subsequent annealing (drive-in). Figure 16.13a shows that the masking oxide is removed once the N wells have been created. In the next stage, thick field oxide is created to surround the active areas. As Fig. 16.13b and 16.13c illustrates, this is achieved by the same LOCOS process used in the NMOS technology.

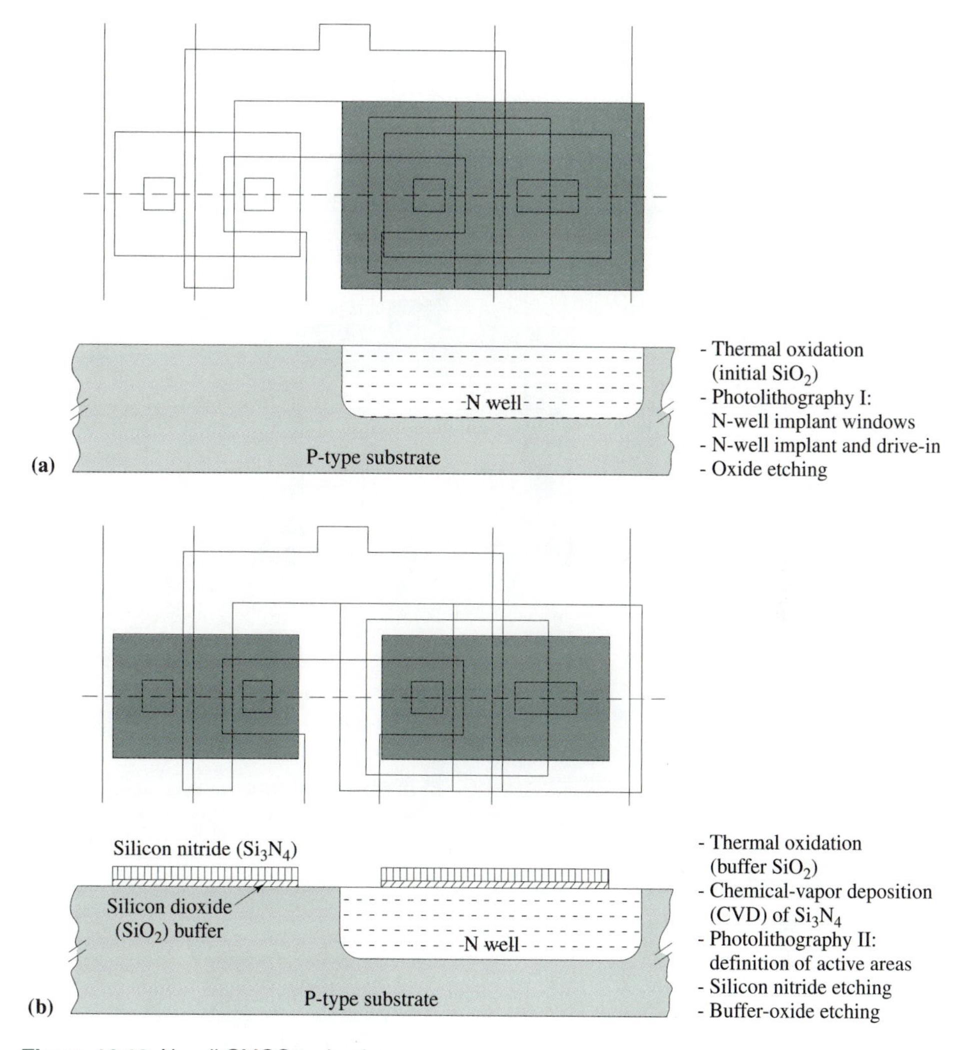

Figure 16.13 N-well CMOS technology process.

The surface donor concentration N_D in the N well is higher than the acceptor concentration N_A in the P-type substrate. This fact as well as the work-function differences cause a lower NMOS threshold voltage and a higher absolute value of the PMOS threshold voltage than the desirable values. The performance of CMOS circuits is maximized with a slight positive NMOS threshold voltage and a negative PMOS threshold voltage (enhancement-type MOSFETs) with the absolute value equal to its NMOS counterpart. To adjust the threshold voltages, boron is implanted into the channel areas of both the NMOS and PMOS transistors. This, the so-called threshold-voltage-adjustment implantation, reduces the effective N-type doping level at the surface of the N well, decreasing the

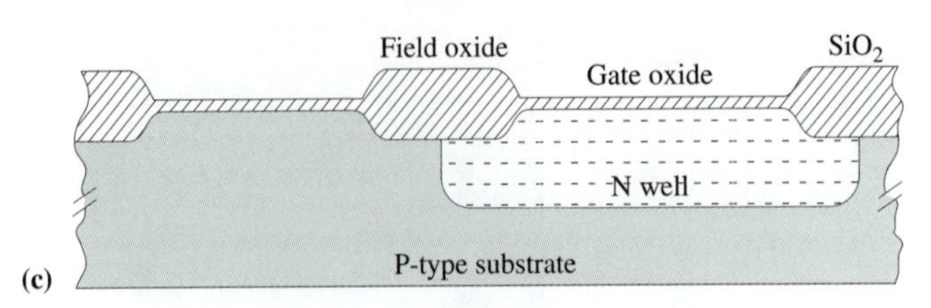

- Thermal oxidation to create field oxide
- Silicon nitride etching
- Buffer oxide etching
- Surface cleaning
- Threshold-voltage adjustment implant
- Gate-oxide growth

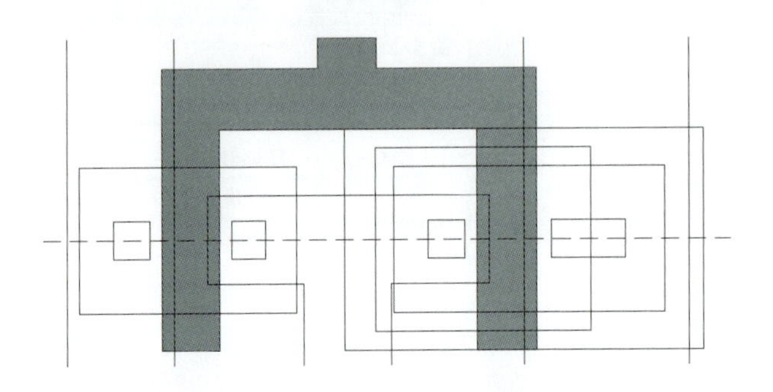

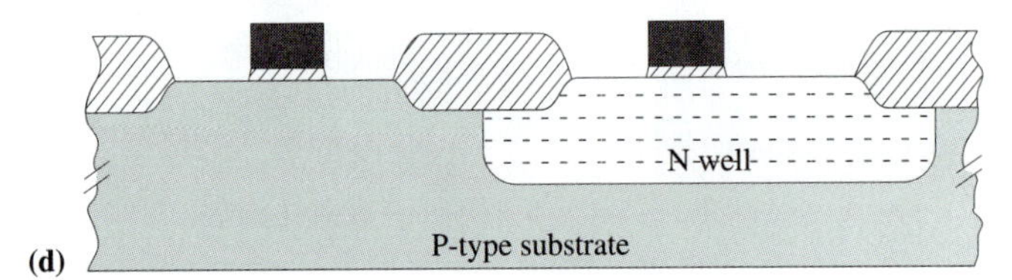

- Chemical-vapor deposition (CVD) of N^+-doped polysilicon
- Photolithography III: polysilicon patterning (gate definition)
- Gate-oxide etching

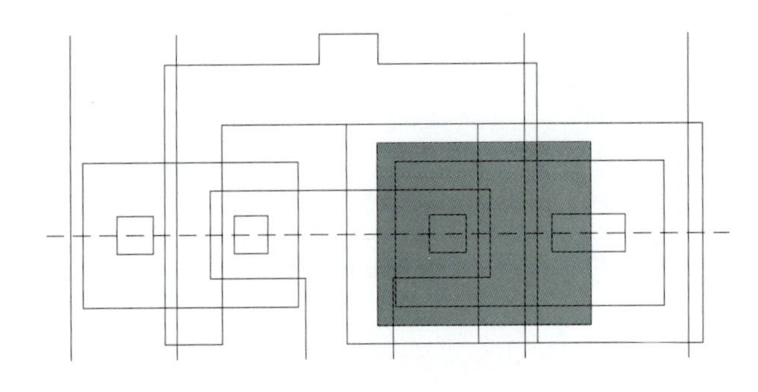

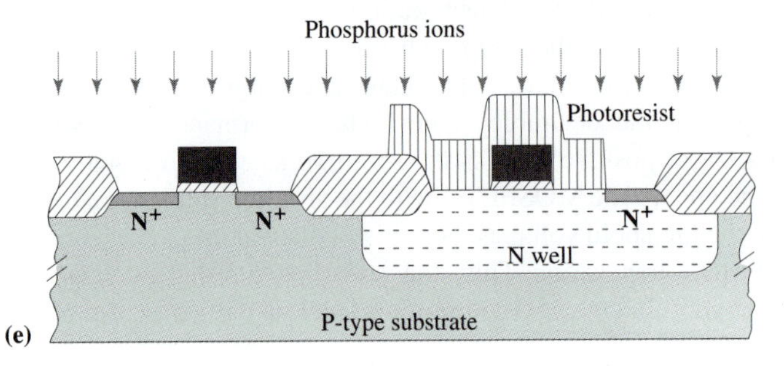

- Photolithography IV: implantation of N^+ regions

Figure 16.13 (Continued)

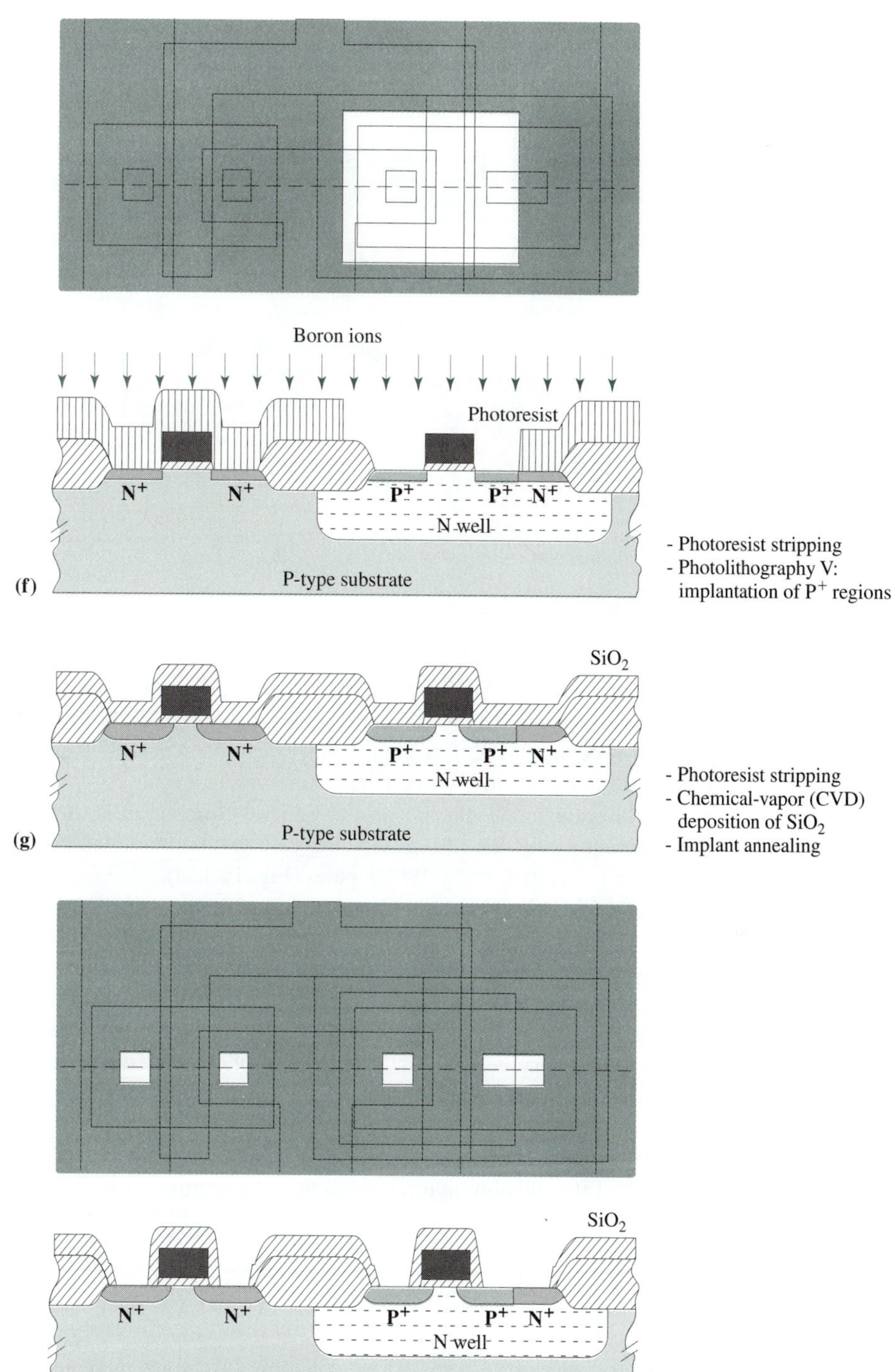

Figure 16.13 (Continued)

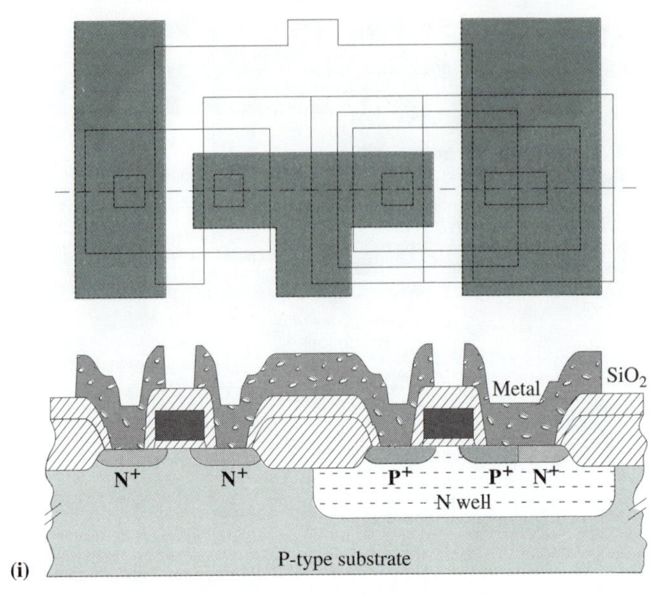

Figure 16.13 (Continued)

absolute value of PMOS threshold voltage. Also, it increases the P-type doping level at the surface of the P-type substrate, increasing the NMOS threshold voltage. In this way, the NMOS and PMOS threshold voltages can be matched by an appropriately determined dose of the threshold-voltage-adjustment implantation.

The gate oxide can be grown either before or after threshold-voltage-adjustment implantation (Fig. 16.13c). Following gate oxidation, doped polysilicon is deposited and patterned to define the MOSFET gates (Fig. 16.13d).

The CMOS technology also makes use of the self-aligning technique, to minimize the overlap between the gate and drain/source areas. The difference here, compared to the NMOS technology, is that the doping is achieved by ion implantation of phosphorus (Fig. 16.13e) and boron (Fig. 16.13f) to obtain the N$^+$ and P$^+$ regions, respectively. An appropriately patterned photoresist is used to mask the areas that are not to be implanted. As Fig. 16.13g shows, the ion implants are followed by oxide deposition (needed to isolate the polysilicon layer from the subsequent metal layer) and annealing, which is needed to activate the implanted doping ions.

The process finishes in the same way as the NMOS technology, by contact hole etching (Fig. 16.13h) and aluminum deposition and patterning (Fig. 16.13i).

EXAMPLE 16.5 Design of Threshold-Voltage-Adjustment Implant

The technological parameters of N-well CMOS technology are given in Table 16.1, together with the values of the relevant physical parameters.

(a) Determine the threshold voltages of the NMOS and PMOS transistors for $\Phi_{implant} = 0$.

(a) Provided the threshold-voltage-adjustment implant is shallow, the threshold-voltage shift $|\Delta V_T|$ due to the implant dose $q\Phi_{implant}$ can be expressed as

$$|\Delta V_T| = \frac{q\Phi_{implant}}{C_{ox}}$$

TABLE 16.1 Technological Parameters of N-Well CMOS Technology

Parameter	Symbol	Value
Substrate doping concentration	N_A	10^{15} cm^{-3}
N-well surface concentration	N_D	5×10^{16} cm^{-3}
Gate-oxide thickness	t_{ox}	15 nm
Dose of threshold-voltage-adjustment implantation	$\Phi_{implant}$	?
Oxide charge density	N_{oc}	10^{10} cm^{-2}
Type of gate		N$^+$-polysilicon
Intrinsic-carrier concentration	n_i	1.02×10^{10} cm^{-3}
Energy gap	E_g	1.12 eV
Thermal voltage at room temperature	$V_t = kT/q$	0.026 V
Oxide permittivity	ε_{ox}	3.45×10^{-11} F/m
Silicon permittivity	ε_s	1.04×10^{-10} F/m

Design the dose of threshold-voltage-adjustment implant so that the NMOS and PMOS threshold voltages are matched ($V_{T-NMOS} = |V_{T-PMOS}|$).

SOLUTION

(a) The threshold voltage of an NMOS transistor is given by Eq. (8.9), whereas the threshold voltage of a PMOS transistor is given by Eq. (8.11). The threshold voltages are calculated using the procedure given in Example 8.1:

- NMOS:

$$\phi_F = V_t \ln \frac{N_A}{n_i} = 0.30 \text{ V}$$

$$C_{ox} = \varepsilon_{ox}/t_{ox} = 2.3 \times 10^{-3} \text{ F/m}^2$$

$$\phi_{ms} = \phi_m - \left(\chi + \frac{E_g}{2q} + \phi_F\right) = -E_g/2q - \phi_F = -0.86 \text{ V}$$

$$V_{FB} = \phi_{ms} - \frac{qN_{oc}}{C_{ox}} = -0.87 \text{ V}$$

$$\gamma = \sqrt{2\varepsilon_s q N_A}/C_{ox} = 0.079 \text{ V}^{1/2}$$

$$V_{T-NMOS} = V_{FB} + 2\phi_F + \gamma\sqrt{2\phi_F} = -0.21 \text{ V}$$

- PMOS:

$$\phi_F = -V_t \ln \frac{N_D}{n_i} = -0.40 \text{ V}$$

$$C_{ox} = \varepsilon_{ox}/t_{ox} = 2.3 \times 10^{-3} \text{ F/m}^2$$

$$\phi_{ms} = \phi_m - \left(\chi + \frac{E_g}{2q} + \phi_F\right) = -E_g/2q - \phi_F = -0.16 \text{ V}$$

$$V_{FB} = \phi_{ms} - \frac{qN_{oc}}{C_{ox}} = -0.17 \text{ V}$$

$$\gamma = \sqrt{2\varepsilon_s q N_D}/C_{ox} = 0.561 \text{ V}^{1/2}$$

$$V_{T-PMOS} = V_{FB} - 2|\phi_F| - \gamma\sqrt{2|\phi_F|} = -1.47 \text{ V}$$

(b) Without the ion-implant adjustment, the NMOS threshold voltage is negative, making it effectively a depletion-type MOSFET. A boron implant will increase this threshold voltage (due to an increase in the surface P-type concentration), and it will simultaneously reduce the absolute value of the PMOS threshold voltage (due to effective reduction in the surface doping level of the N well). Therefore the threshold voltages after the threshold-voltage-adjustment implantation can be expressed as

$$V_{T-NMOS}(\Phi_{implant}) = V_{T-NMOS}(0) + q\Phi_{implant}/C_{ox}$$

$$|V_{T-PMOS}(\Phi_{implant})| = |V_{T-NMOS}(0)| - q\Phi_{implant}/C_{ox}$$

When we choose different values for the implant dose $\Phi_{implant}$, the results in Table 16.2 are obtained. Table 16.2 illustrates the effect of the adjustment-implant dose on the NMOS and PMOS threshold voltages. The implant dose that matches the threshold voltage can be found directly by (1) observing that each of the threshold-voltages should be shifted by $\Delta V_T = |V_{T-PMOS}(\Phi_{implant})| - V_{T-NMOS}(\Phi_{implant})/2 = (1.47+0.21)/2 = 0.84$ V and (2) finding the dose that corresponds to this threshold-voltage shift. As Table 16.2 shows, this dose is $\Phi_{implant} = 1.21 \times 10^{12}$ cm^{-2}.

TABLE 16.2 Iterative Solutions for Example 16.5

$\Phi_{implant}$ (cm^{-2})	ΔV_T (V)	$V_{T-NMOS}(\Phi_{implant})$ (V)	$\|V_{T-PMOS}(\Phi_{implant})\|$ (V)
0	0.00	−0.21	1.47
1.0×10^{11}	0.07	−0.14	1.40
3.0×10^{11}	0.21	0.00	1.26
1.00×10^{12}	0.70	0.49	0.77
1.21×10^{12}	0.84	0.63	0.63

16.2.4 Silicon-on-Insulator (SOI) Technology

A silicon-on-insulator (SOI) structure is very suitable for many applications. A number of improvements can be achieved with SOI CMOS technology. The SOI structure, illustrated in Fig. 16.14a, is typically obtained by one of the following two techniques: (1) **s**eparation of silicon by **im**planted **ox**ygen (SIMOX) and (2) bonded wafers. In the SIMOX process, oxygen is implanted as deeply as possible into the silicon wafer and is then annealed at very high temperature to create the buried oxide (SiO_2) layer. The thickness of the buried oxide obtained by SIMOX is typically around 400 nm, whereas the thickness of the top monocrystalline silicon is around 200 nm. Much thicker buried oxide and more flexible thicknesses of the top silicon layer can be achieved by the bonded wafer process. In this

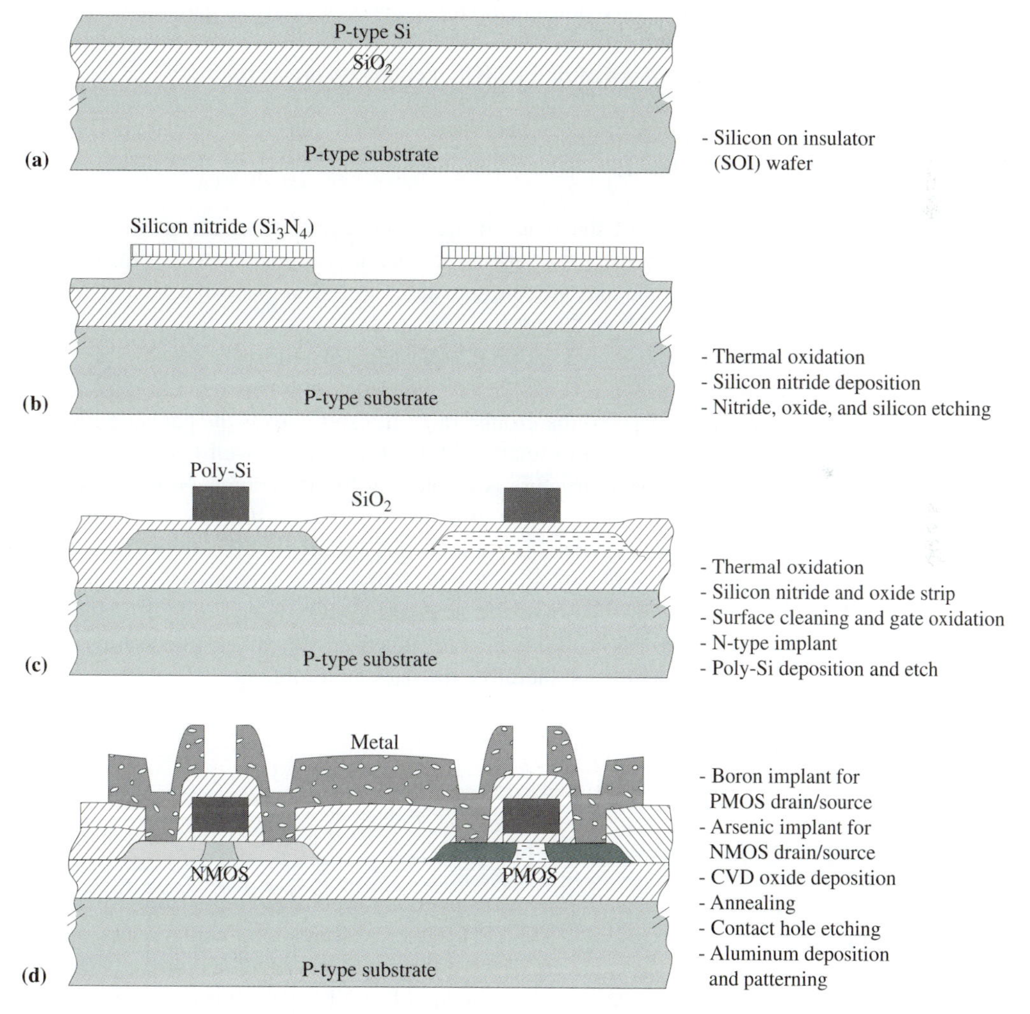

Figure 16.14 Silicon-on-insulator (SOI) CMOS technology.

case, two silicon wafers with pregrown oxides at their surfaces are placed face to face and exposed to further thermal oxidation that bonds them to each other. After this, one of the wafers is thinned down to the desired top-silicon thickness.

To create a CMOS inverter using an SOI substrate, active areas are defined by depositing and patterning silicon nitride (Fig. 16.14b), similar to the basic CMOS technology process. The silicon in the field region (outside the active regions) is etched to approximately half of the original thickness and then exposed to thermal oxidation (LOCOS). As the field oxide grows, it consumes the remaining top silicon in the field region, joining with the buried oxide (Fig. 16.14c). In this way, islands of silicon, isolated from each other by oxide, are created. The remaining processing is very similar to the basic CMOS technology process: polysilicon is deposited and patterned (Fig. 16.14c), P^+ and N^+ areas implanted to create the sources and the drains of the P-channel and N-channel MOSFETs, and isolation oxide and top metal are deposited and patterned to create the final structure (Fig. 16.14d).

The main advantages of SOI CMOS technology are as follows:

1. The top silicon layer can be thinned down to tens of nanometers. This reduces the punch-through problem, which helps in the design of nanoscale CMOS devices (refer to Section 8.4.3).
2. The thickness of the oxide in the field region is significantly increased, which significantly reduces the parasitic capacitances created between the interconnecting metal lines and the silicon substrate. Reduced parasitic capacitances result in improved switching speed.
3. N-channel and P-channel MOSFETs are isolated by a dielectric (as distinct from the isolation by a reverse-biased P–N junction). This reduces the leakage current between the V_+ and the ground rails. It also removes the parasitic PNPN thyristor structure existing in ordinary CMOS ICs. In an N-well CMOS, for example, the parasitic thyristor structure is created by the P^+-source–N-well–P-substrate–N^+-source layers. This thyristor structure is normally *off;* however, if turned *on* under certain unpredicted conditions, it creates a short circuit between the V_+ and ground rails, permanently damaging the IC. This effect, known as *latch-up,* is one of the biggest reliability problems in standard CMOS ICs. The SOI CMOS technology eliminates the N-well–P-substrate junction altogether, eliminating the parasitic thyristor structure and therefore the latch-up problem.

16.3 BIPOLAR IC TECHNOLOGIES

16.3.1 IC Structure of NPN BJT

The doping layers (basically, diffusion layers) of bipolar integrated circuits are designed to optimize the characteristics of the NPN BJT. All other circuit components, including PNP BJTs and resistors, are made out of the diffusion layers designed for the NPN BJT.

Figure 16.15 illustrates the IC structure of an NPN BJT. It is immediately obvious that the active part of the device (N^+PN layers highlighted by the zoom-in rectangle) occupies a small portion of the total cross-sectional area. A large part of the cross-sectional area

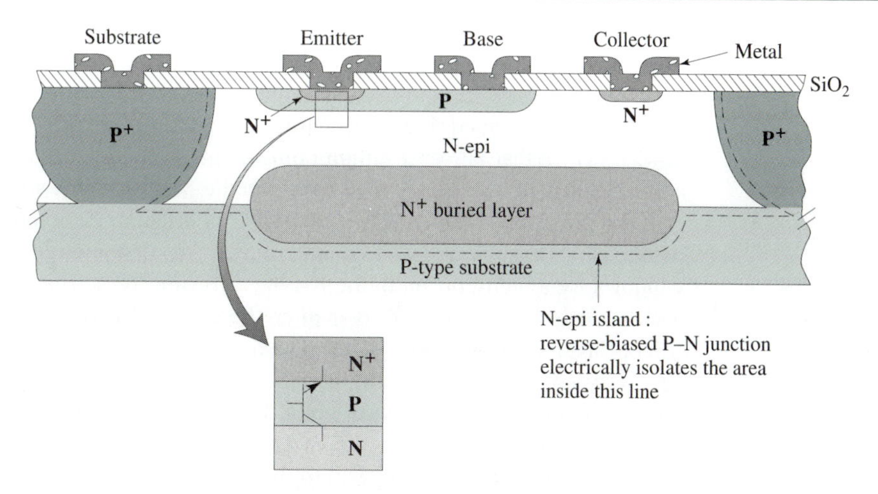

Figure 16.15 Integrated-circuit structure of an NPN BJT.

is taken to satisfy the following two requirements: (1) electrical isolation from the other components of the IC and (2) enabling surface contacts to the three device terminals: base, emitter, and collector.

Electrical isolation is provided by the reverse-biased P–N junction, indicated by the dashed line in Fig. 16.15. To create a P–N junction that encloses the device, a P-type substrate (the fourth layer in addition to the three device layers) is needed. Having the P-type substrate at the bottom, a P^+ ring is diffused around the device to cut the bottom N-type layer into so-called N-epi islands. The isolation P^+ ring takes a large amount of area, not only because it encircles the device but also due to the lateral diffusion that is about 80% of the vertical diffusion, and the vertical diffusion has to be sufficient to penetrate through the whole depth of the N-epi layer. To activate the P–N junction isolation, the P^+ isolation region has to be connected to the lowest potential in the circuit (V_-); this ensures that the isolation P–N junctions are reverse-biased.

To provide surface contacts to the device, the P-type layer is extended beyond the N^+ area to contact the base, and the N-epi layer is even wider to provide room for the collector contact. As the aluminum and low-doped N-type silicon make rectifying Schottky contact rather than ohmic contact, an N^+ region is created at the collector contact. Note that this N^+ region is created by the same diffusion process used for the N^+ emitter layer.

The eye-catching N^+ buried layer may not extend the lateral dimensions of the device, but it certainly makes the technology process more complex. However, if the buried layer had not been introduced, the current collected by the active part of the collector (the N-epi region under the N^+ emitter) would have had to face the resistance of the relatively long and low-doped N-epi layer before it reached the collector contact. The N^+ buried layer provides a low-resistive section between the active part of the collector and the collector contact, significantly reducing the collector parasitic resistance. To further reduce this resistance, some bipolar technology processes introduce an additional diffusion process that provides a deep N^+ region connecting the collector contact and the N^+ buried layer. The following are disadvantages of this so-called deep collector diffusion: (1) increased complexity of the technology process and (2) increased lateral dimensions, especially due to the significant lateral diffusion of this deep layer.

16.3.2 Standard Bipolar Technology Process

In this section, the process sequence used to fabricate the standard NPN structure of Fig. 16.15 is described. Practically, this is the description of *standard bipolar technology process,* given that all other circuit components (resistors, capacitors, diodes, PNP BJTs) are implemented with the layers existing in the standard NPN structure.

The processing sequence begins with (a) thermal oxidation of P-type silicon substrate (wafers) and (b) subsequent oxide patterning by the first photolithography process to create windows for a high-concentration N-type diffusion (N^+). The created N^+ diffusion layer will become the N^+ buried layer after the epitaxial deposition of a low-doped N-type silicon layer (Fig. 16.16b).[1] The epitaxial growth of the low-doped N-type silicon is necessary not only to create the buried layer but also to enable the creation of the four-layer structure: P substrate–N collector–P base–N^+ emitter. If these four layers were to be created by diffusing one layer into another three times (N collector into P substrate, P base into N collector, and N^+ emitter into the P base), a hardly achievable and inconveniently low concentration of the initial P-type substrate would be necessary. With the epitaxial process, the concentration of the N-epitaxial (N-epi) layer is independent of the doping level of the underlying P-type substrate and can be set at an appropriately low level.

After the epitaxial process, P^+ diffusion is employed to define the electrically isolated N-epi islands that carry the individual circuit components. The photolithography mask used for this process, shown in Fig. 16.16c for the case of positive photoresist, illustrates the top view of an N-epi island surrounded by the P^+ isolation diffusion. Vertically, the N-epi islands are still not disconnected at this stage, because the P^+ diffusion still does not reach the underlying P-type substrate as in the final structure shown previously in Fig. 16.15. The reason for doing this diffusion "part way" at this stage is that more high-temperature (diffusion) processes follow, which will cause simultaneous diffusion of the doping atoms in the P^+ isolation areas. It is important to limit the overall P^+ diffusion to the level needed to reach the underlying P-type substrate because any excessive lateral diffusion would mean wasting a significant surface area by the wider P^+ isolation areas, which dominate the surface area of the IC even without any waste.

The definition of the N-epi islands by the P^+ isolation diffusion is followed by the P-type and N^+-type diffusions needed for the P-type base and N^+-type emitter. Of course, the diffusion areas are defined by the associated photolithography processes, labeled as photolithography III and photolithography IV in Fig. 16.16d and 16.16e, respectively. Note that photolithography IV opens two windows for the N diffusion: one for the emitter of the NPN BJT and the other for the collector contact.

At this stage, the NPN BJT is created in the silicon, and what remains is the photolithography V to etch the contact holes through the insulating oxide layer (Fig. 16.16f), then to deposit the aluminum layer, and to pattern the metalization by the sixth photolithography process (Fig. 16.16g).

[1] The deposition of a monocrystalline layer on monocrystalline substrate is called *epitaxial* growth, and the deposited layer is called the epitaxial layer, or epi layer for short.

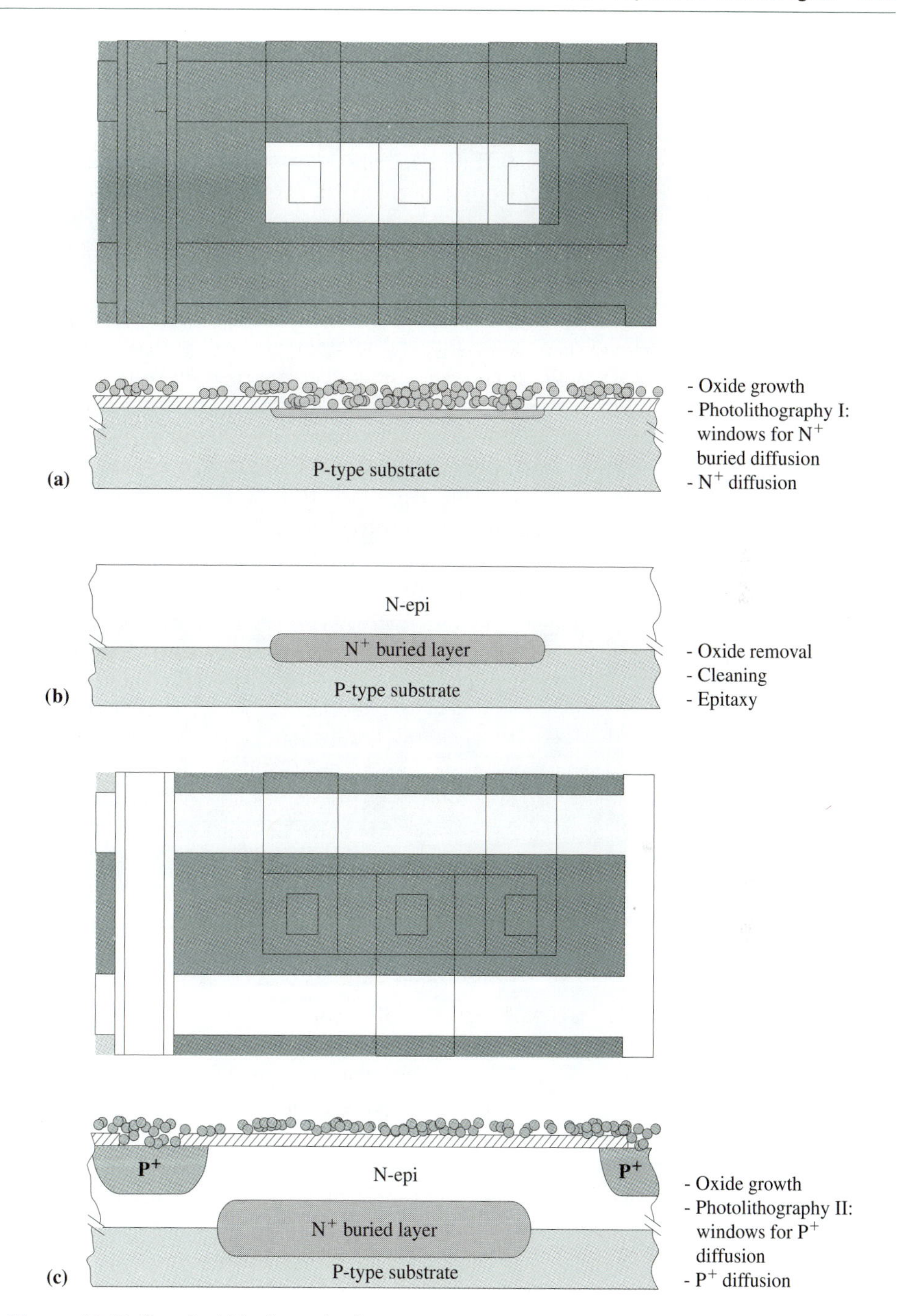

Figure 16.16 Standard bipolar technology.

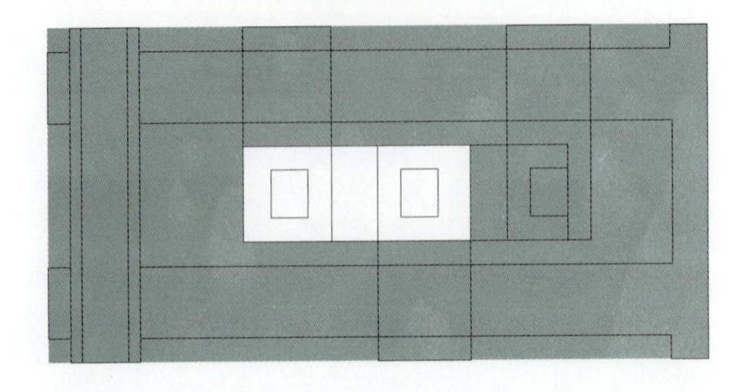

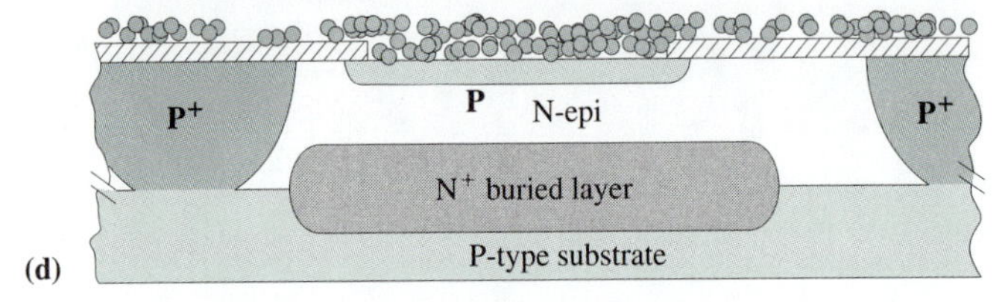

(d)

- Oxide removal and growth
- Photolithography III: windows for P diffusion
- P diffusion

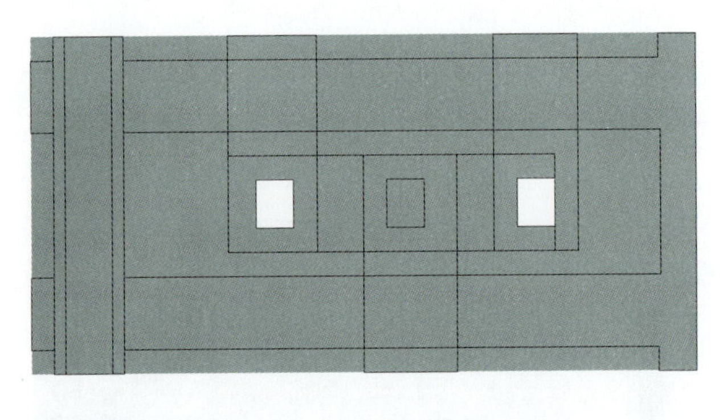

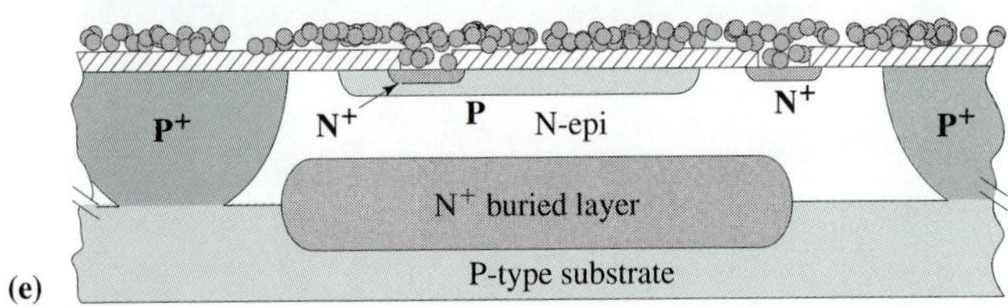

(e)

- Oxide growth
- Photolithography IV: windows for N^+ diffusion
- N^+ diffusion

Figure 16.16 (Continued)

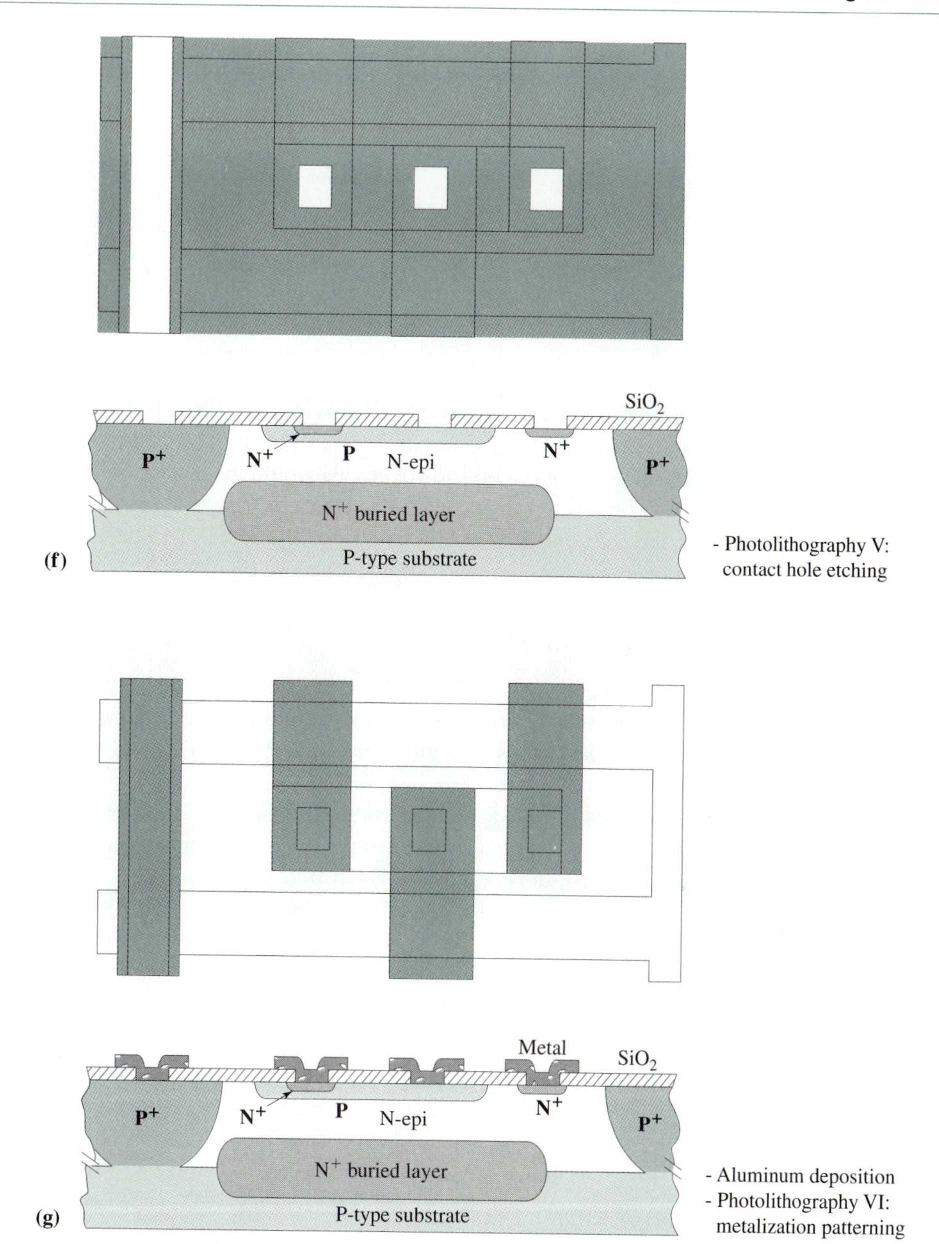

Figure 16.16 (Continued)

16.3.3 Implementation of PNP BJTs, Resistors, Capacitors, and Diodes

As already mentioned, the semiconductor layers of the bipolar ICs are designed to maximize the performance of the main component, the NPN BJT. These layers are labeled

TABLE 16.3 The Semiconductor Layers of Standard Bipolar ICs

Layer Number	Layer Name	Typical Characteristics
1	P-type substrate/P^+ isolation diffusion	
2	N-epi	10^{15} cm^{-3}, 10 μm
3	P-base diffusion	200 $\Omega/\square$, 2–4 μm
4	N^+-emitter diffusion	5 $\Omega/\square$, 1–2 μm

according to the function they perform in the NPN BJT (Table 16.3). Theoretically, it is possible to introduce additional layers to optimize or improve the characteristics of the other circuit components; however, this would mean an increase in the process complexity that was practically unjustifiable. Consequently, all other circuit components are designed from the four existing layers, listed in Table 16.3. This section describes the ways these components are implemented in the standard bipolar ICs.

Substrate PNP BJT

It is possible to make a PNP BJT using both existing P-type layers listed in Table 16.3 (the P-type substrate/P^+ isolation diffusion and the P-type base layers) with the N-epi layer as the base. This configuration of the PNP BJT, called *substrate PNP,* is illustrated in Fig. 16.17.

Electrically, the P-type substrate/P^+ isolation is a single region, connected to the most negative potential in the circuit (V_-) to ensure the isolation of the N-epi islands. Because of that, the collector of a substrate PNP cannot be connected to an arbitrary point in the circuit—it is automatically connected to V_-. This severely limits the application of the substrate PNP to the particular case where the circuit needs a PNP BJT with the collector connected to V_-. Because the substrate PNP provides better characteristics than any other PNP in ICs, it should be used as the first option whenever the collector biasing is appropriate.

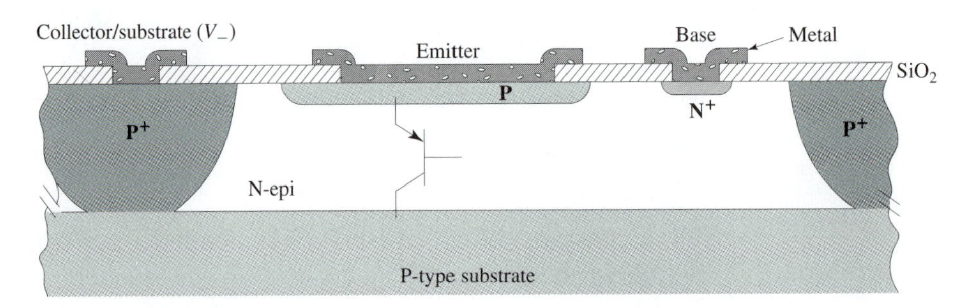

Figure 16.17 Substrate PNP BJT: the collector is connected to the most negative potential in the circuit.

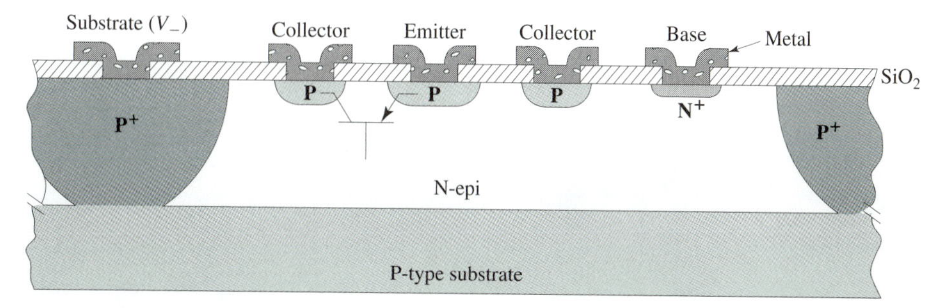

Figure 16.18 Lateral PNP BJT.

Lateral PNP BJT

Obviously, the P-type substrate/P^+ isolation diffusion layer cannot be used if the application limitation regarding the collector biasing is to be removed. The only remaining P-type layer is the P-type base diffusion; hence this layer has to be used for both the emitter and the collector. This is possible by the lateral structure illustrated in Fig. 16.18.

Because the collector of this BJT collects only a small part of laterally emitted holes, the transport factor and consequently the current gains α and β are very small. To minimize the waste of emitted holes, the collector P-type region normally surrounds the emitter P-type region (closed geometry). This is why two P-type collector regions appear in the cross section of Fig. 16.18.

Resistors

Unlike the CMOS ICs, which use only complementary pairs of MOSFETs (an example is the inverter of Fig. 8.6), bipolar ICs generally use resistors and capacitors, and frequently diodes as well. The general structure of an IC resistor is shown in Fig. 3.3. The resistive body is made of one of the available layers (N-epi, base diffusion, or emitter diffusion), and it is isolated from the other resistors and the rest of the IC by a reverse-biased P–N junction. Figure 16.19 illustrates the case of a base-diffusion resistor. The body of the resistor is the P-type base-diffusion layer, which is surrounded by the N-epi layer connected to the most positive potential in the circuit (V_+) to ensure that the resistor current is confined within

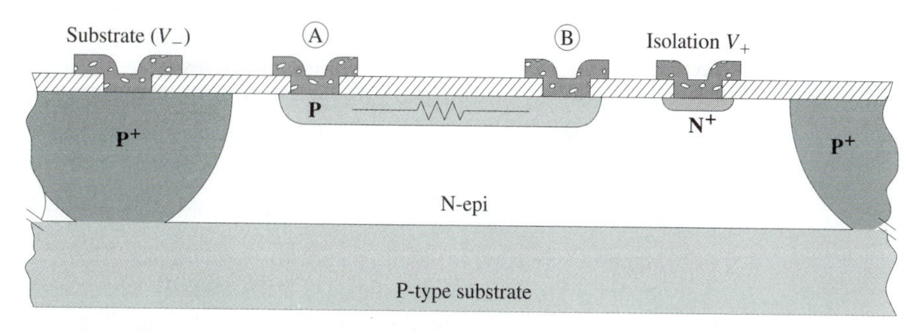

Figure 16.19 The cross section of base-diffusion resistor.

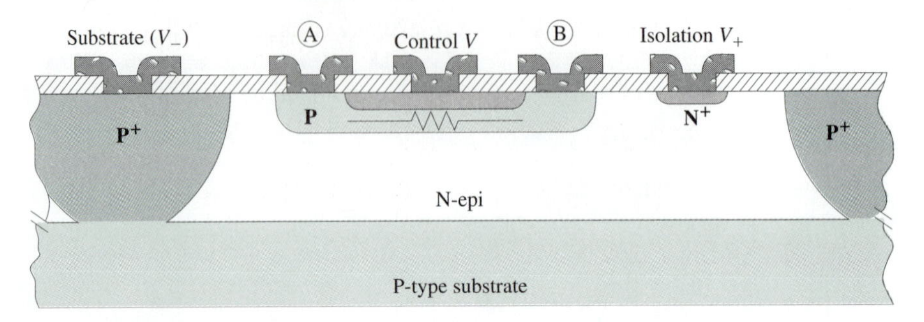

Figure 16.20 Base-diffusion pinch resistor. The emitter region reduces the electrically active cross-sectional area of the resistor.

the P-type region. Connecting the surrounding N-epi layer to V_+ also enables us to place many base-diffusion resistors into a single N-epi island, which is a much better solution than using a separate N-epi for every single resistor. This is an example of *layer merging*. The resistance of the resistor is determined by the surface dimensions L and W and by the sheet resistance of the layer R_S, as explained in Section 3.2.1.

Resistors can analogously be created using the N^+ diffusion layer and the N-epi, noting that an N-epi island can accommodate only a single N-epi resistor. The N^+ layer offers the smallest sheet resistance and is therefore the most suitable for small-value resistors.

Large-value resistors are hard to implement in bipolar ICs. The length of a 100-kΩ base-diffusion resistor can exceed many times the length of the whole IC chip. A solution for large-value resistors is to use a "snake" geometry. Nonetheless, a couple of resistors of this type may occupy a significant part the IC area. This constraint can be relaxed to a certain extent by using "pinch" resistors.

Figure 16.20 illustrates the cross section of a base-diffusion pinch resistor. The idea is not to mask the P-type resistor body from the emitter N^+-diffusion but to use the emitter-diffusion layer to reduce the electrically active cross-sectional area of the resistor. If an appropriate voltage is applied to the N^+ emitter layer, which means more positive than the terminal voltages V_A and V_B, the N^+-emitter–P-base junction is reverse-biased and the resistor current is forced to flow through the reduced cross-sectional area of the remaining P-type body. In addition, the resistance can be changed to some extent by the voltage applied to the N^+ layer because a larger reverse-bias voltage will produce a wider depletion layer, reducing the effective cross section of the resistor.

Pinch resistors can also be created with the N-epi as the resistor body, and the P-type base diffusion layer can be used to reduce the resistor cross section.

Capacitors and Diodes

P–N junctions are used for the IC capacitors and diodes. There are three P–N junctions in the four-layer structure of the standard NPN BJT: P-substrate–N-epi, N-epi–P base, and P base–N^+ emitter. The application of the P-substrate–N-epi junction is limited by the fact that the P substrate is connected to the most negative potential (V_-). The P-base–N^+-

emitter junction has low breakdown voltage, about 6–7 V, because of the heavy doping in the emitter region. This fact makes it useful as a reference diode. Obviously, there is no choice of reference diodes in the standard bipolar technology, and the reference voltages have to be obtained from reverse-biased base–emitter junctions (6–7 V) and forward-biased P–N junctions ($\approx$ 0.7 V). When a diode with a breakdown voltage larger than 6–7 V is needed, the base–collector P–N junction has to be used.

EXAMPLE 16.6 PNP BJT in Standard IC Technology

Draw the cross section of the class B output stage amplifier of Fig. 16.21, implemented in the standard bipolar IC technology.

SOLUTION

The T_{NPN} transistor is a standard NPN BJT, the cross section of which is shown in Fig. 16.15. The transistor T_{PNP} is a PNP BJT, and there are two possible implementations in the standard bipolar technology: the substrate PNP (Fig. 16.17) and the lateral PNP (Fig. 16.18). The characteristics of the substrate PNP are superior; however, its collector has to be connected to the lowest potential in the circuit (V_-). This is the case in the circuit of Fig. 16.21, so the choice is the substrate PNP. The cross section of the whole circuit is given in Fig. 16.22.

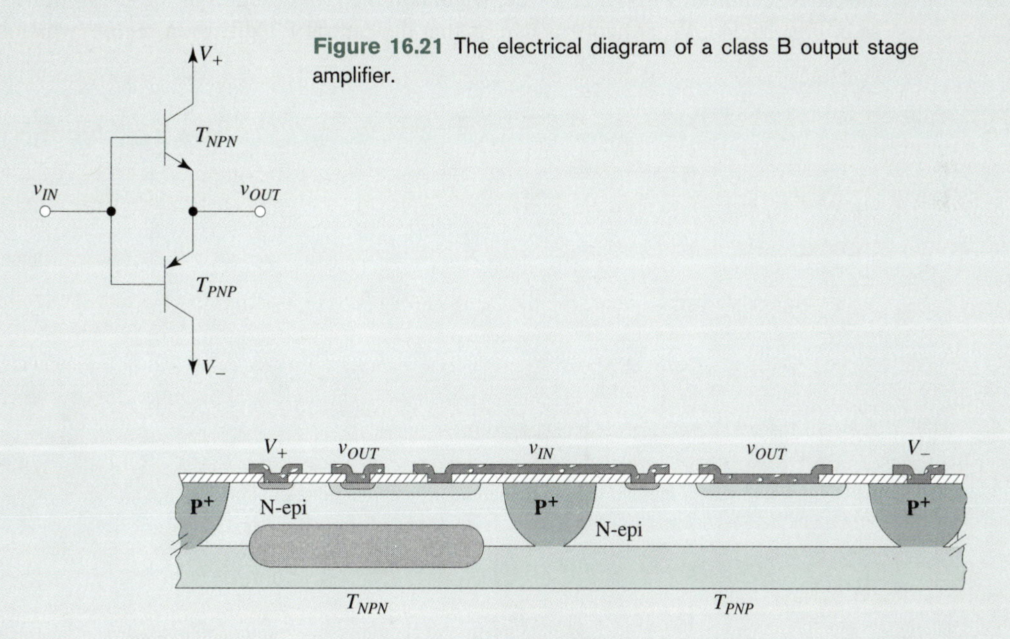

Figure 16.21 The electrical diagram of a class B output stage amplifier.

Figure 16.22 The cross section of a class B output stage amplifier.

16.3.4 Layer Merging

Miniaturization of integrated circuits, within the technology limits, generally improves the IC performance:

- Parasitic components, such as parasitic resistances and capacitances, are reduced as the length of the conductive lines and the area of the P–N junctions is reduced. This is helpful in terms of different types of limitations; in particular, it improves the circuit-response time due to reduced RC time constants.
- Smaller devices mean a smaller active area of the integrated circuit and therefore an improvement in the manufacturing yield. To understand this effect, think of a defect appearing randomly in the silicon crystal; the chance of this defect appearing in the P–N junction region and causing leakage current is higher if the P–N junction area is larger.
- Smaller devices mean that circuits with a larger number of devices can be integrated while avoiding the zero-yield situation, if not even maintaining the yield at the same level.

Miniaturization has proved to be a powerful tool for improving the performance and applicability of electronic systems. Layer merging is a way of reducing the integrated-circuit size by eliminating nonfunctional structures from the IC chip.

The circuit of Fig. 16.23 helps to explain the layer-merging principle. Let us concentrate on the transistors T_1 and T_3. These are standard NPN BJTs; by replicating the cross section of Fig. 16.15 twice, we obtain the cross section of these two transistors, as in Fig. 16.24a. What happens here is that the central P^+ diffusion region, which takes

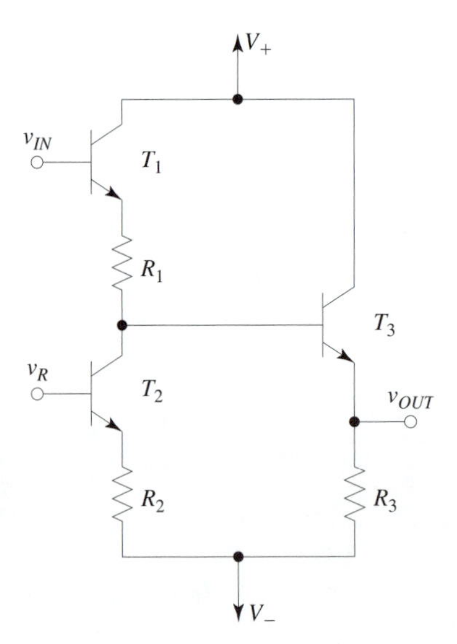

Figure 16.23 A circuit suitable for an illustration of the layer-merging principle (the circuit performs the function of a level shifter).

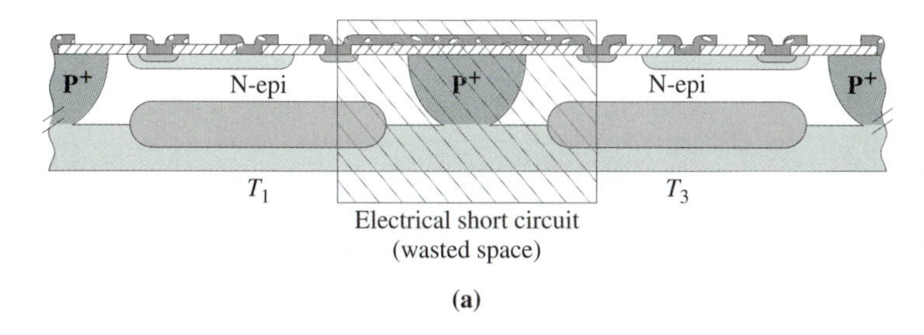

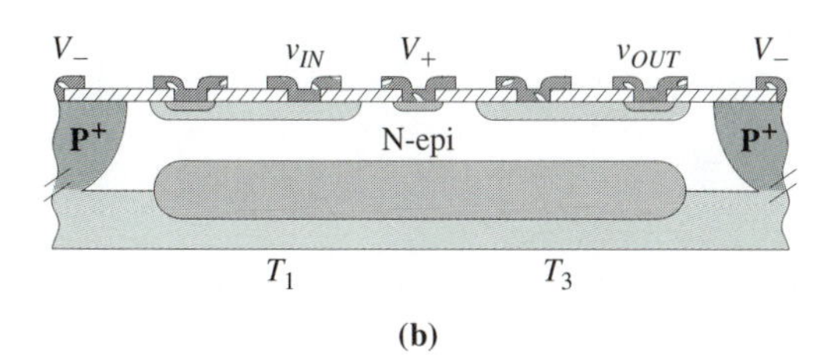

Figure 16.24 Cross section of T_1 and T_3 from Fig. 16.23. (a) Straightforward replication of the NPN BJT cross section (no layer merging) leads to the ineffective region marked by the shaded rectangle. (b) After the layer merging, the two transistors are placed in a single N-epi island.

a lot of space to isolate the two transistors, is electrically short-circuited by the metal line running over the top of the isolation region. There is no functional reason for isolating the two N-epi islands if they have to be short-circuited because the collectors of the two transistors are short-circuited. It may seem that a straightforward layout design of the IC, an easier design automation, as well as clearer IC layout and cross section presentations might be valid arguments against layer merging; but they do not stand against the benefits of IC miniaturization, which is obviously helped by merging the two N-epi islands into one.

EXAMPLE 16.7 IC Layer Merging

Figure 16.25 shows the electrical diagram of a clocked set–reset flip-flop realized in the standard bipolar IC technology, with the resistors being implemented as base diffusion resistors.

(a) Draw the cross section of transistor T_1.
(b) Draw the cross section of transistors T_2 and T_3.
(c) Group the devices into N-epi islands.

SOLUTION

(a) We can think of transistor T_1 as two transistors with short-circuited collector and base terminals. Although this is the electrical equivalent of T_1, the layer-merging principle dictates merging the epi layers first (to short-circuit the collectors) and then merging the P-type base regions (to short-circuit the bases). What is left are two N^+ emitter regions appearing in a single P base and a single N-epi, as illustrated in Fig. 16.26.

(b) The situation with transistors T_2 and T_3 is analogous to transistors T_1 and T_3 in the circuit of Fig. 16.23. The fact that, in this case, the emitters are short-circuited is irrelevant from the layer-merging point of view. The bases of T_2 and T_3 are separate, the P-type base regions cannot be merged, and therefore the two emitters have to appear in two different P-type regions, regardless of the fact that they will have to be electrically short-circuited. Therefore, the cross section is analogous to the one given in Fig. 16.24b.

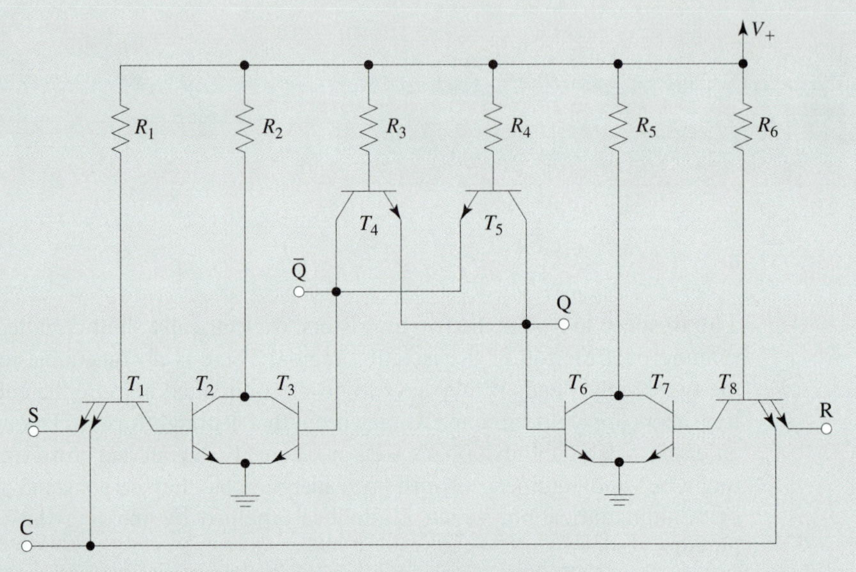

Figure 16.25 The electrical diagram of a clocked set–reset flip-flop.

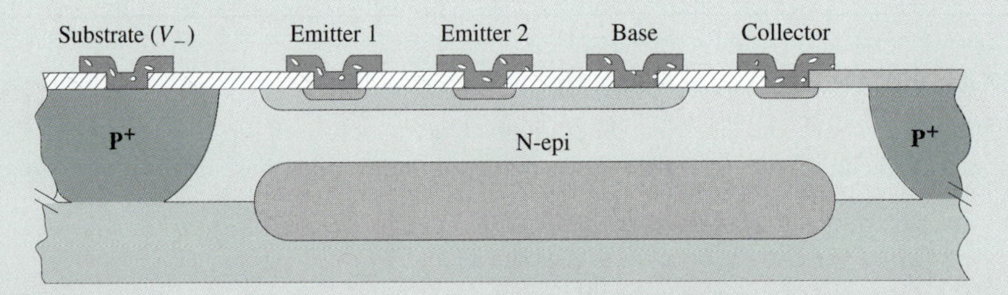

Figure 16.26 Cross-section of double-emitter transistor.

(c) All the resistors can share a single N-epi, which is connected to V_+ to ensure that the P-type region of every single resistor creates a reverse-biased P–N junction with the common N-epi, isolating them electrically from each other. Therefore, the circuit elements can be grouped as follows: N-epi 1, $R_1, \ldots, R_6$; N-epi 2, T_1; N-epi 3, T_2 and T_3; N-epi 4, T_4; N-epi 5, T_5; N-epi 6, T_6 and T_7; and N-epi 7, T_8.

EXAMPLE 16.8 Merging of PNP BJTs

Figure 16.27 shows the electrical diagram of a differential amplifier. Assuming that the resistors are realized as base-diffusion resistors, group the circuit elements into N-epi islands.

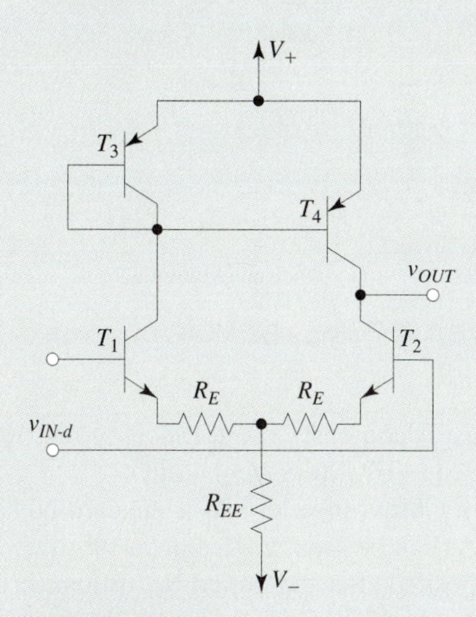

Figure 16.27 The electrical diagram of a differential amplifier.

SOLUTION

This circuit contains PNP BJTs (T_3 and T_4) with separate collectors but a common base. As the N-epi layer plays the role of the base in the case of PNP BJT, the two PNP BJTs (T_3 and T_4) can be placed into a single N-epi that will provide the common base. It is a general statement to say that the merging consideration should start with the N-epi layer but it is not a general rule to check first if there are BJTs with short-circuited collectors. Furthermore, the bases of T_3 and T_4 are connected to the collector of T_1, which means T_1 can be placed into the same N-epi island.

Further consideration may be given to the fact that the N-epi containing T_1, T_3, and T_4 is at a higher potential than any of the resistor terminals. This would suggest that the resistors can be placed into the N-epi of T_3 and T_4. Although this appears as the optimum solution for the circuit of Fig. 16.27 considered in isolation, this may not be the case when a more complex circuit is considered in which the differential amplifier of Fig. 16.27 is only a small part.

Finally, T_2 needs a separate N-epi island.

16.3.5 BiCMOS Technology

The comparison between the BJT and the MOSFET (Section 9.1.6) shows that no device offers the ultimate advantage. BJTs are especially good as output buffers, needed to provide the necessary output power. It is possible to merge the CMOS and standard bipolar IC technologies to have the advantages of both CMOS and bipolar circuits on the same chip. Although this inevitably leads to a more complex, and therefore more expensive,

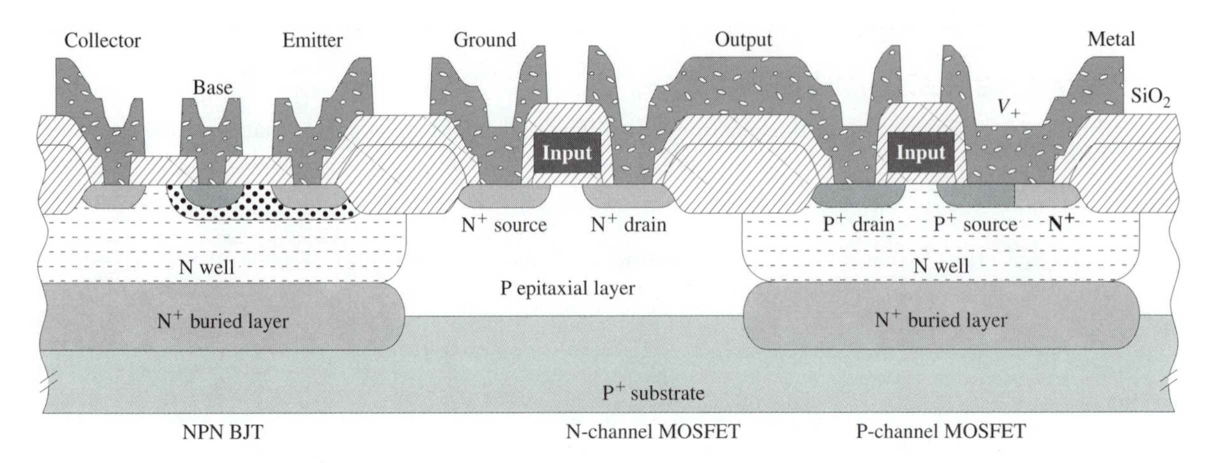

Figure 16.28 The cross section of a CMOS inverter and a standard NPN BJT illustrating a BiCMOS technology.

technology, the benefits gained sometimes justify the cost. The technology that merges bipolar and CMOS IC technologies is referred to as BiCMOS technology.

Figure 16.28 shows the cross section of a CMOS inverter and a standard NPN BJT obtained by a BiCMOS technology process. As can be seen, a P^+ silicon substrate is the starting material and a P-epitaxial layer is deposited after the buried N^+ diffusion. Using the P-epi is an obvious difference from the standard bipolar technology that uses N-epi, but it is needed to provide the substrate for the N-channel MOSFETs. The equivalent of the N-epi islands, appearing in the standard bipolar technology, is achieved by deep N-type diffusion, which creates the N-well regions, needed also for the P-channel MOSFETs. Note that P^+ diffusion, which is necessary for the source and drain of the P-channel MOSFETs, is also used to improve the base contact of the NPN BJT.

Buried N^+ diffusion and the deposition of a P-epi layer are additional process steps in comparison to the standard N-well CMOS technology. Another addition is P-type diffusion, needed for the base of the NPN BJT. This is a more complex process than that required N-well CMOS technology; however, it provides all the devices available in the standard bipolar technology, in addition to the CMOS devices.

SUMMARY

1. An IC process sequence consists of multiple doping steps (different dopant types, surface/peak concentrations, and depths) into selected areas, multiple layer-deposition steps, and selective etching to define the pattern/windows in the deposited layers. Selective layer etching, which also provides the windows for selective doping, is achieved by a set of process steps known as *photolithography*. It is photolithography that enables the transfer of IC layout design onto semiconductor chips; thus, the IC-layout designers create the patterns that appear on the photolithography masks.
2. It is possible to create integrated circuits with N-channel MOSFETs only, using enhancement MOSFETs as voltage-controlled switches and depletion MOSFETs as load resistances—*NMOS technology*. A current flows through the enhancement-type

MOSFETs set in the *on* mode, resulting in considerable power dissipation and limiting the circuit complexity that can be achieved with NMOS technology.

3. *CMOS technology* uses enhancement-type N-channel and P-channel MOSFETs that operate as complementary switches. Given that one of the two complementary switches is *off* for any logic state, the static power dissipation is reduced to leakage-current levels, enabling very complex ICs that perform sophisticated functions. The older *P-well* technology uses N-type substrate for the P-channel MOSFETs, placing the N-channel MOSFETs in P wells. The newer CMOS technologies use a P substrate for the N-channel MOSFETs and use N wells for the P-channel MOSFETs, an approach that requires *threshold-voltage adjustment*. Threshold-voltage adjustment is typically achieved by ion implantation of doping atoms, a process that provides excellent control of the doping dose and profile.
4. *Silicon-on-insulator* (SOI) technology represents an advance that may prove as an important "enabling" technology, in particular for high-speed and high-reliability CMOS ICs.
5. The standard bipolar IC technology is designed to optimize the performance of the NPN BJT as the main device. A low-doped N-epitaxial layer is deposited onto P substrate and "cut" into N-epi islands by deep P^+ diffusion stripes. The boundary N-epi–P junction is reverse-biased to provide electrical isolation of the individual components, placed into the N-epi islands. The N-epi islands become collectors, with P-type base and N^+ emitter regions of increasing doping concentrations sequentially diffused into one another.
6. PNP BJTs can be implemented as either substrate (P-base–N-epi–P-substrate as emitter–base–collector regions) or lateral (P-base–N-epi–P-base regions). The neutral regions of base- and emitter-diffusion regions, as well as the N-epi, are used as resistors. The base–collector, base–emitter, and collector–substrate P–N junctions can be used as diodes and capacitors. The breakdown voltage of the E–B junction is low ($\approx$ 7V) due to the very heavy emitter doping; this junction can be used as a reference diode.
7. The physical merging of layers that belong to different devices but are short-circuited electrically enables significant area reduction and the associated speed/performance and yield improvements.
8. Integration of bipolar and CMOS technologies (BiCMOS) is an advanced technology that makes available the superior characteristics of both MOSFETs and BJTs to the IC designers.

PROBLEMS

16.1 The dimensions of a diffusion resistor on the photolithography mask are $W_m = 1\ \mu$m and $L_m = 20\ \mu$m. The resistor is to be created by boron diffusion to the junction depth of $x_j = 2\ \mu$m. Assuming that the mask dimensions are perfectly transferred onto the masking oxide and that the lateral diffusion is equal to 80% of the vertical diffusion, determine the actual dimensions of the resistor.

16.2 Calculate the diffusion coefficient of phosphorus at $T = 1100°$C, knowing that $D_0 = 19.7$ cm^2/s and $E_A = 3.75$ eV.

TABLE 16.4 Ion Implant Parameters

			Energy (keV)			
			20	50	100	200
B	R_p	(μm)	0.0662	0.1608	0.2994	0.5297
	ΔR_p	(μm)	0.0283	0.0504	0.0710	0.0921
P	R_p	(μm)	0.0253	0.0607	0.1238	0.2539
	ΔR_p	(μm)	0.0119	0.0256	0.0456	0.0775
As	R_p	(μm)	0.0159	0.0322	0.0582	0.1114
	ΔR_p	(μm)	0.0059	0.0118	0.0207	0.0374

16.3 The concentration profile of drive-in diffusion is

$$N(x,t) = \frac{\Phi}{\sqrt{\pi D t}} \exp(-x^2/4Dt)$$

What deposition dose (Φ) and drive-in time (t) are needed to obtain a layer with surface concentration $N_s = 5 \times 10^{16}$ cm^{-3} and junction depth $x_j = 2$ μm? The doping element is phosphorus, which is being diffused into a P-type wafer with uniform concentration $N_A = 10^{15}$ cm^{-3}. The drive-in temperature is $T = 1100°$C.

16.4 Calculate the sheet resistance of the layer designed in Problem 16.3. Assume constant electron mobility $\mu_n = 1250$ cm^2/V·s. It is known that $\int_0^\infty \exp(-u^2/2)\,du = \sqrt{\pi/2}$. **A**

16.5 What would the sheet resistance of the layer designed in Problem 16.3 be if the actual drive-in temperature were 1102°C (2°C higher than the nominal 1100°C)?

16.6 The temperature variation of a drive-in diffusion furnace is ±0.1%. Determine the junction depth tolerance if the nominal drive-in temperature is $T = 1050°$C. Express the result as a percentage. The doping element is

(a) phosphorus ($E_A = 3.75$ eV)

(b) arsenic ($E_A = 3.90$ eV) **A**

16.7 Ion-implantation profiles can be approximated by the Gaussian distribution function:

$$N(x) = \frac{\Phi}{\sqrt{2\pi}\,\Delta R_p} \exp\left[-\frac{1}{2}\left(\frac{x - R_p}{\Delta R_p}\right)^2\right]$$

where Φ is the dose, R_p is the range, and ΔR_p is the straggle of the implanted ions. The range and the straggle depend on the energy of the implantation, as shown in Table 16.4. Sketch the implant profiles of boron for three different energies (20 keV, 50 keV, and 100 keV) and dose of 5×10^{11} cm^{-2}. Using the medium energy, change the dose to 10^{11} cm^{-2} and 10^{12} cm^{-2} and sketch the implant profiles. Comment on the influence of energy and dose on the implant profiles.

16.8 The design dose of a threshold-voltage-adjustment implant is $\Phi_{implant} = 10^{12}$ cm^{-2}. What is the mismatch between the threshold voltages ($V_{T-NMOS} - |V_{T-PMOS}|$) if the actual dose is 1% higher? The doping element of the adjustment implant is boron, and the gate-oxide thickness is 4 nm. Assume that all the implanted atoms are in the depletion layer. **A**

16.9 The threshold voltage of a CMOS inverter (V_{TI}) can be defined as the input voltage corresponding to the output voltage that is equal to half of the supply voltage $V_{DD} = V_H$. V_{TI} can be adjusted by the layout design of the NMOS and PMOS transistors. Ideally, the inverter threshold voltage should be equal to $V_{DD}/2$ (centered transfer characteristic). In the case of CMOS inverter (Fig. 8.6), both NMOS and PMOS are in saturation at $V_{GS} = V_{TI}$.

(a) Neglecting the channel pinch-off in the SPICE LEVEL 3 model ($L_{pinch} = 0$), derive an equation for $r = \frac{W_{PMOS}/L_{PMOS}}{W_{NMOS}/L_{NMOS}}$ when both the NMOS and the PMOS are in saturation.

(b) Determine r so that $V_{TI} = V_{DD}/2$ for the case of the following technological parameters: $|2\phi_{F-NMOS}| = 0.91$ V, $\gamma_{NMOS} = 0.65$ V$^{1/2}$,

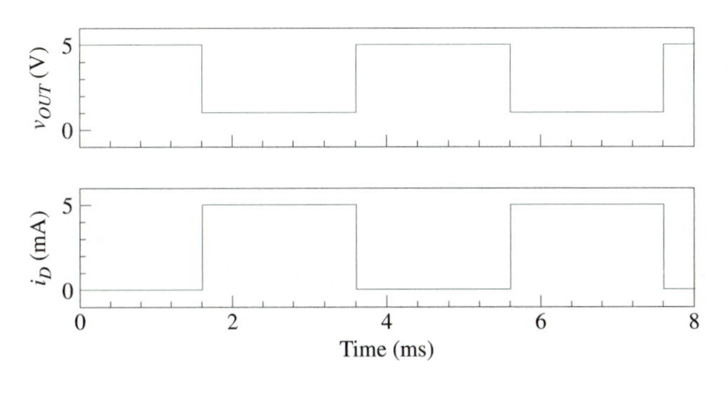

Figure 16.29 Ideal switching response of an NMOS inverter.

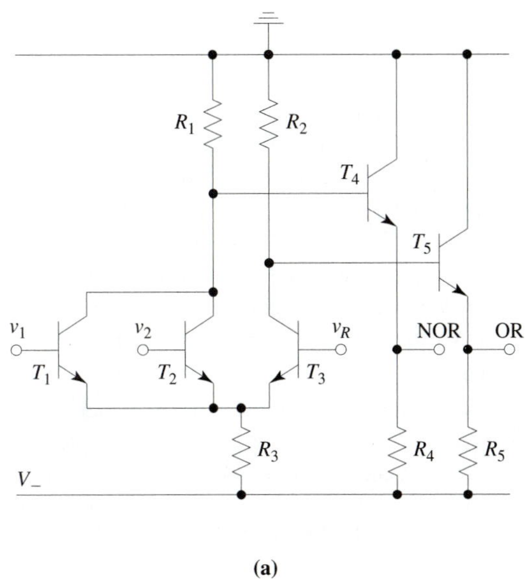

(b)

Figure 16.30 Example circuits.

$\mu_{eff-NMOS} = 380$ cm^2/Vs. $\mu_{eff-PMOS} = 100$ cm^2/V·s, $|2\phi_{F-PMOS}| = 0.58$ V, $\gamma_{PMOS} = 0.03$ V$^{1/2}$. Assume perfectly matched NMOS and PMOS threshold voltages, $V_{T-NMOS} = -V_{T-PMOS} > 0$. **A**

(c) Specify the layout-design parameters of the MOSFETs if the minimum channel dimension is limited to 0.15 μm. The design should minimize the input capacitance and maximize MOSFET currents.

16.10 Figure 16.29 shows the ideal response (no delays and parasitic capacitances to be charged/discharged) of an NMOS inverter. The power-supply voltage is 5 V.

(a) Find the average power dissipated by this inverter.

(b) Find the average dissipated power if the inverter was implemented in CMOS technology.

16.11 The dynamic power dissipation of a CMOS inverter is given by

$$P_{diss} = CV^2 f$$

where V is the power-supply voltage, f is the switching frequency, and C is the relevant capacitance. The technology progress enables design of CMOS ICs with reduced input capacitance and increased switching frequency, although this may require a reduction in the power-supply voltage. Assuming that the average power dissipated by the inverter is $P_{diss} = 5\ \mu$W, find P_{diss} for a new design that

(a) halves the input MOSFET capacitances and quadruples the switching frequency, maintaining the power supply voltage. **A**

(b) reduces the input capacitance five times, increases the switching frequency 25 times, and reduces the power-supply voltage from 5 V to 1.5 V.

16.12 The NPN BJT of Fig. 16.15 is to be designed for a 10-μm-thick N-epitaxial layer. Simplified design rules can be expressed as follows: the depth of the P^+ isolation diffusion should be $x_{j-P+} = 12\ \mu$m, the minimum size of the diffusion and contact windows is $W_w \times W_w = 5\ \mu\text{m} \times 5\ \mu$m, the minimum spacing between two windows is $S_{w-w} = 5\ \mu$m, the minimum spacing between a contact edge and the closest P–N junction is $S_{c-j} = 3\ \mu$m, and the minimum spacing between two P–N junctions is $S_{j-j} = 5\ \mu$m. The lateral diffusion is 80% of the vertical diffusion.

(a) Calculate the minimum size of the NPN BJT, defined as the distance between the center of the left and the center of the right P^+ isolation diffusions in Fig. 16.15.

(b) Assume that the BJT has a square shape, with the side of the square being equal to the minimum distance determined in (a). What percent of the total BJT area is occupied by the isolation P^+ diffusion? **A**

16.13 The average doping levels of a base-diffusion pinch resistor (Fig. 16.20) are $N_{epi} = 10^{14}$ cm^{-3}, $N_{base} = 10^{16}$ cm^{-3}, and $N_{emitter} = 10^{20}$ cm^{-3}. The emitter–base and base–collector junction depths are $x_{jE} = 3\ \mu$m and $x_{jB} = 5\ \mu$m, respectively. Assuming abrupt P–N junctions and a hole mobility of $\mu_p = 400$ cm^2/V · s, calculate the sheet resistance of this resistor for two different control voltages:

(a) $V_{EB} = 0$ V

(b) $V_{EB} = 5$ V **A**

16.14 The base–collector junction is to be used as a 0.1-nF capacitor in a bipolar IC. Calculate the needed area of the capacitor if the doping concentrations for the abrupt-junction model are $N_{epi} = 10^{14}$ cm^{-3} and $N_{base} = 10^{16}$ cm^{-3}. The DC bias is $V_R = 5$ V.

16.15 The resistors of the circuit given in Fig. 16.30a are implemented as base-diffusion resistors.

(a) Draw the cross section of T_1 and T_2.

(b) Draw the cross section of T_4 and T_5.

(c) Group the devices into N-epi islands. **A**

16.16 Assuming that the resistors are implemented as base-diffusion resistors, and the diode as a base–collector diode, group the devices of the circuits given in Fig. 16.30b into N-epi islands.

REVIEW QUESTIONS

R-16.1 What is the difference between dose and concentration of doping atoms, and how are they related? What are the respective units?

R-16.2 How are the devices/gates isolated from each other in NMOS and CMOS ICs?

R-16.3 What are the advantages of IC layer merging?

R-16.4 Why is polysilicon, and not aluminum, used as the gate material in modern MOSFET technologies?

R-16.5 Which technology process is more complex, NMOS or CMOS?

R-16.6 What are the advantages of CMOS ICs compared to NMOS ICs?

R-16.7 Why is threshold-voltage-adjustment implant needed?

R-16.8 How do the characteristics of a lateral PNP compare to the characteristics of the standard NPN BJT?

R-16.9 Why is the PNP BJT in the standard bipolar IC technology not made with different but complementary layers (mirror image)? In other words, why are the same layers used at the expense of PNP performance?

BIBLIOGRAPHY

R. J. Baker, H. W. Li, and D. E. Boyce, *CMOS: Circuit Design, Layout, and Simulation,* IEEE Press, New York, 1998.

S. Brandt and H. D. Dahmen, *The Picture Book of Quantum Mechanics,* Springer-Verlag, New York, 1995.

P. A. M. Dirac, *The Principles of Quantum Mechanics,* 4th ed., Oxford University Press, Oxford, 1958.

D. K. Ferry, *Quantum Mechanics—An Introduction for Device Physicists and Electrical Engineers,* Institute of Physics Publishing, Bristol, 1995.

R. P. Feynman, R. B. Leighton, and M. Sands, *The Feynman Lectures on Physics: Quantum Mechanics,* Addison-Wesley, Reading, MA, 1965.

D. Foty, *MOSFET Modeling with SPICE: Principles and Practice,* Prentice-Hall, Upper Saddle River, NJ, 1997.

W. Heisenberg, *The Physical Principles of the Quantum Theory,* Dover, New York, 1949.

J. R. Hook and H. E. Hall, *Solid State Physics,* 2nd ed., Wiley, Chichester, UK, 1991.

P. S. Kireev, *Semiconductor Physics,* 2nd ed., Mir, Moscow, 1978.

C. Kittel, *Introduction to Solid State Physics,* 7th ed., Wiley, New York, 1996.

J.-M. Lévy-Leblond and F. Balibar, *Quantics—Rudiments of Quantum Physics,* 2nd ed., North-Holland, Amsterdam, 1990.

G. Massobrio and P. Antognetti, *Semiconductor Device Modeling with SPICE,* 2nd ed., McGraw-Hill, New York, 1993.

R. S. Muller and T. I. Kamins, *Device Electronics for Integrated Circuits,* 2nd ed., Wiley, New York, 1986.

D. A. Neamen, *Semiconductor Physics & Devices: Basic Principles,* 3rd ed., McGraw-Hill, Boston, MA, 2003.

R. F. Pierret, *Semiconductor Device Fundamentals,* Addison-Wesley, Reading, MA, 1996.

D. K. Reinhard, *Introduction to Integrated Circuit Engineering,* Houghton Mifflin, Boston, 1987.

E. H. Rhoderick, *Metal–Semiconductor Contacts*, Clarendon Press, Oxford, 1978.

W. R. Runyan and K. E. Bean, *Semiconductor Integrated Circuit Processing Technology,* Addison-Wesley, Reading, 1990.

D. K. Schroder, *Semiconductor Material and Device Characterization*, 2nd ed., Wiley, New York, 1998.

M. Shur, *Introduction to Electronic Devices,* Wiley, New York, 1996.

J. Singh, *Semiconductor Devices: An Introduction,* McGraw-Hill, New York, 1994.

J. Singh, *Semiconductor Devices: Basic Principles,* Wiley, New York, 2001.

H. T. Stokes, *Solid State Physics,* Allyn and Bacon, Boston, MA, 1987.

B. G. Streetman and S. Banerjee, *Solid State Electronic Devices,* 5th ed., Prentice-Hall, Upper Saddle River, NJ, 2000.

S. M. Sze, *Physics of Semiconductor Devices,* 2nd ed., Wiley, New York, 1981.

S. M. Sze, ed., *VLSI Technology,* McGraw-Hill, New York, 1983.

W. C. Till and J. T. Luxon, *Integrated Circuits: Materials, Devices, and Fabrication,* Prentice-Hall, Englewood Cliffs, NJ, 1982.

P. Tuinenga, *SPICE: A Guide to Circuit Simulation Using PSPICE,* Prentice-Hall, Englewood Cliffs, NJ, 1988.

C. T. Wang, *Introduction to Semiconductor Technology: GaAs and Related Compounds,* Wiley, New York, 1990.

R. M. Warner and B. L. Grung, *Semiconductor-Device Electronics,* Oxford University Press, New York, 1991.

ANSWERS TO SELECTED PROBLEMS

CHAPTER 1

1.1 (b) $PF = 0.740$; **1.2** (a) $d_{Ga-As} = 0.245$ nm; **1.3** (c) 7.10 cm^3/mol; **1.6** $N_{Si} = 4.826 \times 10^{22}$ cm^{-3}; **1.9** (c) $N_{\{111\}} = 2.3 \times 10^{14}$ cm^{-2}; **1.10** (a) $N_{\{110\}} = 9.6 \times 10^{14}$ cm^{-2}; **1.13** [100], $[\bar{1}00]$, [010], $[0\bar{1}0]$, [001], $[00\bar{1}]$; **1.18** (a) $n = 2.0 \times 10^{23}$ cm^{-3}; **1.20** (b) (1); **1.21** (b) $p = 4.41 \times 10^{-4}$ cm^{-3}; **1.26** GaAs: $V = 2.27 \times 10^4$ cm^3; **1.27** (b) $n = 5.9 \times 10^{12}$ cm^{-3}; **1.28** (a) $p = 10^{15}$ cm^{-3}

CHAPTER 2

2.3 (b) 4.31×10^{11} m^{-1}, 1.45×10^{-11} m, 2.16×10^{19} s^{-1}, and 2.91×10^{-19} s; **2.5** (b) 1.88×10^7 m/s, 1.62×10^{11} m^{-1}, 3.88×10^{-11} m, and 2.07×10^{-18} s; **2.7** (b) $P = 0.31$; **2.10** (a) 6.1×10^{-5}; **2.12** (b) 4.6×10^{-10} m; **2.13** $m_l^* = 0.16m_0$, $m_h^* = 0.52m_0$; **2.16** $D = \frac{2(2m^*)^{1/2}}{h} E_{kin}^{-1/2}$; **2.17** (b) 2D: $m^* = 0.68m_0$ 1D: $m^* = 1.26m_0$; **2.24** (b) $N_V = 8.1 \times 10^{18}$ cm^{-3}; **2.26** (b) 0.702 eV; **2.30** $E_C - E_F < E_F - E_V$, i.e., $E_F > E_i \Longrightarrow$ N type; **2.33** (b) P type, 1.1×10^{15} cm^{-3}; **2.35** (b) 4.20 eV; **2.37** $n_i = 1.08 \times 10^7$ cm^{-3}; **2.40** (a) 0.069 eV; (b) 0.080 eV

CHAPTER 3

3.5 $L = 246$ μm, $R_1 = 76.5$ Ω; **3.9** $R_S = 1.5$ k$\Omega/\Box$; **3.11** $L = 56.4$ μm, $W = 28.2$ μm; **3.12** $R_S = 180$ $\Omega/\Box$, $R_c = 5$ Ω; **3.13** 2.32 $(\Omega \cdot \text{cm})^{-1}$; **3.16** $N_A = 4.7 \times 10^{15}$ cm^{-3}, $\rho = 3.65$ $\Omega \cdot$cm; **3.20** (b) 27°C: $R_S = 0.625$ $\Omega/\Box$; 700°C: $R_S = 0.625$ $\Omega/\Box$; **3.23** (b) $t_{tr} = 0.65$ ms; **3.24** (c) $R = 53.3$ Ω; **3.27** $l_{sc} = 16.4$ nm; **3.29** $\tau_{sc2} = 0.19$ ps

CHAPTER 4

4.3 (b) 1.96×10^{15} cm^{-3}; **4.10** $E(L_p) = 318.7$ V/m; **4.12** 14.0%; **4.14** (b) $D = 7.3$ cm^2/s; **4.16** (c) -2.74×10^{13} cm^{-3}; **4.17** (b) 9.2×10^{19} cm^{-3} s^{-1}

CHAPTER 5

5.3 (c) $r_h = 2.5 \times 10^{10}$ cm^{-3} s^{-1}; **5.8** $p(t) = G_{ext}\tau_p(1 - e^{-t/\tau_p}) + p_0$; **5.10** (a) $j_{diff}(0) = 8$ A/cm^2; **5.12** (b) $\tau_p = 0.32$ μs; **5.14** (b) $U = -10^{12}$ cm^{-3} s^{-1}; **5.16** (c) $U = -8.7 \times 10^{15}$ cm^{-3} s^{-1}

CHAPTER 6

6.4 (b) $V_{bi} = 0.30$ V; **6.9** (b) $n_p(|w_p|) = 1.26 \times 10^{-7}$ cm^{-3}, $p_n(|w_n|) = 1.26 \times 10^{-11}$ cm^{-3}; **6.10** (b) $j_n = 3.20 \times 10^{-2}$ A/m^2, $j_p = 3.00 \times 10^{-7}$ A/m^2; **6.13** (b) $I_S = 2.01 \times 10^{-44}$ A, $V_D = 2.54$ V; **6.16** $I_{n,p \to n} = 1.25 \times 10^{-16}$ A, $I_{p,n \to p} = 1.25 \times 10^{-13}$ A; **6.20** $r_S = 4.7$ Ω; **6.22** $V_D = 0.835$ V; **6.23** (b) $I_S = 8.10 \times 10^{-12}$ A, $n = 1.575$; **6.28** (c) $Q_C = 0.58$ mC/m^2; **6.29** $Q_C = 2.12$ mC/m^2; **6.30** (b) $\Delta Q_C \approx 0.098$ mC/m^2; **6.33** (c) $C_d(0)/A = 0.108$ mF/m^2; **6.36** (a) $V_{bi} = 0.588$ V; (d) abrupt junction: $E_{max} = 1.04$ V/μm, linear junction: $E_{max} = 0.53$ V/μm; **6.43** (b) $Q_{s-rec} = 2.9 \times 10^{-10}$ C; **6.44** $Q_s = 2.49 \times 10^{-11}$ C; **6.49** (b) $j_{gen} = 4.7 \times 10^{-24}$ A/cm^2, $j_{diff} = 7.2 \times 10^{-41}$ A/cm^2

CHAPTER 7

7.4 (b) 5.4 eV and 0.65 eV; **7.6** (a) $V_F = 0.14$ V; **7.9** (a) $C_d(0) = 0.145$ mF/m^2; **7.16** $E_{ox} = 0.2$ V/nm, toward the substrate; $E_s = 0$; **7.19** (a) $q\phi_{ms} = 0.203$ eV; **7.20** (b) $V_{FB} = 0.87$ V; **7.21** electrons: $Q_I = 9.2$ mC/m^2; **7.23** N-type; $Q_I = 0$ C/m^2; **7.26** $Q_I = 3.22$ mC/m^2; **7.28** $N_A = 1.77 \times 10^{17}$ cm^{-3}; **7.29** $\varphi_s = 0.80$ V, $V_{ox} = 5.16$ V; **7.31** (a) $E_{ox} = E_s = 0$; **7.32** (b) $V_{BRinv} = -78.20$ V

CHAPTER 8

8.4 (b) $\varphi_s = 0.172$ V, $Q_d = 5.36 \times 10^{-4}$ C/m^2 for $V_{GS} = -0.5$ V, $\varphi_s = 0.605$ V, $Q_d = 1.01 \times 10^{-3}$ C/m^2 for $V_{GS} = 0$ V; (d) $\varphi_s = 0.797$ V, $Q_d = 1.15 \times 10^{-3}$ C/m^2; **8.8** 1.05 times; **8.9** (b) $Q_I = 2.17$ mC/m^2 for $V_{GS} = -0.75$ V; $Q_I = 0$ for $V_{GS} = 0$, and 0.75 V; **8.15** $j = 9.0 \times 10^{10}$ A/m^2, $\mu_{eff} = 320$ cm^2/V $\cdot$ s; **8.16** (b) $V_{DSsat} = -0.943$ V, $I_{Dsat} = 2.22$ mA, $I_D = 0.80$ mA; **8.17** (b) $\Delta I_D = 1.49$ mA; **8.20** (a) 100 kΩ

CHAPTER 9

9.8 (b) $I_C = 0.038$ μA; **9.9** (b) $g_m = 10.1$ A/V; **9.11** (e) BJT damaged (impossible); **9.13** $|I_{B-max}| = 14.33$ μA;

9.14 $\beta_R = 0.65$; **9.15** (c) $I_C = -164.88$ mA, $I_E = 82.44$ mA, $I_B = 82.44$ mA; **9.19** $V_{BR} = 618$ V

CHAPTER 10

10.4 99.38% for $N_{el} = 0$, 0.62% for $N_{el} = 1$, and 0.0019% for $N_{el} = 2$; **10.5** (a) $p(1, 1) = 0.0281$; **10.10** (b) $I_D = 367.6$ mA; **10.12** (b) $N = 364$, $d = 0.27$ nm; **10.13** (a) $1.1 k\Omega$; (d) $N = 4$; **10.16** (c) 3.3×10^7 cm/s

CHAPTER 11

11.5 $p_t = 2.5$; **11.6** $p_t = 2$; **11.9** $I_D = 1.81$ mA; **11.11** $C_{GD0} = 345$ pF/m;

CHAPTER 12

12.3 (b) $P = 2.1$ mW; **12.7** (a) $\Phi_{opt} = 2.26 \times 10^{21}$ s^{-1} m^{-2}; (c) $I_{photo} = 361.6\ \mu$A; **12.9** (a) $V_{oc} = V_t \ln(1 + I_{photo}/I_S)$; $V_{oc} = 0.565$ V; (c) $N_{series} = 25$, $N_{parallel} = 5$; **12.11** (a) $I_{photo} = 3.08$ mA; (b) 89.8% is diffusion photocurrent; (c) $n_p = 1.5 \times 10^{11}$ cm^{-3}, $p_n = 5 \times 10^{10}$ cm^{-3}

CHAPTER 13

13.3 $V_{DSsat} = 4.12$ V; $I_{DSsat} = 50.8$ mA; **13.11** $N_D = 1.18 \times 10^{17}$ cm^{-3}, $V_{bi} = 1.0$ V, $R_{ch} = 75.9\ \Omega$

CHAPTER 14

14.3 (a) $V_{BR} = 304$ V, (c) $L_N \geq 1.9\ \mu$m; **14.6** (b) $C_{GS} = 1.86$ nF

CHAPTER 15

15.4 $f = 28.6$ GHz

CHAPTER 16

16.4 $R_s = 1116\ \Omega/\square$; **16.6** (b) $\frac{dx_j}{x_j} \times 100 = 1.70\%$ **16.8** $\Delta V_T = 3.7$ mV; **16.9** (b) $r = 2.89$; **16.11** (a) $P_{diss-a} = 10\ \mu$W; **16.12** (b) 46.7%; **16.13** (b) $R_S = 13.95\ k\Omega/\square$; **16.15** (c) epi 1: T_4, T_5, and the resistors; epi 2: T_1 and T_2; epi 3: T_3

Index